PHOTOVOLTAIC SOLAR ENERGY

PHOTOVOLTAIC SOLAR ENERGY

FROM FUNDAMENTALS TO APPLICATIONS

Edited by

Angèle Reinders
University of Twente
Enschede
The Netherlands

Pierre Verlinden
Trina Solar
Changzhou, Jiangsu
China

Wilfried van Sark
Copernicus Institute
Utrecht University
The Netherlands

and

Alexandre Freundlich
University of Houston
Texas
USA

WILEY

This edition first published 2017

Registered Office
John Wiley & Sons, Ltd, The Atrium, Southern Gate, Chichester, West Sussex, PO19 8SQ, United Kingdom

For details of our global editorial offices, for customer services and for information about how to apply for permission to reuse the copyright material in this book please see our website at www.wiley.com.

Library of Congress Cataloguing-in-Publication Data

Names: Reinders, Angèle, editor. | Verlinden, Pierre, editor. | Sark, Wilfried van, editor. | Freundlich, Alexandre, editor.
Title: Photovoltaic solar energy : from fundamentals to applications / edited by Angèle Reinders, Pierre Verlinden, Wilfried van Sark, Alexandre Freundlich.
Description: Chichester, West Sussex, United Kingdom ; Hoboken, NJ : John Wiley & Sons Ltd, 2017. | Includes bibliographical references and index.
Identifiers: LCCN 2016024802 | ISBN 9781118927465 (cloth) | ISBN 9781118927489 (epub) | ISBN 9781118927472 (Adobe PDF)
Subjects: LCSH: Photovoltaic power generation. | Photovoltaic cells.
Classification: LCC TK1087 .P466 2017 | DDC 621.31/244–dc23
LC record available at https://lccn.loc.gov/2016024802

A catalogue record for this book is available from the British Library.

Cover image: gerenme/Gettyimages
Pedro Castellano/gerenme/Spectral-Design/luchschen/EunikaSopotnicka/Klagyivik/iStock

Set in 10/12pt Times by SPi Global, Pondicherry, India

10 9 8 7 6 5 4 3 2 1

The world has enough for everyone's need, but not enough for everyone's greed

----

Mahatma Gandhi

Contents

List of Contributors

Pietro Altermatt
State Key Laboratory of PV Science and Technology (PVST), Trina Solar,
Changzhou,
Jiangsu, China

Daniel Amkreutz
Institute for Silicon Photovoltaics,
Helmholtz-Zentrum Berlin für Materialien und Energie GmbH,
Germany

Georgia Apostolou
Delft University of Technology,
The Netherlands

Greg J. Ball
Solar City,
Oakland, CA,
USA

Christophe Ballif
Ecole Polytechnique Fédérale de Lausanne,
Institute of Microengineering (IMT),
Photovoltaics and Thin-Film Electronics Laboratory (PV-lab),
Swiss Center for Electronics and Microtechnology (CSEM),
PV-Center, Neuchâtel,
Switzerland

Andrew Blakers
Australian National University,
Canberra,
Australia

Mathieu Boccard
Arizona State University,
Tempe, AZ,
USA

Georg Bopp
Fraunhofer Institute for Solar Energy Systems ISE,
Freiburg,
Germany

Matthew Campbell
SunPower Corporation,
Richmond, CA,
USA

John P.D. Cook
SUNLAB University of Ottawa,
Canada

Michael Debije
Technical University,
Eindhoven,
The Netherlands

Chris Deline
National Renewable Energy Laboratory
Golden, CO,
USA

Frank Dimroth
Fraunhofer Institute for Solar Energy
Systems, ISE,
Freiburg,
Germany

Rhett Evans
The University of New South Wales,
Sydney,
Australia

Simon Fafard
Laboratoire Nanotechnologies
Nanosystèmes,
Institut Interdisciplinaire d'Innovation
Technologique (3IT),
Université de Sherbrooke,
Canada

Andreas Fell
Fraunhofer Institute for Solar Energy
Systems, ISE,
Freiburg,
Germany

Halden Field
PV Measurements, Inc.,
Boulder, CO,
USA

Vasilis Fthenakis
Center of Life Cycle Analysis, Columbia
University, New York,
USA

Gianluca Fulli
European Commission Joint Research Centre,
Institute for Energy and Transport,
Petten,
The Netherlands

Onno Gabriel
PVcomB/Helmholtz-Zentrum Berlin für
Materialien und Energie GmbH,
Germany

Flavia Gangale
European Commission,
Joint Research Centre,
Institute for Energy and Transport,
Petten,
The Netherlands

Chris Geurts
TNO,
Delft/Eindhoven,
The Netherlands

Veronique S. Gevaerts
Energy Research Centre of the
Netherlands – Solliance,
Eindhoven,
The Netherlands

Martin Green
Australian Centre for Advanced
Photovoltaics,
School of Photovoltaic and Renewable
Energy Engineering,
University of New South Wales,
Sydney,
Australia

Jan Haschke
Institute for Silicon Photovoltaics/
Helmholtz-Zentrum Berlin für Materialien
und Energie GmbH,
Germany

Franz Haug
Ecole Polytechnique Fédérale de Lausanne,
Photovoltaics and Thin Film Electronics
Laboratory (PV-Lab),
Neuchâtel,
Switzerland

Martin Hermle
Fraunhofer Institute for Solar Energy Systems, ISE,
Freiburg,
Germany

Karin Hinzer
School of Electrical Engineering and Computer Science,
University of Ottawa,
Canada

Bram Hoex
School of Photovoltaic and Renewable Energy Engineering,
University of New South Wales,
Kensington,
Australia

Zachary Holman
Arizona State University,
Tempe, AZ,
USA

Seth Hubbard
NanoPower Research Laboratories,
Rochester Institute of Technology (RIT),
Rochester, NY,
USA

Phillip Jenkins
U.S. Naval Research Laboratory,
Washington, DC,
USA

Henner Kampwerth
University of New South Wales,
Sydney,
Australia

Michael Kempe
National Renewable Energy Laboratory,
Golden, CO,
USA

Geoffrey S. Kinsey
U.S. Department of Energy,
Washington, DC,
USA

Bernard Kippelen
Center for Organic Photonics and Electronics,
School of Electrical and Computer Engineering,
Georgia Institute of Technology,
Atlanta, GA,
USA

Sarah Kurtz
National Center for Photovoltaics;
U.S. Department of Energy National Renewable Energy Laboratory,
Golden, CO,
USA

Alison Lennon
The University of New South Wales,
Sydney, Australia

Atse Louwen
Copernicus Institute,
Utrecht University,
The Netherlands

Sylvain Marsillac
Virginia Institute of Photovoltaics,
Old Dominion University,
Norfolk, VA,
USA

Barbara van Mierlo
Department of Social Sciences,
Knowledge, Technology and Innovation,
Wageningen University,
The Netherlands

Panos Moraitis
Copernicus Institute,
Utrecht University,
The Netherlands

Etienne Moulin
Ecole Polytechnique Fédérale de Lausanne,
Institute of Microengineering (IMT),
Photovoltaics and Thin-Film Electronics Laboratory (PV-lab),
Neuchâtel,
Switzerland

Negar Naghavi
Centre National de la Recherche Scientifique,
Institute of Research and Development on Photovoltaic Energy (IRDEP),
Chatou,
France

Albert van den Noort
PowerMatching City,
DNV GL,
Arnhem,
The Netherlands

Bernd Rech
Institute for Silicon Photovoltaics/
Helmholtz-Zentrum Berlin für Materialien und Energie GmbH,
Germany

Michiel Ritzen
Zuyd University of Applied Sciences,
Heerlen,
The Netherlands

Angus Rockett
University of Illinois,
Urbana, IL,
USA

Rutger Schlatmann
PVcomB/Helmholtz-Zentrum Berlin für Materialien und Energie GmbH,
Germany

Jan-Willem Schüttauf
Swiss Center for Electronics and Microtechnology (CSEM),
PV-Center,
Neuchâtel,
Switzerland

William Shafarman
Institute of Energy Conversion,
University of Delaware,
Newark, DE,
USA

Susanne Siebentritt
Laboratory for Photovoltaics,
University of Luxembourg,
Luxembourg

Henry J. Snaith
Clarendon Laboratory,
University of Oxford,
Oxford, UK

Joshua S. Stein
Sandia National Laboratories
Albuquerque, NM
USA

Samuel D. Stranks
Research Laboratory of Electronics,
Massachusetts Institute of Technology,
Cambridge, MA,
USA

Thorsten Trupke
ARC Centre of Excellence for Advanced Silicon Photovoltaics,
University of New South Wales,
Sydney,
Australia

Odysseas Tsafarakis
Copernicus Institute,
Utrecht University,
The Netherlands

Christopher E. Valdivia
SUNLAB University of Ottawa,
Canada

Matthias Vetter
Fraunhofer Institute for Solar Energy Systems ISE
Freiburg,
Germany

Zeger Vroon
Zuyd University of Applied Sciences,
Heerlen,
The Netherlands

Rob Walters
US Naval Research Lab,
Washington, DC,
USA

David Wilt
Air Force Research Laboratory,
Ohio,
USA

Woojun Yoon
U.S. Naval Research Laboratory,
Washington, DC,
USA

Ngwe Zin
Australian National University,
Canberra,
Australia

Foreword

Education on photovoltaic solar energy involves a broad range of disciplines, ranging from physics, electrical engineering, material science to design engineering. Moreover, since around 1995, the field of photovoltaics has rapidly been developing due to technological advances, significant cost reductions of solar cells, panels and inverters, and by more beneficial regulatory frameworks, such as feed-in tariffs.

As such, when I was teaching on the Master's course Solar Energy at the University of Twente in The Netherlands, in the framework of the national Master's programme Sustainable Energy Technology, it was quite difficult to find an appropriate, contemporary and affordable textbook that covered the "full spectrum" of knowledge about photovoltaic solar energy.

This situation was confirmed by my co-editors, Pierre Verlinden, Wilfried van Sark and Alex Freundlich, all of whom have vast experience of and long-term commitment to photovoltaics. Also at Utrecht University in The Netherlands, on the Master's course Solar Energy Physics, which is part of the Energy Science Master's programme, and at the University of Houston in the USA on the Electrical and Computer Engineering Program, and for training young engineers in the industry, a book covering the full spectrum of photovoltaic solar energy was lacking.

Therefore, in 2014, we took the initiative to compile and edit this book together with many photovoltaic specialists, our colleagues and good friends. The PV specialists who have written individual chapters of this book are currently active in R&D, covering PV material sciences, solar cell research and application. Our initiative was strongly supported by our publisher, John Wiley & Sons, and the photovoltaic community. You now find in your hands the result of these efforts put together in a book which provides fundamental and contemporary knowledge about various photovoltaic technologies in the framework of material science, device physics of solar cells, engineering of PV modules and the design aspects of photovoltaic applications. The aim of this book is to inform undergraduate and/or post-graduate students, young or experienced engineers, about the basic knowledge of each aspect of photovoltaic technologies and the applications thereof in the context of the most recent advances in science and engineering, and to provide insight into possible future developments in the field of photovoltaics.

If used as a university textbook, this book would be suitable for most universities with technical study programs offering Master's level and graduate courses in renewables, solar energy, photovoltaics, and PV systems.

We do not aim to compete with the typical specialists' books, which go deeper into a particular topic, but, instead, this book provides a broad scope of information, and engages an audience with a more diverse background in science and engineering than physics only.

We hope that students in different engineering disciplines will find the book easy to read and to use during their studies. We tried to keep the size of the book to an affordable level for students, while still maintaining a very broad scope and giving numerous references of recent publications to allow students to go deeper into the selected topics. We aimed to bring the relevant information about all existing photovoltaic technologies, including crystalline silicon devices, organic PV, chalcogenide solar cells, PV modules, concentrator PV, space PV technologies, characterization, modeling, and reliability testing, together in one comprehensive book of a reasonable length.

We hope that, with its structured set-up, this book will reach out to interested students who have already completed a bachelor's degree in exact or natural sciences, or in an engineering study, but also to PV specialists who would like to gain a better understanding of the topic other than their own field in photovoltaics. As such, we wish to reach a wide audience and engage more people in the growing field of photovoltaic technologies.

Angèle Reinders
University of Twente
The Netherlands
June 2016

Acknowledgments

Over the past two years we, the editors of *Photovoltaic Solar Energy*, have had the great pleasure of working with many colleagues to bring this book to life.

Our sincere thanks goes therefore to all contributing authors and respected photovoltaic specialists, such as Seth Hubbard, Andreas Fell, Andrew Blakers, Ngwe Zin, Bram Hoex, Martin Hermle, Zachary Holman, Mathieu Boccard, Pietro Altermatt, Martin Green, Susanne Siebentritt, Sylvain Marsillac, Negar Naghavi, Bill Shafarman, Etienne Moulin, Jan-Willem Schüttauf, Christophe Ballif, Onno Gabriel, Daniel Amkreutz, Jan Haschke, Bernd Rech, Rutger Schlatmann, Franz Haug, Bernard Kippelen, Woojun Yoon, Sam Stranks, Henry Snaith, Veronique Gevaerts, Halden Field, Thorsten Trupke, Henner Kampwerth, Angus Rockett, Geoffrey Kinsey, Frank Dimroth, Simon Fafard, Karin Hinzer, Christopher Valdivia, John Cook, Michael Debije, Rob Walters, Phillip Jenkins, Dave Wilt, Alison Lennon, Rhett Evans, Michael Kempe, Sarah Kurtz, Greg Ball, Chris Deline, Matthias Vetter, Georg Bopp, Atse Louwen, Odysseas Tsafarakis, Panos Moraitis, Joshua Stein, Michiel Ritzen, Zeger Vroon, Chris Geurts, Georgia Apostolou, Albert van den Noort, Gianluca Fulli, Flavia Gangale, Matthew Campbell, Barbara van Mierlo, and Vasilis Fthenakis.

We would also like to thank the Wiley team, Peter Mitchell, Ella Mitchell, Liz Wingett, Shruthe Mothi, Yamuna Jayaraman and Shivana Raj, for their professional support in this endeavor.

Thank you to all the solar deities who through the centuries represented the beneficial powers of the Sun and who give us humans free solar power every day. Thank you, Surya, Ra, Sol, Aryaman, Helios, Doumo, Magec, Tonatiuh.

Our hope continues to be that mankind finally realizes that solar power is powering our precious one and only Earth.

About the Companion Website

This book is accompanied by a companion website:

www.wiley.com/go/reinders/photovoltaic_solar_energy

The website includes:

- Figures
- Tables

Part One

Introduction to Photovoltaics

1.1

Introduction

Angèle Reinders[1], Wilfried van Sark[2], and Pierre Verlinden[3]
[1] *University of Twente, Enschede, The Netherlands*
[2] *Copernicus Institute, Utrecht University, The Netherlands*
[3] *Trina Solar, Changzhou, Jiangsu, China*

1.1.1 Introduction to Photovoltaic Solar Energy

At present, photovoltaic (PV) systems have become an established part of the electrical energy mix in Europe, the United States, Japan, China, Australia and many more countries all around the globe. So far, no single other energy technology has shown such a distributed set-up and modularity as PV systems. Stand-alone and grid-connected applications provide power in an extended range, from tenths of watts up to hundreds of megawatts. At the end of 2015, the total global cumulative capacity of installed PV systems exceeded 227 gigawatts, and this capacity is equivalent to about 280 coal-fired plants. For instance, in 2014, in Germany, Italy and Greece, 6 to 11% of the annual electricity generated originated from PV systems, while across Europe PV systems account for 3.5% of the electricity need (IEA-PVPS, 2015). According to the IEA (2016), electricity generated by PV systems contributed 0.8% of the total electricity production in the USA, in Japan 3.9%, and in China 1.0%.

Originally the development of PV technology was driven by the need for reliable and durable electricity systems for space applications, such as satellites. Nowadays, the implementation of PV systems in our society is driven by the need to reduce CO_2 emissions. Since PV systems have an extremely low CO_2 emission per kWh of electricity generated, namely below 30 g/kWh, see Figure 1.1.1 and Louwen *et al.* (2015), they are considered by policy-makers an important technology to slow down global warming due to the increased greenhouse effect (IPCC, 2013). This should be compared with the amount of more than 800 g CO_2/kWh emitted by coal-fired plants, see Figure 1.1.1. At present, all the major economies have policy targets to reduce greenhouse gas emissions, for instance, a 20% reduction of CO_2 emission is set by 2020 for Europe, 40% by 2030 and a 80–95% cut in greenhouse gases by 2050 compared to 1990

Photovoltaic Solar Energy: From Fundamentals to Applications, First Edition.
Edited by Angèle Reinders, Pierre Verlinden, Wilfried van Sark, and Alexandre Freundlich.

Companion website: www.wiley.com/go/reinders/photovoltaic_solar_energy

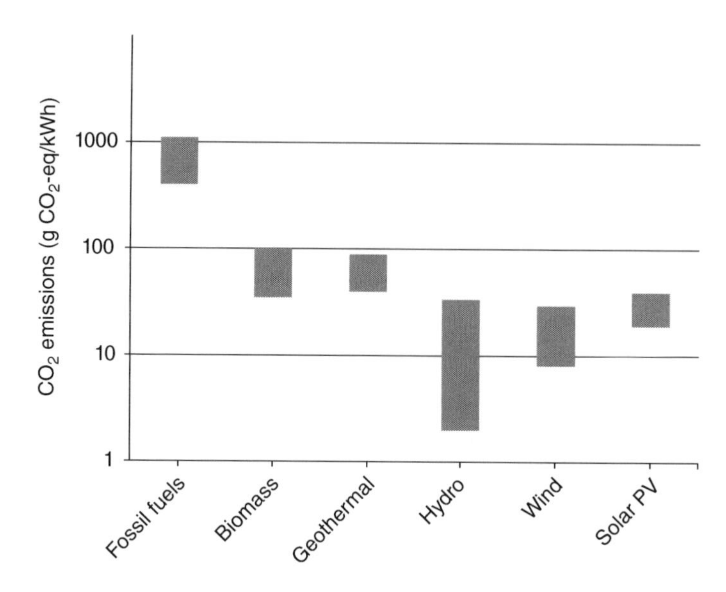

Figure 1.1.1 Comparison of CO_2-equivalent emissions of various energy technologies. Note the logarithmic *y*-scale. Courtesy of A. Reinders, University of Twente

levels (European Commission, 2011). To achieve these targets solar photovoltaic technologies will be unavoidable, as well as other sustainable energy technologies in combination with an increased energy efficiency of society. Therefore, we can expect further growth of the volume of PV systems in our electrical energy mix and in off-grid applications. According to Greenpeace's (2015) updated *Energy Revolution* scenario, we can even expect a 100% sustainable energy supply, which will end global CO_2 emissions, by 2050, which will be achieved by 20% of our electricity demand being produced by PV systems.

The growth of the market for PV technologies brings economies of scales. In the past decade, economies of scales together with technological progress in solar cell efficiencies, standardization of technologies, improved manufacturing and lower costs of production of feedstock materials, such as silicon, have brought down the cost of, for example, silicon PV modules from $4/Wattpeak to less than $1/Wattpeak (see Chapter 13.1). Present record efficiencies of 21% for commercial PV modules will reduce these costs even more in the forthcoming years. Due to these low investment costs and low O&M costs, the price of PV electricity is able to compete with consumer electricity prices in many countries, thus realizing grid parity on the customer side of the meter (Hurtado Muñoz *et al.*, 2014). Though incentives still remain necessary to overcome the hurdle of upfront investment costs at present, in the long run it seems feasible that PV technology will become an affordable self-sustained energy technology within the reach of many consumers, in particular in urbanized areas where the technology's silent operation with zero emissions during use will perfectly fit into a built environment. Additionally, a market for PV systems exists in developing countries, where more than one billion people do not have access to electricity. There, solar electricity can make a difference by being a means to access information, for instance, with the internet on laptops or through television, to communicate, by internet and by phone, to pump water, to cool food and medication, and to generate income.

All the rapid developments in the field, described above, have happened in the past five years. The global PV market is still growing rapidly at about 20–40% per annum, as of 2014. Most PV technologies are still relatively young and going through various innovations. Triggered by serious interest from "newcomers" in the field of solar energy, the PV scientific community faces a great educational challenge to change the educational message from a "physicist-only" audience to an audience with more diverse backgrounds in engineering, economics, and business. To keep the solar PV field growing and to prepare a successful future for photovoltaics, it is our sincere ambition to address a large part of these interested audiences and spread knowledge, enthusiasm, and inspiration about photovoltaic energy.

1.1.2 Properties of Irradiance

Readers of this book will often come across photons and other matters related to irradiance. Therefore, to make sure that we all have the same starting point, in this section we will briefly address a few basic topics related to irradiance.

1.1.2.1 Photons

German physicists Max Planck and Albert Einstein proposed in 1900 and 1905, respectively, that light, or more correctly, irradiance is composed of discrete particles. However, both Max Planck and Albert Einstein never used a specific term for these particles. It took some time, until the late 1920s, until the word "photon" became a synonym for the light quantum. The name photon is derived from the Greek word *φώτο* (= photo), which means light. The suffix "-on" at the end of the word indicates that the photon is an elementary particle belonging to the same class as the proton, the electron, and the neutron.

The energy contained by a photon, E, is given by Equation (1.1.1):

$$E = \frac{hc}{\lambda} \tag{1.1.1}$$

where λ is the wavelength (in m), h is Planck's constant ($6.626 \cdot 10^{-34}$ J · s) and c is the speed of light in vacuum ($2.998 \cdot 10^{8}$ m/s). The energy contained in a photon is rather small and can therefore be better expressed by the unit electron-volt (indicated by eV) than the more common SI unit for energy, which is the joule (J). An electron-volt is the energy gained by one electron when accelerated through 1 volt of electric potential difference. Since the electric charge of one electron is given by the elementary charge, $q = 1.602\ 10^{-19}$ C, in this situation 1 eV is equivalent to $1.602 \cdot 10^{-19}$ J. In practice, this results in the following simplified relationship for daily use: E (in eV) = $1.24/\lambda$ (in μm). From Equation (1.1.1), it follows that photons with a short wavelength, such as ultra-violet and blue light, have a high energy and those with a long wavelength, such as red and infrared light, have a low energy. Figure 1.1.2 shows the colors of light, visible to human eye, with the corresponding wavelengths (nm) and energy (eV).

Since the frequency ν (in Hz), wavelength λ, and speed of light c, are related by $\lambda \cdot \nu = c$, Equation (1.1.1), also called the Planck-Einstein relation, can be represented by:

$$E = h \cdot \nu \tag{1.1.2}$$

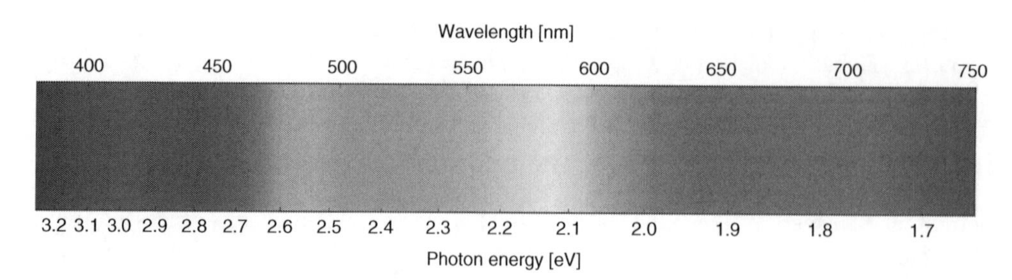

Figure 1.1.2 Relation between color of (visible) light, wavelength and its energy. Courtesy of A. Louwen, Utrecht University. (*See insert for color representation of the figure*)

In 1905, Einstein concluded that "the wave theory of light (Maxwell, 1865) has worked well in the representation of purely optical phenomena," however,

> observations associated with blackbody radiation, fluorescence, the production of cathode rays by ultraviolet light, and other related phenomena connected with the emission or transformation of light are more readily understood if one assumes that the *energy of light is discontinuously distributed in space*. In accordance with the assumption to be considered here, the energy of a light ray spreading out from a point source is not continuously distributed over an increasing space but *consists of a finite number of energy quanta* which are *localized at points in space*, which move without dividing, and *which can only be produced and absorbed as complete units*. (Einstein, 1905)

These assumptions are essential for the understanding of the photovoltaic effect and therefore the functioning of most of the photovoltaic materials applied nowadays in photovoltaic solar cells. In this sense, the classic wave theory of light in which light is represented as a form of electromagnetic radiation did not appear to be sufficient for an understanding of certain phenomena which are relevant to this field of research, though the wave theory is indispensable to the explanation and quantification of transmission, reflection and refraction of irradiance in various media.

Albert Einstein used the principle of discrete energy packages, light quanta, to explain the photoelectric effect in metal surfaces. The photovoltaic effect is the creation of voltage or electric current in a material upon exposure to irradiance. It was observed for the first time in 1839 by the French physicist Alexandre Edmond Becquerel, in a solution that contained two metal electrodes. In due course, despite the widespread occurrence of the photovoltaic effect in liquid and solid state systems, the term became predominantly associated with the functioning of layers of semiconductor materials applied in photovoltaic (PV) devices such as PV solar cells.

Einstein's theories on the photovoltaic effect in metal surfaces were experimentally validated by the American physicist Robert Millikan, who was able to confirm that the energy contained by photons varied according to the frequency – or the reciprocal of the wavelength – of the incident irradiance, and not according to the intensity of the irradiance. For these significant discoveries related to the properties of irradiance, Max Planck received the Nobel Prize in 1918, Albert Einstein received it in 1921 for his discovery of the law of the photoelectric effect, and later Robert Millikan also received it in 1923 for his work on the elementary charge of electricity and on the photoelectric effect. These discoveries were essential for the development of PV solar cells, and in the early 1950s, Daryl Chapin, Calvin Fuller, and Gerald Pearson of Bell Laboratories observed the PV effect in doped silicon layers, by chance. In 1954, they presented the first silicon solar cell, which was reported to have a conversion efficiency of 6% (Chapin *et al.*, 1954). This is considered the start of the modern PV era that has led to a multi-billion market today.

1.1.2.2 Solar Irradiance

Solar irradiance is the name for the spectrum of light originating from the sun. For the purpose of photovoltaic applications, we can distinguish between extraterrestrial solar irradiance, which is available in space, and terrestrial solar irradiance, which is received on Earth. Solar irradiance in space is usually measured at the interface of space and the Earth's atmosphere, for the purpose of space applications of photovoltaics, such as satellites, and is standardized by the so-called *AM0* spectrum, which is shown as the ASTM E-490 spectrum in Figure 1.1.3. *AM* stands for Air Mass. The indicator (0 in *AM0*) is the air mass coefficient, which is the ratio of the direct optical path length of solar irradiance through the Earth's atmosphere, L_θ, relative to the path length with the zenith as an origin, called the zenith path length, L. The air mass coefficient at average sea level can easily be calculated using Equation (1.1.3):

$$AM = \frac{L}{L_\theta} \sim \frac{1}{\cos\theta} \tag{1.1.3}$$

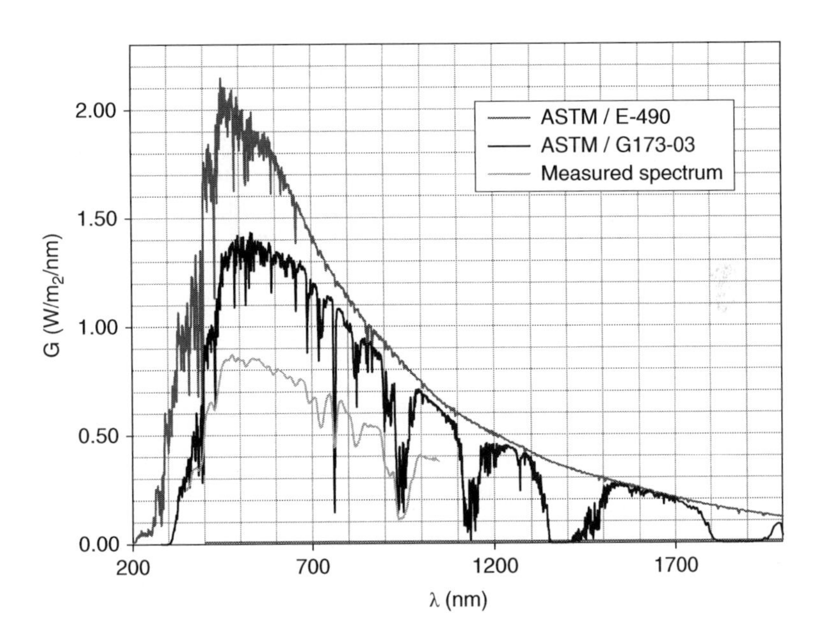

Figure 1.1.3 Solar spectra: ASTM E-490 representing AM0 (black line), ASTM G173-03 representing AM1.5 (red line), and a measured spectrum (green line) showing the differences that can occur in reality. Data from ASTM and University of Twente, The Netherlands. Courtesy of A. Reinders, University of Twente. (*See insert for color representation of the figure*)

where θ is the zenith angle of incidence of irradiance. Equation (1.1.3) provides a good approximation for the air mass coefficient for θ up to around 75°, that is, for air mass up to around 4.

In 2000, the American Society for Testing and Materials (ASTM) developed an *AM0* reference spectrum (ASTM E-490, 2000) for use by the aerospace community, see Figure 1.1.3. The ASTM E-490 solar spectral irradiance is based on data from satellites, space shuttle missions, high-altitude aircraft, rocket soundings, ground-based solar telescopes, and modeled spectral irradiance (NREL, 2015) with an integrated spectral irradiance of 1366.1 W/m^2 representing the value of the solar constant as accepted by the space community.

The photovoltaic industry developed another spectrum for terrestrial applications, which is represented by the standard ASTM G-173-03 containing specific *AM1.5* spectra, and it is referred to as "Standard Tables for Reference Solar Spectral Irradiance at Air Mass 1.5: Direct Normal and Hemispherical for a 37 Degree Tilted Surface" (ASTM, 2012), see Figure 1.1.3. This is a revised version of the earlier single standard, which was available from 1999. The ASTM G-173 spectra represent terrestrial solar spectral irradiance on a surface of specified orientation under one and only one set of specified atmospheric conditions with an air mass of 1.5 (solar zenith angle 48.19°). The receiving surface is defined in the standards as an inclined plane at 37° tilt toward the equator, facing the sun, that is, tilted South on the Northern hemisphere, and tilted North on the Southern hemisphere. The specified atmospheric conditions are defined by a specific temperature, pressure, aerosol density, air density and molecular species density specified in 33 layers (Gueymard, 2004).

Apart from the standardized extraterrestrial and terrestrial solar irradiance spectra a measured solar irradiance spectrum is presented in Figure 1.1.3, to show the difference between reality and the standards. Given the discourse about the effect of the optical path length of solar irradiance through the atmosphere on the solar irradiance spectra received on an earthly surface, it is logical that solar irradiance strongly depends on the moment of the day and the day of the year and hence the related sun positions. Moreover, the irradiance depends on the composition of the atmosphere and, as such, weather, including the cloud formation and the precipitation, the particles and water vapor in the atmosphere, and the gasses that are contained by the atmosphere. These factors all contribute to the absorption and reflectance of irradiance at various wavelengths. For more details about the properties and behavior of solar irradiance in the earthly atmosphere and on terrestrial receiving surfaces, we refer to Sengupta *et al.* (2015).

1.1.2.3 Refraction, Reflection and Transmission

As already mentioned in Section 1.2.1, the wave theory of light (Maxwell, 1865) works very well for the representation of purely optical phenomena. This is a good basis for non-quantum phenomena in photovoltaics, in particular, since the propagation of light waves in matter is an important topic in the field of photovoltaic solar energy. Therefore, in this section, we will provide information about the most important laws and equations regarding the propagation of light in matter.

The speed of light in a vacuum, c, is a physical constant which is exactly 299.792.458 m/s. However if light propagates through a medium, which could be, for instance, air, glass or silicon, its speed is reduced to v, the phase velocity of light in media, where v is lower than c. The refractive index of a material, n, is an indicator of the speed of light in media, since it is equal to the ratio of the speed of light in vacuum, c, and the phase velocity in the medium, v. Logically, in media n is always larger than 1.

The different propagation of light in two different isotropic media results in refraction at the interface of these media, according to Snell's law:

$$\frac{\sin\theta_i}{\sin\theta_t} = \frac{n_2}{n_1} \tag{1.1.4}$$

where θ_i is the angle of incidence (towards the normal of a surface), θ_t is the angle of refraction of the outgoing beam, n_1 is the refractive index of the medium through which the incident beam of light passes and n_2 is the refraction index of the medium through which the refracted beam goes, see Figure 1.1.4. From Snell's law it can be derived that when light propagates from a medium with a higher refraction index to a medium with a lower refractive index (hence, $n_1 > n_2$), total internal reflection occurs for incidence angles greater than the critical angle, θ_{crit}, leading to refraction into the medium with the higher refraction index

$$\theta_{crit} = arsin\frac{n_2}{n_1}$$

The law of refraction states that if light hits an interface between two media with a different refractive index, the angle of reflection. θ_r, is equal to the angle of incidence, θ_i.

Snell's law does not explain what share of the energy contained by the incident irradiance, I_i, is subject to transmission or reflection. However, the law of conservation of energy states that the sum of the transmission coefficient, T, and the reflection coefficient, R, should be 1: $T+R=1$ where $T=I_t/I_i$ and $R=I_r/I_i$ for non-magnetic materials.

The values of T and R depend on the polarization of irradiance. If the vector of the electric field happens to be perpendicularly oriented towards the plane that contains both the incident, refracted and reflected irradiance (this is called s-polarization), then the reflection coefficient is indicated by R_s. If it happens to be in line with the plane of incidence, the so-called p-polarization,

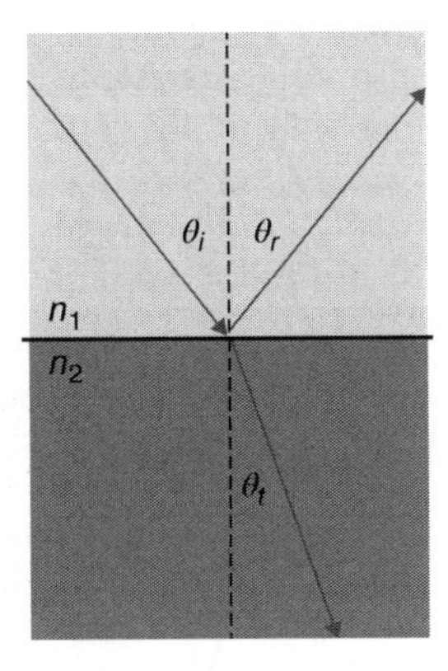

Figure 1.1.4 Scheme representing incident, refracted and reflected irradiance at the interface of two media

we can talk about R_p. The respective transmission coefficients are given by the law of conservation of energy, namely, $T_s + R_s = 1$ and $T_p + R_p = 1$.

R_s and R_p are given by the Fresnel equations for non-magnetic materials:

$$R_s = \left[\frac{\sin(\theta_t - \theta_i)}{\sin(\theta_t + \theta_i)}\right]^2 = \left[\frac{n_1\cos(\theta_i) - n_2\cos(\theta_t)}{n_1\cos(\theta_i) + n_2\cos(\theta_t)}\right]^2 = \left[\frac{n_1\cos(\theta_i) - n_2\sqrt{1-\left(\frac{n_1}{n_2}\sin\theta_i\right)^2}}{n_1\cos(\theta_i) + n_2\sqrt{1-\left(\frac{n_1}{n_2}\sin\theta_i\right)^2}}\right]^2 \tag{1.1.5}$$

$$R_p = \left[\frac{\tan(\theta_t - \theta_i)}{\tan(\theta_t + \theta_i)}\right]^2 = \left[\frac{n_1\cos(\theta_t) - n_2\cos(\theta_i)}{n_1\cos(\theta_t) + n_2\cos(\theta_i)}\right]^2 = \left[\frac{n_1\sqrt{1-\left(\frac{n_1}{n_2}\sin\theta_i\right)^2} - n_2\cos(\theta_i)}{n_1\sqrt{1-\left(\frac{n_1}{n_2}\sin\theta_i\right)^2} + n_2\cos(\theta_i)}\right]^2 \tag{1.1.6}$$

Equations (1.1.5) and (1.1.6) are frequently used to calculate the reflectance and transmittance at the interface between two materials of different refractive indexes at various angles of incidence.

1.1.3 Structure of the Book

After this short introduction to basic physics related to irradiance, Part 2 presents the basic functional principles of photovoltaics, including an introduction to semiconductor materials and several topics related to solar cell device physics in general. Next, in Part 3, design aspects and the actual functioning of crystalline silicon solar cells, the most dominant solar technology in the current PV market, are presented. Subsequent parts will introduce the reader to other material systems for photovoltaic devices, that is, chalcogenide thin film solar cells in Part 4, thin film silicon-based PV technologies in Part 5, organic photovoltaic cells in Part 6 and III-V solar cells, including their concentrator applications, in Part 8. Since photovoltaic research is based on appropriate diagnostics, one Part (Part 7) will be fully devoted to characterization and measurement methods for materials contained by PV cells and PV modules. The second half of the book will focus more on applications of photovoltaic technologies. To start with, Part 9 will explore PV technologies that are applied in space. Next, Part 10 will present PV modules and their manufacturing processes. In Part 11, PV technologies applied in systems, buildings, and various products are placed in the spotlight. The use and issues of PV power production in interaction with conventional electricity grids are discussed in Part 12. Finally, Part 13 will complete this book, covering financial aspects, user aspects, and standards for the PV sector.

List of Symbols

Symbol	Unit	Meaning
E	eV	Energy of a photon
I	W/m^2	Energy contained by irradiance
L	m	Zenith path length of solar irradiance
L_θ	m	Optical path length of solar irradiance
λ	m	Wavelength
n	dimensionless	Refractive index
ν	Hz	Frequency
v	m/s	Phase velocity of light in media
R	dimensionless	Reflection coefficient
θ	$^\circ$	Angle
T	dimensionless	Transmission coefficient

Constants

Symbol	Value	Name
c	$2.998 \cdot 10^8$ m/s	Speed of light in a vacuum
h	$6.626 \cdot 10^{-34}$ J·s	Planck's constant
q	$1.602 \cdot 10^{-19}$ C	Elementary charge

List of Acronyms

Acronym	Meaning
AM	Air Mass
ASTM	American Society for Testing and Materials
CO_2	Carbon dioxide
O&M	Operation and Maintenance
PV	Photovoltaic
SI	Système International d'unités
STC	Standard test conditions

References

ASTM, American Society for Testing and Materials (2010) Standard extraterrestrial spectrum reference E-490-00. Available at: http://www.astm.org/Standards/E490

ASTM, American Society for Testing and Materials (2012) Standard tables for reference solar spectral irradiances: direct normal and hemispherical on 37° tilted surface. Available at: http://www.astm.org/Standards/G173.htm

Becquerel, E. (1839) Mémoire sur les effets électriques produits sous l'influence des rayons solaires. *Comptes Rendus de l'Académie des Sciences*, **9**, 561–567.

Chapin, D.M., Fuller, C.S., and Pearson, G.L. (1954) A new silicon p-n junction photocell for converting solar radiation into electrical power. *Journal of Applied Physics*, **25**, 676.

Chen, Y.F., Feng, Z.Q., and Verlinden, P. (2014) Assessment of module efficiency and manufacturing cost for industrial crystalline silicon and thin film technologies. In *Proceedings of the 6th World Conference on Photovoltaic Energy Conversion (WCPEC6)*. Kyoto, Japan, 2014.

Einstein, A. (1905) Über einen die Erzeugung und Verwandlung des Lichtes betreffenden heuristischen Gesichtspunkt. [On a heuristic viewpoint concerning the production and transformation of light]. *Annalen der Physik*, **322** (6), 132–148.

European Commission (2011) Energy Roadmap 2050. *Communication from the Commission to the European Parliament, the Council, the European Economic and Social Committee and the Committee of the Regions*, COM/2011/0885 final. EC, Brussels.

Greenpeace (2015) *Energy Revolution*, 5th edn, *2015 World Energy Scenario*, Greenpeace International, Global Wind Energy Council, Solar Power Europe (eds S. Teske, S. Sawyer, and O. Schäfer). Greenpeace, New York.

Gueymard, C. (2004) The sun's total and spectral irradiance for solar energy applications and solar radiation models. *Solar Energy*, **76**, 423–453.

Hurtado Muñoz, L.A., Huijben, J.C.C.M., Verhees, B., and Verbong, G.P.J. (2014) The power of grid parity: a discursive approach. *Technology Forecasting & Social Change*, **87**, 179–190.

IEA-PVPS (2016) Snapshot of global PV markets, 1992–2015, Report IEA PVPS T1-29:2016.

IPCC (2013) Summary for policymakers. In *Climate Change 2013: The Physical Science Basis. Contribution of Working Group I to the Fifth Assessment Report of the Intergovernmental Panel on Climate Change* (eds T.F. Stocker, D. Qin, G.-K. Plattner *et al.*). Cambridge University Press, Cambridge.

Louwen, A., Sark, W.G.J.H.M. van, Schropp, R.E.I., *et al.* (2015) Life cycle greenhouse gas emissions and energy payback time of current and prospective silicon heterojunction solar cell designs. *Progress in Photovoltaics*, **23**, 1406–1428.

Maxwell, J.C. (1865) A dynamical theory of the electromagnetic field. *Philosophical Transactions of the Royal Society of London*, **155**, 459–512.

NREL (2015) http://rredc.nrel.gov/solar/spectra/am0/

Planck, M. (1900) Distribution of energy in the normal spectrum. *Verhandlungen der Deutschen Physikalischen Gesellschaft*, **2**, 237–245.

Planck, M. (1901) Über die Elementarquanta der Materie und der Elektrizität. *Annalen der Physik*, **309** (3), 564–566.

Sengupta, M., Habte, A., Kurtz, S. *et al.* (2015) *Best Practices Handbook for the Collection and Use of Solar Resource Data for Solar Energy Applications*. Technical Report NREL/TP-5D00-63112, available at: http://www.nrel.gov/docs/fy15osti/63112.pdf

Part Two

Basic Functional Principles of Photovoltaics

2.1

Semiconductor Materials and their Properties

Angèle Reinders
University of Twente, Enschede, The Netherlands

2.1.1 Semiconductor Materials

Semiconductor materials are the basic materials which are used in photovoltaic (PV) devices. Though often semiconductors are not fully explained in text books about PV technologies (Archer and Hill, 2001; Chen, 2011), it would be helpful for most readers to have a basic knowledge in order to be able to understand important topics in photovoltaics such as doping, photogeneration, and carrier transport, which are discussed later in this book.

The aim of this chapter is to provide an introduction to solid-state physics and semiconductor properties that are relevant to photovoltaics without spending too much time on unnecessary information. Readers wishing a deeper understanding after reading this Part are referred to extensive handbooks such as those edited by Luque and Hegedus (2010) on PV science and engineering, the handbook by Sze and Lee (2012) on semiconductor devices, and even more fundamental works, such as those written by Kittel (2004) on solid-state physics.

A semiconductor material has a conductivity σ (in S/cm) that is used to distinguish conducting materials from insulating materials. This conductivity is particularly sensitive to temperature, impurities in the material, and light. Therefore, the conductivity and therefore also the resistivity, ρ (in Ω cm) can be tuned to the desired values by changing the amount of impurities, called donor and acceptor dopants.

Typical conducting materials, conductors, are the metals, among which are found silver and copper, with conductivity, σ, greater than 10^3 S/cm. Their conductivity can be modified only in a narrow range of values. Insulators have a conductivity, σ, smaller than 10^{-8} S/cm; an example is glass. Between these two extremes, semiconductors have a broad range of conductivity and therefore, depending on the condition, they can also behave as an insulator or a conductor.

Photovoltaic Solar Energy: From Fundamentals to Applications, First Edition.
Edited by Angèle Reinders, Pierre Verlinden, Wilfried van Sark, and Alexandre Freundlich.

Companion website: www.wiley.com/go/reinders/photovoltaic_solar_energy

Table 2.1.1 Part of the Periodic Table of Elements related to semiconductors used in photovoltaic devices

Period	Column II	Column III	Column IV	Column V	Column VI
2		B	C	N	O
3	Mg	Al	Si	P	S
4	Zn	Ga	Ge	As	Se
5	Cd	In	Sn	Sb	Te
6	Hg		Pb		

Since this change in conductivity is controllable, by voltage, electric field, injection, doping, light, temperature, etc., these semiconductor materials can be used in electronic applications.

There are two groups of semiconductors: elemental materials and compound materials. Examples of elemental materials are germanium (Ge) and silicon (Si), also called Group IV materials because they can be found in column IV of the Periodic Table of Elements. Typical compound materials are gallium arsenide (GaAs), which is a III-V compound because it is a combination of gallium from column III and arsenic from column V in the Periodic Table of Elements. Compound materials can also be made from a combination of column II and column VI elements, for instance, cadmium telluride (CdTe), which is then called a II-VI compound made from cadmium and tellurium. In Table 2.1.1, a section of the Periodic Table of Elements is shown, with relevance to a large share of the established semiconductors which are used for photovoltaic devices as presented in this book. In recent years, new semiconductor compounds have been found and applied in research for photovoltaic application. For example, recent progress in the field of perovskite materials is presented in Chapter 6.3 of this book. The inverse design approach (Zakutayev *et al.*, 2013) to semiconductor materials is a generic quest to discover new materials which theoretically could be created for use in different applications, including photovoltaic technology.

2.1.2 Crystalline Structures of Semiconductors

Usually atoms in the group of semiconductor materials form crystalline structures in solid-state conditions. Ideally these structures consist of an infinite repetition of identical groups of atoms. The set of points to which groups of atoms are attached is called the lattice (Kittel, 2004); the group of atoms is called the basis, or the unit cell (Sze and Lee, 2012), that is, by repetition of the unit cell, the entire lattice can be formed.

The bonds between atoms in a semiconductor crystal are single covalent bonds. Covalent bonding is the sharing of valence electrons which are the electrons in the outer electron shell (orbit) of the atom. Due to the fact that, for instance, silicon has four valence electrons in its outer orbit – because it is a Group IV element, see Table 2.1.1 – each silicon atom in a crystalline structure is bound to four neighbouring silicon atoms at an equal distance to the central atom. This tetrahedron organization of the atoms forms the unit cell in the silicon crystal in a diamond lattice structure, see Figure 2.1.1. The same is true for germanium, another Group IV element. The number of valence electrons for a neutral atom is equal to the main group number, Therefore, III-V compound conductors have another lattice than silicon: they have a zincblende lattice. Another crystalline structure is the cubic crystal. In particular, Part 4 of this book will discuss typical crystalline structures of chalcogenide materials, such as $Cu(InGa)Se_2$ and CdTe.

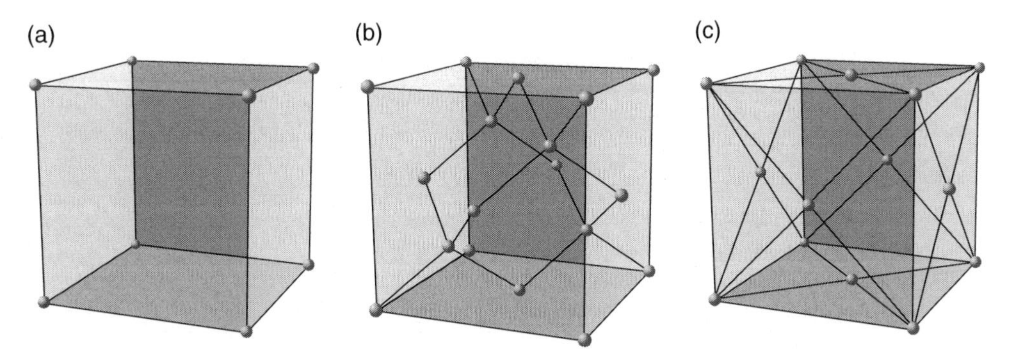

Figure 2.1.1 (a) Simple cubic lattice; (b) Diamond lattice with tetrahedron bonds; (c) Face centered cubic lattice. Source: Courtesy of Boudewijn Elsinga (2015)

2.1.3 Energy Bands in Semiconductors

When individual atoms are brought together in a crystalline lattice, energy bands are formed in a semiconductor material (Kittel, 2004; Sze and Lee, 2012). These energy bands indicate the allowed states of energy of electrons under respective bound conditions and conductive conditions according to the following mechanisms.

At zero Kelvin, electrons in a semiconductor are all bound to their atoms in their lowest energy states; they stay in what is called the valence band. By gaining a discreet amount of energy, an electron can move from the top of the valence band E_v to the bottom of the conduction band E_c where it is free to move to other empty states under the forces of an electric field or diffusion. The discreet amount of energy to move an electron from the valence band to the conduction band must therefore be higher than (E_C-E_V). This energy difference is called the bandgap energy E_g, indicated in Figure 2.1.2. The bandgap energy depends among other things on the composition of the material and the temperature. For semiconductor materials used in photovoltaic devices, E_g is in the range of ~1.1 eV for silicon up to values of ~2.5 eV for III-V compounds.

When an electron leaves the valence band, a 'void' remains in the crystal, which is called a hole. This hole is in principle a deficient electron, for which reason it can be considered an electron with a positive charge. The creation of what are called electron-hole pairs in pure semiconductor materials – also called intrinsic semiconductors – happens at temperatures above zero Kelvin. This process is called thermal excitation.

The density of electrons and holes in intrinsic semiconductors is mainly determined by thermal excitation and therefore by temperature. The density of electrons and holes is equal and is called the intrinsic carrier concentration. For instance, the value of the intrinsic carrier concentration, n_i (in cm^{-3}), in silicon as a function of temperature T (in Kelvin) is given by Equation (2.1.1) (Misiakos and Tsamakis, 1993):

$$n_i(T) = 5.29 \times 10^{19} \times (T/300)^{2.54x} e^{(-6726/T)} \tag{2.1.1}$$

The energy band representation of materials can visually explain the difference between insulators, semiconductors and conductors, see Figure 2.1.3. In Figure 2.1.3, it is shown that in, for instance, a conducting material, the conduction and valence band can overlap so that a bandgap does not exist (another option is that the conduction band is partially

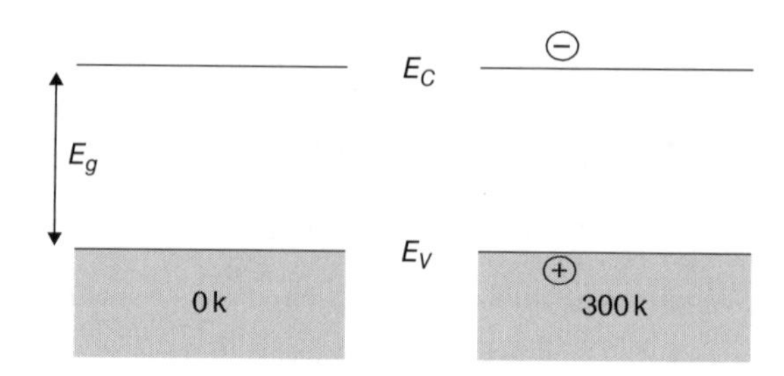

Figure 2.1.2 Energy band diagram at zero Kelvin and at 300 Kelvin

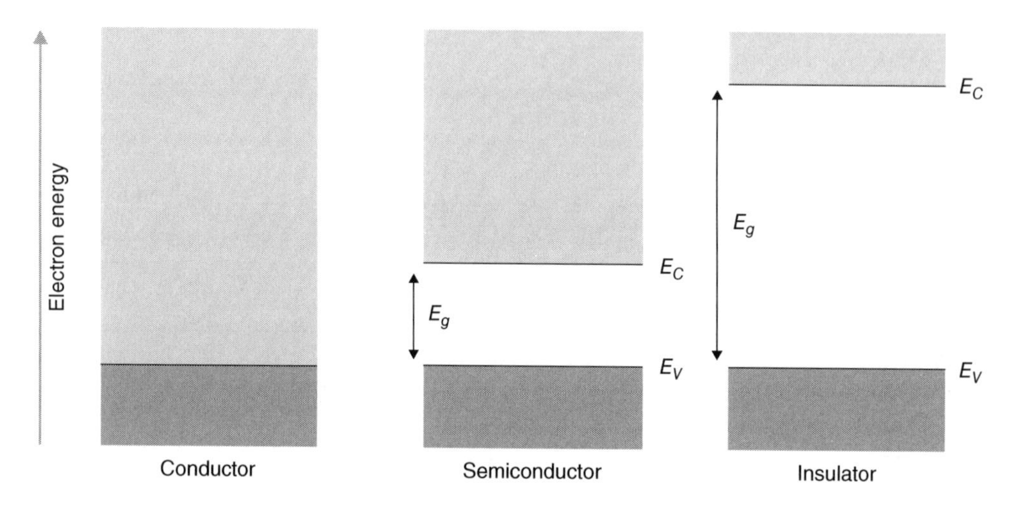

Figure 2.1.3 Energy band diagrams for a conductor, a semiconductor and an insulator

filled, however, this is not visually represented here). Therefore, electrons can freely move even at low temperatures, leading to good conductive properties of the material. In insulators, the valence electrons are very tightly bound to the atoms. Therefore, the thermal energy required to move a valence electron to the conduction band is too high to be achievable. Though a conduction band does exist in insulators, the probability that an electron will surpass the bandgap is therefore too low for good conductive properties. Hence, an insulator has a very high resistivity. On the other hand, a semiconductor's bandgap energy is low enough to achieve conductivity under specific circumstances. Chapter 2.3 will explain this further.

Besides energy band diagrams, other schemes such as energy-momentum diagrams are used to explain the optical-electrical functioning of semiconductor materials, see also Chapter 2.3. In principle, an energy-momentum diagram represents the energy of charge carriers – electrons or holes – E, in relation to their momentum, p, using Equation (2.1.2) for electrons:

$$E = p^2 / 2m_n \tag{2.1.2}$$

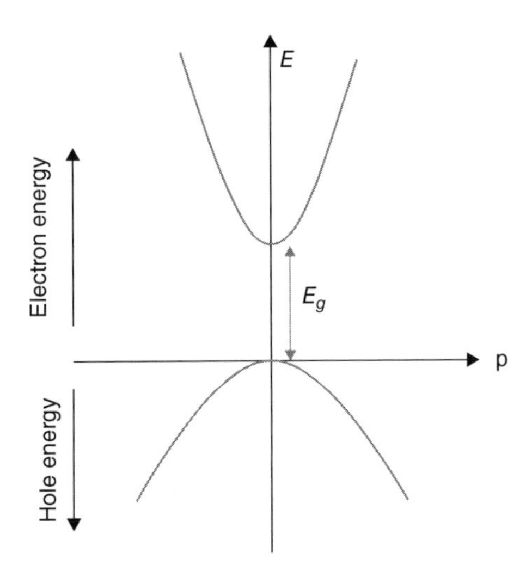

Figure 2.1.4 Energy-momentum diagrams for a semiconductor

where m_n (in g) is the effective mass of an electron and a similar equation for holes where m_n is replaced by m_p the effective mass of an hole. The effective mass depends on the semiconductor material and therefore also on the crystalline structure. Equation (2.1.2) can also be represented by the energy-momentum diagram shown in Figure 2.1.4.

List of Symbols

Symbol	Description	Unit
E	Energy	eV or 1.609×10^{-19} J
E_C	Conduction band	eV
E_g	Band gap energy	eV
E_v	Valence band	eV
m_n	Effective mass of an electron	g
m_p	Effective mass of a hole	g
n_i	Intrinsic carrier concentration	cm^{-3}
p	Momentum	g.m/s
ρ	Resistivity	Ω.cm
T	Temperature	K
σ	Conductivity	S/cm

List of Acronyms

PV	Photovoltaic or Photovoltaics

References

Archer, M.D. and Hill, R. (eds) (2001) *Clean Electricity from Photovoltaics*, Series on Photoconversion of Solar Energy: vol. **1**. Imperial College Press, London.

Chen, C.J. (2011) *Physics of Solar Energy*. John Wiley & Sons, Inc., Hoboken, NJ.

Kittel, C. (2004) *Introduction to Solid State Physics*, 8th edition. John Wiley & Sons, Ltd, Chichester.

Luque, A. and Hegedus, S. (eds) (2010) *Handbook of Photovoltaic Science and Engineering*, 2nd edition. John Wiley & Sons, Ltd, Chichester.

Misiakos, K. and Tsamakis, D. (1993) Accurate measurements of the silicon intrinsic carrier density from 78 to 340 K. *Journal of Applied Physics*, **74** (5), 3293–3297.

Sze, S.M. and Lee, M.-K. (2012) *Semiconductor Devices: Physics and Technology, International Student Version*, 3rd edition. John Wiley & Sons, Inc., New York.

Zakutayev, A., Zhang, X., Nagaraja, A. *et al.* (2013) Theoretical prediction and experimental realization of new stable V-IX-IV semiconductors using the Inverse Design approach. *Journal of the American Chemical Society*, **135** (27), 10048–10054.

2.2

Doping, Diffusion, and Defects in Solar Cells

Pierre Verlinden
Trina Solar, Changzhou, Jiangsu, China

2.2.1 Introduction

Presenting the entire range of techniques used to produce semiconductor substrates, doping and diffusion for photovoltaic (PV) application is beyond the scope of this short chapter. However, before discussing in Chapter 2.3 to 2.6 the physics of solar cells, it is important to introduce the technologies of substrate formation, doping, and diffusion for the most common PV technology, namely, crystalline silicon. In 1954, D. Chapin, C. Fuller and G. Pearson at Bell Laboratories used silicon as semiconductor material to make the first practical solid-state solar cell for power generation. Since then, the PV industry has grown, based not only on the body of knowledge and silicon technologies developed by the semiconductor industry (Beadle *et al.*, 1985), but also benefiting from low-cost silicon feedstock supplied as reject by this "parent" industry. In 2015, crystalline silicon PV still represents more than 90% of the entire global PV market. Silicon is incontestably the most studied element in the Periodic Table and, after oxygen, silicon is the second most abundant element (about 28.2% by weight) present in the Earth's crust. It is clear that the dominance of crystalline silicon in the PV industry is the result of a well-established silicon-based integrated circuit (IC) industry with its supply chain and manufacturing tools. The PV scientific community also developed its own body of knowledge, since its preoccupations differed significantly from the preoccupations of the microelectronics scientific community. Simply speaking, the goal of the IC industry is to make the fastest circuit with the largest amount of transistors, while the goal of the PV industry is to make the most efficient light-to-electricity converter at the lowest cost. Therefore, the science of IC manufacturing focuses on reducing the size of the transistor and reducing the number of particle-generated defects, whereas the science of PV focuses on increasing the size of the solar cell and improving the effective lifetime of minority carriers.

Photovoltaic Solar Energy: From Fundamentals to Applications, First Edition.
Edited by Angèle Reinders, Pierre Verlinden, Wilfried van Sark, and Alexandre Freundlich.

Companion website: www.wiley.com/go/reinders/photovoltaic_solar_energy

It is interesting to note that the PV industry grew much faster than its "parent." By 2002, the use of silicon for PV applications was already greater in surface area than the use of silicon for microelectronics. For the first time in 2004, PV used more silicon feedstock by weight than the semiconductor industry.

2.2.2 Silicon Wafer Fabrication

The fabrication process for crystalline silicon substrates involves five important steps:

1. Reduction of sand to obtain metallurgical-grade silicon.
2. Purification.
3. Production of purified silicon feedstock.
4. Growth of silicon ingots, either mono-crystalline or multi-crystalline.
5. Slicing of silicon ingots into wafers.

2.2.2.1 Metallurgical-grade Silicon

The first step to produce solar-grade silicon substrates is to produce metallurgical-grade silicon (MGS). This is done in a high temperature arc-furnace with submerged electrodes made of carbon. The furnace load is composed of high-purity silica (SiO_2) or quartzite mixed with charcoal, wood chips, and coal. At a temperature above 1900 °C, the carbon reduces the silica, as described in the chemical reaction below, to form silicon, which is liquid at this temperature, and is drained at the bottom of the furnace:

$$SiO_2 + C \xrightarrow{1900\,°C} Si(\text{liquid}) + SiO(\text{gas}) + CO(\text{gas}) \tag{2.2.1}$$

This process requires a large amount of energy, about 13kWh/kg and the purity of silicon is typically around 98%. Other chemical reactions are also present. For example, if there is too much carbon in the furnace, silicon carbide can also be formed, as described in the chemical reaction in Equation (2.2.2):

$$SiO + 2C \xrightarrow{1900\,°C} SiC(\text{solid}) + CO(\text{gas}) \tag{2.2.2}$$

The risk of forming silicon carbide is reduced if the concentration of SiO_2 is kept at a high level:

$$2SiC + SiO_2 \xrightarrow{1900\,°C} 3Si(\text{liquid}) + 2CO(\text{gas}) \tag{2.2.3}$$

2.2.2.2 Purification of Silicon

The second step in producing silicon substrate consists in purifying the MGS material. The most common technique uses a distillation process of trichlorosilane ($SiHCl_3$), but other methods using silicon tetrachloride ($SiCl_4$) or silane (SiH_4) are also applied in practice. Starting from MGS, the trichlorosilane is formed according to the chemical reaction:

$$\mathrm{Si} + 3\mathrm{HCl} \xrightarrow{300\,°\mathrm{C}} \mathrm{SiHCl_3} + \mathrm{H_2} + \mathrm{heat} \tag{2.2.4}$$

Trichlorosilane, which is liquid at room temperature, is purified by fractional distillation.

2.2.2.3 Silicon Feedstock

The production of solar-grade silicon (SoG-Si) consists of reducing the purified trichlorosilane with hydrogen in a chemical vapor deposition (CVD) process at around 1150 °C, the silicon depositing on pure silicon rods is electrically heated, serving as a nucleation surface for the deposition process according to the reaction:

$$2\mathrm{SiHCl_3} + 3\mathrm{H_2} \xrightarrow{1150\,°\mathrm{C}} 2\mathrm{Si} + 6\mathrm{HCl} \tag{2.2.5}$$

The process takes several hours and produces long (several meters in length) rods of purified silicon (in a polysilicon form) up to 20 cm in diameter. The silicon rods are then broken into pieces of useable size for growing ingots. The anhydrous Hydrogen Chloride (HCl) by-product of the reaction in Equation (2.2.5) can be recycled and reused for the production of trichlorosilane in the reaction in Equation (2.2.4). This process, known as the Siemens process, is the best-known technique for producing silicon feedstock for PV application (Figure 2.2.1). Table 2.2.1 compares the typical impurity level in metallurgical-grade silicon, solar-grade silicon, and electronic-grade silicon (EGS) feedstock.

2.2.3 Ingot Formation

2.2.3.1 Mono-crystalline Ingots

Silicon has a diamond-lattice crystal structure, which can be viewed as two interpenetrating face-centered cubic (fcc) lattices, see also Section 2.1.2 in Chapter 2.1. Mono-crystalline silicon ingots for PV applications are grown by the Czochralski (Cz) process in a crystal growth apparatus, called a Cz puller. The Cz method consists of dipping a seed of mono-crystalline silicon of known orientation into molten silicon, then slowly pulling the growing mono-crystal while maintaining a constant rotation, in opposite directions, of both the ingot and the crucible containing the molten silicon. The process of growing a mono-crystalline silicon ingot is very slow and requires a large amount of energy, typically from 40–100 kWh/kg (Jester, 2002). For standard quasi-square 156 mm wafers, ingots of 200 mm or 210 mm in diameter are required. The preferred crystal orientation of the ingot is {100} for two practical reasons: the recombination of minority carriers (see Chapter 2.4) is lower on {100} planes because of a lower density of interface states compared to other planes, and {100} surfaces can easily be textured by an anisotropic alkaline etching process to form random pyramids. In 2015, a typical Cz puller has an annual capacity of 3.3–4.2 MW/year with the following process parameters:

- Ingot diameter: 200–220 mm.
- Ingot length: 1.8–2.1 m.
- Si load: 190 kg.
- Pulling speed: 1.1 mm/min to 1.2 mm/min.
- Total process time per ingot: 56–61 hours.

Figure 2.2.1 Solar-grade crystalline silicon feedstock loaded in a fused silica crucible ready for multi-crystalline ingot production. Courtesy of Trina Solar

Table 2.2.1 Comparison of typical impurity level in metallurgical-grade silicon (MGS), solar-grade silicon (SoG-Si) and electronic-grade silicon (EGS) feedstock (values in ppm except as noted). Data from Sze (1983) and various silicon feedstock suppliers collected by the author

Impurity	MGS	SoG-Si	EGS
Donor impurity		<4.5 ppba to <0.9 ppba	<0.3 ppbw to <0.15 ppbw
Acceptor impurity		<2.67 ppba to <0.53 ppba	<0.1 ppbw to <0.05 ppbw
Al	1570		
B	44		
Fe	2070		
P	28		
Cr	137		
Mn	70		
Ni	4		
Ti	163		
V	100		
Total bulk metals (Fe, Cu, Ni, Cr, Zn, Na)		<200 ppbw to <50 ppbw	<2 ppbw to <1 ppbw
C	80	<0.9 ppma to <0.5 ppma	<0.3 ppma to <0.08 ppma
O		<2 ppma	<0.2 ppma

2.2.3.2 *Multi-crystalline Ingots*

Multi-crystalline silicon (mc-Si) ingots are relatively cheaper to produce than mono-crystalline ingots mostly because of the simplicity of the process and the equipment, the lower energy consumption and the greater size of the ingot, nowadays reaching more than 800 kg per load. The method is based on the growth of large silicon ingots (typically 950 × 950 × 400 mm) by controlling the heat removal from the melt in a fused silica crucible to maintain a flat solid-liquid interface and to promote columnar silicon grains of few millimeters (Ferrazza, 2005). After complete solidification and cooling, the silicon ingot is cut into bricks of 156 × 156 × 350 mm. For example, a G6 furnace would produce 36 bricks.

In 2015, a typical mc-Si furnace (Figure 2.2.2) has an annual capacity of 14 MW/year with the following process parameters:

- mc-Si ingot dimensions: 1000 mm × 1000 mm × 350 mm.
- Number of bricks: 36 bricks of 156 mm × 156 mm × 350 mm.
- Si load: 800–820 kg.
- Growth rate: 0.1 mm/min to 0.4 mm/min.
- Total process time per ingot: 70–74 hours.

Figure 2.2.2 Production facility of multi-crystalline silicon ingots. Courtesy of Trina Solar

2.2.3.3 *Slicing*

Wire saws for slicing silicon ingots into wafers were specifically developed for the PV industry in the early 1990s to replace the old ID (inside diameter) saws. The main reason for this development was to improve productivity by reducing kerf loss and reducing cost. Wires saws are nowadays widely used in the PV and semiconductor industry and allow the slicing of wafers that are typically 180 microns thick and with a kerf loss of about 200 microns. Several hundred kilometers of a bronze-coated stainless steel wire of about 160 microns in diameter is wound on a spool and fed on grooved wire guides, allowing several thousand wafers to be sliced at a time (Figure 2.2.3). Several bricks of mono- or multi-crystalline silicon are glued onto a glass plate, which is mounted on a motor-driven table moving the silicon bricks downward through the mesh of wires at a speed of about 0.3–0.4 mm/min (Figure 2.2.4). During the entire sawing process the wire speed is maintained at about 5–10 m/sec feeding an abrasive slurry containing a fine powder of silicon carbide. When the wires reach the glass plate (total process time of about 6–8 hours for 156 mm wafers), the wafers are detached from the glass and cleaned before processing into cells.

2.2.4 Doping and Diffusion

Doping the semiconductor with impurities is necessary to alter the conductivity and type of the semiconductor material. As discussed in Section 2.1.1 of Chapter 2.1, silicon atoms have four electrons on their external orbit and each silicon atom has four nearest neighboring atoms

Figure 2.2.3 Four bricks in a wire saw ready to be sliced into wafers. Courtesy of Trina Solar

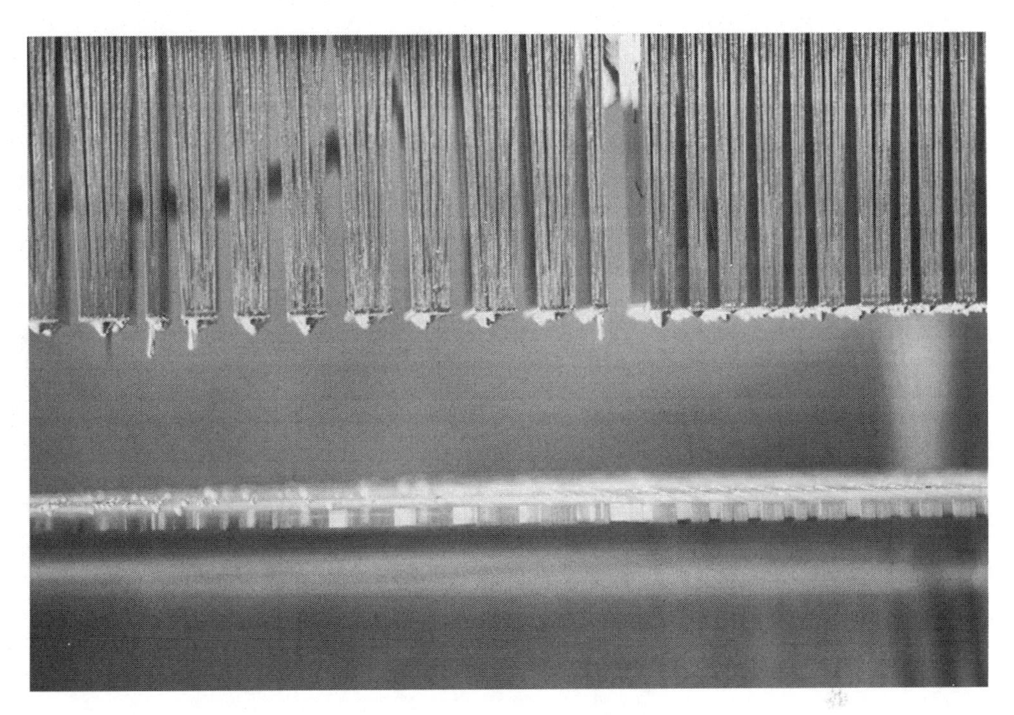

Figure 2.2.4 Bricks of multi-crystalline silicon after wafer slicing in a wire saw. Courtesy of Trina Solar

to which it is covalently bonded. Boron and phosphorus are typical impurities for silicon. Phosphorus in a substitutional position, that is, substituting one silicon atom, is a donor impurity because four of its five external-orbit electrons are covalently bonded to the four silicon neighboring atoms. Its fifth electron is free for electrical conduction. Phosphorus-doped silicon is said to be n-type. Other donor dopants are, for example, Arsenic (As) or Antimony (Sb), but these impurities are not used in PV cell manufacturing. Boron, having three electrons on its external orbit that are covalently bonded to three silicon atoms, is a substitutional acceptor. Boron-doped silicon is said to be p-type because the deficiency of one electron to complete the covalence bonding is the basis for hole conduction. Other acceptor impurities are Aluminum (Al), Gallium (Ga) and Indium (In). Aluminum is used as a dopant for Al-doped back surface field (BSF) and Gallium is sometimes used as a substrate background doping to avoid light-induced degradation (LID) due to the boron-oxygen defect.

Although other structures are also possible, the most common architecture of a silicon solar cell is a homojunction of a p-type and n-type semiconductor. The substrate could have, for example, a background doping with acceptor impurities like boron (p-type) and a p-n junction formed by the diffusion of phosphorus on one surface of the wafer.

Silicon feedstock produced by the Siemens process is relatively pure and undoped. The doping of the silicon substrates is provided during the ingot production by the addition of highly doped silicon chips to the charge of the crucible. Typical p-type silicon substrates for solar cell fabrication have a resistivity of 1–3 Ω.cm, that is, an acceptor dopant density of around $10^{16}\,cm^{-3}$.

The p-n junction formation is realized by the diffusion of impurities through one surface of the wafer. For example, if p-type wafers are used, phosphorus is diffused to form a n-p junction. Dopant atoms can be introduced into silicon wafers in many different ways, but these techniques can be classified into three categories: (1) diffusion from a chemical source (solid, liquid or gaseous) in a vapor form at high temperature; (2) diffusion from a doped silicon dioxide source; or (3) ion implantation followed by annealing.

The theory of dopant diffusion is based on Fick's law, expressing that the local diffusion flux, J (in atoms/cm^2.s), or diffusion rate per unit area, is proportional to the gradient of concentration of impurity, $C(x,t)$ in (atoms/cm^3), and a proportionality constant defined as the diffusion coefficient, D (in cm^2/s) according to Equation (2.2.6):

$$J = -D\frac{\partial C(x,t)}{\partial x} \tag{2.2.6}$$

Fick's second law of diffusion is derived from the fact that if there is a change in the local concentration of impurity, it must be due to a local decrease or increase of diffusion flux:

$$\frac{\partial C(x,t)}{\partial t} = -\frac{\partial J(x,t)}{\partial x} \tag{2.2.7}$$

By substituting Equation (2.2.6) into Equation (2.2.7) and assuming a constant diffusion coefficient, we derive Equation (2.2.8):

$$\frac{\partial C(x,t)}{\partial t} = D\frac{\partial^2 C(x,t)}{\partial x^2} \tag{2.2.8}$$

Two simple solutions exist for Equation (2.2.8). The first is the result of assuming a constant surface concentration, which is the case during diffusion from a chemical source in a vapor form at high temperature, also called a "predeposition diffusion." In this case, the solution of Equation (2.2.8) is given by:

$$C(x,t) = C_s\, erfc\left[\frac{x}{2\sqrt{Dt}}\right] \tag{2.2.9}$$

where C_s is the constant surface concentration of dopant (in atoms/cm^3), x is the depth coordinate (in cm), t is the diffusion time (in sec) and *erfc* is the complementary error function.

The second solution of Equation (2.2.8) is the result of assuming diffusion from a doped silicon dioxide source or after ion implantation of a finite dose of dopant, Φ (in atoms/cm^2). The solution of Equation (2.2.8) is:

$$C(x,t) = \frac{\phi}{\sqrt{\pi Dt}}\, exp\left[-\frac{x^2}{4Dt}\right] \tag{2.2.10}$$

resulting in a Gaussian distribution of dopant with a surface concentration $C_s = \frac{\phi}{\sqrt{\pi Dt}}$

Equations (2.2.9) and (2.2.10) are of course simplified solutions of a very complex problem with, for example, redistribution of dopant during multiple high-temperature steps, diffusion

Table 2.2.2 Intrinsic diffusivity of boron and phosphorus in silicon (Sze, 1983)

		Boron	Phosphorus
D_0	cm^2/s	0.76	3.85
E_A	eV	3.46	3.66

in an oxidizing ambient, diffusion coefficients that are concentration-dependent, all of which require numerical simulation. One important thing to note from Equations (2.2.9 and 2.2.10) is that the depth of diffusion is proportional to $\sqrt{Dt}$.

The diffusion coefficients of dopants are determined experimentally and, as expected, increase with increasing temperature following the general expression:

$$D = D_0\, exp\left[-\frac{E_A}{kT}\right] \quad (2.2.11)$$

where D_0 is called the frequency factor (in cm^2/s), E_A is the activation energy (in eV), T is the absolute temperature (in K) and k is the Boltzmann constant (in eV/K). Table 2.2.2 gives the diffusion coefficients of boron and phosphorus in silicon.

2.2.5 Defects in Silicon

Multi-crystalline wafers are composed of multiple grains of mono-crystals, and mono-crystalline wafers are supposed to be made of a perfect single crystal. In reality, the mono-crystals forming the multi-crystalline wafers or even the mono-crystalline wafers are not perfect. They differ from the ideal crystal in many aspects. First, they are not infinite crystals. They have finite dimensions and, as a consequence, atoms at the surface that are incompletely bonded. Within the crystal lattice, some atoms may be displaced from their ideal crystallographic location. In general, crystallographic defects can be classified into four categories: (1) point defects; (2) line defects; (3) planar defects; and (4) volume defects.

2.2.5.1 Point Defects

Point defects are typically impurities (non-silicon atoms) that are introduced into the silicon lattice at either a substitutional or interstitial site. Even if they are not considered as such, dopant impurities are actually also "point defects." A vacancy, that is, a missing atom in the lattice, is also a point defect and is called a "Schottky defect." If a silicon atom is displaced from its ideal position in the lattice to an interstitial site and is associated with a vacancy, it becomes a point defect called a "Frenkel defect." Point defects generate permitted energy levels in the bandgap that are also called traps because they have a significant impact on the recombination process of generated carriers in the solar cell (see Chapter 2.4). There are several common unintentional (i.e. other than dopants) impurities in a silicon wafers:

- oxygen, coming from the dissolution of the crucible by the molten silicon during the ingot production;

- carbon or silicon carbide, coming from the graphite susceptor supporting the crucible, the cover of the crucible or the crystallization furnace itself during the ingot production;
- iron, present in almost every piece of equipment used for the production of the silicon wafers.

Undesirable impurities in silicon are the ones with traps located close to the middle of the bandgap, like iron or most metallic impurities. Oxygen is also highly undesirable because, in addition to generating thermal donors, oxygen also forms a defect complex with boron which is activated by light. Oxygen can also be caused by SiO_2 precipitates which represent a compressive strain on the lattice that is relieved by the generation of dislocations.

2.2.5.2 Line Defects

Line defects or dislocations often appear in a crystal lattice. Two primary types of dislocations exist: edge (or line) dislocations and screw (or spiral) dislocations. An edge dislocation can be seen as an extra plane of atoms. The closest analogy is a book with, in the middle, a page cut in half. The "edge dislocation" is located at the edge of this extra plane of atoms. A screw dislocation is much more difficult to visualize. Imagine a crystal lattice that you cut half-way along a crystallographic plane. Then you slip one half across the other by one lattice vector. The boundary of the cut is a screw dislocation with a helical path traced around the linear defect, like a spiral-sliced ham. Mixed dislocations consist of dislocations with both screw and edge characters.

Dislocations are introduced, for example, by thermal stress during processing, mechanical stress, excess of point-defect concentration, precipitates of impurities, or substitutional impurities with a covalent radius significantly larger or smaller than silicon. Dislocations are also generated during the solidification of multi-crystalline silicon ingots. To reduce the amount of dislocations, it is preferable to keep the grain size of the mc-Si ingots as small as possible.

2.2.5.3 Planar Defects

Planar defects are a large area discontinuity in the lattice, for example, grain boundaries, that is, a transition between two crystals of different orientations. Twins are another type of planar defect. Like a grain boundary, a twin boundary has different crystal orientations on its two sides, but not randomly. The two orientations are related in a mirror-image way.

Planar defects are surfaces with a high density of dangling bonds, like the surface of a wafer, and represent surfaces with a high recombination rate for excess minority carriers.

2.2.5.4 Volume Defects

The fourth class of crystal defects is caused by precipitates of impurities or dopant atoms. Every impurity introduced, intentionally or unintentionally, into the silicon has a limit of solubility which decreases with decreasing temperature. During the cooling process (during the ingot production or during the solar cell processing), some impurities may reach their limit of solubility. If the cooling rate is high, a supersaturated condition may appear. However, if the cooling rate is slow, the crystal may achieve an equilibrium state by precipitating the impurity atoms, the concentration of which is higher than the solubility. This is generally undesirable because precipitates will create a mechanical stress and would become a site for dislocation generation.

List of Symbols

Symbol	Description	Unit
C	Concentration of impurity	atoms/cm^3
C_s	Surface concentration of dopant	atoms/cm^3
D	Diffusion coefficient	cm^2/sec
D_0	Frequency factor (diffusivity)	cm^2/sec
E_A	Activation Energy	eV
J	Diffusion flux	atoms/cm^2.sec
k	Boltzmann constant	eV/K
T	Temperature	K
Φ	Dopant dose	atoms/cm^2

List of Acronyms

Acronym	Meaning
BSF	Back Surface Field
Cz	Czochralski
fcc	Face-Centred Cubic
IC	Integrated Circuit
LID	Light-Induced Degradation
mc-Si	Multi-Crystalline Silicon
MGS	Metallurgical-Grade Silicon
EGS	Electronic-Grade Silicon
PV	Photovoltaic or Photovoltaics
SoG-Si	Solar-Grade Silicon
ppbw	Part Per Billion by Weight
ppba	Part Per Billion by Atom density
ppma	Part Per Million by Atom density

References

Beadle, W.E., Tsai, J.C.C., and Plummer, R.D. (1985) *Quick Reference Manual for Silicon Integrated Circuit Technology*. John Wiley & Sons, Inc., New York.

Ferrazza, F. (2005) Crystalline silicon: manufacture and properties. In *Solar Cells, Materials, Manufacture and Operation* (eds T. Markvart and L. Castañer). Elsevier Science, Oxford, pp. 71–88.

Jester, T. (2002) Crystalline silicon manufacturing process, *Progress in Photovoltaics: Research to Applications*, **10**, 99–106.

Sze, S.M. (1983) *VLSI Technology*. McGraw-Hill Book Company, New York.

2.3

Absorption and Generation

Seth Hubbard
NanoPower Research Laboratories, Rochester Institute of Technology (RIT), Rochester, NY, USA

2.3.1 Introduction

The most important function of a solar cell is generation of photocurrent and photovoltage. In order to do this, the solar cell must first absorb light from the sun and generate charge carriers. In this chapter, we discuss both the various charge carrier generation processes as well the process of absorption of light by a semiconductor.

2.3.2 Generation of Electron Hole Pairs in Semiconductors

Generation is a process that increases the density of either free electrons, holes or both. This process requires some minimum generation energy, which in most cases except for impurities, is equal to the bandgap energy E_g. The four basic processes that lead to free carrier generation include impact ionization, thermal generation, photogeneration (also called optical generation), and impurity-mediated generation. The processes are explained briefly below:

1. *Impact ionization*: Carriers (electrons or holes) with sufficient kinetic energy can promote a valence electron into the conduction band, basically knocking a bound electron away from the host atom. This process requires significant kinetic energy, for example, under a high electric field, and is typically not encountered in a standard solar cell in forward bias.
2. *Thermal generation*: Occurs due to high energy lattice vibrations, called phonons, interacting with the host atoms. At higher temperatures or for low bandgap materials, thermal vibration can be great enough to free an electron from the host atom and create free carriers.

Photovoltaic Solar Energy: From Fundamentals to Applications, First Edition.
Edited by Angèle Reinders, Pierre Verlinden, Wilfried van Sark, and Alexandre Freundlich.

Companion website: www.wiley.com/go/reinders/photovoltaic_solar_energy

3. *Photogeneration*: The most important process for solar cells is generation from photons, or photogeneration. In this case, the input photon energy must be greater than the bandgap energy in order to create an electron hole pair (EHP).
4. *Impurity-mediated generation*: Localized trap states in the bandgap can generate single free carriers by interaction with photons or phonons.

As well, due to the principle of microscopic reversibility, each of these generation processes discussed above must have an equivalent recombination process, which is discussed in Chapter 2.4. In equilibrium, there is a continuous process of generation and recombination of carriers, while under non-equilibrium conditions, one process may dominate in order to attempt to restore equilibrium conditions.

It should be noted that thermal generation occurs in all semiconductors and will depend strongly on the bandgap and temperature. Since a charge cannot accumulate indefinitely in one band, at equilibrium, thermal generation must be balanced by thermal recombination. This thermally generated intrinsic carrier density (n_i) accounts for the background equilibrium carrier concentration in a semiconductor. For typical semiconductors with bandgaps near 1 eV, the net number of thermally generated carriers at room temperature ($kT \sim 26$ meV) is quite small, allowing for the extrinsic doping that is used to created p-type and n-type materials. Under non-equilibrium conditions, such as illumination, we are thus concerned with the change in carrier populations due to excess generation, or recombination, beyond the thermal equilibrium intrinsic or extrinsic values.

2.3.3 Absorption and Photogeneration

As discussed above, photogeneration is perhaps the most important optical process in solar cells. The rate of photogenerated carriers $G(\text{EHP/cm}^3\text{s})$ must be known when working with solar cells under illumination. In order to calculate G, we need first to understand how the intensity or flux of photons varies with depth and energy within a bulk semiconductor material.

Consider the situation in Figure 2.3.1, showing light incident on one side of a bulk semiconductor with intensity I_0 (mW/cm^2) or flux Γ_0 (photons/cm^2/s). We then define the absorption coefficient, $\alpha(E)$ (cm^{-1}), as the probability that a photon of energy E (eV) is absorbed. Assuming the bulk semiconductor has a uniform absorption coefficient, in a microscopic width dx, then $\alpha(E)dx$ photons are absorbed. Thus, we can relate the microscopic change in intensity to the absorption coefficient as:

$$\frac{dI(E)}{I(E)} = \alpha(E)dx \tag{2.3.1}$$

Solving this equation for intensity gives the well-known Beer-Lambert law, predicting an exponential decay of intensity with distance:

$$I(E,x) = I_0(E)e^{-\alpha(E)x} \tag{2.3.2}$$

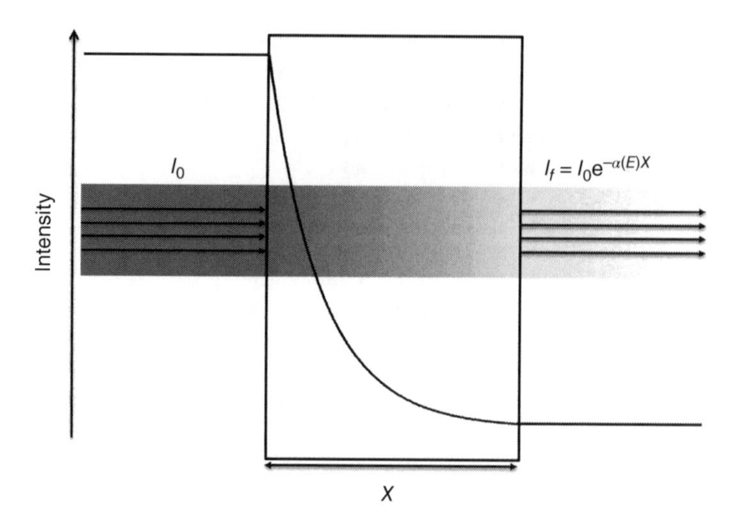

Figure 2.3.1 The absorption process in a semiconductor (the Beer-Lambert law)

This equation is equally valid when considering photon intensity I or photon flux. The absorption coefficient is related to the complex index of refraction through the extinction coefficient κ as:

$$\alpha(\lambda) = \frac{4\pi\kappa(\lambda)}{\lambda} \tag{2.3.3}$$

where λ is the wavelength (in cm) and κ is a unitless factor that is typically wavelength-dependent.

The photogeneration rate, at a specific energy E and at a depth x, can be given as the product of the carrier flux and the absorption coefficient:

$$g(E,x) = \alpha(E)\Gamma(E,x) \tag{2.3.4}$$

Applying the Beer-Lambert law and accounting for any reflective loss from the front surface gives:

$$g(E,x) = \left[1 - R(E)\right]\alpha(E)\Gamma_0(E)e^{-\alpha(E)x} \tag{2.3.5}$$

where $R(E)$ is the reflectance at the surface of the material.

The total generate rate is then found by integration over all energies where α is non-zero (i.e., above the bandgap energy):

$$G(x) = \int_{E_g}^{\infty} \left[1 - R(E)\right]\alpha(E)\Gamma_0(E)e^{-\alpha(E)x}dE \tag{2.3.6}$$

The absorption length is a useful quantity for solar materials. It gives the depth where the intensity of light has dropped by a factor of e^{-1} or ~36%:

$$L_a = \alpha^{-1} \tag{2.3.7}$$

We will discuss this value for various types of semiconductors in the following chapters.

2.3.4 Absorption Coefficient for Direct and Indirect Bandgap Semiconductors

In this section we will discuss the form of absorption and related absorption coefficients for various types of semiconductors.

2.3.4.1 Direct Bandgap

Figure 2.3.2 (a) illustrates the *E-k* diagram for a direct bandgap semiconductor. This plot shows the allowed values of energy as a function of crystal wave vector ($\mathbf{k}(cm^{-1})$) from the zone center (Γ point) along the high symmetry <100> (ending at the X point) and <111> (ending at the L point) crystal directions. The crystal wavevector is related to the crystal momentum through Planck's constant ($\hbar$). In Figure 2.3.2, a photon of angular frequency ω is shown incident on the semiconductor.

In semiconductors when the conduction band minimum and the valence band maximum occur at the same value of the crystal wave vector, then energy and momentum are conserved simply as $\Delta E = \hbar\omega$ (where ω is related to the wavelength using the speed of light c, as $2\pi c/\lambda$). In this case, there is no change in **k** from initial to final states. A single photon of energy $E > E_g$ is sufficient to create an EHP. Due to this, absorption coefficients tend to be rather high for direct bandgap semiconductors. Many III-V and II-VI semiconductors have direct bandgaps, including GaAs, InP and CdTe. As an example, Figure 2.3.3 shows a plot of bandgap energy

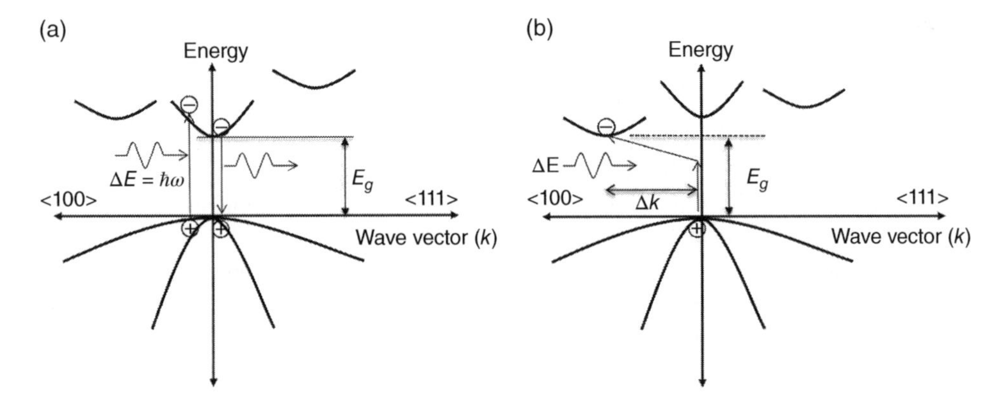

Figure 2.3.2 The *E-k* diagram and absorption process for (a) the direct bandgap semiconductor; and (b) the indirect bandgap semiconductor

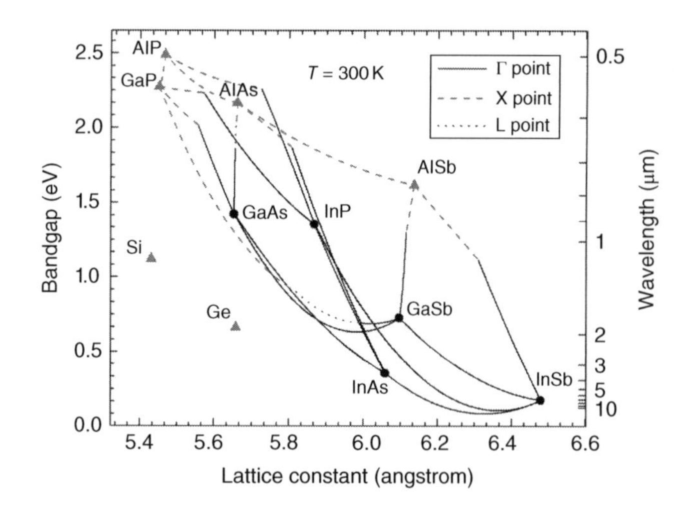

Figure 2.3.3 Plot of bandgaps and lattice constants for a number of semiconductor. Circles indicate direct binary materials and triangles indirect binary materials. The solid, dashed and dotted lines indicate if the ternary bandgap is direct (Γ) or indirect (X, L)

versus lattice constant for a number of direct bandgap (Γ point) and indirect bandgap (X,L) semiconductors. For tertiary and quaternary semiconductors (line and planes in Figure 2.3.3), the exact nature of the bandgap depends strongly on the materials stoichiometry.

The exact value of the absorption coefficient can be derived from the optical matrix element and the density of states (Nelson, 2003), however, in most semiconductors near the band edge, where the parabolic band approximation is valid, the absorption coefficient can be expressed as:

$$\alpha(E) = C(E - E_g)^{1/2} \tag{2.3.8}$$

where C (eV/cm) is a material-dependent constant. Beyond ~100 mV above E_g, this approximation for α is no longer valid. Well above the band edge, the absorption coefficient begins to stabilize, with occasional peaks due to transitions into higher energies of the conduction band or to other bands at higher energies.

2.3.4.2 Indirect Bandgap

Figure 2.3.2 (b) illustrates the E-k diagram for an indirect bandgap semiconductor. In semiconductors where the conduction band minimum and the valance band maximum occur at different values of the crystal wave vector, a single photon of $E > E_g$ is not sufficient to create an EHP. Momentum must also be conserved by a change in wave vector Δk. Since the photon momentum is very small, this can only be accomplished through the participation of a phonon of suitable momentum and with energy $\hbar\Omega$, where Ω is the phonon frequency. The transition can be accomplished through either absorption or generation of a photon of energy $\hbar\Omega$. Since both a photon and a phonon must be absorbed roughly simultaneously, indirect bandgap materials are generally much weaker absorbers than direct bandgap materials. As seen in

Figure 2.3.3, Group IV semiconductors Si and Gi are both indirect bandgap materials, as are some of the III-V materials such as AlAs and GaP. Near the band edge of an indirect bandgap material, the absorption coefficient varies as:

$$\alpha(E) \propto \left(E - E_g \pm \hbar\Omega\right)^2 \tag{2.3.9}$$

where the $\pm\hbar\Omega$ terms indicates either absorption or generation of a phonon, respectively. The phonon energy $\hbar\Omega$ is usually quite small compared to the bandgap energy, on the order of a few kT. At energies well above the band edge for indirect materials, direct optical transitions to the Γ point are observed as abrupt changes in the slope for $\alpha(E)$.

Figure 2.3.4 shows $\alpha(E)$ for various semiconductors based on reported values of the extinction coefficient (Adachi, 2013). As we expect, both GaAs and $InGaP_2$ have very steep absorption edges and maintain a relatively high value beyond the bandgap. However, indirect semiconductors show a slower, more graduate increase in absorption beyond the bandgap, eventually reaching higher values of absorption coefficient due to direct transitions.

Due to the nature of indirect bandgap materials, the absorption coefficient is typically orders of magnitude smaller. Typical values for the indirect absorption coefficients near the bandgap energy are in the range of $\alpha_i \sim 10^{-1}$ to $10^{-2}\,\text{cm}^{-1}$. However, for direct bandgap materials near the bandgap energy, these values can range from $\alpha_d \sim 10^2$ to $10^6\,\text{cm}^{-1}$. This has implications for the absorption length discussed earlier. For indirect bandgap materials, bulk absorber thickness on the order of 100 μm or more are often necessary to absorb all of the light above the bandgap. However, in direct bandgap materials, absorber thicknesses of only ~3 μm are able to fully absorb all the light above the bandgap.

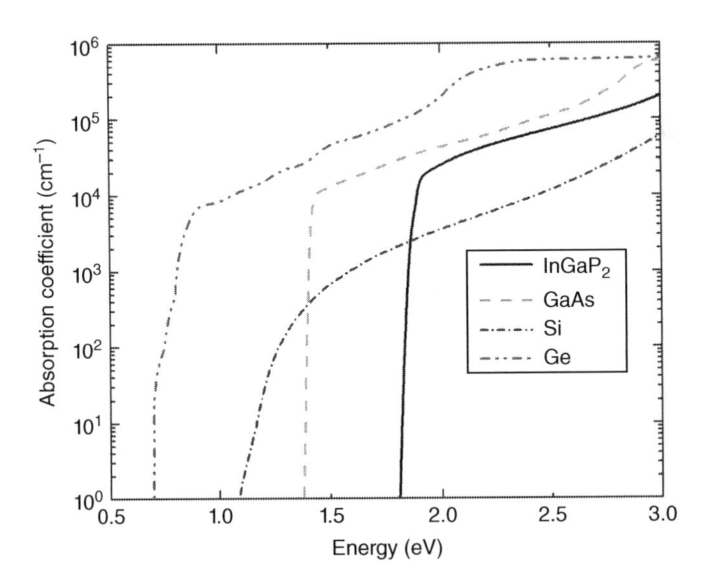

Figure 2.3.4 Plot of the absorption coefficient for a number of direct (GaAs, $InGaP_2$) and indirect (Ge, Si) semiconductors

It should be noted again that the predicted behavior of $\alpha(E)$ is only valid near the band edge where the parabolic band approximation is valid. At $E>>E_g$, changes in $\alpha(E)$ are more gradual due to transitions to other higher energy bands. As well, at $E<E_g$, our model predicts zero absorption. However, in reality, other process can lead to non-zero absorption near the band-edge. For example, the Columbic attractions between electron and holes leads to a lower energy bound state, called an exciton. Absorption and emission from excitons will occur at energies lower than the bandgap. In semiconductors, the exciton binding energy is typically in the order of a few kT, but can be much larger in organic materials. As well, crystal defects and impurities in the semiconductor can lead to absorption below the bandgap, often called band tailing (Urbach, 1953). Better approximation of this behavior can be gathered from empirical models of the absorption coefficient (Aspnes and Studna, 1983; Adachi, 1992).

For a list of symbols we refer to Chapter 2.6.

References

Adachi, S. (1992) *Physical Properties of III-V Semiconductor Compounds: InP, InAs, GaAs, GaP, InGaAs, and InGaAsP*. John Wiley & Sons, Inc., New York.

Adachi, S. (2013) *Optical Constants of Crystalline and Amorphous Semiconductors: Numerical Data and Graphical Information*. Springer Science & Business Media, New York.

Aspnes, D.E. and Studna, A.A. (1983) Dielectric functions and optical parameters of Si, Ge, GaP, GaAs, GaSb, InP, InAs, and InSb from 1.5 to 6.0 eV. *Physical Review B*, **27** (2): 985–1009.

Nelson, J. (2003) *The Physics of Solar Cells*. Imperial College Press, London.

Urbach, F. (1953) The long-wavelength edge of photographic sensitivity and of the electronic absorption of solids. *Physical Review*, **92** (5), 1324.

2.4

Recombination

Seth Hubbard
NanoPower Research Laboratories, Rochester Institute of Technology (RIT), Rochester, NY, USA

2.4.1 Introduction

As discussed in Chapter 2.3, recombination is necessary from the standpoint of microscopic reversibility. In a solar cell, recombination acts to restore the non-equilibrium light generated EHP population to its thermal equilibrium value. The three types of recombination in a bulk semiconductor are:

- *Radiative recombination* resulting in the emission of a photon. This is often called band-to-band recombination.
- *Non-radiative recombination* resulting in phonon emission. This is often called trap-assisted recombination.
- *Auger recombination* resulting when kinetic energy is imparted to a third carrier. This is also a non-radiative process.

Each of these processes is illustrated in Figure 2.4.1 and will be discussed in depth below. Of the above types of recombination, radiative and Auger are a basic property of the semiconductor materials being used, while non-radiative depends on the quality of the material through the density of non-radiative traps. For a general two-particle recombination process, the rate of recombination (U (cm^{-3}s^{-1})) can be expressed as:

$$U = n_1 n_2 \beta_{12} \tag{2.4.1}$$

where n_1 (cm^{-3}) and n_2 (cm^{-3}) are the densities of recombination species and β_{12} (cm^{3}s^{-1}) is the recombination coefficient. The recombination coefficient can be expressed in terms of a

Photovoltaic Solar Energy: From Fundamentals to Applications, First Edition.
Edited by Angèle Reinders, Pierre Verlinden, Wilfried van Sark, and Alexandre Freundlich.

Companion website: www.wiley.com/go/reinders/photovoltaic_solar_energy

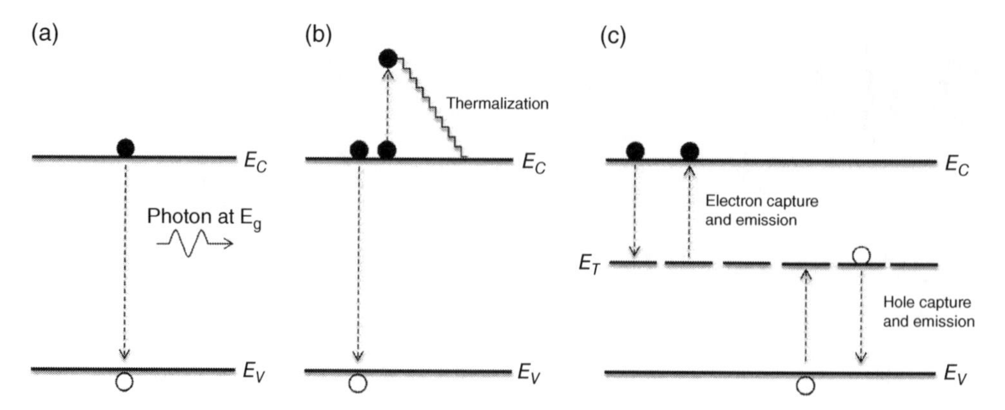

Figure 2.4.1 The three basic types of recombination processes: (a) radiative band-to-band; (b) non-radiative Auger; and (c) non-radiative recombination centers (traps)

cross-section for that process (σ (cm^2)) and the thermal velocity (v_{th} (cm/s)) of the particle as (Blood and Orton, 1992):

$$\beta_{12} = \sigma v_{th} \tag{2.4.2}$$

2.4.2 Radiative Recombination

Radiative recombination results from the recombination of an electron in the conduction band with a valence band hole, resulting in the emission of a photon at the bandgap energy E_g (Figure 2.4.1 (a)). The rate of spontaneous radiative recombination (band-to-band recombination) from a semiconductor will depend physically on the flux of photons from the material (blackbody radiation) as well as the absorption coefficient of the material. As well, the external bias on a pn junction serves to increase the photon flux from the material, as is the case of a light emitting diode. A detailed discussion of the background and derivations for spontaneous radiative recombination can be found in either Nelson (2003) or Battacharaya (1997). A simple expression for band-to-band radiative recombination for a non-degenerate semiconductor under non-equilibrium conditions is given as:

$$U_{rad} = B_r \left(np - n_i^2\right) \tag{2.4.3}$$

where B_r (cm^3/s) is called the *radiative recombination coefficient* and is a property of a material. The first term in the expression corresponds to the non-equilibrium electron-hole population (from illumination or external bias), while the second term contains the intrinsic thermal carrier population ($n_i^2 = n_0 p_0$). The above rate describes the change in radiative recombination from its equilibrium value, which will be useful when considering solar cell operation through the continuity equations discussed in Chapter 2.5.

The radiative recombination coefficient, B_r, depends on a material's bandgap as well as the strength of the absorption coefficient. Values range from 10^{-9}–10^{-11} cm^3/s for direct bandgap materials and from 10^{-13}–10^{-15} cm^3/s for indirect bandgap materials. In a doped material, the radiative recombination rate can be further simplified. For example, in n-type material with $N_D = n_0 >> p_0$ and under low level injection such that the excess EHP population (Δn and Δp) is significantly less than the majority carrier population, the above equation is simplified to:

$$U_{rad} \approx B_r N_D \Delta p = \frac{p - p_0}{\tau_{p,rad}} \tag{2.4.4}$$

where we define the hole minority carrier radiative lifetime, $\tau_{p,rad}$, (s), i.e., the lifetime of the holes in n-type material as:

$$\tau_{p,rad} = \frac{1}{B_r N_D} \tag{2.4.5}$$

The radiative lifetime is a very important parameter in solar cells. When used in context with the continuity equations, it expresses the average time a carrier will spend in the excited state before recombination. This time will have a direct impact on the carrier collection efficiency in solar cells.

Similarly, for p-type materials under low-level injection:

$$U_{rad} \approx B_r N_A \Delta n = \frac{n - n_0}{\tau_{n,rad}} \tag{2.4.6}$$

$$\tau_{n,rad} = \frac{1}{B_r N_A} \tag{2.4.7}$$

where $\tau_{n,rad}$ (s) is the electron minority carrier radiative lifetime, i.e., the lifetime of the electrons in p-type material.

As mentioned above, since B_r is generally much larger for direct bandgap materials, the radiative lifetime τ_{rad} is generally much shorter (10^{-9}–10^{-7} s) in direct bandgap materials versus indirect bandgap materials (10^{-2} s). Therefore, since radiative processes are much less probable in indirect bandgap materials, other non-radiative process will tend to dominate.

For very high quality direct bandgap materials, such as GaAs, the radiative recombination of carriers can provide the ultimate limit to device efficiency. These types of solar cells are said to be operating near the radiative limit. However, most practical solar cells are limited by non-radiative processes as discussed in Sections 2.4.3 and 2.4.4.

2.4.3 Auger Recombination

Auger recombination is a non-radiative process where the energy of the photogenerated carrier is dissipated not by photon emission, but rather by increasing the kinetic energy of another free carrier. For example, the collision of two electrons can result in recombination of one of the electrons with a hole and the resulting energy is given up to increase the kinetic energy of the second electron. This kinetic energy is then very quickly (typically less than a picosecond) dissipated through the process of thermalization, where the electron emits multiple phonons and returns to an energy near the band edge. This is shown graphically in Figure 2.4.1 (b). This process is the inverse process of impact ionization where a high energy carrier can "knock" other free carriers out the crystal lattice. Since this is a three-carrier process, the rates will vary as the square of one of the carrier types. For Auger recombination, accounting for the background thermal Auger processes, the net change in recombination rate under non-equilibrium conditions is:

$$U_{Auger} = B_{Auger,p}\left(n^2 p - n_0^2 p_0\right) \tag{2.4.8}$$

for two electrons, where $B_{Auger,p}$ (cm^6/s) is the hole Auger recombination coefficient, and

$$U_{Auger} = B_{Auger,n}\left(p^2 n - p_0^2 n_0\right) \tag{2.4.9}$$

for two holes, where $B_{Auger,n}$ (cm^6/s) is the electron Auger recombination coefficient.

The Auger lifetime ($\tau_{n,p}(s)$), similar to radiative lifetime, can be found for doped materials under low injection conditions as:

$$\tau_{n,Auger} = \frac{1}{B_{Auger,n} N_A^2} \tag{2.4.10}$$

for electrons in p-type material and

$$\tau_{p,Auger} = \frac{1}{B_{Auger,p} N_D^2} \tag{2.4.11}$$

for holes in n-type materials. Values of B_{Auger} for Si and GaAs are both on the order of 10^{-30} cm^6/s. In direct bandgap semiconductors, due to the short radiative lifetime, Auger processes are typically not observed except at very high injection conditions (e.g., under high solar concentration). However, in indirect bandgap materials, since the Auger processes are also able to conserve momentum, these processes are the dominant recombination pathway, and thus are the efficiency-limiting loss mechanism for high purity Si or Ge solar cells.

2.4.4 Non-Radiative (Shockley-Read-Hall) Recombination

Growing extremely high purity semiconductors can be difficult. Most practical semiconductors will have some degree of impurities or vacancies that are incorporated into the crystal structure during material growth. These impurities lead to spatially localized defect levels within the bandgap of the bulk semiconductor, which are often referred to as trap states due to the ability of the level to spatially capture a delocalized free carrier. This type of trap-assisted non-radiative recombination is often the practical limiting loss mechanism in most solar cells. Figure 2.4.1 (c) shows this type of recombination processes for both electron and holes. As seen in Figure 2.4.1, electron and hole traps can both capture and then thermally emit carriers. Traps located near the middle of the bandgap are sometimes called recombination centers. For example, an empty acceptor-like recombination center can capture an electron followed by a hole capture from the valance band. This effective recombination event is most often non-radiative with the capture energies being emitted as phonons.

We can mathematically derive the net rate of recombination for shallow trap states or recombination centers. This is often referred to as the Shockley-Read-Hall (SRH) recombination, after the individuals who first derived this type of trap-assisted recombination behavior (Hall, 1952; Shockley and Read, 1952). We will briefly show how this rate is derived. As shown in Figure 2.4.1 (c), we have illustrated a trap at energy E_t(eV) with density N_t(cm^{-3}). The traps in this example are acceptor-like and thus are neutral when empty and negative when full. The net rate of electron capture is given by the equation:

$$\frac{dn_t}{dt} = n\left(N_t - n_t\right)B_n - nN_cB_ne^{-(E_c-E_t)/kT} \tag{2.4.12}$$

where n_t(cm^{-3}) is the number of traps filled with electrons and B_n(cm^3/s) is the electron capture coefficient. The first term is just the generalized recombination equation as discussed before, applied to n electrons in the conduction band and $N_t - n_t$ unfilled trap states. It is positive since it adds to the total number of filled traps. The second term is due to the thermal emission of carriers back into the conduction band. It can easily be derived by detailed balance and depends on the conduction band density of states as well as exponentially on the difference in energy between the trap and the conduction band. Not surprisingly, traps closer to the conduction band will have a much higher rate of thermal emission at a given kT.

The electron capture coefficient can be expressed as $B_n = \sigma_n v_{th}$ where the parameter σ_n is the electron capture cross-section. The capture cross-section indicates how effective a trap is for electron capture, with values ranging from 10^{-10} cm^2 for extended complex sets of traps to 10^{-15} cm^2 for point defects. Cross-sections for specific impurity elements in specific semiconductor materials can be found in numerous textbooks and the literature (Mitonneau *et al.*, 1979; Blood and Orton, 1992).

In a similar manner to electrons, we can also find the net rate of hole capture as:

$$\frac{d\left(N_t - n_t\right)}{dt} = pn_tB_p - \left(N_t - n_t\right)N_vB_pe^{-(E_t-E_v)/kT} \tag{2.4.13}$$

where the hole capture coefficient $B_p = \sigma_p v_{th}$ and σ_p is the hole capture cross-section. As with electrons, the first term is just capture of holes into filled trap states, while the section term is due to thermal emission of holes into the valence band. The net rate of recombination under steady state conditions must be balanced or a charge would build up on the traps, therefore:

$$U_{SRH} = \frac{dn_t}{dt} = \frac{d\left(N_t - n_t\right)}{dt} \tag{2.4.14}$$

The above relationship can be used to eliminate n_t and express the SRH recombination rate as:

$$U_{SRH} = \frac{np - n_i^2}{\tau_{p,SRH}\left(n + \hat{n}\right) + \tau_{n,SRH}\left(p + \hat{p}\right)} \tag{2.4.15}$$

where we have defined the following variables:

$$\tau_{p,SRH} = \frac{1}{B_p N_t} \tag{2.4.16}$$

$$\tau_{n,SRH} = \frac{1}{B_n N_t} \tag{2.4.17}$$

$$\hat{n} = n_i e^{(E_t - E_i)/kT} \tag{2.4.18}$$

$$\hat{p} = n_i e^{(E_i - E_t)/kT} \tag{2.4.19}$$

The SRH or non-radiative lifetime $\tau_{p,SRH}(s)$ and $\tau_{n,SRH}(s)$ represent the minimum lifetime for electrons and holes when all of the traps are filled with carriers of the opposite type. The carrier densities $\hat{n}$ and $\hat{p}$ represent the electron and hole densities when the Fermi level is located at the trap energy. In general terms the Fermi level can be described by any energy level having the probability that it is exactly half filled with electrons. Levels of lower energy than the Fermi level tend to be entirely filled with electrons, whereas energy levels higher than the Fermi tend to be empty.

The physical significance of U_{SRH} can be seen by examination of the numerator. When the np product is greater than n_i^2, there is a net positive recombination that attempts to restore the carrier densities toward equilibrium. However, when the np product is less than n_i^2, there is a net generation (negative recombination) that again attempts to restore carrier densities to equilibrium values. As well, by examining the denominator we see that it has a minimum value when $E_t = E_i$, indicating that levels near mid-gap are the most effective recombination centers.

For doped semiconductors under low-level injection conditions, we can simplify the SRH recombination such that:

$$U_{SRH} \approx \frac{n - n_0}{\tau_{n,SRH}} \tag{2.4.20}$$

for p-type doping, and

$$U_{SRH} \approx \frac{p - p_0}{\tau_{p,SRH}} \tag{2.4.21}$$

for n-type doping.

2.4.5 Surface, Interface and Grain Boundary Recombination

At the surface of a solar cell, dangling unpassivated bonds as well as absorbed impurity molecules create a very high density of trap states. This is shown schematically in Figure 2.4.2. These trap states form an almost continuous distribution across the bandgap. Thus, surface

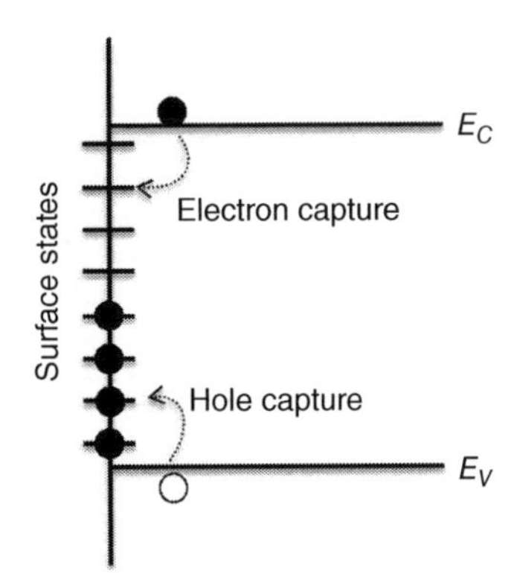

Figure 2.4.2 Electron and hole capture at the surface of a semiconductor

states provide an infinite sink or source for carrier recombination and generation, and, if left unpassivated, can account for a major loss mechanism in the device. To a lesser degree, interfaces between different semiconductors or high-angle grain boundaries in polycrystalline materials also act as sources of recombination and generation due to interface states. We can treat this type of recombination similar at the bulk SRH recombination with a slight modification to account for the areal versus volume quantities.

We can derive the SRH recombination rate for a surface or interface trap with a density of $N_{t,s}$ (cm^{-2}) and energy level $E_{t,s}$ (eV). This can be done similar to Section 2.4.3 and results in:

$$U_{SRH} = \frac{n_s p_s - n_i^2}{\frac{1}{s_n}(n_s + \hat{n}_s) + \frac{1}{s_p}(p_s + \hat{p}_s)} \tag{2.4.22}$$

where n_s, p_s(cm^{-3}) are the surface electron and hole concentrations and $\hat{n}_s, \hat{p}_s$(cm^{-3}) are given by Equations (2.4.18) and (2.4.19) applied at the surface trap energy $E_{t,s}$. We also define the variable $s_{n,p}$(cm/s) $= N_{t,s}\sigma_{n/p,s}v_{th}$, where σ_s is the capture cross-section for the surface states. This variable has units of velocity and is defined as the *surface recombination velocity* for electrons or holes. Note that this velocity has no bearing on the actual velocity of carriers, but rather refers to the rate that excess carriers recombine at the surface. For p-type material near the surface, the SRH recombination expression simplifies to:

$$U_{SRH,s} \approx s_n(n_s - n_0) \tag{2.4.23}$$

The magnitude of s_n (or s_p for n-type materials) is an important factor in most solar cells. Unpassivated surfaces or low quality interfaces can easily have s_n, s_p as high as 10^5–10^6 cm/s. The surface recombination current, $J_s(A/cm^2) = qs_n(n_s - n_0)$, derived using the continuity equations discussed in Chapter 2.5, thus acts to "steal" carriers that would otherwise be collected by the solar cell. Surface passivation using high quality hetero-epitaxial growth is able to significantly reduce the surface recombination velocity, to values as low at 1–10 cm/s.

For a list of symbols we refer to Chapter 2.6.

References

Bhattacharya, P. (1997) *Semiconductor Optoelectronic Devices*. Prentice Hall, Upper Saddle River, NJ.

Blood, P. and Orton, J.W. (1992) *The Electrical Characterization of Semiconductors: Majority Carriers and Electron States*. Academic Press, London.

Hall, R.N. (1952) Electron-hole recombination in germanium, *Physical Review*, **87** (2), 387.

Mitonneau, A., Mircea, A, *et al.* (1979) Electron and hole capture cross-sections at deep centers in gallium-arsenide, *Revue de Physique Appliquée*, **14** (10), 853–886.

Nelson, J. (2003) *The Physics of Solar Cells*. Imperial College Press, London.

Shockley, W. and Read, W.T. (1952) Statistics of the recombinations of holes and electrons. *Physical Review*, **87** (5), 835–842.

2.5

Carrier Transport

Seth Hubbard
NanoPower Research Laboratories, Rochester Institute of Technology (RIT), Rochester, NY, USA

2.5.1 Introduction

In order to analyze the basic pn junction diode in Chapter 2.6, we must have some understanding of the mechanisms that govern carrier transport in a semiconductor. Importantly, we must understand carrier transport under the non-equilibrium conditions that prevail in a solar cell, for example, light illumination and voltage bias. In this section, we will first discuss drift and diffusion of carriers, the two main mechanisms that give rise to current flow. Then we will tie the current flows together with the recombination and generation mechanisms discussed in Chapters 2.3 and 2.4, using the continuity equations. At the end of the chapter, we make a number of assumptions so that we can reformulate the complex continuity equations into simplified minority carrier diffusion equations. These represent the basic transport equations that must be solved, along with Poisson's equation, to calculate carrier density and current within a pn junction diode.

2.5.2 Drift Current

Drift current arises due to the carrier motion inside an electric field. Charged particles (electrons or holes) subjected to an electric field will experience a force ($\mathbf{F} = q\mathbf{E}$). Since electron hole pairs are free carriers, they will also accelerate parallel or anti-parallel to the electric field direction, giving rise to a current.

In order to analyze this current, let us first consider hole (or electron) motion without an applied field. From a microscopic perspective, at thermal equilibrium, carriers are in constant thermal motion v_{th} ($\sim 10^7$ cm/s at 300K). Carriers moving at this thermal velocity are constantly colliding with the vibrating lattice atoms, each other or crystal impurities (Figure 2.5.1 (a)). This motion is random and results in a zero net current. We define a mean scattering time

Photovoltaic Solar Energy: From Fundamentals to Applications, First Edition.
Edited by Angèle Reinders, Pierre Verlinden, Wilfried van Sark, and Alexandre Freundlich.

Companion website: www.wiley.com/go/reinders/photovoltaic_solar_energy

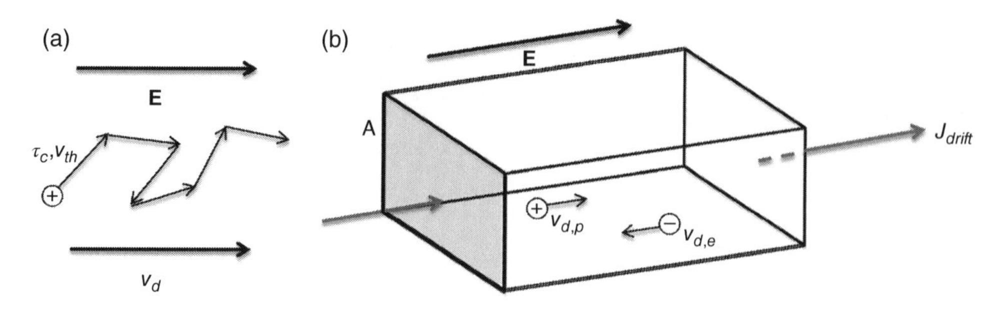

Figure 2.5.1 (a) Random electron scatter and drift velocity due to an electric field; and (b) drift current due to both electrons and holes in a semiconductor of cross-section area A

($\tau_c(s)$) as the average time between collisions. Upon application of an electric field, carriers are accelerated along the field direction between collisions. This will lead to a net carrier velocity component which we call the *drift velocity* $\mathbf{v}_d$(cm/s). The drift velocity is typically many orders of magnitude less than the thermal velocity and can be related to the electric field by considering the carrier acceleration during the scattering time τ_c,

$$\mathbf{v}_d = \frac{q\tau_c}{m^*_{e,h}}\mathbf{E} = \mu_{e,h}\mathbf{E} \tag{2.5.1}$$

where $m^*_{e,h}$(kg) is the electron or hole effective mass and we have defined the *mobility* $\mu_{e,h}$(cm^2/Vs) as the constant of proportionality that relates the drift velocity and electric field. The subscript (e or h) denotes either electron or hole mobility. Therefore, the value of mobility is essentially an indicator of how easily a carrier can move through the semiconductor under an applied electric field.

The drift current density $\mathbf{J}_{drift}$(A/cm^2) is then found by considering the drift velocity of all available carriers (Figure 2.5.1 (b)). Thus:

$$\mathbf{J}_{drift} = \mathbf{J}_{n,drift} + \mathbf{J}_{p.drift} = \left(nq\mu_n + pq\mu_p\right)\mathbf{E} \tag{2.5.2}$$

The constant of proportionality relating current density and electric field is called the *conductivity* σ_s (Siemens/cm) and its inverse is called the *resistivity,* ρ_s (Ω-cm).

$$\sigma_s = \frac{1}{\rho_s} = nq\mu_n + pq\mu_p \tag{2.5.3}$$

The above equations are just another formulation of the well-known Ohm's Law. The mobility of a semiconductor (and ultimately its conductivity) depend on the effective mass of the material, with low effective mass materials having higher mobility. As well, the scattering time factor leads to the dependence of the mobility on the temperature (phonon scattering), doping (carrier-carrier scattering), impurities, defects, and interfaces. For many common semiconductors (e.g., Si and GaAs), there are often empirical relationships that can be used to find mobility dependence on doping and temperature. Most common semiconductor texts list these relationships (Pierret, 1996; Muller *et al.*, 2003; Sze and Ng, 2007). As well, statistical and quantum mechanics can be used to derive scattering times for phonon, impurities and interfaces (Singh, 1993). Standard values for high-purity lightly doped Si are:

$$\mu_n \simeq 1360\frac{cm^2}{Vs}\text{and}\,\mu_p \simeq 460\frac{cm^2}{Vs}\text{at 300K.}$$

For GaAs, these values are $\mu_n \simeq 8000\frac{cm^2}{Vs}$ and $\mu_n \simeq 320\frac{cm^2}{Vs}$, due to its higher electron and lower hole effective mass.

2.5.3 Diffusion Current

The diffusion current arises due to a concentration gradient of free carriers inside a semiconductor. As a result of their random thermal motion, carriers will always tend to flow from regions of high to low concentration. This result is true for any type of mass flux and is captured by Fick's First Law of Diffusion. This states that under steady state condition, particle flux will be proportional to the concentration gradient, where the proportionality constant is called the *diffusion coefficient*. As shown in Figure 2.5.2, if the particles are free charge carriers, the flux will give rise to a current. Thus, the diffusion current for electrons and holes in one dimension can be given by:

$$J_{diff} = J_{n,diff} + J_{p,diff} = qD_n\frac{dn}{dx} - qD_p\frac{dp}{dx} \tag{2.5.4}$$

where D_n and D_p are the electron and hole diffusion coefficients, respectively, and have units of cm²/s. The signs for electron and hole diffusion currents ensure that particle flow will be down the concentration gradient.

In three dimensions we can generalize the above using the gradient as:

$$\mathbf{J}_{diff} = \mathbf{J}_{n,diff} + \mathbf{J}_{p,diff} = qD_n\nabla n - qD_p\nabla p \tag{2.5.5}$$

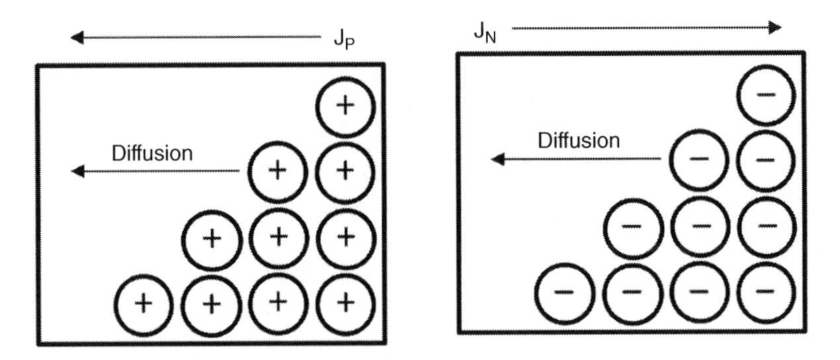

Figure 2.5.2 Diffusion currents due to the concentration gradients of holes and of electrons

The diffusion coefficient is not an independent material property. By considering the mechanics of particles in a solid, it can be shown that the diffusion coefficient and mobility are related. These are known as the *Einstein relationships* and are given as:

$$\frac{D_n}{\mu_n} = \frac{kT}{q} \text{ and } \frac{D_p}{\mu_p} = \frac{kT}{q} \tag{2.5.6}$$

2.5.4 Total Current

The total carrier current is then the linear sum of drift and diffusion components. Breaking these down by particle type, we obtain:

$$\mathbf{J}_n = \mathbf{J}_{n,drift} + \mathbf{J}_{n,diff} = q\mu_n n\mathbf{E} + qD_n \nabla n \tag{2.5.7}$$

$$\mathbf{J}_p = \mathbf{J}_{p,drift} + \mathbf{J}_{p,diff} = q\mu_p p\mathbf{E} - qD_p \nabla p \tag{2.5.8}$$

The total free carrier current is then just given by the sum of each type of particle current:

$$\mathbf{J}_{Total} = \mathbf{J}_p + \mathbf{J}_n \tag{2.5.9}$$

2.5.5 Quasi-Fermi Levels and Current

Solar cells operate under both light illumination and voltage bias, in other words, under non-equilibrium conditions. A useful construct that allows one to use the concepts and equation developed for equilibrium carrier densities, but under non-equilibrium conditions, is the *quasi-Fermi level* (E_{F_n} and E_{F_p}). We assume that under a small perturbation, the individual populations of electrons and holes will quickly relax to some new *quasi-thermal equilibrium* value. This requires fast thermalization times within the band, but slower band-to-band recombination. Typically this is true as thermalization occurs on the order of femtoseconds to picoseconds, while radiative, Auger and SRH recombinations are on the order of nanoseconds to microseconds or longer. Under this assumption we can replace the Fermi energy in our carrier density formulae with the quasi-Fermi energy, such that:

$$n = n_i e^{\left(E_{F_n} - E_i\right)/kT} \tag{2.5.10}$$

$$p = n_i e^{\left(E_i - E_{Fp}\right)/kT} \tag{2.5.11}$$

Thus, the np product can be redefined under non-equilibrium conditions as:

$$np = n_i^2 e^{\left(E_{F_n} - E_{Fp}\right)/kT} \tag{2.5.12}$$

The separation of quasi-Fermi levels is just an indication of the degree of excess electron-hole pairs under non-equilibrium conditions. By applying the above equations to our definition for total electron and hole current, we can express the currents in terms of the quasi-Fermi level as:

$$\mathbf{J}_n = \mu_n n \nabla E_{F_n} \tag{2.5.13}$$

$$\mathbf{J}_p = \mu_p p \nabla E_{F_p} \tag{2.5.14}$$

These relationships represent a compact mathematical form of the semiconductor current equations and are thus often more convenient to use when analyzing semiconductors under light or voltage bias. These equations also give some physical insight into band diagrams showing quasi-Fermi levels. Gradients in quasi-Fermi levels in a band diagram show, at a glance, that current must be flowing through that region of the semiconductor.

2.5.6 Continuity Equations

The carrier processes discussed in the previous sections (current flow, recombination and generation) can be related to one another though the continuity equations. These equations are simply a statement of the conservation of charge for each carrier type. Consider the example in Figure 2.5.3 for electron flow into a one-dimensional region of a semiconductor. The time rate of change of electrons in the slice *dx* is just the number of electrons flowing in from the left, minus the number leaving from the right, plus the difference between any electron generation (G_n) and recombination (U_n) inside the slice. Mathematically we can express this as

$$\frac{\partial n}{\partial t} = \frac{1}{q}\frac{\partial J_n}{\partial x} + \left(G_n - U_n\right) \tag{2.5.15}$$

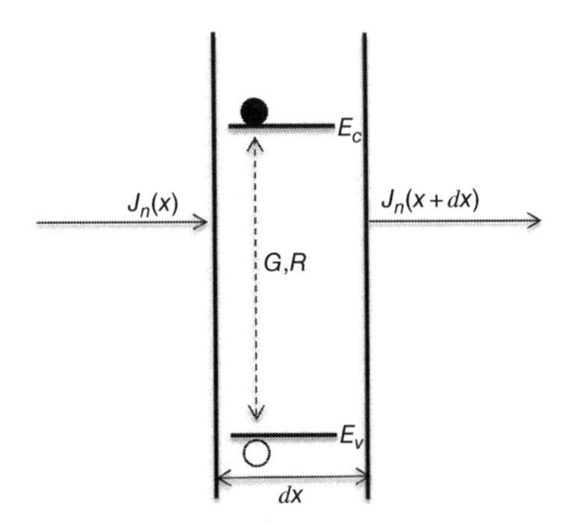

Figure 2.5.3 Illustration of the concept leading to the continuity equation

where the derivative arises by taking a Taylor series expansion of $J_n(x) - J_n(x+dx)$. The analysis can be applied to both electrons and holes and extends to three dimensions using the divergence for electrons:

$$\frac{\partial n}{\partial t} = \frac{1}{q}\nabla \bullet \mathbf{J}_n + (G_n - U_n) \tag{2.5.16}$$

and for holes:

$$\frac{\partial p}{\partial t} = \frac{1}{q}\nabla \bullet \mathbf{J}_p + (G_p - U_p) \tag{2.5.17}$$

These two equations represent very general equations of state for any semiconductor device. These equations are complex, involving both time and space derivatives, as well as spatially varying generation and recombination processes. A closed form solution is often not possible. The continuity equations can of course be used directly in numerical device simulation software. However, in Section 2.5.7 we introduce a number of simplifying assumptions that allow us to apply these equations to derive a closed form solution of the ideal *pn* junction diode.

2.5.7 Minority Carrier Transport Equations

In order to apply the continuity equations to the analysis of a pn junction, we make a set of simplifying assumptions to arrive at a useful set of transport equation that can be solved in closed form. We first restrict our analysis to one dimension so that the gradients can be replaced with partial derivatives along one coordinate. Applying the continuity equations to the definitions of drift and diffusion current in Section 2.5.3 and assuming that the mobility is not spatially varying lead to a set of transport equations applicable to both majority and minority carrier devices, for electrons:

$$\frac{\partial n}{\partial t} = \mu_n n \frac{\partial E}{\partial x} + \mu_n E \frac{\partial n}{\partial x} + D_n \frac{\partial^2 n}{\partial x^2} + G_n - U_n \tag{2.5.18}$$

and for holes:

$$\frac{\partial p}{\partial t} = -\mu_p p \frac{\partial E}{\partial x} - \mu_p E \frac{\partial p}{\partial x} + D_p \frac{\partial^2 p}{\partial x^2} + G_p - U_p \tag{2.5.19}$$

Furthermore, in devices dominated by minority carrier actions, such as the pn junction diode, we can further simplify the equations above by assuming that:

1. Analysis is restricted to minority carriers. We then use a subscript to denote which region the carrier is in, for example, n_p for minority carrier electrons in p-type material.
2. As discussed in Chapter 2.6, analysis will be restricted to the quasi-neutral regions of the diode such that $E \approx 0$. Also, we assume any disturbance from equilibrium is small (low-level injection) so that the thermal equilibrium values are basically unchanged (e.g., $n \simeq n_0$ and $\Delta p \ll n_0$ for n-type material). In this case, both the first and second terms of Equations (2.5.18) and (2.5.19) can be ignored.

3. The thermal carrier concentrations (n_0,p_0) are assumed to be independent of position. Therefore, we are only concerned with derivatives with respect to *excess* electron and hole populations.
4. For solar cells, we are typically only concerned with behavior at a fixed bias voltage and with constant illumination from the sun. Thus, operation in the steady state leads to $\frac{\partial p}{\partial t} = \frac{\partial n}{\partial t} = 0$.
5. Under the low-level injection condition we can replace the net recombination term with the recombination mechanisms discussed in Chapter 2.4, namely:

$$U_n = \Delta n\left(\frac{1}{\tau_{n,r}} + \frac{1}{\tau_{n,Auger}} + \frac{1}{\tau_{n,SRH}}\right) = \frac{\Delta n}{\tau_n} \tag{2.5.20}$$

   where τ_n is the net minority carrier lifetime due to radiative, Auger and SRH recombination.
6. The only source of generation is photogeneration, so that electron and holes are produced in pairs and $G_n = G_p = G_L$.

Applying these assumptions to Equations (2.5.18) and (2.5.19) leads to a pair of second order, ordinary differential equations that can easily be solved given the appropriate boundary conditions:

$$D_n \frac{\partial^2 \Delta n_p}{\partial x^2} - \frac{\Delta n_p}{\tau_n} + G_L = 0 \tag{2.5.21}$$

for minority carrier electrons in a p-type material, and

$$D_p \frac{\partial^2 \Delta p_n}{\partial x^2} - \frac{\Delta p_n}{\tau_p} + G_L = 0 \tag{2.5.22}$$

for minority carrier holes in n-type material. Equations (2.5.21) and (2.5.22) are known as the minority carrier diffusion equations. These represent the working relationships that, in combination with Poisson's equation, are used to model the carrier densities and currents in a minority carrier device, with respect to applied voltage and illumination.

For a list of symbols we refer to Chapter 2.6.

References

Muller, R.S., Kamins, T.I. *et al.* (2003) *Device Electronics for Integrated Circuits*. John Wiley & Sons, Inc., New York.
Pierret, R.F. (1996) *Semiconductor Device Fundamentals*, Addison-Wesley, Reading, MA.
Singh, J. (1993) *Physics of Semiconductors and their Heterostructures*. McGraw-Hill, New York.
Sze, S.M. and Ng, K.K. (2007) *Physics of Semiconductor Devices*. Wiley-Interscience, Hoboken, NJ.

2.6

PN Junctions and the Diode Equation

Seth Hubbard
NanoPower Research Laboratories, Rochester Institute of Technology (RIT), Rochester, NY, USA

The pn junction is the heart of many solar cells (e.g. Si, GaAs and other inorganic devices). Variations on this concept, such as heterojunction and selective contacts devices, account for many additional types of solar cells. In this section, we first analyze the physics of a pn homojunction. We then apply the minority carrier diffusion equations to the pn junction to arrive at the ideal diode equation. Finally, we will present non-ideal behavior in the diode by considering currents arising from the space charge region.

2.6.1 Properties of a pn Homojunction

Junctions can be formed either by diffusion of dopants of an opposite type (e.g., phosphorus diffused into boron doped p-type Si wafers) or epitaxially (using chemical or physical deposition methods). For simplicity of analysis, we consider in this chapter the second case, an abrupt epitaxial pn junction. Additional types of junctions such as the linear and exponential graded junctions and the heterojunction exhibit similar behavior and are discussed extensively in the semiconductor texts (Singh, 1993; Pierret, 1996; Sze and Ng, 2007).

The basic energy band diagram of the pn junction is shown in Figure 2.6.1. When separate, the n-type and p-type regions are described by thermal equilibrium statistics with $p_0 = N_A$ on the p-side and $n_0 = N_D$ on the n-side. After contact, due to the large imbalance in carrier type between the two regions, electrons flow to the left and holes to the right. However, this leaves uncompensated ionized donor (N_D^+) and acceptor ions (N_A^-), i.e., a fixed space charge, on either side of the junction. The fixed charge then creates an electric field across the junction (from right to left) that opposes further carrier motion. At equilibrium the net diffusion current across

Photovoltaic Solar Energy: From Fundamentals to Applications, First Edition.
Edited by Angèle Reinders, Pierre Verlinden, Wilfried van Sark, and Alexandre Freundlich.

Companion website: www.wiley.com/go/reinders/photovoltaic_solar_energy

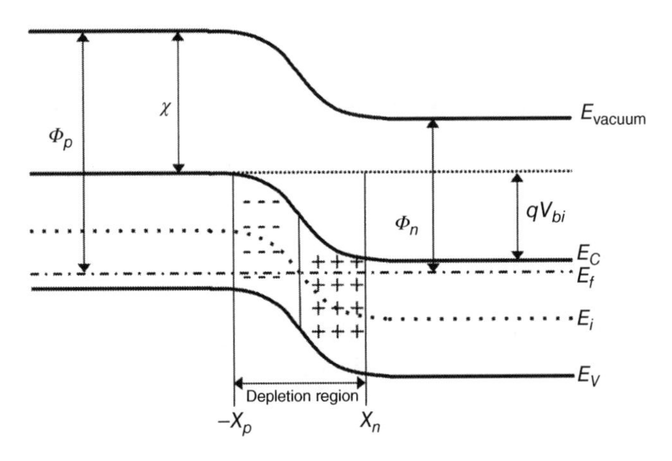

Figure 2.6.1 A pn junction band diagram showing the Fermi level, depletion region and built-in voltage

the junction is exactly balanced by the net drift current. Also, the Fermi level on both sides is now equal and the built-in field created by the fixed space charge leads to an energy barrier (qV_{bi}).

Note that all the potential is dropped across the junction region and that far from the junction, each region has returned to its respective equilibrium carrier density. We can see from Figure 2.6.1 that the free carrier concentration will rapidly decrease in the vicinity of the junction. We call this region the *depletion region* or *space charge region* due to the fact that most of the free carrier population is depleted and only the fixed ionized space charge remains.

The built-in voltage V_{bi}(V) can be found by considering the difference in work function for each side of the junction. This can then be related to the shift in E_i with respect to E_F as:

$$V_{bi} = \frac{1}{q}\left(\phi_n - \phi_p\right) = \frac{1}{q}\left[\left(E_i - E_F\right)_{p-side} + \left(E_F - E_i\right)_{n-side}\right] \tag{2.6.1}$$

where ϕ_n(eV) and ϕ_p(eV) are the work functions for the n-type and p-type materials, respectively.

The Fermi energy difference can then easily be related to the equilibrium doping levels on each side of the junction using Equations (2.5.10) and (2.5.11), giving,

$$V_{bi} = \frac{kT}{q}\ln\left(\frac{N_A N_D}{n_i^2}\right) \tag{2.6.2}$$

If an external voltage is applied to the pn junction, the built-in voltage will be modified accordingly. Under the low-level injection conditions discussed previously, we assume that very little voltage is dropped across the contacts or the bulk of the semiconductor and that the entire applied voltage will appear across the depletion region. For a voltage V_a(V) applied to the p-type region, the junction voltage becomes:

$$V_j = V_{bi} - V_a \tag{2.6.3}$$

Thus, for positive voltage, the built-in potential barrier is lowered from its equilibrium value. This will tend to increase the diffusion of majority carriers across the depletion region, leading to an exponential increase in current. However, for a negative voltage, the built-in potential barrier increases. This effectively eliminates any diffusion of majority carriers across the depletion region. The only current that will flow will be a very small drift current due to minority carrier electrons on the p-side and holes on the n-side. This behavior leads to the rectifying behavior expected from a diode, in other words, large current flow in the forward bias and almost no current under the reverse bias.

In addition to the rectification properties discussed above, the asymmetry to different types of current flow (electron vs. hole) produced by the pn junction leads to its photovoltaic properties. Consider a uniform light illumination of the pn junction that leads to excess electron hole pairs on either side of the junction. The excess minority carrier electrons near the depletion region on the p-side are easily "swept" across the junction by the built-in electric field, while the excess holes are blocked and must remain on the p-side. A similar situation occurs for minority carrier holes on the n-side. In effect, the junction asymmetry has provided a low resistance path for electrons from the p-side to flow to the n-side and for holes from the n-side to flow to the p-side, while majority carriers on each side are forced to the ohmic contacts. This separation of charge pairs will then drive a photocurrent (in short circuit) or a photovoltage (in open circuit).

In order to analyze the pn junction further, we begin with Poisson's equation for a doped semiconductor:

$$\nabla^2\phi = -\frac{\rho}{\varepsilon_s} = -\frac{q}{\varepsilon_s}\left(p - n + N_D - N_A\right) \tag{2.6.4}$$

where ρ(C/cm^3) is the space charge density, ε_s the permittivity of the semiconductor and the dopants are assumed to be fully ionized such that $N_D^+ = N_D$ and $N_A^- = N_A$. In general, for an arbitrary charge distribution, the solution would require a numerical approach. Even in the case of the pn step junction, the actual variations in the space charge near the edges of the depletion region lead to a more involved solution. However, two assumptions can be made to simplify the solution of Poisson's equation:

1. *Quasi-neutrality*: All voltage is assumed to be dropped across the junction, thus $E \simeq 0$ from the p-type contact to $x = -x_p$ and from $x = x_n$ to the n-type contact, where $-x_p\,(\text{cm})$ and x_n(cm) are the edges of the depletion regions as indicated in Figure 2.6.1, with $x=0$ occurring at the metallurgical junction.
2. *Depletion approximation*: The space charge region is assumed to be fully depleted of mobile carriers up to the edges of the depletion region. Thus, $\rho = N_A$ for $-x_p < x < 0$ and $\rho = N_D$ for $0 < x < x_n$.

These assumption are illustrated in Figure 2.6.2 (a), for a pn junction that has $N_A > N_D$. Due to the higher acceptor doping, more of the n-type donor region is depleted. We can now split the solution of Poisson's equation into two regions such that (in one dimension):

$$\frac{d^2\phi}{dx^2} = -\frac{q}{\varepsilon_s}N_A \tag{2.6.5}$$

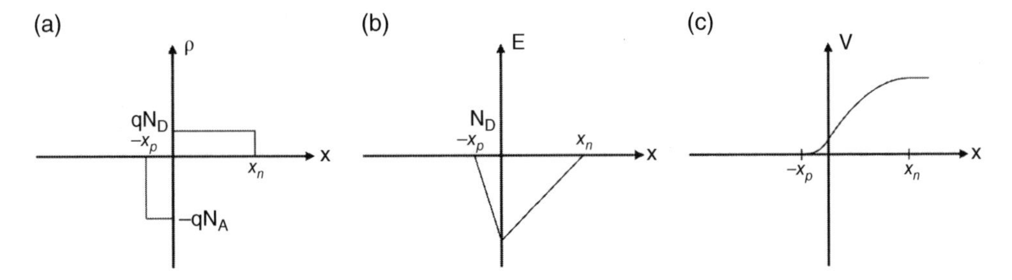

Figure 2.6.2 Plots of (a) charge density; (b) electric field; and (c) potential across a pn junction

for $-x_p < x < 0$, and

$$\frac{d^2\phi}{dx^2} = \frac{q}{\varepsilon_s} N_D \tag{2.6.6}$$

for $0 < x < x_n$. Integration of these two equations once, with the boundary conditions of $E = 0$ at $x = -x_p$ and $x = x_n$, gives the electric field across the depletion region.

$$E(x) = -\frac{qN_A}{\varepsilon_s}(x + x_p) \tag{2.6.7}$$

for $-x_p < x < 0$

$$E(x) = -\frac{qN_D}{\varepsilon_s}(x_n - x) \tag{2.6.8}$$

for $0 < x < x_n$. This result, shown graphically in Figure 2.6.2 (b), gives a negative electric field that is linearly increasing and decreasing on both sides of the junction. As expected, the field direction produces a force to oppose the additional diffusion flow of electrons toward the left and holes to the right. The field will be maximum at the metallurgical junction, x=0. In addition, since our device is a homojunction, the field must be continuous at the junction, which gives a simple condition to relate the depletion widths on each side of the junction, namely, $N_A x_p = N_D x_n$.

A second integration can then be performed to yield the potential with the required boundary conditions that $\phi = 0$ at $x = -x_p$ and $\phi(x_n) = V_{bi}$,

$$\phi(x) = \frac{qN_A}{2\varepsilon_s}(x + x_p)^2 \tag{2.6.9}$$

for $-x_p < x < 0$

$$\phi(x) = V_{bi} - \frac{qN_D}{2\varepsilon_s}(x_n - x)^2 \tag{2.6.10}$$

for $0 < x < x_n$. The form of the potential can be seen in Figure 2.6.2 (c). The potential is quadratic in nature and is just a mirror image of the band diagram, as expected, based on the definitions of potential and potential energy.

The depletion region edges x_n and x_p can be found by forcing both E and ϕ to be continuous at the metallurgical junction. This results in the following set of equations:

$$x_p = \sqrt{\frac{2\varepsilon_s V_{bi}}{q}\left(\frac{N_D}{N_A(N_A + N_D)}\right)} \tag{2.6.11}$$

$$x_n = \sqrt{\frac{2\varepsilon_s V_{bi}}{q}\left(\frac{N_A}{N_D(N_A + N_D)}\right)} \tag{2.6.12}$$

$$x_d = x_p + x_n = \sqrt{\frac{2\varepsilon_s V_{bi}}{q}\left(\frac{1}{N_A} + \frac{1}{N_D}\right)} \tag{2.6.13}$$

The depletion region width is inversely proportional to the doping levels and is only symmetric for $N_A = N_D$. In many solar cells, the half of the junction nearest the illumination source (emitter) is doped by at least a factor of 10 greater than the other half of the junction (base). The majority of the depletion region will then lie in the base of the solar cell, with typically depletion widths on the order of 10–100 nm. In forward bias, the junction voltage is reduced by the factor $V_{bi} - V_a$ and the depletion region width will shrink. In reverse bias, the junction voltage is increased by the factor $V_{bi} + V_a$ and the depletion region width will increase as well.

2.6.2 Ideal pn Diode in the Dark

We are now in a position to analyze the pn junction itself. The assumptions made in the previous section are still valid, namely, that $E \simeq 0$ in the quasi-neutral region and the bias level does not approach V_{bi}, so we stay in a low-level injection condition. As shown in Figure 2.6.3, we will analyze a pn junction diode of cross-sectional area A, with a p-type emitter and n-type base. The emitter ohmic contact is made at $-W_E$ and the base ohmic contact at W_B. The n-region is grounded and a voltage V_a is applied to at the p-region. While the applied voltage is certainly dropped across both the quasi-neutral regions and the junction, the resistivity of the semiconductors and the current density of the diode are typically low enough that any voltage across the quasi-neutral region can effectively be ignored when compared to the junction voltage, lending validity to our initial assumption. In this section, there is no external illumination of the diode.

The relevant equations to describe the carrier density and current on the p-side of the semiconductor are found using the minority carrier diffusion equations from Section 2.5.7 with $G_L = 0$:

$$D_n \frac{d^2 \Delta n_p}{dx^2} - \frac{\Delta n_p}{\tau_n} = 0 \tag{2.6.14}$$

$$J_n \simeq qD_n \frac{d\Delta n}{dx} \tag{2.6.15}$$

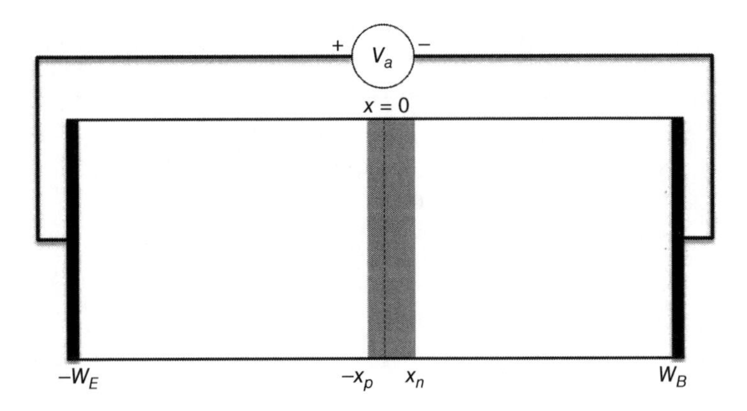

Figure 2.6.3 A pn junction diode under an external bias V_a

where we assume that the only minority carrier current component in the quasi-neutral region is diffusion since $E \simeq 0$. Similar equations can be obtained for the n-side quasi-neutral region. We also assume to first order that the depletion region does not produce excess current (this will be revisited in Section 2.6.3). Thus, J_n and J_p must be constant across the depletion region. Since this is a two-terminal device, the current must be a constant across the entire device, and we can express the total current as the sum of the electron and hole currents at the depletion edges:

$$J = J_N\left(-x_p\right) + J_p\left(x_n\right) \tag{2.6.16}$$

Therefore, in order to obtain a solution for the pn junction diode, we can solve the minority carrier diffusion equations for both electrons and holes, calculate currents, then add the values of these currents at the depletion region edges. The general solution of Equation 2.6.14, for carrier density on the p-side of the device, is a simple exponential of the form

$$\Delta n_p\left(x\right) = Ae^{\frac{x+x_p}{\sqrt{D_n\tau_n}}} + Be^{-\frac{x+x_p}{\sqrt{D_n\tau_n}}} \tag{2.6.17}$$

for $-W_E < x < -x_p$. We can define a new quantity, the minority carrier diffusion length L_n(cm) for electrons as

$$L_n = \sqrt{D_n\tau_n} \tag{2.6.18}$$

This length designates the average distance an electron will diffuse before recombination with a majority carrier hole. A similar solution for the n-side can be found with hole minority carrier diffusion length L_p(cm) as:

$$L_p = \sqrt{D_p\tau_p} \tag{2.6.19}$$

which indicates the average distance a hole will diffuse in n-type material before recombination with an electron. The constants A and B are found by considering the two limiting cases of $x = -W_E$ and $x = -x_p$. In many solar cells, the emitter and base widths are many times greater than L_n and L_p. Therefore, the excess minority carrier density must tend to zero at $-W_E$

and therefore B must be zero. The constant A is determined from knowledge of $\Delta n_p(-x_p)$. This can be found from the concept of the quasi-Fermi level discussed previously. We have assumed that some degree of quasi-thermal equilibrium is established in the depletion region. In this case E_{F_n} and E_{F_p} are constant across the depletion region and $qV_a = E_{F_n} - E_{F_p}$. The excess current density can then be found from the np product under non-equilibrium conditions (Equation 2.5.12) evaluated at $-x_p$, such that:

$$\Delta n_p\left(-x_p\right) = \frac{n_i^2}{N_A}\left(e^{qV/kT} - 1\right) = n_{p0}\left(e^{qV/kT} - 1\right) \tag{2.6.20}$$

where n_{p0} is the equilibrium minority carrier concentration on the p-side. Using this value for our boundary condition, the excess minority carrier density and current density on the n-side are:

$$\Delta n_p\left(x\right) = \frac{n_i^2}{N_A}\left(e^{qV/kT} - 1\right)e^{\frac{x+x_p}{L_n}} \tag{2.6.21}$$

and

$$J_n\left(x\right) = \frac{qn_i^2 D_n}{N_A L_n}\left(e^{qV/kT} - 1\right)e^{\frac{x+x_p}{L_n}} \tag{2.6.22}$$

The injected minority carrier electron current is thus maximum across the depletion region (from x_n to $-x_p$) and decays exponentially (with decay constant L_n) toward the p-type contact as electrons recombine with majority carrier holes. The majority carrier current, based on our definition, is then just $J_p(x) = J(x) - J_N(x)$. Due to our assumption that $L_n >> W_E$, at the contact, the current will be entirely made up of majority carrier holes.

A similar analysis of the minority carrier hole current on the n-side gives:

$$J_p\left(x\right) = \frac{qn_i^2 D_p}{N_D L_p}\left(e^{qV/kT} - 1\right)e^{-\frac{x-x_n}{L_p}} \tag{2.6.23}$$

The total current is then just:

$$J\left(V\right) = J_N\left(x_n\right) + J_p\left(-x_p\right) = qn_i^2\left[\frac{D_n}{N_A L_n} + \frac{D_p}{N_D L_p}\right]\left(e^{qV/kT} - 1\right) \tag{2.6.24}$$

$$J\left(V\right) = J_{0,diff}\left(e^{qV/kT} - 1\right) \tag{2.6.25}$$

where we define $J_{0,diff}$ as the *saturation current density* due to carrier diffusion in the quasi-neutral region. The above equation is often referred to as the *ideal diode equation*. As can be seen, this equation describes the expected rectifying behavior of the diode. The saturation current density is typically a very small number, on the order of $10^{-10}\,\text{mA/cm}^2$ in Si and $10^{-17}\,\text{mA/cm}^2$ in GaAs. In forward bias, the barrier to minority carried diffusion is lower and we expect a large current flow. In the equation, beyond a few *kT/q* of forward bias, the exponential term dominates and we indeed predict a large current flow. In reverse bias, the barrier to minority carrier diffusion is very large and only a small drift current should flow, given by $-J_{0,diff}$. This is shown graphically for an example diode in Figure 2.6.4 (a), showing a linear plot giving the expected exponential turn-on in forward bias. Figure 2.6.4 (b) shows a logarithmic plot of the ideal diode (dotted line), illustrating the large asymmetry between forward and reverse currents.

While the assumptions that lead to the ideal diode equation are many, the analysis presented represents the starting point for more complex devices. For example, in solar cells where the emitter and base width are not significantly greater than the diffusion length, the front and back surface recombination will influence the diode current. However, this leads only to a

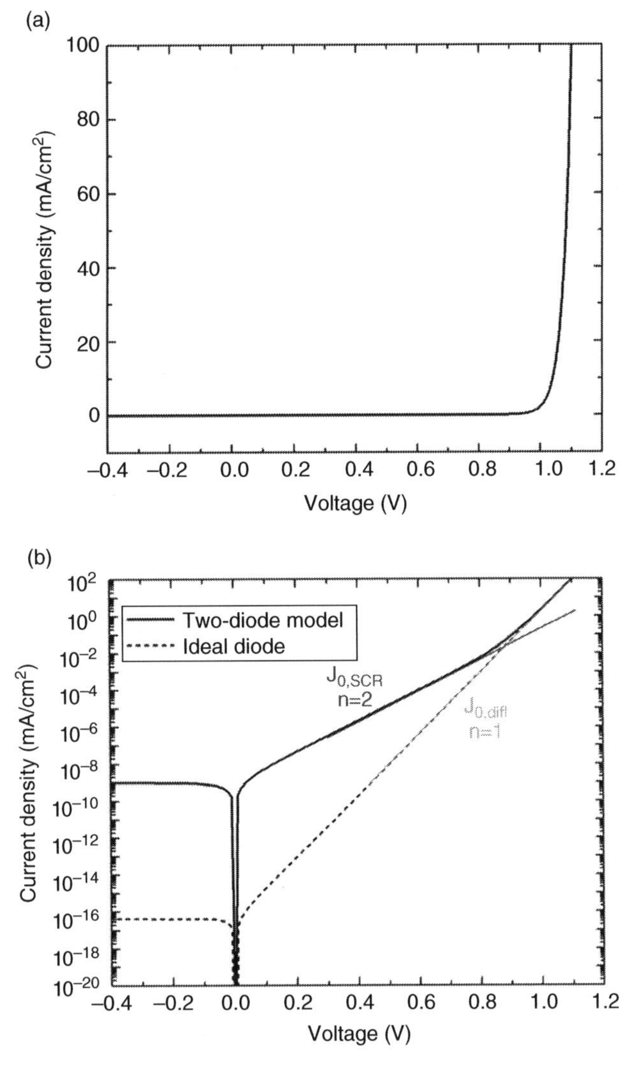

Figure 2.6.4 Examples of dark current density versus voltage (JV) curves showing (a) rectifying behavior when plotted on a linear scale; and (b) reverse saturation current and diode ideality factor when plotted on a logarithmic scale

change in $J_{0,diff}$ and not to the voltage dependence. Diode analysis for this situation as well as for hetero-structures can be found in a number of solar cells textbooks (Fahrenbruch and Bube, 1983; Nelson, 2003).

2.6.3 Depletion Region Effects and the General Diode Equation

In Section 2.6.1, we assumed that generation-recombination in the depletion region did not add any additional current to the overall device. However, in many solar cells this is not the case, especially at low bias or under low illumination conditions. In this case, we can calculate the depletion region current directly from the continuity equation as:

$$J_{SCR} = \int_{-x_p}^{x_n} (U - G)\, dx \tag{2.6.26}$$

An approximate solution of this equation in forward bias is given by Sah *et al.* for SRH recombination centers at mid-gap (Sah, *et al.* 1957). The form of this space charge region (SCR) current is given as:

$$J_{SCR}(V) = \frac{qn_i x_d}{\sqrt{\tau_n \tau_p}}\left(e^{qV/2kT} - 1\right) = J_{0,SCR}\left(e^{qV/2kT} - 1\right) \tag{2.6.27}$$

where we define $J_{0,SCR}$ as the saturate current density due to SCR recombination. This term will depend strongly on the nature of the trap states, the width of the depletion region (which is bias dependent) and both the electron and hole lifetimes. Also, note that $J_{SCR}(V)$ has a factor of $(2kT)^{-1}$ contained in the exponential factor as compared to $(kT)^{-1}$ for the ideal diode. The total diode current for a solar cell in the dark is then expressed as:

$$J(V) = J_{0,diff}\left(e^{qV/kT} - 1\right) + J_{0,SCR}\left(e^{qV/2kT} - 1\right) \tag{2.6.28}$$

Because the saturation current density due to space charge region recombination can be significantly greater ($\sim 10^{-10}\,\text{mA/cm}^2$ or higher) than the diffusion based saturation current density, it will tend to dominate the IV curve at lower voltage bias. Examining a logarithmic plot of current density, as in Figure 2.6.4 (b), we can see a marked change in the slope of the two-diode model curve around the point where $J_{0,diff}$ begins to become a significant contribution to the forward current. In high quality solar cells, the SCR and diffusion-dominated regions of the dark IV curve are easily distinguished and can be modeled using Equation 2.6.28. However, in other devices, more than one process can dominate and separate region are not so easily identified. In this case, we define a general diode equation:

$$J(V) = J_0\left(e^{qV/mkT} - 1\right) \tag{2.6.29}$$

where we define an *ideality factor m* and saturation current J_0. The ideality factor has a value $m = 1$ for the ideal diode and $m = 2$ for a space charge region limited diode. Higher values of ideality can sometimes be measured in non-ideal and heterojunction devices, but its physical significance in that case is limited, and ideality is only used as a fitting factor.

Acknowledgments

The author would like to thank Zachary Bittner for his help in creating the illustrations in chapters 2.3–2.6.

List of Symbols

Symbol	Description	Unit
α	Absorption coefficient	cm^{-1}
β_{12}	Recombination coefficient	cm^3s^{-1}
σ	Conductivity of semiconductor	Siemens/cm
σ_s	Capture cross-section for the surface states	cm^2
σ_n	Electron capture cross-section coefficient	cm^2
σ_p	Hole capture cross-section coefficient	cm^2
Γ	Flux	photons/cm^2/s
$\hbar$	Planck's constant	$6.626*10^{-34}$joule-second
λ	Wavelength	cm
ε_s	Permittivity of semiconductor	unitless
Δn	Deviation of electron carrier density from its equilibrium value	cm^{-3}
Δn_p	Excess electron density in p-type material	cm^{-3}
Δp_n	Excess hole density in n-type material	cm^{-3}
Δp	Deviation of hole carrier density from its equilibrium value	cm^{-3}
ϕ_n	n-type semiconductor work function	eV
ϕ_p	p-type semiconductor work function	eV
ρ	Resistivity of semiconductor	Ω-cm
τ_c	Mean scattering time	s
τ_{rad}	Minority carrier radiative lifetime	s
τ_n	Net electron lifetime	s
$\tau_{n,auger}$	Electron Auger lifetime	s
$\tau_{n,rad}$	Electron radiative lifetime	s
$\tau_{n,SRH}$	Electron Shockley-Read-Hall lifetime	s
τ_p	Net hole lifetime	s
$\tau_{p,auger}$	Hole carrier Auger lifetime	s
$\tau_{p,rad}$	Hole radiative lifetime	s
$\tau_{p,SRH}$	Hole Shockley-Read-Hall lifetime	s
μ_n	Electron mobility	cm^2/Vs
μ_p	Hole mobility	cm^2/Vs
ω	Angular frequency	rad/s
Ω	Phonon frequency	s^{-1}
v_{th}	Thermal velocity	cm/s
B_r	Radiative recombination coefficient	cm^3/s
$B_{Auger,p}$	Hole Auger recombination coefficient	cm^3/s
$B_{Auger,n}$	Electron Auger recombination coefficient	cm^3/s
B_n	Electron capture coefficient	cm^3/s

(*Continued*)

(*Continued*)

Symbol	Description	Unit
B_p	Hole capture coefficient	cm^3/s
D_n	Electron diffusion coefficient	cm^2/s
D_p	Hole diffusion coefficient	cm^2/s
E	Energy	eV
E_c	Conduction band energy	eV
E_{F_n}	Quasi-Fermi energy for electrons	eV
E_{F_p}	Quasi-Fermi energy for holes	eV
E_g	Bandgap energy	eV
E_i	Intrinsic energy level	eV
E_t	Trap energy	eV
$E_{t,s}$	Surface trap energy	eV
E_v	Valence band energy	eV
$g(E,x)$	Specific photogeneration rate	ehp/cm^3s
G	Total generation rate	ehp/cm^3s
G_n	Electron generation	ehp/cm^3s
G_L	Light generation	ehp/cm^3s
G_p	Hole generation	ehp/cm^3s
I	Illumination intensity	mW/cm^2
J_0	Generalized saturation current density	mA/cm^2
$J_{0,diff}$	Saturation current density due to diffusion	mA/cm^2
$J_{0,SCR}$	Saturation current density due to SCR recombination	mA/cm^2
J_{drift}	Drift current	mA/cm^2
J_{diff}	Diffusion current	mA/cm^2
J_n	Total electron current	mA/cm^2
$J_{n,diff}$	Diffusion current for electrons	mA/cm^2
$J_{n,drift}$	Drift current for electrons	mA/cm^2
J_p	Total hole current	mA/cm^2
$J_{p,diff}$	Diffusion current for holes	mA/cm^2
$J_{p,drift}$	Drift current for holes	mA/cm^2
J_s	Surface recombination current	mA/cm^2
J_{SCR}	Depletion region current	mA/cm^2
J_{Total}	Total free carrier current	mA/cm^2
κ	Extinction coefficient	unitless
k	Wavenumber	cm^{-1}
$\mathbf{k}$	Crystal wavevector	cm^{-1}
L_a	Absorption length	cm
L_n	Diffusion length for electrons	cm
L_p	Diffusion length for holes	cm
$m^*_{e,h}$	Electron/hole effective mass (e denotes electron, h denotes hole)	kg
n	Electron carrier density	cm^{-3}
m	Ideality factor	unitless
n_i	Intrinsic carrier density	cm^{-3}
n_{p0}	Equilibrium electron concentration on p-side of diode	cm^{-3}

Symbol	Description	Unit
n_t	Number of traps filled with electrons	cm^{-3}
n_0	Equilibrium electron carrier density	cm^{-3}
n_s	Surface electron density	cm^{-3}
$\hat{n}$	Electron density when the Fermi level is located at the trap energy	cm^{-3}
N_A	Density of acceptors	cm^{-3}
N_A^+	Density of ionized acceptor	cm^{-3}
N_C	Effective density of conduction band states	cm^{-3}
N_D	Density of donors	cm^{-3}
N_D^+	Density of ionized donor	cm^{-3}
$N_{t,s}$	Surface or interface trap density	cm^{-3}
N_t	SRH trap density	cm^{-3}
N_v	Total number of holes	cm^{-3}
p	Hole carrier density	cm^{-3}
p_s	Surface hole density	cm^{-3}
p_0	Equilibrium hole carrier density	cm^{-3}
$\hat{p}$	Hole density when the Fermi level is located at the trap energy	cm^{-3}
q	Charge of an electron	C
s_n	Electron surface recombination velocity	cm/s
s_p	Hole surface recombination velocity	cm/s
T	Temperature	K
R	Reflectance at surface of a material	
U_{Auger}	Auger recombination rate	cm^3/s
U_{rad}	Spontaneous radiative recombination rate	cm^3/s
U_{SRH}	Shockley-Read-Hall recombination rate	cm^3/s
$U_{SRH,s}$	Shockley-Read-Hall surface recombination rate	cm^3/s
V_{bi}	Built-in potential	eV
v_d	Drift velocity	cm/s
v_{th}	Thermal velocity	cm/s
V	Voltage	V
V_a	Applied voltage	V
V_{bi}	Built-in voltage	V
V_j	Junction voltage	V
x_d	Total width of the depletion region	cm
x_n	n-side width of the depletion region	cm
x_p	p-side width of the depletion region	cm

List of Acronyms

EHP	Electron-hole-pair
SCR	Space charge region
SRH	Shockley-Read-Hall

References

Fahrenbruch, A.L. and Bube, R.H. (1983) *Fundamentals of Solar Cells: Photovoltaic Solar Energy Conversion.* Academic Press, New York.

Hall, R.N. (1952) Electron-hole recombination in germanium, *Physical Review*, **87** (2), 387.

Nelson, J. (2003) *The Physics of Solar Cells.* Imperial College Press, London.

Pierret, R.F. (1996) *Semiconductor Device Fundamentals.* Addison-Wesley, Reading, MA.

Sah, C.T., Noyce, R.N. *et al.* (1957) Carrier generation and recombination in p-n junctions and p-n junction characteristics. *Proceedings of the Institute of Radio Engineers*, **45** (9), 1228–1243.

Singh, J. (1993) *Physics of Semiconductors and their Heterostructures.* McGraw-Hill, New York.

Sze, S.M. and Ng, K.K. (2007) *Physics of Semiconductor Devices.* Wiley-Interscience, Hoboken, NJ.

Part Three

Crystalline Silicon Technologies

3.1

Silicon Materials: Electrical and Optical Properties

Andreas Fell
Fraunhofer Institute for Solar Energy Systems, ISE, Freiburg, Germany

3.1.1 Introduction

The aim of this chapter is to present the material properties of silicon which are relevant for an understanding and modelling of the performance of typical silicon solar cell devices. Published models and parameterizations of those properties are reviewed and discussed.

To identify the relevant material properties, it is helpful to look at the fundamental equations which describe the carrier transport in a semiconductor, and which form the basis of any simulation tool or analytical models. In their shortest form they can be written using the chemical potentials or quasi-Fermi potentials φ_{Fn} and φ_{Fp} of the electron and hole densities n and p, respectively, and are equivalent to the popular drift-diffusion model (as can be derived from Würfel and Würfel, 2009):

$$\text{Transport of electrons}: q\frac{\partial n}{\partial t} = q(G-R) - \nabla\left(n\mu_n \nabla\varphi_{Fn}\right) \tag{3.1.1}$$

$$\text{Transport of holes}: q\frac{\partial p}{\partial t} = q(G-R) + \nabla\left(p\mu_p \nabla\varphi_{Fp}\right) \tag{3.1.2}$$

$$\text{Poisson equation for electric potential } \varphi_{el}: \nabla\left(\varepsilon\nabla\varphi_{el}\right) = -q\left(N_D - N_A + p - n\right) \tag{3.1.3}$$

Here q denotes the elementary charge and ε the permittivity of silicon. Essentially the generation and recombination rate G and R, the electron and hole mobilities μ_n and μ_p, the (active)

Photovoltaic Solar Energy: From Fundamentals to Applications, First Edition.
Edited by Angèle Reinders, Pierre Verlinden, Wilfried van Sark, and Alexandre Freundlich.

Companion website: www.wiley.com/go/reinders/photovoltaic_solar_energy

donor and acceptor doping densities N_D and N_A, as well as a relation between carrier densities and quasi-Fermi levels are required to solve this system of equations (plus boundary conditions). The doping densities are commonly known and specific to the investigated cell design. The generation rate is determined by band-to-band absorption of photons generating electron–hole pairs, as described in the optical properties, Section 3.1.3. The mobilities, recombination rate and carrier densities are functions of various other material properties and electrical excitation conditions (the injection level), as described in the electrical properties in Section 3.1.2.

A general comment on the temperature is made at this point. The temperature influences all of the relevant properties to a different extent. Even small temperature changes, for example, between 25 °C (standard testing conditions) and 300 K (28.15 °C, often used for characterization and simulation), can make a significant difference in solar cell characteristics. Understanding silicon properties at varying temperatures is therefore important, given the large temperature variations of solar cells and modules in the field, and deliberate temperature changes in some characterization techniques. Not all relevant material properties are known accurately within a wide range of temperatures, requiring careful assessment of the validity of chosen material properties when investigating varying temperatures.

3.1.2 Electrical Properties

3.1.2.1 Bandgap and Intrinsic Carrier Density

A fundamental property of any semiconductor is its band gap E_g. The bandgap of silicon is close to the optimum regarding the efficiency potential of a single junction solar cell (Würfel and Würfel, 2009). It is related to the intrinsic carrier density n_i by

$$n_i^2 = N_C N_V \exp\left(-\frac{E_g}{k_B T}\right), \quad (3.1.4)$$

where k_B denotes the Boltzmann constant and T the absolute temperature. N_C and N_V are the density of states in the conduction and valance band, respectively, with the widely accepted values published in (Green, 1990):

$$N_C = 2.86\times10^{19}\left(T/300\text{K}\right)^{1.58}\ \text{cm}^{-3} \text{ and } N_V = 3.10\times10^{19}\left(T/300\text{K}\right)^{1.85}\ \text{cm}^{-3}.$$

The value for the intrinsic carrier density n_i has a major influence on modelling solar cell device performance (opposed to as in microelectronics), and several efforts within the PV community have been made to determine its value accurately. The latest published and widely accepted value is $9.65\times10^9\,\text{cm}^{-3}$ at a temperature of 300 K (Altermatt *et al.*, 2003). Note that n_i has a strong temperature dependence, with consequently a large impact on modelled cell characteristics. Temperature dependent measurements of n_i have been published by Misiakos and Tsamakis (1993), which must be multiplied by 0.9953 to match $n_i(300\,\text{K})=9.65\times10^9\,\text{cm}^{-3}$.

While the value for E_g was rather precisely measured, for example, in (Pässler, 2002), it is usually adjusted for modeling purposes to be consistent with the ultimately more important measured intrinsic carrier density. For solar cell modeling, a consistent temperature dependent value of E_g should therefore be calculated from the best known n_i, resulting in $E_g = 1.130\text{eV}$ at 300 K.

3.1.2.2 Bandgap Narrowing

For high carrier densities due to doping and/or injection level, the bandgap of silicon decreases by $\Delta E_{g,BGN}$, where BGN means bandgap narrowing. For typical solar cells, it mainly impacts the performance of highly doped regions (e.g., emitters), and needs to be accounted for when modelling those. The most relevant effect of BGN in this context is an increase of the intrinsic carrier density, resulting in an effective intrinsic carrier density $n_{i,eff}$ defined by:

$$n_{i,eff}^{\;2} = N_C N_V \exp\left(-\frac{E_g - \Delta E_{g,BGN}}{k_B T}\right) = n_i^2 \exp\left(\frac{\Delta E_{g,BGN}}{k_B T}\right). \tag{3.1.5}$$

Several studies on quantifying BGN in silicon can be found in the literature, with a good recent overview given in (Yan and Cuevas, 2013, 2014), indicating that significant uncertainty still exists. Two alternatives for calculating BGN can be considered as being most up-to-date:(1) the model published by Schenk (Schenk, 1998), which includes temperature and injection dependence; and (2) the parameterizations in (Yan and Cuevas, 2013, 2014), which are derived from fits to measurements of total recombination in highly doped regions and are therefore claimed to best reproduce this ultimately most important quantity. However, it is given at 300 K only, does not include injection dependence and, as it is rather new at the time of writing, has not yet been extensively applied and validated within the community.

Figure 3.1.1 shows the dependence of $n_{i,eff}$ on temperature and doping level calculated with the assumption of n_i described in Section 3.1.2.1, and Schenk's BGN model. A strong influence of temperature can be seen, while the influence of doping level (via BGN) is significant primarily for highly doped regions, and not the bulk (where typically $N_{A/D} < 1\times10^{16}\,\text{cm}^{-3}$).

Historically, what is called the "apparent" BGN was prominently used. This modifies the BGN value to effectively include the degeneration from Fermi-Dirac statistics to be able to still use the mathematically simpler Boltzmann statistics (see Section 3.1.2.3). While still useful for analytical modelling, the use of apparent BGN should be avoided if possible, as it is physically less meaningful and in some cases might lead to misinterpretation of experimental results, see, for example, Altermatt *et al.* (2002).

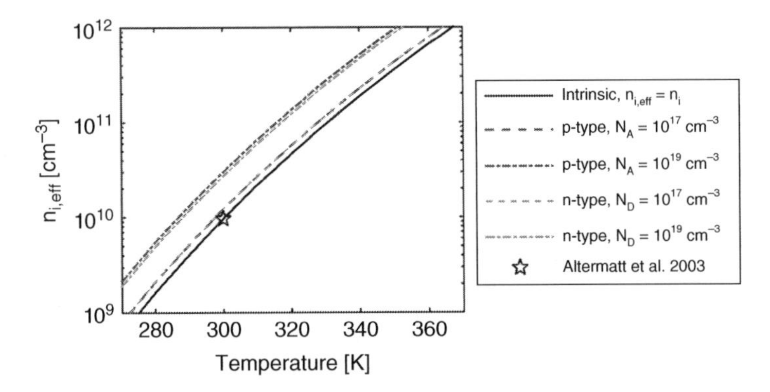

Figure 3.1.1 Effective intrinsic carrier density $n_{i,eff}$ as a function of temperature and doping level and type; lines are calculated using the models and assumptions discussed in the text; the symbol marks the measured value of n_i in. Source: (Altermatt *et al.*, 2003). (*See insert for color representation of the figure*)

3.1.2.3 *Carrier Statistics*

To solve the semiconductor Equations (3.1.1)–(3.1.3), one must relate the quasi-Fermi potentials to the carrier densities. Such a relation is derived from what are called carrier statistics, which describe the concentration of carriers in the valence and conduction band of a semiconductor for a given energy band structure. For silicon, the general applicable Fermi-Dirac statistics can be simplified to the mathematically less complex Boltzmann statistics if doping levels are low ($N_{A/D} < 1\times10^{19}\,\mathrm{cm}^{-3}$). Boltzmann statistics have consequently been popular in the PV community in particular for the derivation of analytical models, while proper Fermi-Dirac statistics can and should be used for numerical simulation of highly doped regions.

An important result from the carrier statistics is the relation of the *pn* product and the intrinsic carrier concentration n_i. For intrinsic (i.e. non-excited and non-doped) silicon, this results simply in $pn = n_i^2$, which expands to

$$pn = n_{i,eff}^{\;2}\gamma_{deg} exp\left(\frac{\varphi_{Fn} - \varphi_{Fp}}{V_t}\right), \tag{3.1.6}$$

to account for effects from doping and excitation (i.e. quasi-Fermi level splitting). Here $V_t = \frac{k_B T}{q}$ denotes the thermal voltage. The degeneracy factor γ_{deg} represents the deviation of Fermi-Dirac statistics from Boltzmann statistics, i.e. $\gamma_{deg} = 1$ represents Boltzmann statistics, see Altermatt *et al.* (2002).

3.1.2.4 *Mobility*

The mobility of electrons and holes μ_n and μ_p in silicon depend on the injection level and the doping densities. The most widely accepted and used mobility model for phosphorus or boron doped silicon is that of Klaassen (1992a, 1992b), which accounts for temperature dependence. It provides a good overall fit to measured data, as most recently comprehensively shown in (Altermatt, 2011). Figure 3.1.2 shows a comparison of measured minority electron and hole mobilities compared with Klaassen's model, indicating some considerable uncertainty of the minority hole mobility. In recent years compensated silicon, which has both phosphorous and boron concentrations in considerable amounts, has attracted interest. Here, measured mobilities differ significantly from the predictions of Klaassen's model, and empirical mobility models have been suggested in (Forster *et al.*, 2013; Schindler *et al.*, 2014).

3.1.2.5 *Intrinsic Recombination*

The intrinsic recombination is the "unavoidable" contribution to the recombination rate *R* related to fundamental physical mechanisms present even in perfect material quality. There are two different intrinsic recombination mechanisms: radiative and Auger recombination.

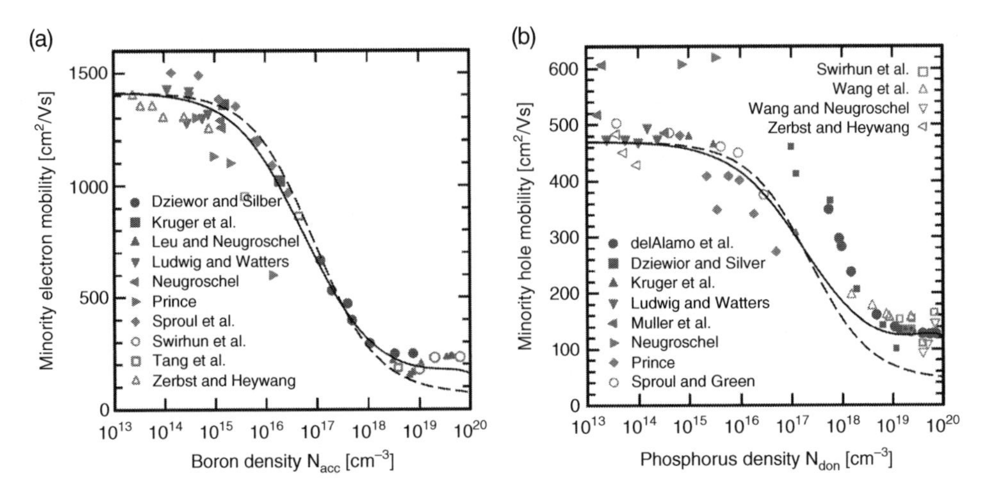

Figure 3.1.2 Minority electron (a) and hole (b) mobility as a function of doping density comparing Klaassen's model (solid line: minority mobility, dashed line: majority mobility) with various measurements (symbols). Source: (Altermatt, 2011), see also references therein

3.1.2.5.1 Radiative Recombination

With silicon being an indirect semiconductor, band-to-band recombination of an electron-hole pair by emitting a photon is a process with low probability. This makes radiative recombination an insignificant loss factor for most silicon solar cells. It is, however, used for popular characterization methods where the photons emitted via radiative recombination are detected, known primarily as electro- and photo-luminescence (EL and PL). The strong dependence on electrical properties and injection level can be used for contactless measurement of decisive solar cell properties. For this purpose, an accurate model for radiative recombination is important.

Commonly, the radiative recombination rate is calculated with the radiative recombination coefficient B_{rad} by $R_{rad} = B_{rad}pn$. B_{rad} can also be calculated from first principles using Planck's law, where it can be derived that the product $B_{rad}n_{i,eff}^{2}$ is solely a function of the complex refractive index of silicon, mainly of the absorption coefficient (Würfel and Würfel, 2009). This relation has been used to derive accurate temperature dependent values for the latter from luminescence measurements in (Nguyen *et al.*, 2014b), which have been shown to be consistent with earlier measurements of B_{rad} (Trupke *et al.*, 2003). Therefore, the temperature-dependent radiative recombination given in (Nguyen *et al.*, 2014a) is considered the best current choice to quantify the radiative recombination.

3.1.2.5.2 The Auger Recombination

The dominant intrinsic recombination process in silicon is the Auger recombination, where the energy is not released via a photon but via an increase of energy of a second electron, which subsequently thermalizes. It is significant for solar cells within highly doped regions, and also within the bulk when approaching efficiencies of 25% and beyond. The latest and most widely accepted parameterization of the Auger recombination as a function of doping and excitation is given in (Richter *et al*,. 2012). Notably, the parameterization is valid at 300 *K* only. However, for small (several *K*) temperature changes the influence of the strongly changing intrinsic carrier concentration dominates over the temperature dependence of the parameters. Significant errors must be expected for larger temperature variations.

3.1.2.6 The Shockley-Read Hall (SRH) Defect Recombination

The main loss contribution in common crystalline silicon solar cells is recombination via defects, mainly impurities within the crystal and surfaces (including grain-boundaries in multi-crystalline silicon). The defects essentially introduce energy states within the band-gap, which provide a low-energy pathway for electron-hole pairs to recombine. Unless trapping becomes significant (Macdonald and Cuevas, 2003), the recombination rate is well quantified by the simplified Shockley-Read-Hall (SRH) formalism (Shockley and Read Jr, 1952). It calculates the minority carrier lifetime τ_{SRH} from the defect energy level E_T and the fundamental electron and hole lifetime parameters τ_{n0} and τ_{p0}, which in turn are functions of the defect density N_T and the electron and hole capture cross-sections σ_n and σ_p. It should be noted that τ_{SRH} has generally a strong dependence on the injection level. To set an arbitrary bulk lifetime within solar cell modeling, it is therefore advisable not to use a constant value for τ_{SRH}, but rather to set $\tau_{n0} = \tau_{p0} = \tau_{bulk}$ and use a defect energy in the center of the bandgap, which will produce an approximately typical injection dependence of the bulk lifetime.

3.1.2.7 Metal Impurities

For the common elemental metal impurities, the fundamental defect properties E_T, σ_n and σ_p have been experimentally derived. A summary of values can be found in (Schmidt *et al.*, 2012).

3.1.2.8 SRH via Complexes

Besides elemental impurities, "complexes" of several elements can form and act as electrical defects. The most prominent one is the boron-oxygen (BO) complex, which is most prominent in boron-doped Czochralski (Cz) silicon. The defect activates only after prolonged exposure to light, resulting in what is called the light-induced degradation (LID) defect. Due to technological and economic reasons, limits exist as to by how much the oxygen content can be reduced. That is why LID was long believed to impose a fundamental limit for the conversion efficiency of industrial p-type solar cells. More recently, the possibility of permanent deactivation has been discovered (Herguth *et al.*, 2006) and current R&D efforts are aiming to gain an understanding and to introduce this into production. A widely used parameterization to quantify the BO-complex SRH recombination rate for different activation states and impurity levels (not yet considering the permanent deactivation state) was presented in (Bothe *et al.*, 2005).

A further defect complex impacting on silicon solar cell performance results from phosphorous precipitation in highly phosphorus-doped regions (emitter), where the phosphorus is incompletely ionized. This has been quantified in (Min *et al.*, 2014). Similarly, Al-O complexes form in aluminium-alloyed regions, where Al is incompletely ionized. A suitable model and applicable parameters were presented in (Rosenits *et al.*, 2011; Rüdiger *et al.*, 2011).

3.1.3 Optical Properties

The generation rate G of electron-hole pairs is equal to the rate of photons absorbed via band-band absorption. G is further indirectly influenced by the reflection, refraction and scattering at surfaces, which depend on the refractive index of silicon. Thus, a knowledge of the optical properties for the entire relevant spectrum (~250 nm – ~1300 nm), and their temperature dependence, is essential.

3.1.3.1 Complex Refractive Index n + ik *of Intrinsic Silicon*

For intrinsic (i.e., undoped and lowly injected) silicon, the band-to-band absorption coefficient α_{BB} largely dominates the extinction coefficient k, where $\alpha_{BB} = 4\pi k / \lambda$. α_{BB} can therefore be derived from fitting the refractive index to reflection and/or ellipsometry measurements. Results from such a fitting approach to various measurements in literature are given in (Green, 2008), including temperature dependence, which constitutes the currently most widely used data for optical properties of silicon. For the near-infrared wavelength region however, some significant differences with values of α_{BB} derived from luminescence measurements were found (Nguyen *et al.*, 2014b). These can be considered to be more accurate, see also Section 3.1.2.5.1. A best set of optical properties data consists therefore by combining α_{BB} from (Nguyen *et al.*, 2014b) with (Green, 2008) for the wavelengths not included in the former, and n from (Green, 2008). The combined data set shown in Figure 3.1.3 is valid for a wide range of temperatures. For an assessment of the uncertainty of the absorption coefficient, see (Schinke *et al.*, 2014).

3.1.3.2 Free Carrier Absorption (FCA)

When carrier densities become sufficiently high due to doping or injection, absorption of photons by free carriers (FCA) can become significant. While for typical silicon solar cells

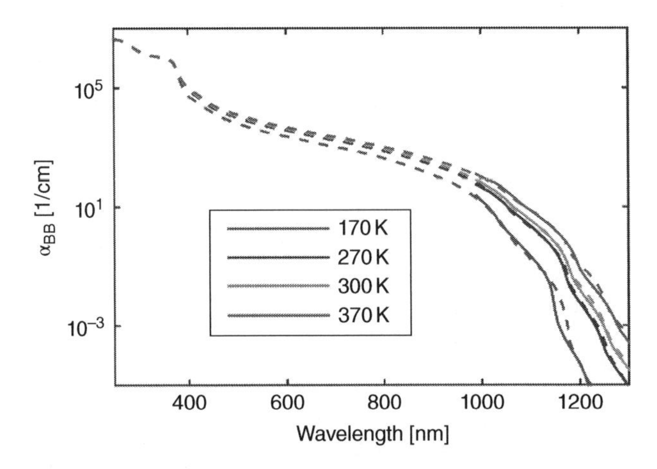

Figure 3.1.3 Absorption coefficient α_{BB} of silicon for several temperatures; *dashed lines*: calculated from (Green, 2008); *solid lines*: data from (Nguyen *et al.*, 2014b); though differences between the different data sets appear minor, they are significant for luminescence spectroscopy. (*See insert for color representation of the figure*)

this is rarely the case within the bulk, FCA within highly doped regions can negatively influence the device performance, as the absorbed photons do not contribute to G. Several parameterizations of the FCA absorption coefficient α_{FCA} exist, with the most recent one providing a good overview (Baker-Finch *et al.*, 2014). Notably, the parameterizations are valid at a temperature of $300\,K$ only.

3.1.4 Conclusion

Table 3.1.1 summarizes the main electrical and optical properties of silicon required to model solar cell performance, including a recommendation for the latest best-known values, models or parameterizations to use.

Table 3.1.1 Main electrical and optical properties of silicon

Property	Recommended value, model or parameterization	Comment
Intrinsic carrier density	(Misiakos and Tsamakis, 1993) multiplied by 0.9953 to match $n_i(300K) = 9.65\times10^9\,\text{cm}^{-3}$ (Altermatt *et al.*, 2003)	
Density of states	$N_C = 2.86\times10^{19}(T/300K)^{1.58}\,\text{cm}^{-3}$ and $N_V = 3.10\times10^{19}(T/300K)^{1.85}\,\text{cm}^{-3}$ (Green, 1990)	
Bandgap	E_g calculated from n_i via $n_i^2 = N_C N_V \exp\left(-\frac{E_g}{kT}\right)$	
Bandgap narrowing BGN	(1) (Schenk, 1998), or (2) (Yan and Cuevas, 2013, 2014)	B) 300 K parameterization, no injection dependence
Carrier statistics	Fermi Dirac recommended, Boltzmann can be used for low doping, or if unavoidable	For Boltzmann statistics and $N_{A/D} > 1\times10^{19}\text{cm}^{-3}$, 'apparent' BGN needs to be used
Mobility	(Klaassen, 1992a, 1992b)	inaccurate for compensated silicon
Radiative recombination	$B_{rad}n_{i,eff}^{\;2}$ from (Nguyen *et al.*, 2014a)	
Auger recombination	(Richter *et al.*, 2012)	300 K parameterization
SRH recombination	metal impurities: (Macdonald and Geerligs, 2004) BO-complex: (Bothe *et al.*, 2005) Al-O complex: (Rosenits *et al.*, 2011; Rüdiger *et al.*, 2011)	
Refractive index	n: (Green, 2008) k/α_{BB}: (Nguyen *et al.*, 2014b), (Green, 2008) for wavelengths not covered in the former	
Free carrier absorption	(Baker-Finch *et al.*, 2014)	300 K parameterization

List of Symbols

Symbol	Description	Unit
q	Elementary charge	1.602×10^{-19} C
n	Electron density	cm^{-3}
p	Hole density	cm^{-3}
t	Time	s
G	Generation rate	$cm^{-3}s^{-1}$
R	Recombination rate	$cm^{-3}s^{-1}$
μ_n	Electron mobility	$cm^2V^{-1}s^{-1}$
μ_p	Hole mobility	$cm^2V^{-1}s^{-1}$
φ_{Fn}	Electron quasi-Fermi potential	V
φ_{Fp}	Hole quasi-Fermi potential	V
ε	Permittivity	$AsV^{-1}m^{-1}$
φ_{el}	Electric potential	V
N_D	Donor density	cm^{-3}
N_A	Acceptor density	cm^{-3}
T	Temperature	K
E_g	Band gap	eV
k_B	Boltzmann constant	1.3806×10^{-23} m^2 kg s^{-2} K^{-1}
n_i	Intrinsic carrier density	cm^{-3}
$n_{i,eff}$	Effective intrinsic carrier density	cm^{-3}
N_C	Density of states in conduction band	cm^{-3}
N_V	Density of states in valence band	cm^{-3}
γ_{deg}	Degeneracy factor	
V_t	Thermal voltage	V
B_{rad}	Radiative recombination coefficient	cm^3s^{-1}
τ_{n0}	Electron lifetime parameter	s
τ_{p0}	Hole lifetime parameter	s
τ_{SRH}	Shockley-Read-Hall lifetime	s
N_T	Defect density	cm^3s^{-1}
E_T	Defect energy level	eV
σ_n	Electron capture cross section	cm^2
σ_p	Hole capture cross section	cm^2
n	(real part of) Refractive index	
k	Extinction coefficient	cm^{-1}
λ	Wavelength	cm

List of Acronyms

α_{BB}	Band-to-band absorption coefficient
BGN	Band Gap Narrowing
EL	Electro-luminescence
PL	Photo-luminescence

(*Continued*)

(*Continued*)

SRH	Shockley-Read-Hall
BO	Boron-Oxygen
Cz	Czochralski
LID	Light Induced Degradation
FCA	Free Carrier Absorption

References

Altermatt, P.P. (2011) Models for numerical device simulations of crystalline silicon solar cells: a review. *Journal of Computational Electronics*, **10** (3), 314–330.

Altermatt, P.P., Schenk, A., Geelhaar, F., and Heiser, G. (2003) Reassessment of the intrinsic carrier density in crystalline silicon in view of band-gap narrowing. *Journal of Applied Physics*, **93** (3), 1598–1604.

Altermatt, P.P., Schumacher, J.O., Cuevas, A. *et al.* (2002) Numerical modeling of highly doped Si:P emitters based on Fermi–Dirac statistics and self-consistent material parameters. *Journal of Applied Physics*, **92** (6), 3187–3197.

Baker-Finch, S.C., McIntosh, K.R., Yan, D. *et al.* (2014) Near-infrared free carrier absorption in heavily doped silicon. *Journal of Applied Physics*, **116** (6).

Bothe, K., Sinton, R., and Schmidt, J. (2005) Fundamental boron–oxygen-related carrier lifetime limit in mono- and multicrystalline silicon. *Progress in Photovoltaics: Research and Applications*, **13** (4), 287–296.

Forster, M., Rougieux, F., Cuevas, A. *et al.* (2013) Incomplete ionization and carrier mobility in compensated p-type and n-type silicon. *IEEE Journal of Photovoltaics*, **3** (1), 108–113.

Green, M.A. (1990) Intrinsic concentration, effective densities of states, and effective mass in silicon. *Journal of Applied Physics*, **67** (6), 2944–2954.

Green, M.A. (2008) Self-consistent optical parameters of intrinsic silicon at 300 K including temperature coefficients. *Solar Energy Materials and Solar Cells*, **92** (11), 1305–1310.

Herguth, A., Schubert, G., Kaes, M., and Hahn, G. (2006) A new approach to prevent the negative impact of the metastable defect in boron doped Cz silicon solar cells. Paper presented at the Photovoltaic Energy Conversion, 2006 IEEE 4th World Conference.

Klaassen, D.B.M. (1992a) A unified mobility model for device simulation: I. Model equations and concentration dependence. *Solid-State Electronics*, **35** (7), 953–959.

Klaassen, D.B.M. (1992b) A unified mobility model for device simulation: II. Temperature dependence of carrier mobility and lifetime. *Solid-State Electronics*, **35** (7), 961–967.

Macdonald, D. and Cuevas, A. (2003) Validity of simplified Shockley-Read-Hall statistics for modeling carrier lifetimes in crystalline silicon. *Physical Review B*, **67** (7), 075203.

Macdonald, D. and Geerligs, L. J. (2004) Recombination activity of interstitial iron and other transition metal point defects in p- and n-type crystalline silicon. *Applied Physics Letters*, **85** (18), 4061–4063.

Min, B., Wagner, H., Dastgheib-Shirazi, A. *et al.* (2014) Heavily doped Si: P emitters of crystalline Si solar cells: recombination due to phosphorus precipitation. *physica status solidi (RRL)-Rapid Research Letters*, **8**, 680–684.

Misiakos, K. and Tsamakis, D. (1993) Accurate measurements of the silicon intrinsic carrier density from 78 to 340 K. *Journal of Applied Physics*, **74** (5), 3293–3297.

Nguyen, H.T., Baker-Finch, S.C., and Macdonald, D. (2014a) Temperature dependence of the radiative recombination coefficient in crystalline silicon from spectral photoluminescence. *Applied Physics Letters*, **104** (11), 112105.

Nguyen, H.T., Rougieux, F.E., Mitchell, B., and Macdonald, D. (2014b) Temperature dependence of the band-band absorption coefficient in crystalline silicon from photoluminescence. *Journal of Applied Physics*, **115** (4).

Pässler, R. (2002) Dispersion-related description of temperature dependencies of band gaps in semiconductors. *Physical Review, B*, **66** (8), 085201.

Richter, A., Glunz, S.W., Werner, F., Schmidt, J., and Cuevas, A. (2012) Improved quantitative description of Auger recombination in crystalline silicon. *Physical Review B*, **86** (16), 165202.

Rosenits, P., Roth, T., and Glunz, S.W. (2011) Erratum on "Determining the defect parameters of the deep aluminum-related defect center in silicon" [*Appl. Phys. Lett.* 91, 122109 (2007)]. *Applied Physics Letters*, **99** (23), 239904.

Rüdiger, M., Rauer, M., Schmiga, C., and Hermle, M. (2011) Effect of incomplete ionization for the description of highly aluminum-doped silicon. *Journal of Applied Physics*, **110** (2), 024508.

Schenk, A. (1998) Finite-temperature full random-phase approximation model of band gap narrowing for silicon device simulation. *Journal of Applied Physics*, **84** (7), 3684–3695.

Schindler, F., Forster, M., Broisch, J. *et al.* (2014) Towards a unified low-field model for carrier mobilities in crystalline silicon. *Solar Energy Materials and Solar Cells*, **131** (0), 92–99.

Schinke, C., Bothe, K., Peest, P. C., Schmidt, J., and Brendel, R. (2014) Uncertainty of the coefficient of band-to-band absorption of crystalline silicon at near-infrared wavelengths. *Applied Physics Letters*, **104** (8), 081915.

Schmidt, J., Lim, B., Walter, D. *et al.* (2012) Impurity-related limitations of next-generation industrial silicon solar cells. Paper presented at the 2012 IEEE 38th Photovoltaic Specialists Conference (PVSC), vol. **2**.

Shockley, W. and Read Jr, W. (1952) Statistics of the recombinations of holes and electrons. *Physical Review*, **87** (5), 835.

Trupke, T., Green, M.A., Würfel, P. *et al.* (2003) Temperature dependence of the radiative recombination coefficient of intrinsic crystalline silicon. *Journal of Applied Physics*, **94** (8), 4930–4937.

Würfel, P. and Würfel, U. (2009) *Physics of Solar Cells: From Basic Principles to Advanced Concepts*. John Wiley & Sons, Inc., New York.

Yan, D. and Cuevas, A. (2013) Empirical determination of the energy band gap narrowing in highly doped n+ silicon. *Journal of Applied Physics*, **114** (4), 044508.

Yan, D. and Cuevas, A. (2014) Empirical determination of the energy band gap narrowing in p+ silicon heavily doped with boron. *Journal of Applied Physics*, **116** (19), 194505.

3.2

Silicon Solar Cell Device Structures

Andrew Blakers and Ngwe Zin
Australian National University, Canberra, Australia

3.2.1 Introduction

About 90% of the world's photovoltaic market is serviced by crystalline silicon solar cells. Silicon has important advantages, including elemental abundance, moderate cost, non-toxicity, high (20–25%) efficiency, device performance stability, simplicity (it is a mono-elemental semiconductor), physical toughness, the highly advanced state of knowledge of silicon material and technology, and the advantages of incumbency. The latter comprises extensively standardized and sophisticated supply chains, large-scale investment in mass production facilities, deep understanding of silicon PV technology and markets, and the presence of thousands of highly trained silicon specialists: scientists, engineers, and technicians.

A silicon solar cell is a large area diode that efficiently absorbs sunlight. Silicon solar cells are made on silicon wafers that have a typical dimension of 125 × 125 or 156 × 156 mm^2 and a thickness of 0.15–0.2 mm. Light with a wavelength of less than 1.1 μm is sufficiently energetic to break a bond in the silicon, creating one free electron and one free hole per incident photon. As described in Part 2, the electrons and holes diffuse through the crystal lattice. Most free electrons created in the p-region reach the edge of the pn junction and are swept across into the n-region; and similarly for holes created in the n-region. The pn junction acts like a one-way membrane. Electrons accumulate in the n-region and holes accumulate in the p-region. This accumulation of charge allows a voltage to build up across the pn junction. If the n-region is connected to the p-region by an external wire with a load (e.g. a battery), then power can be extracted from the solar cell (Figure 3.2.1).

As described in Chapter 10.1, various process steps including texturing, diffusion, passivation and metallization are used to convert a silicon wafer into a solar cell. Groups of 50–100 solar cells are electrically connected and encapsulated within thin layers of plastic (e.g., ethylene vinyl

Photovoltaic Solar Energy: From Fundamentals to Applications, First Edition.
Edited by Angèle Reinders, Pierre Verlinden, Wilfried van Sark, and Alexandre Freundlich.

Companion website: www.wiley.com/go/reinders/photovoltaic_solar_energy

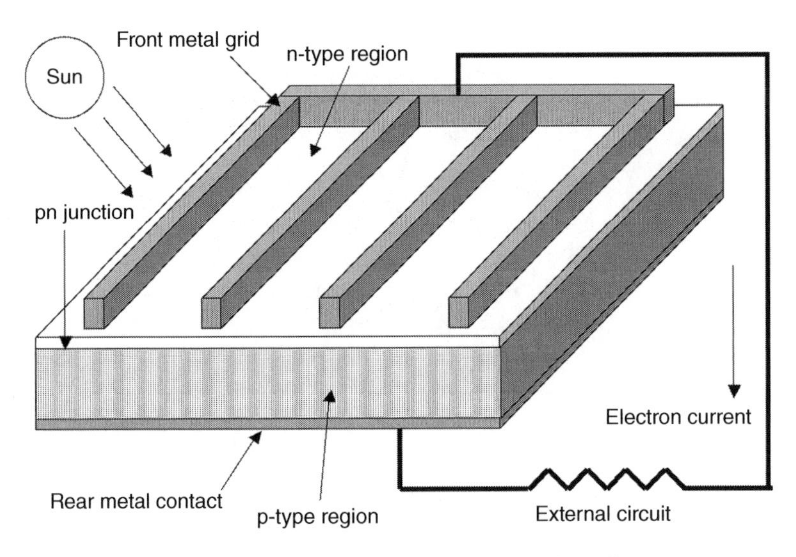

Figure 3.2.1 Schematic of a typical solar cell

acetate and polyvinyl fluoride), see Chapter 10.2, and laminated behind a tough 3 mm-thick glass cover to form solar modules, each with a power of several hundred Watts. Dozens to millions of solar modules are mounted together and electrically connected to form a solar power system.

3.2.2 Solar Cell Optics

The principal optical losses in a silicon solar cell are the reflection of light from the sunward surface, the parasitic absorption of sunlight within the solar cell at the rear metal contact, and the escape of long wavelength, weakly absorbed, infrared light from the silicon wafer.

About one third of the sunlight striking a bare silicon wafer is reflected. Application of an antireflection coating reduces this loss. An antireflection coating is a thin transparent film composed of one or several dielectric materials such as silicon dioxide, silicon nitride, aluminium oxide or titanium dioxide. It works by creating a destructive interference between the incoming and reflected light. The antireflection coating often is useful for other purposes as well, such as reducing the recombination of electrons and holes at the silicon surface and providing electrical isolation between the silicon and the cell metallization.

Almost all silicon solar cells are textured, that is, the sunward surface is roughened, as described in Chapter 10.1. One advantage of texturing is that light that is reflected upon initial impingement on the silicon wafer has a second impingement and hence a second chance of absorption. The second impingement can be directly from one silicon facet to a neighbouring facet, or can result from total internal reflection at the air-glass interface of the solar module. With an ideally textured surface, the amount of incident light that is reflected from the uncoated silicon surface is reduced to approximately 11%, as illustrated in Figure 3.2.2.

The combination of texturing and antireflection coating reduces the reflection from the sunward surface to a few percent. A second advantage of texturing is that infrared light, which is relatively poorly absorbed in silicon, is trapped within the silicon, resulting in improved absorbance. Although silicon absorbs half of the usable photons in the solar spectrum within

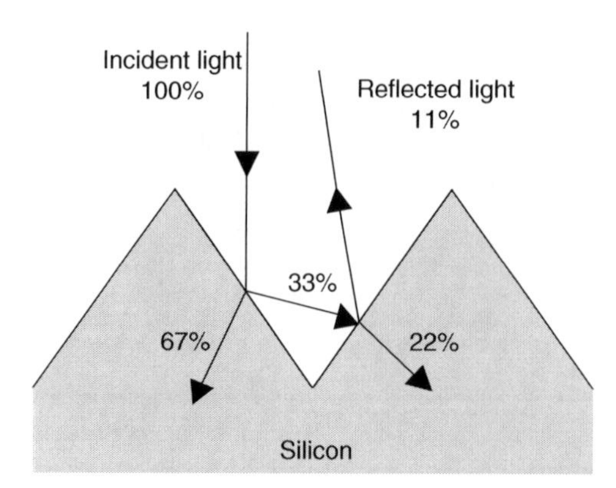

Figure 3.2.2 Effect of a textured surface on the reflectivity of silicon

the first 3 μm, and three-quarters within the first 20 μm, several millimetres of silicon are required to absorb most of the remaining (infrared) photons. Light trapping in wafers with textured surfaces arises from the fact that silicon has a high refractive index (3.6 for infrared light). Only light striking the surface at a small angle relative to the surface normal ($<16°$ in air and $<25°$ in a glass-portant package) can escape. In a textured wafer, most of the light striking a wafer surface will fail to escape.

The metal fingers on the sunward surface of a solar cell (Figure 3.2.1) cause reflection losses. Minimizing the number and width of the metal fingers reduces reflection losses, but at the expense of increased electrical resistance losses. A compromise is necessary.

Most solar cells have a sheet of metal on the rear surface that forms one of the metal contacts. Most light impinging on the rear metallized surface is reflected. However, a substantial fraction is parasitically absorbed within the metal. This absorption loss can be almost eliminated by placing a thin film of transparent dielectric material between the silicon and the metal, although this introduces additional process costs associated both with the deposition of the dielectric film and the formation of electrical contacts at intervals through the film.

3.2.3 Minimizing Electron-Hole Recombination

Absorption of sunlight creates free electrons and holes. It is desirable that most of these electrons and holes contribute to electrical current in the external circuit rather than recombining within the silicon. Most recombination losses occur at imperfections in the silicon crystal. At such sites the regular location of silicon atoms within the crystal is disrupted, leading to defect states that enhance electron-hole recombination.

A major source of recombination is the silicon surface where the silicon crystal is interrupted, that is, the sunward and rear surfaces and the wafer edges. It is possible to passivate silicon surfaces with thin films of a variety of materials, including silicon dioxide, silicon nitride, aluminium oxide and amorphous silicon. These thin films greatly reduce the density of surface defects, and frequently double as antireflection coatings. Additionally, the films may be electrically charged, which electrostatically repels either electrons or holes and hence reduces the recombination rate. Another way of reducing surface recombination rates is to dope the silicon surface with boron or phosphorus, which has the effect of reducing the electron or hole concentration respectively at the surface.

Within the bulk of the silicon wafer common imperfections include foreign atoms, dislocations and other crystal defects, and grain boundaries, which are present in multicrystalline wafers sliced from block-cast ingots. The use of high purity materials and clean process equipment reduces contamination. Some thermal steps remove impurities to reduce their concentration and activity. The electrical activity of crystal defects and grain boundaries can be minimized using specific thermal steps and passivation with hydrogen. There is an economic trade-off between the cost of minimizing crystal imperfections and the value of the resulting increase in solar cell efficiency.

3.2.4 Minimizing Electrical Losses

Electrical series resistance degrades the cell performance by causing voltage drops within a cell or module. Common locations include the lateral conduction of electrons and holes within the silicon to the metal fingers located on the sunward surface of a solar cell; in the electrical contact of cell metallization to the silicon; within the metal fingers on the front surface; and within the copper ribbons that connect the cells together within a module. Trade-offs between resistive losses and other losses are required in the cell design process. For example, the resistive and optical losses of the metal fingers must be balanced. Poor design or fabrication can lead to short-circuits within a solar cell. Shorted cells are normally detected and discarded during the cell screening process prior to module assembly.

3.2.5 Screen-Printed Silicon Solar Cells

The leading commercial solar cell technology over the past four decades uses screen printing of the metallic contacts onto p-type single or multi-crystalline silicon wafers (Figure 3.2.3). Commercial solar cell efficiencies of 15–21% are achieved with this technology.

The first process step is wafer texturing in order to reduce reflection losses and improve light trapping. Phosphorus diffusion into the surfaces follows. In order to avoid a short-circuit between the front and rear surfaces of the finished solar cell, plasma or chemical etching is used to remove the phosphorus doping at the edges or at the back of the wafer. The next step is the deposition of a thin silicon nitride layer on the front surface, which doubles as a passivation and antireflection layer.

Screen printing is then used to deposit silver paste in a finger pattern on the sunward surface and sheets of aluminium and silver paste on the rear surface (Figure 3.2.4). The pastes consist of small metal particles together with solvents and binders. The pastes are dried and then fired by placing the wafer in a furnace to bind the metal particles together and form electrically conductive layers. The firing also drives the metal fingers through the silicon nitride layer. Careful processing is required to avoid the metal fingers penetrating the pn junction and shorting the cell. On the rear surface the fired aluminium paste causes what is called a back surface field (BSF) to form, which reduces recombination rates at the rear surface. Silver, printed in the form of 3–5 linear electrodes, is needed to facilitate soldering of metal interconnect ribbons to the rear surface.

The efficiency of typical industrial monocrystalline cells ranges between 16–19% and for multicrystalline cells between 15–18%. The simple screen-printed solar cell process remains overwhelmingly commercially dominant. However, numerous design compromises are necessary in order to minimize costs. Improved designs and fabrication processes can circumvent many of these compromises.

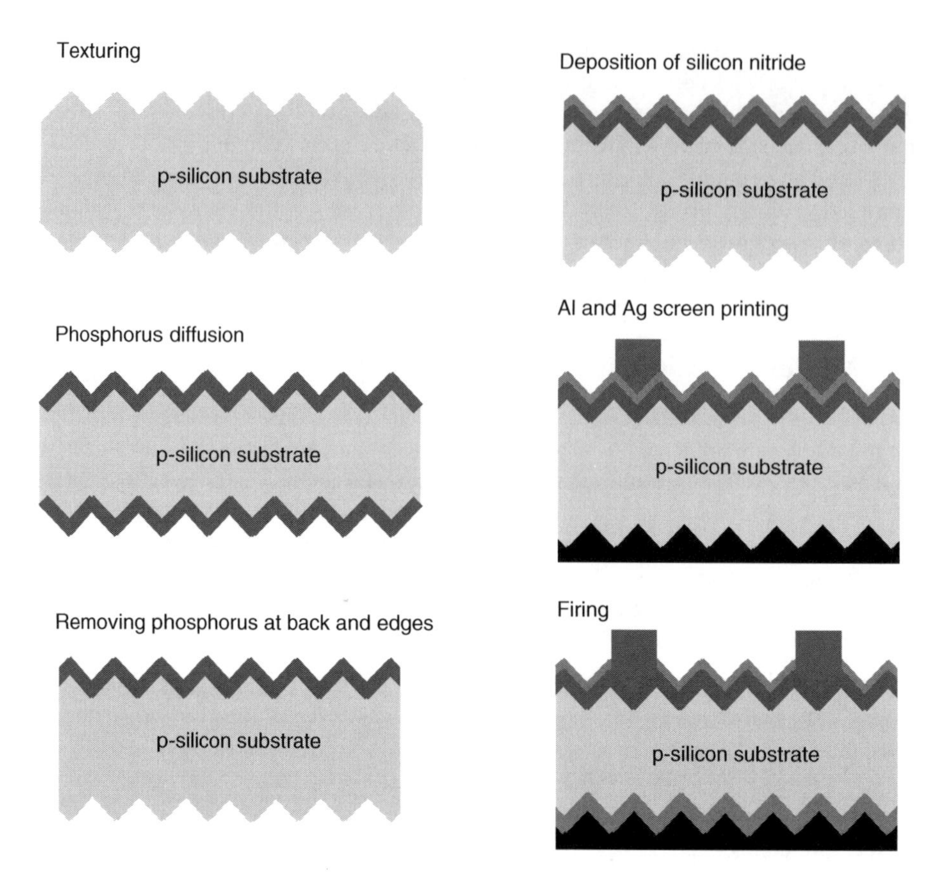

Figure 3.2.3 A typical process of a screen-printed silicon solar cell. (*See insert for color representation of the figure*)

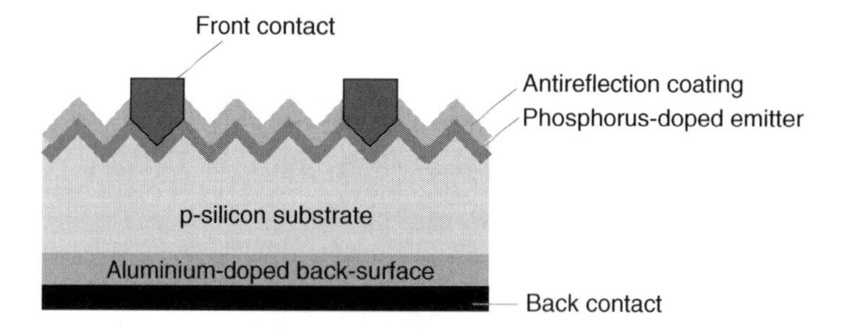

Figure 3.2.4 Screen-printed silicon solar cell

3.2.6 Selective Emitter Solar Cells

Ideally, the silver fingers on the sunward surface are thick and narrow (high aspect ratio) and closely spaced in order to minimize resistive and optical losses. However, there are limitations on the minimum width of the silver fingers, and their aspect ratio, limits to what the fabrication technique can achieve. The contact resistance between the metal fingers and the silicon is reduced if the surface of the silicon is relatively heavily doped with phosphorus. Additionally, the lateral resistance to the electron flow within the phosphorus-doped layer (the emitter) is reduced by heavier doping. However, heavy phosphorus doping in the surface region leads to reduced responsiveness to blue and ultra-violet light, which reduces the solar cell efficiency.

The selective emitter approach entails heavy phosphorus doping beneath the metal contacts (3–10% of the cell's front surface) and relatively light doping over the rest of the surface. In this way the blue response of the solar cell is improved while maintaining low contact resistance and low recombination at the metal contacts. However, additional process steps and cost are required for selective emitter cells (Figure 3.2.5).

Steadily improving printing machines are facilitating thinner fingers with increased aspect ratio. Improved screen printing pastes are allowing low resistance contact to be made to less heavily doped silicon surfaces. Industrial monocrystalline silicon solar cells using improved paste have achieved 19.4% efficiency, and the efficiency is expected to reach and exceed 20% in the next few years (Mikeska *et al.*, 2013). These improvements are reducing the advantage of a selective emitter.

3.2.7 PERC and PERL Solar Cells

PERC (Passivated Emitter and Rear Cells) were developed in 1988 (Blakers *et al.*, 1989) to address several shortcomings of conventional screen printed solar cells, primarily related to the rear surface (Figure 3.2.6). The screen-printed aluminium layer has relatively high

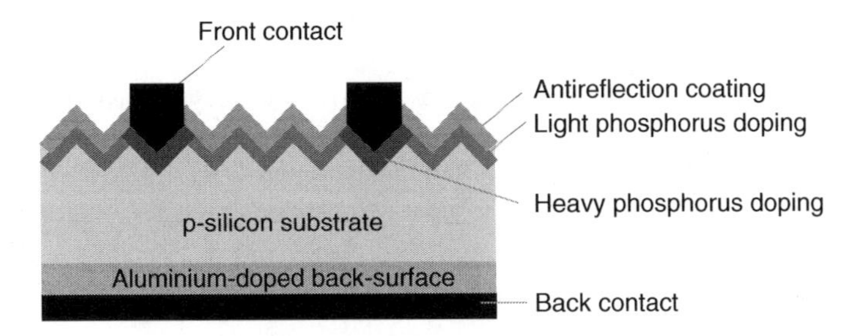

Figure 3.2.5 Screen printed silicon solar cell with selective emitter

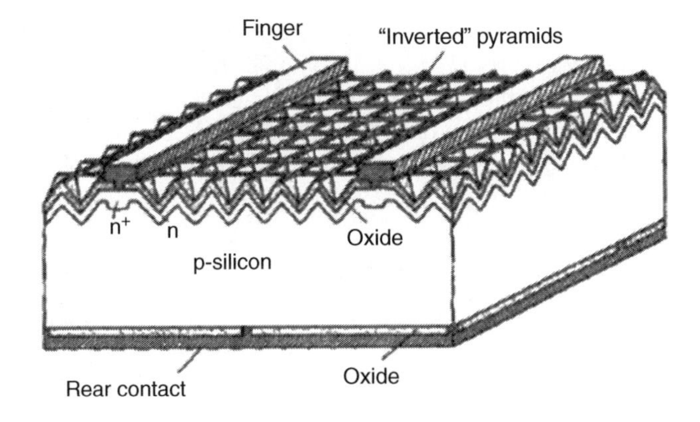

Figure 3.2.6 PERC solar cell. Source: (Blakers *et al.*, 1989)

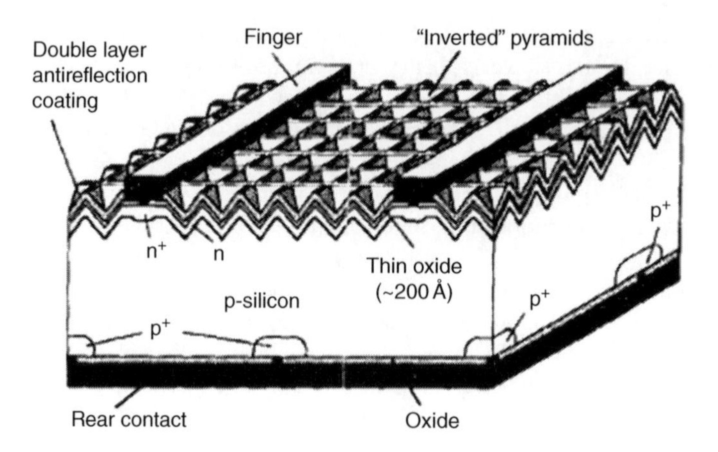

Figure 3.2.7 PERL solar cell. Source: (Zhao *et al.*, 1999; Zhao, 2011)

parasitic absorbance of light and relatively high recombination rate for electrons and holes. A thin dielectric film can be placed between the silicon and the rear metal contact to substantially reduce both optical absorption losses and recombination. An array of holes in the dielectric film covering about 1% of the rear surface and spaced about a millimetre apart allows electrical contact to be made directly to the rear surface of the boron doped silicon wafer. The improved performance of a PERC cell has to be balanced against the slightly higher cost of cell fabrication. Many companies are now introducing PERC cells into production.

PERL (Passivated Emitter and Rear Local diffusion) solar cells have a similar design to PERC solar cells with the addition of boron diffusions at the rear contact points to reduce recombination and resistive losses (Figure 3.2.7). For many years the PERL design held the world silicon solar cell efficiency record of 25% (Zhao *et al.*, 1999; Zhao, 2011). PERL cells are complicated and have not yet entered commercial production.

3.2.8 Switching to Phosphorus-Doped Substrates

Hitherto the photovoltaic industry has been dominated by boron-doped p-type crystalline silicon, partly due to historical reasons. However, it is now well established that p-type silicon suffers from reduced minority carrier lifetime as compared to n-type silicon due to the presence of boron-oxygen defects and greater recombination strengths of most common metallic impurities in p-type silicon (Macdonald and Geerligs, 2004). This can allow solar cells fabricated on phosphorus-doped n-type silicon to have a performance advantage (Glunz *et al.*, 2010).

3.2.9 N-Type Rear Emitter Silicon Solar Cells

The simplest way of implementing phosphorus-doped wafers is to use a similar design and processes as has been very well established for boron-doped wafers. In this case, the pn junction will be located at the rear surface rather than the front surface. The p-type emitter (p^+) is formed by aluminium-alloying the silicon (or by boron diffusion). As the pn junction is

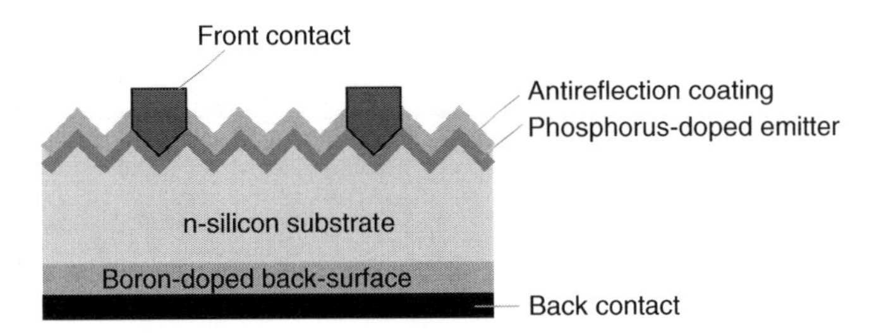

Figure 3.2.8 n-type rear emitter silicon solar cell

located at the rear of the cell, photogenerated carriers (which are mostly generated near the sunward surface) have to diffuse to the rear surface to be collected by the junction, therefore requiring higher quality silicon substrate material as well as effective front passivation and thinner wafers. Solar cells fabricated on phosphorus-doped wafers with full-area aluminium alloyed rear have demonstrated efficiency of 19–20% (Glunz *et al.*, 2010; Schmiga *et al.*, 2010). The performance of the cell shown in Figure 3.2.8 can be further improved to above 20% by incorporating a passivation dielectric over most of the rear surface (analogous to the PERC structure), which reduces rear surface recombination and optical absorption, increasing the cell efficiency to exceed 20% (Schmiga *et al.*, 2009).

Interdigitated back contact silicon solar cells and heterojunction back contact silicon solar cells typically use n-type substrates (lightly doped) because of the high and stable minority carrier lifetimes which can be obtained. In these cells, both the positive and negative electrodes are located on the rear surface, and both electrons and holes (which are mostly created near the front surface) are transported to the rear surface for charge separation and collection. These cell designs are used in the highest efficiency commercial modules, and are discussed in detail in Chapter 3.3.

3.2.10 N-type Front Emitter Silicon Solar Cell

By locating the p^+ emitter at the front and n^+ diffusion at the rear, an n-type front emitter silicon solar cell with p^+nn^+ structure with full rear metal coverage can be formed. Various institutes have reported n-type passivated emitter rear locally diffused (PERL) cells with efficiencies of 23.4% (Benick *et al.*, 2009) and 23.9% (Glunz *et a*l., 2010). Replacing the full area aluminum coverage at the rear with a metal finger grid similar to the front, n-type front emitter (see Figure 3.2.9) cell can take advantage of the bifacial solar cell benefits of collecting light from both sides of the cell. This type of cell also does not require high-quality n-type silicon substrate material since photogenerated carriers can be collected at the front emitter. The p^+ emitter and n^+ back surface field (BSF) are formed by boron and phosphorus diffusion respectively. The p^+ emitter can be effectively passivated by thermally grown oxide (Altermatt *et al.*, 2006; King and Swanson, 1991), low-temperature acid grown oxide (Valentin *et al.*, 2008) and aluminium oxide (Hoex *et al.*, 2006; Hoex *et al.*, 2007; Hoex, *et al.*, 2008), while the n^+ BSF can be passivated by thermally grown oxide and plasma enhanced chemical vapor deposited (PECVD) silicon nitride (Kerr *et al.*, 2001). Low-cost bifacial solar cells with an efficiency greater than 20% were reported by the Energy Research Centre of the Netherlands (ECN) (Romijn *et al.*, 2013), while HIT (Heterostructure with Intrinsic Thin Layer) bifacial solar

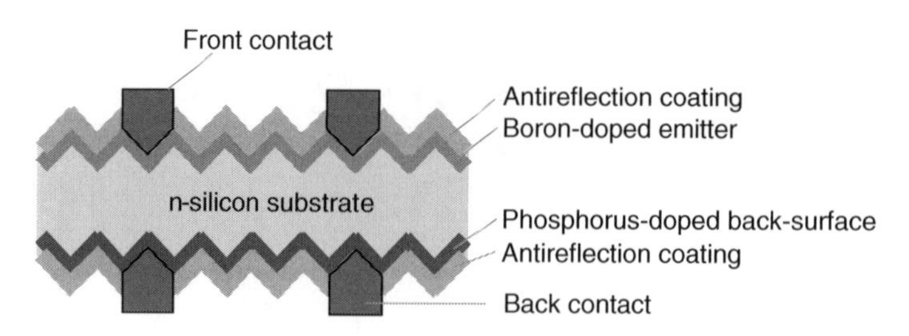

Figure 3.2.9 n-type front emitter silicon solar cell

cells developed by Sanyo Electronic achieved an efficiency of 24.7% (Taguchi *et al.*, 2014). In HIT cells, surface passivation is achieved using amorphous silicon, while passivation of electrical contacts is achieved using a thin layer of amorphous silicon, thin enough to allow electron and hole transport.

3.2.11 Efficiency Improvements for Industrial Solar Cells

3.2.11.1 Passivation Schemes for Boron-Doped Emitter

Switching from p- to n-type silicon requires effective passivation of boron-doped emitters since the traditional method of passivating phosphorus-doped surfaces by PECVD nitride is not particularly effective for boron doped surfaces (Chen *et al.*, 2006). Thermally grown silicon oxide has been the passivation layer used to develop high efficiency silicon solar cells as it passivates both n- and p-diffusion effectively. The oxidation is normally undertaken between 800–1100 °C in oxygen ambient. Additionally, oxide can be used as an etch and diffusion barrier. However, the process of oxidation is relatively slow, and the thermal treatment can result in the bulk degradation of silicon wafers (especially in multicrystalline wafers). Effective passivation of boron-doped surfaces by an ultra-thin (1.5 nm) silicon oxide layer grown in nitric acid at room temperature or higher temperatures is possible. Acid-grown oxide has been successfully implemented industrially (Mihailetchi *et al.*, 2010). Ultra-thin oxide can also be formed by exposure to UV light in the presence of ozone (O_3). Another method of passivating a boron-doped emitter is aluminium oxide, which has a high fixed negative charge density. A very low emitter saturation current density of 10 fA/cm^2 was achieved on a boron-doped emitter when passivated with aluminum oxide synthesized by plasma-assisted atomic layer deposition (ALD) (Hoex *et al.*, 2007). Aluminum oxide deposited by in-line plasma-enhanced chemical vapour deposition (PECVD) systems for industrial environments have achieved outstanding passivation (Saint-Cast *et al.*, 2009).

3.2.11.1.1 Thinner Wafers

Reducing the wafer thickness reduces the volume and cost of the expensive pure silicon used in solar cell manufacture. Wafer slicing technology has improved significantly, resulting in a reduction of substrate thickness from >300 μm to 180 μm. Wafers thinner than 150 μm can be used in solar cell production provided that the wafer breakage rate remains low.

Current wafer slicing technology is dominated by the slurry-assisted wire saw, which cuts silicon wafer blocks with a fine steel wire while being sprayed with abrasive slurry (Möller, 2006). The slurry wire saw technique provides a high throughput, the ability to cut varied block sizes and with reasonable surface quality. However, the technique has some trade-offs: loss of silicon – known as kerf loss – during the sawing process, high wear and tear on the wire, and the requirement to periodically remove silicon debris from the slurry to maintain an effective cutting speed.

An alternative to the slurry wire saw is the diamond wire saw method, in which diamond particles attached to the wire cut the silicon block, and the abrasive slurry is replaced by a cooling liquid (Bidiville *et al.*, 2009). The diamond wire saw method offers higher productivity through fast cutting, reduced kerf loss, longer lifespan of diamond wire, easier recycling of the cooling liquid and similar solar cell efficiency as compared to the slurry wire saw. However, the diamond wire saw technique needs to address the high cost of diamond abrasive in the wire and the capability to cut wafers thinner than 150 μm. Other methods of slicing are being investigated, including ion-beam induced wafer cleaving (Henley *et al.*, 2008; Henley *et al.*, 2009).

3.2.11.1.2 Improved Metallization

The dominant metallization technology for commercial silicon solar cells is screen printing. Improved printers and pastes are allowing higher conversion efficiency to be achieved with screen printing. However, demand for thinner silicon wafers is increasing, and this constrains the metallization technique to minimize wafer breakage. Additionally, the cost of silver in the pastes may become substantial as PV demand grows and squeezes the supply of silver. The International Technology Roadmap for Photovoltaics (2013) envisages new metallization techniques that will provide solar cells with a superior performance with lower usage of silver.

Light-induced plating (LIP) is an attractive alternative to achieve narrower fingers, an improved aspect ratio, a higher finger conductivity and lower shading losses (Mette *et al.*, 2006). The LIP technique exploits the photovoltaic effect of a solar cell to obtain reduced front reflectance and lower finger resistance. Copper metallization using LIP and similar techniques offers lower material costs and superior cell performance (Rehman and Lee, 2014). Various companies and research institutes including Kaneka, SunTech, SunPower, Silevo, Tetrasun, Hyundai, Schott Solar, IMEC, Fraunhofer ISE, and Roth and Rau have achieved solar cells based on Cu metallization with an efficiency above 20%.

3.2.12 Conclusion

The current dominance of the world's photovoltaic markets by silicon technology is likely to continue. The learning associated with mass production, coupled with continued improvements in cell design and production technology, is steadily reducing module manufacturing costs. As PV modules continue to decrease as a proportion of PV system costs, the relative premium for greater efficiency (which leverages cost reductions across all area-related balance of costs) increases. Many and varied routes are available to increase standard commercial cell efficiencies at a modest additional cost and process complexity. Standard commercial cell efficiencies above 20% are likely in the near term, rising over the years to above 25%.

References

Altermatt, P.P., Plagwitz, H., Bock, R. *et al.* (2006) The surface recombination velocity at boron-doped emitters: comparison between various passivation techniques. Paper presented at the 21st European Photovoltaic Solar Energy Conference, Dresden, Germany.

Benick, J., Hoex, B., Dingemans, G. *et al.* (2009) High-efficiency n-type silicon solar cells with front side boron emitter. Paper presented at the 24th European Photovoltaic Solar Energy Conference, Hamburg, Germany.

Bidiville, A., Wasmer, K., Kraft, R., and Ballif, C. (2009) Diamond wire-sawn silicon wafers: from the lab to the cell production. Paper presented at the 24th European Photovoltaic Solar Energy Conference, Hamburg, Germany.

Blakers, A.W., Wang, A., Milne, A.M. *et al.* (1989) 22.8% efficient silicon solar cell. *Applied Physics Letters*, **55** (13), 1363–1365. DOI: http://dx.doi.org/10.1063/1.101596.

Chen, F. W., Li, T.-T.A., and Cotter, J.E. (2006) Passivation of boron emitters on n-type silicon by plasma-enhanced chemical vapor deposited silicon nitride. *Applied Physics Letters*, **88** (26), 263514. DOI:http://dx.doi.org/10.1063/1.2217167.

Glunz, S.W., Benick, J., Biro, D. *et al.* (2010) n-type silicon – enabling efficiencies > 20% in industrial production. Paper presented at the 35th IEEE Photovoltaic Specialists Conference, Hawaii.

Henley, F., Kang, S., Liu, Z. *et al.* (2009) Kerf-free 20-150 µm c-Si wafering for thin PV manufacturing. Paper presented at the 24th European Photovoltaic Solar Energy Conference, Hamburg, Germany.

Henley, F., Lamm, A., Kang, S., *et al.* (2008) Direct film transfer (DFT) technology for kerf-free silicon wafering. Paper presented at the 23rd European Photovoltaic Solar Energy Conference, Valencia, Spain.

Hoex, B., Gielis, J.J.H., van de Sanden, M.C.M., and Kessels, W.M.M. (2008) On the c-Si surface passivation mechanism by the negative-charge-dielectric Al2O3. *Journal of Applied Physics*, **104** (11), 113703. DOI: http://dx.doi.org/10.1063/1.3021091.

Hoex, B., Heil, S.B.S., Langereis, E., *et al.* (2006) Ultralow surface recombination of c-Si substrates passivated by plasma-assisted atomic layer deposited Al2O3. *Applied Physics Letters*, **89** (4), 042112. DOI: http://dx.doi.org/10.1063/1.2240736.

Hoex, B., Schmidt, J., Bock, R., *et al.* (2007) Excellent passivation of highly doped p-type Si surfaces by the negative-charge-dielectric Al2O3. *Applied Physics Letters*, **91** (11), 112107. DOI: doi:http://dx.doi.org/http://dx.doi.org/10.1063/1.2784168.

International Technology Roadmap for Photovoltaics (2013) http://www.itrpv.net/

Kerr, M.J., Schmidt, J., Cuevas, A., and Bultman, J.H. (2001) Surface recombination velocity of phosphorus-diffused silicon solar cell emitters passivated with plasma enhanced chemical vapor deposited silicon nitride and thermal silicon oxide. *Journal of Applied Physics*, **89** (7), 3821–3826. DOI: http://dx.doi.org/10.1063/1.1350633.

King, R.R. and Swanson, R.M. (1991) Studies of diffused boron emitters: saturation current, bandgap narrowing, and surface recombination velocity. *IEEE Transactions on Electron Devices*, **38** (6), 1399–1409. DOI: 10.1109/16.81632.

Macdonald, D. and Geerligs, L.J. (2004) Recombination activity of interstitial iron and other transition metal point defects in p- and n-type crystalline silicon. *Applied Physics Letters*, **85** (18), 4061–4063. DOI: http://dx.doi.org/10.1063/1.1812833.

Mette, A., Schetter, C., Wissen, D. *et al.* (2006) Increasing the efficiency of screen-printed silicon solar cells by light-induced silver plating. Paper presented at the Photovoltaic Energy Conversion, 2006 IEEE 4th World Conference.

Mihailetchi, V.D., Jourdan, J., Edler, A. *et al.* (2010) Screen printed n-type silicon solar cells for industrial application. Paper presented at the 25th European Photovoltaic Solar Energy Conference, Valencia, Spain.

Mihailetchi, V.D., Komatsu, Y., and Geerligs, L.J. (2008) Nitric acid pretreatment for the passivation of boron emitters for n-type base silicon solar cells. *Applied Physics Letters*, **92** (6), 063510. DOI: http://dx.doi.org/10.1063/1.2870202.

Mikeska, K.R., Carroll, A.F., Cheng, L.K., and Li, Z. (2013) Screen-printed silver contact mechanisms. Paper presented at the 28th European Photovoltaic Solar Energy Conference, Paris, France.

Möller, H.J. (2006) Wafering of silicon crystals. *physica status solidi (a)*, **203** (4), 659–669. DOI: 10.1002/pssa.200564508.

Rehman, A. and Lee, S. (2014) Review of the potential of the Ni/Cu plating technique for crystalline silicon solar cells. *Materials*, **7** (2), 1318–1341.

Romijn, I.G., Anker, J., Burgers, A.R. *et al.* (2013) Industrial n-type solar cells with greater 20% cell efficiency. Paper presented at the China PV Technology International Conference (CPTIC), Shanghai, China.

Saint-Cast, P., Kania, D., Hofmann, M. *et al.* (2009) Very low surface recombination velocity on p-type c-Si by high-rate plasma-deposited aluminum oxide. *Applied Physics Letters*, **95** (15), 151502. DOI: http://dx.doi.org/10.1063/1.3250157.

Schmiga, C., Hörteis, M., Rauer, M. *et al.* (2009) Large-area n-type silicon solar cells with printed contacts and aluminium-alloyed rear emitter. Paper presented at the 24th European Photovoltaic Solar Energy Conference, Hamburg, Germany.

Schmiga, C., Rauer, M., Rüdiger, M. *et al.* (2010) Aluminium-doped p+ silicon for rear emitters and back surface fields: Results and potentials of industrial n-and p-type solar cells. Paper presented at the 25th European Photovoltaic Solar Energy Conference and Exhibition/5th World Conference on Photovoltaic Energy Conversion, Valencia, Spain.

Taguchi, M., Yano, A., Tohoda, S. *et al.* (2014) 24.7%; Record Efficiency HIT Solar Cell on Thin Silicon Wafer. *IEEE Journal of Photovoltaics*, **4** (1), 96–99. DOI: 10.1109/JPHOTOV.2013.2282737.

Zhao, J. (2011) Passivated emitter rear locally diffused solar cells. *Bulletin of Advanced Technology Research*, **5**, 41–43.

Zhao, J., Wang, A., and Green, M.A. (1999) 24·5% efficiency silicon PERT cells on MCZ substrates and 24·7% efficiency PERL cells on FZ substrates. *Progress in Photovoltaics: Research and Applications*, **7** (6), 471–474. DOI: 10.1002/(SICI)1099-159X(199911/12)7:6<471::AID-PIP298>3.0.CO;2-7.

3.3

Interdigitated Back Contact Solar Cells

Pierre Verlinden
Trina Solar, Changzhou, Jiangsu, China

3.3.1 Introduction

3.3.1.1 A Different Concept for High-Efficiency Silicon Solar Cells

An alternative to the conventional two-sided contacted silicon solar cells described in Chapter 3.2 is to develop solar cells with both collecting diffused regions, i.e. emitter and back surface field (BSF), and electrodes located on the rear side of the silicon substrate (Schwartz *et al.*, 1975; Lammert and Schwartz, 1977). This solar cell design is commonly called an interdigitated back contact (IBC) solar cell and is schematically represented in Figure 3.3.1. The design has several advantages over conventional two-sided contacted solar cells, in particular:

- No shadow loss from the front electrode, corresponding to a gain of about 5–7% in photogenerated current compared to conventional solar cells.
- Low series resistance due to the fact that each electrode can cover almost 50% of the back surface of the solar cell.
- Decoupling of the optical optimization of the front side and the electrical optimization of the back side of the cell. Unlike conventional cell designs, the efficiency of IBC cells is not limited by a trade-off between the sheet resistance of a front emitter, to keep the series resistance to an acceptable level, and the recombination parameter J_o of the same front emitter, to keep a good quantum efficiency and high open-circuit voltage V_{oc}.
- The front surface, where the largest concentration of minority carriers is located, can be optimized to obtain a very low surface recombination velocity S or a low recombination parameter J_o.

Photovoltaic Solar Energy: From Fundamentals to Applications, First Edition.
Edited by Angèle Reinders, Pierre Verlinden, Wilfried van Sark, and Alexandre Freundlich.

Companion website: www.wiley.com/go/reinders/photovoltaic_solar_energy

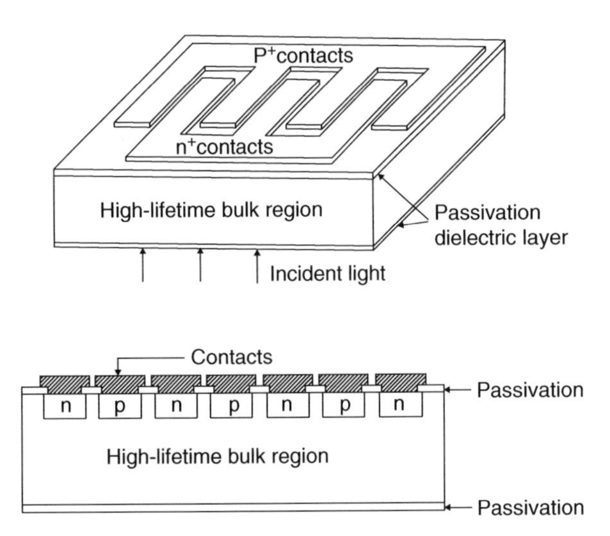

Figure 3.3.1 Interdigitated back contact solar cell. Source: (Verlinden, 1985)

- The IBC cells are easy to assemble into modules with ribbons or interconnects with stress relief since both polarities are on the same side of the cell. Module assembly with a conductive backsheet is also possible but more complicated.

The two-dimensional current flow of carriers and distribution of potential inside the IBC cell (see Figure 3.3.2) are more complicated to understand than that of a conventional cell that is mostly one-dimensional. In its first approximation, the IBC cell can be understood as a rear junction solar cell (Yernaux *et al.*, 1984). In a conventional two-sided contacted solar cell with a front junction, most of the carriers are generated near the junction and the quantum efficiency is only slightly affected by the carrier lifetime, except for long wavelengths. In a rear junction cell, the photo-generated carriers are generated close to the front surface and have to travel all the way through the thickness of the silicon wafer, with the minority carriers toward the emitter and the majority carriers toward the bulk contact or back surface field (BSF). The efficiency of the IBC cell is, therefore, very sensitive to the bulk lifetime and the front surface recombination. It requires a long minority carrier lifetime and a low front surface recombination velocity to achieve a decent efficiency. Practically speaking, the IBC design requires a minority carrier diffusion length much greater (about 5 or 10 times) than the thickness of the solar cell, equivalent to a carrier lifetime greater than 1 ms and a front surface recombination parameter J_o smaller than 10 fA/cm^2. As a result of these requirements and the particular design of the IBC cell, the fabrication process is complicated and more expensive than that of conventional cells, requiring:

- Higher quality silicon wafers with longer minority carrier lifetime. Initially IBC was only implemented on Float Zone (FZ) substrates until high-quality Czochralski (Cz) wafers became available in the 1990s. n-type substrates are generally preferred because they offer a much longer minority carrier lifetime than p-type substrates.

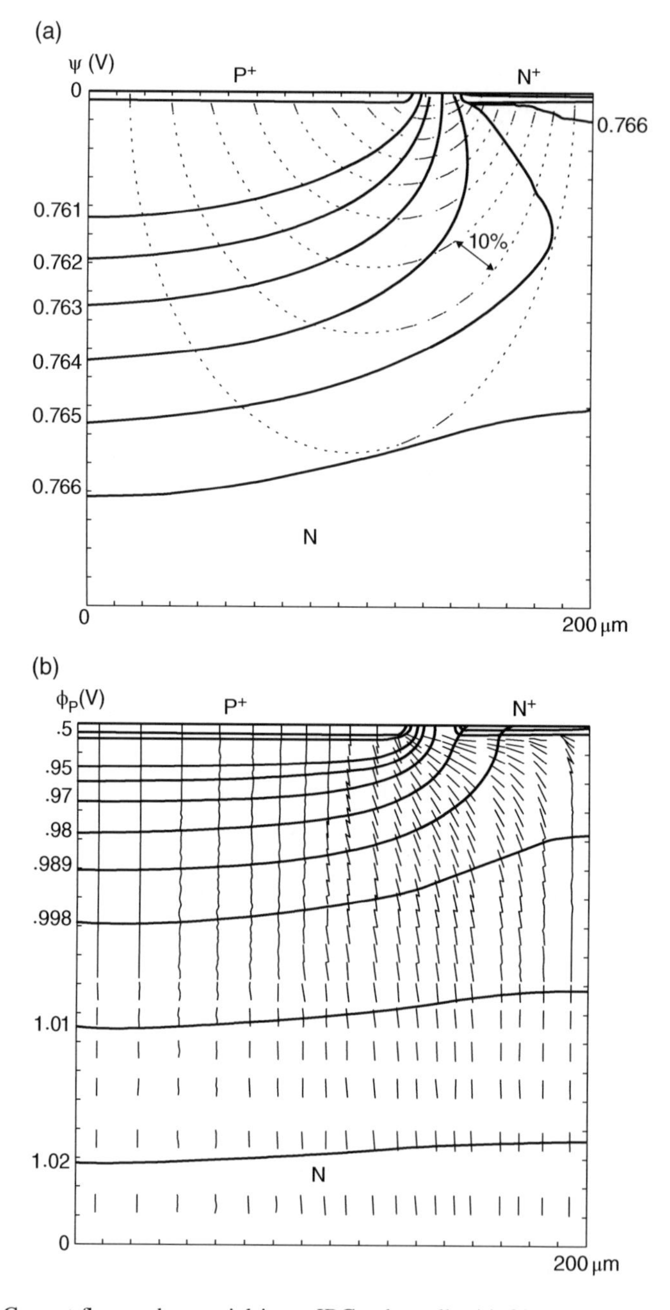

Figure 3.3.2 Current flow and potential in an IBC solar cell with 200 μm half-pitch under 50 suns (5 W/cm^2). (a) Minority carrier current flow J_p and quasi-Fermi level Φ_p. (b) Total current flow J_T=J_p+J_n and potential Ψ. Source: (Verlinden, 1985)

- Patterning and alignment of n-type and p-type regions on the same side of the wafer with their corresponding contact windows and metal contacts.
- Two or three doping steps: emitter, BSF and optionally a front surface field (FSF) or a front floating emitter (FFE).
- Superior contamination control and sophisticated cleaning procedures, for example, RCA cleaning (Kern and Puotinen, 1970), although simpler and cheaper cleaning procedures using solutions of diluted acids with ozone have been proposed (Li, 2014).

The concept of IBC cells was originally introduced for concentrator PV (CPV) applications as it would allow silicon solar cells to be used in several hundred times concentrated sunlight, an application that was out of reach of conventional solar cells (Sinton, 1987). With IBC cells, it is possible to obtain a series resistance in the range of 10^{-3} ohm.cm^2 with a very fine pitch of interdigitated n+and p+regions, or about three orders of magnitude lower than for a conventional silicon solar cell. Because of manufacturing costs, CPV was the only application for IBC cells until around 2004 when the SunPower Corporation introduced the first commercial flat-plate IBC module (McIntosh *et al.*, 2003). The few exceptions were the use of IBC cells for very high-value applications, for example, solar race cars and solar airplanes (Verlinden *et al.*, 1994; Zhou *et al.*, 1997). Despite its very high efficiency, the use of IBC cells for space applications was hindered by the device's poor electron and proton radiation tolerance as the radiation-induced degradation of the bulk carrier lifetime led to much faster efficiency degradation than with conventional front junction cells.

3.3.2 Different Types of IBC Solar Cells

3.3.2.1 Front Surface Field Solar Cells

Over the years, several variants of the IBC design have been introduced. The front surface field (FSF) solar cell has a high-low (n+/n or p+/p) junction on the front side of the IBC solar cell (Figure 3.3.3 (a)). The function of the front surface field is to reduce the effective front-surface recombination velocity for the carriers generated in the bulk of the device. It behaves the same way as a back surface field (BSF) in a conventional n+/p/p+solar cell. However, since the high-low junction is now applied on the front side of the cell, which is not contacted and is passivated with a dielectric layer such as thermal silicon oxide, silicon nitride or aluminum oxide, it can be very lightly doped and optimized for the lowest effective surface recombination velocity or the lowest J_o. For high-efficiency FSF solar cells, the lowest J_o must be targeted, which corresponds to a sheet resistance >350 ohm/sq. and front surface doping concentration around 10^{18} cm^{-3} (King, 1990). It is important to note that, in a high-level injection, the front surface recombination current becomes proportional to the square of the carrier concentration. Therefore, the presence of a FSF creates a sub-linearity of the photo-generated current when the bulk is in a high-level injection, for example, for a CPV application (Verlinden, 2005).

The use of low sheet resistivity FSF was sometimes proposed to reduce the loss in internal bulk series resistance for large pitch IBC cells and therefore to improve the fill factor (FF). This is contradictory to the advantages of the IBC cell design and the loss in J_{sc} or V_{oc} is generally more important than the gain in FF.

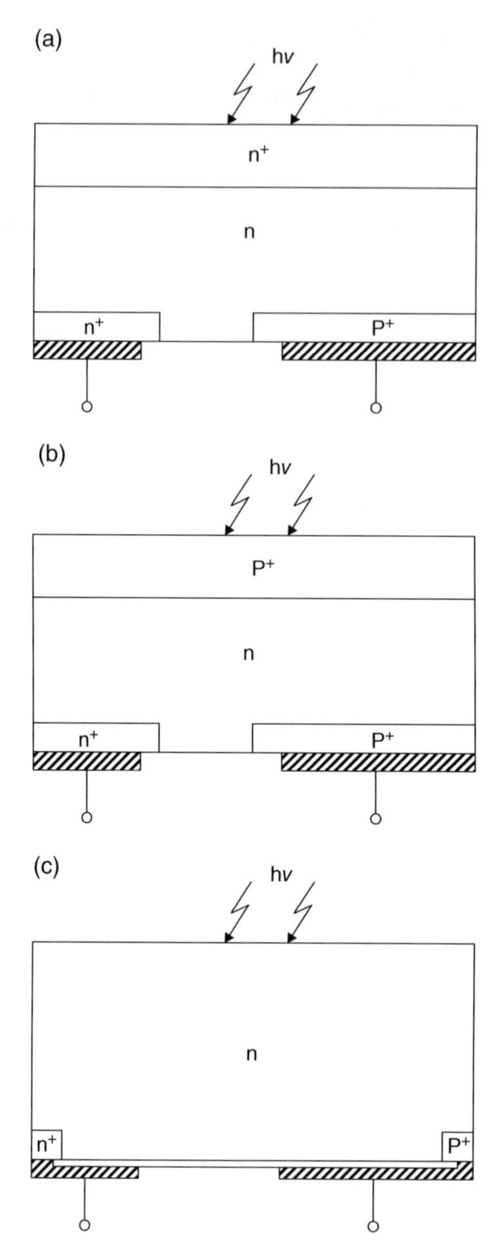

Figure 3.3.3 (a) FSF, (b) FFE and (c) PC solar cell designs

3.3.2.2 Front Floating Emitter Solar Cells

The effective recombination velocity of the front surface can also be reduced with a front floating emitter (FFE) instead of a FSF (Figure 3.3.3 (b)). This type of structure was originally and wrongly called a "tandem junction" because the structure was comparable to a pnp bipolar transistor with a floating emitter, and its analytical modeling used Ebers-Moll equations (Matzen *et al.*, 1979; Goradia, 1980). The analytical modeling of the FFE, FSF and IBC cell was later unified. The emitter recombination parameter J_o is the main characteristics of the FFE, and the behavior of FFE cells is actually identical to the behavior of FSF cells (Fossum *et al.*, 1980; Verlinden, 1985) as long as the light-generated voltage across the FFE is greater than about 400 mV (not the cell voltage, but the voltage between the front surface and the bulk). For example, when measuring the spectral response of FFE cells (with a diffused or field-induced floating emitter), it is important to flood the cell with white light bias (Matzen *et al.*, 1979), which is not necessary for FSF cells or IBC cells with a field-induced accumulation region at the front surface. For the same reason, FFE cells are more sensitive than FSF cells to front surface defects, creating shunts of the front floating junction.

The optimum FFE is obtained with a doping profile giving the lowest possible J_o, which is usually obtained by a shallow lightly doped emitter (King, 1990). The use of low sheet resistance FFE has sometimes been proposed to reduce the loss in J_{sc} due to "electronic shading" (Cesar *et al.*, 2014), a loss due to the fact that carriers generated above the BSF of the IBC cell require a longer distance to reach the rear emitter than carriers generated directly above the emitter, and suffer from additional recombination while approaching the BSF region (see Figure 3.3.3 (a)). This solution of using low resistivity FFE generally results in an improved J_{sc} but low V_{oc} and *FF* because of the high value of the J_o of low-resistivity FFEs.

3.3.2.3 Point-Contact Solar Cells

In order to decrease the emitter recombination which can become significant in IBC cells in the CPV application, Swanson (1985) introduced and proposed a quasi-3D analytical model for an improved design with reduced emitter area and reduced metal contact, called a Point-Contact (PC) solar cell (Figure 3.3.3 (c)). Parrott (Parrott and Al-Juffali, 1987) presented a comparison of point and line contact IBC cells for different concentration ratios. Under high concentration, PC solar cells have achieved the highest efficiencies to date for silicon solar cells. A small (3 mm × 5 mm) laboratory PC cell has reached an efficiency of 28.3% (Swanson *et al.*, 1988) and a 1 cm^2 production PC cell reached 27.6% under 9.7 W/cm^2 (Slade and Garboushian 2005).

3.3.2.4 IBC Solar Cells for Concentrator Application

IBC solar cells with a PC design for high-concentration ratio CPV application (Figure 3.3.4) require thin substrates (100–150 μm) to reduce the Auger recombination, a very fine pitch, typically 50–100 μm, with low emitter coverage, around 10%, and a small contact area, typically around 5% of the cell area (Sinton, 1987). Due the very high current, they also require a sophisticated metallization scheme with a double-level metallization scheme (Verlinden *et al.*, 1988). With very fine pitch and a double-level metallization, series resistances in the range of 10^{-3} ohm.cm^2 can be achieved. Because of the very high injection level, the substrate resistivity is not important. The critical parameter is the bulk carrier lifetime in high injection. Therefore, n-type high resistivity substrates are preferred. The highest efficiency of 27.6% at a concentration ratio of 97X for a production device was reported by Slade and Garboushian (2005). Incorporating passivated contact technology with polysilicon emitters could be a

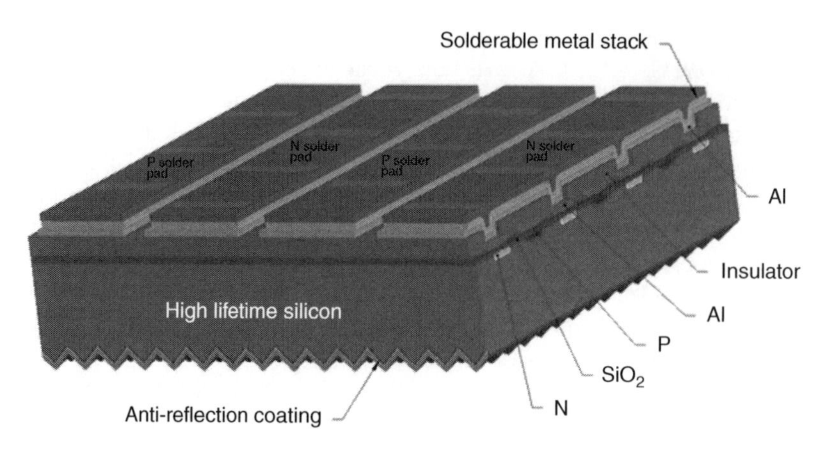

Figure 3.3.4 Structure of PC solar cell for high-concentration CPV application. Source: (Verlinden, 2005). (*See insert for color representation of the figure*)

potential pathway to achieve 30% efficiency (Verlinden, 2005) at a concentration ratio of 275X.

Recently, IBC cells originally developed for one-sun application have also been slightly modified for low concentration PV (LCPV) application (Bunea *et al.*, 2010).

3.3.3 IBC Solar Cells for One-Sun Flat-Plate Modules

Due to cost constraints, IBC solar cells for one-sun flat-plate modules cannot be fabricated with very fine patterning technique such as photolithography. Therefore, the process must be simplified and low-cost patterning, such as screen printing, limits the minimum pitch (the width of the full unit cell, from the center of the BSF to the center of the next BSF region) to about 800–1500 μm. The rear emitter should have a large coverage fraction (>85%) to avoid "electronic shading" and the BSF region should be reduced to the smallest coverage fraction that is technically possible. The pitch of the IBC cell and the substrate resistivity are the determining factors for the bulk series resistance (McIntosh *et al.*, 2003), while the size and the J_o of the BSF region determine the loss in J_{sc} due to "electronic shading." The optimum substrate is typically n-type with a resistivity around 2 ohm.cm, being a trade-off between bulk lifetime and internal series resistance. The metal contact coverage fraction, typically 1–5%, must be reduced to the smallest value but, in practice, this is limited by the contact resistivity achievable with the selected metallization technology.

An important process improvement in the manufacturability of low-cost IBC cells was presented by Sinton (Sinton and Swanson, 1990) consisting of a uniform p+emitter diffusion followed by patterning, etch back and diffusion of the n+BSF regions, reducing by one the number of patterning steps. This technique is still in use for most IBC cell manufacturing processes. Other proposed process simplifications include masked ion implantation (Mo *et al.*, 2012), screen-printed dopant paste (Scardera *et al.*, 2015), atmospheric chemical vapor deposition (APCVD) of doped oxide followed by a single step co-diffusion (Sinton *et al.*, 1989).

There are several examples of small-area laboratory-scale IBC solar cells, using three or more photolithography steps and vacuum-deposited metallization with measured one-sun efficiencies greater than 23%. Aleman *et al.* (2012) reported an efficiency of 23.3% on a 4 cm^2

device. Merkle *et al.* (2014) reported an ion-implanted and laser-structured IBC cell with an independently confirmed efficiency of 23.7% measured on a designated area of 3.96 cm^2. Franklin *et al.* (2014) developed a 4 cm^2 IBC cell with a designated area efficiency of 24.4% and presented ways to achieve efficiencies exceeding 25%.

The difficulty of extending such high efficiencies in low-cost mass production with large substrates certainly should not be underestimated. Contamination control and patterning fine geometries with low-cost high-volume processes are critical to achieve the same efficiencies as with small-area IBC cells. Samsung reported a 22.4% efficient IBC solar cell with an area of 155 cm^2 fabricated with a simplified process using masked ion implantation (Mo *et al.*, 2012). Trina Solar has demonstrated a 156 × 156 mm IBC solar cell fabricated with standard low-cost large-volume manufacturing with a champion cell reaching a total area efficiency of 23.5% (Guanchao *et al.*, 2016). The best results with IBC solar cells have been presented by SunPower, and Panasonic. Smith *et al.* (2016) presented a 25.2% efficient (total area efficiency, 153.5 cm^2) IBC cell. The aperture area (121.1 cm^2) efficiency of the same cell was 25.3%, the difference between the two efficiency values being due to edge losses. The V_{oc} of the SunPower cell is an impressive 737 mV (Smith *et al.*, 2016), suggesting that the emitter has passivated contacts, probably formed by a thin tunnel oxide and a doped polysilicon emitter (Swanson, 2008). Smith *et al.* (2016) give an excellent analysis of the power loss in the cell, estimating that a total area efficiency of 25.7% can be achieved by improving the total J_o of the cell and the rear surface optical performance to increase light trapping. A practical full area efficiency of 26% seems attainable. The IBC cells currently in production by SunPower have two sets of three bonding pads at two opposite edges of the cell, which design limits the acceptable total dimensions of the cell and requires a relatively thick metallization (several tens of microns) formed by copper plating.

Recently Nakamura *et al.* (2014) and Masuko *et al.* (2014) have presented heterojunction interdigitated back contact (HIBC) solar cells with two heterojunctions for both emitter and BSF. The best HIBC cell has a designated area (143.7 cm^2) efficiency of 25.6%. It is to date the highest one-sun efficiency for any single-junction silicon solar cell. This cell is a combination of the IBC solar design and the heterojunction (HJ) solar cell technology previously developed by Panasonic which had reached a previous record of 24.7% efficiency. Both n-type and p-type back side contacts consist of doped and intrinsic amorphous silicon stack similar to the conventional HJ solar cell. The V_{oc} reaches an impressive value of 740 mV (Masuko *et al.*, 2014).

3.3.4 Conclusion

In this chapter we presented an overview of the development of IBC silicon solar cells from their introduction in 1975 to the most recent results in 2014, summarized in Table 3.3.1. Although more complicated to fabricate and more expensive than their conventional counterparts, IBC solar cells have demonstrated the highest efficiencies for single-junction silicon cells, for both CPV (up to 28.3%) and one-sun applications (up to 25.2%). Recently, the combination of IBC design with heterojunctions in HIBC cells has demonstrated the highest efficiency to date with 25.6%. Several avenues have been presented for future development and efficiencies greater than 26% seem achievable. High-quality substrate, contamination control, cleaning procedure, fine geometries, good surface passivation and diffusion regions

Table 3.3.1 Recent performance achievements in IBC solar cell development

Cell	Area (cm^2)	Type of Efficiency	J_{sc} (mA/cm^2)	V_{oc} (mV)	FF (%)	Efficiency (%)	Ref.
Panasonic HIBC	143.7	Designated area	41.8	740	82.7	25.6	Masuko *et al.*, 2014
Sharp HIBC	3.72	Aperture area	41.7	736	81.9	25.1	Nakamura *et al.*, 2014
SunPower IBC	153.5	Total area	41.33	737	82.7	25.2	Smith *et al.*, 2016
ANU IBC	4.0	Designated area	41.95	702.5	82.7	24.37	Franklin *et al.*, 2014
ISFH IBC	3.96	Designated area	41.3	696	82.8	23.7	Merkle *et al.*, 2014
Imec IBC	4.0	Designated area	41.6	696	80.4	23.3	Aleman *et al.*, 2012
Trina Solar IBC	238.6	Total area	42.08	689.9	80.9	23.5	Guanchao *et al.*, 2016
Samsung IBC	155	Total area	40.9	676	81.0	22.4	Mo *et al.*, 2012

with low recombination are key to efficient IBC cells. Process simplification, passivated contacts and a heterojunction will be part of the next generation of IBC cells approaching the efficiency limit for a single-junction silicon solar cell.

List of Symbols

Symbol	Description	Unit
FF	Fill Factor	
J_{sc}	Short-circuit current density	A/cm^2
J_o	Saturation current density or Emitter recombination parameter	A/cm^2
J_n	Electron current density	A/cm^2
J_p	Hole current density	A/cm^2
J_T	Total current density	A/cm^2
V_{oc}	Open-circuit voltage	V
Φ_p	Quasi-Fermi level for holes	V
Ψ	Potential	V

List of Acronyms

Acronym	Meaning
APCVD	Atmospheric Pressure Chemical Vapor Deposition
BSF	Back Surface Field
CPV	Concentrator Photovoltaics
Cz	Czochralski
FFE	Front Floating Emitter
FSF	Front Surface Field
FZ	Float Zone
HIBC	Heterojunction Interdigitated Back Contact
IBC	Interdigitated Back Contact
LCPV	Low Concentration Photovoltaics
PC	Point Contact
RCA	Radio Corporation of America

References

Aleman, M., Das, J., Janssens, T. *et al.* (2012) Development and integration of a high efficiency baseline leading to 23% IBC cells. *2nd Silicon PV Conference, Leuven, 2012, in Energy Procedia*, **27**, 638–645. DOI:10.1016/j.egypro.2012.07.122.

Bunea, M.M., Johnston, K.W., Bonner, C.M. *et al.* (2010) Simulations and characterization of high-efficiency back contact cells for low-concentration photovoltaics. *Proceedings of the 35th IEEE Photovoltaic Specialists Conference (PVSC)*. DOI: 10.1109/PVSC.2010.5617188.

Cesar, I., Guillevin, N., Burgers, A.R. *et al.* (2014) Mercury: a novel design for back junction back contact cell with high efficiency and simplified processing. *Proceedings of the 29th[h] European Photovoltaic Solar Energy Conference*, Amsterdam, NL, 22–26 September.

Fossum, J.G., Neugroschel, A., and Lindholm, F.A. (1980) A unifying study of tandem junction, front-surface field and interdigitated back contact solar cells. *Solid-State Electronics*, **23**, 1127–1138.

Franklin, E., Fong, K., McIntosh, K. *et al.* (2014) Design, fabrication and characterization of a 24.4% efficient interdigitated back contact solar cell. *Progress in Photovoltaics Research and Applications*, 2014. DOI: 10.1002/pip.2556.

Goradia, C. (1980) A one-dimensional theory of high base resistivity tandem junction solar cells in low injection. *IEEE Transactions on Electron Devices*, **27** (4), 777–785.

Guanchao, X., Yang, Y., Zhang, X.L. *et al.* (2016) 6 inch IBC cells with efficiency of 23.5% fabricated with low-cost industrial technologies. *Proceedings of the 43rd IEEE Photovoltaic Specialists Conference*, Portland, OR, June 5–10, 2016.

Kern, W. and Puotinen, D.A. (1970) Cleaning solutions based on hydrogen peroxide for use in silicon semiconductor technology. *RCA Review*, **31** (2), 187–206.

King, R.R. (1990) Studies of oxide-passivated emitters in silicon and applications to solar cells, PhD thesis, Stanford University, Stanford, CA.

Lammert, M.D. and Schwartz, R.J. (1977) The interdigitated back contact solar cell: A silicon solar cell for use in concentrated sunlight. *IEEE Transactions on Electron Devices*, **24** (4), 337–342.

Li, Z., Yang, Y., Zhang, X. *et al.* (2014) High-lifetime wafer cleaning method using ozone dissolved in DIW/HF/HCL solution. *Proceedings of the 29th European Photovoltaic Solar Energy Conference*, Amsterdam.

Masuko, K., Shigematsu, M., Hashiguchi, T. *et al.* (2014) Achievement of more than 25% conversion efficiency with crystalline silicon heterojunction solar cell. *Proceedings of the 40th IEEE Photovoltaic Specialists Conference*, Denver, CO.

Matzen, W.T., Chiang, S.Y., and Garbajal, B.G. (1979) A device model for the tandem junction solar cell. *IEEE Transactions on Electron Devices*, **26** (9), 1365–1368.

McIntosh, K.R., Cudzinovic, M.J., Smith, D.D. *et al.* (2003) The choice of silicon wafer for the production of low-cost rear-contact solar cells, *Proceedings of the 3rd World Conference on Photovoltaic Energy Conversion*, Osaka, 2003, pp. 971–974.

Merkle, A., Peibst, R., and Brendel, R. (2014) High efficient fully ion-implanted, co-annealed and laser-structured back junction back contacted solar cells. *Proceedings of the 29th European Photovoltaic Solar Energy Conference*. WIP-Renewable Energies: Munich, 2014, pp. 954–958. DOI 10.4229/EUPVSEC20142014-2AV.2.61.

Mo, C.B., Park, S.J., Kim, Y.J. *et al.* (2012) High efficiency back contact solar cells via ion implantation. *Proceedings of the 27th European Photovoltaic Solar Energy Conference*, Frankfurt, Germany.

Nakamura, J., Asano, N., Hieda, T. *et al.* (2014) Development of heterojunction back contact Si solar cells. *Proceedings of the 29th European Photovoltaic Solar Energy Conference*, Amsterdam, pp. 373–375.

Parrott, J.E. and Al-Juffali, A.A.(1987) Comparison of the predicted performance of IBC and point contact solar cells. *Proceedings of the 19th IEEE Photovoltaic Specialists Conference*. New Orleans, LA, May 4–8, pp. 1520, 1521.

Scardera, G., Inns, D., Wang, G. *et al.* (2015) Screen-printed dopant paste interdigitated back contact solar cells. *Proceedings of the 42nd IEEE Photovoltaic Specialists Conference*, New Orleans.

Schwartz, R.J. and Lammert, M.D. (1975) Silicon solar cells for high-concentration application. *IEEE International Electron Devices Meeting*, Washington, DC, pp. 350–351.

Sinton, R.A. (1987) Device physics and characterization of silicon point-contact solar cells. PhD thesis, Stanford University, Stanford, CA.

Sinton, R.A., King, R.R., and Swanson, R.M. (1989) Novel implementation of backside-contact silicon solar cell designs in one-sun and concentrator applications. *Proceedings of the 4th International Photovoltaic Science and Engineering Conference*, Sydney, Feb. 14–17, 1989.

Sinton, R.A. and Swanson, R.M. (1990) Simplified backside-contact solar cells. *IEEE Transactions on Electron Devices*, **37** (2), 348–352.

Slade, A. and Garboushian, V. (2005) 27.6% efficient silicon concentrator cell for mass production. *Technical Digest of the 15th International Photovoltaic Science and Engineering Conference*, Shanghai, October, p. 701.

Smith, D.D., Reich, G., Baldrias, M. *et al.* (2016) Silicon solar cells with total area efficiency above 25%. *Proceedings of the 43rd IEEE Photovoltaic Specialists Conference*, Portland, OR, June 5–10, 2016.

Swanson, R.M. (1985) Point contact silicon solar cells: theory and modeling. *Proceedings of the 18th IEEE Photovoltaic Specialists Conference*, Las Vegas, pp. 604–610.

Swanson, R.M. (1988) Point contact solar cells: modeling and experiment. *Solar Cells*, **7** (1), 85–118.

Swanson, R.M. (2008) Back side contact solar cell with doped polysilicon regions. US Patent US 7,468,485.

Swanson, R.M., Sinton, R.A., Midkiff, N., and Kane, D.E. (1988) Simplified designs for high-efficiency concentrator solar cells. *Sandia Report SAND88-0522*, Sandia National Laboratories, Albuquerque, NM, July.

Van Kerschaver, E. and Beaucarne, G. (2006) Back-contact solar cells: a review. *Progress in Photovoltaics Research and Applications*, **14** (2), 107–123.

Verlinden, P. (1985) Modélisation de cellules solaires pour lumière concentrée. Analyse des cellules à contacts sur la face arrière, PhD thesis. Université Catholique de Louvain, Belgium.

Verlinden, P.J. (2005) High-efficiency concentrator silicon solar cells. In *Solar Cells, Materials, Manufacture and Operation* (eds T. Markvart and L. Castañer). Elsevier Science, Oxford, pp. 371–391.

Verlinden, P.J., Sinton, R.A., and Swanson, R.M. (1988) High-efficiency large-area back contact concentrator solar cells with a multilevel interconnection. *International Journal of Solar Energy*, **6**, 347–366.

Verlinden, P.J., Swanson, R.M., and Crane R.A. (1994) 7000 High-efficiency cells for a dream. *Progress in Photovoltaics: Research and Applications*, Special Issue, April 1994, **2**, 143–152.

Yernaux, M.I., Battochio, C., Verlinden, P., and Van de Wiele, F. (1984) A one-dimensional model for the quantum efficiency of front-surface field solar cells. *Solar Cells*, **13**, 83–97.

Zhou C.Z., Verlinden P.J., Crane R.A., and Swanson R.M. (1997) 21.9% efficient silicon bifacial solar cells. *26th IEEE Photovoltaic Specialists Conference*, Anaheim, September 29–October 3, pp. 287–290.

3.4

Heterojunction Silicon Solar Cells

Wilfried van Sark
Copernicus Institute, Utrecht University, The Netherlands

Silicon-based solar cells are predominantly of the so-called homojunction type, i.e., the junction is made so that the material on either side of the junction is the same, albeit that doping levels and type can be different, see also Part 2 in this book. When two different semiconductor materials form a junction, where the materials differ in bandgap, it is denoted as a heterojunction, and this results in band offsets and band discontinuities. Heterojunctions were first discussed in the 1950s (Sze, 1981; Kroemer, 2000) and they have found widespread applications in, among others, lasers, light emitting diodes, and solar cells. The latter will be discussed here.

Heterojunctions can be found in a wide variety of solar cells. Examples include the Heterojunction with Intrinsic Thin-layer (HIT) solar cell, which is based on a thin amorphous silicon p-type doped layer on top of an n-type doped c-Si wafer (Van Sark *et al.*, 2012; De Wolf, 2012), the cadmium telluride (CdTe/CdS) and copper indium (Ga) diselenide (CIGS/CdS) solar cell (Scheer and Schock, 2011; see also Part 4) and the bulk heterojunction organic PV (Part 6). Heterojunctions are also encountered in multijunction solar cells (Part 8). High efficiency cells have been demonstrated with >40% efficiency using material alloys from Groups III and V of the Periodic Table (Philipps and Bett, 2014).

The silicon heterojunction cell (SHJ), also commonly termed the HIT cell (a trademark of the Panasonic Company), can best be characterized as a device structure that combines the best of thin film (Part 5) and bulk silicon technology. It has been referred to as a marriage between these two technologies and was first announced as a "honeymoon" cell (Hamakawa *et al.*, 1983; Tanaka *et al.*, 1992). Early patenting has given Sanyo a head start in commercializing the SHJ technology, which is now marketed by Panasonic. When the SHJ technology is combined with a backside contact structure, it has been demonstrated to result in 25.6% efficiency, the highest "one-sun" efficiency to date for a crystalline silicon solar cell (Masuko

Photovoltaic Solar Energy: From Fundamentals to Applications, First Edition.
Edited by Angèle Reinders, Pierre Verlinden, Wilfried van Sark, and Alexandre Freundlich.

Companion website: www.wiley.com/go/reinders/photovoltaic_solar_energy

et al., 2014). Interestingly, another record efficiency of 25.1% for a two-sided contacted SHJ cell was reported by Kaneka (Yamamoto *et al.*, 2015).

In this chapter, first, the basics of heterojunctions will be presented, drawing on the basics of homojunctions presented in Part 2. The subsequent sections will present the development of the SHJ cell.

3.4.1 Basic Principles

We consider the energy band diagram of a *pn* heterojunction in Figure 3.4.1 (Muller *et al.*, 2003). An isotype heterojunction, where doping type of both materials is the same, i.e., an n^+n or p^+p junction, is not treated here. Figure 3.4.1 (a) shows the energy band diagram of the two separated semiconductors that will form the *pn* heterojunction, i.e., a wide bandgap *n*-type semiconductor on the left and a narrow bandgap *p*-type semiconductor on the right with bandgaps E_{g1} and E_{g2}, respectively, with $E_{g1} \neq E_{g2}$. In the general case, the electron affinities of the materials can be different, i.e., $\chi_1 \neq \chi_2$. Doping densities are N_{d1} and N_{a2}. Also note that the conduction band edge of the wide bandgap material is higher in energy than that of the narrow bandgap material, and vice versa for the valence band edges.

When the two materials are brought together, electrons and holes will flow, leaving a depletion region at the junction between the materials just as in the homojunction case (Chapter 2.6), and thermal equilibrium is reached, which is clear from the Fermi level E_F that is the same at both sides of the junction (Figure 3.4.1 (b)). Unlike with the homojunction, this leads to band edge discontinuities of the conduction and valence band at the junction. In the case shown in Figure 3.4.1 (b), we notice what is called a "spike" and a "notch" in the conduction band, in which free charges could accumulate. The magnitude of ΔE_c is determined by the doping levels on either side of the junction.

The resulting built-in voltage Φ_i (also denoted as V_{bi}) can be found as follows:

$$q\Phi_i = q\Phi_{p2} - q\Phi_{n1} = \left(q\chi_2 + E_{c2} - E_F\right) - \left(q\chi_1 + E_{c1} - E_F\right) \tag{3.4.1}$$

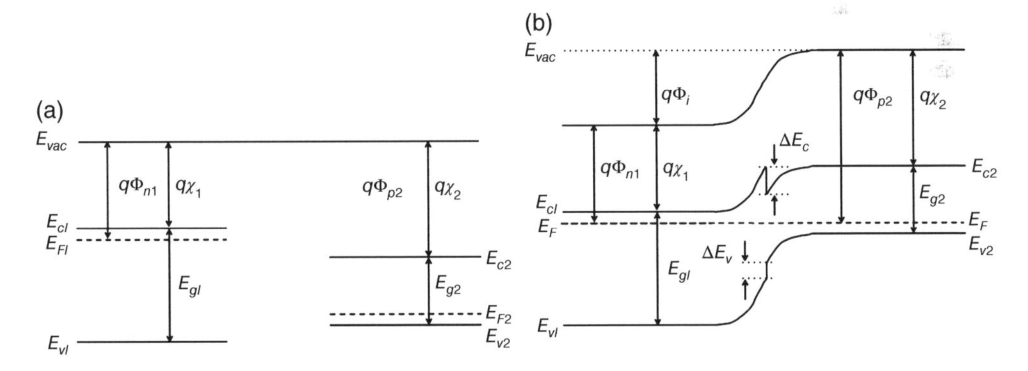

Figure 3.4.1 Energy band diagrams for a heterojunction. (a) Two separated semiconductors, with a wide bandgap n-type semiconductor on the left and a narrow band gap p-type semiconductor on the right. (b) Energy levels after bringing the materials into contact and reaching an equilibrium state. The vacuum level remains continuous, while the conduction and valence band are discontinuous. Source: Adapted from Muller *et al.* (2003)

with

$$E_{c1} - E_F = kT \ln\left(\frac{N_{c1}}{N_{d1}}\right) \tag{3.4.2}$$

$$E_{c2} - E_F = E_{g2} - (E_F - E_{v2}) = E_{g2} - kT \ln\left(\frac{N_{v2}}{N_{a2}}\right) \tag{3.4.3}$$

with N_{c1} and N_{v2} being the effective density of states at the conduction and valence band edges of the two materials.

Combined with Equations (3.4.2) and (3.4.3), Equation (3.4.1) becomes:

$$\Phi_i = \chi_2 - \chi_1 + \frac{1}{q}\left(E_{g2} - kT \ln\left(\frac{N_{c1}N_{v2}}{N_{d1}N_{a2}}\right)\right) \tag{3.4.4}$$

For the case that the electron affinities of both materials are equal, the discontinuity in the conduction band disappears and Equation (3.4.4) is reduced to a similar form as was derived for a homojunction, except that a term with E_{g2} remains:

$$\Phi_i = \frac{1}{q}\left(E_{g2} + kT \ln\left(\frac{N_{d1}N_{a2}}{N_{c1}N_{v2}}\right)\right) \tag{3.4.5}$$

where it should be noted that the product $N_{c1}N_{v2}$ would equal the intrinsic carrier density squared (n_i^2) if both materials were the same (the homojunction case, see Chapter 2.6).

The potential drop is divided over the two sides as follows (Muller *et al.*, 2003):

$$\Phi_{i1} = \Phi_i \frac{\varepsilon_2 N_{a2}}{\varepsilon_1 N_{d1} + \varepsilon_2 N_{a2}} \tag{3.4.6}$$

$$\Phi_{i2} = \Phi_i \frac{\varepsilon_1 N_{d1}}{\varepsilon_1 N_{d1} + \varepsilon_2 N_{a2}} \tag{3.4.7}$$

with ε_1 and ε_2 the permittivities of the two materials. As with the homojunction, the built-in voltage drop is larger across the material with the lightest doping concentration. This is important in a heterojunction due to the presence of a spike in the conduction band edges. For example, when the doping in material 1 is higher than that in material 2, most of the band bending occurs in material 2, leading to the fact that the top of the spike is lower than the conduction band edge of material 2 (see Figure 3.4.1(b)) (Muller *et al.*, 2003). This influences the operational performance of heterojunction devices, such as solar cells. Proper analysis of band diagrams thus is necessary to understand device performance. The current across the junction can be expressed in a similar way as for homojunctions (Chapter 2.6), albeit that electron and hole currents are dependent on the band discontinuities in the conduction band (Muller *et al.*, 2003).

3.4.2 a-Si:H/c-Si Cell Development

The silicon heterojunction solar cell is based on a device structure that combines thin film and bulk silicon technology. Of particular interest is the heterojunction cell with an intrinsic thin layer (HIT) (Tanaka *et al.*, 1992) with which Sanyo had achieved a very high efficiency of 23% in 2009 (Taguchi *et al.*, 2009), and of 25.6% efficiency most recently (Masuko *et al.*, 2014). This strongly challenges the standard high efficiency c-Si homojunction cell (Green *et al.*, 2015). The emitter can be an amorphous (Taguchi *et al.*, 2009) or microcrystalline (Van Kleef *et al.*, 1997) silicon type. It is claimed by Sanyo (Tanaka *et al.*, 1992) that an excellent surface passivation of c-Si at the c-Si/a-Si interface is most important for high efficiency.

In practical heterojunction solar cells, such as the a-Si:H/c-Si one, the band offsets between different materials are accommodated in both the conduction band and the valence band. This results in barriers for carrier transport. Figure 3.4.2 shows the band diagram of *p*-type a-Si:H (bandgap 1.7 eV) and *n*-type c-Si (bandgap 1.1 eV). As indicated in Figure 3.4.2, the holes experience a spike which acts as a barrier to hole transport. Holes can move from c-Si to a-Si:H by tunnelling, trap-assisted tunnelling and/or thermionic emission (Zeman and Zhang, 2012).

Figure 3.4.3 shows the development of the structure of SHJ cells, illustrating increased complexity. The basic structure consists of a *p*-type a-Si:H layer on top of an *n*-type c-Si wafer. With this solar cell Sanyo achieved an efficiency of 12.3% (Tanaka *et al.*, 1992). It was shown that the optimal *p*-type a-Si:H layer thickness in terms of efficiency is 10 nm. It was found that a high defect-state density at the a-Si:H/c-Si interface is limiting the open-circuit voltage (V_{oc}) and fill factor (*FF*) of the device, i.e., 0.56 V and 0.71, respectively. The short-circuit current (J_{sc}) was 30 mA/cm^2 at the maximum, and increasing the p-type a-Si:H layer thickness beyond 10 nm resulted in a decrease of J_{sc} due to absorption losses in the a-Si:H material.

To reduce the defect density at the *p-n* interface, a thin intrinsic a-Si:H layer between the n-type c-Si wafer and the *p*-type a-Si:H was introduced, which was denoted as ACJ-HIT (artificially constructed junction-heterojunction with intrinsic thin film) solar cell by Sanyo (Tanaka *et al.*, 1992). The thin intrinsic a-Si:H layer was to passivate the dangling bonds on the c-Si surface, and the a-Si:H/c-Si interface defect-state density was indeed reduced. This resulted in an improvement of V_{oc} and *FF* to 0.6 V and 0.79, respectively, with J_{sc} of 33 mA/cm^2

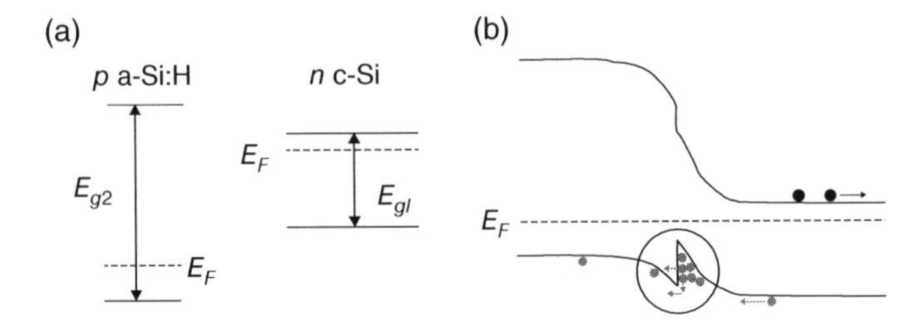

Figure 3.4.2 Energy band diagrams for (a) the separate materials p-type a-Si:H and n-type c-Si; and (b) the a-Si:H/c-Si heterojunction. Black and grey dots represent electrons and holes, respectively. The circle indicates the carrier (hole) transport through the energy barrier at the interface. Source: After Zeman and Zhang (2012)

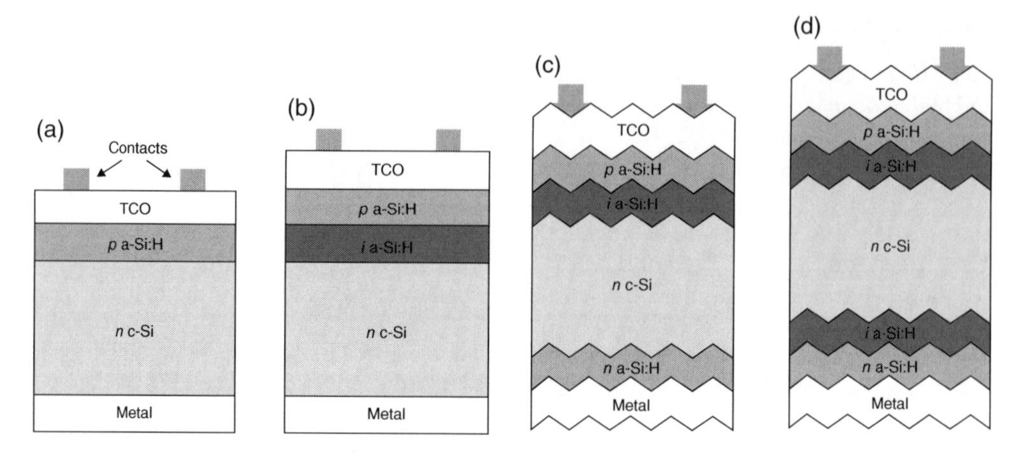

Figure 3.4.3 Schematic structure of heterojunction a-Si:H/c-Si solar cells. (a) basic structure with transparent conductive oxide and metal as top and back contact; (b) idem, with intrinsic a-Si:H layer sandwiched between p a-Si:H and n c-Si; (c) idem, with textured interfaces and back-surface field layer of n a-Si:H, (d) idem, with additional *i* a-Si:H. Note, drawings are not to scale

for an optimal intrinsic layer thickness of 4 nm. The highest efficiency was 14.8%. With increasing *i*-layer thickness, the absorption losses increase and as a consequence J_{sc} decreases. Also, the *FF* decreases due to the high resistivity of the intrinsic layer.

Further improvements have been achieved by introducing random texturing of the c-Si wafer to achieve optimal light trapping. At the same time, a back surface field (BSF) is introduced, which is formed by an *n*-type a-Si:H (Figure 3.4.3 (c)). The best device performance was J_{sc} = 37.9 mA/cm^2, V_{oc} = 0.614 V, *FF*=0.78, and an efficiency of 18.6% (Tanaka *et al.*, 1992). A further step is to use intrinsic a-Si:H on both sides of the c-Si wafer for passivation. This yielded J_{sc} = 38.6 mA/cm^2, V_{oc} = 0.717 V, *FF*=0.77, and efficiency of 21.3% (Tanaka *et al.*, 2003). In all device designs, transparent conductive oxides (TCOs) are necessary to facilitate carrier transport to the contacts as the a-Si:H are highly resistive. Usually, indium tin oxide (ITO) is used as the TCO.

The highest efficiency reported to date is 25.6%, using an interdigitated back contact structure to reduce the optical loss from absorption in the TCO layer and reflection by the front grid electrode. Silicon nitride was used as an antireflection layer, instead of the usual TCO layer. It was reported that the short-circuit current was improved to 41.8 mA/cm^2, and V_{oc} = 0.74 V, *FF*=0.827 (Masuko *et al.*, 2014).

3.4.3 Key Issues in a-Si:H/c-Si Cells

3.4.3.1 Surface Passivation

Surface passivation of the c-Si wafer is a key requirement to optimize the performance of SHJ solar cells. The design of SHJ solar cells allows for the deposition of a surface passivating thin intrinsic layer consisting of a-Si:H before depositing the doped emitter and BSF layers, which can increase V_{oc}. The passivating properties of intrinsic a-Si:H have been widely studied

(Olibet *et al.*, 2007; Illiberi *et al.*, 2010; Schulze *et al.*, 2010; Schüttauf *et al.*, 2011), and excellent results have been obtained by various laboratories using different deposition techniques. For example, Schüttauf *et al.* (2011) compared different chemical vapor deposition (CVD) techniques, i.e., standard radio frequency plasma-enhanced CVD (rf-PECVD) at 13.56 MHz, very high frequency PECVD (VHF PECVD) at 50 MHz, and hot-wire CVD (HWCVD). It was shown that minority carrier lifetimes of 34–87 μs could be obtained, which could be increased to values > 10 ms after prolonged annealing at ~200 °C. Annealing is crucial as it enables the movement of hydrogen in the a-Si:H layer toward the a-Si:H/c-Si interface, which strongly reduces the dangling bond density at the interface (Olibet *et al.*, 2007; De Wolf *et al.*, 2012).

The excellent minority carrier lifetimes that can be obtained with a-Si:H can be explained by distinguishing different passivation mechanisms that govern the lifetime characteristics. Generally, surface passivation is ascribed to two different phenomena, namely, chemical passivation and field effect passivation. The former is caused by a reduction in defect states at the a-Si:H/c-Si interface, the latter by the repulsion of minority charge carriers from the interface by fixed charges. At high injection levels (light intensity) the Auger recombination also has to be taken into account (Schüttauf *et al.*, 2011). It was reported that the high quality of surface passivation, measured as surface recombination velocity (SRV), is mainly caused by a reduction in dangling bond density at the interface (N_{DB}) due to a saturation of dangling bonds by atomic hydrogen. The excellent c-Si surface passivation implies that the interface dangling bond density is on the order of 10^8 cm^{-2}. An SRV as low as 0.93 cm/s is calculated at an injection level of 10^{15} cm^{-3} (Schüttauf *et al.*, 2011).

In SHJ cells, the influence of the doped layers is also crucial (Schüttauf *et al.*, 2011; De Wolf *et al.*, 2012). The BSF creates additional field effect passivation, repelling minority charge carriers from the a-Si:H/c-Si interface at the back side, thereby further reducing recombination losses. On the emitter side, minority carriers are passing through the a-Si:H/c-Si interface region in the front side, which enhances the recombination losses. Furthermore, doped a-Si:H contains high levels of defect density, as the formation enthalpy for defects counteracting the active dopants is reduced when the Fermi level approaches one of the band edges (Street, 1991).

3.4.3.2 Wafer Cleaning

Before passivation layers are deposited, the c-Si wafer needs to be cleaned (Angermann *et al.*, 2009). Obviously, this is needed to remove any contamination present, but also the cleaning step in the processing sequence is used to partially passivate dangling bonds at the surface with hydrogen. One generally determines the quality of the cleaning step by measuring the carrier lifetimes using a Sinton contactless lifetime tester (Sinton and Cuevas, 1996). The RCA (Radio Corporation of America) cleaning method (Kern, 1970) is often used for high-lifetime processing of silicon wafers, which combines the removal of organic and inorganic metallic contaminants by a diluted solution of hydrogen peroxide with ammonia, followed by a diluted solution of hydrogen peroxide and hydrochloric acid, followed by an oxide removal step by hydrofluoric acid (HF) etching. This produces contamination-free c-Si surfaces that are temporarily chemically stable for further processing (Grundner and Jacob, 1986). Note that after cleaning, the wafers need to be transferred quickly to deposition systems to prevent fast native oxide growth on the c-Si surface.

3.4.3.3 *Texturing*

To lower material costs, c-Si wafers should be as thin as possible, however, this limits the absorption of light. This has prompted the development of textured surfaces as a means to increase light trapping. For c-Si wafers one generally uses a random pyramid texture. This can be attained using Si (100) substrates which are etched in alkaline solutions: the anisotropic etching yields Si (111) faceted pyramids (Bean, 1978), which lowers external reflection and at the same time improves internal reflection. Subsequent deposition of layers that are uniform in thickness is possible on the flat facets of the pyramids using CVD techniques.

3.4.4 Advantages Compared to c-Si Cells

In conventional c-Si solar cells, the p–n junction is formed by a thermal diffusion for which temperatures around 900 °C are required. In case of a-Si:H/c-Si-based heterojunction solar cell, processing, temperatures below 200 °C are typically used, as the p–n junction is formed by depositing a thin doped a-Si:H layer on a c-Si wafer (Van Sark *et al.*, 2012). This lower temperature processing (1) decreases the thermal budget of the production process; (2) limits the thermal degradation of the c-Si wafer quality; (3) enables the use of thinner wafers (<100 μm), avoiding warping issues associated with traditional high temperature processing; and (4) results in reduced environmental impact and production costs (Louwen *et al.*, 2015). Low-temperature (<200 °C) processing of a-Si:H/c-Si-based heterojunction solar cell is actually required after the a-Si:H film silicon deposition to avoid recrystallization by epitaxy of the amorphous film. This requirement, however, complicates the formation of low-resistivity metallized contacts for SHJ solar cells.

Light-induced degradation of performance that is usually seen in a-Si:H devices (Part 5) is not observed for a-Si:H/c-Si cells. This is due to the fact that the amorphous layers are very thin (5–10 nm) and their contribution to power generation can be ignored.

The temperature coefficient of a-Si:H/c-Si cells is lower than that of c-Si cells: the efficiency decreases as the function of the temperature is less severe for a-Si:H/c-Si cells. Although the a-Si:H layer is thin, its effect on the improvement of the surface passivation and as a consequence the higher V_{oc} has been shown to lead to this improved temperature dependence (Taguchi *et al.*, 2008). The lower temperature coefficient explains the higher annual yields for SHJ modules compared to conventional c-Si technology.

List of Symbols

Symbol	Description	Unit
ΔE_c	Conduction band discontinuity	eV
E_{c1}	Conduction band material 1	eV
E_{c2}	Conduction band material 2	eV
ε_1	Permittivity of material 1	–
ε_2	Permittivity of material 2	–
E_F	Fermi level	eV
E_{F1}	Fermi level of material 1	eV
E_{F2}	Fermi level of material 2	eV
E_{g1}	Bandgap material 1	eV

Symbol	Description	Unit
E_{g2}	Bandgap material 2	eV
E_{vac}	Vacuum energy level	eV
FF	Fill factor	–
J_{sc}	Short-circuit density	mA/cm^2
k	Boltzmann constant ($1.38064852 \times 10^{-23}$)	$m^2\,kg\,s^{-2}\,K^{-1}$
n_i	Intrinsic carrier density	cm^{-3}
N_{a2}	Acceptor density of material 2	cm^{-3}
N_{c1}	Effective density of states at the conduction band edge of material 1	cm^{-3}
N_{d1}	Donor density of material 1	cm^{-3}
N_{DB}	Dangling bond density at interface	cm^{-2}
N_{v2}	Effective density of states at the valence band edge of material 2	cm^{-3}
q	Elementary charge ($1.60217662 \times 10^{-19}$)	C
χ_1	Electron affinity of material 1	V
χ_2	Electron affinity of material 2	V
Φ_i	Built-in voltage	V
Φ_{n1}	Work function of material 1	V
Φ_{p2}	Work function of material 2	V
T	Temperature	K
V_{bi}	Built-in voltage	V
V_{oc}	Open-circuit voltage	V

List of Acronyms

Acronym	Meaning
ACJ-HIT	Artificially Constructed Junction-Heterojunction with Intrinsic Thin film
BSF	Back Surface Field
CVD	Chemical Vapor Deposition
CIGS	Copper Indium Gallium diSelenide
HF	Hydrofluoric acid
HIT	Heterojunction with Intrinsic Thin layer
HWCVD	Hot-Wire Chemical Vapor Deposition
ITO	Indium Tin Oxide
RCA	Radio Corporation of America
rf-PECVD	radio frequency Plasma-Enhanced Chemical Vapor Deposition
SHJ	Silicon HeteroJunction
SRV	Surface Recombination Velocity
TCO	Transparent Conductive Oxide
VHF	Very High Frequency

References

Angermann, H., Conrad, E., Korte, L. *et al.* (2009) Passivation of textured substrates for a-Si:H/c-Si hetero-junction solar cells: Effect of wet-chemical smoothing and intrinsic a-Si:H interlayer. *Materials Science and Engineering B*, **159–160**, 219–223.

Bean, K.E. (1978) Anisotropic etching of silicon. *IEEE Transactions on Electron Devices*, **25**, 1185–1193.

De Wolf, S., Descoeudres, A., Holman, Z.C., and Ballif, C. (2012) High-efficiency silicon heterojunction solar cells: a review. *Green*, **2**, 7–25.

Green, M.A., Emery, K., Hishikawa, Y. *et al.* (2015) Solar cell efficiency tables (version 46). *Progress in Photovoltaics: Research and Applications*, **23**, 805–812.

Grundner, M. and Jacob, H. (1996) Investigations on hydrophilic and hydrophobic silicon (100) wafer surfaces by X-ray photoelectron and high-resolution electron energy loss-spectroscopy. *Applied Physics A: Solids and Surfaces*, **39**, 73–82.

Hamakawa, Y., Fujimoto, K. Okuda, K. *et al.* (1983) New types of high efficiency solar cells based on a-Si. *Applied Physics Letters*, **43**, 644.

Illiberi, A., Sharma, K., Creatore, M., and Van de Sanden, M.C.M. (2010) Role of a-Si: H bulk in surface passivation of c-Si wafers. *Physica Status Solidi RRL*, **4**, 172–174.

Kern, W. and Puotinen, D.A. (1970) Cleaning solutions based on hydrogen peroxide for use in silicon semiconductor technology. *RCA Review*, **31**, 187–206.

Kroemer, H. (2000) Quasi-electric fields and band offsets: teaching electrons new tricks, Nobel Prize laureate lecture (and references therein). http://www.nobelprize.org/nobel_prizes/physics/laureates/2000/kroemer-lecture.pdf (accessed October 30, 2015).

Louwen, A., Van Sark, W.G.J.H.M., Schropp, R.E.I. *et al.* (2015) Life cycle greenhouse gas emissions and energy payback time of current and prospective silicon heterojunction solar cell designs. *Progress in Photovoltaics*, **23**, 1406–1428.

Masuko, K., Shigematsu, M., Hashiguchi, T. *et al.* (2014) Achievement of more than 25% conversion efficiency with crystalline silicon heterojunction solar cell. *IEEE Journal of Photovoltaics*, **4**, 1433–1435.

Muller, R.S., Kamins, T.I., and Chan, M. (2003) *Device Electronics for Integrated Circuits*. John Wiley & Sons, Inc., New York.

Olibet, S., Vallat-Sauvain, E., and Ballif, C. (2007) Model for a-Si:H/c-Si interface recombination based on the amphoteric nature of silicon dangling bonds. *Physical Review B*, **76**, 035326.

Philipps, S.P. and Bett, A.W. (2014) III-V Multi-junction solar cells and concentrating photovoltaic (CPV) systems. *Advanced Optical Technologies*, **3**, 469–478.

Scheer, R. and Schock, H.W. (2011) *Chalcogenide Photovoltaics: Physics, Technologies, and Thin Film Devices*. Wiley-VCH, Weinheim.

Schulze, T.F., Beushausen, H.N., Leendertz, C. *et al.* (2010) Interplay of amorphous silicon disorder and hydrogen content with interface defects in amorphous/crystalline silicon heterojunctions. *Applied Physics Letters*, **96**, 252102.

Schüttauf, J.W.A., Van der Werf, C.H.M., Kielen, I.M. *et al.* (2011) Excellent crystalline silicon surface passivation by amorphous silicon irrespective of the technique used for chemical vapor deposition. *Applied Physics Letters*, **98**, 153514.

Schüttauf, J.W.A., Van der Werf, C.H.M., Kielen, I.M. *et al.* (2012) Improving the performance of amorphous and crystalline silicon heterojunction solar cells by monitoring surface passivation. *Journal of Non-Crystalline Solids*, **358**, 2245–2248.

Sinton, R. and Cuevas, A. (1996) Contactless determination of current-voltage characteristics and minority-carrier lifetimes in semiconductors from quasi-steady-state photoconductance data. *Applied Physics Letters*, **69**, 2510.

Street, R.A. (1991) *Hydrogenated Amorphous Silicon*. Cambridge University Press, Cambridge.

Sze, S.M. (1981) *Physics of Semiconductor Devices*. John Wiley & Sons, Inc., New York.

Taguchi, M., Maruyama, E., and Tanaka, M. (2008) Temperature dependence of amorphous/crystalline silicon heterojunction solar cells. *Japanese Journal of Applied Physics*, **47**, 814–818.

Taguchi, M., Tsunomura, Y., Inoue, H. *et al.* (2009) High efficiency HIT solar cell on thin (<100 μm) silicon wafer. *Proceedings of the 24th European Photovoltaic Solar Energy Conference*, Hamburg, WIP-Munich, p. 1690.

Tanaka, M., Okamoto, S., Tsuge, S., and Kiyama, S. (2003) Development of HIT solar cells with more than 21% conversion efficiency and commercialization of highest performance HIT modules. *Photovoltaic Energy Conversion*, **1**, 955.

Tanaka, M., Taguchi, M., Matsuyama, T. *et al.* (1992) Development of new a-Si/c-Si heterojunction solar cells: ACJ-HIT (Artificially Constructed Junction-Heterojunction with Intrinsic Thin-layer). *Japanese Journal of Applied Physics*, **31**, 3518–3522.

Van Cleef, M.W.M., Rath, J.K., Rubinelli, F.A. *et al.* (1997) Performance of heterojunction p^+ microcrystalline silicon in crystalline silicon solar cells. *Journal of Applied Physics*, **82**, 6089.

Van Sark, W.G.J.H.M., Korte, L. and Roca, F. (eds) (2012) *Physics and Technology of Amorphous-Crystalline Heterostructure Silicon Solar Cells*. Springer Verlag, Heidelberg.

Yamamoto, K., Adachi, D., Uzu, H. *et al.* (2015) Progress and challenges in thin-film silicon photovoltaics: heterojunctions and multijunctions. *Proceedings of the 31st European Photovoltaic Solar Energy Conference*, Hamburg, WIP-Munich, pp. 1003–1005.

Zeman, M. and Zhang, D. (2012) Heterojunction silicon based solar cells. In *Physics and Technology of Amorphous-Crystalline Heterostructure Silicon Solar Cells* (eds W. Van Sark, L. Korte, and F. Roca). Springer Verlag, Heidelberg, pp. 13–43.

3.5

Surface Passivation and Emitter Recombination Parameters

Bram Hoex
School of Photovoltaic and Renewable Energy Engineering, University of New South Wales, Kensington, Australia

3.5.1 Introduction

The surface of c-Si is a severe interruption from its crystallographic structure. A high density of defect states is therefore present in the c-Si bandgap due to the presence of this surface. Recombination at a c-Si surface can be described by Shockley-Read-Hall (SRH) statistics (Hall, 1952; Shockley and Read, 1952), similar to bulk recombination, and is quantified by what is called effective surface recombination velocity S_{eff}. As shown in Figure 3.5.1, surface recombination on the rear side of a c-Si solar cell has a tremendous impact on the solar cell efficiency, especially when the solar cell thickness is reduced further. The current industry standard aluminum back surface field (Al-BSF) has a S_{eff} of ~200–600 cm/s (Narasinha and Rohatgi, 1997; Peters, 2004) and it can be seen that, by lowering this value to 10–100 cm/s, a significant gain in the solar cell conversion efficiency can be obtained. These values can actually be obtained on c-Si, as will be shown in this chapter.

3.5.2 Surface Passivation Mechanisms

3.5.2.1 Chemical and Field-Effect Passivation

Recombination at the c-Si surface can be reduced by two different strategies. As the recombination rate is directly proportional to the interface defect density, the first strategy is based on reducing the number of defects at the interface. This is typically achieved by the deposition of a dielectric film on the c-Si surface, as will be discussed in detail in Part 3. Typically the interface defect density is not sufficiently low directly after the deposition of the dielectric film and a subsequent high-temperature treatment (e.g., an anneal) is required to reduce the interface

Photovoltaic Solar Energy: From Fundamentals to Applications, First Edition.
Edited by Angèle Reinders, Pierre Verlinden, Wilfried van Sark, and Alexandre Freundlich.

Companion website: www.wiley.com/go/reinders/photovoltaic_solar_energy

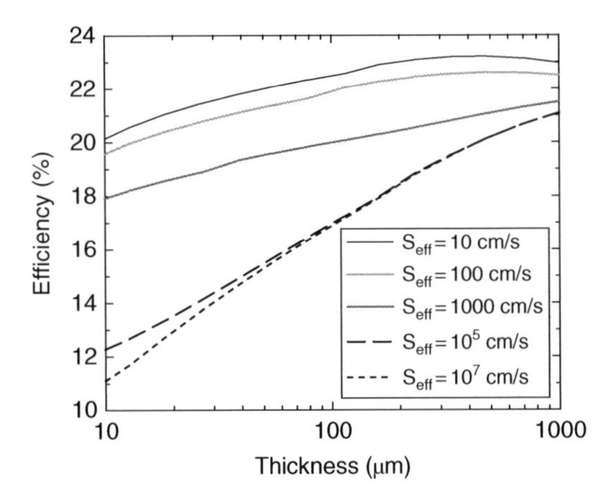

Figure 3.5.1 Simulated solar cell for a high-efficiency silicon wafer solar cell as a function of the solar cell thickness for various values of the effective surface recombination velocity at the rear side of the solar cell. The values that can be obtained by the standard aluminum back surface field are in the range of 200–600 cm/s while more advanced solar cell architectures can achieve values well below 100 cm/s. These simulations were conducted in the software package PC1D. Source: (Basore, 1990). (*See insert for color representation of the figure*)

defect density further by hydrogen passivation. The interface state density D_{it} can be reduced by several orders of magnitude to values of $1 \times 10^9 \, eV^{-1}cm^{-2}$ by growing a thermal SiO_2 followed by an anneal in a H_2 containing gas, for example, a forming gas anneal or an *alneal*, which consists of a forming gas anneal with the presence of a thin evaporated aluminum layer, assumed to be a catalyst for the production of atomic hydrogen (Eades and Swanson, 1985). As will be discussed in detail in Part 3, most commonly applied dielectrics have D_{it} values in the range of 10^9–$10^{12} \, eV^{-1}cm^{-2}$. As can be deduced from the SRH statistics, the recombination rate scales linearly with the D_{it} value. Hence, a tenfold reduction in D_{it} reduces the surface recombination rate by a factor of ten as well (assuming that the defect type that dominates the surface recombination rate remains the same). When comparing the passivation performance of various dielectric films, one should also consider the capture cross-section of electrons and holes for the dominant defect in conjunction with the D_{it} value, as these can be significantly different for the various dielectric films (Aberle, 1999).

The second surface passivation strategy is based on the significant reduction of the surface electron or hole concentration by an internal field below the surface. This can easily be understood from the SRH statistics as well, as the surface recombination process requires the presence of both electrons and holes. This internal field can be obtained by the application of a doping profile below the interface or by the presence of positive or negative fixed charge density, $Q_{f,}$ at the semiconductor interface as schematically shown in Figure 3.5.2. As discussed in detail in Part 3, most dielectric films exhibit either a positive or negative Q_f in the range of 10^{10}–$10^{13} \, cm^{-2}$ (unit elementary charges).

3.5.2.2 *Passivation of Undiffused Silicon Surfaces*

There has been extensive work to model surface passivation using quantities that can be determined experimentally (Girisch *et al.*, 1988; Aberle *et al.*, 1992; Kuhlmann *et al.* 1995;

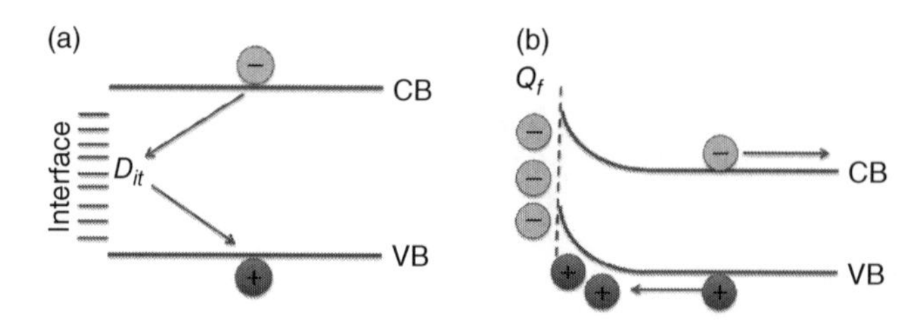

Figure 3.5.2 Schematic illustration of (a) chemical passivation; and (b) field-effect passivation

Table 3.5.1 Analytical relations for S_{eff} and J_{0s} for *n*-type c-Si for certain boundary conditions. S_{n0} and holes S_{p0} are the fundamental surface recombination velocity for electrons and holes, q is the elementary charge, n_d is the electron concentration at the position nearest to the surface where the energy bands are unaffected by the presence of fixed charge ($n_d = N_d + \Delta n_d$, with Δn_d the injection level at position *d*), n_i is the intrinsic carrier density, k is the Boltzmann constant, T is the temperature, ε_{si} is the permittivity of silicon, N_D is the bulk doping density, and Q_f is the fixed charge density

Boundary conditions	S_{eff}	J_{0s}
Negligible Q_f	S_{p0}	$q\frac{S_{p0}}{n_d}n_i^2$
Low Q_f, low injection	$S_{p0}\exp\left(\frac{-Q_f}{\sqrt{kT\varepsilon_{si}N_d}}\right)$	$q\frac{S_{p0}}{n_d}\exp\left(\frac{-Q_f}{\sqrt{kT\varepsilon_{si}N_d}}\right)n_i^2$
Large Q_f resulting in accumulation	$S_{p0}n_d\frac{2kT\varepsilon_{si}}{Q_f^2}$	$qS_{p0}\frac{2kT\varepsilon_{si}}{Q_f^2}n_i^2$
Large Q_f resulting in inversion	$S_{n0}n_d\frac{2kT\varepsilon_{si}}{Q_f^2}$	$qS_{n0}\frac{2kT\varepsilon_{si}}{Q_f^2}n_i^2$

Source: adapted from McIntosh and Black (2014).

Aberle, 1999; Brody and Rohatgi, 2001; McIntosh and Black, 2014). The approaches by Girisch *et al.* and Aberle *et al.* were done numerically which, although more precise, hide some of the useful "fundamental relations" that can be applied in most application-relevant cases. McIntosh and Black derived the analytical expressions for the most common cases that allow a better comparison between the results obtained for various surface-passivating films on various types of samples. The main parameters are the fundamental surface recombination velocity for electrons S_{n0} and holes S_{p0}. They are given by $S_{n0} = v_{thn}N_{it}\sigma_n$ and $S_{p0} = v_{thh}N_{it}\sigma_n$ where v_{thn} and v_{thh} are the thermal velocities of electrons and holes, σ_n and σ_p are the electron and hole capture cross-section, and N_{it} is the interface defect density in cm^{-2}.[1] The effective surface recombination velocity S_{eff} is most commonly used to express the level of surface passivation in the literature and, as can be seen in Table 3.5.1, it can be expressed as a function of the fundamental recombination velocities S_{n0} and S_{p0}. McIntosh and Black advocate the use of what is called the surface recombination current J_{0s} in most cases, as this property is not

[1] As actual interfaces typically have multiple defects it is more general to consider the energy-dependent interface trap density $D_{it}(E)$ (in cm^{-2} eV^{-1}) and capture constant for electrons and holes $\sigma_{n,p}(E)$ (in cm^2).

dependent on the bulk or surface dopant density and, just as important, can relatively easily be extracted from lifetime measurements (Kane and Swanson, 1985).

3.5.2.3 Passivation of Diffused Silicon Surfaces

The recombination in emitters is governed by a combination of bulk and surface recombination. The doping concentration in emitters varies over various orders of magnitude over very short distances and this results in significant differences in the minority carrier concentration, the Auger recombination, the SRH recombination, carrier mobility, free-carrier absorption, electron and hole effective mass, and the bandgap (McIntosh *et al.*, 2013). In recent years there has been good progress in modeling recombination in emitters (Aberle *et al.*, 1995; Altermatt *et al.*, 2002, 2006; Hoex *et al.*, 2007; McIntosh and Altermatt, 2010). Traditionally the recombination in emitters could only properly be assessed by solving the complete set of semiconductor equations numerically in programs such as Sentaurus from Synopsys. Recently McIntosh and Altermatt released the program EDNA (currently version 2 is available on www.pvlighthouse.com.au) that enables researchers who are not necessarily an expert in advanced computer simulation to accurately model n-type and p-type emitters, for example, to extract the *effective* surface recombination velocity S_{eff} at the highly doped n-type and p-type surface (McIntosh and Altermatt, 2010; McIntosh *et al.*, 2013). In Figure 3.5.3 (b), the simulated emitter saturation current density J_{0e} is shown as a function of S_{n0} and Q_f as simulated in Sentaurus. It can be seen that the impact of Q_f is not as strong for highly doped surfaces as in the case of lowly diffused surfaces. This dependence is exploited in recent work where it is shown that moderately negatively charged dielectrics can be used to passivate n^+ emitters and moderately positively charged dielectrics to passivate p^+ emitters (Lin *et al.*, 2012; Duttagupta *et al.*, 2013b; Duttagupta *et al.*, 2014c). In the specific case shown in Figure 3.5.3, a negative

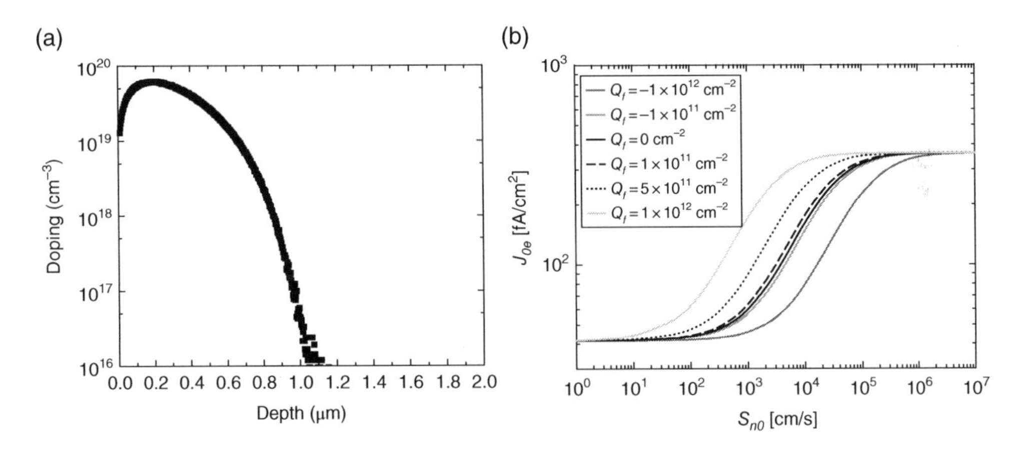

Figure 3.5.3 (a) Active boron doping depth profile as determined by electrochemical capacitance-voltage (ECV) profiling; and (b) resulting simulated J_{0e} value as a function of S_{n0} for various values of Q_f (in unit of elementary charges). The sheet resistance of the boron emitter was determined to be 30 Ω/□ by means of a four-point-probe measurement. The simulations were performed by Fajun Ma from the Solar Energy Institute of Singapore (SERIS) in the software package Sentaurus of Synopsys, assuming only intrinsic recombination in the emitter and an SRH recombination at the surface. (*See insert for color representation of the figure*)

or positive Q_f value of $1 \times 10^{11}\,cm^{-2}$ (in units of elementary charges) has barely an impact on the J_{0e} value and the recombination rate is mainly determined by the value of S_{n0}. From this simulation it can also clearly be seen that a positive fixed charge density of $1 \times 10^{12}\,cm^{-2}$ is strongly increasing the emitter recombination rate as it increases the minority carrier concentration at the surface. A negative Q_f of $-1 \times 10^{12}\,cm^{-2}$ allows roughly a six times higher value of S_{n0} for a similar level of J_{0e} in this case. It should be mentioned that the relationship shown in Figure 3.5.3 would be different for every diffusion profile. When a proper analysis is done using the latest physical models, it can actually be shown that, at least for the case of SiN_x on n^+ emitters and AlO_x on p^+ emitters, the S_{p0} or S_{n0} is actually to be found independent of the surface doping density in the experimentally investigated range (Ma *et al.*, 2012; McIntosh *et al.*, 2013 ; Black *et al.*, 2014).

3.5.3 Commonly Used Surface Passivating Films

In this section we will summarize the experimental results that have been reported for the industrially relevant silicon surfaces (i.e. lightly and highly doped n-type and p-type c-Si) by various dielectric thin films. We divide the various dielectrics into two groups with an identical polarity of the fixed charge. It can be seen that very low S_{eff} and J_{0e} values can be obtained for all industrially relevant c-Si surfaces and some dielectric films (such as SiO_x or AlO_x) can provide an excellent level of surface passivation on all relevant surfaces. This is, for example, very beneficial in the case of all-back-contact solar cells where both highly doped n^+ and p^+ regions are present on the same side of sample. In the next sub-sections we will look into the various dielectrics in some more details.

The wafer resistivity, in the case of undiffused c-Si, and sheet resistivity, in the case of diffused c-Si, is given in Table 3.5.2 as well as the reported values that strongly depend on these experimental conditions, as is also apparent from Table 3.5.1. The upper-limit $S_{eff,UL}$ values were calculated assuming an infinite bulk minority carrier lifetime. For some dielectric films both results in the lab as well as on industrially relevant equipment are given, while for Al_2O_3 the best results for ALD (atomic layer deposition) and PECVD (plasma-enhanced chemical vapor deposition) films are shown which are both industrially applied. Most of these dielectric films were capped with a silicon nitride film that can act as a source of hydrogen in a high temperature activation step.

3.5.3.1 Silicon Nitride (SiN_x)

SiN_x is currently the most widely applied surface passivation film in the field of silicon wafer solar cells. Its success can largely be attributed to the fact that SiN_x can simultaneously act as an antireflection coating while providing a high level of surface and bulk passivation. The advantages of a-SiN_x:H were already reported in the literature in the early 1980s (Hezel and Schorner, 1981) but the large-scale implementation in the c-Si solar cell manufacturing started in 2000 when high-throughput production equipment became available. As can be seen from Table 3.5.2, SiN_x is a positive-charge dielectric and, hence, it is particularly effective in passivating n-type c-Si surfaces. Due to the increased interest in *n*-type silicon wafer solar cells, various researchers have tried to improve the passivation level of SiN_x films on p^+ silicon surfaces. This can, for example, be achieved by prolonged annealing (F. W. Chen *et al.*, 2006) or by the insertion of a thin SiO_x layer between c-Si and the SiN_x film (Mihailetchi *et al.*, 2008; Duttagupta *et al.*, 2013b).

Table 3.5.2 Non-exhaustive summary of the experimentally obtained properties for the most commonly used dielectric films for passivating c-Si surfaces

Charge polarity	Surface passivation parameters on c-Si					Electronic properties	
	Material	*n-type* $S_{eff,\,UL}$ [cm/s]	*n*⁺⁺ *emitter* J_{0e} [fA/cm^2]	*p-type* $S_{eff,\,UL}$ [cm/s]	*p*⁺⁺ *emitter* J_{0e} [fA/cm^2]	Mid-gap D_{it} [10^{10}eV^{-1}cm^{-2}]	Q_f [10^{11} cm^{-2}]
Positive	Thermal SiO_x	2.4 (1.5 Ω cm)[a]	10 (180 Ω/+)[b]	11.8 (1.0 Ω cm)[a]	20 (160 Ω/+)[c]	0.1[d]	0.1-1[e]
	PECVD SiO_x (lab)	2 (1.0 Ω cm)[f]	210[1],[g]	11 (2.2 Ω cm)[h]	–[2]	1-10[i]	5-10[k]
	PECVD SiO_x (industrial)	7 (1.5 Ω cm)[l]	10 (150 Ω/+)[m]	8 (1.5 Ω cm)[l]	25 (75 Ω/+)[n]	3-20[n]	0.8-10[n]
	PECVD SiN_x (lab)	<1.0 (0.47 Ω cm)[m]	30 (120 Ω/+)[b]	1.6 (0.85 Ω cm)[o]	25 (100 Ω/+)[p]	30-500[q]	1-70[q]
	PECVD SiN_x (industrial)	8.2 (1-2 Ω cm)[r]	40 (150 Ω/+)[l]	11.1 (1-2 Ω cm)[r]	26[3] (75 Ω/+)[l]	50-90[r]	40-80[r]
Negative	AlO_x (ALD)	1.2 (2.5 Ω cm)[s]	60 (100 Ω/+)[t]	5 (2 Ω cm)[s]	10 (120 Ω/+)[u]	4-10[v]	-(10-130)[v]
	AlO_x (PECVD)	1.0 (3.5 Ω cm)[w]	15 (145 Ω/+)[x]	10 (1 Ω cm)[y]	7 (135 Ω/+)[z]	2-5[y]	-(10-70)[y]
	TiO_2	3 (2.5 Ω cm)[aa]	–[4]	8 (2.0 Ω cm)[aa]	27 (80 Ω/+)[bb]	–[4]	–[4]

[a](Kerr and Cuevas, 2002); [b](Kerr *et al.*, 2001); [c](Kerr, 2002); [d](Eades and Swanson, 1985); [e] (Aberle, 1999); [f](Mueller *et al.*, 2008); [g](Z. Chen, Rohatgi, and Ruby, 1994); [h](Dingemans, *et al.*, 2011a); [i](Z. Chen *et al.*, 1993b); [k](Z. Chen, 1993a); [l](Duttagupta, 2014); [m](Duttagupta, Hoex, and Aberle, 2013); [n](Duttagupta, *et al.*, 2014b); [o](Wan *et al.*, 2013); [p](F.W. Chen, Li, and Cotter, 2006); [q] (Aberle, 1999); [r](Duttagupta *et al.*, 2014a); [s](Liao *et al.*, 2013); [t](Hoex *et al.*, 2012); [u](Hoex *et al.*, 2007); [v](Dingemans *et al.*, 2011b); [w](Dingemans, van de Sanden, and Kessels, 2012); [x](Duttagupta *et al.*, 2013c); [y](Saint-Cast *et al.*, 2009); [z](Saint-Cast *et al.*, 2012); [aa](Liao *et al.*2014); [bb](Liao *et al.*, 2015).

[1]Not mentioned in the publication.

[2]Not available at this time but expected to be similar to the industrial results.

[3]The c-Si surface contained a chemical oxide prior to the deposition of the SiN_x film.

[4]Still under investigation.

3.5.3.2 *Silicon Oxide (SiO_x)*

Thermally grown SiO_x has traditionally been the best surface passivation layer for silicon wafer solar cells and has also been used in the 25.0% passivated emitter rear locally diffused (PERL) solar cell from UNSW, Australia, which held the world record for the most efficient silicon solar cell for more than a decade (Zhao, Wang, and Green, 1999). The surface passivation mechanism of thermal SiO_x mainly relies on its unravelled low interface defect density that can be achieved on c-Si, as can be seen from Table 3.5.2. This excellent interface is predominantly the result of the fact that the SiO_x layer grows into the silicon wafer, hence, is almost unaffected by the initial surface condition. It should be noted that the interface defect density and the positive fixed charge density strongly depend on the process conditions used. Additionally, extensive post-growth processing is required to obtain the highest quality interfaces. Typically extensive anneals in H_2 or *alneal* processes are used for this purpose (Aberle, 1999). The formation of high-quality thermal oxides requires high process temperatures that are not compatible with lower-quality silicon material such as multicrystalline silicon. As a consequence, there has been extensive work to develop low temperature SiO_x films typically grown by PECVD. The interface quality is typically found to be slightly poorer for these kinds of films. The positive fixed charge density is usually found to be higher in the films that provide a good level of surface passivation. Also it is found to be critical to use a capping (typically SiN_x) layer for these films to ensure that they are thermally stable e.g. during the high-temperature firing step. As can be seen from Table 3.5.2, very good results have recently been reported on lifetime samples symmetrically passivated by PECVD SiO_x films capped with PECVD SiN_x.

3.5.3.3 *Aluminum Oxide (AlO_x)*

AlO_x is a material that has recently gained considerable interest for application in the field of c-Si solar cells. In the late 1980s, Al_2O_3 was already applied for c-Si surface passivation in metal-insulator-semiconductor (MIS) solar cells (Hezel and Jaeger, 1989). Hezel and Jaeger demonstrated that AlO_x could provide a reasonable level of surface passivation with an effective surface recombination velocity of ~200 cm/s on 2 Ω cm *p*-type c-Si. The interest in AlO_x re-emerged when excellent results were reported for films synthesized by atomic layer deposition (ALD) (Agostinelli *et al.*, 2006; Hoex *et al.*, 2006). As can be seen in Table 3.5.2, AlO_x is a negative charge dielectric and this is particularly beneficial in passivating *p*-type c-Si surfaces. It has been shown that AlO_x can also passivate n^+ c-Si surfaces as long as the magnitude from the negative fixed charge is sufficiently low (Duttagupta *et al.*, 2013a). AlO_x is seen as the main candidate for application at the rear of passivated emitter rear contact (PERC) and aluminum local back surface field (Al-LSBF) silicon wafer solar cells, which are widely expected to dominate the crystalline silicon solar cell industry in the coming years.

3.5.3.4 *Titanium Oxide (TiO_x)*

TiO_x was the standard antireflection coating for silicon solar cells until 2000. At that time SiN_x surpassed TiO_x as TiO_x did not provide a sufficiently high level of surface and bulk passivation (Richards, 2004). Very recently there has been very good progress reported for surface passivation by TiO_x. It was shown that the level of surface passivation provided by TiO_x could be significantly improved and J_{0e} values of 90 fA/cm^2 were reported on a 200 Ω/+ p^+ emitter (Thomson and McIntosh, 2012). More recently, excellent results were demonstrated on lightly doped n-type and p-type c-Si and p^+ emitters for TiO_x films synthesized by ALD (Liao *et al.*, 2014, 2015). Both studies seem to indicate that TiO_x is most likely negatively charged, which is consistent with the excellent results on p-type c-Si.

3.5.4 Conclusion

In this chapter we have given a short overview of the basics of surface and emitter recombination and summarized the results obtained for the most commonly used dielectrics. There has been tremendous progress in the recent years and the most industrially relevant silicon surfaces can now all be passivated to levels which are sufficient for very high efficiency silicon wafer solar cells. There has also been a significant progress in the modeling of the highly doped regions of silicon wafer solar cells with the availability of accessible (or even free) simulators such as EDNA that allow researchers to analyse their experimental results without necessarily be a modeling expert.

List of symbols

Symbol	Description	Unit
D_{it}	Interface state density	eV^{-1}cm^{-2}
J_{0e}	Emitter saturation current density	A/cm^2
J_{0s}	Surface recombination current density	A/cm^2
k	Boltzman constant	J/K
n_d	Electron concentration at the position nearest to the surface where the energy bands are unaffected by the presence of fixed charge	cm^{-3}
N_D	Bulk donor density	cm^{-3}
n_i	Intrinsic carrier density	cm^{-3}
N_{it}	Density of defects at the c-Si/dielectric interface	cm^{-2}
q	Elementary charge density	C
Q_f	Fixed charge density	cm^{-2} [unit elementary charges]
S_{eff}	Effective surface recombination velocity	cm/s
S_{n0} (S_{p0})	Surface recombination velocity for electrons (holes)	cm/s
T	Temperature	K
v_{thn} (v_{thh})	Thermal velocity for electrons (holes)	cm/s
Δn_d	The injection level at position d	cm^{-3}
ε_{si}	Permittivity of silicon	F/m
σ_n (σ_p)	Capture cross-section for electrons (holes)	cm^2

List of acronyms

Acronym	Description
Al-BSF	Aluminum Back Surface Field
ALD	Atomic Layer Deposition
Al-LBSF	Aluminium Local Back Surface Field
PECVD	Plasma Enhanced Chemical Vapor Deposition
PERC	Passivated Emitter Rear Contact
UNSW	University of New South Wales
SERIS	Solar Energy Institute of Singapore
SRH	Shockley-Read-Hall

References

Aberle, A.G. (1999) *Crystalline Silicon Solar Cells: Advanced Surface Passivation and Analysis*. UNSW Publishing and Printing Services, Sydney.

Aberle, A.G., Altermatt, P.P., Heiser, G. *et al.* (1995) Limiting loss mechanisms in 23-percent efficient silicon solar-cells. *Journal of Applied Physics*, **77** (7), 3491–3504.

Aberle, A.G., Glunz, S., and Warta, W. (1992) Impact of illumination level and oxide parameters recombination at the Si-Si02 interface. *Journal of Applied Physics*, **71** (9), 4422–4431.

Agostinelli, G., Delabie, A., Vitanov, P. *et al.* (2006) Very low surface recombination velocities on p-type silicon wafers passivated with a dielectric with fixed negative charge. *Solar Energy Materials and Solar Cells*, **90** (18–19), 3438–3443.

Altermatt, P.P., Plagwitz, H., Bock, R. *et al.* (2006) The surface recombination velocity at boron-doped emitters: comparison between various passivation techniques. *Proceedings of the 21st EU-PVSEC*, 647–651.

Altermatt, P.P., Schumacher, J.O., Cuevas, A. *et al.* (2002) Numerical modeling of highly doped Si : P emitters based on Fermi-Dirac statistics and self-consistent material parameters. *Journal of Applied Physics*, **92** (6), 3187–3197.

Basore, P.A. (1990) Numerical modeling of textured silicon solar-cells using PC-1D. *IEEE Transactions on Electron Devices*, **37** (2), 337–343.

Black, L.E., Allen, T., McIntosh, K.R., and Cuevas, A. (2014) Effect of boron concentration on recombination at the p-Si–Al2O3 interface. *Journal of Applied Physics*, **115** (9), 093707.

Brody, J. and Rohatgi, A. (2001) Analytical approximation of effective surface recombination velocity of dielectric-passivated *p*-type silicon. *Solid-State Electronics*, **45** (9), 1549–1557.

Chen, F.W., Li, T.T.A., and Cotter, J.E. (2006) Passivation of boron emitters on n-type silicon by plasma-enhanced chemical vapor deposited silicon nitride. *Applied Physics Letters*, **88** (26), 263514.

Chen, Z., Pang, S.K., Yasutake, K., and Rohatgi, A. (1993) Plasma-enhanced chemical-vapor-deposited oxide for low surface recombination velocity and high effective lifetime in silicon. *Journal of Applied Physics*, **74** (4), 2856–2859.

Chen, Z., Rohatgi, A., and Ruby, D. (1994) Silicon surface and bulk defect passivation by low temperature PECVD oxides and nitrides. *Proceedings of the 24th IEEE PVSC*, 1331–1334.

Chen, Z., Yasutake, K., Doolittle, A., and Rohatgi, A. (1993) Record low SiO2/Si interface state density for low temperature oxides prepared by direct plasma-enhanced chemical vapor deposition. *Applied Physics Letters*, **63** (15), 2117–2119.

Dingemans, G., Mandoc, M.M., Bordihn, S. *et al.* (2011a) Effective passivation of Si surfaces by plasma deposited SiO_x/a-SiN_x:H stacks. *Applied Physics Letters*, **98** (22), 222102.

Dingemans, G., Terlinden, N.M., Pierreux, D. *et al.* (2011b) Influence of the oxidant on the chemical and field-effect passivation of si by ALD Al_2O_3. *Electrochemical and Solid State Letters*, **14** (1), H1–H4.

Dingemans, G., van de Sanden, M.C.M., and Kessels, W.M.M. (2012) Plasma-enhanced chemical vapor deposition of aluminum oxide using ultrashort precursor injection pulses. *Plasma Processes and Polymers*, **9** (8), 761–771.

Duttagupta, S. (2014) Advanced surface passivation of crystalline silicon for solar cell applications. PhD thesis, National University of Singapore.

Duttagupta, S., Hoex, B., and Aberle, A.G. (2013a) Progress with industrially-feasible excellent surface passivation of heavily-doped *p*-type and *n*-type crystalline silicon by PECVD SiO_x/SiN_x with optimised anti-reflective performance. *Proceedings of the 28th European PVSEC*, 993–996.

Duttagupta, S., Lin, F., Wilson, M. *et al.* (2014a) Extremely low surface recombination velocities on low-resistivity *n*-type and *p*-type crystalline silicon using dynamically deposited remote plasma silicon nitride films. *Progress in Photovoltaics*, **22** (6), 641–647.

Duttagupta, S., Ma, F.J., Hoex, B., and Aberle, A.G. (2013b) Extremely low surface recombination velocities on heavily doped planar and textured p^+ silicon using low-temperature positively-Charged PECVD SiO_x/SiN_x dielectric stacks with optimised antireflective properties. *Proceedings of the 39th IEEE PVSC*, pp. 1776–1780.

Duttagupta, S., Ma, F.J., Hoex, B., and Aberle, A.G. (2014b) Excellent surface passivation of heavily doped p^+ silicon by low-temperature plasma-deposited SiO_x/SiN_y dielectric stacks with optimised antireflective performance for solar cell application. *Solar Energy Materials and Solar Cells*, **120**, 204–208.

Duttagupta, S., Ma, F.J., Lin, S.F. *et al.* (2013c) Progress in surface passivation of heavily doped *n*-Type and *p*-type silicon by plasma-deposited AlO_x/SiN_x dielectric stacks. *IEEE Journal of Photovoltaics*, **3** (4), 1163–1169.

Eades, W.D. and Swanson, R.M. (1985) Calculation of surface generation and recombination velocities at the Si-SiO_2 interface. *Journal of Applied Physics*, **58** (11), 4267–4276.

Girisch, R.B.M., Mertens, R.P., and Dekeersmaecker, R.F. (1988) Determination of Si-SiO_2 interface recombination parameters using a gate-controlled point-junction diode under illumination. *IEEE Transactions on Electron Devices*, **35** (2), 203–222.

Hall, R. (1952) Electron-hole recombination in germanium. *Physical Review*, **87** (2), 387.

Hezel, R. and Jaeger, K. (1989) Low-temperature surface passivation of silicon for solar-cells. *Journal of the Electrochemical Society*, **136** (2), 518–523.

Hezel, R. and Schorner, R. (1981) Plasma si nitride: a promising dielectric to achieve high-quality silicon mis-il solar-cells. *Journal of Applied Physics*, **52** (4), 3076–3079.

Hoex, B., Heil, S.B.S., Langereis, E. *et al.* (2006) Ultralow surface recombination of c-Si substrates passivated by plasma-assisted atomic layer deposited Al_2O_3. *Applied Physics Letters*, **89** (4), 042112.

Hoex, B., van de Sanden, M.C.M., Schmidt, J. et al. (2012) Surface passivation of phosphorus-diffused n^+-type emitters by plasma-assisted atomic-layer deposited Al2O3. *Physica Status Solidi-Rapid Research Letters*, **6** (1), 4–6.

Hoex, B., Schmidt, J., Bock, R., Altermatt, P.P. *et al.* (2007) Excellent passivation of highly doped *p*-type Si surfaces by the negative-charge-dielectric Al_2O_3. *Applied Physics Letters*, **91** (11), 112107.

Kane, R.M., and Swanson, D.R. (1985) Measurement of the emitter saturation current by a contactless photoconductivity decay method. *Proceedings of the 18th IEEE Photovoltaic Specialists Conference*, **578**.

Kerr, M.J. (2002) Surface, emitter and bulk recombination in silicon and development of silicon nitride passivated solar cells. PhD thesis, Australian National University.

Kerr, M.J. and Cuevas, A. (2002) Very low bulk and surface recombination in oxidized silicon wafers. *Semiconductor Science and Technology*, **17** (1), 35–38.

Kerr, M.J., Schmidt, J., Cuevas, A., and Bultman, J.H. (2001) Surface recombination velocity of phosphorus-diffused silicon solar cell emitters passivated with plasma enhanced chemical vapor deposited silicon nitride and thermal silicon oxide. *Journal of Applied Physics*, **89** (7), 3821–3826.

Kuhlmann, B., Aberle, A.G., and Hezel, R. (1995) Two-dimensional numerical optimisation study of inversion layer emitters of silicon solar cells. *Proceedings of the 13th European PVSEC*, **1209**.

Liao, B.C., Hoex, B., Aberle, A.G., *et al.* (2014) Excellent c-Si surface passivation by low-temperature atomic layer deposited titanium oxide. *Applied Physics Letters*, **104** (25), 253903.

Liao, B.C., Hoex, B., Shetty, K.D. *et al.* (forthcoming) Excellent passivation of boron doped industrial silicon emitters by thermal atomic layer deposited titanium oxide, *IEEE Journal of Photovoltaics*.

Liao, B.C., Stangl, R., Ma, F. *et al.* (2013) Excellent c-Si surface passivation by thermal atomic layer deposited aluminum oxide after industrial firing activation. *Journal of Physics D-Applied Physics*, **46** (38), 385102.

Lin, F., Duttagupta, S., Aberle, A.G., and Hoex, B. (2012) Excellent passivation of n^+ and p^+ silicon by PECVD SiO_x/ AlOx stacks. *Proceedings of the 27th European PVSEC*, 1251–1254.

Ma, F.J., Duttagupta, S., Peters, M. *et al.* (2012) Numerical modelling of silicon p^+ emitters passivated by a PECVD AlO_x/SiN_x stack. *Energy Procedia*, **33**, 104–109.

McIntosh, K.R. and Altermatt, P.P. (2010) A freeware 1D emitter model for silicon solar cells. *Proceedings of the 35th IEEE Photovoltaic Specialists Conference*, 2188–2193.

McIntosh, K.R., Altermatt, P.P., Ratcliff, T.J. *et al.* (2013) An examination of three common assumptions used to simulate recombination in heavily doped silicon. *Proceedings of the 28th European Photovoltaic Solar Energy Conference and Exhibition*, 1672–1679.

McIntosh, K.R. and Black, L.E. (2014) On effective surface recombination parameters. *Journal of Applied Physics*, **116** (1), 014503.

Mihailetchi, V.D., Komatsu, Y., and Geerligs, L.J. (2008) Nitric acid pretreatment for the passivation of boron emitters for n-type base silicon solar cells. *Applied Physics Letters*, **92** (6), 63510.

Mueller, T., Schwertheim, S., Scherff, M., and Fahrner, W.R. (2008) High quality passivation for heterojunction solar cells by hydrogenated amorphous silicon suboxide films. *Applied Physics Letters*, **92** (3), 033504.

Narasinha, S. and Rohatgi, A. (1997) Optimized aluminum back surface field techniques for silicon solar cells. *Proceedings of the 26th IEEE Photovoltaic Specialists Conference*, 63–66.

Peters, S. (2004) Rapid thermal processing of crystalline silicon materials and solar cells. PhD thesis, University of Konstanz.

Richards, B.S. (2004) Comparison of TiO_2 and other dielectric coatings for buried-contact solar cells: a review. *Progress in Photovoltaics*, **12** (4), 253–281.

Saint-Cast, P., Kania, D., Hofmann, M., Benick, J. *et al.* (2009) Very low surface recombination velocity on *p*-type c-Si by high-rate plasma-deposited aluminum oxide. *Applied Physics Letters*, **95** (15), 151502.

Saint-Cast, P., Richter, A., Billot, E. *et al.* (2012) Very low surface recombination velocity of boron doped emitter passivated with plasma-enhanced chemical-vapor-deposited AlO_x layers. *Thin Solid Films*, **522**, 336–339.

Shockley, W. and Read, W. (1952) Statistics of the recombinations of holes and electrons. *Physical Review*, **87** (5), 835–842.

Thomson, A.F. and McIntosh, K.R. (2012) Light-enhanced surface passivation of TiO_2-coated silicon. *Progress in Photovoltaics: Research and Applications*, **20** (3), 343–349.

Wan, Y., McIntosh, K.R., Thomson, A.F., and Cuevas, A. (2013) Low Surface recombination velocity by low-absorption silicon nitride on c-Si. *IEEE Journal of Photovoltaics*, **3** (1), 554–559.

Zhao, J.H., Wang, A.H., and Green, M.A. (1999) 24 center dot 5% efficiency silicon PERT cells on MCZ substrates and 24 center dot 7% efficiency PERL cells on FZ substrates. *Progress in Photovoltaics*, **7** (6), 471–474.

3.6

Passivated Contacts

Martin Hermle
Fraunhofer Institute for Solar Energy Systems, ISE, Freiburg, Germany

3.6.1 Introduction

Metal contacts are an essential element of a solar cell to extract the generated power from the device. However, the direct contact of metal to a semiconductor interface leads to quasi-continuous defect distribution in the semiconductor bandgap and thus to an enormous sink for minority charge carriers (Würfel, 2005). The traditional way to reduce the recombination rate at such a highly recombinative metal/semiconductor interface is the introduction of a highly doped region, which reduces the number of minority carriers at the interface and thus reduces the recombination rate (see also Chapter 3.5). Figure 3.6.1 shows the simulated dark saturation current density J_0 which is a good measure of the recombination rate as a function of the semiconductor surface concentration. It can be seen that for oxide-passivated surfaces with low surface defect densities, lower doped silicon leads to lower J_0 values, where in contrast the lowest J_0 values for metal contacted surfaces can be achieved for highly doped surfaces. However, such highly doped region leads to an increase of non-avoidable Auger recombination, which finally limits the efficiency of the silicon solar cells (Richter, Hermle, and Glunz, 2013).

In the past, the classical way to overcome this opposed requirement between a passivated and a metallized surface was to reduce the area of the metallized regions and restrict the highly doped regions only locally beyond the metal contacts. This approach leads to the long-standing 25% PERL (Passivated Emitter and Rear Locally Diffused) efficiency record for silicon solar cells, featuring a passivated emitter with a selective emitter underneath the front contacts and a passivated rear side with locally diffused regions underneath the metal contacts at the rear side (Zhao, Wang, Green, and Ferrazza, 1998). However, even though the fraction of metallized area in this cell is less than 1%, these areas still dominate the overall recombination of this kind of solar cell device (Benick *et al.*, 2008). Furthermore, shrinking the metal contacts

Photovoltaic Solar Energy: From Fundamentals to Applications, First Edition.
Edited by Angèle Reinders, Pierre Verlinden, Wilfried van Sark, and Alexandre Freundlich.

Companion website: www.wiley.com/go/reinders/photovoltaic_solar_energy

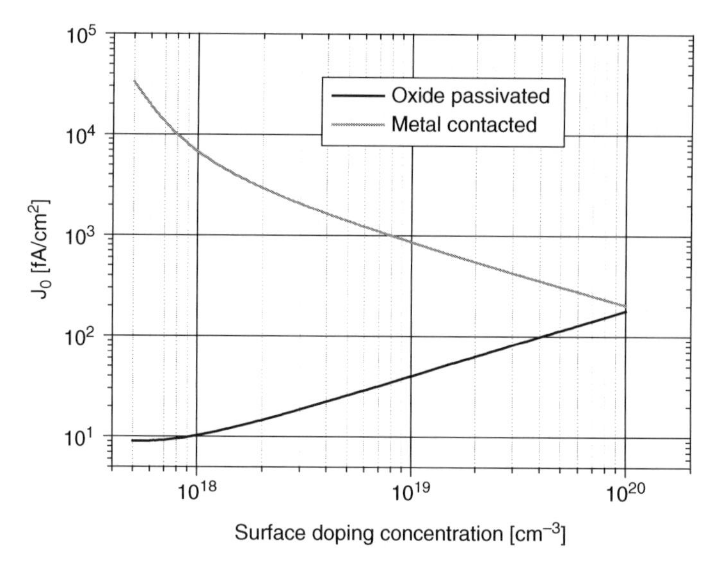

Figure 3.6.1 Simulated dark saturation current density for a phosphorus doped surface either passivated with SiO_2 (surface recombination velocity, SRV, is a function of N_A after (Altermatt, 2011)) or contacted with metal (SRV = 10^7 cm/s) as a function of the surface doping concentration. Low saturation current densities can be achieved with passivated lowly doped surfaces. The metallization of such lowly doped surfaces leads to very high J_0 values. The J_0 can be decreased by increasing the surface doping concentration. However, values of at least one order of magnitude higher than the passivated areas are still present. The simulations were conducted using the software package EDNA (McIntosh and Altermatt, 2015)

to local point contacts leads to an additional loss mechanism, as the majority carrier's transport path is increased and current crowding occurs at the local contacts which leads to an additional power loss (Catchpole and Blakers, 2002).

Thus, to approach the theoretical efficiency limit of silicon solar cells of 29.4% (Richter *et al.*, 2013), alternative contacts are necessary (Swanson, 2005), which suppress the minority carrier recombination and simultaneously allow a barrier-free transport for the majority carrier. Such contacts are called "passivated contacts" (Cousins *et al.*, 2010). In this chapter, the theoretical background and an overview of experimentally realized passivated contacts are presented.

3.6.2 Theory of Passivated Contacts

The fundamental requirement of electrical contacts in solar cells is the extraction of electrons and holes from the device. A crystalline silicon solar cell consists of a silicon wafer which acts as the absorber material to generate electron–hole pairs by absorbing photons with energies larger than the bandgap. To supply power to an external electric circuit, the electrons and holes have to be separated and transported to their respective contacts, the electron and hole contacts. These contacts are connected to the external circuit by electrical conductive electrodes. In the literature, the term "contact" mainly refers to the direct metal/semiconductor

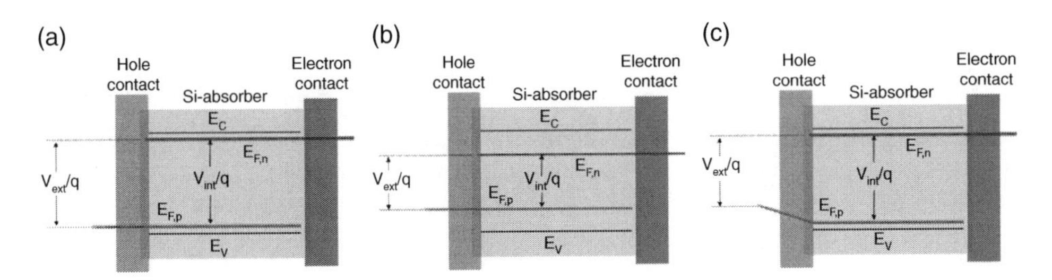

Figure 3.6.2 Schematic silicon band diagram with different electron and hole contact systems, leading to either an ideal solar cell (a) or recombination or transport limited solar cells (b) and (c) respectively

interface. In this chapter, a more general definition of "contacts" is used. It includes the metal/semiconductor interface itself and any induced or diffused regions inside of the silicon wafer. Thus, the contacts in this definition are responsible for the extraction of electrons and holes from the silicon absorber.

In the following, the fundamental requirements of passivated electron and hole contacts will be discussed. In Figure 3.6.2, a schematic silicon band diagram with a silicon absorber and three different electron and hole contacts is shown, in order to explain these requirements. In the ideal case (a), the internal voltage (V_{int}) which is defined by the separation of the quasi Fermi levels ($E_{F,n}$-$E_{F,p}$) in the absorber, is only limited by the recombination losses of the silicon absorber itself and the external voltage (V_{ext}) measured at the metal electrodes corresponds to the internal voltage. In this case, the contacts can be referred to as passivated contacts. In the minority carrier recombination limited case (b), the external voltage still corresponds to the internal voltage ($V_{ext}=V_{int}$), however, the internal voltage is reduced by the additional recombination at the contacts (as is the case, e.g., for the industrial standard silicon solar cell with a full area Al-BSF rear contact). In the third case (c), the implied voltage of the cell is similar to the ideal case (a), but the external voltage is reduced ($V_{ext}<V_{int}$), which results from a voltage drop in the contact region's majority Fermi energy. Such a drop in the majority Fermi energy (in this case, $E_{F,p}$) occurs, if the majority carrier faces any kind of transport losses. In this case, a voltage drop occurs and the internal voltage cannot be fully extracted at the external electrodes.

From this, the following two fundamental prerequisites of a passivated contact system can be deduced:

1. The recombination in the vicinity of the contact has to be suppressed to allow for a high internal voltage (V_{int}).
2. A drop in the majority Fermi energy in the contact region has to be avoided. This means, that any voltage losses in the contact region have to be avoided ($V_{ext}=V_{int}$) and no barrier is present which would hinder the carrier transport of the majority carrier.

Requirement 2 can be realized by a highly doped region being directly in contact with an appropriate metal layer, forming an ohmic contact. However, this leads to a relevant recombination in the highly doped region by the Auger recombination (see Figure 3.6.1) and at the metal-semiconductor interfaces and thus contradicts Requirement 1. To fulfill Requirement 1, highly doped regions have to be avoided and the surface has to be passivated. That is why ideal contacts are called "passivated contacts."

In principle, the two requirements stated above can be combined into one statement proposed by Würfel *et al.* (2015): A passivated contact "is achieved by differences in the conductivities of electrons and holes in two distinct regions of the device, which, for one charge carrier, allows transport to one contact and block transport to the other contact." This difference in conductivity leads to a suppression of recombination of one carrier type (Requirement 1) because the carriers with the low conductivity cannot reach the metal contact and the carriers with the higher conductivity can be extracted without losses (Requirement 2).

An often-used synonym for "passivated contact" is "carrier selective contact." However, a highly phosphorus-doped region under a metal contact leads to an electron-selective contact, which, however, is poorly passivated. A passivated contact has good selectivity for one specific carrier type and furthermore reduces the recombination rate values which cannot be achieved by metal/semiconductor contacts with underlying highly doped regions. Thus, a passivated contact is always a carrier-selective contact, but not all carrier-selective contacts are passivated contacts.

In the following section, different technological approaches for passivated contacts will be presented.

3.6.3 Experimentally Realized Passivated Contacts

In this section the experimental results for passivated contacts that have been reported in the literature will be summarized. As stated above, one critical requirement for passivated contacts is the suppression of recombination to allow for high internal voltages. As described in Chapter 3.5, the minimization of the surface recombination rate can be achieved by: (1) a chemical passivation to reduce the number of surface states by passivating dangling bonds (characterized by an interface trap density (D_{it})); and (2) the reduction of the number of electrons or holes at the surface. In all realized passivated contacts, both mechanisms are used to ensure a low surface recombination rate. In principle, two categories of passivated contacts can be distinguished: (1) the heterojunction concept; and (2) the conductor-insulator-semiconductor (CIS) concept (Singh, Green, and Rajkanan, 1981).

A heterojunction is formed, when two semiconductors with different band gaps and/or band offsets are brought into contact. If the heterojunction is formed with a highly doped semiconductor layer with a larger bandgap on top of a lower doped absorber, band bending is induced into the absorber leading to a reduction of one carrier species at the heterojunction interface, which reduces the carrier recombination rate. If a crystallographic perfect interface can be realized, as can be done, e.g. in epitaxial grown III/V solar cells (Kayes *et al.*, 2011), the interface trap density is low and a well-passivated contact can be realized.

The CIS concept summarizes the group of metal insulator (MIS) and semiconductor insulator semiconductor (SIS) junctions, where a thin oxide layer is sandwiched between the silicon absorber and a metal or doped semiconductor (i.e. transparent conductive oxides (TCO) or highly doped polycrystalline silicon), respectively. These contacts have in common that they establish a well-passivated surface using the thin oxide layer and reduce the number of minority carriers at the surface via induced band bending or a very shallow diffused doping profile in the silicon absorber.

In the following, experimentally realized approaches will be discussed in more details.

3.6.3.1 Silicon Heterojunction

The most prominent heterojunction for silicon solar cells is the a-Si:H/c-Si heterojunction. Silicon heterojunction (SHJ) solar cells have attracted a lot of attention due to the very high conversion efficiencies achieved so far (25.6%) (Masuko *et al.*, 2014) and their lean process flow, which is based solely on low-temperature processes. The way in which the two fundamental prerequisites of a passivated contact are solved by a SHJ will be discussed using the example of the TCO/a-Si:H(p)/a-Si:H(i)/c-Si(n) heterojunction. The corresponding band diagram in thermal equilibrium is given in Figure 3.6.3. Due to the differences in the bandgap E_g and the electron affinity χ between a-Si:H ($E_{g,a\text{-}Si:H} \approx 1.7$, eV, $\chi_{a\text{-}Si:H} = 3.9\,\text{eV}$) and c-Si ($E_{g,c\text{-}Si} = 1.12\,\text{eV}$, $\chi_{c\text{-}Si} = 4.05\,\text{eV}$) a heterojunction is formed, which leads to a strong band bending in the c-Si absorber.

The chemical passivation of the silicon surface is realized by the intrinsic a-Si:H(i) buffer layer, which has a low defect concentration and reduces the number of defect states at the a-Si/c-Si interface. The overlaying doped a-Si:H(p) layer has a twofold function: (1) it leads to an additional band bending in the crystalline silicon and thus to an induced junction; and (2) if doped high enough, it guarantees an unhindered majority carrier transport to the TCO (Bivour *et al.*, 2014; Pysch *et al.*, 2011). The induced junction further reduces the carrier recombination at the hetero interface, as it leads to a reduction of one carrier species (either inversion forming a pn-junction or accumulation). With this contact, the highest voltages for silicon solar cells have been achieved so far (Taguchi *et al.*, 2014).

An alternative approach is the GaP/Si heterojunction (Beck, Blakeslee, and Gessert, 1988; Wagner *et al.*, 2014). The band diagram of a GaP/Si heterojunction looks quite similar to that of an ideal solar cell structure as, for example, shown in Würfel and Würfel (2009). The technological advantage of GaP as a wide bandgap ($E_g = 2.26\,\text{eV}$) (Lorenz, Pettit, and Taylor, 1968) contact layer is that it has a similar lattice constant to silicon (5,450 A and 5.431 A, respectively). Thus, a direct growth with low interface defect density is in principle possible. First, processed GaP/Si solar cells represent a proof of concept, but still exhibit low open-circuit voltages (Grassman *et al.*, 2014). One reason for the growing interest in a GaP/Si heterojunction as a passivated contact for silicon solar cells is that GaP can be used as a

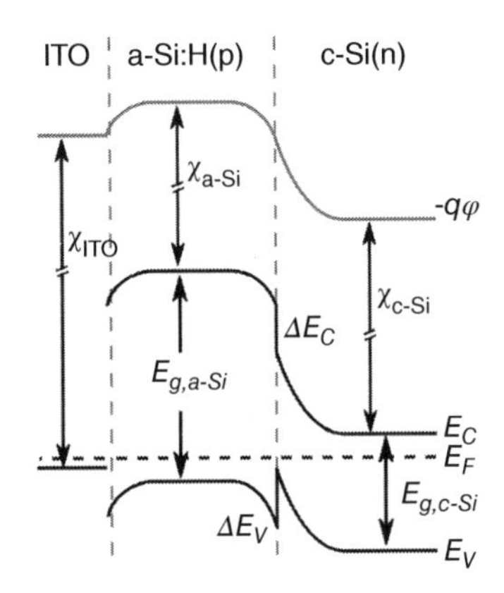

Figure 3.6.3 Band diagram of TCO/a Si:H(p)/c-Si structure (not to scale)

nucleation layer for the growth of other III/V layers such as GaAs or GaInP on top of a silicon layer to realize a tandem solar cell with an efficiency potential higher than 45% for one sun (Grassman *et al.*, 2014).

3.6.3.2 Conductor-Insulator-Semiconductor (CIS) Based Passivated Contacts

Metal insulator (MIS) and semiconductor insulator-semiconductor (SIS) junctions were investigated in the 1970s and were proposed as an alternative to diffused pn-junctions for photovoltaic applications(Shewchun, Singh, and Green, 1977; Green, King, and Shewchun, 1974). The structure of an MIS or SIS contact based on a thin high bandgap dielectric, e.g. SiO, is placed in-between the silicon wafer and either a metal or a metal-like structure (MIS) or another doped semiconductor (SIS). The working principle of an MIS or SIS contact relies on: (1) the passivation of the surface by the insulator; and (2) the depletion of one type of carrier at the surface by using a metal with a suitable work function (MIS, see Figure 3.6.4) or a doped semiconductor (SIS, see Figure 3.6.4).

The dielectric insulator reduces interface traps which would otherwise lead to an unfavorable pinning of the Fermi level at the silicon surface. By deactivating most surface states the metal work function or the Fermi energy of the semiconductor determine the surface band bending in the c-Si.

The influence of the work function or Fermi energy on the selectivity of such MIS/SIS structures can be seen in Figure 3.6.5. Shown are the simulated internal voltage (dotted lines) and the external voltage (straight lines) as a function of the applied work function and the surface recombination velocity (SRV) at the absorber surface. It can be seen, that for very low SRV (1 cm/s) the internal voltage (V_{int}, dotted lines) is independent of the work function and very high internal voltages can be achieved. To achieve a high external voltage (V_{ext}, straight lines), a material with an appropriate work function has to be added on top of the oxide layer. To get a hole contact (solid lines), a high work function material has to be used (e.g. p-doped a-Si or MoO_3). For an electron contact (dashed lines), a low work function material (e.g. n-doped a-Si or Cs_2CO_3). For higher SRV (5000 cm/s), the work function is even more important, as it lead to a depletion or accumulation at the interface of the absorber and thus also influences the internal voltage.

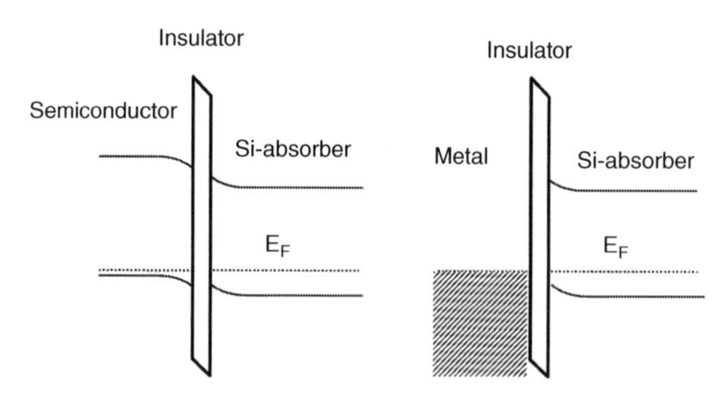

Figure 3.6.4 Band diagram of hole SIS contact (left) and MIS contact (right)

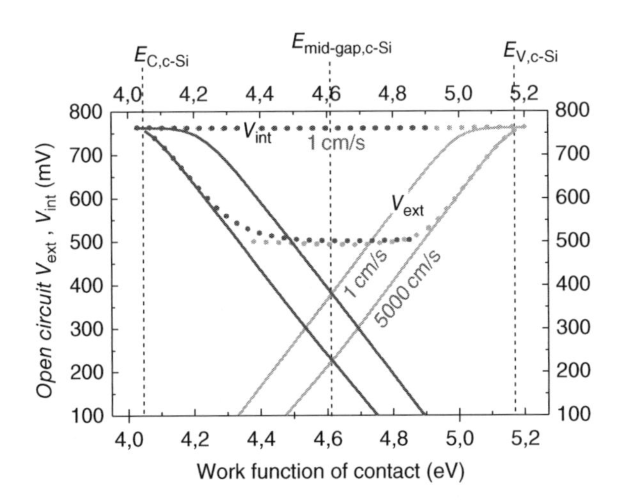

Figure 3.6.5 Calculated internal open circuit voltage (V_{int}) and external open circuit voltage (V_{ext}) as a function of the work function of the contact material for different interface recombination velocities

Although a dielectric with a large energy barrier to minority carriers and a small one to majority carriers is conceivable, it should be noted that a typical dielectric always poses an energy barrier to both majority and minority carriers. If tunnelling is assumed as the transport mechanism, the thickness of the dielectric is a very critical parameter which determines the efficiency of such a solar cell and it must be minimized without degrading the surface passivation. According to Shewchun, the oxide thickness must not exceed 15 Å for sufficient tunnelling probability (Shewchun *et al.*, 1977). Beside the majority carrier transport via tunnelling, also alternative transport mechanisms are discussed in the literature. For example, Peibst *et al.* (2014) proposed that small pinholes in the thin oxide layers could be responsible for the majority carrier transport rather than tunnelling through the oxide layer. Both mechanisms, tunnelling and transport through pinholes or a combination of both, are possible and can also depend on the detailed properties of the thin oxide layer and the additional layers on top of it.

The most direct way to realize a passivated CIS contact is the MIS concept, because the metal contact is placed directly on top of the dielectric passivation layer. The early MIS solar cells were intended as an alternative to diffused pn-junctions (Green and Godfrey, 1976) to avoid the high temperature diffusion step. In these cells, the work function of the applied metal layer leads to the inversion of one carrier species forming the pn junction. An enhancement of the classical MIS contact was proposed by Blakers and Green (1981), where the pn-junction is formed by diffusion and the metal grid was separated from the diffused region by a thin tunnel oxide. In this MINP junction (metal-insulator-np junction), the MIS contact could be described as a passivated contact reducing the minority carrier recombination at the metal contact. As a result, a V_{oc} of 678 mV was achieved with such a simple solar cell structure. Metz *et al.* even demonstrated a 21.1% efficient solar cell (Metz and Hezel, 1997). Recently, this concept has been revisited by several research groups. Due to its success in PV, Al_2O_3 was given special consideration for realizing passivated contacts (ZnO/Al_2O_3/c-Si stacks) (Smit *et al.*, 2014); (Al/Al_2O_3/c-Si stacks) (Deckers *et al.*, 2014); and (Al/a-Si:H(i)/Al_2O_3/c-Si stacks) (Bullock *et al.*, 2014).

Different groups have realized very good results for SIS contacts based on silicon oxide layers capped with a highly doped silicon layer. One of the early works on silicon solar cells was performed by Gan and Swanson (1990). They used a technology developed for bipolar transistors (BJTs) (Post, Ashburn, and Wolstenholme, 1992; Wolstenholme, Jorgensen, Ashburn, and Booker, 1987). In this approach, the oxide layer is thermally grown and a poly-silicon layer is deposited using LPCVD. After additional high temperature doping and annealing steps, they achieved low J_0 values between 10 and 50 fA/cm^2 and $\rho c \sim 10^{-5}\,\Omega\,\text{cm}^2$ for phosphorus doped electron contacts on test samples (Gan and Swanson, 1990). Recently, Römer *et al* (Peibst *et al.*, 2014) achieved even lower results for electron and hole contacts $J_0 < 5$ fA/cm^2 with $\rho_c < 10\,\text{m}\Omega\,\text{cm}^2$ on test structures using the same technology approach. Similar results were also achieved by other groups using different oxides and deposition techniques for the doped silicon layer (Di Yan *et al.*, 2015; Nemeth *et al.*, 2014). An alternative technology approach was proposed by Feldmann *et al.* (2014a). They used a wet chemically grown oxide in combination with a PECVD deposited silicon layer. After a high temperature annealing in the range of 800 to 900 °C, the amorphous silicon layer is partly recrystallized (Feldmann *et al.*, 2014b). A TEM picture of such a structure is shown in Figure 3.6.6.

After a hydrogen passivation step they achieved J_{0e} values of down to 8 fA/cm^2 and $\rho_c < 10\,\text{m}\Omega\,\text{cm}^2$ (Feldmann *et al.*, 2014a). They integrated their passivated contact system, called tunnel oxide passivated contact (TOPCon), as a full area electron rear contact into a both side contacted n-type silicon solar cell and achieved an efficiency of 24.9% (Hermle, IEEE, 2015) showing the excellent performance of the passivated contact on cell level.

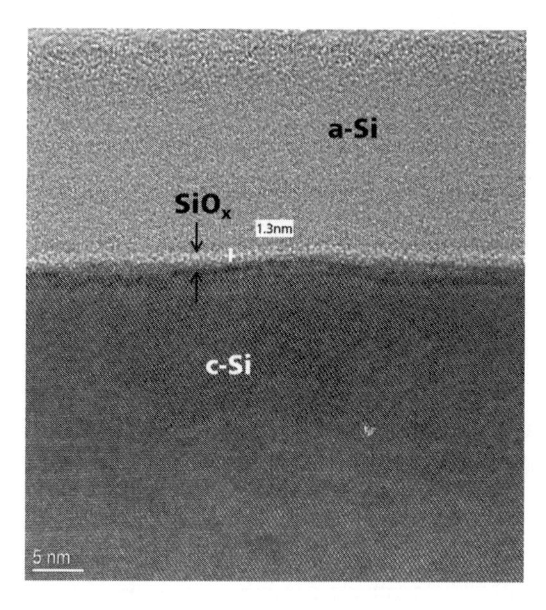

Figure 3.6.6 TEM image of a cross-section of a TOPCon contact. Between the crystalline Si absorber and the amorphous/nanocrystalline silicon layer, the SiOx tunnel layer is clearly seen

3.6.4 Conclusion

Passivated contacts are a relevant technology to push the solar efficiency to the practical limits. In this chapter, the theoretical background and experimental results are presented. The fundamental requirements, suppression of recombination and lossless majority carrier transport, are discussed and the experimental realizations like heterojunctions and the different SIS and MIS concepts are presented.

List of Symbols

Symbol	Description	Unit
J_0	Dark saturation current density	fA/cm^2
N_A	Acceptor density	cm^{-3}
V_{int}	Internal or implied Voltage	V
V_{ext}	External Voltage	V
$E_{F,n}$	Electron quasi Fermi level	eV
$E_{F,p}$	Hole quasi Fermi level	eV
D_{it}	Interface trap density	$eV^{-1}\ cm^{-2}$
E_g	Bandgap	eV
χ	Electron affinity	eV
ρ_c	Contact resistance	Ωcm^2

List of Acronyms

SRV	Surface recombination velocity
PERL	Passivated Emitter and Rear Locally Diffused
Al-BSF	Aluminum Back Surface Field
CIS	Conductor-insulator-semiconductor
MIS	Metal-insulator-semiconductor
SIS	Semiconductor-insulator-semiconductor
TCO	Transparent conductive oxides
SHJ	Silicon heterojunction
GaP	Gallium Phosphide
Si	Silicon
GaAs	Gallium Arsenide
GaInP	Gallium Indium Phosphide
MINP	Metal-insulator-np
LPCVD	Low Pressure Chemical Vapor Deposition
PECVD	Plasma Enhanced Chemical Vapor Deposition
TEM	Transmission Electron Microscopy
SiO	Silicon Oxide

References

Altermatt, P.P. (2011) Models for numerical device simulations of crystalline silicon solar cells: a review. *Journal of Computational Electronics*, **10** (3), 314–330.

Beck, E.E., Blakeslee, A.E., and Gessert, T.A. (1988). Application of GaP/Si heteroepitaxy to cascade solar cells. *Solar Cells*, **24** (1–2), 205–209.

Benick, J., Hoex, B., van de Sanden, M.C.M. *et al.* (2008) High efficiency n-type Si solar cells on Al_2O_3-passivated boron emitters. *Applied Physics Letters*, **92** (25), 253504.

Bivour, M., Reusch, M., Schroer, S. *et al.* (2014) Doped layer optimization for silicon heterojunctions by injection-level-dependent open-circuit voltage measurements. *IEEE Journal of Photovoltaics*, **4** (2), 566–574.

Blakers, A.W. and Green, M.A. (1981) 678-mV open-circuit voltage silicon solar cells. *Applied Physics Letters*, **39** (6), 483–485.

Bullock, J., Cuevas, A., Yan, D. *et al.* (2014) Amorphous silicon enhanced metal-insulator-semiconductor contacts for silicon solar cells. *Journal of Applied Physics*, **116** (16), 163706.

Catchpole, K.R. and Blakers, AW. (2002) Modelling the PERC structure for industrial quality silicon. *Solar Energy Materials and Solar Cells*, **73** (2), 189–202.

Cousins, P.J., Smith, D.D., Luan, H.-C., *et al.* (2010) Generation 3: Improved performance at lower cost. In *35th IEEE Photovoltaic Specialists Conference Honolulu, Hawaii, 2010*, pp. 275–278.

Deckers, J., Cornagliotti, E., Debucquoy, M. *et al.* (2014) Aluminum oxide-aluminum stacks for contact passivation in silicon solar cells. *Energy Procedia*, **55**, 656–664.

Di Yan, Cuevas, A., Bullock, J., *et al.* (2015) Phosphorus-diffused polysilicon contacts for solar cells. *Solar Energy Materials and Solar Cells* **142**, 71–82.

Feldmann, F., Bivour, M., Reichel, C *et al.* (2014a) Passivated rear contacts for high-efficiency n-type Si solar cells providing high interface passivation quality and excellent transport characteristics. *Solar Energy Materials and Solar Cells*, **120**, 270–274.

Feldmann, F., Simon, M., Bivour, M. *et al.* (2014b) Efficient carrier-selective p- and n-contacts for Si solar cells. *Solar Energy Materials and Solar Cells*, **131**, 100–104.

Gan, J. Y. and Swanson, R. M. (1990) Polysilicon emitters for silicon concentrator solar cells. In *Polysilicon Emitters for Silicon Concentrator Solar Cells* (eds J.Y. Gan and R.M. Swanson), IEEE Explore, pp. 245–250.

Grassman, T.J., Carlin, J. A., Galiana, B. *et al.* (2014) MOCVD-Grown GaP/Si subcells for integrated III–V/Si multi-junction photovoltaics. *IEEE Journal of Photovoltaics*, **4** (3), 972–980.

Green, M.A. and Godfrey, R.B. (1976) MIS solar cell: General theory and new experimental results for silicon. *Applied Physics Letters*, **29** (9), 610–612.

Green, M.A., King, F.D., and Shewchun, J. (1974) Minority carrier MIS tunnel diodes and their application to electron- and photo-voltaic energy conversion—I: Theory. *Solid-State Electronics*, **17** (6), 551–561.

Kayes, B.M., Nie, H., Twist, R., *et al.* (2011) 27.6% Conversion efficiency, a new record for single-junction solar cells under 1 sun illumination. In *2011 37th IEEE Photovoltaic Specialists Conference (PVSC)*, pp. 4–8.

Lorenz, M.R., Pettit, G.D., and Taylor, R.C. (1968) Band gap of gallium phosphide from 0 to 900°k and light emission from diodes at high temperatures. *Physical Review*, **171** (3), 876–881.

Masuko, K., Shigematsu, M., Hashiguchi, T. *et al.* (2014) Achievement of more than 25% conversion efficiency with crystalline silicon heterojunction solar cell. *IEEE Journal of Photovoltaics*, **4**(6), 1433–1435.

McIntosh, C. and Altermatt, P.P. (2015) EDNA 2: www.pvlighthouse.com.au.

Metz, A. and Hezel, R. (1997) Record efficiencies above 21% for MIS-contacted diffused junction silicon solar cells. *Proceedings of IEEE Photovoltaic Specialist Conference*, Anaheim, CA, Sept, pp. 283–286.

Nemeth, B., Young, D. L., Yuan, H.-C., *et al.* (2014) Low temperature Si/SiOx/pc-Si passivated contacts to n-type Si solar cells. In *2014 IEEE 40th Photovoltaic Specialists Conference (PVSC)* pp. 3448–3452.

Peibst, R., Romer, U., Hofmann, K.R., *et al.* (2014) A simple model describing the symmetric I–V characteristics of p polycrystalline Si/n monocrystalline Si, and n polycrystalline Si/p monocrystalline Si junctions. *IEEE Journal of Photovoltaics*, **4** (3), 841–850.

Peibst, R., Romer, U., Larionova, Y. *et al.* Building blocks for back-junction back-contacted cells and modules with ion-implanted poly-Si junctions. In *2014 IEEE 40th Photovoltaic Specialists Conference (PVSC)*, pp. 852–856.

Post, I.R.C., Ashburn, P., and Wolstenholme, G.R. (1992) Polysilicon emitters for bipolar transistors: a review and re-evaluation of theory and experiment. *IEEE Transactions on Electron Devices*, **39** (7), 1717–1731.

Pysch, D., Meinhard, C., Harder, N.-P. *et al.*(2011) Analysis and optimization approach for the doped amorphous layers of silicon heterojunction solar cells. *Journal of Applied Physics*, **110** (9), 94516.

Richter, A., Hermle, M., and Glunz, S. W. (2013) Reassessment of the limiting efficiency for crystalline silicon solar cells. *IEEE Journal of Photovoltaics*, **3** (4), 1184–1191.

Shewchun, J., Singh, R., and Green, M. A. (1977) Theory of metal-insulator-semiconductor solar cells. *Journal of Applied Physics*, **48** (2), 765.

Singh, R., Green, M.A., and Rajkanan, K. (1981) Review of conductor-insulator-semiconductor (CIS) solar cells. *Solar Cells*, **3** (2), 95–148.

Smit, S., Garcia-Alonso, D., Bordihn, S. *et al.* (2014) Metal-oxide-based hole-selective tunneling contacts for crystalline silicon solar cells. *Solar Energy Materials and Solar Cells*, **120**, 376–382.

Swanson, R.M. (2005) Approaching the 29% limit efficiency of silicon solar cells. In *31st IEEE Photovoltaic Specialists Conference Lake Buena Vista, Conference Record*, pp. 889–894.

Taguchi, M., Yano, A., Tohoda, S., *et al.* (2014) 24.7% record efficiency HIT solar cell on thin silicon wafer. *IEEE Journal of Photovoltaics*, **4** (1), 96–99.

Wagner, H., Ohrdes, T., Dastgheib-Shirazi, A. *et al.* (2014) A numerical simulation study of gallium-phosphide/silicon heterojunction passivated emitter and rear solar cells. *Journal of Applied Physics*, **115** (4), 44508.

Wolstenholme, G.R., Jorgensen, N., Ashburn, P., and Booker, G.R. (1987) An investigation of the thermal stability of the interfacial oxide in polycrystalline silicon emitter bipolar transistors by comparing device results with high-resolution electron microscopy observations. *Journal of Applied Physics*, **61** (1), 225.

Würfel, P. (2005) *Physics of Solar Cells: From Principles to New Concepts*, Wiley-VCH, Weinheim.

Würfel, P., and Würfel, U. (2009) *Physics of Solar Cells: From Basic Principles to Advanced Concepts* (2nd, updated and expanded ed.). Wiley-VCH, Weinheim.

Würfel, U., Cuevas, A., and Würfel, P. (2015) Charge carrier separation in solar cells. *IEEE Journal of Photovoltaics*, **5** (1), 461–469.

Zhao, J., Wang, A., Green, M.A., and Ferrazza, F. (1998) 19.8% efficient "honeycomb" textured multicrystalline and 24.4% monocrystalline silicon solar cells. *Applied Physics Letters*, **73** (14), 1991.

3.7

Light Management in Silicon Solar Cells

Zachary Holman and Mathieu Boccard
Arizona State University, Tempe, AZ, USA

3.7.1 Introduction

This chapter is concerned with light management, which is – very generally – the business of tuning the optical properties of the layers of a solar cell so as to best utilize the incident spectrum in generating electrical power. More specifically, light management is primarily associated with maximizing the short-circuit current density (J_{sc}), though there are instances in which the optical properties of a solar cell influence, for example, its open-circuit voltage (V_{oc}). Light management is particularly important in silicon solar cells because silicon is an indirect-bandgap material with poor absorption near its bandgap. The problem is compounded in thin-film silicon solar cells because the double-pass path length is so short; in response, thin-film silicon researchers have pioneered many absorption-enhancement techniques. Recently, light management has also gained importance in crystalline, wafer-based silicon cells for two reasons: First, whereas improving the V_{oc} of silicon cells has been a major research focus in past decades, several device structures have now reached V_{oc} close to the theoretical limit using passivating contacts. This makes optical losses the new lowest-hanging fruit. Second, there is a continued trend towards thinner wafers, which allow for higher V_{oc} than their thicker counterparts. Maintaining high J_{sc} in thin cells requires still further attention to light management.

This chapter begins with the theory of light management and an experimental approach to access the key light management characteristics. The three ingredients in light management – front-surface reflection, parasitic absorption, and light trapping – are then developed further, and the chapter concludes with an example.

Photovoltaic Solar Energy: From Fundamentals to Applications, First Edition.
Edited by Angèle Reinders, Pierre Verlinden, Wilfried van Sark, and Alexandre Freundlich.

Companion website: www.wiley.com/go/reinders/photovoltaic_solar_energy

3.7.2 Theory and Experiment

High J_{sc} is achieved by maximizing interband (band-to-band) absorption within the absorber: the wafer, in the case of silicon solar cells.[1] That is, the fraction of the incident light absorbed in the wafer, A_{Si}, should be made to approach unity for all photons with energies greater than the bandgap energy of silicon ($h\nu > E_g$).[2] To calculate A_{Si}, we can follow a representative incident ray as it makes its way through a solar cell, as illustrated Figure 3.7.1. Noting that all light must be reflected, transmitted, or absorbed somewhere in the cell, we have:

$$1 = \underbrace{R_0 + A_{Front_0}}_{\text{Entry}} + \underbrace{A_{Si_1} + A_{Rear_1} + T_1}_{\text{First pass}} + \underbrace{A_{Si_2} + A_{Front_2} + T_2}_{\text{Second pass}} + \underbrace{A_{Si_3} + A_{Rear_3} + T_3}_{\text{Third pass}} + \cdots \tag{3.7.1}$$

where R_m, A_{Front_m}, A_{Si_m}, A_{Rear_m}, and T_m are the fractions of the light incident on the solar cell that are reflected, absorbed in the layers on the front (sunward) side of the wafer, absorbed within

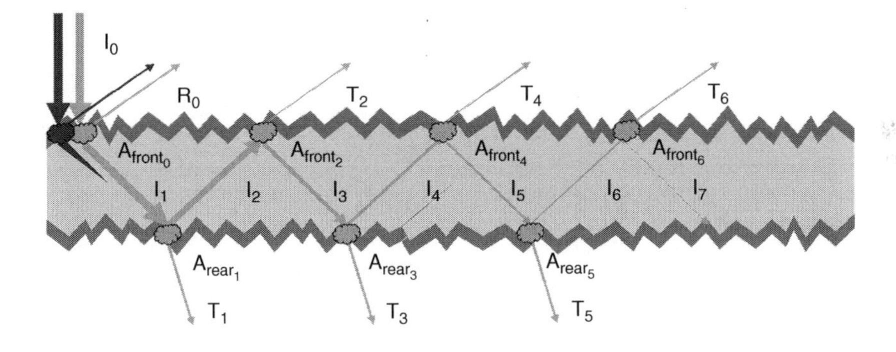

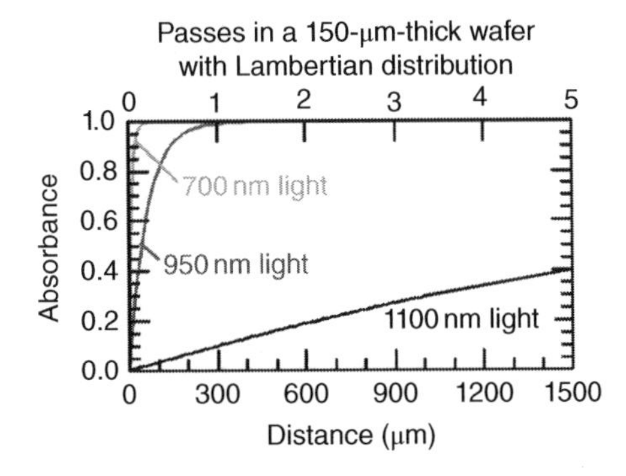

Figure 3.7.1 Schematic depiction of the path taken by visible and infrared light in a silicon solar cell and penetration depth of red and infrared light in a silicon wafer. (*See insert for color representation of the figure*)

---

[1] J_{sc} is the product of the spectrally weighted interband absorbance and carrier collection, integrated over the spectrum. However, we ignore the carrier collection in this discussion since it is outside the realm of optics.

[2] Hereafter, the discussion will refer exclusively to these photons unless otherwise stated.

the wafer, absorbed in the layers on the rear side of the wafer, and transmitted out of the wafer on the mth pass, respectively. Each A_{Si_m} can be written in terms of the more familiar reflectances, single-pass absorbances, and transmittances,[3] for example:

$$A_{Si_2} = \frac{I_2}{I_0} a_{Si} = \underbrace{(1-r)(1-a_{Front})(1-a_{Si})(1-a_{Rear})(1-t_{Rear})}_{\text{Attenuation before second pass}} \underbrace{a_{Si}}_{\text{Single-pass absorbance}} \tag{3.7.2}$$

Here, $\frac{I_2}{I_0}$ is the remaining fraction of light when the ray begins its second pass through the wafer, and lowercase letters are used to denote reflectance, absorbance, and transmittance. The single-pass absorbance in the wafer is given by

$$a_{Si} = 1 - e^{-\alpha y d} \tag{3.7.3}$$

with α the absorption coefficient in the wafer, d the wafer thickness, and

$$y = \left\langle \frac{1}{\cos\theta} \right\rangle \tag{3.7.4}$$

the average path length enhancement during a single pass through the wafer (θ is the angle light travels with respect to the wafer normal).[4]

Finally, summing the absorbance in each pass through the wafer yields the desired light-management metric:

$$A_{Si} = \sum_{m=1}^{\infty} A_{Si_m} \tag{3.7.5}$$

$$\begin{aligned}&(1-r)(1-a_{Front})\sum_{k=1}^{\infty} a_{Si} \\ &\cdot\Big\{(1-a_{Rear})^{k-1}(1-t_{Rear})^{k-1}(1-a_{Front})^{k-1}(1-t_{Front})^{k-1}(1-a_{Si})^{2k-2} \\ &+(1-a_{Rear})^{k}(1-t_{Rear})^{k}(1-a_{Front})^{k-1}(1-t_{Front})^{k-1}(1-a_{Si})^{2k-1}\Big\}\end{aligned} \tag{3.7.6}$$

Though it is conceptually simple to arrive at Equation (3.7.6), the equation is cumbersome and tempting to skip over. Nevertheless, it provides us with all we need to know to understand light management. Every term in Equation (3.7.6) except a_{Si} (following the summation) is a loss. These are grouped into three categories that can be attacked individually when trying to enhance J_{sc}: (1) front-surface reflection, which is the r term out front; (2) "parasitic" absorption, which refers to all a_{Front} and a_{Rear} terms; and (3) light trapping, which encompasses the t_{Front}

[3] These are the fractions of the *remaining* light reflected, absorbed, or transmitted, rather than the fractions of the incident light on the cell.

[4] This angle can change from pass to pass. Most typically it is small for the first pass and evolves to 60°. The angle changes not only y and thus the absorbance in wafer, but also the transmittance and parasitic absorption.

and t_{Rear} terms, as well as y hidden inside a_{Si} (see Equation 3.7.3). It can be summarized as: successful light management requires that the incident light be transmitted into the solar cell, not be absorbed in layers other than the wafer, and have a long path length through the wafer before escaping. Note, however, that the third criterion (and, to some extent, the second) are irrelevant for wavelengths that are completely absorbed within a single pass through the wafer ($\lambda \lesssim 950$ nm for standard silicon solar wafers). In this case, the summation collapses to unity and the only losses are front-surface reflection and parasitic absorption in the front layers. (This is easier to see in Equation 3.7.1.).

While Equation (3.7.6) tells us which losses to investigate, there is no easy way to directly measure, e.g., t_{Front}, a_{Rear}, y, or even a_{Si} in a completed solar cell; the individual terms are experimentally inaccessible. Fortunately, the three loss *categories* can be parsed by measuring the reflectance, transmittance, and external quantum efficiency (*EQE*) of a finished cell. Referring to Equation (3.7.1) and Figure 3.7.1, these measurements yield

$$R = R_0 + R_{Escape} = R_0 + T_2 + T_4 + T_6 + \ldots \quad (3.7.7)$$

$$T = T_1 + T_3 + T_5 + \ldots \quad (3.7.8)$$

$$EQE \cong A_{Si} = A_{Si_1} + A_{Si_2} + A_{Si_3} + \ldots \quad (3.7.9)$$

Note that Equation (3.7.9) is valid only when the effective diffusion length of carriers substantially exceeds the wafer thickness so that the probability of collecting a photogenerated electron–hole pair approaches unity. This is a good assumption at short circuit for most of today's commercial silicon solar cells (those with $V_{oc} \gtrsim 650$ mV, for a rough guide), and an even better assumption at reverse bias.[5]

Rewriting Equation (3.7.1) in terms of Equations (3.7.7)–(3.7.9),

$$1 = R + T + EQE + A_{Parasitic} \quad (3.7.10)$$

with

$$A_{Parasitic} = A_{Front_0} + A_{Rear_1} + A_{Front_2} + A_{Rear_3} + \ldots \quad (3.7.11)$$

For all cells and modules with full backside metal coverage, $T=0$ (as will hereafter be assumed), and it is instructive to plot *EQE* and $1-R$ together, as shown in Figure 3.7.2 (a) for an amorphous silicon/crystalline silicon heterojunction solar cell. With this display of the spectra, the losses are immediately apparent for $\lambda \lesssim 950$ nm: The area above the $1-R$ curve corresponds to the front-surface reflectance, the area between the *EQE* and $1-R$ curves corresponds to

[5] Reverse bias helps for cells with significant junction recombination but not necessarily for cells with a poor collection of carriers generated near the rear of the wafer, since a very large bias would be needed to extend the depletion region to within a diffusion length of these carriers. Note that implicit in reverse-bias measurements (and even short-circuit measurements) is the assumption that absorption is independent of the cell's operating condition. *EQE* is measured under conditions for which the collection probability is near unity, and the resulting spectrum is taken to be representative of A_{Si} at short circuit, open circuit, and maximum power point, which is the operating condition that truly matters. This should not be assumed for absorbers with high internal radiative efficiency high-quality III-V materials.

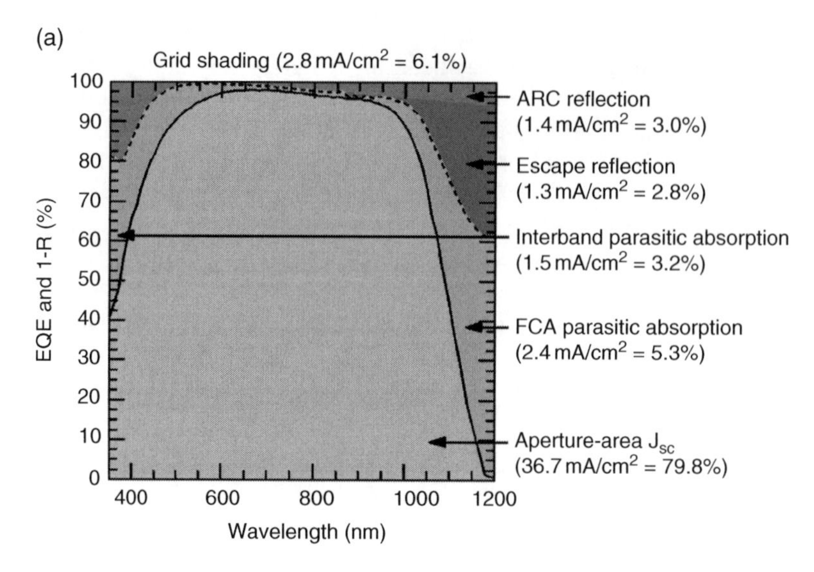

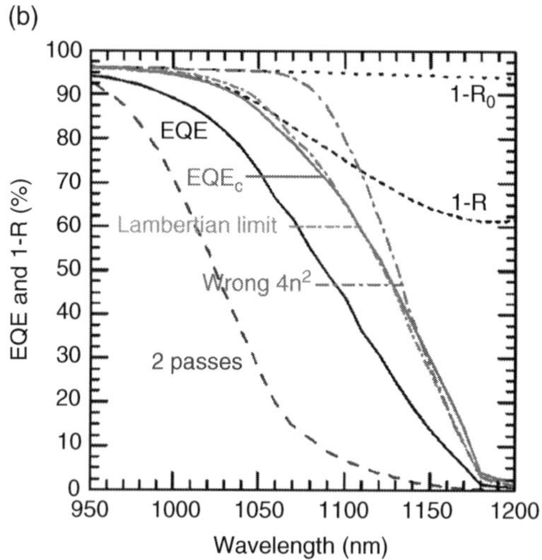

Figure 3.7.2 (a) External quantum efficiency and 1 – reflection spectra for a representative silicon heterojunction solar cell on a 110-μm-thick wafer. (b) Measured and corrected (free of parasitic absorption) external quantum efficiency of the same cell, and light-trapping limits to which the corrected spectra may be compared

parasitic absorption in the layers at the front of the wafer, and there are no losses caused by imperfect light trapping (refer to Equations (3.7.7)–(3.7.11) for the case of full absorption in a single pass).

For longer wavelengths, the interpretation is complicated by the fact that R, EQE, and $A_{Parasitic}$ are all sensitive to light trapping. The front-surface reflectance loss can be found by linearly extrapolating R at shorter wavelengths out to 1200 nm (as was done in Figure 3.7.2 (a)) or, if more accuracy is needed, fitting it with a thin-film simulator such as OPAL 2 (McIntosh and Baker-Finch, 2012). The parasitic absorption loss is again the difference between the $1-R$ and *EQE* curves. To assess the light-trapping loss, we must first artificially remove the (long-wavelength) parasitic absorption. "Light trapping" is used throughout this chapter expressly to refer to path-length enhancement *in the absence of absorption* – either in the wafer or in other layers – which means that it is a function entirely of scattering. This is consistent with its above association with escape from the wafer and wafer traverse angle (t_{Front}, t_{Rear}, and y), and it allows the disambiguation of the effects of, e.g., a poor rear reflector and poor texture. Some 950–1200 nm light that is absorbed parasitically would, in the absence of that absorption, either escape out the front of the wafer or be absorbed within the wafer; that is, it would contribute to R or EQE. Rearranging Equation (3.7.10) and using Equation (3.7.7):

$$1-R_0 = R_{Escape} + EQE + A_{Parasitic} \tag{3.7.12}$$

In the hypothetical case of no parasitic absorption, Equation (3.7.12) would become:

$$1-R_0 = R_{Escape,c} + EQE_c \tag{3.7.13}$$

with $R_{Escape,c}$ and EQE_c the new, "corrected" quantities. To enforce that $A_{Parasitic}$ be split equitably between $R_{Escape,c}$ and EQE_c, we require that

$$\frac{EQE}{R_{Escape} + EQE} = \frac{EQE_c}{R_{Escape,c} + EQE_c} \tag{3.7.14}$$

and thus

$$EQE_c = \frac{1-R_0}{1-R_0-A_{Parasitic}} \cdot EQE = \frac{1-R_0}{R_{Escape} + EQE} \cdot EQE \tag{3.7.15}$$

It is unnecessary to separately calculate $1-R_c = 1-(R_0 + R_{Escape,c})$ because it is equivalent to EQE_c; the two curves now collapse to one. In addition, for a typical antireflection coating with a reflectance minimum near 600 nm, $R_0 < 10\%$ and it is acceptable to ignore it in Equation (3.7.15).

Figure 3.7.2 (b) shows the same spectra as in Figure 3.7.2 (a), as well as EQE_c for this solar cell and three relevant light-trapping "limits." To assess the light-trapping loss, EQE_c is to be compared with the Lambertian limit, which, as described in Section 3.7.5, is a practical limit that has yet to be surpassed. Integrating any of the three losses, weighted by the incident spectrum, over the 350–1200 nm wavelength range yields the associated J_{sc} loss. This must be done with care, however, as the current lost is not always equivalent to the current that could be gained if the loss were eliminated (see example in Section 3.7.6).

3.7.3 Front-Surface Reflection

The primary source of front-surface reflection in silicon solar cells is the large mismatch in the refractive indices of air and silicon ($n = 3.5$–6 in the wavelength range of interest) Two approaches are used to combat this loss: wafer texturing and antireflection coatings.

Monocrystalline wafers are textured in alkaline solutions of, e.g., KOH, that preferentially etch the {100} crystalline planes, leaving behind upright, randomly placed pyramids with nominally {111} facets. From silicon's crystal structure, we expect these pyramids to have a base angle of 54.7° and to reflect normally incident light downwards so that 70% of rays hit two facets and the remaining 30% hit three or more (Baker-Finch *et al.*, 2011). Each reflection at the silicon surface results in a fraction of light being transmitted into the wafer according to the Fresnel equations, and thus more bounces per ray reduces the front-surface reflectance. Baker-Finch and McIntosh found that real random pyramids are in fact "hillocks" with octagonal bases and smaller base angles of 50–52°; these are slightly less effective than ideal 54.7° pyramids at scattering light downwards and increase the simulated reflectance of a bare wafer at 800 nm from approximately 8.5–10% (Baker-Finch *et al.*, 2012a). Nevertheless, the naturally forming random "pyramid" texture is still excellent at reducing front-surface reflectance; it performs nearly as well as textures that are significantly harder to achieve, including inverted pyramids (Baker-Finch *et al.*, 2011). In addition, with appropriate chemistry, it is possible to make random pyramids with characteristic sizes as small as 1 μm, which makes the texture suitable for cells as thin as 10 μm. Multi-crystalline wafers are typically textured in (crystallographic-orientation-agnostic) acidic solutions that leave a pitted surface. The texture is sensitive to the chemistry and not easy to describe analytically, but most measurements show that the features are not as effective as random pyramids at reducing front-surface reflection (Baker-Finch *et al.*, 2012).

Antireflection coatings (ARCs) – the second approach to reducing front-surface reflectance – are most commonly formed by requiring an existing layer to serve a second, optical role. In diffused-junction solar cells, the dielectric front passivation layer doubles as an ARC; in silicon heterojunction solar cells, it is the front transparent conductive oxide (TCO) lateral transport layer. For these single-layer ARCs, the AM1.5G-weighted front-surface reflectance is minimized for normally incident light when the reflectance minimum occurs at 500–600 nm. The layers typically have a refractive index of approximately 2 – roughly the geometric mean of the refractive indices of air and silicon, fortuitously, since this makes for the broadest reflectance minimum – and thus the layers should be approximately 75 nm thick. Note that this thickness takes into account the fact that (normally incident) light traverses the ARC at an angle because of the front surface texture.

An additional, secondary source of front-surface reflectance is "shading" of the cell by the metal fingers and busbars. A typical commercial solar cell has screen-printed fingers that are 50–70 μm wide and spaced apart by 1.5–2 mm, resulting in 3–5% loss in J_{sc}. The busbars account for an additional loss of approximately 2–3%. While screen-printed fingers as narrow as 50 μm or less have been reported (Hannebauer *et al.*, 2013), the width reduction is accompanied by an undesirable height reduction that increases the finger resistance. Approaches to fabricate high-aspect-ratio metallic lines that reduce both shading and series resistance include double print, copper plating, and stencil printing. These are becoming commonplace in research cells and high-efficiency commercial cells but not yet in the majority of commercial cells.

Figure 3.7.2 (a) shows the front-surface reflectance for a silicon heterojunction solar cell that has a random pyramid texture and a TCO ARC. The shading loss is also given, and, together, these comprise nearly a 10% loss. However, what we are ultimately interested in is the light management in a *module*, not a cell, and while parasitic absorption and light trapping are similar in cells and modules, front-surface reflectance is not. In particular, shading decreases because light scattered off the fingers and busbars at high angles is totally internally reflected at the glass-air interface and is again incident upon the cell, the Fresnel reflection at the cell surface decreases because the refractive-index transition at the encapsulant/ARC interface is smaller than at the air/ARC interface, and 4% Fresnel reflection is introduced at the new air/glass interface. This last loss is significant and has spurred some module manufacturers to begin using glass with its own ARC.

A final note on reflection: Front-surface reflection and light trapping are often confused because of their shared dependence on the front texture. They are not, however, the same. A planar wafer with a multi-layer ARC coating can have very low front-surface reflectance but no light trapping (a path-length enhancement of twice the cell thickness). Conversely, a bare wafer with a polished front surface and textured rear surface can be excellent at trapping light once it has entered the wafer, but have more than 30% front-surface reflectance.

3.7.4 Parasitic Absorption

There are two common sources of parasitic absorption in silicon solar cells: interband absorption and free-carrier absorption. Interband parasitic absorption occurs exclusively at the front of the wafer and at short wavelengths since the layers at the rear are illuminated only with $\lambda \gtrsim 950\,\text{nm}$ light and silicon solar cells do not typically contain layers with bandgaps small enough to absorb this light. The absorbed light creates an excited electron-hole pair that in most cases cannot be collected. To avoid incurring losses, only wide-bandgap or small-amplitude-oscillator (e.g., indirect-bandgap) layers should be used on the sunward side of the wafer. Diffused-junction solar cells have dielectric passivation layers with $E_g > 3\,\text{eV}$, such as silicon nitride or silicon oxide, on their front sides and thus have only minor losses ($<0.5\,\text{mA/cm}^2$). Silicon heterojunction solar cells like that in Figure 3.7.2 (a), on the other hand, suffer interband parasitic absorption losses of up to $2\,\text{mA/cm}^2$ because the amorphous silicon passivation and carrier-collection layers, though only 5–10 nm thick each, have a narrow bandgap of $E_g = 1.7\,\text{eV}$.

Free-carrier absorption (FCA) is a resonant phenomenon in which free electrons or holes absorb photons with energies near their plasma frequency through their collective oscillation. It occurs in doped semiconductor, TCO, and metal layers in silicon solar cells (including in the wafer itself). For doped semiconductor and TCO layers, the plasma resonance is in the infrared and the resulting FCA is observed in silicon solar cells as an *EQE* loss at long wavelengths. The absorption can occur either at the front of the solar cell (e.g., in a heavily doped emitter or TCO layer) or, for $\lambda \gtrsim 950\,\text{nm}$, at the rear. At wavelengths shorter than the plasma wavelength, the absorption coefficient for these materials has the form $\alpha \propto N\lambda^2/\mu$, where N is the free-carrier density and μ is their mobility. Consequently, FCA is most severe at long wavelengths and can be mitigated by choosing materials with lighter doping or higher mobility (more precisely, longer scattering times). These choices typically have other repercussions in solar cells, of course, since the doping density of an emitter determines junction formation and the sheet resistance of a TCO layer influences lumped series resistance.

For metals, the plasma resonance is in the ultraviolet or visible and the absorption coefficient at wavelengths longer than the plasma wavelength has the form $\alpha \propto \sqrt{N\mu/\lambda}$. This indicates that FCA is strongest in metals at short wavelengths; nevertheless, it is seen in solar cells only for $\lambda \gtrsim 950\,\text{nm}$ because metals with large area coverage are used exclusively at the rear of the wafer. Silver, aluminum, and copper are the three most common metals used as rear reflectors in silicon solar cells. Silver is an excellent reflector (low FCA) but is expensive, aluminum is a comparatively poor reflector but is cheap, and copper offers the best tradeoff (despite its color in the visible, it is a good infrared reflector) but it must not be in direct contact with silicon or it will diffuse at room temperature and form recombination-active defects.

In determining which source of parasitic absorption to address first, note that all wavelengths pass through the front layers at their full intensity, and thus any lossy layers on the sunward side of the wafer are particularly detrimental. These can be tricky to identify depending on the wavelength: parasitic absorption at $\lambda \lesssim 950\,\text{nm}$ (as seen from a plot of EQE and $1-R$) necessarily occurs at the front of the wafer, but parasitic absorption at longer wavelengths could occur at either surface. After maximizing the transparency of the front layers, effort should be dedicated to the rear layers in proportion to the thinness of the wafer. Infrared light interacts with the rear surface (and indeed, the front, too) more times per path length through the wafer as the wafer is thinned, increasing the importance of the associated loss.

When we imagine parasitic absorption losses, we most commonly think of light traversing the thickness of a layer, losing intensity as it does so. However, weakly absorbed infrared light that is trapped in a wafer often arrives at the front or rear surface outside the escape cone as it bounces around. The resulting internal reflection is total only if no absorbing layer is within the penetration depth of the evanescent wave that decays outwards from the interface; if a lossy layer is present, it will absorb light approximately in proportion to its absorption coefficient and thickness, and the process is termed *attenuated* total reflectance. (The same physics is put to use in the popular Fourier transform infrared spectroscopy sampling method of the same name.) A specific, but common, instance of this phenomenon occurs at the metallic rear reflector and is the dominant parasitic absorption mechanism in many silicon cells with full-area aluminum rear reflectors (Holman *et al.*, 2013a). The penetration depth of the evanescent wave is only on the order of 100 nm at the wavelengths of interest, depending on the angle of incidence and the refractive index of the layer into which it penetrates. The loss can thus be largely mitigated by displacing the metal from the wafer surface with a thin, transparent dielectric layer. This has been shown to boost the J_{sc} of a cell with an aluminum reflector by up to $1.5\,\text{mA/cm}^2$ (Holman *et al.*, 2013a).

Thus far in this chapter, we have concerned ourselves only with "super-bandgap" photons ($h\nu > E_g$) since these are the photons that have the potential to contribute to J_{sc}. Though less critical to cell performance, light management, and, in particular, parasitic absorption, also matters for "sub-bandgap" photons ($h\nu < E_g$). To see this, consider two solar cells that are identical except that the rear reflector of one absorbs sub-bandgap photons whereas the other reflects them. At short circuit, this difference will go unnoticed: the cells will have the same absorption in the wafer and thus J_{sc} since they are strictly indistinguishable with respect to super-bandgap photons. However, what if we consider the role of optics in a cell at maximum power point in outdoor conditions – the operating condition that ultimately matters? The cells will have the same current density at maximum power point (J_{mpp}),[6] but the cell with the rear reflector that absorbs sub-bandgap photons will heat up in the field and have a lower *voltage* at maximum power point V_{mpp} (Vogt, 2015). Apparently, light management is not exclusively

[6] J_{sc} is a good predictor of J_{mpp} in the absence of high series resistance or low shunt resistance.

about current, and infrared parasitic absorption – a very common phenomenon in silicon solar cells, as can be seen in Figure 3.7.2 (a) at 1200 nm – can reduce the cell voltage if the cell temperature is not artificially controlled. Thus, a solar cell with perfect light management will not only appear "black" at super-bandgap wavelengths, it will also appear "white" or transparent at sub-bandgap wavelengths.

3.7.5 Light Trapping

The high refractive index n of silicon makes it possible to have a higher concentration of photons in an isotropically illuminated silicon wafer than in air, by a factor of n^2, due to the increase in the density of photonic states with refractive index (Yablonovitch, 1982). This is true by transitivity even in a complete module that has an ARC with a refractive index close to 2 at the front surface and encapsulant and glass with a refractive index of 1.5 between this coating and air.

For cells with a rear reflector that are illuminated from the front, the concentration of photons is double, due to the two passes between entry and exit, and for randomized light in a wafer, the average distance traveled by photons between the front and rear surfaces is enhanced by a factor of two compared to photons traveling normal to the wafer (i.e., $\theta = 60°$ and $y = 2$ in Equation 3.7.4). This $2 \cdot 2 \cdot n^2$ is the source of the so-called "$4n^2$ limit" for light trapping and means that, on average, *non-absorbed* light will travel a distance $4n^2d$ in the wafer before escaping, where d is again the wafer thickness. This also applies to weakly absorbed light, which thus suffers $4n^2$ times more absorption than during a single pass in the wafer[7]. We emphasize that, for this random-scattering limit to be met, isotropic illumination, weak absorption in the wafer, no parasitic absorption, and no front-surface reflection are all required (all terms in Equation 3.7.1 are zero except A_{Sim} and T_m).

The constraint of isotropic illumination is relaxed if light is scattered into a Lambertian distribution when entering the cell. The limit can also be revised to account for non-weak absorption and non-zero parasitic absorption, as was first done by Deckman *et al.* (1983) by tracing an average ray of light through the wafer. Boccard *et al.* (2012) recently derived an even more general expression for the random-scattering absorption limit that includes parasitic absorption in the front of the cell, non-Lambertian light scattering upon entry, and front-surface reflection. As described in Section 3.7.2, the way to assess the loss associated with imperfect light trapping in a completed solar cell is to "correct" the *EQE* (which stands in for A_{Si}) by removing infrared parasitic absorption and to compare this to the Lambertian light-trapping limit. This limit is found from Boccard's expression by eliminating the parasitic absorption terms:

$$A_{Si} = (1 - R_0)\frac{a_{Si_{double-pass}}}{1 - Attenuation_{double-pass}} = (1 - R_0)\frac{1 - e^{-4\alpha d}}{1 - e^{-4\alpha d}\left[1 - (1 - R_0)/n^2\right]} \tag{3.7.16}$$

Equation (3.7.16) is the result of performing the summation of the geometric series in Equation (3.7.6) with a_{Front}, a_{Rear}, and t_{Rear} set to zero, in which case attenuation comes only from

[7] Absorption is linear in path length for very small absorption coefficients: when $4n^2\alpha d \ll 1$, $A_{Si} \approx 4n^2\alpha d$.

absorption in the wafer or transmission at the front interface. Transmission at the front interface occurs only for light in the escape cone ($1/n^2$ for a Lambertian distribution) and with probability equal to the transmission of this interface ($1-R_0$). Note that for perfect transmission ($R_0=0$) and weak absorption ($e^{-4\alpha d} \approx 1-4\alpha d$) the first-order approximation of Equation (3.7.16) is $A_{Si} = 4n^2\alpha d$ as expected.

Equation (3.7.16) is plotted in Figure 3.7.2 (b) (labeled the "Lambertian limit") and closely matches EQE_c, indicating that the light trapping afforded by random pyramids is within experimental error of Lambertian light trapping, at least for normal incidence (under which the *EQE* was measured). Early ray-tracing analyses of various textures on silicon wafers predicted this (Campbell and Green, 1987). $A_{Si} = e^{-4n^2\alpha d}$ is also plotted (labeled the "Wrong 4n^2") for comparison; this is sometimes mistakenly taken as the Lambertian limit. It converges with Equation (3.7.16) only for complete absorption in one double-pass (both expressions equal one) and in the weak-absorption limit (where both expressions yield $A_{Si} < 20\%$); in all other cases, the true Lambertian limit yields lower absorption.

Light trapping is achieved in practice by diverting photons away from normal (compared to the plane of the wafer) before they reach an interface through which they can escape. The degree of diversion, which, when averaged, is given by θ in Equation (3.7.4), affects all three quantities in Equation (3.7.6) linked to light trapping: The t_{Front} and t_{Rear} escape terms are modified by total internal reflection and the y thickness-enhancement factor hidden inside Equation (3.7.3) is enlarged by making light travel obliquely through the wafer. The micrometer-sized random pyramids and etch pits described in Section 3.7.3 scatter light according to the laws of geometric optics and are commonly and successfully used for light trapping in silicon solar cells. Much of the infrared light in the wafer arrives at the textured front surface (and the rear surface, in bifacial cells) at angles beyond the critical angle of approximately 16° and is thus internally reflected. Note that the surface texture must allow some light to be coupled out of the wafer (by reciprocity, since the front-surface reflectance is low), unlike a perfect waveguide in which light between two perfectly flat and parallel interfaces will remain trapped forever if it is above the critical angle. Nevertheless, textured surfaces are substantially better at light trapping than planar surfaces since there is no mechanism to couple light into the waveguide modes of the latter, resulting in a maximum path length of twice the wafer thickness.

Other, non-geometric-optics phenomena can also lead to light trapping, including plasmonic resonances in the vicinity of metal nanoparticles and diffraction from wavelength-sized random textures or periodic gratings (Haug and Ballif, 2015). None of these, however, has outperformed random pyramids. The parasitic absorption in metal nanoparticles tends to heavily outweigh any light-trapping gain, making the overall J_{sc} lower than for traditional textures. Diffraction at random, rough interfaces has been successfully used to approach the Lambertian limit in thin-film silicon solar cells, but small features are challenging to passivate. Diffraction at periodic gratings has, in the best case, just matched the light-trapping performance of random diffractive textures. In some instances, these light-trapping phenomena have been claimed to exceed the $4n^2$ limit. Recall, however, that $4n^2$ is the incorrect limit for comparison because the specific assumptions used in its derivation are rarely realized in experiments. Even geometric textures have been shown to exceed $4n^2$ path-length enhancement trapping in the case of normal incidence (Campbell and Green, 1987; Brendel, 1995), yet no experimental or simulated light-trapping scheme has surpassed this limit when the path-length enhancement is averaged over all angles of incidence.

3.7.6 Conclusion

This chapter prescribed a methodology for assessing light management in silicon solar cells and, in particular, determining the importance of the three J_{sc} loss categories (front-surface reflection, parasitic absorption, and light trapping). We conclude by employing this approach in an example: the silicon heterojunction solar cell from Figure 3.7.2. Multiplying by the AM1.5G spectrum and integrating the shaded areas in Figure 3.7.2 (a) over the 350–1200 nm wavelength range gives the J_{sc} losses listed in Figure 3.7.2. (The shading loss was determined by comparing the active-area J_{sc} and the aperture-area J_{sc}.) However, these values are not identical to the amount of current that could be *gained* by eliminating each loss, which is what we are actually interested in. For example, if front-surface reflection were eliminated, most of the newly transmitted light would result in collected electron–hole pairs, but some would be absorbed parasitically; if infrared parasitic absorption were removed, roughly half of the saved light would escape instead of being absorbed in the wafer. Calculating the potential gains, we find that front-surface reflection is the most severe loss (2.2 mA/cm^2 available from shading and 1.1 mA/cm^2 available from ARC reflection), parasitic absorption is nearly as important (1.5 mA/cm^2 available from interband absorption and 1.2 mA/cm^2 available from FCA), and light trapping is as good as Lambertian.

List of Symbols

Symbols	Description	Unit
α	Absorption coefficient in the wafer	cm^{-1}
a_{Front}	Single-pass absorbance in the layers on the front of the solar cell	Unitless or %
a_{Rear}	Single-pass absorbance in the layers on the rear of the solar cell	Unitless or %
a_{Si}	Single-pass absorbance in the wafer	Unitless or %
A_{Front_m}	Fraction of the incident light on the solar cell that is absorbed in layers on the front of the cell on the mth pass of light through the cell	Unitless or %
$A_{Parasitic}$	Fraction of the incident light on the solar cell that is absorbed not in the wafer	Unitless or %
A_{Rear_m}	Fraction of the incident light on the solar cell that is absorbed in layers on the rear of the cell on the mth pass of light through the cell	Unitless or %
A_{Si}	Fraction of the incident light on the solar cell that is absorbed in the wafer	Unitless or %
A_{Si_m}	Fraction of the incident light on the solar cell that is absorbed in the wafer on the mth pass of light through the cell	Unitless or %
d	Wafer thickness	cm
E_g	Bandgap energy of silicon	eV
EQE	External quantum efficiency	Unitless or %
EQE_c	External quantum efficiency in the absence of parasitic absorption	Unitless or %

(*Continued*)

(*Continued*)

Symbols	Description	Unit
h	Planck's constant	eV s
I_m	Light intensity at the beginning of the mth pass of light through the cell	W m^{-2}
J_{mpp}	Current density at maximum power point	mA cm^{-2}
J_{sc}	Short-circuit current density	mA cm^{-2}
λ	Wavelength	nm
μ	Mobility of free charge carriers	cm^{-3}
n	Refractive index	Unitless
N	Density of free charge carriers	cm^{-3}
r	Front-surface reflectance of the solar cell	Unitless or %
R	Total, measured reflectance of the solar cell	Unitless or %
R_c	Total reflectance of the solar cell in the absence of parasitic absorption	Unitless or %
R_{Escape}	Fraction of the incident light on the solar cell that enters the solar cell and subsequently escapes out the front surface	Unitless or %
$R_{Escape,c}$	Fraction of the incident light on the solar cell that would enter the solar cell and subsequently escape out the front surface in the absence of parasitic absorption	Unitless or %
R_0	Fraction of the incident light on the solar cell that is reflected at the front surface	Unitless or %
t_{Front}	Transmittance out of the front of the solar cell	Unitless or %
t_{Rear}	Transmittance out of the rear of the solar cell	Unitless or %
T	Total, measured transmittance of the solar cell	Unitless or %
T_m	Fraction of the incident light on the solar cell that is transmitted out of the cell on the mth pass of light through the cell	Unitless or %
θ	Average angle light travels through the wafer with respect to the wafer normal	Radians or degrees
v	Photon frequency	s^{-1}
V_{mpp}	Voltage at maximum power point	V
V_{oc}	Open-circuit voltage	V
y	Average path length enhancement of light during a single pass through the wafer	Unitless

List of Acronyms

AM1.5G	Reference global solar spectral irradiance for an air mass of 1.5
ARC	Antireflection Coating
FCA	Free-carrier Absorption
KOH	Potassium Hydroxide
TCO	Transparent Conductive Oxide

References

Baker-Finch, S.C. and McIntosh, K.R. (2011) Reflection of normally incident light from silicon solar cells with pyramidal texture. *Progress in Photovoltaics: Research and Applications*, **19** (4), 406–416.

Baker-Finch, S.C. and McIntosh, K.R. (2012) Reflection distributions of textured monocrystalline silicon: implications for silicon solar cells. *Progress in Photovoltaics: Research and Applications*, **21** (5), 960–971.

Baker-Finch, S.C., McIntosh, K.R., and Terry, M.L. (2012) Isotextured silicon solar cell analysis and modeling 1: Optics. *IEEE Journal of Photovoltaics*, **2** (4), 457–464.

Boccard, M., Battaglia, C., Haug, F.J. *et al.* (2012) Light trapping in solar cells: Analytical modeling. *Applied Physics Letters*, **101**, 151105.

Brendel, R. (1995) Coupling of light into mechanically textured silicon solar cells: a ray-tracing study. *Progress in Photovoltaics*, **3** (1), 25–38.

Campbell, P. and Green, M.A. (1987) Light trapping properties of pyramidally textured surfaces. *Journal of Applied Physics*, **62** (1), 243–249.

Deckman, H.W., Wronski, C.R., Witzke, H., and Yablonovitch, E. (1983) Optically enhanced amorphous silicon solar cells. *Applied Physics Letters*, **42** (11), 968–970.

Hannebauer, H., Dullweber, T., Falcon, T., and Brendel, R. (2013) Fineline printing options for high efficiencies and low Ag paste consumption. *Energy Procedia*, **38**, 725–731.

Haug, F.J. and Ballif, C. (2015) Light management in thin film silicon solar cells. *Energy and Environmental Science*, **8** (3), 824–837.

Holman, Z.C., Descoeudres, A., De Wolf, S., and Ballif, C. (2013a) Record infrared internal quantum efficiency in silicon heterojunction solar cells with dielectric/metal rear reflectors. *IEEE Journal of Photovoltaics*, **3** (4), 1243–1249.

Holman, Z.C., De Wolf, S., and Ballif, C. (2013b) Improving metal reflectors by suppressing surface plasmon polaritons: a priori calculation of the internal reflectance of a solar cell. *Light: Science and Applications*, **2**, e106.

McIntosh, K.R. and Baker-Finch, S.C. (2012) OPAL 2: Rapid optical simulation of silicon solar cells. *Proceedings of the 38th IEEE Photovoltaic Specialists Conference*, pp. 265–271.

Vogt, M.R. (2015) Numerical modeling of C-Si modules by coupling the semiconductor with the thermal conduction and radiation equations. Paper presented at the 5th International Conference on Crystalline Silicon Photovoltaics, Konstanz, Germany.

Yablonovitch, E. (1982) Statistics ray optics. *Journal of the Optical Society of America*, **72** (7), 899–907.

3.8

Numerical Simulation of Crystalline Silicon Solar Cells

Pietro Altermatt

State Key Laboratory of PV Science and Technology (PVST), Trina Solar, Changzhou, Jiangsu, China

3.8.1 Introduction

In the following, a brief outline is given to assist in deciding why to use numerical simulations, and which software is suitable. Also, a detailed example provides an insight into how numerical simulations may be performed.

3.8.2 Why Numerical Simulations?

Numerical modeling is defined in this chapter as solving the following system of semiconductor differential equations fully coupled and self-consistently:

$$\nabla \cdot (\varepsilon \nabla \Psi) = -q\left(p - n + N_{don}^{+} - N_{acc}^{-}\right)$$
$$\frac{\partial n}{\partial t} = \frac{1}{q}\nabla \cdot \vec{J}_n + G - R \qquad \frac{\partial p}{\partial t} = -\frac{1}{q}\nabla \cdot \vec{J}_p + G - R \tag{3.8.1}$$
$$\vec{J}_n = -q\mu_n n \nabla \Psi + qD_n \nabla n \qquad \vec{J}_p = -q\mu_p p \nabla \Psi - qD_p \nabla p$$

The free parameters are the electric potential Ψ, the electron density n, and the hole density p. The constants are the permittivity ε and the elementary charge q, while the remaining parameters are the density of the donor and acceptor ions N_{don}^{+} and N_{acc}^{-}, the electron and hole currents J, the electron and hole mobilities μ or diffusivities D, and finally the photo-generation and recombination rates G and R. For a more detailed explanation, see Part 2, "Basic functional principles of photovoltaics" and Chapter 3.2, "Silicon solar cell device structures."

Photovoltaic Solar Energy: From Fundamentals to Applications, First Edition.
Edited by Angèle Reinders, Pierre Verlinden, Wilfried van Sark, and Alexandre Freundlich.

Companion website: www.wiley.com/go/reinders/photovoltaic_solar_energy

Alternatively, analytical theories solve these equations only under specific conditions and with certain assumptions. For example, these equations may be de-coupled (e.g. without the Poisson equation in quasi-neutral regions), or inputs may be simplified (such as constant quasi-Fermi levels).

Simulations, whether numerical or analytical, are generally used (1) to analyze experiments; (2) to quantify losses in test samples or fabricated cells; (3) for parameter studies; and (4) to provide a road map for further experiments. In deciding whether to use an analytical theory or numerical modeling to fulfill one of these four tasks, the following considerations may help:

1. The device dynamics may easily be conveyed by an elegant set of analytical equations. Hence, analytical theories may lead to a straightforward *understanding* of effects, while numerical device simulations may merely *quantify* effects in various circumstances. Examples of the BSF are Godlewski *et al.* (1973) vs. Harder *et al.* (2004).
2. There are many analytical theories that describe the losses in Si solar cells. Examples of PERC cells are Cuevas (2013), Plagwitz *et al.* (2006), Saint-Cast *et al.* (2010), and Wolf *et al.* (2010). However, the precision of such models needs to be carefully tested. For example, the analytical theory of Wolf *et al.* (2010) was compared to numerical modeling by Kimmerle *et al.* (2012) and was found to be rather imprecise if intermediate-injection or even high-injection conditions prevail in the base region at the maximum power point (mpp). However, the base region is almost always under intermediate-injection conditions at mpp, because the optimum base resistivity is rather high in PERC cells. Hence, analytical models need either to be extended to account for injection effects, such as in Kimmerle *et al.* (2012); or one needs to resort to numerical device simulation, as in Rüdiger and Hermle (2012) to arrive at a precise loss analysis of PERC cells. Similarly, an analytical theory underestimated the resistive losses in the base region of PERC cells and caused an overestimation of the contact resistivity at the rear in Gatz *et al.* (2011), affecting the predicted optimum rear metallization, as shown by numerical simulations in Müller (2014).
3. Numerical device simulations are usually simpler to perform and more precise than analytical theories:
   - if there is additional recombination occurring via defects having a specific defect profile, such as inactive phosphorus in emitters, Al-O complexes in the Al-BSF, and contamination under metal contacts, as is the case in most standard, mass-produced Si solar cells;
   - if there are non-homogeneities other than those mentioned above, for example, the edge of small test samples, grains and grain boundaries in mc-Si material, lateral non-homogeneity of the emitter, etc.;
   - if there is an interplay between various effects, as, for example, in shunt currents occurring near a contact of a SiN_x-passivated rear surface of a p-type cell. There, the electrostatics of the interface are strongly coupled to the high recombination rates at the metal contacts, which causes a gradient in the quasi-Fermi levels.

3.8.3 Commonly Used Software for the Numerical Simulation of Si Solar Cells

Table 3.8.1 gives an overview of the most commonly used software for the numerical modeling of crystalline Si solar cells.

Table 3.8.1 Simulation software most commonly used for simulating crystalline Si solar cells

Criteria	Software	Strengths	Weaknesses
User-friendly and simple	PC1D (Clugston and Basore, 1997)	Easy to use, yet powerful; visual output well suited to solar cells; high educational value; well accepted and understood in the PV community.	One-dimensional; no full electrostatics (no MIS); Boltzmann statistics;[a] outdated software.
Most detailed	Sentaurus (Synopsis Inc., 2014)	Vast range of models (which may be implemented by users as plug-ins); one- to three-dimensional; interface to other tools (e.g. process, optical, and statistical simulations); detailed input; the standard in the IC industry.	Learning curve takes effort; complex software; a multitude of models need to be mutually consolidated; output may need to be processed.
Most detailed	Atlas (Silvaco Inc., 2014)	Similar to Sentaurus	Similar to Sentaurus
Midway between the simple PC1D and the complex Sentaurus or Atlas software.	Quokka (Fell, 2013)	Rather simple software because the diffused regions are included as conductive boundaries (CBs), so 2D and 3D problems are solved very quickly; (measured) J_0 values serve as input; good visual output that is well suited to solar cells.	CBs make it difficult to quantify photo-generation and collection efficiency in the front emitter; e.g. resistive losses in the space-charge region of the p-n junction are ignored; the base is assumed to be quasi-neutral in all circumstances.

[a] Fermi-Dirac statistics are included in a derivate by Haug *et al.* (2014).

Which software is most suitable depends on the simulation task:

1. PC1D does a perfect job in one-dimensional problems, where the dopant density does not exceed 10^{19} cm^{-3}, and where no metal-insulator-semiconductor (MIS) structures exist. An example of this is the modeling of the lifetime measurements of wafers by means of photoconductance decay (Cuevas and Sinton, 1997). More sophisticated software is necessary if the sample's edge affects the measurements (Kessler *et al.*, 2012), or if surface passivation leads to lifetime degradation below the surface (Steingrube *et al.*, 2010). At dopant densities higher than 10^{19} cm^{-3}, Boltzmann statistics may be compensated by effective band gap narrowing, but not fully consistently (Altermatt *et al.*, 2002).
2. Sentaurus or Atlas is suitable when a quantitative base needs to be provided for the industry to decide, for example, which fabrication tools to buy and which processes to develop for mass production. Although solar cells have a rather simple structure and are fabricated as simply as possible, their behavior is clearly complex. To help the industry effectively, this complexity should be embraced in the simulation tools, as has been successfully done in the IC industry for decades.
3. Software like Quokka (Brendel, 2012; Fell, 2013) has medium-level complexity. It is not as simple or restricted as PC1D, because it is both two- and three-dimensional. On the other hand, it does not require as detailed an input as Sentaurus or Atlas. For example, the diffused regions do not require dopant profiles and surface recombination velocities, but

rather J_0 values that are either measured or calculated with EDNA (McIntosh and Altermatt, 2010). Because the diffused regions are included as conductive boundaries, 3D problems are solved very quickly (Fell, 2013). Such software is very well suited for a quick estimate and analysis in an R&D environment, but its predictive power may be weaker than that of the more detailed Sentaurus and Atlas models.

An example follows of how to choose the optimum software. Assume that you have collected a detailed set of inputs, such as dopant profiles, the bulk lifetime derived from test samples, etc. but Sentaurus or Atlas does not reproduce the measured IV parameters with satisfactory precision. You decide to use Quokka, so you measure the J_0 values from test samples, and with Quokka you find precise agreement with the measured IV parameters. However, which software is the better choice depends on what you aim to achieve with your simulations. Obviously, parasitic losses occur, because the initial Sentaurus or Atlas simulations do not deliver satisfactory results. If you want to pinpoint these parasitic losses, Sentaurus or Atlas gives you the means to do so because, in Sentaurus and Atlas, J_0 is a result of dopant profiles, surface recombination, etc., and therefore gives you a more detailed insight into what causes a higher than expected J_0. However, if you are not primarily interested in reducing these losses, but instead prefer to optimize the geometry of the contacts and the existing local diffusions, Quokka is the easier choice, because the optimum geometry depends on J_0, and the measured J_0 values serve as an input to Quokka, regardless of which losses cause these J_0 values. Note that most common measurement procedures tend to underestimate J_0, see Mäckel and Varner (2013) and Min *et al.* (2014), and may limit the precision of the Quokka simulations.

3.8.4 General Simulation Approach

Figure 3.8.1 illustrates the general simulation approach. Various measurement techniques provide input to various aspects of modeling (indicated as boxes), which may then be combined to provide a multitude of outputs.

In concrete terms, the usual document scanners are suitable for measuring the device geometry, while light and electron microscopy imaging techniques may show more details, such as the front metal finger profiles and the geometry of local Al-BSFs. Part of these geometries may be an important input for either ray tracing to calculate the optical generation profile or, in the case of local Al-BSFs, for modeling the recombination and resistive losses. The use of ECV and SIMS dopant profiles is discussed in the example below. Injection dependent lifetime measurements of the wafer material are an important input, because underestimated or overestimated losses at the front or back may easily be compensated with an unknown base lifetime. To obtain the lifetime after processing, a cell is usually etched back at both sides and re-passivated, optimally with Al_2O_3 or SiO_2, because SiN_x may dominate the recombination of the samples at injection densities near 10^{14}cm^{-3}, which prevail near mpp in the cell. If specific defects are limiting the lifetimes, CV and DLTS measurements may reveal the defect distribution and capture cross-sections, e.g., at passivated surfaces.

So far, the input has been device-specific. A model also has general material properties as input, such as the band structure of Si, including band gap narrowing, intrinsic recombination, transport properties, etc. An overview of consistent models is given in Altermatt (2011). Last but not least, the boundary conditions are needed to solve the semiconductor equations, such as ohmic or Schottky contacts, charges at interfaces, tunneling, etc.

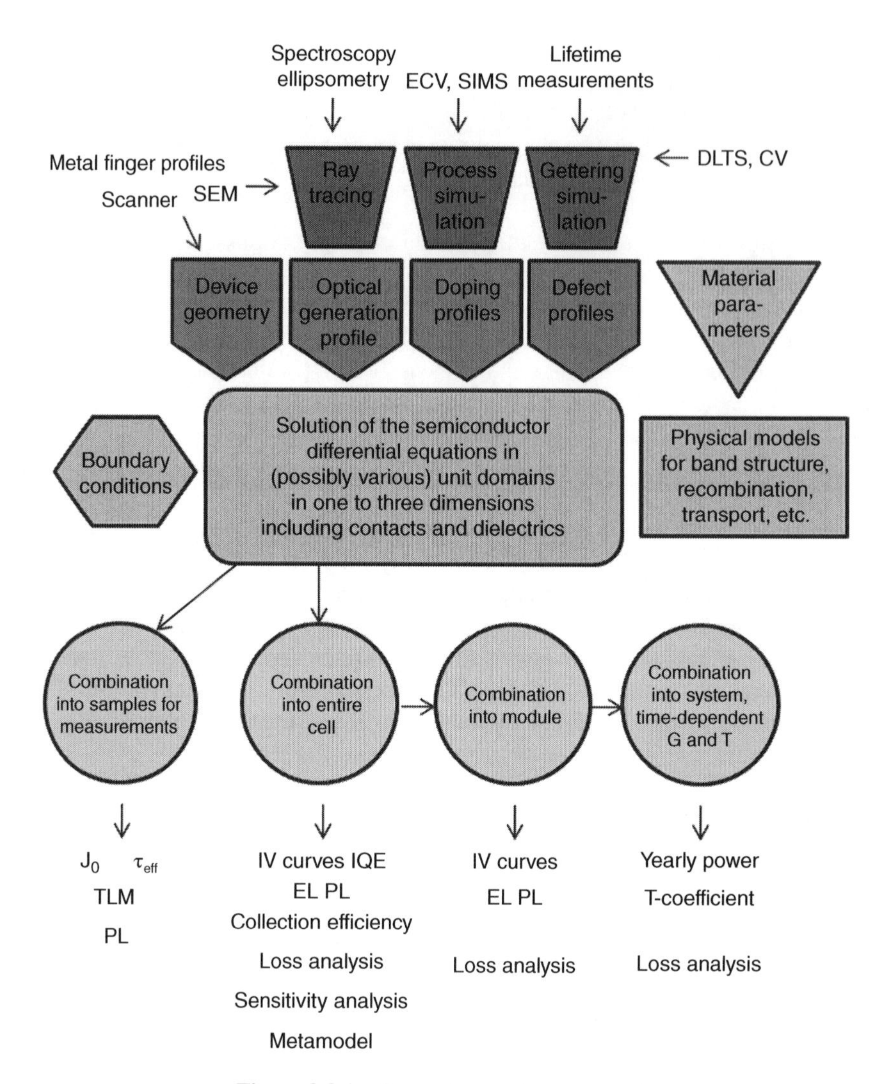

Figure 3.8.1 General simulation approach

Solving the fully coupled set of semiconductor equations self-consistently is numerically demanding. The finite-difference method has been applied most successfully in two and three dimensions, such as in Sentaurus and Atlas. The finite-element or finite-volume method requires particular basis functions, especially in two and three dimensions (Angermann and Wang, 2003; Comsol Multiphysics, 2014). This is one of the reasons why PC1D, a finite-element code, is difficult to expand to two dimensions. Quokka does not solve the p-n junction space charge regions, where the carrier densities change rather abruptly, and is therefore numerically less demanding.

Finally, the obtained solution usually needs some post-processing and combination of output values to serve a specific goal. Some examples are: the extraction of the effective lifetime or J_0 from simulations of test samples (Mäckel and Varner, 2013); the derivation of the IQE from the simulation of collection efficiency and absorbance; the simulation of IV curves at three different yet similar illumination intensities for the extraction of R_s (Fong *et al.*, 2013), etc.

3.8.5 Detailed Numerical Simulation of an mc-Si Solar Cell

In the following example from Fell (2014), industrial mc-Si solar cells are simulated such that most major optical, recombination and resistance losses are included. In this way, a variation of these losses with parameter changes is described as realistically as possible. This gives the model a high degree of predictive power, not only for the average IV parameters, but also for the distribution of the IV parameters in mass production and for designing roadmaps for further cell development.

The optical generation and parasitic losses are calculated with standard ray tracing (Greulich, unpublished work, 2015). For the emitter, a measured ECV profile for the active dopants, and a SIMS profile for the total phosphorus density are used from Dastgheib-Shirazi *et al.* (2013). The difference between SIMS and ECV is assigned to inactive phosphorus. The following recombination models are chosen:

- SRH recombination at the front surface in dependence of the total phosphorus concentration according to Min *et al.* (2014).
- SRH recombination via the inactive phosphorus in the emitter according to Min *et al.* (2014).
- SRH recombination at the contacts with thermal velocities.
- SRH recombination of the wafer material with symmetrical lifetime parameters (FeB-limited).
- Auger recombination (Richter *et al.*, 2012) and radiative recombination (Altermatt, 2011).

Iso-texturing increases the surface area and, hence, the J_0 of the emitter by a factor of about 1.2. This means that, in this simulation example, J_0 needs to be increased from 167 to 200 fA/cm^2. This increase in J_0 cannot be mimicked in the planar simulation domain simply by increasing the S_p of the front surface and the σ_p of the inactive phosphorus by a factor of 1.2, because the limited carrier collection efficiency causes a non-linear dependency between these input parameters and the resulting J_0. In many industrial emitters, these two parameters must be increased by a factor of about 1.35. In our example, the multiplication factor turned out to be 1.73 (which incidentally is the surface enhancement factor of KOH texturing), resulting in $S_p = 1.64 \times 10^7$ cm/s and $\sigma_p = 1.3 \times 10^{-17}$ cm^2. This slightly underestimates the IQE in the UV and the blue parts of the spectrum. However, due to the above-mentioned nonlinearities, it is not possible to represent a textured surface correctly with a planar simulation domain, and some compromises must be expected.

The base is assumed to be homogeneous (2 Ωcm) with no grain boundaries, and with homogeneous SRH lifetime parameters of $\tau_n = \tau_p = 75\,\mu$s (FeB-limited), i.e. neglecting injection dependencies. The approximation of homogeneous lifetime was investigated in Wagner *et al.* (2013). The Auger recombination and the radiative recombination are again taken from Richter *et al.* (2012) and Altermatt (2011).

For the Al-BSF, a measured ECV profile is used from Bock *et al.* (2010), and is etched so that its depth is approximately 7 um. The following effects are taken into account:

- SRH recombination via the Al-O complexes using the model of Rüdiger *et al.* (2011) and the defect parameters of Rosenits (2011).
- The assumption that a fraction of only 6×10^{-4} Al-O defects is created compared to the Cz crystallization process of an Al-doped ingot (Altermatt, 2011).

- Incomplete ionization, neglecting its weakening of band gap narrowing (Steinkemper *et al.*, 2015).
- Again, the Auger recombination (Richter *et al.*, 2012) and the radiative recombination (Altermatt, 2011).
- SRH recombination at the contacts with thermal velocities, neglecting the silver islands and silver crystallites.

The major recombination losses in this Al-BSF region are SRH both in the bulk (due to the Al-O complexes) and at the contact. Depending on the depth of the dopant profile, only one of the two may dominate (Chen, 2012).

The resistive losses in bulk silicon are taken into account with the simulated carrier densities, currents and Klaassen's mobility model (Altermatt, 2011). The contact resistivities chosen are 2 m Ωcm^2 at the screen-printed Ag metal fingers at the front, and 5 m Ωcm^2 at the full-area screen-printed Al contact at the rear. The resistive losses in the metallization are calculated using a spice model. The 75 uniform front fingers have a resistivity of 4.5 $\mu\,\Omega$cm. The geometrical parameters of the front metallization vary greatly among manufacturers. The rear Al layer is 30 μm thick and has a resistivity of 35 $\mu\,\Omega$cm. Ag pads are neglected. The resulting lumped series resistance at MPP is extracted from three IV curves at different illumination intensities (Fong *et al.*, 2013) and is 0.73 Ωcm^2, with a contribution from the silicon and the silicon/metal contacts of 0.36 Ωcm^2.

The cell is well optimized: Figure 3.8.2 shows that the recombination currents of all three major cell regions have similar magnitudes at mpp. Note that recombination in the emitter has a different dependency on external bias compared to both the base and the BSF, contributing to a reduction of the fill factor and the pseudo fill factor. This example shows that the fill factor is influenced not only by R_s, but also by recombination-rate-saturation mechanisms such as, in this case, lateral potential drops along the emitter. A careful distinction between resistive and recombination losses is particularly helpful in the development of PERC cells, where the emitter has an increased sheet resistivity.

Also note that the recombination rates are dominated by SRH recombination throughout the whole device: via inactive phosphorus in the emitter; via B-O complexes, crystal defects, Fe or other contaminants in the base; and via Al-O complexes in the Al-BSF.

A more detailed model may include recombination losses in the perimeter and busbar regions, contamination due to screen printing underneath the front metal fingers, grain boundaries and variations of bulk lifetime in the various grains, as well as inhomogeneity in the front metal fingers. To design roadmaps and to calculate the efficiency distribution in mass production, the metamodeling approach of Müller *et al.* (2013) may be used.

3.8.6 Conclusion

While analytical theories may lead to a straightforward understanding of device dynamics, numerical modeling quantifies device behavior in various circumstances with a minimal set of assumptions and approximations. Among the commonly used software for simulating crystalline Si solar cells, PC1D is suitable for one-dimensional problems (care needs to be taken with MIS structures and regions with dopant densities above $10^{19}\,cm^{-3}$), while Sentaurus or Atlas (which require effort to become acquainted with) is suitable for most detailed models

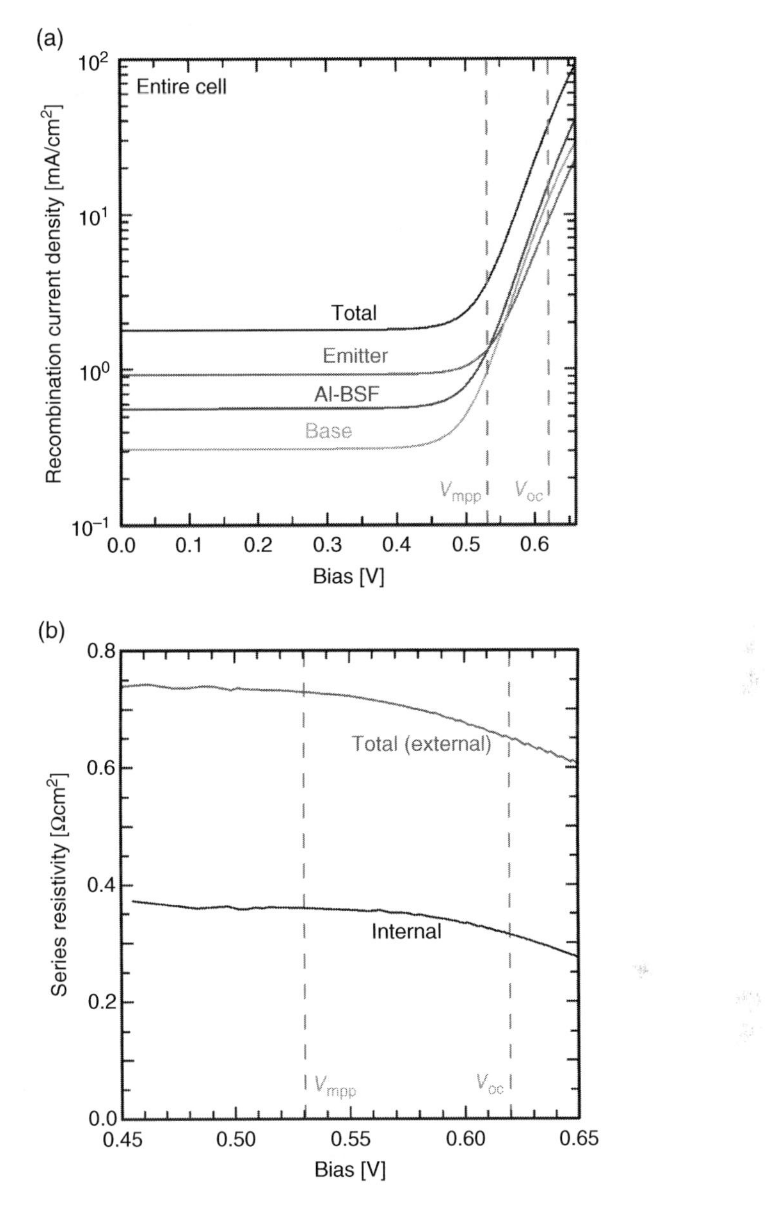

Figure 3.8.2 (a) Simulated recombination currents; and (b) simulated series resistivity of an mc-Si solar cell, under one-sun standard conditions. (*See insert for color representation of the figure*)

and input. Quokka (which models the diffused regions as conductive boundaries, with accompanying approximations) is suitable for cases that fall between these scenarios. A detailed simulation example of a mc-Si cell is also given, with reference to the most recent literature.

References

Altermatt, P.P. (2011) Models for numerical device simulations of crystalline silicon solar cells—a review. *Journal of Computational Electronics*, **10**, 314–330.

Altermatt, P P., Schumacher, J.O., Cuevas, A. *et al.* (2002) Numerical modeling of highly doped Si:P emitters based on Fermi-Dirac statistics and self-consistent material parameters. *Journal of Applied Physics*, **92**, 3187–3197.

Angermann, L. and Wang, S. (2003) Three-dimensional exponentially fitted conforming tetrahedral finite elements for the semiconductor continuity equations. *Applied Numerical Mathematics*, **46**, 19–43.

Bock, R., Altermatt, P.P., Schmidt, J., and Brendel, R. (2010) Formation of aluminum-oxygen complexes in highly aluminum-doped silicon. *Semiconductor Science and Technology*, **25**, 105007.

Brendel, R. (2012) Modeling solar cells with the dopant-diffused layers treated as conductive boundaries. *Progress in Photovoltaics* **20**, 31–43.

Clugston, D.A. and Basore, P.A. (1997) PC1D version 5: 32-bit solar cell modelling on personal computers. *26th IEEE Photovoltaic Specialists Conference*, 207–210.

Comsol Multiphysics (2014) *Semiconductor Module User Guide*. Stockholm, Sweden.

Cuevas, A. (2013) Physical model of back line-contact front-junction solar cells. *Journal of Applied Physics*, **113**, 164502.

Cuevas, A. and Sinton, R.A. (1997) Prediction of the open-circuit voltage of solar cells from the steady-state photoconductance. *Progress in Photovoltaics*, **5**, 79–90.

Dastgheib-Shirazi, A., Steyer, M., Micard, G. *et al.* (2013) Relationships between diffusion parameters and phosphorus precipitation during the $POCl_3$ diffusion process. *Energy Procedia*, **38**, 254–262.

Fell, A. (2013) A free and fast three-dimensional/two-dimensional solar cell simulator featuring conductive boundary and quasi-neutrality approximations. *IEEE Transactions on Electron Devices*, **60**, 733–738.

Fell, A., McIntosh, K.R., Altermatt, P.P. *et al.* (2014) Input parameters for the simulation of silicon solar cells in 2014. *Digest of the 6th World Conference on Photovoltaic Energy Conversion (WCPEC-6)*, Kyoto, November 2014.

Fong, K.C., McIntosh, K.R., and Blakers, A.W. (2013) Accurate series resistance measurement of solar cells. *Progress in Photovoltaics*, **21**, 490–499.

Gatz, S., Dullweber, T., and Brendel, R. (2011) Evaluation of series resistance losses in screen-printed solar cells with local rear contacts. *IEEE Journal of Photovoltaics*, **1**, 37–42.

Godlewski, M.P., Baraona, C.R., and Brandhorst, H.W. (1973) Low-high junction theory applied to solar cells. *10th IEEE Photovoltaic Specialists Conference*, 40–49.

Greulich, J. (2015) Unpublished results.

Harder, N.P., Altermatt, P.P., Cuevas, A., and Heiser, G. (2004) Numerical modeling of rear junction Si solar cells using Fermi-Dirac statistics. *19th European Photovoltaic Solar Energy Conference*, 828–831.

Haug, H., Kimmerle, A., Greulich, J. *et al.* (2014) Implementation of Fermi-Dirac statistics and advanced models in PC1D for precise simulations of silicon solar cells. *Solar Energy Materials and Solar Cells*, **131**, 30–36.

Kessler, M., Ohrdes, T., Altermatt, P.P., and Brendel, R. (2012) The effect of sample edge recombination on the averaged injection-dependent carrier lifetime in silicon. *Journal of Applied Physics*, **111**, 054508.

Kimmerle, A., Rüdiger, M., Wolf, A. *et al.* (2012) Validation of analytical modelling of locally contacted solar cells by numerical simulations. *Energy Procedia*, **27**, 219–226.

Mäckel, H. and Varner, K. (2013) On the determination of the emitter saturation current density from lifetime measurements of silicon devices. *Progress in Photovoltaics*, **21**, 850–866.

McIntosh, K.R. and Altermatt, P.P. (2010) A freeware 1D emitter model for silicon solar cells. *35th IEEE Photovoltaic Specialists Conference*, pp. 2188–2193.

Min, B., Dastgheib-Shirazi, A., Altermatt, P.P., and Kurz, H. (2014) Accurate determination of the emitter saturation current density for industrial P-diffused emitters. *29th European Photovoltaic Solar Energy Conference*, Amsterdam, pp. 463–466.

Min, B., Wagner, H., Dastgheib-Shirazi, A. *et al.* (2014) Heavily doped Si:P emitters of crystalline Si solar cells: recombination due to phosphorus precipitation. *Physica Status Solidi RRL*, **8**, 680–684.

Müller, M. (2014) Sensitivity of solar cells. PhD thesis, Institute of Solid-State Physics, Leibniz, University of Hannover, Germany.

Müller, M., Altermatt, P.P., Wagner, H., and Fischer, G. (2013) Sensitivity analysis of industrial multicrystalline PERC silicon solar cells by means of 3-D device simulation and metamodeling. *IEEE Journal of Photovoltaics*, **4**, 107–113.

Plagwitz, H. and Brendel, R. (2006) Analytical model for the diode saturation current of point-contacted solar cells. *Progress in Photovoltaics*, **14**, 1–12.

Richter, A., Glunz, S.W., Weener, F. *et al.* (2012) Improved quantitative description of Auger recombination in crystalline silicon. *Physical Review B*, **86**, 165202.

Rosenits,P., Roth, T., and Glunz, S.W. (2011) Erratum on "Determining the defect parameters of the deep aluminum-related defect center in silicon" [Appl. Phys. Lett. 91, 122109 (2007)]. *Journal of Applied Physics* **99**, 239904.

Rüdiger, M. and Hermle, M. (2012) Numerical analysis of locally contacted rear surface passivated silicon solar cells. *Japanese Journal of Applied Physics*, **51**, 10NA07.

Rüdiger, M., Rauer, M., Schmiga, C., and Hermle, M. (2011) Effect of incomplete ionization for the description of highly aluminum-doped silicon. *Journal of Applied Physics*, **110**, 024508.

Saint-Cast, P., Rüdiger, M., Wolf, A. *et al.* (2010) Advanced analytical model for the effective recombination velocity of locally contacted surfaces. *Journal of Applied Physics*, **108**, 013705.

Silvaco Inc. (2014) *Atlas User Guide*. Santa Clara, CA.

Steingrube, S., Altermatt, P.P., Steingrube, D.S., Schmidt, J., and Brendel, R. (2010) Interpretation of recombination at c-Si/SiN_x interfaces by surface damage. *Journal of Applied Physics*, **108**, 014506.

Steinkemper, H., Rauer, M., Altermatt, P.P. *et al.* (2015) Adapted parameterization of incomplete ionization in aluminum-doped silicon and impact on numerical device simulation. *Journal of Applied Physics* (to appear).

Synopsys Inc. (2014) *Sentaurus User Guide*. Mountain View, CA.

Wagner, H., Müller, M., Fischer, G., and Altermatt, P.P. (2013) A simple criterion for predicting multicrystalline Si solar cell performance from lifetime images of wafers prior to cell production. *Journal of Applied Physics*, **114**, 054501.

Wolf, A., Biro, D., Nekarda, J.F. *et al.* (2010) Comprehensive analytical model for locally contacted rear surface passivated solar cells. *Journal of Applied Physics*, **108**, 124510.

3.9

Advanced Concepts

Martin Green
Australian Centre for Advanced Photovoltaics, School of Photovoltaic and Renewable Energy Engineering, University of New South Wales, Sydney, Australia

3.9.1 Introduction

From the mid-1980s until recently, silicon solar cell manufacturing was almost completely dominated by the screen-printed, aluminium "Back Surface Field" (Al-BSF) processes applied to p-type mono-crystalline and multi-crystalline wafers, described in Parts 3 and 10. Several alternative silicon device structures giving improved efficiency have been developed during this period as described in Chapters 3.3 to 3.6 with some of these now finding their way into production.

A key attraction of these alternatives is improved energy conversion efficiency. Over the last decade, the efficiency of the average silicon solar module has improved by about 0.3% absolute/year (2% relative/year), providing an increasingly important contribution to the reduction in module manufacturing costs demonstrated over this period. This has been by leveraging the costs of all materials used in module fabrication, such as glass cover sheets, encapsulants, junction boxes, interconnects and the like. Additional leveraging occurs when the modules are put to use, by reducing transport and installation costs, including the costs of site preparation, mounting structures and wiring. Improved efficiency will therefore always be an attractive option for the industry, provided it can be obtained without adding excessively to cell processing costs.

Apart from increased efficiency, other potentially attractive features offered by some advanced technologies are bifacial response, whereby the solar module responds to light incident on both its front and rear. A second is the ability to contact both cell polarities from the cell rear. When these additional desirable attributes are taken into account, this makes the four advanced technologies of Figure 3.9.1 of near-term interest.

Photovoltaic Solar Energy: From Fundamentals to Applications, First Edition.
Edited by Angèle Reinders, Pierre Verlinden, Wilfried van Sark, and Alexandre Freundlich.

Companion website: www.wiley.com/go/reinders/photovoltaic_solar_energy

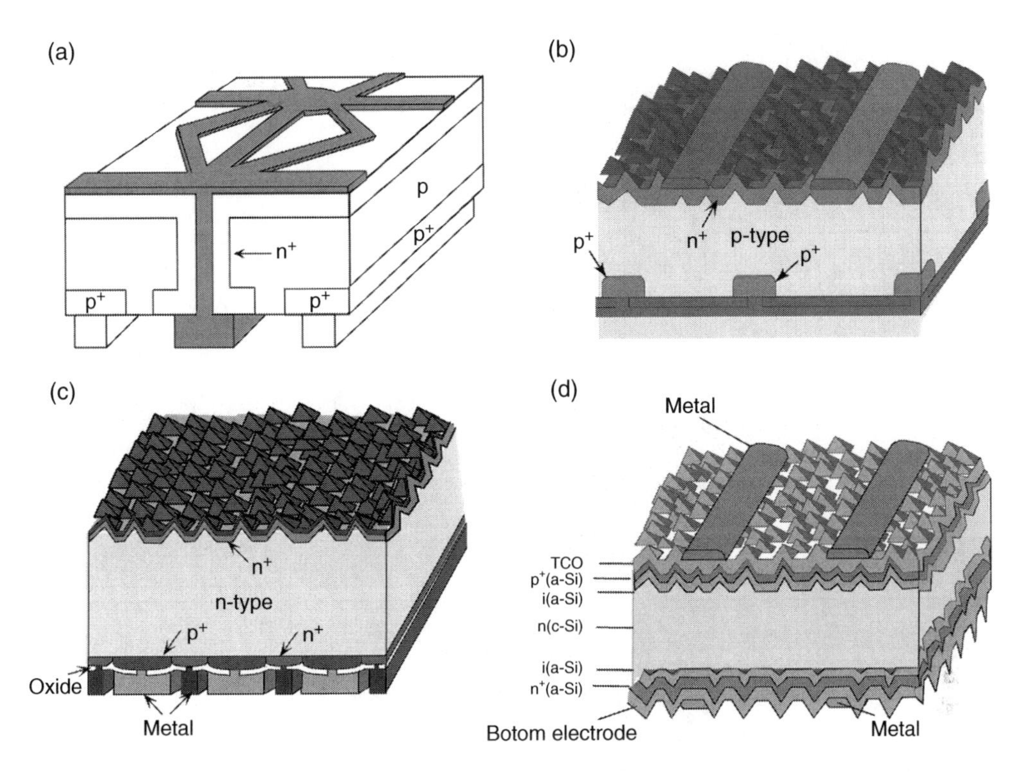

Figure 3.9.1 Four advanced silicon cell technologies: (a) Metal Wrap Through (MWT); (b) Passivated Emitter and Rear Cell (PERC); (c) Interdigitated Back Junction (IBJ); (d) Heterojunction Cell (HJT). *(See insert for color representation of the figure)*

3.9.2 Near-Term Advanced Options

The "Metal Wrap Through" (MWT) cell shown in Figure 3.9.1 (a) has the advantages of reduced coverage of the cell's top surface by metal and the potential for improved contacting, by bringing both polarity contacts to the cell rear (Walter *et al.*, 2013). This can simplify module assembly and reduce stresses on the cell in operation, improving reliability. MWT is a generic approach that can also be used in conjunction with both multi-crystalline and mono-crystalline wafers of either polarity and with different cell technologies, including the advanced Passivated Emitter and Rear (PERC) and Heterojunction (HJT) technologies of Figure 3.9.1. Although MWT cells have been used in a limited number of commercial installations and have been demonstrated in pilot production by a number of manufacturers, market penetration is small and may remain so while the other approaches are still developing. Competing approaches such as the use of metal meshes for the top cell contact may also decrease interest in this approach.

The PERC cell (Green, 2015) has been the most rapidly adopted of the advanced technologies over recent years with this trend expected to continue as shown in Figure 3.9.2 (a). By 2020, PERC is expected to become the solar cell technology manufactured in the largest volume as shown in Figure 3.9.2 (b).

Note that Figure 3.9.2 (a) and Figure 3.9.2 (b) are from different sources. Although both agree on the prime role to be played by PERC technology over the coming decade, there is

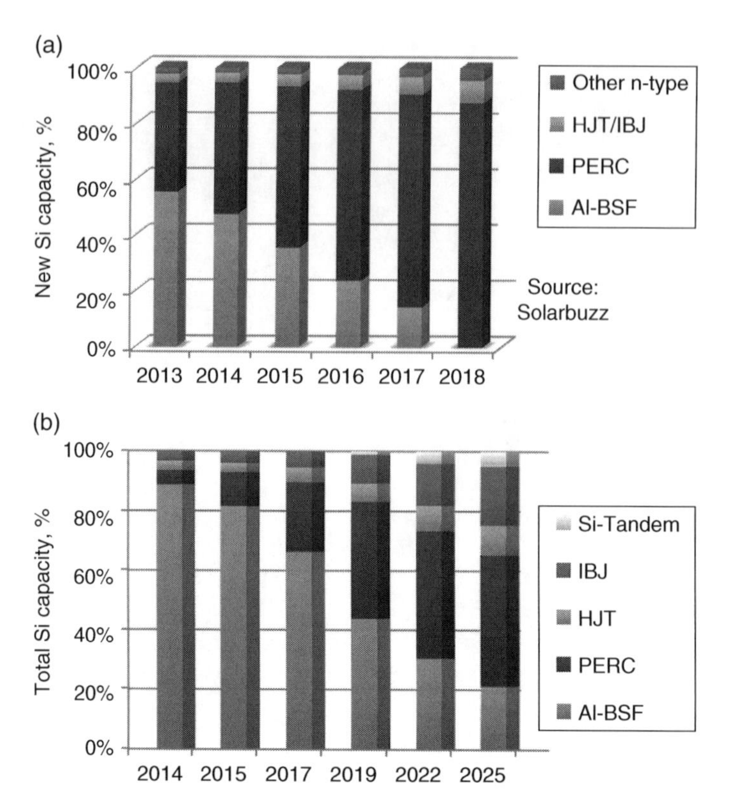

Figure 3.9.2 (a) Share of new silicon-based manufacturing capacity from different cell approaches. Source: Solarbuzz (2014); (b) Expected total market share of different silicon cell technologies. Source: ITRPV (2015)

some disagreement on the adoption rates of other advanced technologies, specifically of the IBJ and HJT technologies of Figure 3.9.1. This can be traced to the different groups providing input to the two graphs. An industrial consulting group prepared the first, while the second represents the consensual view of a cross-section of both industry and academic researchers. The latter may be more optimistic about introduction rates of new technologies, since they are not always as familiar with the commercial hurdles opposing such introduction.

PERC technology is suitable for multicrystalline and monocrystalline wafers of either polarity and is capable of high efficiency on good quality wafers, being the first silicon technology to achieve 25% efficiency on such wafers. Key to its rapid adoption has been the minimal changes required from the previously standard "Al BSF" technology, with production lines able to be upgraded to PERC simply by the addition of the extra equipment required for the additional processing of the cell rear (Green, 2015).

The strengths of the IBJ and HJT approaches have been discussed in earlier chapters. As well as a high efficiency potential, with champion cells also demonstrating 25% efficiency, up to 25.6% when in combination, both have additional advantages. The IBJ cell automatically has both contacts on the rear allowing simplified cell interconnection, producing modules more tolerant of environmental stress. On the other hand, the HJT approach, because of its structural symmetry, is well suited to bifacial operation, allowing an energy yield up to 20% higher than standard modules when mounted in open-back configurations (Chieng and Green, 1993).

Offsetting these considerable advantages is the need to use high quality, monocrystalline wafers to obtain the desired efficiency advantage. For IBJ cells, this is because most photogeneration occurs near the top surface, requiring carrier diffusion across the wafer for their photocurrent contribution. Diffusion lengths more than five times larger than the wafer thickness are required for less than 2% loss of such carriers. For the HJT cell, high quality wafers are required to get the open-circuit voltage advantage producing high efficiency. On moderate or low quality wafers where this voltage is limited by bulk wafer properties rather than surface properties, performance would be lower than with alternative approaches. This is due to reduced current output, caused by extra parasitic absorption in the amorphous silicon and conducting oxide layers involved, and lower fill factors, due to the increased difficulty in making low resistance cell contact due to the low allowable processing temperatures in the final stages of cell processing.

Notwithstanding these challenges, two companies, SunPower and Panasonic, have established strong market positions as the sole commercial proponents to date of IBJ and HJT technology, respectively. The high efficiency monocrystalline modules produced by these companies command a market premium. According to SPV Market Research, in 2014, the average selling price (ASP) of such premium modules increased to US$1.64/Watt, 2–2.5 times higher than the market average ASP of US$0.71/Watt (Mints, 2015). Although this price premium would compensate for the additional wafer and processing costs associated with these technologies, it is a premium that can be accommodated in niche applications, rather than by the market as a whole. This makes the high penetration rates projected in Figure 3.9.2 (b) for these technologies unlikely in practice. This would require: (1) a reversal in the 25-year trend towards an increasing market share of low cost p-type multicrystalline wafers; (2) large investments in new production capacity for high-lifetime, n-type monocrystalline wafers and a significant reduction in the costs of such wafers; and (3) a much larger investment in new IBJ and HJT cell processing lines than anticipated in Figure 3.9.2 (a).

Instead, it is believed that in the near to medium term, IBJ and HJT technologies will continue to produce premium products for the high-performance end of the market, maintaining a combined market share similar to historical values. The performance margin over mainstream production will diminish with time as the full potential of PERC processing is realized routinely in production. In the longer term, more disruptive technology may influence the way the industry evolves.

3.9.3 Longer-Term Advanced Options

At present, the dominant commercial position of silicon cell technology seems likely to be maintained well into the future, with no obvious threat to this position identified. The combined market share of alternative thin film options has been declining over the past 5 years, with this trend appearing likely to continue.

At present, silicon has a clear energy conversion efficiency advantage over these alternatives. A potentially disruptive development would be the reversal of this situation, probably only possible in the near term if a tandem stacked cell version of one of the present thin-film

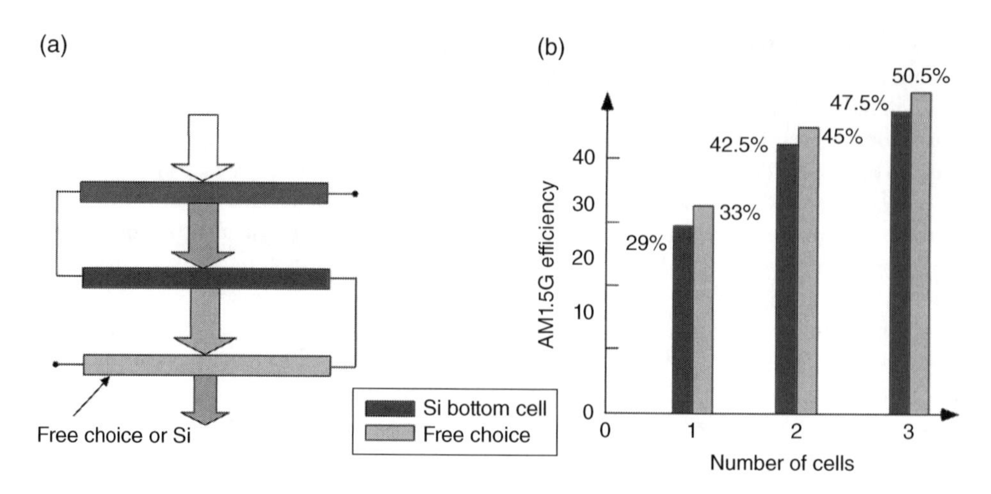

Figure 3.9.3 Limiting energy conversion efficiency under the AM1.5G spectrum for tandem stacks both with silicon as the bottom cell and with an unrestricted choice of bottom cell material

contenders were developed. Alternatively, if a tandem cell stack based on silicon were developed first, this would consolidate silicon technology into a position from which it would be difficult to dislodge it.

The theoretical efficiency limit on a silicon cell by itself is 29.4% (Richter *et al.*, 2013). The best laboratory devices with PERC, IBJ and HJT technologies all lie in the 25–26% range, quite close to this theoretical limit suggesting there is little scope for further gains. However, if a cell of higher bandgap is stacked on top of the silicon cell and incoming photons are shared allowing them to be connected in series, the limiting efficiency increases to 42.5%, a 45–50% relative increase (Figure 3.9.3). If the corresponding increase could be obtained experimentally, this would take experimental efficiencies to above 35%.

Adding more cells increases the limiting efficiency, to 47.5% for a 3-cell stack (Figure 3.9.3), and above 50% for a 4-cell stack. Adding even more cells gives diminishing returns, although each additional cell increases the limiting efficiency marginally, approaching a fundamental limit for an infinite stack of 59.1%.

If the tandem cell is implemented on silicon as a thin-film, this may not add significantly to cell processing costs. Standard "Al BSF" cells use a Si_3N_4 antireflection coating, deposited by plasma-enhanced chemical vapor deposition (PECVD), a more sophisticated deposition approach than required for some thin film cells. HJT cells have four separate layers deposited by PECVD plus two layers of conducting oxide, all absorbed within the cell-processing budget. Adding additional cells on top of silicon will add additional processing costs but, since thin layers would be required, these costs are amenable to reduction by increased production volumes. Hence, the increased efficiency is likely to significantly reduce costs per Watt, particularly as a mature technology. Moreover, increased cell numbers in the stack could be added incrementally, making this a viable approach to prolonged ongoing efficiency improvement.

The challenge is in finding thin-film cells that work well on silicon. Since the efficiency improvement in a tandem stack comes from the higher voltage of the uppermost cells, these cells need to give a good voltage output relative to their bandgap. This requirement can be quantified in terms of the cell's external radiative efficiency, quantified elsewhere (Green, 2012). Cells with radiative efficiency comparable to or higher than silicon are required.

The benefits of cell stacking have been best demonstrated in the III-V semiconductor alloy system, where high efficiency 3-cell stacks are routinely used for high value applications on spacecraft and in concentrating photovoltaic systems. For high radiative efficiency, III-V cells need to be single crystalline. Although single crystalline wafers are widely used in silicon photovoltaics, silicon atoms are more tightly packed than in the III-V materials of interest. The challenge in producing efficient III-V cell stacks on silicon is to find a way of overcoming this atomic packing mismatch.

Other thin-film technologies have displayed high radiative efficiency in polycrystalline form, avoiding this issue. The chalcogenide semiconductors, notably CIGS and CdTe, have the required radiative efficiency, but have too low a bandgap for use with silicon and also involve toxic and scarce elements. A closely related material Cu_2ZnSnS_4 (CZTS) avoids the latter problem and belongs to a large alloy system that should allow the required bandgap range to be accommodated. However, radiative efficiencies to date have been low, although steadily improving.

A recently emerging prospect is to stack cells based on organic-inorganic halide perovskites (Green *et al.*, 2014) on silicon. Perovskites have recently joined the select group of materials producing cells of over 20% efficiency and furthermore have high radiative efficiency. The most successful cells have been based on the compound formamidinium lead iodide ($HC(NH_2)_2PbI_3$) with a bandgap around 1.5 eV, but higher bandgaps needed for tandem cells can be obtained most readily by replacing the iodine anion by a lighter halogen (Br or Cl). Experimentally a monolithic tandem cell of 13.7% has been demonstrated on silicon (Mailoa *et al.*, 2015), but the full potential has been shown in a split-spectrum approach, where 28% efficiency has been demonstrated (Uzu *et al.*, 2015). These perovskites can be deposited inexpensively at low temperature, low enough to be suitable for use even with HJT silicon cells, but are very sensitive to moisture and have demonstrated a range of other instabilities. The challenge with these materials is to massively increase their ruggedness, so they can be used in conjunction with silicon cells without compromising silicon's exceptional durability.

3.9.4 Conclusion

The push to ever increasing energy conversion efficiency is providing the opportunity for advanced silicon cell technologies to make their mark, commercially. In the near term, the traditional "Al BSF" approach is being replaced by higher efficiency PERC sequences, with the transition expected to be largely completed by 2020. Other advanced approaches, particularly IBJ and HJT cells are expected to at least maintain market share in a rapidly growing market. All three approaches will take silicon cell efficiency in production close to the 25–26% range demonstrated in the laboratory, not far from silicon's fundamental limit of 29%.

In the longer term, tandem cell stacks on silicon offer the potential for much higher efficiency, more than doubling the fundamental limit to 59%, although the challenge is find a compatible materials system for these overlying cells. Crystalline III-V cells are probably too costly, earth-abundant chalcogenides have still not demonstrated the required performance levels, while organic-inorganic halide perovskites are moisture sensitive and plagued by a host of other stability issues. Nonetheless, the drive to ever-increasing efficiency is likely to intensify interest in these tandem cell options over the coming years. Successful implementation may well consolidate the dominant position of silicon cell technology indefinitely.

List of Acronyms

Acronym	Meaning
Al-BSF	Aluminum Back Surface Field
ASP	Average Selling Price
CZTS	Cu_2ZnSnS_4
HJT	Heterojunction Cell
IBJ	Interdigitated Back Junction
MWT	Metal Wrap Through
PECVD	Plasma-Enhanced Chemical Vapor Deposition
PERC	Passivated Emitter and Rear Cell

References

Chieng, C. and Green, M.A. (1993) Computer simulation of enhanced output from bifacial photovoltaic modules. *Progress in Photovoltaics*, **1** (4), 293–299.

Green, M.A. (2012) Radiative efficiency of state-of-the-art photovoltaic cells. *Short Communication, Progress in Photovoltaics*, **20** (4), 472–476.

Green, M.A. (2015) The Passivated Emitter and Rear Cell (PERC): from conception to mass production. *Solar Energy Materials and Solar Cells*, **143**, 190–197.

Green, M.A., Ho-Baillie, A., and Snaith, H.J. (2014) The emergence of perovskite solar cells. *Nature Photonics*, **8**, 506–514.

ITRPV (2015) *International Technology Roadmap for Photovoltaic*, Sixth edition, April 2015. www.itrpv.net.

Mailoa, J.P., Bailie, C.D., Johlin, E.C. *et al.* (2015) A 2-terminal perovskite/silicon multijunction solar cell enabled by a silicon tunnel junction. *Applied Physics Letters*, **106**, 121105.

Mints, P. (2015) Priced to sell, *PV Magazine*, **6**, 108–109.

Richter, A., Hermle, M., and Glunz, S.W. (2013) Reassessment of the limiting efficiency for crystalline silicon solar cells. *IEEE Journal of Photovoltaics*, **3** (4), 1184–1191.

Solarbuzz (2014) Efficiency enhancements to define solar PV technology roadmap for the next five years, 14 October, 2014.

Uzu, H., Ichikawa, M., Hino, M *et al.* (2015) *Applied Physics Letters*, **106**, 013506.

Walter, J., Hendrichs, M., Clement, F. *et al.* (2013) Evaluation of solder resists for module integration of MWT solar cells. *Energy Procedia*, **38**, 395–403.

Part Four

Chalcogenide Thin Film Solar Cells

4.1

Basics of Chalcogenide Thin Film Solar Cells

Susanne Siebentritt
Laboratory for Photovoltaics, University of Luxembourg, Luxembourg

4.1.1 Introduction

Thin film solar cells are considered the second generation of PV, based on their lower material and energy consumption during production. Chalcogenide thin film solar cells are based on chalcogenide absorbers, like CdTe, $Cu(In,Ga)Se_2$, or $Cu_2ZnSn(S,Se)_4$. The highest efficiency of all thin film technologies at the time of writing has been obtained with $Cu(In,Ga)Se_2$ for small lab cells with 22.6% (Jackson *et al.*, 2016), higher than polycrystalline Si cells efficiencies (Green *et al.*, 2016). $Cu(InGa)Se_2$ modules are now produced in GW volumes (Kushiya, 2014). The largest production volume has been established for CdTe solar modules, with nearly 2 GW produced worldwide (PVtech, 2014). Lab cells have achieved an efficiency of 22.1% (First Solar, 2016). An emerging material is $Cu_2ZnSn(S,Se)_4$ which is being seriously investigated because its raw materials are abundant. So far efficiencies for lab cells of 12.6% have been achieved (Wang *et al.*, 2014), which is not yet high enough to consider production. A further advantage of thin film solar cells is that they can be made flexible by using a polymer foil substrate. The best efficiency for flexible solar cells has been achieved with $Cu(InGa)Se_2$ absorbers and has achieved 20.4% (Chirila *et al.*, 2013). In addition, flexible solar cells are lightweight, which offers further advantages for applications.

In this chapter, the electronic band structure of a typical thin film solar cell will be discussed, which is based on a p/n heterojunction, together with the role and some fundamental design rules of the different layers and their interfaces. This is followed by a description of the

Photovoltaic Solar Energy: From Fundamentals to Applications, First Edition.
Edited by Angèle Reinders, Pierre Verlinden, Wilfried van Sark, and Alexandre Freundlich.

Companion website: www.wiley.com/go/reinders/photovoltaic_solar_energy

various recombination paths which can be found in a thin film solar cell and which can limit the open circuit voltage as well as the fill factor. The standard methods of analysing these are discussed.

4.1.2 The Electronic Structure of Thin Film Solar Cells

A typical cell structure which is used with little variation in thin film solar cells for all absorber materials is shown in Figure 4.1.1 (a), together with the corresponding band structure of a $Cu(InGa)Se_2$ cell (Figure 4.1.1 (b)).

The main structural difference between $Cu(InGa)Se_2$ and $Cu_2ZnSn(S,Se)_4$ cells, on the one hand, and CdTe cells, on the other, is the growth direction: the former are grown in substrate configuration, i.e. the glass or foil substrate is on the back contact side and the cell is grown starting with the back contact, whereas CdTe cells are grown in superstrate configuration with the glass on the window side and the growth starting with the TCO (transparent conductive oxide) layer.

All these cells are based on a p/n junction which is formed between the p-type absorber and the n-type window. The n-type window consists of a TCO layer and one or more buffer layers. Figure 4.1.1 (b) shows also the band diagram of a $Cu(InGa)Se_2$ solar cell. The band diagram of a CdTe solar cell is similar but there are two differences: (i) the interface between the absorber and the buffer (usually CdS) is much less sharp due to interdiffusion between the buffer and the absorber during the high temperature preparation of the absorber (McCandless, Moulton, and Birkmire, 1997; Herndon *et al.*, 1999). This difference is a consequence of the superstrate structure in CdTe as opposed to the substrate structure in $Cu(InGa)Se_2$; and (ii) the back contact in CdTe is not just a metal contact, but requires the formation of a p+layer by selective etching and/or deposition of another telluride material like ZnTe or HgTe, otherwise the wider bandgap of CdTe would lead to a Schottky barrier at the back contact (McCandless and Sites, 2011).

The p/n junction is basically formed by the n-type TCO and the p-type absorber. Since the doping level of the TCO (above 10^{20} cm^{-3}, see e.g. Cebulla, Wendt, and Ellmer (1998)) is much

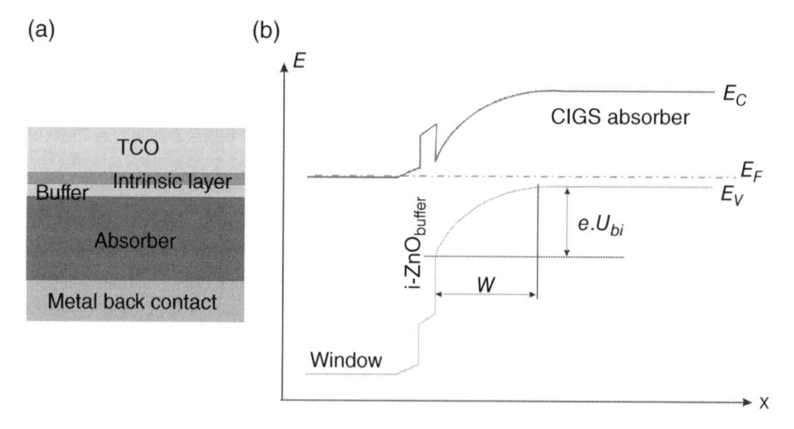

Figure 4.1.1 (a) Physical structure of chalcogenide solar cells; (b) electronic band diagram of a $Cu(InGa)Se_2$ cell

higher than the doping level of the absorber (typically around or below 10^{16} cm^{-3} in Cu(InGa)Se_2 e.g. (Scheer and Schock, 2011; Pianezzi *et al.*, 2014) and in the 10^{14} cm^{-3} range for CdTe (McCandless and Sites, 2011), the junction is one-sided and the space charge region extends only in the absorber layer (plus the buffer layers). This is a major advantage since the absorption starts only where the space charge region is, which is the main region, where the short circuit current is generated (Klenk, 2001). Regarding the role of the buffer layer, a number of reasons have been discussed: (1) protection against sputter damage by the window layer deposition, which has been disproven by the observation of high efficiency solar cells with sputtered buffer layers (Romeo *et al.*, 2010); (2) reduction of interface defects by better lattice match, which is unlikely since a large number of materials with different lattice constants have been used successfully (Hariskos, Spiering, and Powalla, 2005; Siebentritt *et al.*, 2002); (3) formation of a spike, i.e. the conduction band of the buffer is higher than the conduction band of the absorber, which reduces the interface recombination (see Section 4.1.3) (Turner, Schwartz, and Gray, 1988); and (4) forming interface states, which enable an inverted absorber surface, i.e. the Fermi level at the interface is close to the conduction band of the absorber, which suppresses interface recombination even in the case of a high density of interface states (Klenk, 2001). In a one-dimensional device model, the role of the resistive intrinsic-ZnO layer cannot be understood. Its main role is to isolate areas of bad diode behaviour, i.e. low voltage, from the rest of the cell or module (Rau, Grabitz, and Werner, 2004).

The absorber layer is the main part of the solar cell. It absorbs the light and collects the charge carriers, therefore, it needs to exhibit: (1) a suitable bandgap; (2) the proper doping level; and (3) a sufficient diffusion length. First, the bandgap needs to be between about 1 and 1.5 eV (Siebentritt, 2011), where Cu(InGa)Se_2 with 1.15eV is close to the secondary maximum of the theoretical Shockley-Queisser limit, when calculated considering an AM1.5 spectrum, and CdTe with a bandgap of 1.5eV is not far from the maximum of the Shockley-Queisser limit. Second, the absorber needs to have a suitable doping level. The net doping level N_A determines the width W of the space charge region (see Figure 4.1.1) in principle by (Sze, 1981):

$$W = \sqrt{\frac{2\varepsilon\varepsilon_0}{q}\frac{U_{bi}}{N_A}}, \tag{4.1.1}$$

where ε is the static dielectric constant of the absorber material, ε_0 the permittivity of vacuum, q the unit charge and U_{bi} the built-in potential, which is in first approximation defined by the difference in the Fermi level between the window and the absorber and thus increases with the doping level, but only logarithmically. For an exact description, Equation (4.1.1) has to be modified to take localized charges in the buffer layer and at the interfaces into account (Scheer and Schock, 2011). If the doping level is too low (typically below 10^{15} cm^{-3}) the built-in voltage becomes rather low, which in the first approximation limits the open circuit voltage.[1] Additionally the space charge region becomes very wide (several μm), thereby increasing the recombination rate in the space charge region, which further decreases the open circuit voltage (see Section 4.1.3). Further, to avoid depleting the absorber, the absorber layer has to be

[1] In principle, a solar cell does not need to have a built-in voltage (Würfel, 2005), since the current is driven by the gradient in the Fermi level. However, in general, in a p/n junction solar cell the open circuit voltage is limited by flat band conditions and thus by the built-in voltage. Würfel's ideal solar cell has flat bands at the open circuit voltage.

thicker than the space charge region, which causes longer deposition times and higher material cost. If the doping level becomes too high (typically above a few times 10^{16} cm^{-3}), the band bending in the space charge region becomes very steep, thereby increasing tunnelling assisted recombination. This reduces the open circuit voltage (see Section 4.1.3) and in severe cases can even reduce the short circuit current (Depredurand *et al.*, 2014). Furthermore, a high doping level makes the space charge region short and thus reduces the collection length. Finally, the collection length can be approximated as the sum of the width of the space charge region and the diffusion length (Klenk, Schock, and Bloss, 1994). This describes the depth in the absorber from which charge carriers are collected. Thus, for an optimized short circuit current, it has to be as long as possible.

A critical quantity in the band diagram of Figure 4.1.1 is the band alignments. Although the alignment at the back contact has to be considered, the most important band alignments are the ones between the absorber and the buffer and between the buffer and the window. As discussed above, a small spike in the conduction band of the absorber/buffer interface is needed. However, it was shown that the buffer/window interface is also critical to avoid interface recombination (Nguyen *et al.*, 2003; Wilhelm, Schock, and Scheer, 2011). It should not display a large cliff in the conduction band. Experimentally the band offset is determined from the photoemission spectroscopy for the valence band and by inverse photoemission spectroscopy for the conduction band. The valence band offset is basically given by the difference in energy of the valence band edges of the two materials. To avoid the influence of electrostatic shifts by interface charges and/or band bending, the energies have to be referenced to atomic core levels in each material (Kraut *et al.*, 1980). The most reliable way to measure band offsets is by stepwise deposition of one semiconductor on top of the other in the same order as in the solar cell, since this will form the same interface as in the solar cell (Klein and Schulmeyer, 2006). A common way to determine the conduction band offset from the valence band offset is by simply adding the bulk bandgap on either side. However, since the surface bandgap is not necessarily the same as the bulk bandgap (Morkel *et al.*, 2001), the conduction band offset has to be determined separately by inverse photoemission. Since diffusion processes can change the band alignment, these measurements measure the band offsets directly at the interface. However, to construct a band diagram as in Figure 4.1.1, the band alignments between the bulk materials are needed as well. Therefore, these measurements have to be treated with caution.

The band offsets are also determined theoretically. The Anderson approach (Anderson, 1962) determines band offsets simply from the differences in electron affinity and ionization energy, thus using the vacuum level as a reference level. This can lead to dramatically wrong results, particularly in cases where the two materials involved have hugely different electronegativities (Mönch, 2004; 2014). The failure of the vacuum level as a reference level is based on the fact, that at the interface between two semiconductors there will always be interface-induced gap states, basically states of one semiconductor that decay exponentially into the gap of the other semiconductor (Tersoff, 1984). These states change from being valence band-like to conduction band-like at the branch point, i.e. their charge state changes from being a donor to being an acceptor at this energy, which is also called the charge neutrality level. Thus, an interface can only be without dipoles if the two branch points coincide. Therefore, a more suitable reference level is the branch point, which can be calculated as the average energy between the valence and the conduction band (Schleife *et al.*, 2009). Another approach is similar to the experimental one: using atomic core-level energies as reference levels (Zhang *et al.*, 1998).

An average electrostatic potential level is also used to determine the band offsets (Van de Walle and Martin, 1986; Hinuma *et al.*, 2013).

4.1.3 The Current-Voltage Characteristics of Thin Film Solar Cells

The critical parameters for the efficiency of any solar cell are the open circuit voltage, i.e. the voltage at the open terminals under illumination, the short circuit current, i.e., the current that flows under illumination when the terminals are shorted, and the fill factor, which gives the ratio of the power at maximum power point to the product of open circuit voltage and short circuit current. Table 4.1.1 presents a comparison between the best rigid thin film solar cell (Jackson *et al.*, 2016), the best flexible thin film solar cell (Chirila *et al.*, 2013), the best multicrystalline Si cell (Green *et al.*, 2016), and the best monocrystalline Si solar cell (Masuko *et al.*, 2014).

Like any other p/n junction the jV characteristic of thin film solar cells is determined by the Shockley equation, taking parasitic resistances into account:

$$j = j_0\left(e^{q(V-R_S j)/nkT} - 1\right) + \frac{V - R_S j}{R_P} - j_{ph}, \tag{4.1.2}$$

where j is the current density, j_0 is the reverse saturation current density, V the applied voltage, R_S the series resistance, R_P the parallel or shunt resistance, n the diode ideality factor, k the Boltzmann constant, T the temperature and j_{ph} the photocurrent density, which is approximately equal to j_{SC}, the short circuit current density, only if R_S is sufficiently low and R_P sufficiently high. Good thin film solar cells show typically values of or below $0.5\,\Omega\,\text{cm}^2$ for R_S and above $1\,\text{k}\Omega\,\text{cm}^2$ for R_P (Contreras *et al.*, 1999). It should be noted that the approximation of a voltage independent photocurrent, which is made in Equation (4.1.2), is often not valid in thin film solar cells, because the collection length is given in about equal contributions by the space charge width and the diffusion length, where the space charge width is voltage-dependent. This makes the following analyses more approximate. The efficiency of any solar cell depends on the short circuit current, the open circuit voltage, and the fill factor. The short circuit current density in good solar cells is given by j_{ph}, which essentially depends on the bandgap of the absorber, its space charge width and its transport properties (see Section 4.1.2). The open circuit voltage is obtained by setting the current density in Equation (4.1.2) to zero, in which case the series resistance does not play a role. In any decent solar cell the shunt resistance is

Table 4.1.1 Comparison of solar cell parameters for thin film solar cells and crystalline Si solar cells

	Best thin film cell, rigid	Best thin film cell, flexible	Best Si cell, multicrystalline	Best Si cell, monocrystalline
Efficiency (%)	22.6	20.4	21.3	25.6
V_{OC}/mV	741	736	668	740
j_{SC}/mAcm^{-2}	37.8	35.1	39.8	41.8
FF (%)	80.6	78.9	80.0	82.7
Area/cm^2	0.5	0.5	243.9	143.7

high enough that the second term in Equation (4.1.2) can be ignored. The open circuit voltage is then given by:

$$V_{OC} \approx \frac{nkT}{q} \ln\left(\frac{j_{ph}}{j_0}\right) \tag{4.1.3}$$

The fill factor obviously depends on the forward and reverse slopes of the *jV* curve, i.e., on the parasitic resistances. For the case, where those can be ignored, the fill factor can be expressed by the voltage at maximum power point V_{MPP}, which depends in turn almost linearly on V_{OC}:

$$FF^{-1} \approx \frac{1}{V^{*2}}\left(1+V^*\right)\cdot\left(V^*+\ln V^*\right) \text{ with } V^* = \frac{qV_{MPP}}{kT} \tag{4.1.4}$$

Thus, the open circuit voltage and the fill factor depend critically on the diode ideality factor and the reverse saturation current. These two quantities are determined by the dominating recombination mechanism. Generally j_0 is determined by a thermally activated recombination rate with an activation energy E_a:

$$j_0 = j_{00} e^{-E_a/nkT} \tag{4.1.5}$$

where j_{00} is a temperature-independent pre-factor. Usually the diode factor is determined from a direct fit of the *jV* curve, ideally using orthogonal distance regression (Burgers *et al.*, 1996) or by a stepwise analysis of the *jV* curve, using linear fits in each plot (Hegedus and Shafarman, 2004). The activation energy can be determined by the extrapolation to 0K of a temperature-dependent V_{OC} measurement by plugging Equation (4.1.5) into Equation (4.1.3):

$$V_{oc} = \frac{E_a}{q} - \frac{nkT}{q} \ln\left(\frac{j_{00}}{j_{sc}}\right) \tag{4.1.6}$$

or by an Arrhenius plot of j_0, which is also obtained by fitting the *jV* curves. However, care has to be taken since the diode factor might be temperature-dependent. A plot over $1/nkT$ is more advisable than a plot over $1/kT$.

For the ideal case of neutral zone recombination, the diode ideality factor n equals 1 and the activation energy equals the bandgap E_g (Sze, 1981), whereas for the Shockley-Read-Hall recombination in the space charge region the diode factor = 2 and the activation energy $E_a = E_g$ in the simplest model (Sze, 1981). If there are no mid-gap states the diode factor will be found between 1 and 2 (Fahrenbruch and Bube, 1983), since it basically depends on the ratio of voltage drop on the n-side and the p-side from the main recombination location. This ratio depends also on the details of the band structure and can in special cases lead to a diode factor >2 (Fahrenbruch and Bube, 1983). In the case of recombination via tail states it was shown that the diode factor becomes temperature-dependent and lies also between 1 and 2 (Walter, Herberholz, and Schock, 1996). In all cases the activation energy stays equal to E_g, since the recombination rate depends on the presence of free electrons and holes in the bulk, which is activated by the bandgap energy. If the doping in the absorber is high, the band bending within the space charge region becomes steep and a tunnelling-assisted recombination via interface or gap states becomes the most likely recombination path. Tunnelling-assisted recombination is usually discussed in the context of interface recombination (Rau *et al.*, 2000; Scheer and Schock, 2011), but the carriers need not tunnel into interface states, they can also tunnel into bulk defects within the space charge region. Since tunnelling is not a thermally activated process, the (effective) diode factor becomes ~1/T (Riben and Feucht, 1966).

In thin film solar cells additional recombination paths can occur at the interfaces. In the case of Fermi level pinning at the interface, the device can be modeled as two opposite Schottky junctions (Dolega, 1963), which leads to a diode ideality factor between 1 and 2, determined again by the ratio of voltage drop on the n-side and the p-side of the interface and the activation energy is given by the distance of the (pinned) Fermi level from the valence band edge at the interface (assuming an inverted junction), i.e. the activation energy is smaller than the bandgap (Rau *et al.*, 2000). It has been argued that in the case of Fermi level pinning, the diode ideality factor should be 1 (Scheer and Schock, 2011), however, this is only true if the voltage drop over the buffer and the i-layer can be ignored. Without Fermi level pinning, the activation energy is determined by the "interface bandgap," which is given by the lowest conduction band (either of the buffer or the absorber) and the highest valence band (generally of the absorber) (Hengel, *et al.*, 2000; Scheer and Schock, 2011). Thus, for a spike configuration, the interface bandgap is given by the absorber bandgap, which makes the activation energy of interface recombination the same as the bulk recombination. It was demonstrated that in this case the interface recombination can be ignored, even with a very high interface recombination velocity, i.e. a high density of interface states, if the interface is inverted, meaning that the Fermi level at the interface is close to the absorber conduction band (Klenk, 2001). If there is a cliff, i.e., a negative conduction band offset when going from the absorber to the buffer, the interface bandgap will be smaller than the absorber bandgap, so will be the activation energy of the recombination path, making interface recombination the dominant recombination path. A low activation energy of the main recombination path will in general lead to a high saturation current density and therefore limit V_{OC} according to Equation (4.1.3).

Current best devices are limited by the space charge region recombination (Scheer and Schock, 2011; Siebentritt, 2011), possibly enhanced by the band bending at charged grain boundaries. Further optimization of thin film solar cells will require the defect density to be reduced and the production of devices which are limited by radiative recombination in the neutral zone.

4.1.4 Conclusion

Laboratory-scale thin film solar cells are in the same range as the efficiencies of multicrystalline Si solar cells. They are based on a hetero p/n junction, which offers the advantage that the regions of highest absorption are right in the space charge region, where the collection is highest. On the other hand, this makes for a complex multi-layer structure and introduces interfaces as a potential source of recombination. Important design rules are: (1) a considerable higher doping of the window than the absorber; (2) a positive conduction band offset (spike), when going from the absorber to the buffer layer; (3) a band offset that is not too negative, when going from the buffer to the window layer; (4) inversion at the absorber surface; (5) absorber doping around 10^{16} cm^{-3}; and (6) good transport properties, with diffusion lengths on the order of absorber thickness.

List of Symbols

Symbol	Discription	Units
E_a	Activation energy of recombination rate	eV
E_g	Band gap	eV
ε	Static dielectric constant	
ε_0	Permittivity of vacuum	$8.85 . 10^{-12}$ As/Vm
FF	Fill factor	
j	Current density	$mAcm^{-2}$
j_0	Reverse saturation current density	$mAcm^{-2}$
j_{00}	Prefactor of reverse saturation current density	$mAcm^{-2}$
j_{ph}	Photo current density	$mAcm^{-2}$
j_{SC}	Short circuit current density	$mAcm^{-2}$
k	Boltzmann constant	$1.38 \cdot 10^{-23}$ J/K
n	Diode ideality factor	
N_A	Acceptor concentration	cm^{-3}
q	Charge of the electron	$1.6 \cdot 10^{-19}$ As
R_S	Series resistance	Ω
R_P	Parallel or shunt resistance	Ω
T	Temperature	K
V	Applied voltage	V
V_{MPP}	Voltage at maximum power point	V
V_{OC}	Open circuit voltage	V
U_{bi}	Built-in voltage	V
W	Width of the space charge region	cm

List of Acronyms

Acronym	Definition
AM1.5	Air mass 1.5 spectrum
PV	Photovoltaics
TCO	Transparent Conducting Oxide

References

Anderson, R.L. (1962) Experiments on Ge-GaAs heterojunctions. *Solid State Electronics*, **5**, 341–351.

Burgers, A.R., Eikelboom, J.A., Schonecker, A., and Sinke, W.C. (1996) Improved treatment of the strongly varying slope in fitting solar cell I-V curves. Paper presented at the 25th IEEE Photovoltaic Specialists Conference.

Cebulla, R., Wendt, R., and Ellmer, K. (1998) Al-doped zinc oxide films deposited by simultaneous rf and dc excitation of a magnetron plasma: relationships between plasma parameters and structural and electrical film properties. *Journal of Applied Physics*, **83** (2), 1087–1095.

Chirila, A., Reinhard, P., Pianezzi, F., *et al.* (2013) Potassium-induced surface modification of Cu(In,Ga)Se_2 thin films for high-efficiency solar cells. *Nature Materials*, **12** (12), 1107–1111.

Contreras, M., Egaas, B., Ramanathan, K. *et al.* (1999) Progress towards 20% efficiency in Cu(In,Ga)Se_2 polycrystalline thin-film solar cells. *Progress in Photovoltaic Research Applications*, **7**, 311–316.

Depredurand, V., Tanaka, D., Aida, Y. *et al.* (2014) Current loss due to recombination in Cu-rich $CuInSe_2$ solar cells. *Journal of Applied Physics*, **115**, 044503.

Dolega, U. (1963) Theorie des pn-Kontaktes zwischen Halbleitern mit verschiedenen Kristallgittern. *Zeitung für Naturforschung A*, **18**, 653–666.

Fahrenbruch, A.L. and Bube, R.H. (1983) *Fundamentals of Solar Cells*. Academic Press, New York.

First Solar. (2016) http://investor.firstsolar.com/releasedetail.cfm?ReleaseID=956479. Retrieved August 2016, August 2016.

Green, M.A., Emery, K., Hishikawa, Y. *et al.* (2016) Solar cell efficiency tables (version 48) *Progress in Photovoltaics: Research and Applications*, **24** (7), 905–913.

Hariskos, D., Spiering, S. and Powalla, M. (2005) Buffer layers in Cu(In,Ga)Se_2 solar cells and modules. *Thin Solid Films*, **480–481**, 99–109.

Hegedus, S.S. and Shafarman, W.N. (2004) Thin-film solar cells: device measurements and analysis. *Progress in Photovoltaics: Research and Applications*, **12** (2–3), 155.

Hengel, I., Neisser, A., Klenk, R., and Lux-Steiner, M.-C. (2000) Current transport in $CuInS_2$:Ga/CdS/ZnO-solar cells. *Thin Solid Films*, **361–362**, 458–462.

Herndon, M.K., Gupta, A., Kaydanov, V., and Collins, R.T. (1999) Evidence for grain-boundary-assisted diffusion of sulfur in polycrystalline CdS/CdTe heterojunctions. *Applied Physics Letters*, **75** (22), 3503–3505.

Hinuma, Y., Oba, F., Kumagai, Y., and Tanaka, I. (2013) Band offsets of $CuInSe_2$/CdS and $CuInSe_2$/ZnS (110) interfaces: a hybrid density functional theory study. *Physical Review B*, **88** (3), 035305.

Jackson, P., Wuerz, R., Hariskos, *et al.* (2016) Effects of heavy alkali elements in Cu(In,Ga)Se_2 solar cells with efficiencies up to 22.6%. *Physica Status Solidi (RRL) – Rapid Research Letters*, DOI: 10.1002/pssr.201600199.

Klein, A. and Schulmeyer, T. (2006) Interfaces of Cu-chalcopyrites. In *Wide Gap Chalcopyrites* (eds S. Siebentritt and U. Rau). Springer, Berlin.

Klenk, R. (2001) Characterization and modelling of chalcopyrite solar cells. *Thin Solid Films*, **387**, 135–140.

Klenk, R., Schock, H.-W., and Bloss, W.H. (1994) Photocurrent collection in thin film solar cells - caluculation and characterization for $CuGaSe_2$/(Zn,Cd)S. Paper presented at the 12th European Photovoltaic Solar Energy Conference.

Kraut, E.A., Grant, R.W., Waldrop, J.R., and Kowalczyk, S.P. (1980) Precise determination of the valence-band edge in x-ray photoemission spectra: application to measurement of semiconductor interface potentials. *Physical Review Letters*, **44** (24), 1620–1623.

Kushiya, K. (2014) CIS-based thin-film PV technology in solar frontier K.K. *Solar Energy Materials and Solar Cells*, **122** (0), 309–313.

Masuko, K., Shigematsu, M., Hashiguchi, T. *et al.* (2014) Achievement of more than 25% conversion efficiency with crystalline silicon heterojunction solar cell. *IEEE Journal of Photovoltaics*, **4** (6), 1433–1435.

McCandless, B.E., Moulton, L.V., and Birkmire, R.W. (1997) Recrystallization and sulfur diffusion in $CdCl_2$-treated CdTe/CdS thin films. *Progress in Photovoltaics: Research and Applications*, **5**(4), 249–260.

McCandless, B. and Sites, J.R. (2011) Cadmium telluride solar cells. In *Handbook of Photovoltaic Science and Engineering* (eds A. Luque and S. Hegedus), (pp. 600–641). John Wiley and Sons, Ltd, Chichester.

Miasolé (2014) http://miasole.com/en/product/modules/(accessed October 2014).

Mönch, W. (2004) *Electronic Properties of Semiconductor Interfaces*. Springer, Berlin.

Mönch, W. (2014) On the band-structure lineup at Schottky contacts and semiconductor heterostructures. *Materials Science in Semiconductor Processing*, **28**, 2–12.

Morkel, M., Weinhardt, L., Lohmüller, B. *et al.* (2001) Flat conduction-band alignment at the CdS/$CuInSe_2$ thin-fillm solar-cell heterojunction. *Applied Physics Letters*, **79**, 4482–4484.

Nguyen, Q., Orgassa, K., Koetschau, I. *et al.* (2003) Influence of heterointerfaces on the performance of Cu(In,Ga)Se_2 solar cells with CdS AND In(OHx, Sy) buffer layers. *Thin Solid Films*, **431–432**, 330–334.

Pianezzi, F., Reinhard, P., Chirila, A. *et al.* (2014) Unveiling the effects of post-deposition treatment with different alkaline elements on the electronic properties of CIGS thin film solar cells. [10.1039/C4CP00614C]. *Physical Chemistry Chemical Physics*, **16** (19), 8843–8851.

PVtech. (2014) http://www.pv-tech.org/news/first_solar_analyst_day_next_major_production_capacity_expansion_in_2015. (accessed October 2014).

Rau, U., Grabitz, P.O., and Werner, J.H. (2004) Resistive limitations to spatially inhomogeneous electronic losses in solar cells. *Applied Physics Letters*, **85** (24), 6010–6012.

Rau, U., Jasenek, A., Schock, H.W. *et al.* (2000) Electronic loss mechanisms in chalcopyrite based heterojunction solar cells. *Thin Solid Films*, **361–362**, 298–302.

Riben, A.R. and Feucht, D.L. (1966) nGe-pGaAs Heterojunctions. *Solid State Electronics*, **9**, 1055–1065.

Romeo, N., Bosio, A., Mazzamuto, S. *et al.* (2010) CIGS thin films prepared by sputtering and selenization by using In_2Se_3, Ga_2Se_3 and Cu as sputtering targets. Paper presented at the 35th IEEE. Photovoltaic Specialists Conference (PVSC).

Scheer, R. and Schock, H.W. (2011) *Chalcogenide Photovoltaics: Physics, Technologies, and Thin Film Devices*. Wiley-VCH, Weinheim.

Schleife, A., Fuchs, F., Rödl, C. *et al.* (2009) Branch-point energies and band discontinuities of III-nitrides and III-/II-oxides from quasiparticle band-structure calculations. *Applied Physics Letters*, **94**(1), 012104.

Siebentritt, S. (2011) What limits the efficiency of chalcopyrite solar cells? *Solar Energy Materials and Solar Cells*, **95**(6), 1471–1476.

Siebentritt, S., Kampschulte, T., Bauknecht, A. *et al.* (2002) Cd-free buffer layers for CIGS solar cells prepared by a dry process. *Solar Energy Materials and Solar Cells*, **70**, 447–457.

Sze, S.M. (1981) *Physics of Semiconductor Devices*. John Wiley and Sons, Inc., New York.

Tersoff, J. (1984) Theory of semiconductor heterojunctions: the role of quantum dipoles. *Physical Review B*, **30** (8), 4874–4877.

Turner, G.B., Schwartz, R.J., and Gray, J.L. (1988) Band discontinuity and bulk vs. interface recombination in CdS/CuInSe2 solar cells. Paper presented at the 20th IEEE PV Specialist Conference, Las Vegas.

Van de Walle, C.G. and Martin, R.M. (1986) Theoretical calculations of heterojunction discontinuities in the Si/Ge system. *Physical Review B*, **34** (8), 5621–5634.

Walter, T., Herberholz, R., and Schock, H.W. (1996) Distribution of defects in polycristalline chalcopyrite films. *Solid State Phenomena*, **51**/52, 309–316.

Wang, W., Winkler, M.T., Gunawan, O. *et al.* (2014) Device characteristics of CZTSSe thin-film solar cells with 12.6% efficiency. *Advanced Energy Materials*, **4**, 1301465.

Wilhelm, H., Schock, H.-W., and Scheer, R. (2011) Interface recombination in heterojunction solar cells: Influence of buffer layer thickness. *Journal of Applied Physics*, **109** (8), 084514.

Würfel, P. (2005) *Physics of Solar Cells*. Wiley-VCH, Weinheim.

Zhang, S.B., Wei, S.-H., Zunger, A., and Katayama-Yoshida, H. (1998) Defect physics of the $CuInSe_2$ chalcopyrite semiconductor. *Physical Review B*, **57** (16), 9642–9656.

4.2

$Cu(In,Ga)Se_2$ and CdTe Absorber Materials and their Properties

Sylvain Marsillac
Virginia Institute of Photovoltaics, Old Dominion University, Norfolk, VA, USA

4.2.1 Introduction

The success of both $Cu(In,Ga)Se_2$ and CdTe in reaching conversion efficiencies above 20% is due as much to the talent and ingenuity of the scientists and engineers who built the devices, as to the properties of these absorber materials. Understanding and, eventually, controlling these properties are key to the success of these chalcogenide technologies. Table 4.2.1 summarizes the main properties of $CuInSe_2$ and CdTe at 300 K. In the following sections, we will introduce their structural properties, their phase diagram, their electronic properties, their optical properties, and, finally, the evolution of these properties in their related alloys.

4.2.2 Structural Properties

To understand the crystal structure of both $CuInSe_2$ and CdTe, first, one has to understand the crystal structure of silicon (Figure 4.2.1 (a)), as seen previously in Part 2. To build a silicon crystal, one starts with a face-centered cubic (FCC) structure, made up of silicon atoms. Another interpenetrating FCC structure, again made of silicon atoms, is then positioned with a translation of a quarter along all axis. It looks therefore as if the second FCC has shifted from the original position by one quarter along the diagonal direction. To build the CdTe crystal, one simply replaces the silicon atoms from the first FCC structure with Cd atoms, and the silicon atoms from the second FCC structure with Te atoms (note that the same substitution can apply to GaAs). This is still a cubic crystal (Figure 4.2.1 (b)). To modify this structure and build a $CuInSe_2$ crystal, the size of the unit cell has first to be doubled along the c-axis (effectively adding another cube along the c-axis), where the c-axis is in the vertical direction

Photovoltaic Solar Energy: From Fundamentals to Applications, First Edition.
Edited by Angèle Reinders, Pierre Verlinden, Wilfried van Sark, and Alexandre Freundlich.

Companion website: www.wiley.com/go/reinders/photovoltaic_solar_energy

Table 4.2.1 $CuInSe_2$ and CdTe selected materials' properties

Property	$CuInSe_2$	CdTe	Reference $CuInSe_2$	Reference CdTe
Space Group	$I\bar{4}2d$	$F\bar{4}3m$	(Shahidi *et al.*, 1985)	(ICDD 15-770)
Lattice constant	a 0.578 nm c 1.162 nm	a_0 0.648 nm	(Suri *et al.*, 1989)	(ICDD 15-770)
Thermal expansion coefficient	(a axis) 11.23 10^{-6}/K (c axis) 7.90 10^{-6}/K	5.9 10^{-6}/K	(Bondar and Orlova, 1985)	(Fonash, 1981)
Optical band gap Eg	1.02 eV	1.50 eV	(Neumann, 1986a)	(Rakhshani, 1997)
Temperature dep.: dEg/dT	−0.2 meV/K	−0.3 meV/K	(Neumann, 1986a)	(Laurenti *et al.*, 1990)
Static Dielectric const.: $\varepsilon(\theta)$	13.6	10.0	(Wasim S, 1986)	(Madelung, 1992)
High Freq. Dielectric const.: $\varepsilon(\infty)$	8.1	7.1	(Wasim S, 1986)	(Madelung, 1992)
Melting point	1259 K	1365 K	(Ciszek, 1984)	(Aven and Prener, 1967)
Density	5.75 g/cm^3	5.3 g/cm^3	(Suri, *et al.* 1989)	(Hartmann *et al*,, 1982)
Electron affinity χe	4.6 eV	4.28 eV	(Romeo, 1980)	(Hinuma *et al.*, 2014)
Effective mass m_e*	0.09 m_0	0.096 m_0	(Neumann, 1986a)	(Aven and Prener, 1967)
Effective mass m_h*	0.71 m_0	0.35 m_0	(Neumann, 1986a)	(Aven and Prener, 1967)
Mobility μ_e	90–900 cm^2/V.s	500–1000 cm^2/V.s	(Neumann and Tomlinson, 1990)	(Aven and Prener, 1967)
Mobility μ_h	15–150 cm^2/V.s	50–80 cm^2/V.s	(Neumann and Tomlinson, 1990)	(Aven and Prener, 1967)
Thermal conductivity	0.086 W.cm^{-1}.K^{-1}	0.075 W.cm^{-1}.K^{-1}	(Neumann, 1986b)	(Glen, 1964)

in Figure 4.2.1. The anions' positions in both structure remain the same, so Te atoms are replaced by Se atoms. For the cations, the Cd is orderly replaced by either Cu or In, so that each Se (anion) is bonded to two Cu atoms and two In atoms, and each Cu atom and In atom is tetrahedrally bonded to four Se atoms (Figure 4.2.1 (c)). According to the Grimm-Sommerfeld rule, if the average number of valence electrons per atom is four in an atomic structure, then that structure will be a tetragonal structure (Grimm and Sommerfeld, 1926; Bohm *et al.*, 1985). For the I-III-VI_2 compounds, the average number of valence electrons per atom is $(1\times1+1\times3+2\times6)/4=4$; they therefore crystallize into a tetragonal structure (Jaffe and Zunger, 1993). Due to the Grimm-Sommerfeld rule, it is also possible to substitute In by Ga without losing the tetragonal lattice configuration. Due to the unequal bond length between the Cu-Se and the In-Se, the tetragonal ratio ($\eta=c/2a$) is not equal to 1 and the anion displacement (u) is not equal to 1/4 but is given by (Jaffe, 1993):

$$u-\frac{1}{4}=\frac{R^2_{Cu-Se}-R^2_{In-Se}}{a} \tag{4.2.1}$$

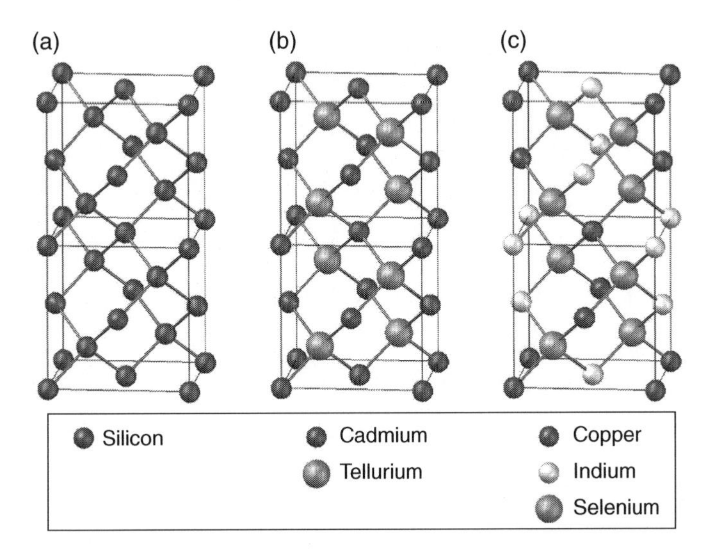

Figure 4.2.1 (a) FCC structure of a Si crystal with two identical atoms basis; (b) CdTe structure derived from the Si structure by replacing identical atoms basis by two different atoms; (c) $CuInSe_2$ structure derived from the CdTe structure by doubling the CdTe structure and replacing the cations by two different cations. (*See insert for color representation of the figure*)

Table 4.2.2 Silicon, CdTe and $CuInSe_2$ associated denominations

Material	Crystal system	Associated mineral	Space group (SH)	Space group (HM)
Silicon	Cubic	Diamond	O_h^7	$Fd\bar{3}m$
CdTe	Cubic	Sphalerite/ZincBlende	T_d^2	$F\bar{4}3m$
	Hexagonal	Wurtzite	C_{6v}^4	$P6_3mc$
$CuInSe_2$	Tetragonal	Chalcopyrite	D_{2d}^{12}	$I\bar{4}2d$

where $R_{Cu\text{-}Se}$ is the bond length of the Cu-Se and $R_{In\text{-}Se}$ is the bond length of the In-Se. These crystals are identified by various nomenclatures, which can be confusing at times. One also should note that CdTe exists in two main polytypes. A summary of these names is provided in Table 4.2.2 using crystal systems, associated minerals and space groups (both Schönflies (SH) notation and Hermann-Mauguin (HM) notation).

The diamond is obviously made of carbon hybridized in the sp^3 configuration. The Sphalerite (or ZincBlende) as well as the Wurtzite are originally minerals made of (Zn,Fe)S. The Chalcopyrite is originally a mineral made of $CuFeS_2$ (not to be confused with chalcogenide, which simply states that the material contains a chalcogen anion (element from the group VI_B) and another more electropositive element).

4.2.3 Phase Diagram

Because CdTe is a binary compound and not a ternary like $CuInSe_2$, its phase diagram is simpler to start with. There are effectively only three materials present in the solid phase (Cd, Te and CdTe) as can be seen in Figure 4.2.2.

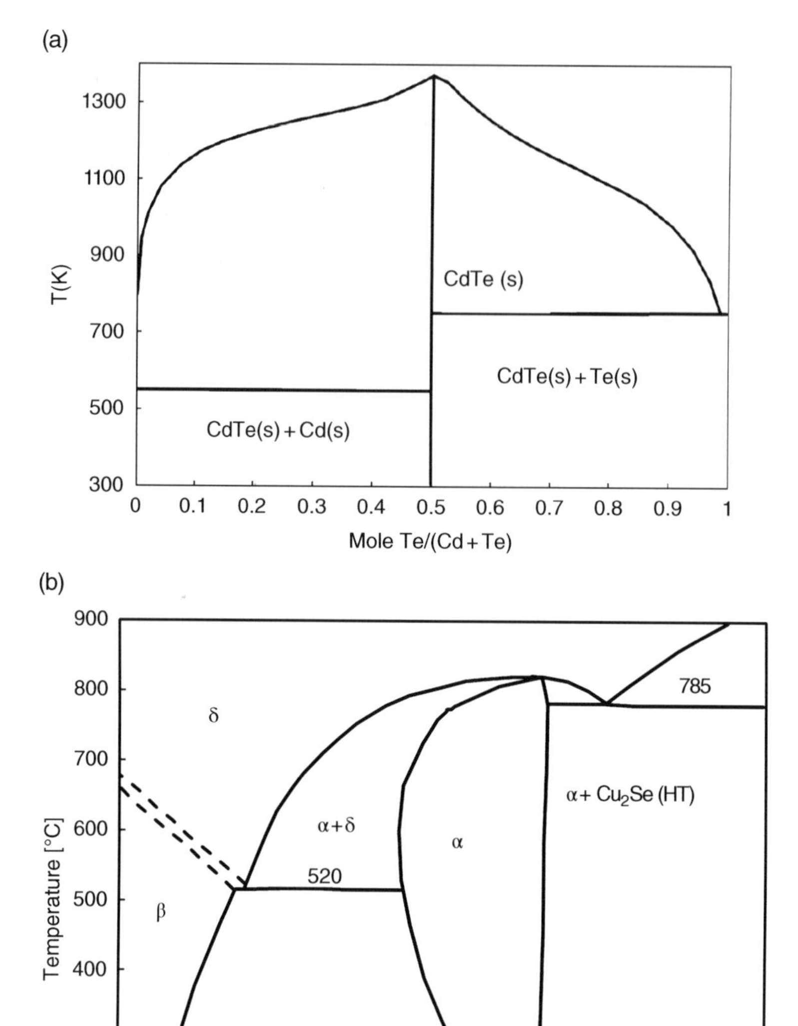

Figure 4.2.2 (a) CdTe phase diagram. Source: after Jianrong, Silk, Watson and Bryant (1995); (b) quasi-binary phase diagram of $CuInSe_2$ along the In_2Se_3-Cu_2Se tie-line. Source: after Godecke (2000)

The left side of the CdTe phase diagram represents pure Cd ($x = 0$) while the right side represents pure Te ($x = 1$). The only other solid compound is CdTe, which occupies a very narrow region (~10^{-6} at. %) at temperature below 500°C. For higher temperatures, this region widens slightly toward the Cd-rich side at first (up to 700°C), then toward the Te-rich side after

22% (Aven and Prener, 1967). One can also note that the melting point for CdTe (T_m = 1092°C) is much higher than that for either Cd (Tm = 321°C) or Te (T_m = 450°C) (Hultgren, 1971). Another important property of CdTe is its capacity to be congruently evaporated, while both Cd and Te have high sublimation pressure. This permits relatively easy single-phase deposition by several physical vapor deposition processes.

In the quasi-binary In_2Se_3-Cu_2Se phase diagram shown in Figure 4.2.2, four different compounds can exist depending on the Cu at.%: three phases for $CuInSe_2$ (α, β, and δ) and Cu_2Se. The α-phase is the chalcopyrite $CuInSe_2$ phase and exists in the narrow range of 24.0–24.5 at.% of Cu at room temperature (note that 25 at.% is not included). This range is maximum around 600°C, but vanishes above 800°C. For pure $CuInSe_2$, the most suitable growth temperature is often around 400–600 °C. At the Cu-poor boundary, the α-phase coexists with another phase called the β-phase, which represents a number of compounds such as $CuIn_3Se_5$ or $CuIn_5Se_8$. Since the building blocks of these compounds are copper vacancies (V_{Cu}) and indium substituted for copper (In_{Cu}), these compounds are called ordered defect compounds. Fractional replacement of Ga and small addition of Na, hinders the ordering of defects (Schock, 2004) and so widens the α-phase towards the Cu-poor boundary. The addition of Ga and Na is therefore crucial for high efficiency solar cell fabrication, as they give more flexibility to the deposition conditions. It is important to note that very high efficiency solar cells (above 19%) have been deposited with a Cu/(In + Ga) ratio as low as 0.7 and as high as 0.95 and a Ga/(In + Ga) ratio ranging from 0.2–0.4 (Jackson *et al.*, 2007), while deposition temperatures below 500°C have been recorded (Chirila *et al.*, 2011).

4.2.4 Electronic Properties and Defects

Many intrinsic defects exist in both $CuInSe_2$ and CdTe, whether in the form of vacancy, interstitial or substitutional defects or even as complex defects. A summary of the main defects for both materials is given in Figure 4.2.3. Extrinsic defects also exist and play a critical role in the overall cell efficiency, but both materials (in their thin film forms) are thought to be intrinsically doped. In both cases, it is also important to distinguish between bulk-crystal properties and thin-film properties, as the grain boundary defects, but also potentially the number of intrinsic defects, will vary quite a lot from one to another. Post-deposition treatments (such as $CdCl_2$ or KF treatment) or addition of extrinsic elements (such as S or Na) are therefore key parameters that will control the ultimate solar cell efficiency by modifying the defects present. Generally, deep states (which lie close to the middle of the bandgap) are defects that will increase the carriers' recombination and are therefore detrimental to the solar cell performance, whereas shallow states (which lie either close to the valence band, i.e., shallow acceptors, or to the conduction band i.e. shallow donors) can contribute positively to the carrier density.

For example, in the case of CdTe, Cd in an interstitial position (Cd_i) is a shallow donor state (0.33 eV) whereas Te in the same position (Te_i) is a deep state (0.74 eV from CB) On the other hand, a vacancy in Cd (V_{Cd}) is a shallow acceptor state (0.21 eV), and a Cd substituting on a Te (Cd_{Te}) is a shallow donor state (0.35 eV). These two defects (V_{Cd} and Cd_{Te}) form one of the main complexes in CdTe, which is a shallow acceptor state (0.10 eV). These various defects depend in part on the deposition process (Close Space Sublimation, Vapor Transport

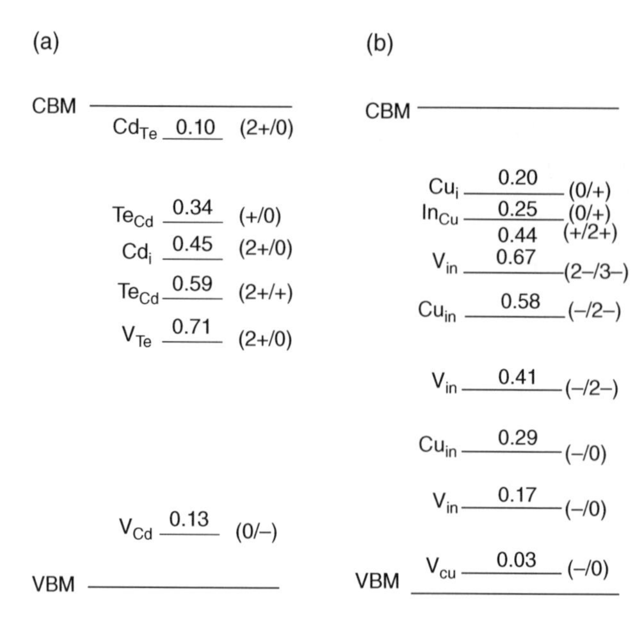

Figure 4.2.3 Main theoretical electronic levels of defects: (a) in CdTe; and (b) $CuInSe_2$. Source: after Wei and Zhang (2002), Zhang *et al.* (1998)

Deposition, Sputtering …), on the post-deposition treatments ($CdCl_2$, O_2) or even on the nature of the back contact (e.g., Cu) (Basol¸1992). Ultimately, the state-of-the-art CdTe solar cells are made of p-type CdTe and n-type CdS. As mentioned previously, the interdiffusion is another critical parameter to take into account when trying to understand the properties of the thin film stack. Device quality CdTe films have majority carrier concentration up to $6 \times 10^{14}\,cm^{-3}$, with minority carrier lifetime of 10 to 15 ns (Mao *et al.*, 2014).

In the case of Cu(In,Ga)Se_2, the deposition conditions are also critical in knowing what type of majority carrier is going to be predominant. If Cu(In,Ga)Se_2 is grown under Cu-rich flux but low Se pressure, Cu can either occupy interstitial sites (Cu_i) or promote Se vacancies (V_{Se}) which leads to a n-type conductivity (Noufi *et al.*, 1984). In this case, V_{Se} is the dominant donor. If Cu(In,Ga)Se_2 is grown under Cu-poor flux and high Se pressure, Cu vacancies (V_{Cu}) lead to a p-type conductivity and represent the dominant acceptor, while vacancies in Se (V_{Se}) are the compensating donors (Schock, 2004). Preparing these materials with a Cu-poor composition gives rise to several defects, which have to stay at a low density or be electronically inactive to yield high quality materials (Schock, 2004). In Cu-poor films, this is achieved naturally by the formation of a primary complex, whereby the formation of $2V_{Cu}^{-}$ is compensated by In_{Cu}^{2+}, i.e. the defect pair $2V_{Cu} + In_{Cu}$ is electrically neutral and has no energy level within the bandgap (Schock, 2004). Another important defect is the Cu replacing an In (Cu_{In}), which leads to a recombination center 0.29 eV from the Valence band. The last critical defect, located around 0.8 eV from the Valence band, is a deep defect that is probably responsible for the major recombination mechanism observed in Cu(In,Ga)Se_2, and has been observed experimentally not to move with the gallium content. This type of deposition process, leading to p-type Cu(In,Ga)Se_2, is used for the preparation of absorber layers in high efficiency solar cells. The typical carrier concentration (holes) in this device of the quality p-type Cu(In,Ga) Se_2 is around 10^{16}–10^{17} cm^{-3}. The minority carrier (electron) lifetime is around 1–9 ns and the diffusion length is around 1 μm (Rau and Schock, 2001).

4.2.5 Optical Properties and Alloys

As with any other materials, the optical and electrical properties of both $CuInSe_2$ and CdTe come from their energy band structure, which is obtained by resolving the time-independent Schrodinger's equation.

In the case of CdTe, the top of the valence band corresponds to the 5p level of Te, while the bottom of the conduction band corresponds to the 5 s level of Cd. This leads to a bandgap of 1.5 eV for CdTe. In the case of $CuInSe_2$, the top of the valence band corresponds to a hybrid between the Cu *d*-states and the Se *p*-states, while the bottom of the conduction band corresponds to the 5p level of In. The bandgap of $CuInSe_2$ varies between 0.81 eV and 1.01 eV for single crystals (Neumann, 1986a) whereas for thin films, this value is very narrow, i.e., E_g of the thin film $CuInSe_2$ is 1.02 ± 0.02 eV (Begou *et al.*, 2011).

In both cases, the maximum of the valence band and the minimum of the conduction band occur at the same value of the crystal momentum, leading therefore to direct bandgap semiconductors. The absorption coefficient can therefore be written as:

$$\alpha = \frac{A}{h\upsilon}\left(h\upsilon - E_g\right)^{1/2} \tag{4.2.2}$$

where α is the absorption coefficient (in cm^{-1}) at photon energy $h\upsilon$, A is a constant (in cm^{-1}.eV$^{1/2}$) and E_g is the bandgap (in eV).

Due to their direct bandgap, the absorption coefficients in both CdTe and $Cu(In,Ga)Se_2$ are very high (10^5 cm^{-1} at 2.5 eV (Paulson *et al.*, 2004), and 10^5 cm^{-1} at 1.4 eV (Kazmerski, 1983), respectively) and most of the photons are absorbed within a few microns. State-of-the-art devices typically use a 3 μm thick absorber for $Cu(In,Ga)Se_2$ (Repins *et al.*, 2008; Chirila *et al.*, 2011; Jackson *et al.*, 2015) and 3 μm thick absorber for CdTe (Mao *et al.*, 2014).

Another important property for both materials is their capacity to form alloys with other elements. In the case of I-III-VI materials, they tend to keep the same crystal structure without any miscibility gap. For the II-VI materials, they can be present as Wurtzite or Zinc Blende, and have miscibility gaps for some of them (e.g. Cd(Te,S)). These alloys are critical for both current applications (allowing device engineering such as bandgap gradient) but also future developments (such as tandem cells fabrication).

For the I-III-VI_2 family, the groups I, III and VI represent the groups of elements in the periodic table from the column I_A, III_B and VI_B respectively. "I" can therefore be either Cu, Ag or Au, "III" be Al, Ga, or In (B has been used in some rare cases) and "VI" be S, Se, or Te (oxygen is also possible but leads to very wide bandgap materials not often used in solar cells). A summary of the lattice constants versus bandgap for some of these materials is presented in Figure 4.2.4.

For the II-VI family, the groups II and VI represent generally elements from the group II_B and VI_B, respectively, even if elements from the group II_A (Mg) have also been used. "II" can therefore be either Zn, Cd, or Hg and "VI" can be S, Se, or Te (oxygen can also be present, but

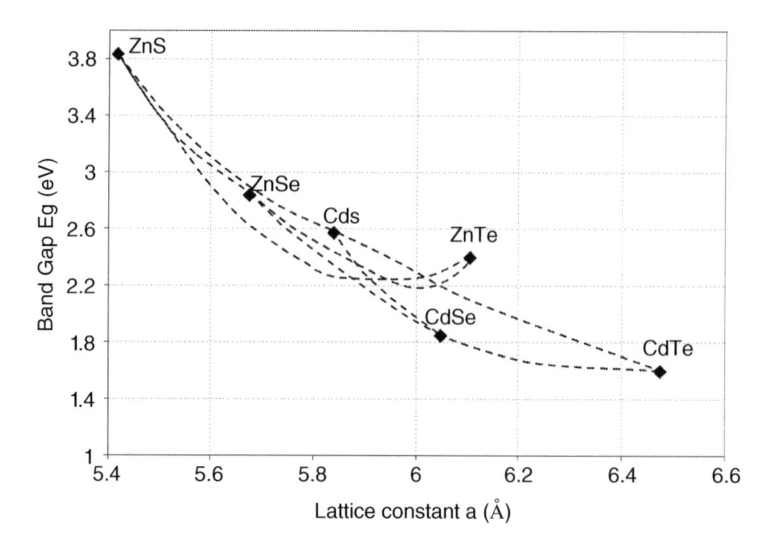

Figure 4.2.4 Summary of the lattice constants versus band gap for some II-VI materials

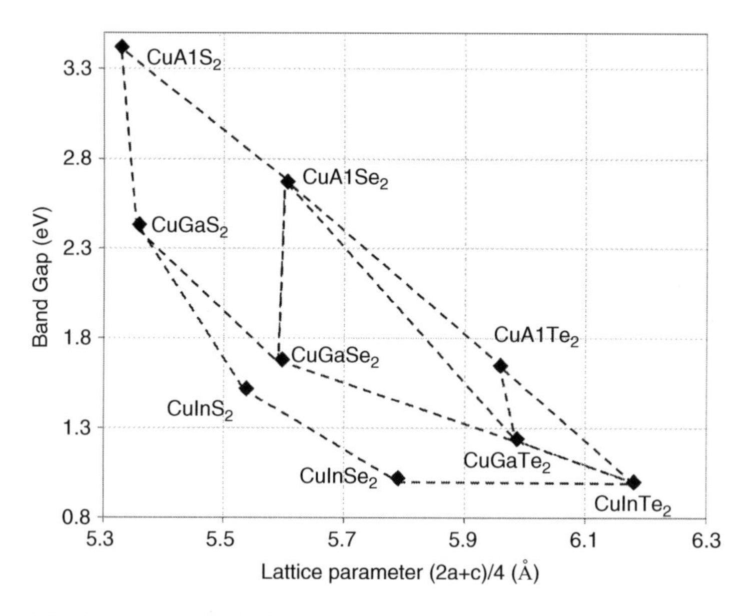

Figure 4.2.5 Summary of the lattice constants versus band gap for some I-III-VI_2 materials

more as a by-product of treatment than as a primary element for the semiconductor). A summary of the lattice constants versus bandgap for some of these materials is also presented in Figure 4.2.4 for II-VI and Figure 4.2.5 for I-III-VI_2.

Some properties of these alloys are easy to extrapolate. For example, since the alloys for the I-III-VI_2 family have no miscibility gap and no phase change, one can extract their lattice constant simply by doing a linear extrapolation between the two end points. However, for the optical properties, one has to take into account some bowing parameter. In the case of $Cu(In,Ga)Se_2$, the bandgap is tunable depending on the Ga substitution and is described by (Albin *et al.*, 1991):

$$E_g = (1-x)E_g(CIS) + xE_g(CGS) - bx(1-x) \tag{4.2.3}$$

where x is the Ga/(In+Ga) ratio, E_g(CIS) is the bandgap of $CuInSe_2$ (in eV), E_g(CGS) is the bandgap of $CuGaSe_2$ (in eV) and b is the bowing parameter (in eV). Generally the bowing parameter lies between 0.15–0.24 eV (Wei, 1998).

In the case of Cd(S,Te), the same type of equation applies and a bowing parameter of 1.88 eV is found (Lane, 2006).

List of Symbols

Symbol	Description	Unit
a	Lattice constant	Nm
c	Lattice constant	nm
E_g	Optical band gap	eV
$\varepsilon(\theta)$	Static Dielectric constant	–
$\varepsilon(\infty)$	High Frequency Dielectric constant	–
χ_e	Electron affinity	eV
m_e^*	Effective mass electron	m_0
m_h^*	Effective mass hole	m_0
μ_e	Electrical Mobility electron	$cm^2/V\,s$
μ_h	Electrical Mobility hole	$cm^2/V\,s$
η	Tetragonal ratio	–
U	Anion displacement	–
Tm	Melting point	K
α	Optical absorption coefficient	cm^{-1}

List of Acronyms

Acronym	Definition
FCC	Face-Centered Cubic
SH	Schönflies Notation
HM	Hermann-Mauguin Notation
Cd_i	Interstitial Cd
Te_i	Interstitial Te
Cu_i	Interstitial Cu
V_{Cd}	Cd vacancies
V_{Se}	Se vacancies

(*Continued*)

(*Continued*)

Acronym	Definition
V_{Cu}	Cu vacancies
Cd_{Te}	Cd substituting to Te
Cu_{In}	Cu substituting to In
In_{Cu}	In substituting to Cu

References

Albin, D.S., Carapella, J.J., Tuttle, J.R., and Noufi, R. (1991) The effect of copper vacancies on the optical bowing of chalcopyrite Cu(In,Ga)Se_2 alloys. *Materials Research Society Symposium Proceedings*, **228**, 267.

Aven, M., and Prener. J. (eds) (1967) *Physics and Chemistry of II-VI Compounds*. John Wiley and Sons, Inc., New York, pp. 211–212.

Basol, B.M. (1992) Processing high efficiency CdTe solar cells. *International Journal Solar Energy*, **12**, 25–35.

Begou, T., Walker, J.D., Attygalle, D. *et al.* (2011) Real time spectroscopic ellipsometry of $CuInSe_2$: Growth dynamics, dielectric function, and its dependence on temperature. *physica status solidi (RRL) - Rapid Research Letters*, **5** (7), 217–219.

Bohm, M., Huber, G., MacKinnon, A. *et al.* (1985) *Physics of Ternary Compounds/Physik der ternären Verbindungen (Numerical Data and Functional Relationships in Science and Technology)*. Springer, New York.

Bondar, I. and Orlova, N. (1985) Thermal expansion of copper aluminum selenide ($CuAlSe_2$), copper gallium selenide ($GuGaSe_2$), and copper indium selenide ($CuInSe_2$) *Inorganic Materials*, **21**, 967–970.

Chirila, A., Buecheler, S., Pianezzi, F. *et al.* (2011) Highly efficient Cu(In,Ga)Se_2 solar cells grown on flexible polymer films. *Nature Materials*, **10**, 857–861.

Ciszek, T. (1984) Growth and properties of $CuInSe_2$ crystals produced by chemical vapor transport with iodine. *Journal of Crystal Growth*, **70**, 405–410.

Fonash, S.J. (1981) *Solar Cell Device Physics*. Academic Press, New York.

Godecke, T.Z. (2000) Phase equilibria of Cu-In-Se I. Stable states and nonequilibrium states of the In_2Se_3-Cu_2Se subsystem. *Zeitschrift für Metallkunde*, **91**, 622–634.

Grimm, H.G. and Sommerfeld, A. (1926) Über den Zusammenhang des Abschlusses der Elektronengruppen im Atom mit den chemischen Valenzzahlen. *Zeitschrift für Physik*, **36** (1), 36–59.

Hartmann, H., Mach, R., and Selle, B. (1982) *Wide Gap II-VI Compounds as Electronic Materials*. North-Holland Publishing Company, New York.

Hinuma, Y., Gruneis, A., Kresse, G., and Oba, F. (2014) Band alignment of semiconductors from density-functional theory and many-body perturbation theory. *Physics Review B*, **90** (15), 155405–155420.

Hultgren, R. R. (1971) *Selected Values of the Thermodynamic Properties of Binary Alloys*. American Society for Metals, Russell Township, Ohio: International Committee for Diffraction Data, Card Number 15–770.

Jackson, P., Hariskos, D., Wurz, R. *et al.* (2015) Properties of Cu(In,Ga)Se_2 solar cells with new record efficiencies up to 21.7%. *Physica status solidi: Rapid Research Letters*, **9** (1), 28–31.

Jackson, P., Wurz, R., Rau, U. *et al.* (2007) High quality baseline for high efficiency, $Cu(In_{1-x},Ga_x)Se_2$ solar cells. *Progress in Photovoltaics: Research and Applications*, **15**, 507–519.

Jaffe, J.E. and Zunger, A. (1993) Electronic-structure of the ternary chalcopyrite semiconductors $CuAlS_2$, $CuGaS_2$, $CuInS_2$, $CuAlSe_2$, $CuGaSe_2$, and $CuInSe_2$. *Physical Review B*, **28** (10), 5822.

Jianrong, Y., Silk, N.J., Watson, A. *et al.* (1995) Thermodynamic and phase diagram assessment of the Cd-Te and Hg-Te systems. *Calphad*, **19** (3), 399–414.

Kazmerski, L. (1983) Optical properties and grain boundary effects in $CuInSe_2$, *Journal of Vacuum Science and Technology A*, (**1**), 395–398.

Lane, D.W. (2006) A review of the optical band gap of thin film CdS_xTe_{1-x}. *Solar Energy Materials and Solar Cells*, **90** (9), 1169–1175.

Laurenti, J.P., Camassel, J., Bouhemadou, A. *et al.* (1990) Temperature dependence of the fundamental absorption edge of mercury cadmium telluride. *Journal of Applied Physics*, **67**, 6454.

Madelung, O. (1992) *Semiconductors Other than Group IV Elements and III-V Compounds*. Springer, New York.

Mao, D., Wickersham, C.E., and Gloeckler, M. (2014) Measurement of chlorine concentrations at CdTe grain boundaries. *IEEE Journal of Photovoltaics*, **4** (6), 1655–1658.

Neumann, H. (1986a) Optical properties and electronic band structure of $CuInSe_2$. *Solar Cells*, **16**, 317.

Neumann, H. (1986b) Lattice vibrational, thermal and mechanical properties of $CuInSe_2$. *Solar Cells*, **16**, 399–418.

Neumann, H. and Tomlinson R.D. (1990) Relation between electrical properties and composition in $CuInSe_2$ single crystals. *Solar Cells*, **28**, 301–313.

Noufi, R., Axton, R., Herrington, C., and Deb, S. (1984) Electronic properties versus composition of thin film $CuInSe_2$, *Applied Physics Letter*, **45**, 668–670.

Paulson, P.D., McCandless, B.E., and Birkmire, R.W. (2004) Optical properties of $Cd_{1-x}Zn_xTe$ films in a device structure using variable angle spectroscopic ellipsometry. *Journal of Applied Physics*, **95** (6), 3010–3019.

Rakhshani, A. (1997) Electrodeposited CdTe-optical properties. *Journal of Applied Physics*, **81** (12), 7988–7993.

Rau, U. and Schock, U.W. (2001) *$Cu(In,Ga)Se_2$ Solar Cells*. Imperial College Press, London.

Repins, I., Contreras, M.A., Egaas, B. *et al.* (2008) 19.9%-efficient ZnO/CdS/$CuInGaSe_2$ solar cell with 81.2% fill factor. *Progress in Photovoltaics: Research and Applications*, **16**, 235–239.

Romeo, N. (1980) Solar-cells made by chalcopyrite materials. *Japanese Journal of Applied Physics*, **19** (3), 5–13.

Schock, H. (2004) *Properties of Chalcopyrite-based Materials and Film Deposition for Thin-Film Solar Cells*. Springer, Berlin.

Shahidi, A.V., Shih, I., Araki, T., and Champness, C.H. (1985) Structural and electronic properties of $CuInSe_2$. *Journal of Electronic Materials*, **14** (3), 297–310.

Slack, G.A., and Galginaitis, S. (1964) Thermal conductivity and phonon scattering by magnetic impurities in CdTe. *Physical Review*, **133**, A253.

Suri, D., Nagpal, K., and Chadha, G. (1989) X-ray study of $CuGa_xIn_{1-x}Se_2$ solid solutions. *Journal of Applied Crystallography*, **22** (578), JCPDS 40-1487.

Wasim, S. (1986) Transport properties of $CuInSe_2$. *Solar Cells*, **16**, 289–316.

Wei, S.H., and Zhang, S.B. (2002) The case of CdTe Chemical trends of defect formation and doping limit in II-VI semiconductors. *Physical Review B*, **66**, 155211.

Wei, S.H., Zhang, S.B., and Zunger, A. (1998) Effects of Ga addition to $CuInSe_2$ on its electronic, structural, and defect properties. *Applied Physics Letters*, **72**, 3199.

Zhang, S.B., Wei, S.H., and Zunger, A. (1998) Defect physics of the $CuInSe_2$ chalcopyrite semiconductor. *Physical Review B*, **57** (16).

4.3

Contacts, Buffers, Substrates, and Interfaces

Negar Naghavi
Centre National de la Recherche Scientifique, Institute of Research and Development on Photovoltaic Energy (IRDEP), Chatou, France

4.3.1 Introduction

In chalcogenide technologies, the nature of the substrate, and the composition and processing of buffers and contacts have a significant effect on the final device performance and its stability. Making progress in understanding the properties of these materials and their interfaces is key to exploring new directions for more efficient device and module production. The aim of this chapter is to provide an overview of the state of the art and recent developments for the most common materials used as contacts, buffers, substrates, and their impact on the performance of $Cu(In,Ga)(S,Se)_2$ (CIGS) and $Cu_2(Zn,Sn)(S,Se)_2$ (CZTS) solar cells.

4.3.2 Substrates

The preparation of CIGS and CZTS-based solar cells starts with the deposition of a Mo back contact layer on a glass or a flexible substrate. The choice of the substrate and properties of the Mo back contact are of prime importance to the final device quality. In these solar cells, the substrate material has to be chosen taking into account its thermo-mechanical properties, to offer compatibility with the subsequent deposition and manufacturing steps of the solar module. Stability and compatibility of the substrate material through the whole production process and operational lifetime of solar modules are mandatory requirements.

Photovoltaic Solar Energy: From Fundamentals to Applications, First Edition.
Edited by Angèle Reinders, Pierre Verlinden, Wilfried van Sark, and Alexandre Freundlich.

Companion website: www.wiley.com/go/reinders/photovoltaic_solar_energy

Table 4.3.1 Typical properties of substrate materials used for CIGS solar cells (Reinhard *et al.*, 2013)

	CTE (10^{-6} K^{-1})	Thickness (μm)	Max T (°C)	Density (g/cm^3_)
Soda-lime glass (SLG)	9	2000–5000	600	2.4–2.5
Borosilicate glass	3.3	2000–5000	>600	2.2
Stainless steel	10–11	25–200	>600	8
Mild steel	13	25–200	>600	7.9
Cu	16.6	50–100	>600	8.9
Ni/Fe alloys	5–11		>600	8.3
Ti	8.6	25–100	>600	4.5
Mo	4.8–5.9	100	>600	10.2
Al	23	100	600	2.7
ZrO_2	5.7	50–300	>600	5.7
Polyimide (PI), i.e. Kapton™ or Upilex™	12–24	12.5–75	<500	1.4
polyethylene terephthalate (PET)	60	300–400	<500	1.4

The substrate material for CIGS and CZTS-based solar cells has to fulfill several physical and chemical requirements (Kessler and Rudmann, 2004; Reinhard *et al.*, 2013). These include:

- vacuum compatibility;
- thermal stability at temperatures higher than 500 °C;
- chemical inertness and no reaction with Mo nor absorber layer;
- moisture barrier and substantial stability on aggressive environments;
- surface smoothness;
- mechanical strength;
- similar thermal expansion coefficient to Mo ($4.8–5.9 \times 10^{-6}$/K) and CIGS ($7–11 \times 10^{-6}$/K).

And for flexible solar cells:

- low cost, and lightweight substrates.

Table 4.3.1 gives an overview of the relevan t properties of materials used as substrates for CIGS deposition processes (Reinhard *et al.*, 2013).

4.3.2.1 Glass Substrates

Glass is the most common substrate material used by manufacturers of CIGS modules. It is available in large quantities at low cost and fulfills all the above requirements. However, the choice of a given glass is of prime importance for the final device quality. As an example Na, often present in common glass, diffuses through the Mo film into the growing absorber, and plays a crucial role in the device performance. It has been shown that it diffuses to the grain boundaries, catalyzes oxygenation and passivation of Se vacancies (Guillemoles, 1999), favors the formation of $MoSe_2$ (or MoS_2) (Kohara *et al.*, 2001; Kessler *et al.*, 2005), and plays a role in the growth of $Cu(In,Ga)Se_2$.

Two types of glass are mainly used:

1. Soda-lime glass (SLG) which contains 73% silica, 14% Na_2O, 9% CaO (by mass) as well as other elements in smaller quantities (<1%). They provide sources of alkali impurities such as Na, which diffuse into the Mo and CIGS films during processing. Sodium plays an important role in the growth and final properties of CIGS solar cells. Moreover soda lime glass has a thermal expansion coefficient of 9×10^{-6}/K (Boyd and Thompson, 1980), which provides a good match to the CIGS and Mo films.
2. Borosilicate glass which contains 70% silica, 10% B_2O_3, 8% Na_2O, 8% K_2O, 1% CaO. The borosilicate glass has the advantage of containing less sodium, which helps to control the amount of Na doping of CIGS absorbers. A process that provides a more controllable supply of Na than diffusion from the glass substrate could be preferred for stability. However, borosilicate glass is more expensive than soda-lime glass and its coefficient of thermal expansion (3.3×10^{-6}/K) is two to three times lower than that of CIGS.

4.3.2.2 Flexible Substrates

Thin film CIGS technologies must evolve towards specific high potential markets where they will be more competitive than standard silicon technology. This is the case for the applications where lightweight modules are needed, in particular for building integrated photovoltaics (BIPV) or space applications with constraints on the weight of the panels. Although glass is the most common substrate material used by manufacturers of CIGS modules and leads to the best solar cell efficiencies, the glass substrate has the disadvantage of being heavy and not suitable for roll-to-roll processes that are used in the manufacturing of flexible panels. Consequently, CIGS technologies using substrates made of lightweight metallic or plastic foils are being investigated specifically for the flexibility of the panel.

4.3.2.3 Metallic Substrates

Metallic substrates have a heat-resistance advantage at the temperatures of synthesis of CIGS films (higher than 500°C). However, they have essentially two disadvantages. First, unpolished metal foils generally are quite rough and need special treatment before deposition of an additional layer can be performed. Moreover metallic substrates require a barrier layer to prevent diffusion of metal atoms into the absorber (Martinez *et al.*, 2001), which has negative effects on the photovoltaic properties of the solar cells.

The most common metal foils used for CIGS solar cells are made of stainless steel, Ti, Cu, Mo, or nickel alloys (Kovar). However, upon annealing, only titanium and Kovar have led to crack-free molybdenum layers (Table 4.3.1).

The barrier layers should also withstand high temperature processing and if they are not conductive, these can also serve as an electrical isolation of the individual devices from the substrate and thus present a benefit for monolithical integration schemes for modules. SiO_2 is the most common barrier layer used in CIGS solar cells. Other types of barrier layers can be used, such as Al_2O_3 and Cr, but the inconvenience is that they can also diffuse into the CIGS material (Martinez *et al.*, 2001). Thus far, CIGS efficiencies of 17.6% (Wuerz *et al.*, 2012) on stainless steel and 17.7–17.9% on titanium have been achieved. ZSW (Powalla *et al.*, 2003), Flisom, Solopower, and other manufacturers have already produced complete modules on flexible metallic substrates (Reinhard *et al.*, 2013).

4.3.2.4 Polymer Substrates

Polymer substrates used in conjunction with CIGS are usually made of special heat-resistant polyimides (PI), like Upilex™ or Kapton™. Nevertheless the highest temperatures they can withstand are around 450–500 °C which remain lower than usual deposition temperatures on glass. However, these substrates have the advantage of allowing a process with fewer steps (i.e. no electrical isolation barrier layer required) and with well-defined surfaces (Chirila *et al.*, 2011a). CIGS devices with efficiencies of up to 20.4% (Chirila *et al.*, 2011b) have been achieved on polyimide substrates. The deposition of Mo tends to bend this type of substrates, but this problem can be circumvented by depositing an additional strain compensation layer of material (insulation, titanium or molybdenum) on the back of the substrate (Wada *et al.*, 1996). In these solar cells, the Na is added either via evaporation of a NaF layer prior to the absorber deposition (Powalla and Dimmler, 2000) or by post deposition and annealing of a NaF layer on the top of the absorber layer (Rudmann *et al.*, 2004; Chirila *et al.*, 2011a).

4.3.3 Back Contact

4.3.3.1 Mo Back Contact

CIGS and CZTS solar cells are mainly grown on molybdenum-coated substrates. Mo is usually grown by sputtering or electron-beam evaporation. The intensive usage of Mo as a back contact, compared to other alternatives is due to the *in-situ* formation at the Mo/Cu(In,Ga)Se_2 interface of a $MoSe_2$ layer that favors the formation of an ohmic contact (Shafarman and Phillips, 1996; Assmann *et al.*, 2005; Zhu *et al.*, 2012). Mo deposition is also crucial to the control of the incorporation of Na that diffuses from the soda-lime glass into the absorber (Bommerbach *et al.*, 2013). The thickness of the Mo layer is determined by the electrical resistance requirements that depend on the specific cell or module configuration. Typically, the Mo back contact presents a sheet resistance of about 0.2 ohm/sq, and provides good adhesion of the semiconductor to the substrate (Scofield *et al.*, 1995).

The two modes of sputter deposition of Mo most commonly used are depositions assisted by radio-frequency sputtering (Assmann *et al.*, 2005) or by DC magnetron (Powalla and Dimmler, 2000) under an argon atmosphere. To achieve both a good adhesion and a good conductivity, Mo multilayers are commonly applied as back contact. First, a thin adhesion layer is deposited at high argon pressure and/or high sputter power, followed by a second thicker and low-resistance layer that is deposited at low sputter power.

4.3.3.2 Alternative Back Contacts

Various alternative metal contacts to p-type CIGS were examined by Matson *et al.* (Au, Ni, Al, Ag) concluding that only Au and Ni ensure the formation of ohmic contacts (Matson *et al.*, 1984). Orgassa *et al.* (2003) fabricated CIGS solar cells with different back contact materials (W, Mo, Cr, Ta, Nb, V, Ti, Mn) and found that W and Mo contacts provide the best CIGS/back-contact interface passivation. For CZTS-based solar cells, numerical simulations using the SCAPS program have shown that metals with higher work functions than Mo, such as Au, In, Ni, Co, Mo, Pd, Pt, Re and W, could improve their V_{oc} (Patel and Ray, 2012).

In the case of very thin absorber layers (<1 μm), using a back reflector, such as Ag, Au or white paint, can lead to significantly increased short circuit current (Malmstrom, 2005; Erfurth *et al.*, 2011). The replacement of Mo by an Au back contact after a lift-off process of CIGS from Mo does not degrade the properties of the cell, as good ohmicity on CIGS is achieved (Jehl *et al.*, 2012; Fleutot *et al.*, 2014). For a standard CIGS layer, an efficiency of 14.3% is measured, which is higher than the initial reference CIGS solar cell efficiency (13.8%). For a CIGS thickness of 0.4 μm, an improvement of photocurrent is achieved by the enhanced light-trapping effect due to the Au back reflector, leading to an absolute efficiency increase of +2.5% (Jehl *et al.*, 2012).

In the case where a transparent back contact is necessary, such as for a superstrate configuration or bifacial solar cells, transparent conductive oxides such as ZnO:Al, SnO_2:F or indium tin oxide (ITO) are used instead of molybdenum (Nakada *et al.*, 2004; Rostan *et al.*, 2005; Heinemann *et al.*, 2014). However, in these cases, it is necessary to deposit a thin layer of Mo, $MoSe_2$ or MoO_3 on top to ensure a good contact ohmicity (Nakada, 2005; Simchi, 2014). CIGS solar cells with efficiencies of up to 15.2% were obtained using a thin interfacial layer of $MoSe_2$. More recently, a back-wall superstrate device structure that outperforms CIGS devices made with conventional substrates has been demonstrated using absorber thickness range of 0.1–0.5 μm. It should be noted that in these back-wall cells a thin layer of MoO_3 was deposited on top of the glass/ITO back contact (Simchi, 2014).

4.3.3.3 Nanostructured Back Contact

Back contact recombination is the dominant path if the diffusion length of minority carriers is large compared to the quasi-neutral region thickness. This mechanism is scaled with the inverse of the absorber thickness, thus it becomes predominant for thin and ultrathin samples. Reduced recombination at the standard rear-contact/CIGS interface is an issue for the optimization of open circuit voltage (V_{oc}) in ultrathin solar cells.

Rear surface recombination in highly efficient CIGS solar cells is mostly limited by using Ga composition grading that is used to create a back surface field (BSF) (Lundberg *et al.*, 2003). Stemming from the Si solar cell industry, the introduction of a rear surface passivation layer with nano-sized contacts was explored (Vermang *et al.*, 2013). Atomic layer deposition or sputtered Al_2O_3 is used to passivate the CIGS surface (Vermang *et al.*, 2014). Such a passivation layer is known to reduce the interface recombination by chemical passivation (due to a reduction in interface trap density) and field-effect passivation (due to a fixed charge density in the passivation layer that reduces the surface minority or majority charge carrier concentration), while the point openings allow for contacting. These nanostructured back contacts have led to significant improvements in V_{oc} of standard CIGS solar cells and result in an improvement of both V_{oc} and J_{sc}, when ultrathin (<500 nm) CIGS absorbers are used (Vermang *et al.*, 2014).

4.3.3.4 $MoSe_2$ Interface Layer

During selenized absorber deposition, an interfacial $MoSe_2$ layer is formed (Wada *et al.*, 2001) at the Mo surface, which facilitates a quasi-ohmic contact between CIGS and Mo (Niki *et al.*, 2010). $MoSe_2$ can be considered a buffer layer between Mo and CIGS, which leads to a better energy-band alignment between these two layers and thus reduces recombination at the interface (Scofield *et al.*, 1995), thus favoring the contact ohmicity (Shafarman and Phillips, 1995), which stresses the importance of the $MoSe_2$ interface formation. $MoSe_2$ is a lamellar semiconductor with p-type conduction, a bandgap of 1.3 eV and weak van der Waals bonding along the *c*-axis. In fact, $MoSe_2$ is regularly found in a hexagonal phase, which consists of Se-Mo-Se sheaths oriented perpendicular to the c-axis, and its anisotropic crystal structure has led to it use as a lubricant (Kubart *et al.*, 2005)

The crystal orientation of the $MoSe_2$ grains is important for the fabrication of solar cells. Due to the lamellar-structure of the $MoSe_2$ compound, if the *c*-axis is oriented perpendicular to the surface of the Mo layer, the CIGS film would tend to delaminate from the Mo layer, leading to cleavage of the interfacial bonds and an unfavorable electronic transport. The formation and orientation of $MoSe_2$ are highly dependent on the substrate and the deposition conditions (temperature, Se activity) (Abou-Ras *et al.*, 2005; Zhu *et al.*, 2012). Fortunately, in most standard CIGS solar cells, the *c*-axis is found to be parallel, and the Van der Waals planes are thus perpendicular to the interface.

A detailed study and in-depth characterization of the Mo/CIGS interface for samples leading to excellent lift-off properties have shown a specific gallium-rich CIGS graded interface structure according to the interfacial sequence glass/Mo/$MoSe_2$/Ga_xSe_y/Ga-rich-CIGS (Fleutot *et al.*, 2014). The excellent lift-off was associated with the formation of a Ga_xSe_y interfacial layer and not only with the presence of $MoSe_2$. The composition gradients of the CIGS resulting from the deposition conditions seem to play a key role in the formation of the Ga_xSe_y interfacial layer. The lamellar Ga_xSe_y structure with Van der Waals planes parallel to the substrate thus appears to ease the lift-off process at the interface Ga_xSe_y/CIGS films.

4.3.4 Front-Contact Layers

In solar cells, an n-type layer is required to complete the p-n junction with CIGS. Historically this layer consisted of only one material (CdS) (Kazmerski *et al.*, 1967), but later evolved into three separate layers (CdS/i-ZnO/ZnO:Al) (Naghavi *et al.*, 2010), with the collective name window layer stack or front contact layers. The reason why it is called window is that all of the materials in the stack are transparent or almost transparent to sunlight, since it is preferable to absorb all light in the CIGS. In the classical chalcopyrite cell, the CdS buffer layer establishes the interface properties, the undoped ZnO (i-ZnO) decouples poorly performing locations, and the highly doped ZnO:Al is responsible for lateral current transport, as is explained below.

4.3.4.1 Buffer Layers

The name buffer layer is derived from the function of acting as an electrical and optical buffer between the p-type CIGS and the n-type ZnO layers. This layer is critical since it defines the p-n junction quality and the device performance when it is deposited on CIGS or CZTS (Naghavi *et al.*, 2010). CdS is the most commonly used buffer layer for CIGS and CZTS thin-film solar cells grown by the so-called chemical bath deposition (CBD) technique. All CIGS record devices on laboratory or even on industrial scale use CBD CdS (Chirila, 2013; Jackson *et al.*, 2015). Nevertheless, recently, efficiencies higher than 20% with CIGS absorbers have also been obtained for CBD Zn(S,O) buffer layers (ZSW 2015) (Osborne, 2014).The CIGS/buffer interface plays a key role in terms of recombination, inter-diffusion, formation of the p-n junction, and the overall device performance (Malm, 2008; Rau *et al.*, 2009).

These buffer-layers must fulfill the following functions:

- The material should be n-type in order to form a p-n junction with the absorber layer.
- The bandgap should be large enough (>2.5 eV) to prevent excessive light absorption.
- The process and material choice of the buffer layer should provide an alignment of the conduction band with the CIGS-based absorber and with the window layers.
- The process for deposition must have the capability to passivate the surface states of the absorber layer.
- The deposition should be highly conformal.

4.3.4.1.1 CdS Buffer Layers

The highest and most reproducible solar-cell efficiencies are obtained by use of the classical chemical-bath deposited (CBD) CdS buffer layer. It should be noted that CIGS cells made with CdS buffers deposited by physical vapor deposition (PVD) have always shown significantly lower efficiencies than cells fabricated with CBD-CdS buffers.

CBD CdS involves the precipitation from the solution of CdS on the absorber. With this method, a thin and conformal buffer layer (30–60 nm) is usually obtained after 5–10 min deposition at a temperature ranging from 60–70 °C, by submerging the substrate into an aqueous ammonia solution in the presence of thiourea ($SC(NH_2)_2$), which acts as a sulfur donor, and cadmium salt (such as $Cd(C_2H_3O_2)_2)_4$, or $CdSO_4$).

The overall reaction of CdS formation can be described as (Ortega-Borges and Lincot, 1993):

$$Cd\left(NH_3\right)_2^{+4} + SC\left(NH_2\right)_2 + 2OH^- \rightarrow CdS + CN_2H_2 + {}_4NH_3 + 2H_2O \qquad (4.3.1)$$

It has been suggested that not only buffer layers deposited by the CBD process function as a protection of the p-n junction from plasma damage during subsequent ZnO sputtering and prevent undesirable shunt paths, but also that chemically deposited buffer layers allow for a conformal and pinhole-free coating of the absorber by the CdS layer. Moreover, when the glass/Mo/absorbers stack is immersed into the chemical bath for the buffer deposition, the CIGS layer surface is probably also subjected to beneficial chemical etching of the surface, e.g., the ammonia used in most CBD recipes is thought to be crucial for the cleaning of the absorber surface by removal of oxides and other impurities (Kessler *et al.*, 1992). In addition, n-type doping of the CIGS region close to the surface by Cd from the CBD solution has been reported (Nakada and Kunioka, 1999; Hiepko *et al.*, 2011). It should be noted that, despite the extreme experimental simplicity, understanding the mechanisms involved in the deposition and the ability to widen the range of deposits obtained (both in composition and the control of other properties) are not so simple.

4.3.4.1.2 Cd-Free Buffer Layers

Because of both cadmium's adverse environmental impact and the somewhat narrow bandgap of CdS (2.4–2.5 eV), one of the major objectives in the field of CIGS technology remains the development and implementation in the production line of Cd-free alternative buffer layers.

The development of Cd-free devices started in 1992 and intensively continued, leading to the current efficiency level of 20.9–21% for cells (Osborne, 2014; ZSW, 2015) and of 17.5% for 30x30 cm^2 sub-modules (Nakamura *et al.*, 2012), using a CBD-Zn(S,O) buffer. However, the Japanese company Solar Frontier is the only one offering Cd-free CIGS modules with a record efficiency of 13.5% (Green *et al.*, 2013). Unfortunately, the material properties of CBD-CdS and the way that it forms the p-n junction with the CIGS seem to be very crucial for good solar cell performance and have so far been hard to reproduce with other materials. Thus, alternative buffer layers need to mimic the properties of CBD-CdS or compensate for them by other means in order to perform as well as CdS.

A general conclusion of the investigations during the last decade suggests that the most relevant Cd-free materials are films based on In_2S_3, Zn(S,O), $Zn_{1-x}Mg_xO$, and $Zn_{1-x}Sn_xO$ (Table 4.3.2) (Witte *et al.*, 2014). All these materials have the potential to replace CdS with solar cell efficiencies approaching or even exceeding 20%. However, whereas for CBD-CdS

Table 4.3.2 Small-area Cu(In,Ga)Se_2 based champion cells with alternative buffer layers which exhibit conversion efficiencies > 16 % in comparison to the world record CIGS cell with a CdS buffer: All denoted efficiencies refer to total-area values with anti-reflective coating if not stated otherwise. Typical cell area A is around 0.5 cm^2

Buffer	Method	Window	η [%]	Institute/Company
CdS	CBD	sputt. i-ZnO/ZnO:Al	21.7	ZSW
Zn(S,O)	CBD	sputt. i-ZnO/ZnO:Al	18.6	NREL/AGU
Zn(S,O)	CBD	sputt. ZnO:Al	18.5	NREL
Zn(S,O)	CBD	sputt. $Zn_{1-x}Mg_xO$/ZnO:Al	21.0	ZSW
Zn(S,O)	CBD	MOCVD-ZnO:B	20.9[2]	Solar Frontier
Zn(S,O)	ALD	sputt. i-ZnO/ZnO:Al	18.5	ASC
Zn(S,O)	ALD	sputt. $Zn_{1-x}Mg_xO$/ZnO:Al	18.1	ZSW/Beneq
Zn(S,O)	ALD	ALD-i-ZnO/MOCVD-ZnO:B	18.7	TUS
Zn(S,O)	Sputt.	sputt. ZnO:Al	18.3	HZB
In_2S_3	ALD	sputt. i-ZnO/ZnO:Al	16.4[1] *(A = 0.1 cm^2)*	ENSCP/ZSW
In_2S_3	ILGAR	sputt. i-ZnO/ZnO:Al	16.8[1,2]	HZB
In_2S_3	Evap.	sputt. i-ZnO/ZnO:Al	17.1	ZSW
$Zn_{1-x}Mg_xO$	ALD	sputt. ZnO:Al	18.1	ASC
$Zn_{1-x}Mg_xO$	Sputt.	sputt. ITO	16.2	Mat. El.
$Zn_{1-x}Sn_xO_y$	ALD	sputt. i-ZnO/ZnO:Al	18.2	ASC

Notes: evap. = evaporation, sputt. = sputtered, [1]Without antireflective coating, [2]CIGS absorber with sulfur in addition to selenium.
Source: Osborne (2014); Witte *et al.* (2014); Jackson *et al.* (2015); ZSW (2015).

buffers, high and reproducible efficiencies are obtained regardless of the absorber used, Cd-free cells often reach their highest efficiencies after light soaking and/or heat treatments.

As shown in Table 4.3.2, the Cd-free materials are deposited using different deposition techniques such as chemical-bath deposition (CBD), atomic-layer deposition (ALD), ion-layer gas reaction (ILGAR), sputtering or evaporation (PVD). Likewise, a comparison of the efficiencies of Cd-free-based solar cells shows that for the most relevant materials mentioned above, the ability to reach efficiencies equivalent to or higher than the corresponding CdS reference cells is highly dependent on the deposition technique used (Table 4.3.2). Often, photovoltaic performance of solar cells made with vacuum-deposited buffer layers is found to be inferior to that of solar cells made with chemical-based buffer layer such as CBD or atomic-layer deposition (ALD). However, the "dry" deposition methods such as sputtering and thermal evaporation could be favorable to the scale-up of manufacturing and processes using industrial scale roll-to-roll coaters. In contrast to the application of CBD in solar cell processing, the PVD techniques can easily be implemented in an in-line system after the absorber deposition without the need to break the vacuum between both process steps.

4.3.4.2 Front Contact and Window Layers

At the front of a solar cell, the window layer acts as an electrical contact as well as a window to allow light through to enter the PV absorber layer. The most common window layers applied in CIGS and CZTS solar cells in fact consist of two layers. The first layer is a thin and highly resistive layer of an i-ZnO film, which is mostly deposited by radio frequency (rf)-magnetron sputtering and normally has a thickness of about 50–100 nm. The second layer is the actual front contact and consists of a transparent conducting ZnO:Al layer, also deposited by rf magnetron sputtering. Typical ZnO:Al thicknesses are 300–500 nm.

As depositing a film with low lateral resistance directly onto the buffer increases the negative influence of local defects, such as pinholes and local fluctuations of absorber properties (e.g., bandgap) (Rau and Schmidt, 2001; Grabitz *et al.*, 2005), first depositing a thin (100 nm) ZnO film with lower conductivity, designated as i-ZnO avoids this. On top of that i-ZnO layer the transparent conducting oxide (TCO) is deposited. The conducting window layer must be highly transparent to move the absorption maximum into the cell, create an interface to the absorber with low recombination losses, and limit the influence of locally distributed shunts and performance fluctuations. In addition, it must have a low sheet resistance to laterally transport the current over macroscopic distances to the nearest metal contact finger or interconnect. Window layers are selected according to their conductivity, transparency, moisture stability, and their compatibility with further processing steps. Reflection, absorption, resistive losses and lost active area, either from the scribed interconnect region in monolithically integrated modules or from the shadow losses of a metal grid in standard modules, typically can reduce the efficiency of the solar cells by 10–25% (Rowell and McGehee, 2011).

For economic and ecological reasons, magnetron sputtering of ZnO doped with Al (ZnO:Al, AZO) or Ga (GZO) is the industrial standard TCO for CIGS-based solar cells today, allowing for high efficiencies at moderate cost. However, some groups use ZnO:B (BZO) deposited by metal organic chemical vapor deposition (Nakamura *et al.*, 2013) as TCO, which often has a higher mobility and a lower absorption compared to ZnO:Al.

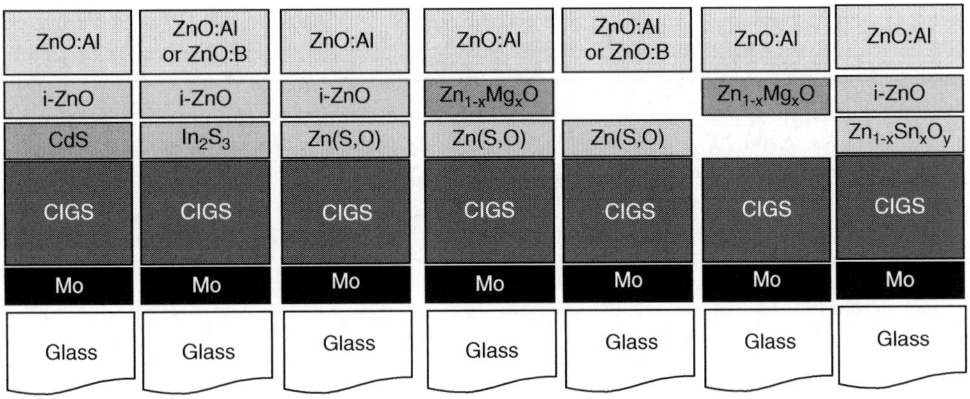

Figure 4.3.1 Scheme of different CIGS cells stacks on glass/Mo substrates with alternative buffer layer systems compared with the commonly used CdS/i-ZnO. Source: adapted from Witte *et al.* (2014)

In the case of some Cd-free buffer layers, the nature and deposition condition of the undoped window layer can strongly influence the performance of cells (Naghavi *et al.*, 2011). For example, for CBD-Zn(S,O) buffer layers, the replacement of i-ZnO by $Zn_{1-x}Mg_xO$ can improve the efficiency and stability of the cells (Hariskos *et al.*, 2009). The Mg addition to ZnO increases the bandgap energy and results in a possibly more favorable band alignment. A more radical approach is to omit the i-ZnO or $Zn_{1-x}Mg_xO$ completely with a direct deposition of the ZnO:B (Nakamura, 2013) onto the Zn(O,S) buffer. Figure 4.3.1 shows the various options discussed above.

Recent investigations have also focused on the development of TCO material by scalable solution processing such as CBD or electrodeposition, which could translate into the reduction of CIGS module costs, lower the equipment complexity, and increase material utilization. Electrodeposition of conductive ZnO:Cl has successfully been applied for co-evaporated CIGS solar cells with efficiencies of up to 15.8% (Rousset *et al.*, 2011). A promising efficiency of 14.8% (Hagendorfer *et al.*, 2014) has been obtained using a novel solution approach based on CBD for highly conductive ZnO:Al.

After the deposition of the TCO layer an anti-reflection layer of MgF_2 is usually deposited on top of the cells in order to limit the reflection losses in these solar cells. Finally, the current collection by the window layer in cells is commonly supported by an Ni-Al metal grid deposited on the front contact. Ni is applied to reduce the formation of a resistive Al_2O_3 barrier. The Ni and Al layers are deposited by electron-beam evaporation with thicknesses of about 50 nm and 1 μm. Although the application of a front contact grid is beneficial to the solar cell performance, it also contributes to a reduction of the active cell area and thus detrimentally affects the overall cell/module efficiency.

4.3.5 Conclusion

As shown in this chapter, the material used and the composition and processing of the buffers and contacts have significant effects on the chalcogenide device performance and their stability. Remarkable progress has been made in the development and understanding of contacts, windows and buffer layers and the generated interfaces in these solar cells, thus opening new perspectives. For the next generation of chalcogenide solar cells such as flexible and lightweight modules, ultrathin or indium free absorbers, or tandem or bifacial devices, the choice of contacts, buffers and the design of interfaces will have a large impact on the device performance. The interfaces will have to be redesigned and re-adapted to further decrease losses, in particular, losses associated with interface recombination.

List of Acronyms

Acronym	Definition
CIGS	$Cu(In,Ga)(S,Se)_2$
CZTS	$Cu_2(Zn,Sn)(S,Se)_2$
Mo	Molybdenum
SLG	Soda-Lime Glass
BIPV	Building Integrated Photovoltaic
NaF	Sodium fluoride
TCO	Transparent Conducting Oxide
ZnO:Al	Aluminum-doped Zinc Oxide
SnO_2:F	Fluorine-doped Tin Oxide
ITO	Tin-doped Indium Oxide
ZnO:B	Boron-doped Zinc Oxide
ZnO:Cl	Chlorine-doped Zinc Oxide
i-ZnO	Intrinsic Zinc Oxide
BSF	Back Surface Field
V_{oc}	Open Circuit Voltage
CdS	Cadmium Sulfide
Cd-Free	Cadmium free
CBD	Chemical-bath deposited
PVD	Physical vapor deposition
ALD	Atomic layer deposition
MOCVD	Metalorganic chemical vapor deposition
ILGAR	Ion layer gas reaction
evap.	Evaporation
sputt.	Sputtered

References

Abou-Ras, D., Kostorz, G., Bremaud, D. *et al.* (2005) Formation and characterisation of $MoSe_2$ for $Cu(In,Ga)Se_2$ based solar cells. *Thin Solid Films*, **480/**481, 433–438. DOI:10.1016/j.tsf.2004.11.098.

Assmann, L., Bernède, J.C., Drici, A. *et al.* (2005) Study of the Mo thin films and Mo/CIGS interface properties, *Applied Surface Science*, **246** (1–3), 159–166.

Bommersbach, P., Arzel, L., Tomassini, M. *et al.* (2013) Influence of Mo back contact porosity on co-evaporated $Cu(In,Ga)Se_2$ thin film properties and related solar cell, *Progress in Photovoltaics: Research and Applications*, **21** (3), 332–343.

Boyd, D. and Thompson, D. (1980) *Kirk-Othmer Encyclopaedia of Chemical Technology*, 3rd edition, John Wiley & Sons Inc, New York, **11**, 807–880.

Chirila, A., Bloesch, P., Pianezzi, F. *et al.* (2011a) Development of high efficiency flexible CIGS solar cells on different substrates. *Proceedings of 21st International Photovoltaic Science and Engineering Conference*, Fukuoka, Japan, 4B-3T-02.

Chirila, A., Buecheler, S., Pianezzi, F. *et al.* (2011b) Highly efficient Cu(In,Ga)Se_2 solar cells grown on flexible polymer films. *Natural Materials*, **10** (11), 857–861.

Chirila, A., Reinhard, P.,Pianezzi, F. *et al.* (2013) Potassium-induced surface modification of Cu(In,Ga)Se_2 thin films for high-efficiency solar cells. *Nature Materials* **12**, 1107–1111. DOI: 10.1038/NMAT3789.

Erfurth, F., Jehl, Z., Bouttemy, M. *et al.* (2011) Mo/Cu(In,Ga)Se_2 back interface chemical and optical properties for ultrathin CIGSe solar cells. *Applied Surface Science*, **258** (7), 3058–3061.

Fleutot, B., Lincot, D., Jubault, M. *et al.* (2014) Easily transferable CIGS thin film solar cells from improved Van-der-Waals mechanical lift-off, *Advanced Materials Interfaces*, **1/4**, 1400044, DOI: 10.1002/admi.201400044.

Grabitz, O., Rau, U. and Werner, J.H. (2005) Electronic inhomogeneities and the role of the intrinsic ZnO layer in Cu(In,Ga)Se_2 thin film solar cells. *Proceedings of the 20th European Photovoltaic Solar Energy Conference*, pp. 1771–1774.

Green, M.A., Emery, K., Hishikawa, Y. *et al.* (2013) Solar cell efficiency tables (version 42) *Progress in Photovoltaic Research Applications*, **21**, 827. DOI: 10.1002/pip.2404.

Guillemoles, J.F., Rau, U., Kronik, L. *et al.* (1999) Cu(In,Ga)Se_2 solar cells: stability based on chemical flexibility, *Advanced Materials*, **11** (11), 957–961.

Hagendorfer, H., Lienau, K., Nishiwaki, S. *et al.* (2014) Highly transparent and conductive ZnO: Al thin films from a low temperature aqueous solution approach. *Advanced Materials*, **26**, 632–636. DOI: 10.1002/adma201303186.

Hariskos, D., Fuchs, B., Menner, R. *et al.* (2009) The Zn(S,O,OH)/ZnMgO buffer in thin-film Cu(In,Ga)(Se,S)$_2$-based solar cells part ii: magnetron sputtering of the ZnMgO buffer layer for in-line co-evaporated Cu(In,Ga)Se_2 solar cells. *Progress in Photovoltaics: Research Applications*, **17**, 479–488.

Heinemann, M.D., Efimova, V., Klenk, R. *et al.* (2014) Cu(In,Ga)Se_2 superstrate solar cells: prospects and limitations. *Progress in Photovoltaics: Research Applications*. DOI: 10.1002/pip2536.

Hiepko, K., Bastek, J., Schlesiger, R. *et al.* (2011) Diffusion and incorporation of Cd in solar-grade Cu(In,Ga)Se_2 layers. *Applied Physics Letters*, **99** (23) 234101.

Jackson, P., Hariskos, D., Wuerz, R. *et al.* (2015) Properties of Cu(In,Ga)Se_2 solar cells with new record efficiencies up to 21.7%, *Physica Status Solidi-Rapid Research Letters*, **9** (1), 28–31, DOI: 10.1002/pssr.201409520.

Jehl Li Kao, Z., Naghavi, N., Erfurth, F. *et al.* (2012) Towards ultrathin copper indium gallium diselenide solar cells: proof of concept study by chemical etching and gold back contact engineering. *Progress in Photovoltaics: Research and Applications*, **20**, 582–587.

Kazmerski, L.L., White, F.R.. and Morgan, G.K. (1976) Thin film $CuInSe_2$/CdS heterojunction solar cells. *Applied Physical Letters*, **29**, 268. DOI:10.1063/1.89041.

Kessler, F., Herrmann, D. and Powalla, M. (2005) Approaches to flexible CIGS thin-film solar cells. *Thin Solid Films*, **480/**481, 491–498.

Kessler, F. and Rudmann, D. (2004) Technological aspects of flexible CIGS solar cells and modules. *Solar Energy*, **77** (6), 685–695.

Kessler, J., Velthaus, K.O., Ruckh, M. *et al.* (1992) Chemical bath deposition of CdS on CIS, etching effects and growth kinetics. *Proceedings of the 6th International Photovoltaic Science and Engineering Conference (PVSEC-6)*, pp. 1005–1010.

Kohara, N., Nishiwaki, S., Hashimoto, Y. *et al.* (2001) Electrical properties of the Cu(In,Ga)Se_2/$MoSe_2$/Mo structure. *Solar Energy Materials and Solar Cells*, **67** (1–4), 209–215.

Kubart, T., Polcar, T., Kopecký, L., *et al.* (2005) Temperature dependence of tribological properties of MoS_2 and $MoSe_2$ coatings. *Surface and Coatings Technology*, **193** (1–3), 230–233.

Lundberg, O., Bodegard, M., Malmstrom, J., and Stolt, L. (2003) Influence of the Cu(In,Ga)Se_2 thickness and Ga grading on solar cell performance. *Progress in Photovoltaics: Research and Applications*, **11**, 77.

Malm, U. (2008) Modelling and degradation characteristics of thin-film CIGS solar cells. PhD thesis, Uppsala University, Uppsala, Sweden.

Malmstrom, J. (2005) On generation and recombination in Cu(In,Ga)Se_2 thin film solar cells. PhD thesis, Uppsala University, Uppsala, Sweden.

Martínez, M., Guillén, C., Morales, A., and Herrero, J. (2001) Arrangement of flexible foil substrates for $CuInSe_2$-based solar cells. *Surface and Coatings Technology*, **148** (1), 61–64.

Matson, R.J., Jamjoum, O., Buonaquisti, A.D. *et al.* (1984) Metal contacts to $CuInSe_2$. *Solar Cells*, **11** (3), 301–305.

Naghavi, N., Abou-Ras, D., Allsop, N. *et al.* (2010) Buffer layers and transparent conducting oxides for chalcopyrite Cu(In,Ga)(S,Se)$_2$ based thin film photovoltaics: present status and current developments. *Progress in Photovoltaics: Research and Applications*, **18**, 411–433.

Naghavi, N., Renou, G., Bockelee, V. *et al.* (2011) Chemical deposition methods for Cd-free buffer layers in CI(G)S solar cells: role of window layers. *Thin Solid Films*, **519** (21), 7600–7605.

Nakada, T. (2005) Microstructural and diffusion properties of CIGS thin film solar cells fabricated using transparent conducting oxide back contacts. *Thin Solid Films*, **480/**481, 419–425.

Nakada, T., Hirabayashi, Y., Tokado, T. *et al.* (2004) Novel device structure for Cu(In,Ga)Se$_2$ thin film solar cells using transparent conducting oxide back and front contacts, *Solar Energy*, **77** (6), 739–747.

Nakada T. and Kunioka, A. (1999) Direct evidence of Cd diffusion into Cu(In,Ga)Se$_2$ thin films during chemical-bath deposition process of CdS films. *Applied Physics Letters*, **74** (17), 2444.

Nakada, T., Yagioka, T., Horiguchi, K. *et al.* (2009) T. CIGS thin film solar cells on flexible foils. *Proceedings of the 24th European Photovoltaic Solar Energy Conference*, Hamburg, Germany, pp. 2425–2428.

Nakamura, M., Chiba, Y., Kijima, S. *et al.* (2012) Achievement of 17.5% efficiency with 30x30cm$_2$-sized Cu(In,Ga)(Se,S)$_2$ submodules, *38th IEEE Photovoltaic Specialists Conference (PVSC)*, pp. 001807–001810.

Nakamura, M., Kouji, Y., Chiba, Y. *et al.* (2013) Achievement of 19.7% efficiency with a small-sized Cu(InGa)(SeS)$_2$ solar cells prepared by sulfurization after selenization process with Zn-based buffer. *Proceedings of the 39th IEEE Photovoltaic Specialists Conference*, pp. 0849–0852.

Niki, S., Contreras, M., Repins, I., *et al.* (2018) CIGS absorbers and processes. *Progress in Photovoltaics: Research and Applications*, **18** (6), 453–466.

Orgassa, K., Schock, H.W. and Werner, J.H. (2003) Alternative back contact materials for thin film Cu (In,Ga)Se$_2$ solar cells. *Thin Solid Films*, **431**, 387–391.

Ortega-Borges R. and Lincot, D. (1993) Mechanism of chemical bath deposition of cadmium sulfide thin films in the ammonia-thiourea system. *Journal of the Electrochemical Society*, **140** (12), 3464–3473.

Osborne (2014) *PVTech*, http://www.pv-tech.org, April 2014.

Patel, M. and Ray, A. (2012) Enhancement of output performance of Cu$_2$ZnSnS$_4$ thin film solar cells: a numerical simulation approach and comparison to experiments. *Physica B:Condensed Matter*, **407** (21), 4391–4397.

Pianezzi, F., Chirilă, A.P., Tiwari, A.N. *et al.* (2012) Electronic properties of Cu(In,Ga)Se2 solar cells on stainless steel foils without diffusion barrier. *Progress in Photovoltaics: Research and Applications*, **20**, 253–259.

Powalla M. and Dimmler, B. (2000) Scaling up issues of CIGS solar cells. *Thin Solid Films*, **361**, 540–546.

Powalla, M., Hariskos, D., Lotter, E. *et al.* (2003) Large-area CIGS modules: processes and properties, *Thin Solid Films*, **431/**432, 523–533.

Rau, U. and Schmidt, M. (2001) Electronic properties of ZnO/CdS/Cu(In,Ga)Se$_2$ solar cells aspects of heterojunction formation. *Thin Solid Films*, **387** (1–2), 141–146.

Rau, U. Taretto, K. and Siebentritt, S. (2009) Grain boundaries in Cu(In,Ga)(Se,S)$_2$ thin-film solar cells. *Applied Physics A*, **96**, 221–234.

Reinhard, P., Chirila, A., Blosch, P. *et al.* (2013) Review of progress toward 20% efficiency flexible cigs solar cells and manufacturing issues of solar modules. *IEEE Journal of Photovoltaics*, **3** (1), 572–580.

Rostan, P.J., Mattheis, J., Bilger, G. *et al.* (2005) Formation of transparent and ohmic ZnO:Al/MoSe$_2$ contacts for bifacial Cu(In,Ga)Se$_2$ solar cells and tandem structures. *Thin Solid Films*, **480/**481, 67–70.

Rousset, J., Saucedo, E., Herz, K. and Lincot, D. (2011) High efficiency CIGS based solar cells with electrodeposited ZnO:Cl as transparent conducting front contact. *Progress in Photovoltaics: Research Applications*, **19**, 537.

Rowell M.W. and McGehee, M.D. (2011). Transparent electrode requirements for thin film solar cell modules. *Energy & Environmental Science*, **4**, 131–134.

Rudmann, D., da Cunha, A.F., Kaelin, M. *et al.* (2004) Efficiency enhancement of Cu(In,Ga)Se$_2$ solar cells due to post-deposition Na incorporation. *Applied Physics Letters*, **84** (7), 1129–1131.

Scofield, J.H., Duda, A., Albin, D. *et al.* (1995) Sputtered molybdenum bilayer back contact for copper indium diselenide-based polycrystalline thin-film solar cells. *Thin Solid Films*, **260** (1), 26–31.

Shafarman, W.N. and Phillips, J.E. (1996) Direct current-voltage measurements of the Mo/CuInSe$_2$ contact on operating solar cells. *Photovoltaic Specialists Conference, Conference Record of the Twenty-Fifth IEEE*, pp. 917–919.

Simchi, H., McCandless, B.E., Meng, T. and Shafarman, W.N. (2014) Structure and interface chemistry of MoO$_3$ back contacts in Cu(In,Ga)Se$_2$ thin film solar cells. *Journal of Applied Physics*, **115** (3), 033514.

Vermang, B., Fjällström, V., Gao, X. and Edoff, M. (2014) Improved rear surface passivation of Cu(In,Ga)Se$_2$ solar cells: a combination of an Al$_2$O$_3$ rear surface passivation layer and nanosized local rear point contacts. *IEEE Journal of Photovoltaics*, **4** (1), 486–492.

Vermang, B., Fjällström, V., Pettersson, J. *et al.* (2013) Development of rear surface passivated Cu(In,Ga)Se_2 thin film solar cells with nano-sized local rear point contacts. *Solar Energy Materials Solar Cells*, **117**, 505–511.

Wada, T., Kohara, N., Negami, T. and Nishitani, M. (2014) Chemical and structural characterization of Cu(In,Ga)Se_2/ Mo Interface in Cu(In,Ga)Se_2 solar cells. *Japanese Journal of Applied Physics*, **35** (10A), L1253.

Wada, T., Kohara, N., Nishiwaki, S. and Negami, T. (2001) Characterization of the Cu(In,Ga)Se_2/Mo interface in CIGS solar cells. *Thin Solid Films*, **387** (1–2), 118–122.

Witte, W., Spiering, S. and Hariskos, D. (2014) Substitution of the CdS buffer layer in CIGS thin-film solar cells: status of current research and record cell efficiencies. *Vakuum in Forschung und Praxis*, **26/**1, 23–27. DOI:10.1002/ vipr.201400546.

Wuerz, R., Eicke, A., Kessler, F. *et al.* (2012) CIGS thin-film solar cells and modules on enamelled steel substrates. *Solar Energy Materials and Solar Cells*, **100**, 132–137.

Zhu, X., Zhou, Z,. Wang, Y. *et al.* (2012) Determining factor of $MoSe_2$ formation in Cu(In,Ga)Se_2 solar cells. *Solar Energy Materials and Solar Cells*, **101**, 57–61.ZSW boosts efficiency of cadmium-free thin-film cells to world record level, *PV Magazine* Photovoltaic markets and technology, 24 Feb. 2015. Available at: http://www.pv-magazine.com/ news/details/beitrag/zsw-boosts-efficiency-of-cadmium-free-thin-film-cells-to-world-record-level_100018323/.

4.4

CIGS Module Design and Manufacturing

William Shafarman
Institute of Energy Conversion, University of Delaware, Newark, DE, USA

4.4.1 Introduction

CIGS technology provides a number of module manufacturing options in the processes for fabrication and the form of the final module and large-scale commercial production, ~ 1 GW per year, is currently underway by Solar Frontier. Two examples of commercially produced CIGS modules that illustrate the range of options are shown in Figure 4.4.1. The Solar Frontier module shown is manufactured on a glass substrate with total area 1.2 m^2 (Sugimoto, 2014). The CIGS-alloy absorber layer in this product is formed by sputtering metal precursors which are then reacted in hydride gasses using a batch process. The module uses a Zn(O,S) buffer layer so it is free of cadmium. The module is fabricated using a monolithic interconnection process that offers one of the significant potential advantages of thin film PV for low cost manufacturing.

A flexible module manufactured by Global Solar Energy (Wiedeman, 2010) is also shown in Figure 4.4.1. In this case, the CIGS is deposited on a stainless steel foil using a continuous roll-to-roll co-evaporation process. Individual cells are fabricated with grids after the deposition of all active layers and interconnected using methods described below. The roll-to-roll processing, flexible product, and versatility in assembly to allow production of different module size and form factor (a 2 m-long module is shown) all provide potential advantages for manufacturing and deployment.

Thin film PV modules have the possibility for very low manufacturing costs. Whether CIGS module production will be able to achieve this low-cost potential will depend on how well the process technology can meet goals for low material costs with high throughput and yield. The issues for the manufacturability of different process options, including batch versus continuous processing, glass versus flexible substrates, and different approaches for module integration are reviewed below.

Photovoltaic Solar Energy: From Fundamentals to Applications, First Edition.
Edited by Angèle Reinders, Pierre Verlinden, Wilfried van Sark, and Alexandre Freundlich.

Companion website: www.wiley.com/go/reinders/photovoltaic_solar_energy

Figure 4.4.1 Large area CIGS-based modules manufactured in 2014 by Solar Frontier on a glass substrate (left) and Global Solar Energy on flexible foil (right)

4.4.2 Deposition Processes and Equipment

The prominent approaches for CIGS deposition being used in manufacturing can be characterized as either a two-step process in which the metals Cu, In and Ga are deposited in one step and then separately reacted to convert them to the chalcopyrite material, or a single-step process where the metals and chalcogen, Se, and in some cases S, are all deposited together, directly forming the absorber film. Selenization or sulfization after selenization (often referred to as SAS) is a two-step process that enables the method for depositing the metals to be chosen to maximize its compatibility with low-cost manufacturing. DC sputtering is commonly selected because it is scalable using commercially available deposition equipment. It can provide good uniformity over large areas with high deposition rates, though materials utilization can be a concern. Typically, the Cu-Ga-In precursor film is sputtered from Cu-Ga and In targets, where the former alleviates the problem of the Ga target melting at just above room temperature. Electrodeposition can provide very high materials utilization at low cost but the deposition-rate is typically slow. Application of particles in an ink or spray can also provide high utilization and uniformity though the inks require solvents which can add impurities to the film.

Reaction of the precursor films in H_2Se and H_2S can take several hours when controlled heat-up and cool-down times are included (Nagoya *et al.*, 2001). As a result, this is done as a batch process in which a large number of sample plates are processed in parallel and each process step can be independently optimized. Reaction is done in one or more large furnaces which are custom designed to withstand corrosive properties of the hydride gases while providing uniform temperature and delivery of the reactive species. An alternative process adds a Se layer to the precursor film, and then reaction can be done in a relatively short process with plates continuously passed through a hot reaction zone (Probst *et al*,. 2001). In practice, it has been found that the reaction requires additional S.

Co-evaporation from elemental sources or sputtering to deposit the CIGS absorber in a single process step is amenable to continuous processing in which the substrate is translated over an array of sources. Throughput can be controlled by the number of sources and flux rate for each element. The most unique and critical aspect for manufacturing CIGS using co-evaporation is the design and control of the sources. Obtaining sufficiently high effusion rate with fluxes that are uniform spatially and stable over many hours of operation has been a difficult issue that many companies have worked hard to solve, developing proprietary designs. Process control incorporating in-situ or in-line tools to measure fluxes or film composition is critical. The three-stage evaporation process (Gabor *et al.*, 1994) provides a well-verified approach to achieve high performance efficiency and allows process monitoring through transitions in film composition that cause changes in film emissivity through the process (Nishitani, Negami, and Wada, 1995).

Since its early development, CIGS PV technology has been implemented on flexible substrates, facilitated by its preferred cell configuration which is compatible with an optically opaque substrate. Roll-to-roll processing, first demonstrated for semiconductor thin films with evaporation of CdS for solar cells (Russell *et al.*, 1982), can be utilized for manufacturing in a combined batch-continuous mode. A single process step can be operated continuously to coat rolls of substrate material that may be on the order of 1000 m long. Then the coated roll can be moved to the next module fabrication step as in a batch mode. Finally, a seemingly ideal process implemented by MiaSolé has deposition rates for all steps coordinated so that a single substrate roll moves continuously through the complete module deposition sequence.

Extrinsic Na is added to the CIGS in many manufacturing processes including when Na-free substrates are used or when films are reacted in hydride gas reactions which non-uniformly react with the soda lime glass necessitating an alkali-blocking barrier layer. Options for Na incorporation include Na incorporation in Mo (Yun *et al.*, 2007) or precursor sputter targets, evaporation of a precursor layer on the Mo or during co-evaporation (Bodegard *et al.*, 2000), or a post-deposition Na treatment (Rudmann *et al.*, 2005).

While the CIGS deposition is rightly viewed as the most critical deposition step, materials and deposition choices must also be made for other deposition steps. An exception is the Mo back contact, which is deposited by dc sputtering in all commercial processes since processing equipment is available for continuous translation of glass plates or a roll-to-roll flexible foil. For the buffer layer, the chemical bath deposition of CdS or Cd-free buffer layers such as Zn(O,S) is suitable for low-cost batch processing, in that it is a surface-controlled process that allows a limited solution volume. The equipment for dipping batches of CIGS-coated substrate plates is relatively simple and can be custom-made or in-line processes using similar bath chemistry can be developed. High utilization or continuous approaches for chemical bath deposition have also been proposed (McCandless and Shaferman, 2003). Alternative materials and processes have been developed at the cell level with one promising approach using sputtered Zn(O,S) (Klenk *et al.*, 2014) which could provide an in-line deposition option for a Cd-free buffer, avoiding toxicity issues.

4.4.3 Module Fabrication

As seen in Figure 4.4.1, CIGS deposited on glass or metal foil substrates leads to very different approaches for module fabrication. On glass, or other insulating substrate including polyimide or an insulator-coated metal foil, monolithic integration is used. This provides an essential opportunity for cost advantage with thin film PV modules as compared to silicon wafer-based PV modules which require substantial extra handling for tabbing and stringing to make the series interconnection. Using a typical monolithic interconnection scheme, a process flow schematic for module fabrication is shown in Figure 4.4.2. This includes five deposition steps, although with a precursor reaction process for absorber formation the CIGS deposition includes separate precursor deposition and reaction steps.

The process flow also includes the three scribe steps used to complete the typical monolithic interconnection shown in Figure 4.4.2. In this scheme, the first scribe step uses using laser ablation for the Mo patterning (P1). The P2 scribe through the CIGS and buffer layers and P3 scribe through the TCO layer commonly use mechanical scribing though recent results have demonstrated the use of laser scribing for P2 and P3 potentially giving greater accuracy and reduced dead area (Heise *et al.*, 2014).

The cell width determined by the scribe spacing depends on the current density of the cell (determined largely by the CIGs band gap or relative Ga content) and the sheet resistance of the TCO layer since the Mo sheet resistance is typically much smaller. In typical configurations sheet resistance on the order of 5 Ω/□ is needed which will require an Al-doped ZnO layer with thickness ≈ 1 μm. Optimization of the TCO and interconnects including cell width, voltage and overall module dimensions then minimize losses and determine the module voltage (Delahoy *et al.*, 2011). Alternative monolithic schemes have been proposed (Wiedeman,

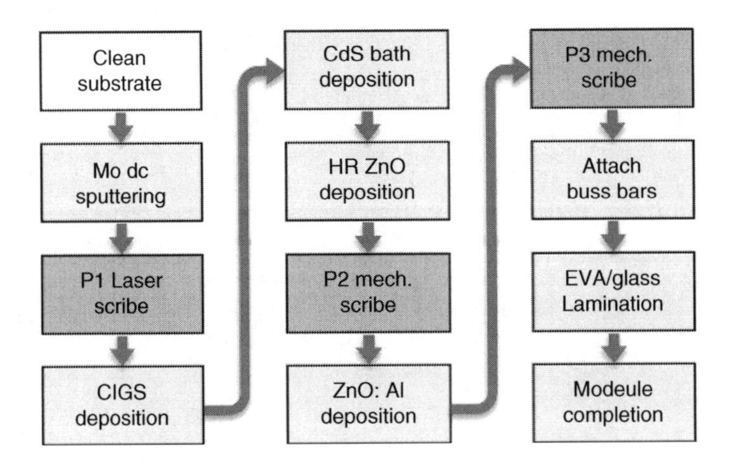

Figure 4.4.2 Process sequence for manufacturing CIGS modules on a glass substrate

Wendt, and Britt, 1999). For example, one approach allows all the scribes to be performed after all depositions are completed if printed insulating and conducting lines are utilized. Also, monolithic integration using both printed or deposited grids with scribe interconnects may be able to reduce losses in some cases (Kessler *et al.*, 2001).

As seen in Figure 4.4.3, final fabrication steps for a module on a glass substrate include attachment of electrical wires and busbars. These are metal stripes that can be soldered, welded, or glued to contact areas near the edges of the substrate plates. Before lamination with a front cover glass, the thin film layers are removed from the outer rim of the substrate plate in order to improve the adhesion to the lamination material, which is usually ethylene vinyl acetate (EVA). Edge sealing, attachment of leads and a junction box, and framing finish the product, though a frame can be omitted for some applications.

Module fabrication follows a different process flow when the cells are deposited on a metal foil substrate since the electrical conductivity of the substrate precludes using the monolithic cell interconnection. Instead, pieces cut from a large web are used to fabricate cells with grids applied to the top TCO layer to assist with current collection, similar to Si wafer solar cells. Two approaches for interconnecting the cells are illustrated in Figure 4.4.4. In a shingle-lapping configuration, the back of one cell at its edge overlaps the top of the next cell and they are connected with a solder or conductive epoxy. Or a stringing and tabbing process, similar

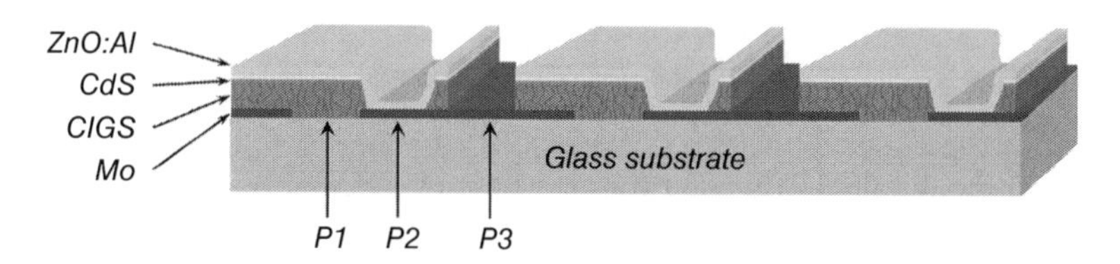

Figure 4.4.3 Schematic for typical monolithic integration including P1, P2, and P3 scribes

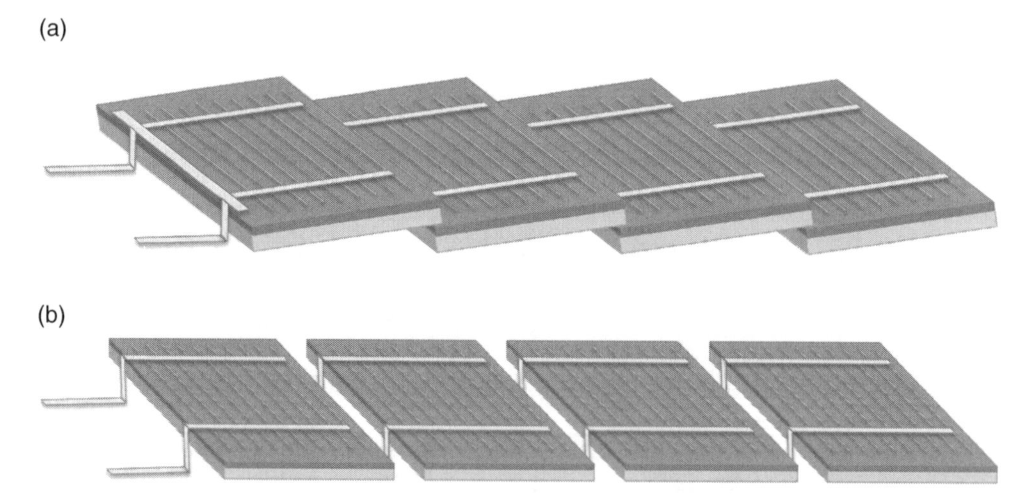

Figure 4.4.4 Approaches for interconnection of CIGS cells using foil substrates with (a) a shingle overlapping connection or (b) stringing and tabbing connection

to that used for connection of Si wafer solar cells, can be used for cell connection. In either case, extra handling of individual cells increases the number of processing steps and adds to the manufacturing cost. However, the ability to measure each cell and bin them according to their performance can relax constraints on production yield.

For a flexible module, of course no front cover glass is used. Instead the module needs an encapsulation layer whose requirements are very demanding. This encapsulant must maintain flexibility, transparency, and keep moisture ingress to a very low level for the full rated lifetime of the module. For a lifetime of 20–30 years no such material is readily available at low cost. Cells manufactured on flexible substrates can also be incorporated into a more standard glass-encased module by laying strings of connected cells as in Figure 4.4.4 together and encapsulating them between sheets of glass. This configuration, however, may be costly since it includes all of the glass, foil, and interconnect costs.

4.4.4 Cost and Materials

The total cost of a PV technology measured in $/W or levelized cost of electricity includes its manufacturing cost, performance, and installation costs. These are affected in different ways by the manufacturing approaches taken. Co-evaporation and precursor reaction approaches to CIGS deposition have demonstrated similar high efficiencies in both laboratory-scale devices and large area modules so they will be distinguished by their cost potential. While efficiency of flexible PV might be a little lower than that of glass-based PV, it could have manufacturing advantages due to the benefits of roll-to-roll processing and the installation and transportation costs may be lower.

While detailed cost comparisons are not available for specific commercial approaches due to the proprietary nature of processes and equipment, a general understanding of cost considerations can be discussed. Following the analysis of Candelise *et al.* (2012), costs for module manufacturing will include:

1. *Capital costs including equipment depreciation*: Many of the processes including glass cleaning and sputtering of Mo or TCO layers can be completed with standard semiconductor tools but the CIGS deposition requires custom equipment regardless of the process. The fractional cost will also depend on the scale of the manufacturing line.
2. *Active materials* used to deposit all layers include high purity metals for sputter targets or evaporation sources, gases, and oxide sputter targets. These costs will depend on layer thicknesses, material utilization and process yields and there may be tradeoffs in the choices of some materials and processes. For example, rotary magnetron sputter targets will cost substantially more than planar targets but will give much higher utilization. For the TCO material, a sputter target for indium tin oxide (ITO) will cost more than for Al-doped ZnO but the ITO film will be more conductive so thinner layers will be needed.
3. *Substrate materials*: while conventional soda lime glass is in many ways ideal for CIGS processing, it is still a significant cost contributor. For flexible cells, stainless steel is most commonly used but foils with sufficiently smooth finish and appropriate composition may be difficult to procure. Polyimide is also a possibility but limits the deposition temperature and is costly.

4. *Other inactive materials* include the top glass sheet which should be costly, hardened glass to withstand hail tests or encapsulation materials for flexible modules. This also includes materials for framing, junction boxes, leads, etc.
5. *Other factory costs* will include direct labor costs, energy, overhead and maintenance.

Cost analysis studies (Candelise *et al.*, 2012; Jimenez, 2013) have found that active materials contribute about 25% of the total materials cost with the greater balance roughly split between the glass and inactive materials, so glass is in fact the single most expensive material in a CIGS module. Overall, the costs of the materials may be around 60% of the total manufacturing cost.

Related to materials' costs, attention must be paid to the issues of toxicity and availability. The CdS buffer layer is very thin, ~ 50 nm, so the total amount of cadmium in a module is small and does not pose any significant environmental risks (Moskowitz and Fthenakis, 1990; Steinberger, 1998). Nevertheless, it does present drawbacks, including the generation of liquid waste which must be disposed of or reprocessed at added expense. Also, the inclusion of Cd can limit the deployability in some markets. Selenium is less toxic and does not present any special concerns in the module or manufacturing process. The exception for precursor reaction processes is the use of hydrogen selenide gas which is highly toxic and, therefore, adds manufacturing expense for containment and monitoring using similar technology to that used in Si or other semiconductor processing.

Materials' availability is primarily a concern with respect to In, the known abundance of which in the Earth's crust indicates that it could potentially constrain the production of CIGS modules, with Ga a lesser concern. Studies to determine the potential limit to production imposed by materials' constraints must make many assumptions regarding module performance, thickness, utilization, and materials' production and recovery, resulting in a wide range of outcomes (Candelisa *et al.*, 2012). One study, assuming efficiency and thickness values comparable to the record modules currently demonstrated and noting the importance of material recovery from recycled spent modules, calculated an annual production potential of up to 106 GW in 2050 (Fthenakis, 2009). Another study found that a growth rate for the production of In continuing at the historical high would allow for the 340 GW annual production of CIGS PV (Kavlak *et al.*, 2015). Currently In is produced as a by-product from the mining of zinc. Achieving annual production levels needed for maximum growth will likely require direct mining of In which will increase its cost, though the impact on the CIGS module should not be prohibitive (Woodhouse *et al.*, 2013).

In addition to materials, the other main production cost for thin film modules is the capital cost of equipment. To first order, any large-scale automated deposition equipment will have a comparable price. Therefore, the throughput or production capacity will be important for determining the capital cost. In an in-line process, throughput will depend on the substrate width and linear speed, which are determined by the deposition rate and desired thickness of the layer. If the deposition rate is relatively low, it can be compensated by having a long deposition zone in the system, for example, by having multiple targets in a sputtering system, with only a relatively small increase in capital cost. A scale-up advantage with respect to production equipment will come when multiple process lines are cloned from a single line after all engineering and equipment design have been optimized.

All cost advantages for thin films are lost if the production is not completed with high yield. The overall manufacturing yield can be broken down into electrical yield and mechanical yield. The mechanical yield is simply the fraction of the substrates entering the production line that make it to the end without losses from broken glass substrates or malfunctioning equipment. The electrical yield, the fraction of the modules produced which fulfill minimum performance criteria, reflects the process reproducibility. In general, the overall yield must be well over 90%.

Energy is another cost consideration worth mentioning. A comprehensive evaluation of a PV system based on CIGS modules with 11.7% efficiency determined an energy payback time of 1 year (de Wild-Scholten, 2013).

4.4.5 Conclusion and Future Considerations

CIGS PV is a highly promising technology based on performance at both cell and module levels but is far from being a mature technology with respect to module manufacturing. In nearly all its production aspects, multiple fabrication approaches provide different advantages and tradeoffs. Large volume production is underway and many companies are pursuing different manufacturing pathways. Included in these is the production of flexible modules for which CIGS is uniquely suited among existing PV technologies to take advantage of higher value applications such as building integrated PV.

A number of pathways are available to reduce module cost and increase production capacity. Perhaps most important is the need to continue increasing the module efficiency with both new innovations developed through research on laboratory scale devices and processes and incorporation of cell level advances to large area production. Such advances include: reduction of the cell thickness to lower materials costs and increase production throughput; increase of the absorber band gap to enhance open circuit voltage, lower series resistance, and hence reduce losses in TCO layers and interconnects and improved TCO materials and cell interconnects. For flexible modules, there is a critical need for low cost encapsulation and packaging materials.

Finally, materials constraints should not limit CIGS from becoming a major contributor to the future mix of PV generation so, with lower cost and higher performance, the manufacture of CIGS modules has a promising future.

List of Acronyms

Acronym	Definition
CIGS	$Cu(In,Ga)(S,Se)_2$
EVA	Ethylene Vinyl Acetate
ITO	Indium Tin Oxide
PV	Photovoltaic
SAS	Selenization or sulfization after selenization
TCO	Transparent Conducting Oxide

References

Bodegård, M., Granath, K., and Stolt, L. (2000) Growth of $Cu(In,GA)Se_2$ thin films by coevaporation using alkaline precursors. *Thin Solid Films*, **361–362**, 9–16.

Candelisa, C., Speirs, J., and Gross, R. (2012) Materials availability for thin film (TF) PV technologies development: a real concern? *Renewable and Sustainable Energy Reviews*, **15**, 4972–4981.

Candelisa, C., Winskel, M., and Gross, R. (2012) Implications for CdTe and CIGS technologies production costs of indium and tellurium scarcity. *Progress in Photovoltaics: Research and Applications*, **20**, 816–831.

de Wild-Scholten, M. (2013) Energy payback time and carbon footprint of commercial photovoltaic systems. *Solar Energy Materials and Solar Cells*, **119**, 296–305.

Delahoy, A. and Guo, S. (2011) Transparent conducting oxides for photovoltaics. In *Handbook of Photovoltaic Science Engineering*, 2nd edition (eds A. Luque and S. Hegedus). John Wiley & Sons, Ltd, Chichester, pp. 716–796.

Fthenakis,V. (2009) Sustainability of photovoltaics: the case for thin-film solar cells. *Renewable and Sustainable Energy Reviews*, **13**, 2746–2750.

Gabor, A. *et al.* (1994) High efficiency CuInxGa1-Se_2 solar cells made from (Inx, Ga1-x) 2 Se. *Applied Physics Letters*, **65**, 198–200.

Heise, G., Heiss, A., Hellwig, C. *et al.* (2013) Optimization of picosecond laser structuring for the monolithic serial interconnection of CIS solar cell. *Progress in Photovoltaics: Research and Applications*, **21**, 681–692.

Jimenez, D. (2013) *Photovoltaics International*, 21st edition, Solar Media Ltd, New York, pp. 76–85.

Kavlak, G. *et al.* (2015) Metal production requirements for rapid photovoltaics deployment. *Energy & Environmental Science*, **8**, 1651–1659.

Kessler, J., Wennerberg, J., Bodegård, M., and Stolt, L. (2001) Design of gridded Cu(In, Ga)Se_2 thin-film modules. *Solar Energy Materials & Solar Cells*, **67**, 59–65.

Klenk, R., Steigert, A., Rissom, T. *et al.* (2014) Junction formation by Zn(O.S) sputtering yields CISSe-based cells with efficiencies exceeding 18%. *Progress in Photovoltaics: Research and Applications*, **22**, 161–165.

McCandless, B. and Shafarman, W. (2003) Chemical surface deposition of ultra-thin semiconductors. *Proceedings of 3rd World Conference Photovoltaic Solar Energy Conversion*, pp. 562–565.

Moskowitz, P. and Fthenakis, V. (1990) Toxic materials released from photovoltaic modules during fires: health risks. *Solar Cells*, **29**, 63–71.

Nagoya, Y. *et al.* (2001) Role of incorporated sulfur into the surface of Cu(InGa)Se_2 thin-film absorber. *Solar Engineering and Material Solar Cells*, **67**, 247–253.

Nishitani, M., Negami, T., and Wada, T. (1995) Composition monitoring method in $CuInSe_2$ thin-film preparation. *Thin Solid Films*, **258**, 313–316.

Probst, V. *et al.* (2001) Rapid CIS process for high efficiency PV-modules: development towards large area processing. *Thin Solid Films*, **387**, 262–267.

Rudmann, D. *et al.* (2005) Efficiency enhancement of Cu (In, Ga) Se_2 solar cells due to post-deposition Na incorporation, *Journal of Applied Physics*, **97**, 084903 1–5.

Russell, T. *et al.* (1982) *Proceedings of 15th IEEE Photovoltaic Specialist Conference*, pp. 743–748.

Steinberger, H. (1998) Health, safety and environmental risks from the operation of CdTe and CIS thin-film modules. *Progress in Photovoltaics: Research and Applications*, **6**, 99–103.

Sugimoto, H. (2014) *Proceedings of the 40th IEEE Photovoltaic Specialist Conference*, pp. 2767–2770.

Wiedeman, S. (2010) *Proceedings of the 35th IEEE Photovoltaic Specialist Conference*, pp. 3485–3490.

Wiedeman, S., Wendt, R. and Britt, J. (1999) *AIP Conference Proceedings* **462**, 17–22.

Woodhouse, M. *et al.* (2013) Supply-chain dynamics of tellurium, indium and gallium within the context of PV module manufacturing costs. *IEEE Journal of Photovoltaics*, **3**, 833–837.

Yun, J., Kim, K.H., Kim, M.S. *et al.* (2007) Fabrication of CIGS solar cells with a Na-doped Molayer on a Na-free substrate. *Thin Solid Films*, **515**, 5876–5879.

Part Five

Thin Film Silicon-Based PV Technologies

5.1

Amorphous and Nanocrystalline Silicon Solar Cells

Etienne Moulin[1], Jan-Willem Schüttauf[2], and Christophe Ballif[1,2]
[1]*Ecole Polytechnique Fédérale de Lausanne, Institute of Microengineering (IMT), Photovoltaics and Thin-Film Electronics Laboratory (PV-lab), Neuchâtel, Switzerland*
[2]*Swiss Center for Electronics and Microtechnology (CSEM), PV-Center, Neuchâtel, Switzerland*

5.1.1 Introduction

In view of the large-scale deployment of solar energy occurring today and expected in the coming decades, silicon (Si) remains the most suitable and privileged material for photovoltaic applications, due to its abundancy, low cost, and non-toxicity. The first wafer-based crystalline silicon (c-Si) solar cells were developed in the 1950s, and so far this technology has dominated the photovoltaic market. Despite the impressive cost reduction of the c-Si technology, manufacturing of high-quality wafers remains an important cost factor driven by Si purification, ingot growth, and wafer cutting; the costs of a finished module are dominated by the material itself and a considerable amount of c-Si is wasted in the kerf losses during sawing. Already in the mid-1950s, concerns about the costs of c-Si wafers, which were prohibitive at that time, inspired the development of alternatives based on *thin films*, a broad class of photovoltaic technologies applying materials such as CuS/CdS, CdTe (entering the scene in the late 1960s), $CuInSe_2$ (in 1975), and Si (in 1976).

Thin silicon is an umbrella term that encompasses a large variety of Si materials and alloys that can be found in the form of monocrystalline, polycrystalline, nanocrystalline and also amorphous silicon, obtained by various deposition or crystal-growth methods. The thin film Si family employs thin "active" layers, with thicknesses typically between 0.1 μm and 50 μm usually deposited on low-cost substrates, such as e.g., glass, graphite, stainless steel, ceramics, foils, or even plastics.

This chapter aims to review some of the major thin Si technologies, with emphasis on the amorphous silicon (a-Si:H) and nano-crystalline silicon (nc-Si:H) technology. We broaden the

Photovoltaic Solar Energy: From Fundamentals to Applications, First Edition.
Edited by Angèle Reinders, Pierre Verlinden, Wilfried van Sark, and Alexandre Freundlich.

Companion website: www.wiley.com/go/reinders/photovoltaic_solar_energy

description of thin film Si to more recent concepts, leading to the highest contemporary conversion efficiencies, in particular to recrystallized polycrystalline silicon (poly-Si) on glass and mono-crystalline silicon (mono-Si) films on foreign substrates fabricated by layer transfer.

5.1.2 Amorphous and Nano-Crystalline Solar Technology

5.1.2.1 General Aspects

The a-Si:H and nc-Si:H photovoltaic technology uses a large variety of electronically active Si materials, grown at low temperature (~200 °C), mostly by plasma-enhanced chemical vapor deposition (PECVD) from different precursor gases, such as silane (SiH_4), germane (GeH_4), methane (CH_4), H_2, or CO_2. The structure, bandgap and opto-electrical properties of the material are to some extent able to be tailored by the gas mixture and deposition conditions that are applied. Plasma excitation conventionally occurs at 13.56 MHz but higher frequencies, up to above 100 MHz, are also used to increase the deposition rates or modify the dissociation chemistry of the precursor gases. Typically, n-type doping is achieved by adding phosphine (PH_3) to the gas mixture, while p-type doping is obtained by using di-borane (B_2H_6), BF_3 or tri-methylboron (TMB) (Chittick and Sterling, 1969; Spear and Le Comber, 1976; Usui and Kikuchi, 1979). In such disordered materials, including in nc-Si:H, the mobility is lower than in ordered materials, but the lifetime is sufficient to ensure a good photoresponse in undoped layers (i-layer). Doping, however, creates electronic defects and the doped layers (p- or n-type) are not photoactive. As a result, in order to realize a working device, a drift zone is created in the intrinsic layer using the doped layers; a *p-i-n* device configuration is therefore used (or *n-i-p*, depending on the deposition sequence). For multijunction devices, made of several junctions, the *p-i-n* (or *n-i-p*) layer sequence is repeated. Figure 5.1.1 illustrates different device structures.

A particular property of a-Si:H is that it degrades under prolonged light exposure; this light-induced degradation (LID) phenomenon – also referred to as the Staebler-Wronski effect

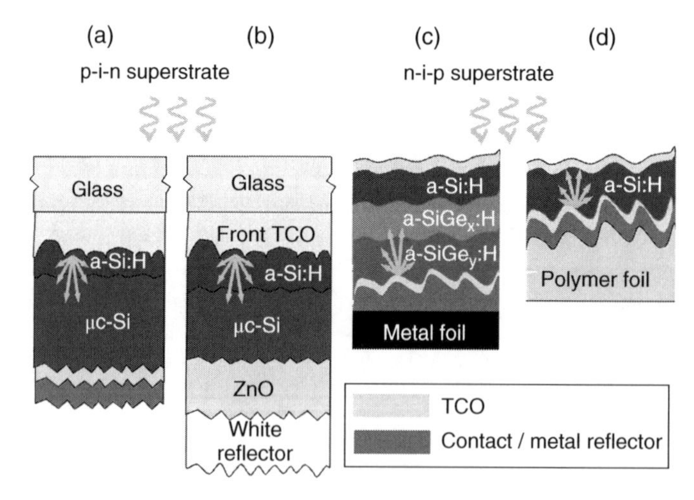

Figure 5.1.1 Schematics of four different device configurations. (*See insert for color representation of the figure*)

(Staebler and Wronski, 1977) – causes an increase in the dangling-bond defect density within the a-Si:H. These dangling bonds act as recombination centers in the material, and result in an enhanced carrier recombination and short drift length of the photo-exited carriers. The Staebler-Wronski effect is, in principle, a reversible process: the light-induced dangling bonds can be (re-)passivated by atomic hydrogen in the material, for instance, by thermal annealing. As light-induced degradation scales with the thickness, the a-Si:H active layer is generally kept thin (usually below 250 nm) to limit LID. For complete devices, using one or several junctions, the total thickness of the active layers ranges from ~200 nm for single-junction a-Si:H cells to a few microns for more complex device configurations incorporating nc-Si:H. Owing to the small thicknesses involved, an advanced light-trapping scheme has to be engineered in order to increase the optical path within the cells and consequently optimize the photocurrent. In the *p-i-n* configuration, this is traditionally realized by depositing the cells on textured electrodes made of transparent and conductive oxide (TCO) providing optical scattering; these textures, with feature sizes in the submicron range, are generally formed at the TCO surface by self-texturization in the case of boron-doped zinc oxide (ZnO:B) (Meier *et al.*, 2012) and fluorine-doped tin oxide (SnO_2:F) (Kambe *et al.*, 2003); for aluminum-doped zinc oxide (ZnO:Al) (Stannowski *et al.*, 2013), textures are usually obtained by chemical etching. Fundamentals of light-trapping concepts will be provided in Chapter 5.2. Typical examples of device structures are illustrated in Figure 5.1.1.

Because of the reduced mobilities of holes, the p-type layers should be on the side where the sunlight enters the device. Thus, p-i-n and n-i-p indicate the order of sequence in the coating process: (a) tandem a-Si:H/nc-Si:H cell with a thin TCO/metal stack at the rear; (b) similar to (a) but with a thick TCO at the rear covered by a white reflector; (c) a triple-junction with a-$SiGe_x$:H material; and (d) a single junction a-Si:H on a polymer foil. In the superstrate configuration, a simple monolithic integration scheme is possible, whereas on an opaque substrate with metal coating, special interconnection schemes have to be applied.

5.1.2.2 *Key Technological Steps Towards Highest Efficiencies in Research and Production*

The first crucial step enabling the fabrication of a-Si:H solar cells was made by Walter Spear and Peter Le Comber in the early 1970s: using PECVD from Si- and H-containing precursor gases, they demonstrated that hydrogen atoms could efficiently passivate a large part of the dangling bond defects present in a-Si (hydrogen-free amorphous silicon). The second step was the discovery that a-Si:H deposited from silane by PECVD could be efficiently doped by adding either phosphine (PH_3) or diborane (B_2H_6) to the gas mixture (Spear and Le Comber, 1975; 1976). The first properly working a-Si:H solar cell was presented in 1976 (Carlson and Wronski, 1976). Throughout the 1980s and 1990s, major improvements were made regarding a-Si:H material stability and deposition processes. The first large-area modules were produced in the late 1980s. However, at that time, most of the a-Si:H devices were used for consumer applications such as pocket calculators due to their good performance under indoor illumination.

Veprek and his co-workers were the first to describe in detail nc-Si:H (also known as microcrystalline silicon, μc-Si:H) (Veprek and Marecek, 1968). From the late 1970s, nc-Si:H was first used for internal junctions and contacts to external circuitry (Usui and Kikuchi, 1979).

nc-Si:H ws recognized as an active absorber material in the mid-1990s, when Meier *et al* (1996) demonstrated that, by reducing the oxygen concentration, the carrier lifetime in nc-Si:H was strongly enhanced. Device grade nc-Si:H is also deposited by PECVD but it requires a higher power density and higher hydrogen dilution than a-Si:H (a hydrogen content typically over 80% in the input gas mixture). This mixed-phase material is composed of Si nanocrystals (with sizes typically in the range of 10–50 nm) embedded in a passivating a-Si:H matrix but with percolating paths between the Si nanocrystals; this specific structure makes it almost unaffected by LID.

In 1996, Meier *et al.* consequently introduced the "micromorph" concept, a tandem configuration based on the stacking of a-Si:H and nc-Si:H junctions, aimed at better exploiting the solar spectrum (Meier *et al.*, 1996); while a-Si:H with its typical bandgap of ~1.7 eV was applied to the top cell, absorbing high-energy radiation, nc-Si:H with a bandgap of 1.1 eV was used for the bottom cell, ensuring photon absorption towards the near infrared wavelength range.

The emergence of nc-Si:H and the possibility of combining it in tandem micromorph or triple-junction devices sparked a strong interest in industrial manufacturers; starting from the late 1990s, mass production grew, initiated by Japanese companies (e.g. Sharp, Kaneka, Mitsubishi Heavy Industries), followed by European companies in the years 2000 to 2010 (e.g. Bosch, Inventux, Schott Solar, Sunfilm, T-Solar Global, Applied Materials, Oerlikon Solar), companies from other Asian countries (e.g. Nexpower, Sunwell, Trony, Hanergy, Baoding), and other international companies. Also the fact that germane or methane could be added to the precursor gas mix to narrow or widen the bandgap of a-Si:H enabled the possibility of generating new multi-junction device configurations. In this respect, the US company United Solar Ovonic was the first to commercialize flexible a-Si/a-SiGe/a-SiGe modules. During the time frame 2010–2015, the situation became more challenging for many thin film Si companies, as at the same time they faced the aggressive cost decline of crystalline Si cells favored by higher volumes, and their low volume and higher initial cost per watt. This has forced several companies to stop module production.

Despite the difficult market conditions, remarkable progress was made regarding cell/module performance and manufacturing processes, leading to contemporary record efficiencies in academic and industrial laboratories. In 2014, TEL Solar AG reported the highest stabilized efficiency ever achieved of 12.34% for Gen-5 ($1.1 \times 1.3\,m^2$) micromorph tandem modules (Cashmore *et al.*, 2015). This impressive result was made possible, in particular, by designing a new plasma box configuration adapted for the use of a narrow gap PECVD reactor technology. The reduction of the inter-electrode distance (down to 7 mm) allows very high process pressures to be supported in the plasma discharge (up to 20 mbar for nc-Si:H). High-pressure regimes ensure a proper growth and high-quality Si material even on highly scattering light-trapping surface textures. Among the highest stable module efficiencies achieved worldwide (>10.5%), we should also mention the results from LG-Electronics, Unisolar, Panasonic/SANYO, PVcomB/Masdar (Shah, 2013). Also on flexible substrates, major improvements have been made in the past few years, with initial module efficiencies above 10% (Jäger *et al.*, 2013). Several reports announced that true fully-loaded module manufacturing costs in the range of 0.35€/W were achievable at moderate volume of 120–250 MW, making it an attractive technology if area-related balance of system costs can be kept low, compensating for the lower efficiency compared to c-Si.

At the cell level (~1 cm^2), the National Institute of Advanced Industrial Science and Technology (AIST) achieved an efficiency of 13.6% (Sai *et al.*, 2015), so far the highest efficiency for triple-junction solar cells (with the a-Si:H/nc-Si:H/nc-Si:H configuration) on honeycomb textures, surpassing the certified 13.4% reported by LG-Electronics in 2013 (Kim *et al.*, 2013). Using a triode PECVD reactor, AIST developed high-quality a-Si:H at a low deposition rate (~1-3 $\times$ 10^{-2} nm/s), exhibiting a low amount of metastable defects and thereby a low light-induced degradation (Matsui *et al.*, 2015). Thanks to that, AIST also achieved the highest certified efficiency for a-Si:H cells with 10.22% (Matsui *et al.*, 2015), surpassing the previous record held by TEL-Solar (Neuchâtel) (Benagli *et al.*, 2009); finally, AIST also managed to obtain a 12.7% efficiency with the micromorph tandem concept (Matsui *et al.*, 2015), thereby exceeding the former results by Kaneka (Green, 2013) and IMT-Neuchâtel (Boccard *et al.*, 2014).

Several upgrades (some are already listed above) have significantly contributed to improving conversion efficiencies over the last few years: in this regard, we should mention the incorporation of silicon oxide (SiO_x) layers used as doped layers and intermediate reflector in multi-junction devices. At the moment, SiO_x has almost been systematically implemented in solar cells and modules, because of its high transparency, low refractive index and shunt-quenching ability (Buehlmann *et al.*, 2007); and buffer layers have also been applied by most laboratories still active in the field (Bugnon *et al.*, 2014; Stückelberger *et al.*, 2014); as compared to SiO_x doped layers, they additionally provide an anti-reflecting effect and prevent defect formation near the p/i interface. Nano-imprinting has revealed itself as a versatile and reliable technique to replicate any surface texture on a large scale; it has been used to produce anti-reflective coatings (Escarre *et al.*, 2012) and advanced light-trapping textures (Steltenpool *et al.*, 2013; Moulin, *et al.*, 2014). In the future, combining the most promising approaches adopted by the different academic and industrial laboratories should potentially lead to new record efficiencies. Also, the use of multi-junctions with three or even four sub-cells is one of the keys to overcoming the barrier of 15%. This subject is treated in detail in Chapter 5.4.

5.1.2.3 Unique Potential for Up-scalability of the a-Si:H and nc-Si:H Technology

The previously listed efficiencies for small-area cells and large-area modules illustrate the great adaptability of thin film Si technology regarding up-scalability. Among the most popular technologies, thin film Si shows the lowest efficiency "loss," when going from 1-cm^2 laboratory cells to > 1 m^2 industrial modules. This is illustrated in Figure 5.1.2 for the case of a-Si:H/nc-Si:H tandem devices. This small difference underlines the scalability of the underlying plasma processes, deposition regimes and manufacturing steps, and also the benefit from an optimized interconnection by means of laser scribing. For other technologies, such as wafer-based c-Si, the presence of area losses between the cells – necessary for series connection and thermal expansion of the ribbons – is responsible for a higher efficiency loss (indicated by the dashed lines). The small difference in efficiency between thin-film Si cells and modules applying the tandem configuration suggests that one can optimistically foresee efficiencies higher than 13% for triple-junction modules (keeping in mind that AIST reached 13.6% at the cell level with the triple-junction configuration).

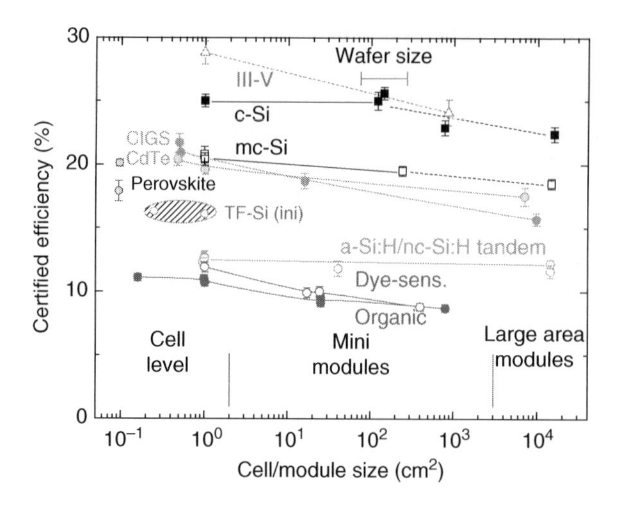

Figure 5.1.2 Independently certified efficiencies of different photovoltaic technologies, plotted with respect to the cell or module size. Source: adapted from Haug (2015) with permission of The Royal Society of Chemistry. (*See insert for color representation of the figure*)

5.1.3 Thin Poly-crystalline and Mono-crystalline Silicon on Glass

One way to improve thin film Si solar cell performance is to use high-quality Si material for the active layer, such as poly-Si or mono-Si. The good correlation between the average size of the Si crystal grains constituting the active-layer material and the solar cell performance was revealed by Bergmann *et al.* (1999). In the literature, thin film Si solar cells made of poly-Si or mono-Si are often regrouped under the term thin (film) *crystalline* solar cells, although a-Si:H or nc-Si:H deposited by PECVD at low temperature (typically below 250 °C) can be used as an intermediate material in the first fabrication stages, i.e., before crystallization. This technology involves film thicknesses typically below 50 μm and it uses a pn junction configuration.

To make this technology economically viable, high-quality Si should be obtained by means of low-cost deposition or crystal-growth techniques, and should involve the utilization of low-cost substrates. A first approach to obtain large-grained poly-Si or mono-Si on such substrates is the direct growth of Si at high temperature, usually above 700 °C. Several techniques are conventionally used to obtain high-quality thin films, e.g., chemical vapor deposition (CVD) (including rapid thermal CVD, atmospheric pressure CVD and low-pressure CVD), liquid phase epitaxy, and ion-assisted deposition. Deposition rates up to 10 μm/min can be achieved through these methods (Faller *et al.*, 1998). High-temperature-resistant substrates can be used: either "native" substrates such as specifically low-cost or reusable c-Si wafers (for the fabrication of the so-called epitaxial cells) or foreign substrates such as ceramics, graphite, and high-temperature glasses.

High-quality Si can also be fabricated by post-deposition treatments of lower-quality Si such as zone melting recrystallization (Atwater *et al.*, 1984), thermal annealing (solid-phase crystallization) (Keevers, *et al.*, 2007; Becker *et al.*, 2014), electron beam or laser radiation (liquid-phase crystallization) (Andrä *et al.*, 2013; Varlamov *et al.*, 2013; Amkreutz *et al.*, 2014; Haschke *et al.*, 2014). The substrate should thus fulfill several requirements: it should

Table 5.1.1 Best solar efficiencies for thin Si solar cells

Cell design, and laboratory	Deposition method	Grain size (μm)	Film thickness (μm)	Max η (%)	Reference
Triple-junction *p-i-n* (a-Si:H/μc-Si:H/ μc-Si:H) AIST	PECVD	A few tens of nm	<5	13.6	Sai *et al.*, 2015
Micromorph *p-i-n* (a-Si:H/μc-Si:H) TEL Solar AG	PECVD	A few tens of nm	~2.2	13.2*	Cashmore *et al.*, 2015
Micromorph *p-i-n* AIST	PECVD	A few tens of nm	~2.5	12.7	Matsui *et al.*, 2015
pn junction Poly-Si HZB	PECVD + liquid phase crystallization	10^4	5–20	12.1	Frijnts *et al.*, 2015
pn junction Mono-Si Solexel	Epitaxy +lift-off	mono c-Si	35	21.2	Solexel, 2014

sustain high-temperature deposition or recrystallization steps, and provide mechanical strength and a thermal coefficient of expansion matching. In view of further reducing costs, production of high-quality Si at low temperature on cost-effective substrates, such as borosilicate glass, is still being intensively studied; epitaxial growth or post-deposition treatments at low or moderate temperatures have shown promising results in the case where the proper seed layer was applied to the substrate (Bergmann *et al.*, 1999). The highest efficiency reported to date for thin-film poly-Si cells on inexpensive borosilicate glass is 12.1% (Frijnts *et al.*, 2015), see above. Chapter 5.2 is dedicated to the detailed description of the preparation method and characteristics of such solar cells.

Another promising approach is based on the utilization of an intermediate reusable substrate to form mono-Si by epitaxy and the subsequent bonding of the final films on a superstrate such as glass or even plastic foils. The company Solexel, for instance, reported on the fabrication of 35-μm-thick mono-Si films attached to a low-cost, flexible superstrate. Thanks to this method, the company could reach a certified efficiency as high as 21.2% in 2014 (Solexel, 2014). The company's target is to produce 20% efficient photovoltaic modules with costs below $0.40 per watt by 2017. Remarkable results were also achieved by Kobayashi *et al.* (2015), applying a heterojunction process on epitaxial wafers of Crystal Solar, reaching an open circuit voltage V_{oc} of 735 mV and efficiencies of 22.5% on 6" wafers (efficiency reported without the shadow loss of the busbar). Even though these were thick epitaxial wafers (>100 microns), this could be transposed effectively onto the thin-film Si category, as is presented in Chapter 5.4.

Table 5.1.1 summarizes the best efficiencies η reported for thin Si solar cells on glass with the corresponding fabrication methods. The * indicates that the listed efficiency corresponds to the efficiency generated "from each 1 cm^2 of active area" of the record module of TEL Solar AG (Cashmore *et al.*, 2015).

5.1.4 Perspectives for Thin Silicon Solar Technology

Thin Si solar technology has demonstrated a high potential of adaptability in the past few decades, as a result of continuous innovation and improvements. Solar cells based on a-Si:H and nc-Si:H have been the subject of intense research and development. Major progress has been made regarding conversion efficiency and manufacturing processes leading to recent world records obtained by AIST and TEL Solar AG with stable efficiencies close to 14% at cell level and 12.3% at module level. Combining the best building blocks available from the different academic and industrial laboratories in triple-junction or even quadruple-junction devices should lead to even higher efficiencies. This would include the use of: (1) advanced light-trapping schemes, such as honeycombs (Sai *et al.*, 2015) or double-textures (Moulin *et al.*, 2014; Tan *et al.*, 2015); (2) deposition regimes and processes allowing for high-quality material on highly textured substrates using a triode reactor (Matsui *et al.*, 2015) or reduced inter-electrode distance (Cashmore *et al.*, 2015); (3) the use of proper supporting layers (Despeisse *et al.*, 2011; Lambertz *et al.*, 2013) and buffer layers (Bugnon *et al.*, 2014; Stückelberger *et al.*, 2014); and (4) an optimized photocurrent matching (Ulbrich *et al.*, 2012; Bonnet-Eymard *et al.*, 2013). Finally, in light of the recent world records from AIST and the ability for up-scalability of the a-Si:H and nc-Si:H photovoltaic technology, one can estimate a short-term module efficiency goal of 13.5–14% if all these strategies are followed by the remaining actors.

Without targeting record efficiencies, the low-cost manufacturing (cost/m^2) and aesthetics of the thin film modules, should still make it strongly attractive for the building market and for specific niche applications. Figure 5.1.3 illustrates three variations of the use of such modules: one as full cladding of an old building, one with a roof in "orange, terracotta-like" thin film modules, and a house with semi-transparent module, where the films have been patterned to the desired degree of transparency.

Regarding thin poly-Si cells on glass (obtained by recrystallization), major progress has been made recently with efficiencies reaching values exceeding 12%. Further improvements need to be made with respect to efficiency and costs in order to have a chance to make the technology really competitive (see Chapter 5.2 for more details). In contrast, some thin-film mono-Si technologies (combining e.g., epitaxy and lift-off processes) have demonstrated efficiency levels such that they could compete with the conventional c-Si wafer-based technology. Thin Si technologies are thus expected to play a significant role in the future, especially when considering the predicted trends of the technology roadmaps, favoring thinner devices (i.e., with reduced costs) applying high-quality material and passivated contacts.

5.1.5 Conclusion

In this chapter, a brief overview of thin film Si-based photovoltaics has been presented, with the main emphasis on a-Si:H- and nc-Si:H-based solar cells, in both single-junction and in multijunction configurations. In the beginning of the section, the most important aspects related to fabrication processes, material properties, cell structure and stability are discussed. Subsequently, recent improvements that have been made in terms of material quality, light management, cell configuration, and thereby device performance that have led to the current state-of-the-art cells and modules are outlined. At the moment, a stabilized cell efficiency of 13.6% (area > 1 cm^2, a-Si:H/μc-Si/μc-Si triple-junction configuration) and a stabilized module

Figure 5.1.3 Examples of thin film silicon modules in the built environment. (Left top and bottom) semi-transparent a-Si:H modules. (Right top) Terracotta orange modules based on thin film Si (courtesy P. Heinstein, EPFL PV-lab). (Bottom right). Old building clad with thin film Si module (Courtesy: F. Frontini, Supsi)

efficiency of 12.3% (area = 1.1 × 1.3 m^2, a-Si:H/μc-Si tandem configuration) are the best results for thin-film Si solar cells based on a-Si:H and nc-Si:H. So-called thin film crystalline Si solar cells form another type of thin film Si-based devices. This technology involves film thicknesses typically below 50 μm fabricated at very high deposition rates and it uses a *pn* junction configuration. Currently, the highest efficiency reported to date for thin film poly-Si cells directly deposited on inexpensive borosilicate glass is 12.1%. Finally, several pathways towards further improvements are briefly discussed. Next to the suggested possible improvements and new approaches, the excellent aesthetics and potentially low-cost manufacturing make thin film Si a very interesting technology for building integration as well as for specific niche applications.

List of Symbols

Symbol	Description
a-Si	Hydrogen-free amorphous silicon
a-Si:H	Hydrogenated amorphous silicon
c-Si	Crystalline silicon
η	Conversion efficiency (%)

(*Continued*)

(*Continued*)

Symbol	Description
CVD	Chemical vapor deposition
LID	Light-induced degradation
μc-Si :H	Hydrogenated microcrystalline silicon
mono-Si	Mono-crystalline silicon
nc-Si:H	Hydrogenated nano-crystalline silicon
PECVD	Plasma-enhanced chemical vapor deposition
Poly-Si	Poly-crystalline silicon
TCO	Transparent conductive oxide
TMB	Tri-methylboron

References

Amkreutz, D.,Haschke, J,. Kühnapfel, S. *et al.* (2014) Silicon thin-film solar cells on glass with open-circuit voltages above 620 mV formed by liquid-phase crystallization. *IEEE Journal of Photovoltaics*, **4**, 1496–1501.

Andrä, G,. Gawlik, A., Hoger, I. *et al.* (2013) Multicrystalline silicon thin film solar cells based on a two-step liquid phase laser crystallization process. *IEEE 39th Photovoltaic Specialists Conference (PVSC)*, pp. 1330–1333.

Atwater, H.A., Smith, H.I., Thompson, C.V. and Geis, M.W. (1984) Zone-melting recrystallization of thick silicon on insulator films. *Materials Letters*, **2** (4), 269–273.

Becker, C., Preidel, V., Amkreutz, D. *et al.* (2014) Double-side textured liquid phase crystallized silicon thin-film solar cells on imprinted glass, online in *Solar Energy Materials and Solar Cells.*

Benagli, S., Borrello, D., Vallat, E. *et al.* (2009) High-efficiency amorphous silicon devices on LPCVD-ZNO TCO prepared in industrial KAI-M R&D reactor. *Proceedings of the 24th European Photovoltaic Solar Energy Conference, Hamburg*, pp. 2293–2298.

Bergmann, R.B. (1999) Crystalline Si thin-film solar cells: a review. *Applied Physics A* **69**, 187–194,

Boccard, M., Despeisse, M., Escarre, J. *et al.* (2014) High-stable-efficiency tandem thin-film silicon solar cell with low-refractive-index silicon-oxide interlayer. *IEEE Journal of Photovoltaics*, **4** (6), 1368–1373.

Bonnet-Eymard, M., Boccard, M., Bugnon, G. *et al.* (2013) Optimized short-circuit current mismatch in multi-junction solar cells. *Solar Energy Materials and Solar Cells*, **117**, 120–125.

Buehlmann, P., Bailat, J., Dominé, D. *et al.* (2007) In situ silicon oxide based intermediate reflector for thin-film silicon micromorph solar cells. *Applied Physics Letters*, **91** 143505.

Bugnon, G., Parascandolo, S., Hänni, M. *et al.* (2014) Silicon oxide buffer layer at the p-i interface in amorphous and microcrystalline silicon solar cells. *Solar Energy Materials and Solar Cells*, **120**, 143–150.

Carlson, D.E. and Wronski, C.R. (1976) Amorphous silicon solar cells. *Applied Physics Letters*, **28**, 671–673.

Cashmore, J.S., Apolloni, M., Braga, A. *et al.* (2015) Record 12.34% stabilized conversion efficiency in a large area thin-film silicon tandem MICROMORPH™) module. *Progress in Photovoltaics* (accepted).

Chittick, R., Alexander, J. and Sterling, H. (1969) The *preparation* and *properties* of amorphous silicon. *Journal of Electrochemical Society*, **116**(1), 77–81.

Despeisse, M., Battaglia, C., Boccard, M. *et al.* (2011) Optimization of thin film silicon solar cells on highly textured substrates. *Physica Status Solidi A*, **208**, 1863–1868.

Escarre, J., Söderstroem, K. Despeisse, M. *et al.* (2012) Geometric light trapping for high efficiency thin film silicon solar cells. *Solar Energy Materials and Solar Cells*, **98**, 185–190.

Faller, F.R., Henninger, V., Hurrle, A. and Schillinger, N. (1998) Optimization of the CVD-process for low-cost crystalline-silicon thin-film solar cells. *2nd World Conference on Photovoltaic Solar Energy Conversion, Vienna*, **2**, 1278–1283.

Frijnts, T., Kühnapfel, S., Ring, S. *et al.* (2015) Analysis of photo-current potentials and losses in thin film crystalline silicon solar cells. *Original Research Article, Solar Energy Materials and Solar Cells*, **143**, 457–466.

Green, M.A., Emery, K. Hishikawa, Y. *et al.* (2013) Solar cell efficiency tables (version 41). *Progress in Photovoltaics: Research and Applications*, **21**, 1–11 (referring to the company Internet site: http://www.kaneka-solar.com).

Haschke, J., Amkreutz, D., Korte, L. *et al.* (2014) Towards wafer quality crystalline silicon thin-film solar cells on glass. *Solar Energy Materials and Solar Cells*, **128**, 190–197.

Haug, F.-J. and Ballif, C. (2015) Light management in thin film silicon solar cells. *Energy Environmental Science*, **8**, 824–837.

Jaeger, K., Lenssen, J., Veltman, P. and Hamers, E. (2013) Large-area production of highly efficient flexible light-weight thin-film silicon PV modules. in *Proceedings of the 28th European Photovoltaiac Solar Energy Conference, Paris, France*, pp. 2164–2169.

Kambe, M., Fukawa, M. Taneda, N. *et al.* (2003) Improvement of light-trapping effect on microcrystalline silicon solar cells by using high haze transparent conductive oxide films. *Proceedings of 3rd World Conference of Photovoltaic Energy Conversion, Osaka, Japan*, **2**, 1812–1815.

Keevers, M., Young, T., Schubert, U. and Green, M. (2007) 10% efficient CSG minimodules. *22nd European Photovoltaic Solar Energy Conference, Milan, Italy*, pp. 1783–1790.

Kim, S., Chung, J.W., Lee, H. *et al.* (2013) Remarkable progress in thin-film silicon solar cells using high-efficiency triple-junction technology. *Solar Energy Materials and Solar Cells*, **119**, 26–35.

Kobayashi, Watabe, Y., Hao, R., and Ravi, T.S. (2015) High efficiency heterojunction solar cells on n-type kerfless mono crystalline silicon wafers by epitaxial growth. *Applied Physics Letters*, **106**, 223504. DOI: 10.1063/1.4922196.

Lambertz, A., Smirnov, V., Merdzhanova, T. *et al.* (2013) Microcrystalline silicon-oxygen alloys for application in silicon solar cells and modules. *Solar Energy Materials and Solar Cells*, **119**, 134–143.

Masuko, K. Shigematsu, M., Hashiguchi, T. *et al.* (2014) Achievement of more than 25% conversion efficiency with crystalline silicon heterojunction solar cell. *IEEE Journal of Photovoltaics*, **4**, 1433–1435.

Matsui, T., Bidiville, A., Maejima, K. *et al.* (2015) High-efficiency amorphous silicon solar cells: Impact of deposition rate on metastability. *Applied Physics Letters*, **106**, 053901.

Meier, J., Torres, P., Platz, R. *et al.* (1996) On the way towards high efficiency thin film silicon solar cells by the "micromorph" concept. *Materials Research Society Symposia Proceedings*, p. 3.

Meier, J., Kroll, U., Benagli, S. *et al.* (2012) From R&D to mass production of micromorph thin film silicon PV. *Energy Procedia*, **15**, 179–188.

Moulin, E., Steltenpool, M., Boccard, M. *et al.* (2014) 2-D periodic and random-on-periodic front textures for tandem thin-film silicon solar cells. *IEEE Journal of Photovoltaics*, **4** (5), 1177–1184.

Sai, H., Maejima, K., Matsui, T. *et al.* (2015) High-efficiency thin-film silicon solar cells on honeycomb textures. (submitted).

Shah, A., Moulin, E. and Ballif, C. (2013) Technological status of plasma-deposited thin-film silicon photovoltaics. *Solar Energy Materials and Solar Cells*, **119**, 311–316.Solexel Press release, 21 July 2014; http://www.solexel.com

Spear, W.E. and Le Comber, P.G. (1975) Substitutional doping of amorphous silicon. *Solid State Communications*, **17**, 1193–1196.

Spear, W.E. and Le Comber, P.G. (1976).Electronic properties of substitutionally doped amorphous Si and Ge. *Philosophical Magazine*, **33**, 935–949.

Staebler, D.L. and Wronski, C. (1977) Reversible conductivity changes in discharge-produced amorphous Si. *Applied Physics Letters*, **31**, 292–294.

Stannowski, B., Gabriel, O., Calnan, S. *et al.* (2013) Achievements and challenges in thin film silicon module production. *Solar Energy Materials and Solar Cells*, **119**, 196–203.

Steltenpool, M., Moulin, E., Haug, F-J. *et al.* (2013) Nano-imprint technology combined with rough TCO morphology as double textured light-trapping superstrate for thin film solar cells. *Proceedings of 28th European Photovoltaics Solar Energy Conference*, pp. 2175–2178.

Stuckelberger, M., Riesen, Y., Despeisse, J.W. *et al.* (2014) Light-induced Voc increase and decrease in high-efficiency amorphous silicon solar cells. *Journal of Applied Physics*, **116**, 094503.

Tan, H., Moulin, E., Si, F.T. *et al.* (2015) Highly transparent modulated surface textured front electrodes for high-efficiency multijunction thin-film silicon solar cells. *Progress in Photovoltaics* (accepted).

Ulbrich, C., Zahren, C., Gerber, A. *et al.* (2013) Matching of silicon thin-film tandem solar cells for maximum power output. *International Journal of Photoenergy*, **2013**, 314097 1–7 .

Usui, S. and Kikuchi, M. (1979) Properties of heavily doped GD-Si with low resistivity. *Journal of Non-Crystalline Solids*, **34**, 1–11.

Varlamov, S., Dore, J., Evans, R. *et al.* (2013) Polycrystalline silicon on glass thin-film solar cells: A transition from solid-phase to liquid-phase crystallized silicon. *Solar Energy Materials and Solar Cells*, **119**, 246–255.

Veprek, S. and Marecek, V. (1986) The preparation of thin layers of Ge and Si by chemical hydrogen plasma transport, *Solid State Electronics*, **11**, 683–684.

5.2

Thin Crystalline Silicon Solar Cells on Glass

Onno Gabriel[1], Daniel Amkreutz[2], Jan Haschke[2], Bernd Rech[2], and Rutger Schlatmann[1]
[1] *PVcomB/Helmholtz-Zentrum Berlin für Materialien und Energie GmbH, Germany*
[2] *Institute for Silicon Photovoltaics/Helmholtz-Zentrum Berlin für Materialien und Energie GmbH, Germany*

5.2.1 Introduction

Photovoltaic (PV) solar cell manufacturing on the Terawatt scale requires abundant, low-cost, non-toxic, and environmentally-friendly materials. Today, the only technology matching these requirements is based on silicon as the absorber material, either in the application of crystalline silicon (c-Si) wafer solar cells or as thin-film silicon solar cells directly deposited on low cost substrates such as large area glass panels. The highest conversion efficiency for c-Si cells exceeded 25% in 2014 with a wafer thickness of only 150 µm (Masuko *et al.*, 2014; Smith *et al.*, 2014). A wafer thickness reduction has been predicted with current or alternative wafer cell technologies from today's limit of about 150–200 µm down to about 100 µm in 2024 (see Figure 5.2.1) (SEMI Solar, 2014). For c-Si wafer-based solar cell manufacturing, two obstacles must be overcome for further significant cost reductions: a strong reduction of kerf losses, which are about 150 µm today and will still be about 100 µm in 2024 with known technologies (SEMI Solar, 2014), and new methods to handle very thin highly flexible wafers. Further reduction of the c-Si absorber thickness below 100 µm, however, would require revolutionary new concepts and cannot be achieved by evolutionary developments in the current c-Si-based PV industry.

Crystalline silicon absorbers with a thickness of less than100 µm, deposited directly on glass substrates is the next generation technology meeting the requirements of the technology roadmap. In this concept, glass is not only a low-cost substrate for the thin absorbers, but also enables large-area fabrication of the solar cells with the potential for a monolithic interconnection and

Photovoltaic Solar Energy: From Fundamentals to Applications, First Edition.
Edited by Angèle Reinders, Pierre Verlinden, Wilfried van Sark, and Alexandre Freundlich.

Companion website: www.wiley.com/go/reinders/photovoltaic_solar_energy

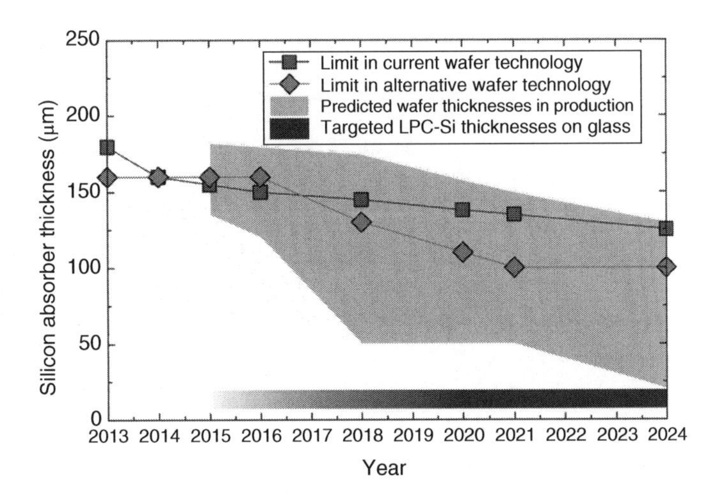

Figure 5.2.1 Predicted c-Si wafer thicknesses in solar cell production within the next decade based on SEMI's technology roadmap for photovoltaic (SEMI Solar, 2014) in comparison with targeted thickness range for thin crystalline silicon on glass technology

polymer-free front-side encapsulation. To achieve high quality thin silicon on glass, two courses have been pursued in the past and are followed in current R&D schemes: First, the direct growth of mono- or multi-crystalline silicon and, second, post-deposition treatments such as annealing or re-recrystallization of lower quality silicon.

Direct deposition is possible by means of plasma-enhanced chemical vapor deposition (PECVD) (Reehal *et al.*, 1996) or by atmospheric pressure chemical vapor deposition (APCVD). However, both methods require temperatures of more than 700 °C, which makes the use of low-cost float glass substrates impossible. To circumvent this problem, crystalline seed layers have been applied, which are subsequently thickened by epitaxial growth at lower temperatures (Rau *et al.*, 2004; Teplin *et al.*, 2007; Gall, 2009; Gestel *et al.*, 2013; Plentz *et al.*, 2014). Another approach is the use of a high-temperature stable intermediate substrate and bonding of the final crystalline silicon films on glass in a succeeding step (Radhakrishnan *et al.*, 2013).

Post-deposition treatment of amorphous or poly-crystalline silicon is another route towards high quality crystalline silicon films. Solid-phase crystallization (SPC) treatment by thermal annealing of amorphous silicon (a-Si:H) has been followed for several years (Keevers *et al.*, 2007; Becker *et al.*, 2013), but the open circuit voltage V_{oc} of resulting solar cells was always limited to values of about 500 mV (Matsuyama *et al.*, 1996; Becker *et al.*, 2013; Gall and Rech, 2013). Recently, liquid-phase crystallization (LPC) of silicon by means of continuous wave (cw) laser radiation or electron beams showed very promising results (Amkreutz *et al.*, 2011; Andrä *et al.*, 2013; Varlamov *et al.*, 2013) with V_{oc} values above 650 mV (Amkreutz *et al.*, 2014a; Haschke *et al.*, 2014), which are comparable to results achieved using conventional multi-crystalline silicon (mc-Si) wafer cells, i.e., 664 mV with a 100 μm-thick mc-Si wafer (Schultz *et al.*, 2004).

Table 5.2.1 summarizes selected material properties of various thin (about 10 μm) crystalline layers and the resulting best solar cell parameters V_{oc} and conversion efficiency η. The following sections focus on a more detailed discussion on the preparation and the properties of solar cells based on thin liquid-phase crystallized silicon absorbers.

Table 5.2.1 Material properties of various thin crystalline silicon films of 2–10 μm thickness and published best open circuit voltages V_{oc} and conversion efficiencies η of cell based on these absorbers

Method	Grain size (μm)	Major. carrier mobility[1] ($cm^2V^{-1}s^{-1}$)	max V_{oc} (mV)	max η (%)	Substrate	Reference
Direct growth	≤1	<20	400	5.3	graphite	(Reehal *et al.*, 1996)
			536	5.0	oxidized c-Si	(Carnel *et al.*, 2006)
Solid phase crystallization	1–5	20–100	492	10.5	borosilicate glass	(Keevers *et al.*, 2007)
			553	9.2	metal	(Matsuyama *et al.*, 1996)
Seed layer thickening	ca. 10	20–500	522	8.54	Al_2O_3 ceramic	(Qiu *et al.*, 2014)
			540	5.2	borosilicate glass	(Plentz *et al.*, 2014)
Liquid phase crystallization	10^4	>500	656	13.2	borosilicate glass	(Frijnts *et al.*, 2015; Sontag *et al.*, 2016)
10 μm silicon-on-insulator	mono c-Si	unknown[2]	623	13.7	wafer	(Jeong *et al.*, 2013)

[1] Mobilities for device grade material with typical doping levels in the range 10^{15} – 10^{16} cm^{-3}.
[2] Not published in given reference. The material is made from a Czochralski wafer, i.e. the value probably in the order of 10^3 $cm^2V^{-1}s^{-1}$.

5.2.2 Solar Cells Based on Liquid-Phase Crystallized Silicon on Glass

In the process of liquid-phase crystallization, an amorphous or nano-crystalline silicon absorber is molten and recrystallized to maximize the grain size. Commonly, millimeter- to centimeter-long grains are formed for silicon layers with a thickness of about 10 μm by scanning a line-shaped energy source at a constant velocity over the entire absorber (Figure 5.2.2). The major challenge during LPC is the stability and integrity of the chosen substrate during crystallization due to the high process temperature, which has inhibited the use of glasses in the past. LPC was generally performed on high temperature ceramics, metallurgical silicon powders or thermally oxidized silicon wafers (Ishihara *et al.*, 1993; Hebling *et al.*, 1996;) Auer *et al.*, 1997; Eyer *et al.*, 2001) and crystallized at scanning speeds around 0.15–0.35 mm/s using graphite strip heaters or focused halogen lamps. An increase of the scanning speed to 1–5 mm/s resulted in a strong reduction of the crystal quality (Ishihara *et al.*, 1993; Reber *et al.*, 2006). These approaches required high steady temperatures of more than 900 °C for precursor deposition or preheating during LPC imposed by the low energy density of the heat sources. In order to transfer the LPC process on glass substrates, the scanning speed must be increased significantly (2–20 mm/s), the steady temperature has to be reduced below 650 °C, and a tailored interlayer stack has to be introduced between the glass substrate and the silicon absorber.

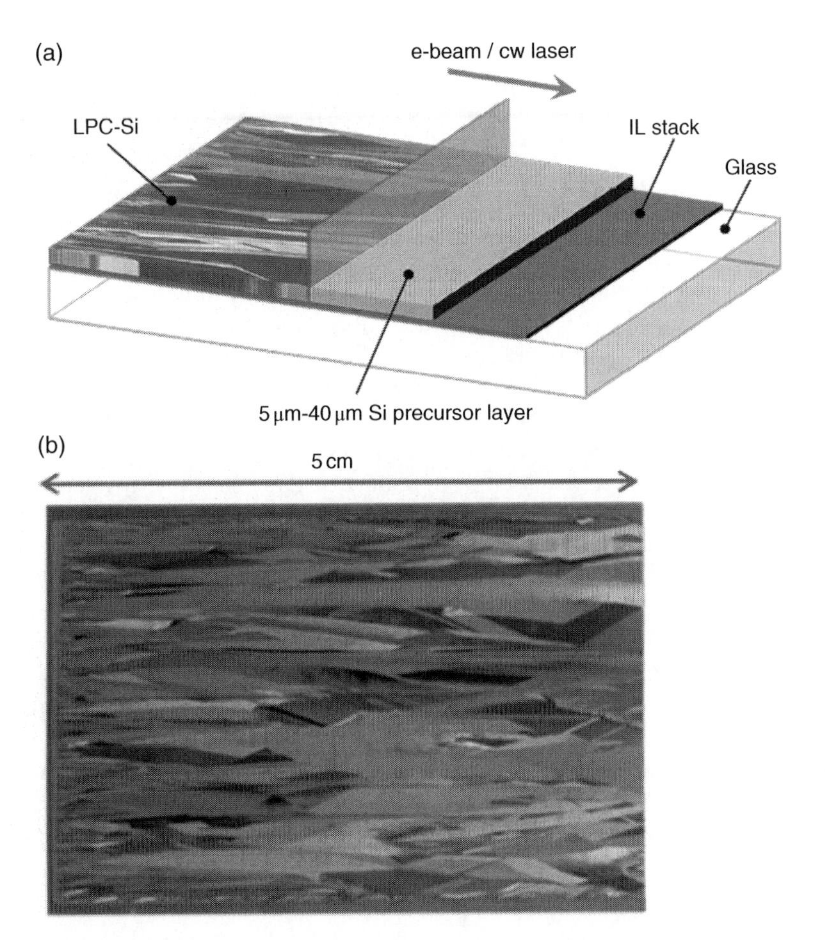

Figure 5.2.2 Scheme of the liquid-phase crystallization process of a silicon precursor layer on glass: (a) coated with an intermediate layer (IL) stack; and (b) the resulting grain structure of the LPC-Si absorber layer

5.2.2.1 Energy Sources for LPC

To achieve the desired increase in scanning speed, energy sources exhibiting energy densities around 150 J/cm^2 are necessary, which can be achieved using line-shaped cw diode lasers. These are a laser diode array by LIMO Lissotschenko Mikrooptik GmbH with a wavelength of 808 nm, a line shape of 30 mm length and 0.17 mm full-width-half-maximum, and a maximum optical power of 1 kW. or electron beams. Upscaling to substrate sizes > 1 m^2 is possible by simply increasing the length of the energy source. Depending on the applied energy and scanning speed, the morphology changes from SPC-like material (grains with random orientation and diameters of several micrometers) to large-grain material with macroscopic grains of several centimeters (right micrograph of Figure 5.2.2). As shown in the literature (Amkreutz *et al.*, 2011; Kühnapfel *et al.*, 2015), the (surface) density of grain boundaries increases with increasing the scanning speed, and for scanning speeds exceeding 20 mm/s dendritic solidification was observed. On the other hand, reducing the scanning velocity causes increasing thermal load on the substrate which results in surface damage or cracking of the glass or the glass/interlayer interface (Kühnapfel *et al.*, 2015). A stable process range was found for scanning speed values comprised between 3 mm/s and 10 mm/s (Amkreutz *et al.*, 2011; Amkreutz *et al.*, 2014a; Kühnapfel *et al.*, 2015).

5.2.2.2 Substrates, Interlayer and Absorber Deposition

As a general rule, the suitability of a specific substrate increases with its temperature stability and match to the thermal expansion coefficient of silicon. Most work up to now has been done on commercially available boro- and/or alumo-silicate glasses such as Schott Borofloat 33, Corning 1737 or Corning Eagle XG. These glasses combine a coefficient of thermal expansion close to that of silicon (3.5 ppm K^{-1}) over a wide temperature region and have a softening point around 1000 K. However, for cost-reduction reasons, a process on soda-lime float glass would be preferable and is currently under development. After cleaning, an interlayer stack is deposited onto the substrate which acts as a diffusion barrier, a wetting layer, and an anti-reflection coating as well as a passivation layer (Figure 5.2.3). Common materials are silicon-oxides, silicon-nitrides, or silicon-carbides and combinations, which can be deposited by either chemical or physical vapor deposition processes. The best results in terms of cell performance were obtained using a 200–300 nm thin layer of SiO_2 as a diffusion barrier, followed by an anti-reflection layer comprised of 70–80 nm SiN_x (Dore *et al.*, 2014). However, the final device quality is governed by the material that is in direct contact with the silicon during LPC. Due to the extraordinary wetting properties of liquid silicon, amorphous, sputtered silicon carbide was used initially (Amkreutz *et al.*, 2011), as it enables a large energy range that can be used for LPC, but it resulted in a poor electronic absorber quality due to out-diffusion of carbon and poor interface stoichiometry (Amkreutz *et al.*, 2014b). So far, the best results in terms of device efficiency have been obtained using thin (<20 nm) SiO_2 or

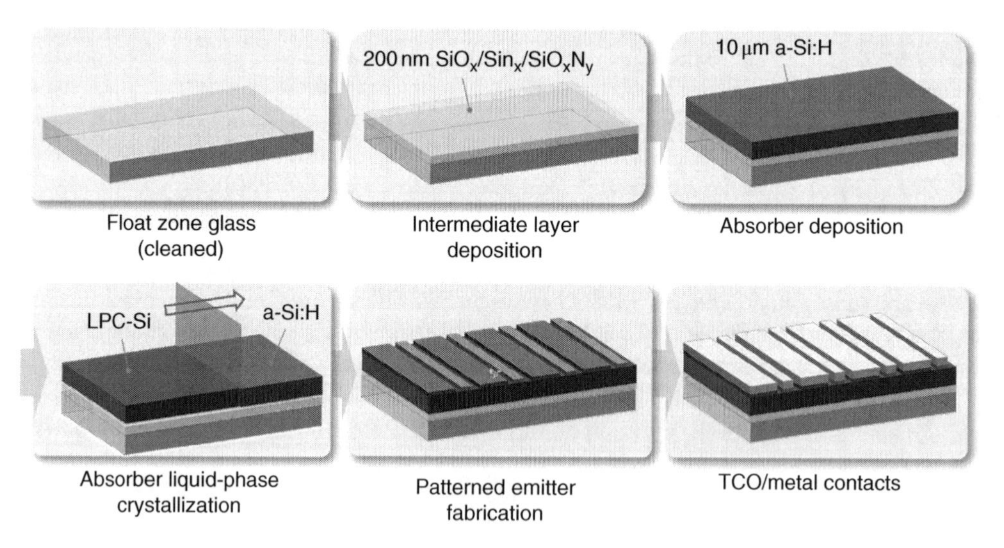

Figure 5.2.3 Scheme of the fabrication process for LPC-Si-based thin-film solar cells: deposition of IL and absorber layers (Section 5.2.2.2), crystallization (Section 5.2.2.1) and cell contacting (Section 5.2.3)

SiO_xN_y layers to minimize surface recombinations (Amkreutz *et al.*, 2014a; Gabriel *et al.*, 2014; Dore *et al.*, 2015).

On top of the interlayer stack, amorphous or nano-crystalline silicon is deposited by means of PECVD, low pressure chemical vapor deposition or high rate evaporation, up to the desired thickness between 5 and 20 micrometers (Figure 5.2.3) (Amkreutz *et al.*, 2011; Becker *et al.*, 2013; Gabriel *et al.*, 2014). For PECVD deposition, it is imperative to effuse hydrogen from all layers before crystallization, because residual hydrogen in the layers would effuse during crystallization and cause the layer stack to blister or delaminate (Gabriel *et al.*, 2014). Depending on the process environment and/or the energy source used for crystallization, the process stability can be significantly increased by the deposition of a silicon dioxide capping layer before crystallization (Amkreutz *et al.*, 2014b). Remaining defects in the form of open silicon bonds within the absorber and at its interfaces can be saturated by means of a hydrogen plasma treatment (Steffens *et al.*, 2014). This treatment not only led to an improvement in V_{oc} of some 10 mV to values of 600 mV and higher, but also led to increased current densities J_{sc} values due to a reduction of the carrier recombination velocities at the interfaces, in particular, at the buried border layer of the absorber (Amkreutz *et al.*, 2014b; Gabriel *et al.*, 2014).

5.2.2.3 *Morphological, Chemical and Electronic Properties*

Figure 5.2.2 shows the grain morphology of a liquid phase crystallized absorber. The morphology is dominated by large grains that extend up to a few cm in length and mm in width. Due to the directional solidification caused by the moving energy source, the preceding crystallized regions act as seed. A detailed investigation with respect to the total crystallographic defect concentration was conducted using transmission electron microscopy (Varlamov *et al.*, 2013; Huang *et al.*, 2015). It was shown that LPC-Si consists of almost perfect up to centimeters-wide monocrystalline regions with dislocation densities below $10^5\,cm^{-2}$ that are separated by mostly electronically inactive twin boundaries. However, in between those grains, highly dislocated regions are present with dislocation densities of $10^7\,cm^{-2}$ to $10^8\,cm^{-2}$ (Seifert *et al.*, 2011; Huang *et al.*, 2015;) and strong recombination activity as shown by electron beam-induced current (EBIC) mappings (Seifert *et al.*, 2011). Due to the high-temperature processing on glass, impurity-diffusion plays an important role as glasses contain different elements that act as a shallow or deep trap in silicon. Secondary ion mass spectroscopy (SIMS) of LPC absorbers on glass showed metallic contaminations below the detection limit and an oxygen concentration between $7\times10^{17}\,cm^{-3}$ and $10^{18}\,cm^{-3}$ (Becker *et al.*, 2013; Varlamov *et al.*, 2013; Dore *et al.*, 2014; Gabriel *et al.*, 2014). Nominally undoped silicon layers exhibit a p-type doping concentration around $10^{15}\,cm^{-3}$ determined by SIMS (Dore *et al.*, 2015) and Hall mobility or four-point-resistivity measurements. To determine the total concentration of electronically active defects on large-scale samples (several cm^2), electron paramagnetic resonance was used (Sontheimer *et al.*, 2013). It was shown that LPC absorbers exhibit an averaged defect density below $10^{15}\,cm^{-3}$ which is at least one order of magnitude lower than the best SPC material (Steffens *et al.*, 2014). For comparison, typical defect densities in mc-Si are very similar (2×10^{15}–$10^{16}\,cm^{-3}$), while in mono-crystalline silicon the defect density is in the $10^{10}\,cm^{-3}$ range or even lower. Hall mobility data for p-type and n-type LPC-Si material showed that a hole mobility up to $400\,cm^2/Vs$ (single grain) (Huang *et al.*, 2015) and an electron mobility up to $620\,cm^2/Vs$ (multiple grains) (Haschke *et al.*, 2014) can be achieved for device-relevant dopant concentrations of 10^{16} cm^{-3}. Based on the applied method (EBIC, numerical device simulations, fitting of available internal quantum efficiency data) the diffusion length is in the range of 2–10 times the absorber thickness (Amkreutz *et al.*, 2014a; Haschke *et al.*, 2014) but significantly reduced to <1 μm or less in regions with a high dislocation density (Seifert *et al.*, 2011).

5.2.3 Cell Concepts for Thin Crystalline Silicon Absorbers

Over 90% of c-Si-based solar cell manufacturing was dominated by p-type absorbers and screen printed back surface field (BSF) cells in 2014. A continuously increasing share of n-type absorbers has been predicted for the next decade, reaching almost 40% in 2024 (SEMI Solar, 2014). The main reason is the higher minority carrier lifetime in n-type c-Si absorbers compared to p-type material (Brendel *et al.*, 2013), allowing alternative and superior cell concepts. In 2024, the share of BSF cell schemes is expected to be reduced to 15%, while passivated emitter rear cells (PERC) will dominate with almost 50%. The remaining 35% will be shared by HIT (heterojunction with intrinsic thin layer) and interdigitated back contacted (IBC) solar cells utilizing n-type c-Si absorbers (SEMI Solar, 2014). These concepts are also applied in today's record wafer solar cells, where Sunpower's 25.0% was achieved using IBC (Smith *et al.*, 2014) and Panasonic's 25.6% cell is based on the HIT-IBC concept (Masuko *et al.*, 2014).

Back contacted solar cells benefit from the advantage of a front side with no shadow losses that come along with non-transparent metal contacts. The front side of the c-Si absorber has to be passivated, e.g. by amorphous a-Si:H/SiN_x:H or Al_2O_3 layers, and can additionally be coated by anti-reflex layers and/or be textured to increase the coupling of light into the absorber. All metal contacts are implemented on the back side, where absorber and emitter contacts are alternatingly fabricated close to each other.

The back contact approach is also pursued in the case of solar cells based on thin crystalline silicon on glass, where only the back side is directly accessible, while the front side is buried between the absorber and the glass substrate. Figure 5.2.4 shows a schematic view of a typical back-contacted thin crystalline silicon solar cells on glass in what is called the superstrate configuration (illumination from the top through the glass). On the front side of the absorber, all materials and layer systems have to be transparent in combination with a minimum of reflection losses. Additionally, the layer in contact with the c-Si absorber has to ensure sufficient front side passivation. The back side can be passivated after absorber fabrication, e.g., by deposition of a thin intrinsic amorphous silicon layer (de Wolf *et al.*, 2012). It is followed by the fabrication of patterned doped regions of n^+ and p^+ amorphous silicon (heterojunction concept), which are finally contacted using transparent conductive oxides (TCO) and metals such as silver, titanium or aluminum ("metallization"), see also Figure 5.2.3.

The requirements for back-contacted solar cells on the absorber material are very similar in the case of thin crystalline silicon on glass compared to conventional silicon wafers. A high minority carrier diffusion length is required, because the minority carriers have to diffuse not only transversally through the absorber towards the emitter, but also laterally over a distance of at least half the absorber contact width. The placement of both contacts on the back side of the solar cell leads also to a higher complexity of the fabrication of IBC cells. The patterned doped regions require several additional process steps to create the pattern. In contrast to BSF cells, the metallization is a separate step and, therefore, challenging in terms of alignment.

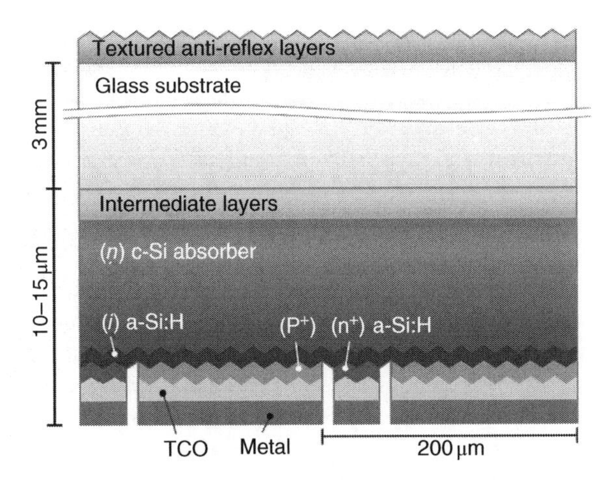

Figure 5.2.4 Schematic representation of an n-type IBC hetero-junction solar cell-based thin crystalline silicon absorbers on glass

Several industrially-compatible approaches for back contacting thin crystalline absorbers have been developed and still are undergoing further optimizations to account for their specific requirements, i.e. lower diffusion length and lower thermal stability compared to wafer silicon. A back contact monolithic interconnection has previously been shown in production by *CSG solar™* for large area SPC-based solar cells on glass (Green *et al.*, 2004; Keevers *et al.*, 2007). The latest approach used an inkjet printing for the formation of contact holes through a white reflective insulating resin layer and a laser scribing technique to define the cell areas. Recently, a similar contact scheme has been developed, where laser beams are applied for cell definition, contact opening and firing (Weizman *et al.*, 2014; Weizman *et al.*, 2015). All these developments have the common objective of a large area monolithic interconnection scheme, fully fabricated from the back side of the LPC-Si-based solar cell.

At the research level, a single-sided contact scheme has been developed that allows illumination from both sides and, therefore, enables sophisticated evaluation of the solar cells (Haschke *et al.*, 2013; Haschke *et al.*, 2014). Using this contact system, an efficiency of 11.8% has been reached using an n-type crystalline silicon absorber with 10 μm thickness (Amkreutz *et al.*, 2014a). The solar cell exhibits an open circuit voltage of 632 mV, a short circuit current density of 27.8 mA/cm^2, with an internal quantum efficiency of 90% over a wide wavelength range. The main limit to the efficiency of this device is the highly resistive metal/semiconductor contact. As an absorber contact, titanium is used, but no titanium silicide is formed due to the low temperature processing, which usually provides a poor ohmic contact. Furthermore, parasitic absorption in the applied diffusive back reflector as well as the Ti contacts is limiting the short circuit current density. To circumvent these limitations, an IBC contact system is currently under investigation, incorporating a textured Si surface and an ITO/Ag back reflector. In this full heterojunction approach, the absorber contact is passivated, which allows current collection also at the absorber contact area, given a sufficiently high diffusion length in the absorber (Sontag *et al.*, 2016). Another concept could be a hybrid approach with a large area passivated minority contact and a small area locally diffused absorber contact as proposed in (Haschke *et al.*, 2016).

5.2.4 Future Outlook

Based on current cell results for silicon on glass technology, one can estimate a short-term efficiency goal of 15% for the present absorber quality and cell concept. However, in order to develop into a competitive technology, two major problems must be tackled. First, the efficiency has to be increased significantly by further improvements of the absorber bulk quality to increase the open circuit voltage in combination with a sophisticated light trapping to enhance the short circuit current density, and, second, the substrate costs must be reduced, ideally by transferring the process to soda lime glasses. The first objective can be achieved by further optimizing the interlayer stack to improve front-surface passivation and to further reduce strain during cool down which so far has resulted in dislocations and isolated cracks. In addition, the total number of electronically active grain boundaries can be reduced and finally even a preferential (100) orientation of the grains can be achieved by further improvement of the crystallization process (Kühnapfel, 2015). Especially for the heterojunction concept a significant boost in V_{oc} can be expected by adapted surface pretreatments and further optimization of the a-Si:H(i) layer.

To enable efficiencies of 18% and higher, a very efficient way to increase the short circuit current density is required, by implementing an adapted light-trapping concept for thin LPC-Si absorbers. As shown by loss analysis (Haschke, 2014), a significant amount of current is lost due to front surface reflection and transmission in the long wavelength region (>700 nm). Different methods have already been tested such as diffuse rear reflectors (white paints or random pyramid texturing) resulting in higher current densities. To approach the Yablonovitch limit (Yablonovitch and Cody, 1982) and, thus, current densities above 35 mA/cm^2 with an absorber thickness of 10 μm, the final concept has to include two-sided textures based on a periodically or randomly textured glass/intermediate layer/silicon and silicon/air interfaces in combination with a high quality reflector. Implementing a glass/silicon texture is a challenging task, since the texture has to withstand the following deposition and the crystallization process. Nevertheless, proof that the concept works for a single- and double-sided textured LPC absorber can be found in (Amkreutz *et al.*, 2013; Preidel *et al.*, 2013; Becker *et al.*, 2014), showing that the total absorption can be significantly increased, especially for double-sided textured samples (Preidel *et al.*, 2013). Unfortunately, the dislocation density of the crystallized absorber was increased in those experiments, so that the gain due to increased absorption is lost by increased recombination losses (Preidel *et al.*, 2013; Becker *et al.*, 2014). Another texturing approach is based on 3D texturing after LPC by means of silicon nanowires (Schmidt *et al.*, 2012). Here, a periodic highly light-trapping structure is etched into the absorber after crystallization. It was shown that by texturing 1/3 of a 6 μm thin absorber, an absorption enhancement of up to 50% can be achieved in the infrared part of the spectrum.

5.2.5 Conclusion

Future PV market developments will increase the need to reduce the material and energy consumption in PV manufacturing. According to the predicted trends of international technology roadmaps, there will be strong pressure to introduce new concepts to grow and handle very thin silicon absorbers while maintaining a high electronic quality. Crystalline silicon thin film technology is one technology that offers a significant potential with regards to material and energy and, therefore, cost-cutting and is in line with predicted industry trends. The recently developed liquid-phase crystallization method is a highly promising technology as it offers thin absorber layers between 5 and 40 μm thickness and wafer-grade electronic quality, which is reflected in the achieved open circuit voltages of about 630–650 mV, on a par with commercial mc-Si wafer cells, and internal quantum efficiencies of 90% over a wide spectral range. Nevertheless, contacting LPC-Si films is a complex process and the ideal contact system has not yet been found. Due to the high temperatures during crystallization, back-sided contact systems in combination with a heterojunction are currently the best choice, but require excellent passivation of the front side interlayer/silicon interface. As the electronic quality of the interface is governed by the material in contact with the silicon absorber, the choice of the interlayer stack is critical, especially as this combination of different materials has to fulfill a variety of tasks. Combinations of silicon oxides, and silicon nitrides deposited by PECVD or sputtering have enabled laboratory-scale efficiencies well above 11%, which are currently limited by series resistance losses in the contacting scheme. The high material quality of liquid phase crystallized silicon in combination with the introduction of a sophisticated light trapping schemes and further improvement of the contact system has the potential to achieve efficiencies towards 18–20%. Research in this field is currently quite active and the solution to the issues addressed could be a game changer for the PV industry.

Acknowledgments

This work was partly supported by the Federal Ministry of Education and Research (BMBF) and the state government of Berlin (SENBWF) in the framework of the program "Spitzenforschung und Innovation in den Neuen Ländern" (Grant no. 03IS2151), and by the Federal Ministry for Economic Affairs and Energy (BMWi) in the projects "DEMO14" (Grant no. 0325237) and "Globe-Si" (Grant no. 0325446A).

List of Symbols and Acronyms

Symbol	Description
APCVD	Atmospheric Pressure Chemical Vapor Deposition
a-Si:H	Hydrogenated amorphous silicon
BSF	Back Surface Field
c-Si	(Mono) crystalline silicon
cw	Continuous wave
EBIC	Electron Beam Induced Current
η	Conversion efficiency (%)
IBC	Interdigitated Back Contact
IL	Intermediate Layer
J_{sc}	Current densities (ma/cm^2)
LPC	Liquid-Phase Crystallization
LPC-Si	Liquid-Phase Crystallized Silicon
mc-Si	Multi-crystalline silicon
PECVD	Plasma-Enhanced Chemical Vapor Deposition
PERC	Passivated Emitter Rear Cell
SIMS	Secondary Ion Mass Spectroscopy
SPC	Solid-Phase Crystallization
TCO	Transparent Conductive Oxide
V_{oc}	Open circuit voltage (V)

References

Amkreutz, D., Haschke, J., Häring, T. *et al.* (2014) Conversion efficiency and process stability improvement of electron beam crystallized thin film silicon solar cells on glass. *Solar Energy Material and Solar Cells*, **123**, 13–16.

Amkreutz, D., Haschke, J., Kühnapfel, S. *et al.* (2014) Silicon thin-film solar cells on glass with open-circuit voltages above 620 mv formed by liquid-phase crystallization. *IEEE Journal of Photovoltaics*, **4**, 1496–1501.

Amkreutz, D., Haschke, J., Schönau, S. *et al.* (2013) Light trapping in polycrystalline silicon thin-film solar cells based on liquid pase crystallization on textured substrates", *Proceedings of the 39th IEEE PVSC*, pp. 1326–1329.

Amkreutz, D., Müller, J., Schmidt, M. *et al.* (2011) Electron-beam crystallized large grained silicon solar cell on glass substrate. *Progress in Photovoltaics: Research and Application*, **19**, 937.

Andrä, D., Gawlik, A., Hoger, I. *et al.* (2013) Multicrystalline silicon thin film solar cells based on a two-step liquid phase laser crystallization process. *IEEE 39th Photovoltaic Specialists Conference (PVSC)*, pp. 1330–1333.

Auer, R., Zettner, J., Krinke, J. *et al.* (1997) Improved performance of thin-film silicon solar cells on graphite substrates. *Proceedings of 26th IEEE PVSC*, pp. 739–742.

Becker, C., Amkreutz, D., Sontheimer, T. *et al.* (2103) Polycrystalline silicon thin-film solar cells: Status and perspectives. *Solar Energy Materials and Solar Cells*, **119**, 112–123.

Becker, C., Preidel, V., Amkreutz, D. *et al.* (2014) Double-side textured liquid phase crystallized silicon thin-film solar cells on imprinted glass. Recently published online in *Solar Energy Materials and Solar Cells.*

Brendel, R., Dullweber, T., Gogolin, R. *et al.* (2013) Recent progress and options for future crystalline silicon solar cells. *Proceedings of 28th European Photovoltaic Solar Energy Conference*, Paris, pp. 676–690.

Carnel, L., Gordon, I., Van Gestel, D. *et al.* (2006) High open-circuit voltage values on fine-grained thin-film polysilicon solar cells. *Journal of Applied Physics*, **100**, 063702.

Dore, J., Ong, D., Varlamov, S. *et al.* (2014) Progress in laser-crystallized thin-film polycrystalline silicon solar cells: intermediate layers, light trapping and metallization. *IEEE Journal of Photovoltaics*, **4**, 33–39.

Dore, J., Varlamov, S., and Green, M.A. (2015) Intermediate layer development for laser-crystallized thin-film silicon solar cells on glass. *IEEE Journal of Photovoltaics*, **5**, 9–16.

Eyer, A., Haas, F., Kieliba, T. *et al.* (2001) Crystalline silicon thin-film (CSiTF) solar cells on SSP and on ceramic substrates. *Journal of Crystal Growth*, **225**, 340–347.

Frijnts, T., Kühnapfel, S., Ring, S. *et al.* (2015) Analysis of photo-current potentials and losses in thin film crystalline silicon solar cells. *Original Research Article, Solar Energy Materials and Solar Cells*, **143**, 457–466.

Gabriel, O., Frijnts, T., Calnan, S. *et al.* (2014) PECVD intermediate and absorber layers applied in liquid-phase crystallized silicon solar cells on glass substrates. *IEEE Journal of Photovoltaics*, **4**, 1343–1348.

Gall, S. (2009) Polycrystalline silicon thin-films formed by the aluminum-induced layer exchange (ALILE) process. In *Crystal Growth of Si for Solar Cells* (eds K. Nakajima and N. Usami), Springer, Berlin, pp. 193–218.

Gall, S. and Rech, B. (2013) Technological status of polycrystalline silicon thin-film solar cells on glass. *Solar Energy Materials and Solar Cells*, **119**, 306–308.

Gestel, D,V., Gordon, I., and Poortmans, J. (2013) Aluminum-induced crystallization for thin-film polycrystalline silicon solar cells: achievements and perspective. *Solar Energy Materials and Solar Cells*, **119**, 261–270.

Green, M., Basore, P., Chang, N. *et al.* (2004) Crystalline silicon on glass (CSG) thin-film solar cell modules. *Solar Energy*, **77**, 857–863.

Haschke, J., Amkreutz, D., Korte, L. *et al.* (2014) Towards wafer quality crystalline silicon thin-film solar cells on glass. *Solar Energy Materials and Solar Cells*, **128**, 190–197.

Haschke, J., Amkreutz, D. and Rech, B. (2016) Liquid phase crystallized silicon on glass: Technology, material quality and back contacted heterojunction solar cells, *Japanese Journal of Applied Physics*, **55**, 04EA04.

Haschke, J., Jogschies, L., Amkreutz, D. *et al.* (2013) Polycrystalline silicon heterojunction thin-film solar cells on glass exhibiting 582 mV open-circuit voltage. *Solar Energy Materials and Solar Cells*, **115**, 7–10.

Hebling, C., Gaffke, R., Lanyi, P. *et al.* (1996) Recrystallized silicon on SiO2-layers for thin film solar cells. *Proceedings of the 25th IEEE Photovoltaic Specialists Conference*, Washington, DC, 13–17 May, pp. 649–652.

Huang, J., Varlamov, S., Dore, J. *et al.* (2015) Micro-structural defects in polycrystalline silicon thin-film solar cells on glass by solid-phase crystallization and laser-induced liquid-phase crystallization. *Solar Energy Materials and Solar Cells*, **132**, 282–288.

Ishihara, T., Arimoto, S., Morikawa, H. *et al.* (1993) High efficiency thin film silicon solar cells prepared by zone-melting recrystallization. *Applied Physics Letters*, **63**, 3604–3606.

Jeong, S., McGehee, M.D. and Cui, Y. (2013) All-back-contact ultra-thin silicon nanocone solar cells with 13.7% power conversion efficiency. *Nature Communications*, **4**, 2950.

Keevers, M., Young, T., Schubert, U., and Green, M. (2007) 10% efficient CSG minimodules. *22nd European Photovoltaic Solar Energy Conference*, Milan, Italy, 3–7 September, pp. 1783–1790.

Kühnapfel, S., Nickel, N.H., Gall, S. *et al.* (2015) Preferential grain orientation in laser crystallized silicon on glass. accepted for publication in *Thin Solid Films*. DOI:10.1016/j.tsf.2015.01.006.

Masuko, K., Shigematsu, M., Hashiguchi, T. *et al.* (2014) Achievement of more than 25% conversion efficiency with crystalline silicon heterojunction solar cell. *IEEE Journal of Photovoltaics*, **4**, 1433–1435.

Matsuyama, T., Terada, N., Baba, T. *et al.* (1996) High-quality polycrystalline silicon thin film prepared by a solid phase crystallization method. *Journal of Non-Crystalline Solids*, **198–200**, Part 2, 940–944.

Plentz, J., Andrä, G., Gawlik, A. *et al.* (2014) Polycrystalline silicon thin-film solar cells prepared by layered laser crystallization with 540 mV open circuit voltage. *Thin Solid Films*, **562**, 430–434.

Preidel, V., Amkreutz, D., Sontheimer, T. *et al.* (2013) A novel light trapping concept for liquid phase crystallized poly-Si thin-film solar cells on periodically nanoimprinted glass substrates. *Proceedings of SPIE*, **8823**, 882307.

Qiu, Y., Kunz, O., Fejfar, A. *et al.* (2014) On the effects of hydrogenation of thin film polycrystalline silicon: A key factor to improve heterojunction solar cells. *Solar Energy Materials and Solar Cells*, **122**, 31–39.

Radhakrishnan, H.S., Martini, R., Depauw, V. *et al.* (2013) Improving the quality of epitaxial foils produced using a porous silicon-based layer transfer process for high-efficiency thin-film crystalline silicon solar cells. *IEEE Journal of Photovoltaics*, **4**, 70–77.

Rau, B., Sieber, I., Schneider, J. *et al.* (2004) Low-temperature Si epitaxy on large-grained polycrystalline seed layers by electron-cyclotron resonance chemical vapor deposition. *Journal of Crystal Growth*, **270**, 396–401.

Reber, S., Eyer, A., and Haas, F. (2006) High-throughput zone-melting recrystallization for crystalline silicon thin-film solar cells. *Journal of Crystal Growth*, **287**, 391–396.

Reehal, R., Thwaites, M., and Bruton, T. (1996) Thin film polycrystalline silicon solar cells prepared by plasma CVD. *physica status solidi (a)*, **154**, 623–633.

Schmidt, S.W., Schechtel, F., Amkreutz, D. *et al.* (2012) Nanowire arrays in multicrystalline silicon thin films on glass: a promising material for research and application in nanotechnology. *Nano Letters* **12**, 4050–4054.

Schultz, O., Glunz, S., and Willeke, G. (2004) Multicrystalline silicon solar cells exceeding 20% efficiency. *Progress in Photovoltaics: Research and Applications*, **12**, 553–558.

Seifert, W., Amkreutz, D., Arguirov, T. *et al.* (2011) Analysis of electron-beam crystallized large grained Si films on glass substrate by EBIC, EBSD and PL. *Solid State Phenomena*, **178-179**, 116–121.

SEMI Solar/PV (2014). International Technology Roadmap for Photovoltaic (ITRPV) 2013 Results.

Smith, D., Cousins, P., Westerberg, S. *et al.* (2014) Toward the practical limits of silicon solar cells. *IEEE Journal of Photovoltaics*, **4**, 1465–1469.

Sonntag, P., Haschke, J., Kühnapfel, S. *et al.* (2016) Interdigitated back-contact heterojunction solar cell concept for liquid phase crystallized thin-film silicon on glass. *Progress in Photovoltaics*, **24**, 716–724.

Sontheimer, T., Schnegg, A., Steffens, S. *et al.* (2013) Identification of intra-grain and grain boundary defects in polycrystalline Si thin films by electron paramagnetic resonance. *Physica Status Solidi Rapid Research Letters*, **7**, 959–962.

Steffens, S., Becker, C., Amkreutz, D. *et al.* (2014) Impact of dislocations and dangling bond defects on the electrical performance of crystalline silicon thin films. *Applied Physics Letters*, **105**, 022108.

Teplin, C.W., Branz, H.M., Jones, K.M. *et al.* (2007) Hot-wire chemical vapor deposition epitaxy on polycrystalline silicon seeds on glass. *MRS Proceedings*, **989**, 0989-A06-16.

Varlamov, S., Dore, J., Evans, R. *et al.* (2013) Polycrystalline silicon on glass thin-film solar cells: a transition from solid-phase to liquid-phase crystallized silicon. *Solar Energy Materials and Solar Cells*, **119**, 246–255.

Weizman, M., Rhein, H., Dore, J. *et al.* (2014) Efficiency and stability enhancement of laser-crystallized polycrystalline silicon thin-film solar cells by laser firing of the absorber contacts. *Solar Energy Materials and Solar Cells*, **120**, 521–525.

Weizman, M. *et al.* (2015) article submitted to *Solar Energy Materials and Solar Cells*.

Wolf, de S., Descoeudres, A., Holman, Z.C., and Ballif, C. (2012) High-efficiency silicon heterojunction solar cells: a review. *Green*, **2**, 7–24.

Yablonovitch, E. and Cody, G.D. (1982) Intensity enhancement in textured optical sheets for solar cells. *IEEE Transactions on Electron Devices*, **29**, 300–305.

5.3

Light Management in Crystalline and Thin Film Silicon Solar Cells

Franz Haug
Ecole Polytechnique Fédérale de Lausanne, Photovoltaics and Thin Film Electronics Laboratory (PV-Lab), Neuchâtel, Switzerland

5.3.1 Introduction

Light management covers a wide range of optical effects that aim at the optimum use of the incoming radiation within the solar cell. The first part of this chapter focusses on simplified, yet powerful, descriptions of how light is directed to the place where it is absorbed, leaving aside functionalities such as spectral conversion or luminescent shifting. The second part discusses parasitic absorption effects and discusses ways to limit their impact on solar cell performance.

The need for light management arises for two reasons; first, most semiconductors have a high index of refraction, which would yield an intolerable amount of reflection. The second reason is weak absorption, which emanates from a low absorption coefficient α or from a low absorber thickness d. Or, as it is the case for thin film silicon, from a combination of both: The absorption coefficients of amorphous silicon as well as that of micro-crystalline silicon show weak onsets, indicative of indirect bandgaps with values of 1.7 and 1.1 eV, respectively. Moreover, material quality limits the absorber layers in solar cells to typical thicknesses of 200 nm and 2 µm, respectively for amorphous and micro-crystalline silicon (Benagli *et al.*, 2009; Hänni *et al.*, 2013). The full symbols in Figure 5.3.1 illustrate the absorption upon a single pass through these thicknesses, $A = 1 - e^{-\alpha d}$. In both cases, the absorption tails off at wavelengths much shorter than those corresponding to the respective bandgaps, i.e. 750 nm and 1100 nm. In the case of weak absorption, the single-pass absorption is often approximated by αd. For the chosen thicknesses, the dashed lines in

Photovoltaic Solar Energy: From Fundamentals to Applications, First Edition.
Edited by Angèle Reinders, Pierre Verlinden, Wilfried van Sark, and Alexandre Freundlich.

Companion website: www.wiley.com/go/reinders/photovoltaic_solar_energy

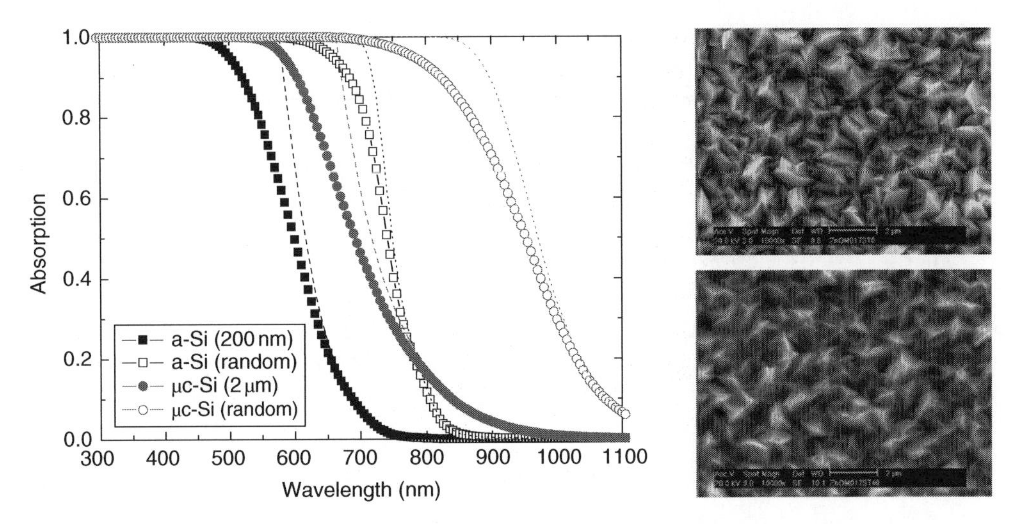

Figure 5.3.1 Absorption in amorphous (squares) and microcrystalline silicon films (circles). Full symbols denote $A = 1 - e^{-\alpha d}$, the single pass absorption with zero reflection at the front and ideal transmission at the back, dashed lines represent the approximation αd. Open symbols show the enhancement according to eq. (1) with scattering and perfect back-reflector, dotted lines illustrate path enhancement by $4n^2$. The images to the right show surface textures of 5 μm thick ZnO films used in microcrystalline solar cells before (upper) and after (lower) surface treatment, Source: (Bailat *et al.*, 2006)

Figure 5.3.1 illustrate that the approximation holds beyond 650 and 750 nm for amorphous and micro-crystalline material, respectively.

For solar cells, the issue of weak absorption was first noted for devices based on crystalline silicon, and texturing of the surface was proposed as a remedy because light is refracted into high angles at the surface facets, undergoing thus a much prolonged absorption path (Redfield, 1974). The effect is often erroneously described as "light trapping by total internal reflection," disregarding the fact that the very same facets will eventually provide a way of refracting weakly absorbed light out of the absorber. A description in terms of an equilibrium between in- and out-coupling is therefore more adequate. Assuming weakly absorbed light and a perfect back reflector, statistical considerations suggest an upper limit of $4n^2$ for the path-length enhancement, where n is the refractive index (Yablonovitch and Cody, 1982). This result is often referred to as "ergodic limit" since it assumes that light is randomized by Lambertian scattering at the interfaces. The intensity is thus equally distributed among all modes. For thick films the modal density of the black body is used, for thin films, the discrete nature of waveguide modes has to be taken into account (Stuart and Hall, 1997; Yu, Raman, and Fan, 2010; Haug *et al.*, 2011).

The treatment for arbitrary absorption as well as non-ideal back reflectors is obtained by summing up the path-lengths in a ray-tracing approach (Deckman *et al.*, 1983).

$$A = \frac{1 - \eta e^{-2\alpha d} - (1-\eta)e^{-4\alpha d}}{1 - (1-\eta)e^{-4\alpha d} + \left[(1-\eta)/n^2\right]e^{-4\alpha d}} \tag{5.3.1}$$

Here, η varies between 0 and 1 to describe either an ideal back reflector ($\eta = 0$), or a complete loss of light at the back by absorption or transmission out of the device ($\eta = 1$). The open symbols in Figure 5.3.1 illustrate again the absorption by amorphous and microcrystalline films with respective thicknesses of 200 nm and 2 μm, but now enhanced by Lambertian scattering and a perfect back reflector. For the case of very weak absorption, Equation (5.3.1) approximates to $A = 1 - e^{-4n^2 \alpha d}$, i.e., a prolongation of the absorption length by a factor of $4n^2$. This is illustrated by the dotted lines in Figure 5.3.1, and it becomes clear that this approximation does not hold for $\alpha d \ll 1$ (weak absorption), but only for the stronger case of $4(n^2 - 1) \cdot \alpha d \ll 1$ (very weak absorption).

Other than statistical treatments, also rigorous electromagnetic modelling is increasingly used to assess absorption enhancement (Rockstuhl *et al.*, 2008; Bittkau and Beckers, 2010; Dewan *et al.*, 2011; Isabella *et al.*, 2012), but the description of typical device thicknesses and random textures requires computationally intensive domain sizes and care to avoid artefacts of periodic supercells. Therefore, simplified models are likely to remain in use (Krc, Smole, and Topic, 2003; Jäger *et al.*, 2012).

5.3.2 Light Scattering Interfaces

Lambertian scattering is a central concept in the preceding section. It bears the name of H. L. Lambert, who described it in his book for the case of parallel light rays incident on a surface (Lambert, 1892). As the surface is tilted to an angle, "the perpendicular illumination relates to the oblique one like unity to the sine of the incident angle". Here, the term *illumination* is understood as the incident flux of energy. Lambert defines the angle of incidence with respect to the plane, but if θ is measured with respect to the surface normal, $\cos\theta$ must be used instead of the sine. Likewise, from the reciprocity of the light path, it follows that the energy received by an observer from a radiating surface will show the same dependence, provided that the observer receives parallel rays from every point of the radiating surface. Generally, this means that the emitter as well as the observer should be small with respect to the distance between them.

The angular dependence of the scattered light intensity enters Equation (5.3.1) via the average path length enhancement. Considering an absorber with a thickness d, scattering into an angle θ with respect to the surface normal yields a prolonged path $d' = d / \cos\theta$. For Lambertian scattering, light is scattered equally into all angles and is described by the angle resolved scattering $ARS_L = 1/\pi \cdot \cos\theta$. This expression includes only $\cos\theta$ that accounts for the projection of the scattering surface and a normalization factor. The average path length enhancement is obtained by weighting the path prolongation with the *ARS* and subsequently integrating over all angles of the hemisphere. For Lambertian scattering this yields a factor of two, i.e., a twofold increase of the light path. In Equation (5.3.1), this is seen by setting η equal to unity, representing thus a complete loss at back electrode that permits only a single passage through the film. The resulting absorption $A = 1 - e^{-2\alpha d}$ nevertheless represents a doubling of the light path. Lambertian scattering is illustrated by the dashed line in the right panel of Figure 5.3.2. Note that the polar diagram is only a 2D cross-section through the hemisphere above the scattering surface. This approach leads to a gross over-representation of the contributions at small angles. The right panel of Figure 5.3.2 represents the same data but multiplied with $\sin\theta$, the Jacobian of spherical surface coordinates, making it clearer that all characteristics integrate to unity.

Whereas the angular distribution may approach a Lambertian one after several bounces within the film, it is unlikely that this will be the case already for the first scattering event.

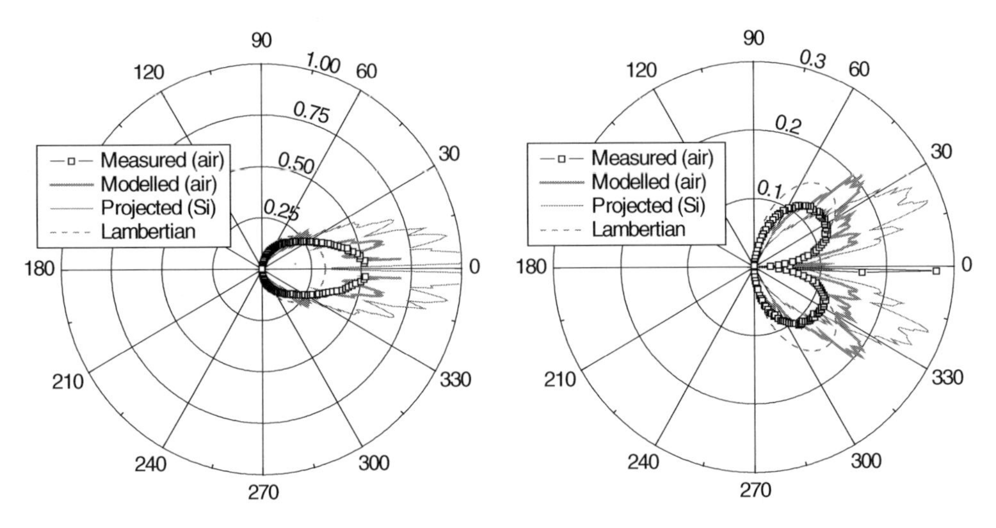

Figure 5.3.2 The left panel shows the angular resolved scattering of the upper ZnO texture in Figure 1; symbols denote scattering into air measured with $\lambda = 543$ nm, the thick line denotes data of a Fourier model on the basis of the atomic force microscopy surface profile, the thin line represents a projection of scattering into silicon. The dashed circle illustrates the Lambertian cosine dependence. The right panel shows the same data after weighting with the spherical Jacobian

For example, the *ARS* in Figure 5.3.2 suggests strong forward scattering if measured in air. However, the actual angular distribution within a complete device is not easily accessible (Schulte *et al.*, 2011) and an assessment must rely on modelling. Fourier theory was successfully applied to this situation, using only measured data like refractive indices and the surface profile (Dominé *et al.*, 2010; Bittkau *et al.*, 2011; Jäger *et al.*, 2012). It describes light with perpendicular incidence as a plane wave whose phase is constant in air or bulk material. Light scattering by the textured interface is described by the phase change that is imparted on the plane wave by the roughness zone, i.e. the depth between the minimum and maximum of the surface morphology (Goodman, 2005). The result was found to agree well with results of rigorous electromagnetic theory (Rockstuhl *et al.*, 2011; Schulte *et al.*, 2011). Since this model does not contain any adjustable parameters, it can be used to predict scattering from the textured ZnO surface into an adjacent silicon film. Figure 5.3.2 suggests that in this case an even larger contribution is scattered into small angles. For the ZnO surfaces shown in Figure 5.3.1, the path-length enhancement of the first bounce averages about 1.25 for the untreated surface (Battaglia, Boccard, Haug, and Ballif, 2012), and it is only 1.1 for the treated texture (Boccard *et al.*, 2012). The latter type is normally used in solar cells (Bailat *et al.*, 2006).

The discrepancy to the factor of 2 for Lambertian scattering inspired a modification of the ray-tracing procedure (Boccard *et al.*, 2012); arguing that a Lambertian distribution will not be achieved immediately but only after several scattering events, the first two of them are treated separately. The resulting model reproduces the external quantum efficiency of cells with an absorber thickness ranging between 1 and 5 μm on the two types of front electrodes shown in Figure 5.3.1. For the example of a 1 μm-thick cell, that yields a photocurrent of 23.5 mA/cm^2, potential improvements can be summarized as follows:

- An anti-reflective functionality can be applied to the front in order to suppress the primary reflection of 4% associated with the flat glass surface. Wide band anti-reflection coatings based on interference were shown to suppress this primary reflection (Meier, 2004), anti-reflective coatings based on textures can gain more because they additionally scatter light (Escarré *et al.*, 2011; Ulbrich *et al.*, 2013).
- Primary reflection also occurs on the other side of the glass substrate at the flat interface to the TCO. For typical refractive indices around 1.9, this amounts to another 2% which can be suppressed by an adequate texture (Battaglia *et al.*, 2011). Likewise, inserting 50 nm of TiO_2 was found to reduce reflection at the interface between the SnO_2 front electrode and Si (Fujibayashi, Matsui, and Kondo, 2006). Note that this might not be necessary for rougher front electrodes since their texture can already provide an anti-reflective effect.

With these experimentally proven modifications, a gain of about 6% can be realistically assumed to yield 25.0 mA/cm^2 for 1 μm-thick absorbers. This value was also demonstrated with the inversed n-i-p cell design where the ITO front electrode doubles as anti-reflective coating (Sai *et al.*, 2013). Then, the following gains related to light scattering can be projected (Boccard *et al.*, 2012):

- 26.1 mA/cm^2 is expected if already the first two scattering events are assumed to be fully Lambertian. Stronger scattering may be obtained on textures with higher aspect ratio, but working on such textures can easily yield absorber-material of inferior quality (Nasuno, Kondo and Matsuda, 2001; Bailat *et al.*, 2006; Li *et al.*, 2008; Python *et al.*, 2008).
- 32.1 mA/cm^2, i.e. a gain of more than 7 mA/cm^2, is expected by suppressing parasitic absorption in the doped layers of the p-i-n junction and in the back electrode. The doped layers are characterized by a high defect density and all minority carriers that are photo-generated in these layers are normally lost by recombination.
- 33.2 mA/cm^2 is projected for the combination of both effects, i.e. Lambertian scattering and the absence of parasitic absorption in the doped layers.

5.3.3 Parasitic Losses

Section 5.3.2 showed that a gain of almost 30% could come from suppressed parasitic absorption. In the past, a substantial amount of work was already devoted to this goal and various routes are still being actively pursued.

The main target was normally the p-doped layer at the front of the device. Alloying with carbon was shown to increase the optical bandgap of amorphous p-layers up to about 2.1 eV and thus to make them more transparent (Li *et al.*, 1994); the photo-current in early devices based on amorphous silicon could thus be improved by as much as 2 mA/cm^2 (Tawada *et al.*, 1982). A similar improvement was later found by depositing the p-layer under conditions of microcrystalline material, i.e., with high hydrogen dilution that yields films with high optical bandgap of about 2 eV (Guha *et al.*, 1986). The parasitic absorption in micro-crystalline p-layers is further reduced by adding oxygen, pushing the optical bandgap as far as 2.3 eV (Sichanugrist *et al.*, 1994). The fabrication of these films requires a very strong dilution, otherwise no micro-crystalline phase is formed and films exhibit poor electrical conductivity (Janotta, 2004; Cuony *et al.*, 2010; Lambertz, Grundler, and Finger, 2011). Since the current transport in micro-crystalline SiOx relies on silicon crystallites that are embedded into an amorphous matrix of SiO_x (Cuony *et al.*, 2012), it is difficult to maintain conductivity beyond a certain oxygen content and the bandgap is therefore limited to about 2.4 eV (Lambertz, Grundler and Finger, 2011). Higher bandgaps and suitable conductivity have been obtained on micro-crystalline SiC, however, its deposition is so far limited to hot-wire chemical vapour deposition (CVD) and it was challenging to demonstrate p-type material (Chen *et al.*, 2014). Note that the definitions of the bandgap cited in this section may differ and that there is a debate on how to relate them among each other (Yan *et al.*, 2013).

A second source of optical losses is associated with the contact layers. Frequently used contact layers are transparent conducting oxides such as indium-tin oxide (ITO), F-doped tin oxide (FTO), or ZnO doped with elements of Group III. Thanks to their high gaps, ITO and FTO are transparent for UV light up to about 4 eV. However, they are usually highly doped which can create defect states below the conduction band edge and thus result in some parasitic UV-absorption. More importantly, strong doping leads to higher free carrier absorption and thus reduces the transmission in the near-IR. ITO films can be made more transparent by reducing the content of the tin dopant; for tin-free In_2O_3 films, doping with hydrogen was demonstrated to yield films that combine excellent transmission and very high conductivity (Koida, Sai, and Kondo, 2010). In ZnO (bandgap of 3.3 eV), the UV-absorption extends further towards the visible range, and in case of heavy doping the absorption in the near-IR is similar to ITO or FTO (Agashe *et al.*, 2004; Steinhauser *et al.*, 2007). An exception may be B-doped ZnO deposited by low-pressure CVD (Wenas *et al.*, 1991; Faÿ *et al.*, 2005), which is usually grown to a thickness of several micrometres in order to develop a natural surface texture. For example, 5 μm are needed for textures shown in Figure 5.3.1. It should be noted that for such a thick ZnO layer, the desired conductivity can then be obtained already with relatively low doping concentrations. Lower doping does not only shift the onset of free carrier absorption further into the IR, but indirectly it also reduces its magnitude; this is related to the fact that a reduced amount of ionized impurity scattering yields to higher mobility (Ellmer, 2001), and higher mobility translates into reduced absorption (Coutts, Young, and Li, 2000). Finally, because most TCOs are degenerate n-type semiconductors, reduced doping shifts the Fermi level closer to the conduction band and thus increases the work function. This aspect is desirable for forming Ohmic contacts to the p-layer (Haug *et al.*, 2012).

In terms of parasitic absorption, the back electrode is just as important. If it is made from metals such as aluminium or silver, the back electrode also serves also as a back-reflector (Banerjee and Guha, 1991). Since the interfaces between silicon and metals are notoriously defective and pose the risk of contamination by metal ions that diffuse into silicon, dielectric buffers are normally inserted between the metal and silicon. The buffer layers can be made from ITO (Deckman *et al.*, 1983), ZnO (Kothandaraman *et al.*, 1991) or n-type SiO_x (Delli Veneri *et al.*, 2010). Note that these conductive oxides can also be made thicker and be used in combination with a diffuse white reflector (Khazaka *et al.*, 2014). Dielectric buffer layers serve the additional benefit of suppressing parasitic absorption by the excitation of surface plasmon polaritons; these are propagating waves of the charge density which are confined to the interface between metals and dielectrics, and excitation is mediated by surface textures. The existence of a buffer layer with a refractive index lower than silicon suppresses their existence, provided the buffer layer is thicker than the optical tunnelling length, i.e. some tens of nanometres (Otto *et al.*, 1968; Haug *et al.*, 2009; Holman *et al.*, 2011). Figure 5.3.3 shows the measured photocurrents of amorphous n-i-p cells as function of the ZnO buffer layer thickness. Purely based on the varying thickness of the buffer layer, one would expect an

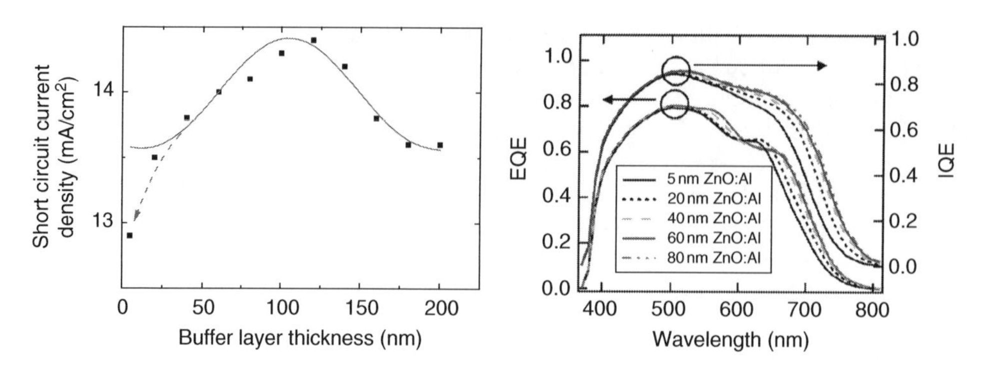

Figure 5.3.3 Photocurrent of n-i-p cells on textured silver electrode with varying buffer layer thickness (left panel). The right panel shows the underlying external quantum efficiency (EQE) and the EQE after correction for reflected light. Source: (Söderström, 2013). (*See insert for color representation of the figure*)

interference effect with a period equal to a wavelength that is representative of the light trapping region, e.g. $700\,\text{nm} / (2 \cdot n_{\text{TCO}}) \approx 175\,\text{nm}$ (note the factor of 2 because reflection doubles the light path). However, the arrow in Figure 5.3.3 shows that the current density in the cell with the thinnest buffer layer is reduced compared to the trend of the sinusoidal (Söderström, 2013). Additional evidence comes from the internal quantum efficiency, i.e. the external quantum efficiency after correction for the reflected light and thus is much less sensitive to interference effects; this indicates that the cells with the thinnest buffer layer absorb substantially less over a wide range of wavelengths from 550 to almost 800 nm. Once the buffer layer thickness exceeds 40 nm, the internal quantum efficiencies are almost identical and suggest an improved absorption in the silicon absorber. Based on the position of the interference maxima, here an optimum current density is found for buffer layer thickness of about 120 nm which is also applicable to microcrystalline absorbers (Sai *et al.*, 2014).

5.3.4 Conclusion

The development of silicon-based solar cells, crystalline as well as thin film, has always been closely accompanied by efforts to enhance the light absorption. Over the years, the optical system in state-of-the-art cells has reached a very mature level and the margin for further improvements is moderate if compared to the amount of absorption losses even in the best devices. Light management should thus not only be understood in terms of directing light into the cell, but it should also include efforts to discover novel materials and device configurations that reduce wasting energy by parasitic absorption effects.

List of Symbols

Symbol	Description
A	Absorptance
α	Cm^{-1} absorption coefficient
ARS	Angle resolved scattering
d	Nm, µm Sample thickness

Symbol	Description
η	Reflector efficiency (only in this section, could be changed by, e.g. ρ if in conflict)
n	Refractive index
θ	Deg incident angle

List of Acronyms

Acronym	Description
CVD	Chemical vapor deposition
EQE	External quantum efficiency
FTO	Fluorine-doped tin oxide
IR	Infra-red
ITO	Indium-tin oxide
TCO	Transparent conducting oxide
UV	Ultra-violet

References

Agashe, C., Kluth, O., Hüpkes, J. *et al.* (2004) Efforts to improve carrier mobility in radio frequency sputtered aluminum doped zinc oxide films. *Journal of Applied Physics*, **95** (4), 1911–1917.

Bailat, J., Dominé, D., Schlüchter, R. *et al.* (2006) High efficiency pin microcrystalline and micromorph thin film silicon solar cells deposited on LPCVD ZnO coated glass substrates. *4th World PVSEC*, Hawaii.

Banerjee, A. and Guha, S. (1991) Study of back reflectors for amorphous-silicon alloy solar-cell application. *Journal of Applied Physics*, **69** (2), 1030–1035.

Battaglia, C., Boccard, M., Haug, F.-J., and Ballif, C. (2012) Light trapping in solar cells: When does a Lambertian scatterer scatter Lambertianly? *Journal of Applied Physics*, **112** (9), 094504.

Battaglia, C., Escarré, J., Söderström, K. *et al.* (2011) Nanomoulding of transparent zinc oxide electrodes for efficient light trapping in solar cells. *Nature Photonics*, **5** (9), 535–538.

Benagli, S., Borrello, D., Vallat-Sauvain, E. *et al.* (2009) High efficiency amorphous silicon devices on LP-CVD TCO prepared in industrial Kai-M R&D reactor. *24th European PVSEC*, Hamburg, Germany.

Bittkau, K. and Beckers, T. (2010) Near-field study of light scattering at rough interfaces of a-Si:H/μc-Si:H tandem solar cells. *physica status solidi (a)* **207** (3), 661–666.

Bittkau, K., Schulte, M., Klein, M. *et al.* (2011) Modeling of light scattering properties from surface profile in thin-film solar cells by Fourier transform techniques. *Thin Solid Films* **519** (19), 6538–6543.

Boccard, M., Battaglia, C., Haug, F.-J. *et al.* (2012) Light trapping in solar cells: Analytical modeling. *Applied Physics Letters*, **101** (15), 151105.

Chen, T., Köhler, F., Heidt, A. *et al.* (2014) Hot-wire chemical vapor deposition prepared aluminum doped p-type microcrystalline silicon carbide window layers for thin film silicon solar cells. *Japanese Journal of Applied Physics*, **53** (5S1), 05FM04.

Coutts, T.J., Young, D. L. and Li, X. (2000) Characterization of transparent conducting oxides. *MRS Bulletin*, **25** (08), 58–65.

Cuony, P., Alexander, D.T.L., and Perez-Wurfl, I. (2012) Silicon filaments in silicon oxide for next-generation photovoltaics. *Advanced Materials*, **24** (9), 1182.

Cuony, P., Marending, M., Alexander, D. *et al.* (2010) Mixed-phase p-type silicon oxide containing silicon nanocrystals and its role in thin-film silicon solar cells. *Applied Physics Letters*, **97** (21), 213502–213503.

Deckman, H.W., Wronski, C.R., Witzke, H., and Yablonovitch, E. (1983) Optically enhanced amorphous silicon solar cells. *Applied Physics Letters*, **42** (11), 968–970.

Delli Veneri, P., Mercaldo, L.V. and Usatii, I. (2010) Silicon oxide based n-doped layer for improved performance of thin film silicon solar cells. *Applied Physics Letters*, **97**, 023512.

Dewan, R., Vasilev, I., Jovanov, V., and Knipp, D. (2011) Optical enhancement and losses of pyramid textured thin-film silicon solar cells. *Journal of Applied Physics*, **110** (1), 013101-013110.

Dominé, D., Haug, F.J., Battaglia, C., and Ballif, C. (2010) Modeling of light scattering from micro- and nanotextured surfaces. *Journal of Applied Physics*, **107**, 044504-8.

Ellmer, K. (2001) Resistivity of polycrystalline zinc oxide films: current status and physical limit. *Journal of Physics D: Applied Physics*, **34** (21), 3097.

Escarré, J., Söderström, K., Battaglia, C. *et al.* (2011) High fidelity transfer of nanometric random textures by UV embossing for thin film solar cells applications. *Solar Energy Materials and Solar Cells*, **95** (3), 881–886.

Faÿ, S., Kroll, U., Bucher, C. *et al.* (2005) Low pressure chemical vapour deposition of ZnO layers for thin-film solar cells: temperature-induced morphological changes. *Solar Energy Materials and Solar Cells*, **86** (3), 385–397.

Fujibayashi, T., Matsui, T., and Kondo, M. (2006) Improvement in quantum efficiency of thin film Si solar cells due to the suppression of optical reflectance at transparent conducting oxide/Si interface by TiO2âˆ• ZnO antireflection coating. *Applied Physics Letters*, **88** (18), 183508.

Goodman, J. (2005) *Introduction to Fourier Optics*, Roberts & Company Publishers, New York.

Guha, S., Yang, J., Nath, P., and Hack, M. (1986) Enhancement of open circuit voltage in high efficiency amorphous silicon alloy solar cells. *Applied Physics Letters*, **49**, 218.

Hänni, S., Bugnon, G., Parascandolo, G. *et al.* (2013) High-efficiency microcrystalline silicon single-junction solar cells. *Progress in Photovoltaics: Research and Applications*, **21** (5), 821–826.

Haug, F.-J., Biron, R., Kratzer, G. *et al.* (2012) Improvement of the open circuit voltage by modifying the transparent indium–tin oxide front electrode in amorphous n–i–p solar cells. *Progress in Photovoltaics: Research and Applications*, **20** (6), 727.

Haug, F.-J., Söderström, T., Cubero, O. *et al.* (2009) *Influence of the ZnO buffer on the guided mode structure in Si/ZnO/Ag multilayers Journal of Applied Physics*, **106**, 044502.

Haug, F.-J., Söderström, K., Naqavi, A., and Ballif, C. (2011) Resonances and absorption enhancement in thin film silicon solar cells. *Journal of Applied Physics*, **107**, 044504.

Holman, Z. C., De Wolf, S., and Ballif, C. (2011) Improving metal reflectors by suppressing surface plasmon polaritons: a priori calculation of the internal reflectance of a solar cell. *Light: Science & Applications*, **2** (10), e106.

Isabella, O., Sai, H., Kondo, M., and Zeman, M. (2012) Full-wave optoelectrical modeling of optimized flattened light-scattering substrate for high efficiency thin-film silicon solar cells. *Progress in Photovoltaics: Research and Applications*, **22** (6), 671–689.

Jäger, K., Fischer, M., van Swaaij, R., and Zeman, M. (2012) A scattering model for nano-textured interfaes and its application in opto-electrical simulations of thin-film silicon solar cells. *Journal of Applied Physics*, **111**, 083108.

Janotta, A., Janssen, R., Schmidt, M. *et al.* (2004) Doping and its efficiency in a-SiOx:H. *Physical Review B*, **69** (11), 115206.

Khazaka, R., Moulin, E., Boccard, M. *et al.* (2014) Silver versus white sheet as a back reflector for microcrystalline silicon solar cells deposited on LPCVD-ZnO electrodes of various textures. *Progress in Photovoltaics: Research and Applications*, n/a.

Koida, T., Sai, H., Kondo, M. (2010) Application of hydrogen-doped In_2O_3 transparent conductive oxide to thin-film microcrystalline Si solar cells. *Thin Solid Films*, **518** (11), 2930–2933.

Kothandaraman, C., Tonon, T., Huang, C., and Delahoy, A.E. (1991) Improvement of a-Si:H p-i-n devices using zinc oxide based back reflectors. *MRS Spring Meeting*, San Francisco, MRS.

Krc, J., Smole, F., and Topic, M. (2003) Analysis of light scattering in a-Si:H solar cells by a one-dimensional semi-coherent optical model. *Progress in Photovoltaics: Research and Applications*, **11** (1), 15–26.

Lambert, J.H. (1892) *Lambert's Photometrie: (Photometria, sive de mensura et gradibus luminis, colorum et umbrae)*. W. Engelmann, Leipzig.

Lambertz, A., Grundler, T., and Finger, F. (2011) Hydrogenated amorphous silicon oxide containing a microcrystalline silicon phase and usage as an intermediate reflector in thin-film silicon solar cells. *Journal of Applied Physics*, **109** (11), 113109–113110.

Li, H., Franken, R.H., Stolk, R.L. *et al.* (2008) Mechanism of shunting of nanocrystalline silicon solar cells deposited on rough Ag/ZnO substrates. *Solid State Phenomena*, **131**/133, 27–32.

Li, Y.-M., Jackson, F., Yang, L. *et al.* (1994) An exploratory survey of p-layers for a-Si:H solar cells. Paper presented at MRS Spring Meeting, San Francisco.

Meier, J., Spitznagel, J., Kroll, U. *et al.* (2004) Potential of amorphous and microcrystalline silicon solar cells. *Thin Solid Films*, **451**, 518–524.

Nasuno, Y., Kondo, M., and Matsuda, A. (2001) Effects of substrate surface morphology on microcrystalline silicon solar cells. *Japanese Journal of Applied Physics*, **2**, 40.

Otto, A. (1968) Excitation of nonradiative surface plasma waves in silver by method of frustrated total reflection. *Zeitschrift für Physik*, **216** (4), 398–410.

Python, M., Vallat-Sauvain, E., Bailat, J. *et al.* (2008) Relation between substrate surface morphology and microcrystalline silicon solar cell performance. *Journal of Non-Crystalline Solids*, **354** (19–25), 2258–2262.

Redfield, D. (1974) Multiple pass thin film silicon solar cell. *Applied Physics Letters*, **25**, 647.

Rockstuhl, C., Fahr, S., Lederer, F. *et al.* (2008) Local versus global absorption in thin-film solar cells with randomly textured surfaces. *Applied Physics Letters*, **93**, 061105.

Rockstuhl, C., Fahr, S., Lederer, F. *et al.* (2011) Light absorption in textured thin film silicon solar cells: A simple scalar scattering approach versus rigorous simulation. *Applied Physics Letters*, **98** (5), 051102–051102-3.

Sai, H., Matsui, T., Matsubara, K. *et al.* (2014) 11.0%-Efficient thin-film microcrystalline silicon solar cells with honeycomb textured substrates. *IEEE Journal of Photovoltaics*, **4** (6), 1349.

Sai, H., Saito, K., Hozuki, N., and Kondo, M. (2013) Relationship between the cell thickness and the optimum period of textured back reflectors in thin-film microcrystalline silicon solar cells. *Applied Physics Letters*, **102** (5), 053509.

Schulte, M., Bittkau, K., Jäger, K. *et al.* (2011) Angular resolved scattering by a nano-textured ZnO/silicon interface. *Applied Physics Letters*, **99**, 111107.

Sichanugrist, P., Sasaki, T., Asano, A. *et al.* (1994) Amorphous silicon oxide and its application to metal/nip/ITO type a-Si solar cells. *Solar Energy Materials and Solar Cells*, **34** (1–4), 415–422.

Söderström, K. (2013) Coupling light into thin silicon layers for high-efficiency solar cells. PhD thesis.

Steinhauser, J., Fay, S., Oliveira, N. *et al.* (2007) Transition between grain boundary and intragrain scattering transport mechanisms in boron-doped zinc oxide thin films. *Applied Physics Letters*, **90** (14), 142107.

Stuart, H.R. and Hall, D.G. (1997) Thermodynamic limit to light trapping in thin planar structures. *Journal of the Optical Society of America A*, **14** (11), 3001–3008.

Tawada, Y., Kondo, M., Okamoto, H., and Hamakawa, Y. (1982) Hydrogenated amorphous silicon carbide as a window material for high efficiency a-Si solar cells. *Solar Energy Materials*, **6** (3), 299–315.

Ulbrich, C., Gerber, A., Hermans, K. *et al.* (2013) Analysis of short circuit current gains by an anti-reflective textured cover on silicon thin film solar cells. *Progress in Photovoltaics: Research and Applications*, **21** (8), 1672–1681.

Wenas, W.W., Yamada, A., Konagai, M., and Takahashi, K. (1991) Textured ZnO thin films for solar cells grown by metalorganic chemical vapor deposition. *Japanese Journal of Applied Physics*, **30** (3B), L441.

Yablonovitch, E. and Cody, G.D. (1982) Intensity enhancement in textured optical sheets for solar cells. *IEEE Transactions on Electron Devices*, **29** (2), 300–305.

Yan, B., Yue, G., Yang, J., and Guha, S. (2013) On the bandgap of hydrogenated nanocrystalline silicon intrinsic materials used in thin film silicon solar cells. *Solar Energy Materials and Solar Cells*, **111**, 90–96.

Yu, Z., Raman, A., and Fan, S. (2010) Fundamental limit of nanophotonic light trapping in solar cells. *Proceedings of the National Academy of Science*, **107** (41), 17491–17496.

5.4

New Future Concepts

Jan-Willem Schüttauf[1], Etienne Moulin[2], and Christophe Ballif[1,2]
[1]*Swiss Center for Electronics and Microtechnology (CSEM), PV-Center, Neuchâtel, Switzerland*
[2]*Ecole Polytechnique Fédérale de Lausanne, Institute of Microengineering (IMT), Photovoltaics and Thin-Film Electronics Laboratory (PV-Lab), Neuchâtel, Switzerland*

5.4.1 Introduction

In this chapter, we review some possible improvements and future pathways for thin-film Si photovoltaics. Currently, most trends in thin film Si photovoltaics (PV) are towards higher conversion efficiencies, rather than towards decreasing production costs. In the past few years, the cell conversion efficiency has increased up to 16.3% initial (Yan *et al.*, 2011) and 13.6% stable (Sai *et al.*, 2015), both obtained in a triple-junction configuration. The record stabilized efficiency of 13.6% was obtained with one a-Si:H and two nc-Si:H component cells. This device configuration has already been extensively studied by several laboratories over the past years, but its potential has not yet fully been explored. In the second part of this chapter we will give an overview of the different aspects that have led to such excellent results being obtained over the past years, and also discuss the building blocks that will allow further improvements.

Since around 2012, quadruple-junction devices have also been attracting more and more attention: by increasing the number of junctions in a device, the theoretical efficiency limit of the device concept increases due to better use of the solar spectrum, and the possibility of using thinner a-Si:H component cells, leading to a reduced light-induced degradation. So far, the efficiencies of quadruple-junction cells have not yet exceeded those of triple-junction cells, but recent simulations and estimations have shown high possible efficiencies for this cell configuration using available thin-film Si-based absorber materials and related alloys (Isabella *et al.*, 2014; Schüttauf *et al.*, 2015). First results, as well as a further outlook on this approach, will be presented in the third section of this chapter.

In the fourth and final part, discussion of further improvements regarding thin c-Si cells will be presented, both in single-junction and in multijunction device configuration. For this device configuration, even higher efficiencies should be within reach. Furthermore, the knowledge

Photovoltaic Solar Energy: From Fundamentals to Applications, First Edition.
Edited by Angèle Reinders, Pierre Verlinden, Wilfried van Sark, and Alexandre Freundlich.

Companion website: www.wiley.com/go/reinders/photovoltaic_solar_energy

obtained from research and development on a-Si:H/c-Si silicon heterojunction (SHJ) solar cells could be beneficial for proper interface engineering, thereby reducing recombination losses and further improving conversion efficiencies.

In theory, Si is a perfect material to create a tailored bandgap with quantum size effects. Indeed, it has been shown that the bandgap of Si can be tailored in such a way (Canham, 1990; Godefroo *et al.*, 2008). However, the presented approaches have so far not been successful in demonstrating high efficiencies (Janz *et al.*, 2013). The major challenges can be summarized as follows: for a sufficiently high effective bandgap, the Si nanoparticles need to be in a wide-bandgap matrix resulting in charge carrier confinement. With a SiO_x matrix, charge carrier confinement has been demonstrated by luminescence, but the high band offsets for both the valence and the conduction band impede electrical transport. Alternatively, a conductive matrix material such as SiC_x can be chosen as confining material (Löper *et al.*, 2013). However, this material system exhibits a high defect density and relatively poor electrical properties (Schnabel *et al.*, 2015). Also the absorptance of thin films including such nanostructures could not really be enhanced, as would have been anticipated by studies which show that the bandgap of c-Si becomes direct at sizes below 3 nm (Lee *et al.*, 2012).

5.4.2 Thin Film Silicon Triple-Junction Solar Cells

Thin film Si triple-junction solar cells have so far been studied in different configurations, of which only the two most important ones will be addressed in this chapter. The first configuration features an a-Si:H top cell, an amorphous silicon-germanium (a-SiGe:H) middle cell and a nc-Si:H bottom cell. The main advantage of this configuration is the use of three different absorber layer materials with different bandgaps, enabling excellent use of the solar spectrum. The main challenge is the limited stability due to the use of two absorber materials that suffer from light-induced degradation, i.e., a-Si:H and a-SiGe:H. This configuration has been extensively studied and commercialized by the company United Solar. In 2011, they showed initial cell efficiency as high as 16.3% in n-i-p substrate configuration (Yan *et al.*, 2011). For small modules (400 cm^2), efficiencies of 12.4% (initial) and 11.85% (after light-soaking) were obtained (Guha *et al.*, 2013). In the p-i-n superstrate configuration, LG Electronics obtained an initial cell efficiency of 16.1% (Kim *et al.*, 2013), and a stabilized module efficiency of 11.2% for Gen5 (1.1 x 1.3 m^2) modules (You *et al.*, 2014).

The highest stabilized conversion efficiencies have so far been obtained with a-Si:H top cells and nc-Si:H middle and bottom cells, respectively. AIST from Japan very recently demonstrated a new record efficiency of 13.6% (Sai, 2015), closely followed by LG Electronics at 13.44% (Kim *et al.*, 2013). The high stabilized efficiencies for this configuration are obtained thanks to the very low light-induced degradation, with values in the range of only 5–7% (Schüttauf *et al.*, 2015). For this reason, this latter configuration is the one mostly studied at this moment. The main challenges when fabricating high-quality triple-junction devices in this configuration are: (1) the fabrication of high bandgap top cells with a low degradation rate (Yan *et al.*, 2012; Stuckelberger *et al.*, 2014); (2) the development of nc-Si:H middle cells with a high open circuit voltage (V_{OC}) (Hänni *et al.*, 2015); (3) the incorporation of nc-Si:H with a high current density; (4) the choice and application of suitable intermediate reflector layers; and (5) the fabrication of excellent front and back electrodes. In the p-i-n configuration, an important challenge is to find the right combination of transparency, conductivity, and morphology for ideal bottom cell growth. The n-i-p configuration might allow for slightly higher efficiencies as one has the opportunity to partially de-couple electrical and optical properties by the application of a so-called flat light-scattering substrate (Sai *et al.*, 2011; Söderström *et al.*, 2012). Further exploration of novel deposition regimes such as the triode reactor configuration developed by AIST and the newly adapted plasma box configuration with narrow inter-electrode distance developed by TEL Solar (see Chapter 5.1), could also help to improve (absorber) material quality and thereby the device performance.

To obtain a major breakthrough in stabilized efficiency compared to the excellent value of 13.6%, the study and incorporation of new absorber materials seem indispensable. For the top cell, the most important challenge is to further increase the open circuit voltage. This is most straightforwardly done by increasing the bandgap, but a reduction of defect densities, especially at the interfaces with the doped layers (in particular, the p-layer structure), could also lead to an increased V_{OC}. Similar approaches could be used for nc-Si:H middle and bottom cells: also here, the band gap of 1.1 eV should allow for higher open circuit voltages than obtained so far. Especially high-quality passivated interfaces – similar to the ones in SHJ solar cells (Van Sark *et al.*, 2012) – could significantly help to this end. Rather than going for the highest efficiency, one could also go for ultrathin triple-junction devices. In this context, very thin (~10–30 nm) 1-D optical cavities of amorphous Germanium (a-Ge:H) form a very interesting option, recently investigated by NEXT Energy in Oldenburg, Germany (Steenhoff *et al.*, 2014). So far, this approach has not yet led to very high efficiencies, but the use of extremely thin a-Ge:H films could enable: (1) a strong cost reduction and (2) reduced recombination.

5.4.3 Thin Film Silicon Quadruple-Junction Solar Cells

Very recently, quadruple-junction devices have attracted significant interest in the thin film Si community. In 2013, Delft University published an optical simulation paper on quadruple-junction devices, showing possible efficiencies approaching 20% (Isabella *et al.*, 2014). Experimental results, however, have not yet exceeded the best efficiencies obtained with triple-junction devices.

For quadruple-junctions, many different configurations are possible. In the simulation paper by Isabella *et al.* (2014), a device in a-Si:H/a-SiGe:H/nc-Si:H/μc-SiGe:H (top to bottom) configuration is proposed. In terms of spectral splitting, this is the optimal configuration for thin-film silicon PV, but material requirements are rather challenging, especially regarding microcrystalline silicon-germanium (μc-SiGe:H) (Matsui *et al.*, 2010; Ni *et al.*, 2014). So far, no experimental devices with a significant response beyond 1100 nm have been presented in the literature. With increasing absorber layer thickness and germanium content, the absorber layer quality decreases very rapidly. To obtain quadruple-junction cells with efficiencies in the 15–20% range, high-quality μc-SiGe:H films with a high germanium content are indispensable.

As long as this challenge remains, it seems more logical to replace the μc-SiGe:H bottom cell by standard μc-Si:H cells with a high crystallinity and thickness, thereby generating high current densities in the infrared. Recent experimental results have indeed followed this approach, using several configurations such as (1) a-SiO:H/a-Si:H/nc-Si:H/nc-Si:H (Si *et al.*, 2014); (2) a-Si:H/a-SiGe:H/μc-Si:H/μc-Si:H (Schüttauf *et al.*, 2014); and (3) a-Si:H/a-Si:H/

nc-Si:H/nc-Si:H (Kirner *et al.*, 2015). In terms of stabilized efficiency, the third approach has shown the best performance so far. Cell efficiencies of 14.0% (initial) and 12.4% (stable) were obtained (Kirner *et al.*, 2015), whereas for small modules (61.44 cm^2), efficiencies of 13.4% (initial) and 12.0% (stable) were presented by the same authors. For both a-Si:H and nc-Si:H, it de-couples high voltage and high current and therefore should allow for higher efficiencies than a-Si:H/nc-Si:H tandem devices and a-Si:H/nc-Si:H/nc-Si:H triple-junction devices. One could typically use an a-Si:H top cell with a V_{OC} above 1 V (Yan *et al.*, 2012; Stuckelberger *et al.*, 2014), and a second cell that delivers sufficient current for the best possible matching, in combination with a SiO_x:H-based intermediate reflector (Buehlmann *et al.*, 2007; Lambertz *et al.*, 2011; Boccard *et al.*, 2014). For the third cell, a high V_{OC} nc-Si:H cell should be used and for the bottom cell a nc-Si:H cell that gives a high current density.

With currently existing building blocks, this device configuration should lead to a stabilized efficiency between 14% and 15% without the necessity of further breakthroughs (Schüttauf *et al.*, 2015). A very interesting approach to fabricating nc-Si:H films at very high crystallinities with excellent electrical properties is to deposit them using an SiF_4 plasma chemistry (Dornstetter *et al.*, 2013).

5.4.4 Further Improvements in Thin Film Crystalline Silicon Solar Cells

Another strategy could be the use of very thin, thin film crystalline silicon absorbers (see Chapters 5.1 and 5.2 for further details) with high current densities and excellent passivated interfaces, such as in silicon heterojunction (SHJ) solar cells (Van Sark *et al.*, 2012). These devices can either be used in single-junction configuration, or as bottom cell in multijunction devices, either completely thin film Si-based, or with top cells based on for instance III-V semiconductors (Essig *et al.*, 2015) or on perovskite component cells (Löper *et al.*, 2015).

Recently, a theoretical reassessment of the limiting efficiency for c-Si based devices was proposed by Richter *et al.* (2013), who predicted that efficiencies well above 24% and 28% are within reach with Si absorber layers of 1 μm and 10 μm, respectively (see Figure 5.4.1). Finally, according to calculations, maximum efficiencies up to 29.43% should – at least in theory – be achievable for 110-μm-thick Si-based solar cells.

The results presented in Figure 5.4.1 imply that perfect contacts applied to high quality thin crystalline layers (with thicknesses up to at most a few tens of microns) can lead to high efficiency devices. Ultrathin cells would benefit more than thick ones from high quality passivating contacts. Indeed, thin c-Si cells with a thickness of 1 μm could reach V_{OC} values over 800 mV with such contacts. Possible materials of choice – in terms of the highest expected quality – are the following:

1. Dense nc-Si:H with a high crystallinity. Indeed, V_{OC} values as high as 608 mV were reported when doing a one-sided passivation with a-Si:H on such layers (Hänni *et al.*, 2014).
2. Recrystallized Si layers, as discussed in Chapter 5.2.
3. Layers obtained by epitaxial and lift-off processes.
4. Other processes such as spallation, in which a region of a c-Si wafer is weakened, e.g., by hydrogen implantation and where a subsequent separation process can take place by applying for instance mechanical stress.

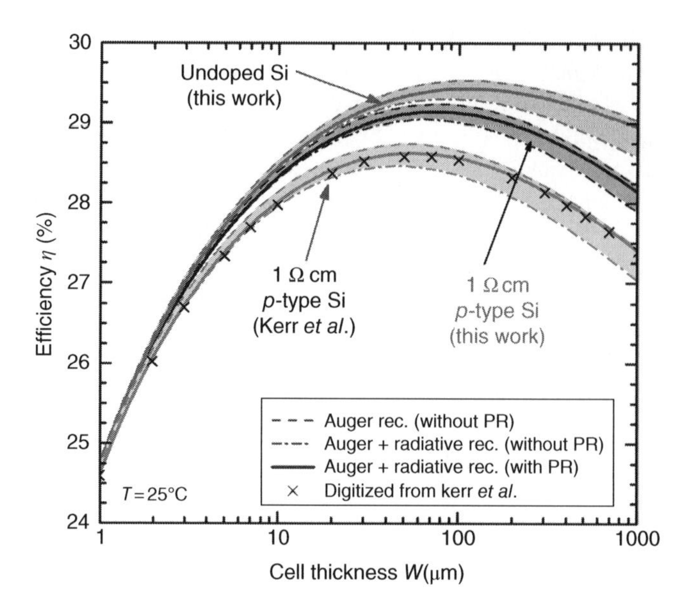

Figure 5.4.1 Efficiency versus thickness for c-Si solar cells under one-sun illumination at 25°C. The solid lines represent the efficiency for a cell made of 1 Ω·cm p-type Si and undoped Si, taking radiative and Auger recombination, as well as photon recycling (PR) into account. The additional curves show cells constrained by either only Auger recombination or by Auger and radiative recombination. Both curves were calculated without taking PR into account and, thus, represent an upper limit assuming complete photon recycling and a lower limit assuming no photon recycling, respectively. Additionally, the curves are also shown for 1 Ω·cm p-type Si calculated with the modeling parameters used by Kerr *et al.* (2003). The cross symbols are the corresponding data points digitized from Kerr *et al.* (2003). Source: Reproduced with permission from *IEEE Journal of Photovoltaics*, licence number: 3665260753169

For approaches where layers are obtained by epitaxial and lift-off processes, 21.7% 6-inch cells on 50 μm-thick wafers have been reported by the company Solexel, without further details of the fabrication process.

For ultra-thin devices with excellent passivation quality, SHJ-like contacting schemes are likely to be a good choice, especially if all the attempts made in the past few years (Holman *et al.*, 2012; Holman *et al.*, 2013) to reduce parasitic absorption can be implemented. For instance, Kobayashi *et al.* (2015) demonstrated 22.5% on a total cell area of 243.4 cm^2 epitaxial cells using a state-of-the art SHJ process. Their total wafer thickness of epitaxially grown Si and Cz c-Si together is 130 μm. Considering the high V_{OC} values achieved on such materials, it is likely that ultrathin epitaxial layers will be able to achieve remarkably high efficiencies in the future.

5.4.5 Conclusion

In this chapter we have presented several approaches that offer the possibility of further increasing the efficiency of thin film Si-based solar cells and modules. Current trends in photovoltaics in general and thin film silicon devices in particular are rather towards higher efficiencies than towards reduced costs. In this context, we have presented three pathways to higher conversion efficiencies in the near future: (1) triple-junction devices; (2) quadruple-junction devices; and (3) thin film crystalline silicon solar cells, possibly with passivated interfaces to be used either in single- or multi-junction configuration. For these three device structures, several approaches towards further possible improvements, as well as practical and theoretical efficiency limits have been assessed and discussed. In particular, approaches using passivating contacts on crystallized thin layers seem the most promising to break the 20% barrier using less than 10 μm of silicon.

List of Symbols

Symbol	Description
a-Si:H	Hydrogenated amorphous silicon
a-SiGe:H	Hydrogenated amorphous silicon-germanium
a-Ge:H	Hydrogenated amorphous germanium
c-Si	Crystalline silicon
μc-SiGe:H	Hydrogenated microcrystalline silicon-germanium
nc-Si:H	Hydrogenated nanocrystalline silicon
PR	Photon recycling
SHJ	Silicon heterojunction
V_{OC}	Open-circuit voltage

References

Boccard, M. Despeisse, M., Escarre, J. *et al.* (2014) High-stable-efficiency tandem thin-film silicon solar cell with low-refractive-index silicon-oxide interlayer. *IEEE Journal of Photovoltaics*, **4**, 1368–1373.

Buehlmann, P., Bailat, J., Dominé, D. *et al.* (2007). In situ silicon oxide based intermediate reflector for thin-film silicon micromorph solar cells. *Applied Physics Letters*, **91**, 143505.

Canham. L.T. (1990) Silicon quantum wire array fabrication by electrochemical and chemical dissolution of wafers. *Applied Physics Letters*, **57**, 1046–1048.

Dornstetter, J.-C., Kasouit, S., and Roca i Cabarrocas, P. (2013) Deposition of high-efficiency microcrystalline silicon solar cells using SiF_4/H_2/Ar mixtures. *IEEE Journal of Photovoltaics*, **3**, 581–586.

Essig, S., Benick, J., Schachtner, M. *et al.* (2015) Wafer-bonded GaInP/GaAs//Si solar cells with 30% efficiency under concentrated sunlight. *IEEE Journal of Photovoltaics*, **5**, 977–981.

Godefroo, S., Hayne, M., Jivanescu, M. *et al.* (2008) Classification and control of the origin of photoluminescence from Si nanocrystals. *Nature Nanotechnology*, **3**, 174–178.

Guha, S., Yang, J., and Yan, B. (2013) High efficiency multi-junction thin film silicon cells incorporating nanocrystalline silicon. *Solar Energy Materials and Solar Cells*, **119**, 1–11.

Hänni, S., Boccard, M., Bugnon, G. *et al.* (2015) Microcrystalline silicon solar cells with passivated interfaces for high open-circuit voltage. *Physica Status Solidi A*, **212**, 840–845.

Holman, Z.C., Descoeudres, A., Barraud, L. *et al.* (2012) Current losses at the front of silicon heterojunction solar cells. *IEEE Journal of Photovoltaics*, **2**, 7–15.

Holman, Z.C., Descoeudres, A., De Wolf, S., and Ballif, C. (2013) Record infrared internal quantum efficiency in silicon heterojunction solar cells with dielectric/metal rear reflectors. *IEEE Journal of Photovoltaics*, **3**, 1243–1249.

Isabella, O., Smets, A.H.M., and Zeman, M. (2014) Thin-film silicon-based quadruple junction solar cells approaching 20% conversion efficiency. *Solar Energy Materials and Solar Cells*, **129**, 82–89.

Janz, S., Löper, P., and Schnabel, M. (2013) Silicon nanocrystals produced by solid phase crystallisation of superlattices for photovoltaic applications. *Materials Science and Engineering B*, **178**, 542–550.

Kerr, M.J., Cuevas, A., and Campbell, P. (2003) Limiting efficiency of crystalline silicon solar cells due to Coulomb-enhanced Auger recombination. *Progress in Photovoltaics: Research Applications*, **11**, 97–104.

Kim, S., Chung, J.-W., Lee, H. *et al.* (2013) Remarkable progress in thin-film silicon solar cells using high-efficiency triple-junction technology. *Solar Energy Materials and Solar Cells*, **119**, 26–35.

Kirner, S., Neubert, S., Schultz, C. *et al.* (2015) Quadruple-junction solar cells and modules based on amorphous and microcrystalline silicon with high stable efficiencies. *Japanese Journal of Applied Physics*, **54**, 08KB03.

Kobayashi, E., Watabe, Y., Hao, R., and Ravi, T.S. (2015) High efficiency heterojunction solar cells on n-type kerfless mono crystalline silicon wafers by epitaxial growth. *Applied Physics Letters*, **106**, 223504.

Lambertz, A., Grundler, T., and Finger, F. (2011) Hydrogenated amorphous silicon oxide containing a microcrystalline silicon phase and usage as an intermediate reflector in thin-film silicon solar cells. *Journal of Applied Physics*, **109**, 113109.

Lee, B. G., Hiller, D., Luo, J.-W. *et al.* (2012) Strained interface defects in silicon nanocrystals. *Advanced Functional Materials*, **22**, 3223–3232.

Löper, P., Canino, M., Qazzazie, D. *et al.* (2013) Silicon nanocrystals embedded in silicon carbide: Investigation of charge carrier transport and recombination. *Applied Physics Letters*, **102**, 033507.

Löper, P., Moon, S.J., Martin de Nicolas, S. *et al.* (2015) Organic-inorganic halide perovskite/crystalline silicon four-terminal tandem solar cells. *Physical Chemistry, Chemical Physics*, **17**, 1619–1629.

Matsui, T., Jia, H., and Kondo, M. (2010) Thin film solar cells incorporating microcrystalline $Si_{1-x}Ge_x$ as efficient infrared absorber: an application to double junction tandem solar cells. *Progress in Photovoltaics: Research Applications*, **18**, 48–53.

Ni, J., Liu, Q., Zhang, J. *et al.* (2014) Microcrystalline silicon-germanium solar cells with spectral sensitivities extending into 1300 nm. *Solar Energy Materials and Solar Cells*, **126**, 6–10.

Richter, A., Hermle, M., and Glunz, S.W. (2013) Reassessment of the limiting efficiency for cyrstalline silicon solar cells. *IEEE Journal of Photovoltaics*, **3**, 1184–1191.

Sai, H., Kanamori, Y., and Kondo, M. (2011) Flattened light-scattering substrate in thin film silicon solar cells for improved infrared response, *Applied Physics Letters*, **98**, 113502.

Sai, H., Matsui, T., Koida, T. *et al.* (2015) High-efficiency thin-film silicon solar cells on honeycomb textures. *Applied Physics Letters*, **106**, 213902.

Schnabel, M., Weiss, C., Löper, P. *et al.* (2015) Self-assembled silicon nanocrystal arrays for photovoltaics. *Physica Status Solidi A*, DOI: 10.1002/pssa.201431764.

Schüttauf, J.-W., Bugnon, G., Stuckelberger, M. *et al.* (2014) Thin-film silicon triple-junction solar cells on highly transparent front electrodes with stabilized efficiencies up to 12.8%. *IEEE Journal of Photovoltaics*, **4**, 757–762.

Schüttauf, J.-W., Niesen, B., Löfgren, L. *et al.* (2015) Amorphous silicon-germanium for triple and quadruple junction thin-film silicon based solar cells. *Solar Energy Materials and Solar Cells*, **133**, 163–169.

Si, F.T., Kim, D.Y., Santbergen, R. *et al.* (2014) Quadruple-junction thin-film silicon-based solar cells with high open-circuit voltage. *Applied Physics Letters*, **105**, 063902.

Söderström, K., Bugnon, G., Haug, F.-J. *et al.* (2012) Experimental study of flat light-scattering substrates in thin-film silicon solar cells. *Solar Energy Materials and Solar Cells*, **101**, 193–199.

Solexel Press Release, 21 July 2014, available at: http://www.solexel.com

Steenhoff, V., Theuring, M., Vehse, M. *et al.* (2014) Ultrathin resonant-cavity-enhanced solar cells with amorphous germanium absorbers. *Advanced Optical Materials*, **3**, 182–186.

Stuckelberger, M., Billet, A., Riesen, Y. *et al.* (2014) Comparison of amorphous silicon absorber materials: kinetics of light-induced degradation. *Progress in Photovoltaics: Research Applications*. DOI: 10.1002/pip.2559.

Van Sark, W.G.J.H.M., Korte, L., and Roca, F. (eds) (2012) *Physics and Technology Of Amorphous-Crystalline Heterostructure Silicon Solar Cells*, Springer Verlag, Heidelberg.

Yan, B., Yang, J., and Guha, S. (2012) Amorphous and nanocrystalline silicon thin film photovoltaic technology on flexible substrates. *Journal Vacuum Science Technology A* **30**, 04D108.

Yan, B., Yue, G., Sivec, L. *et al.* (2011) Innovative dual function nc-SiOx:H layer leading to a >16% efficient multi-junction thin-film silicon solar cell. *Applied Physics Letters*, **99**, 113512.

You, D.J., Kim, S.H., Lee, H. *et al.* (2014) Recent progress of high efficiency Si thin-film solar cells in large area. *Progress in Photovoltaics: Research Applications*, **23**, 973–988.

Part Six

Organic Photovoltaics

6.1

Solid-State Organic Photovoltaics

Bernard Kippelen
Center for Organic Photonics and Electronics, School of Electrical and Computer Engineering, Georgia Institute of Technology, Atlanta, GA, USA

6.1.1 Introduction

Organic photovoltaics, the technology that employs thin films of organic semiconductors to convert sunlight into electricity, has been the subject of active research over the past decades. Today, the power conversion efficiency in laboratory and industrial cells has reached 10% and the field continues to make a steady progress. This technology is receiving strong interest due to its potential to spawn in the near term a new generation of low-cost, solar-powered products with thin and flexible form factors. In the longer term, organic photovoltaic technologies could reach some of the lowest levelized cost of electricity (LCOE). Here, we provide a brief introduction to the basic concepts of organic photovoltaics and discuss some recent science and engineering results.

The chapter is organized as follows. First, we will clarify the definition of organic photovoltaics as used in the context of this book chapter and provide a brief history of the development of this field of research. Then, we will introduce the reader to the basic concepts of organic semiconductors, followed by a description of the design and operation of organic solar cells. Finally, we will discuss the state-of-the-art and future opportunities.

6.1.2 Definition

The photovoltaic effect – the conversion of light into electrical power – can be traced back to Becquerel's 1839 pioneering studies in liquid electrolytes (Becquerel, 1839) and has since then been studied in a wide range of solid-state materials. In the modern era, many solar cell technologies involve the *n*- and *p*-doping of semiconductors allowing for the formation of *pn* junctions such as in the silicon solar cells reported in 1954 by Chapin, Fuller, and Pearson (Chapin *et al.*, 1954). The cell reported by the scientists at Bell Labs can be considered the

Photovoltaic Solar Energy: From Fundamentals to Applications, First Edition.
Edited by Angèle Reinders, Pierre Verlinden, Wilfried van Sark, and Alexandre Freundlich.

Companion website: www.wiley.com/go/reinders/photovoltaic_solar_energy

tipping point that transformed photovoltaics into a technology. Organic photovoltaic devices considered in this chapter are comprised of light absorbers that consist of organic semiconductors that are mainly intrinsic in nature. These intrinsic semiconductors have low conductivity compared to doped inorganic semiconductors used in other photovoltaic technologies, but the low thickness of the absorber films, typically in the range of 100–200 nm, leads to an overall electrical resistance that is sufficiently low to allow for efficient device operation. However, organic semiconductors can also be extrinsically doped and used as electrodes. Doping of organic semiconductors can be employed to build organic solar cells with a *p-i-n* architecture.

Organic compounds are also used as light absorbers in dye-sensitized solar cells (DSSC), however, the operation of such devices involves the use of electrolytes and mesoscale metal-oxide layers. Their design and operation are sufficiently different from that of all solid-state devices that they constitute a class of solar cells on their own. Hence, we will limit the discussion in this chapter to all solid-state organic solar cells.

A schematic representation of an organic solar cell is shown in Figure 6.1.1. It is comprised of an organic absorber sandwiched between a hole-collecting interlayer with a high work function in contact with the organic absorber of one side, and an electron-collecting interlayer with a low work function in contact with the opposite side of the absorber. The organic absorber has the following functions: (1) it harvests a fraction of the sunlight and converting optical radiation into mobile charge carriers; (2) it stores holes and electrons in spatially-separated layers or domains, to yield a photovoltage; and (3) it provides charge transport pathways for carriers towards the hole- and electron-collecting interlayers to yield a photocurrent. The hole- and electron-collecting interlayers are chosen to have a large difference in work function in order to build the necessary asymmetry in the device to force the flow of

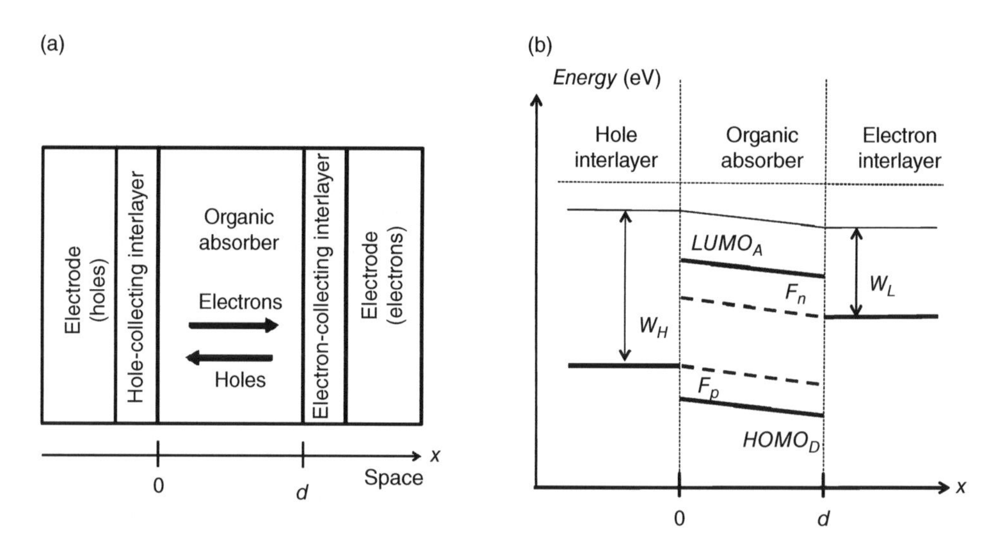

Figure 6.1.1 (a) Cross-sectional schematic representation of an organic solar cell; (b) Corresponding energy level diagram of the solar cell under illumination. W_H: high work function of the hole-collecting interlayer; W_L: low work function of the electron- collecting layer. $LUMO_A$: lowest unoccupied molecular orbital energy of the acceptor material in the organic absorber; $HOMO_D$ highest occupied molecular orbital energy of the donor in the absorber; F_n quasi-Fermi level energy for electrons; F_p quasi-Fermi level energy for holes

electrons towards the low work function interlayer and the holes into the opposite direction (Green, 1982; Wurfel, 2005). The organic absorber is either comprised of a single layer (as in Schottky-type devices), multiple layers, or a blend of several organic materials (as in bulk heterojunction devices). It has evolved over time as described in Section 6.1.3.

6.1.3 Brief History

Early reports on the study of the photovoltaic effect in organic solid-state materials date back to the late 1950s, only a few years after the discovery of the silicon *pn* junction solar cell (Chapin *et al.*, 1954). The use of thick laminated organic films (Kearns and Calvin, 1958) or thick organic crystals such as anthracene (Kallmann and Pope, 1959), resulted in devices with a large internal resistance, yielding to low power conversion efficiency. The dawn of organic photovoltaics as an emerging technology for power generation dates back to the late 1970s, early 1980s with the report of thin-film solar cells with power conversion efficiencies approaching 1% (Morel *et al.*, 1978; Loutfy *et al.*, 1981; Tang, 1986). The geometry of these early devices consisted of an organic semiconductor layer such as merocyanine (Ghosh and Feng, 1978) sandwiched between two metal electrodes with different work function, or a layer of phthalocyanine molecules sandwiched between a conductive oxide and a metal (Loutfy *et al.*, 1981). The operation of these organic solar cells was described in terms of the sequential steps of light absorption, the creation of excitons,[1] the diffusion of excitons and their dissociation into electrons and holes pairs at Schottky-type metal-organic interfaces, carrier transport, and electrical power delivery to an external circuit. An alternative was to use two layers of different organic semiconductors, one made from electron donor-like molecules such as cupper-phthalocyanine or carbazole with a low ionization energy that had the function of a hole transport layer, another comprised of electron acceptor-like molecules that served as an electron transport layer such as perylene diimides molecules (Tang, 1986) or fullerenes. In such bilayer devices, the generation of free carriers is assigned to the dissociation of the excitons at the interface formed between the two organic semiconductors with large differences in ionization energy and electron affinity. In all these device architectures, the thickness of the organic semiconductor layers that absorb the sunlight cannot be much larger than the diffusion length of the excitons (a few tens of nm) to allow them to reach the interface where they dissociate in order to contribute to the current. Organic molecules absorb light very efficiently and typically exhibit absorption coefficients in the range of $10^5\,cm^{-1}$. Hence, light absorption can be rather efficient even in layers of a few tens of nm thickness when considering the incident light that is not fully absorbed after its propagation in the film is reflected back by a reflecting metal electrode. However, the rather small exciton diffusion length compared to the optimal absorption depth remained a challenge and led to the development of polymeric organic photovoltaic cells based on an interpenetrating network formed from a phase-segregated mixture of either two semiconducting polymers (Halls *et al.*, 1995) or a donor-like hole transport polymer mixed with a soluble electron-acceptor fullerene (Yu *et al.*, 1995). Such phase-segregated blends provide both the spatially distributed interfaces necessary for efficient exciton dissociation for charge generation, and the means for separate electron and hole transport in opposite directions towards the charge collecting electrodes. While early

[1] An exciton is bound state of an electron and a hole attracted to each other by the Coulomb force.

studies pointed towards binary mixtures with pure donor- and acceptor-like nanoscale domains, more recent studies using advanced X-ray scattering experiments (such as grazing wide angle X-ray scattering (GWAXS) experiments) have revealed more complex morphologies with the presence of mixed phases (Rivnay *et al.*, 2012). Control and optimization of the morphology of these blends have played an important role in the evolution of organic photovoltaics during the past two decades and the gradual improvement in efficiency from values of 1%, to values greater than 10% reported in single junction cells (Liu *et al.*, 2014) and in tandem cells (You *et al.*, 2013) by academic teams in champion cells, and values of 11% in cells and modules originating from industrial laboratories (Green *et al.*, 2015).

6.1.4 Organic Semiconductors

Organic molecular and polymeric semiconductors can form films with complex morphologies and varying degrees of order and packing modes through the interplay of a variety of non-covalent interactions. Their molecular structure consistently presents a backbone along which the carbon (or nitrogen, oxygen, sulfur) atoms are sp^2-hybridized and thus possess a π-atomic orbital. The conjugation (overlap) of these π orbitals along the backbone results in the formation of delocalized π molecular orbitals (see Figure 6.1.2), which define the frontier (HOMO: highest occupied molecular orbital, and LUMO: lowest unoccupied molecular orbital) electronic levels and determine the optical and electrical properties of the (macro) molecules. The overlap of the frontier π molecular orbitals between adjacent molecules or polymer chains characterizes the strength of the intermolecular electronic couplings (also called transfer integrals or tunneling matrix elements), which represent the key parameter governing charge carrier mobility. Seminal work on conducting polymers dates back to 1970s and was recognized through the Nobel Prize in Chemistry awarded to Alan Heeger, Alan MacDiarmid, and Hideki Shirakawa in 2000.

In crystalline inorganic semiconductors, the three-dimensional character and rigidity of the lattice ensure wide valence and conduction bands and large charge carrier mobility values (typically on the order of several 10^2 to $10^3\,cm^2/Vs$). In contrast, in organic semiconductors, the weakness of the electronic couplings (due to their intermolecular character), the large electron-vibration coupling (leading to pronounced geometry relaxation), and the disorder effects all conspire to produce more modest carrier mobility values due to charge-carrier

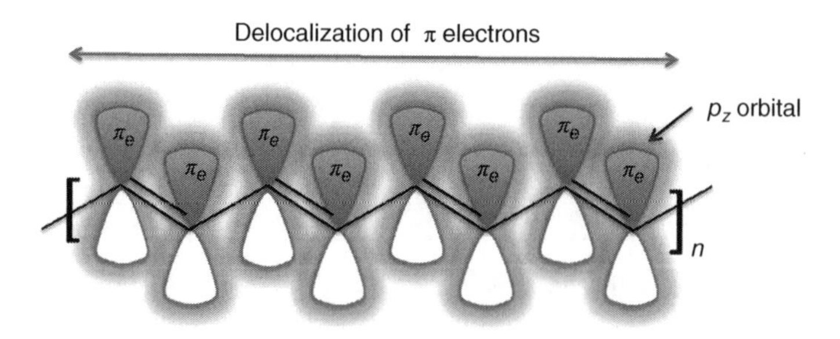

Figure 6.1.2 Schematics of the π electrons (π_e) in the p_z orbitals on each carbon atom and their delocalization over the conjugated polymer chain due to electronic coupling between neighboring carbon atoms. (*See insert for color representation of the figure*)

localization and formation of polarons[2]; transport then relies on polaron-hopping from molecule to molecule or between polymer chain segments (Bredas *et al.*, 2004). As a result, charge carrier mobility values strongly depend on the morphology and can vary over several orders of magnitude when going from highly disordered amorphous films (typically, 10^{-6} to 10^{-3} cm^2/Vs) to highly ordered crystalline materials (>1 cm^2/Vs). One challenge in organic photovoltaics is to design polymers and molecules that yield solid-state charge mobility values in films that are sufficiently high (>10^{-2} cm^2/Vs) to translate into carrier diffusion lengths that are of the order of the thickness of the light absorber (typically 100–200 nm).

Conjugated oligomers and polymers display absorption bands that are usually very intense and broad. The high extinction coefficients are a result of the large wave function overlap between the ground state and the lowest excited states. The width of the absorption spectrum (>1 eV) is due to the significant geometry relaxations that take place in the excited states in π-conjugated systems (Bredas *et al.*, 2004). Once promoted to an excited state, the π-conjugated system relaxes down to the bottom of the potential energy surface of the lowest excited state, the excited state reaches its equilibrium geometry, and an exciton forms. In organic semiconductors the binding energy of an exciton is on the order of a few tenths of an eV, much larger than thermal energy. This high value is a reflection not only of the rather low dielectric constant of π-conjugated organic materials ($\varepsilon_r = 3 - 4$) but also of the significant electron correlation and geometry relaxation effects present in these materials. Thus, in contrast to inorganic semiconductors, the absorption of a photon at room temperature in conjugated materials does not lead to free charge carriers but to neutral, bound electron-hole pairs (excitons). The dissociation of these excitons into mobile free carriers plays a central role in the design and architecture of organic solar cells as will be discussed in more detail in the following sections.

6.1.5 Processing of Organic Semiconductors

Organic semiconductors are processed, depending on their solubility, into thin films by either vacuum deposition techniques or by wet processing. Small molecular compounds with low molecular weight (typically < 1,000 amu) have limited solubility in common solvents, which makes their incorporation into inks difficult. But they can be deposited into thin films using a sublimation process that involves heating the material in a high vacuum (typically $< 1.3 \times 10^{-5}$ Pa). While the need for a high vacuum can be considered a drawback, it is also a blessing since the same sublimation process can be used to purify these materials using thermal-gradient zone sublimation. Larger molecular weight compounds or polymers, on the other hand, show higher solubility and can be processed using standard coating and printing techniques. During the formulation of inks, their viscosity and surface tension are adjusted to comply with the requirements of the selected wet-processing technique.

6.1.6 Physics of Organic Solar Cells

In this section, we now return to the design and operation of organic solar cells. As in any other photovoltaic device, under steady-state illumination the solar cells yields electrical power resulting from the simultaneous production of a photovoltage and a photocurrent.

[2] Landau (1933) introduced the concept of polarons in 1933 in order to consider the polarization induced by the lattice surrounding a charge.

As discussed above, with organic semiconductors, several design options exist. The mechanism of dissociating optically generated excitons into mobile holes and electrons can be achieved in bilayer architectures consisting of separate layers of different semiconductors by analogy of planar *pn*-junction solar cells, or in so-called bulk heterojunction geometries consisting of blends of semiconductors. For simplicity, we will limit the discussion here to the latter case and assume that the absorber consists of a blend of two semiconductors that form three different phases: a mixed phase of the two semiconductors and two pristine phases of each of the semiconductor spatially distributed in the bulk of the blend with each pure phase having interconnected domains that allows for the transport of carriers through the thickness of the absorber. Recent morphology studies suggest that this is the case in efficient cells (Rivnay *et al.*, 2012). In the blend, one of the semiconductors is chosen to have high hole mobility and the other one high electron mobility. The values of the charge mobility need to be sufficiently high to allow for a diffusion length of the carriers that is comparable or preferably larger than the thickness of the absorber, for a given carrier lifetime.

In all organic solar cells, the degree of electronic coupling between the donor-like material used for hole transport, and the acceptor-like material used for electron transport, plays a critical role. This electronic coupling at donor-acceptor interfaces, or in donor-acceptor blends, leads to the formation of an intermolecular charge-transfer complex (Perez *et al.*, 2009; Potscavage *et al.*, 2008; Vandewal *et al.*, 2009) that simultaneously influences the photogeneration of mobile charge carriers and the dark current due to thermal generation. The latter correlates with the value for the open-circuit voltage, and the larger the photon energy of this charge-transfer complex, the larger the resulting open-circuit voltage. This charge-transfer complex is generally spectrally resolved in the low photon energy tail in the power spectrum of the external quantum efficiency (EQE) using highly sensitive Fourier transform photocurrent spectroscopy (Vandewal *et al.*, 2010) or directly in the EQE spectrum of solar cells with large shunt resistance and high dynamic range (Zhou *et al.*, 2014). The photon energy of the charge-transfer complex depends on the ionization energy of the donor material, the electron affinity of the acceptor, and their degree of electronic interaction. It is therefore determined by the choice of donor and acceptor, and also by the morphology of the films.

To describe the operation of organic solar cells, energy-level diagrams can be drawn in which the valence and conduction band energies are replaced by the HOMO and LUMO band energies, respectively, as shown in Figure 6.1.3. Under illumination, the organic semiconductor layers absorb light, creating excitons that dissociate at donor/acceptor interfaces to form mobile holes and electrons. As in any solar cell, the densities of hole and electrons at steady-state under quasi-equilibrium conditions are described by their quasi-Fermi level energies. The gradient in the quasi-Fermi level energy across the absorber provides the driving force for the current. The difference between the quasi-Fermi level energy for electrons (F_n) at the low work function interlayer W_L (see Figure 6.1.1) and the quasi-Fermi level energy for holes (F_p) at the high work function electrode W_H, determines the photovoltage.

6.1.7 State-of-the-Art and Current Trends

Since organic chemistry enables the synthesis of a large variety of chemical compounds, in theory, an infinite number of combinations of donor- and acceptor-like materials can be tested. Some examples of molecules and polymers that have been widely studied are shown in

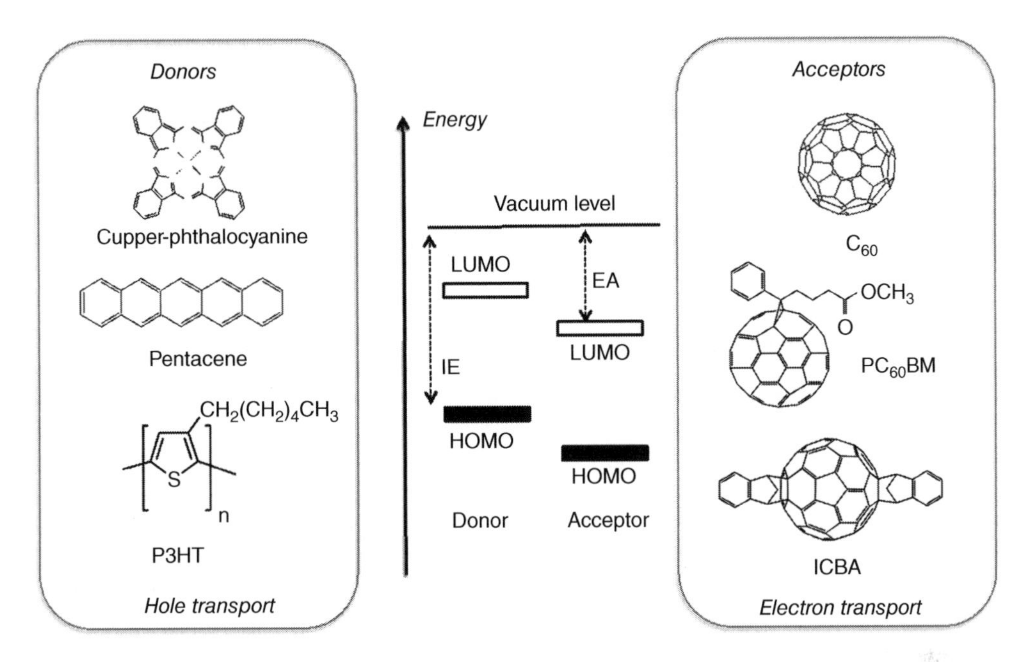

Figure 6.1.3 Energy-level diagram of organic semiconductors (center) showing the relative position of the HOMO and LUMO bands with respect to the vacuum level used as reference. Selected examples of the chemical structure of electron-donor like materials used as hole transport materials in organic photovoltaics are shown on the left. Such materials have a low ionization energy (IE). Selected examples of the chemical structure of electron-acceptor-like materials used as electron transport materials are shown on the right. Such materials have a large electron affinity (EA). P3HT: poly(3-hexylthiophene); $PC_{60}BM$: 6,6-phenyl-C 61-butyric acid methyl ester; ICBA: indene-C 60 bisadduct

Figure 6.1.3. Early organic polymeric cells were fabricated with regio-regular poly(3-hexylthiophene) (P3HT). During the last decade, in an effort to synthesize new materials with optimized optical and electrical properties, the polymer structure has evolved towards more complex structures. To date, the most efficient solar cells are comprised of polymers that are built from building blocks with varying degrees of electron-rich and electron-poor moieties that can extend their optical absorption into the near infrared part of the spectrum through intramolecular charge-transfer. These polymers generally serve as hole-transport materials and they are mixed with soluble fullerenes such as $PC_{60}BM$ or ICBA that serve as electron transport materials. Recently, polymer blends have started to yield devices with increasing power conversion efficiencies (5%) (Mu *et al.*, 2014). The design of new materials continues to be driven by the simultaneous optimization of the intrinsic molecular properties of the constituents as well as the morphology of the films formed by blending them. For details, the reader is referred to few recent reviews found in the scientific literature on the subject (Brabec *et al.*, 2010; Li *et al.*, 2012; Mishra and Bauerle, 2012; Thompson and Frechet, 2008). The number of publications on new materials that report power conversion efficiencies in single cells above 10% is likely to continue to increase.

While new materials for the absorber layer have been the focus of active research, the overall device architecture and the processing techniques used to fabricate cells and modules have evolved drastically over the years. Organic solar cells can also be fabricated in stacked multiple junction geometries (Ameri *et al.*, 2009). In early devices, the hole-collecting interlayer and electrode were comprised of ITO-coated glass (Xue *et al.*, 2004), and the electron-collecting interlayer comprised of highly reactive materials such as Ca capped with a metal electrode (Al) deposited using vacuum thermal evaporation. Gradually, such expensive or reactive materials have been replaced with lower-cost conducting polymers such as PEDOT:PSS combined with solution processed metal-containing inks, and/or metal-oxide inks such as ZnO using roll-to-roll manufacturing techniques (Krebs, 2009). These device architectures, free from ITO and vacuum coating steps, offer cost advantages and can also lead to semitransparent solar cells (Zhou *et al.*, 2010) that can easily be attached to windows. More recently, the strong electronic interaction between materials at interfaces has been used to develop low work function conducting polymers, enabling the demonstration of all organic semitransparent solar cells (Zhou *et al.*, 2012).

The future of organic solar cells as a pervasive technology for portable power will largely rely on their economic potential in terms of levelized cost of energy. This depends on a number of intricate factors such as efficiency, manufacturing cost, lifetime, form factor, weight, scalability, and manufacturing using low-energy (and consequently low carbon footprint) processing. At this point, two main manufacturing techniques can be foreseen, vacuum processing and wet processing. Both present advantages and challenges and undergo advances towards lower cost and lower environmental impact.

If organic photovoltaic technologies mature beyond niche consumer market applications and become players in power generation, their composition must be based on materials available on large scales at low cost. The feedstock for most (or all) (Zhou *et al.*, 2012) of the materials used being oil, the technology can be considered highly scalable. Beyond petroleum chemistry, natural materials such as wood can be used to produce cellulose-based materials that can also become part of organic photovoltaics (Zhou *et al.*, 2013). Last, but not least, organic solar cells must demonstrate lifetimes of several years. While organic cells might not show twenty years or more of operational lifetime like crystalline silicon cells, the recent demonstration of 100,000 h operational lifetime in organic light-emitting diodes (Chu *et al.*, 2006) is indicative that long lifetimes are within reach with organic semiconductors, if encapsulation is adequate.

6.1.8 Conclusion

In summary, photovoltaic technologies are still under development and have not yet been commercialized on a large scale. However, they could become the cleanest renewable energy source as their manufacturing, deployment, operation, and recovery at the end of life-cycle have the potential to leave the lowest carbon footprint. Full life-cycle assessment becomes an important method when the scale of deployment of a technology such as photovoltaics is expected to grow exponentially. Too little attention has been devoted to this aspect so far, but it will become more important. The amount of energy needed for manufacturing will play a critical role when comparing the economics and potential of different photovoltaic platforms in the future. With their device architecture simplicity and low temperature processing. organic photovoltaics are expected to become very competitive as their performance in terms of power conversion efficiency continues to rise. The field of organic photovoltaics provides an exciting playground at the frontiers of science, engineering, and technology. Advances in the near term are likely to lead to solar cells with efficiencies close to 20% in single heterojunction geometries and efficiencies up to 25% in tandem-cell geometries. If organic photovoltaics keep their promise, they can soon become an ubiquitous, clean and sustainable technology for portable power and potentially provide large-scale energy production for future generations.

Acknowledgments

This book chapter is the result of research on organic solar cells that is funded in part by several programs of the Office of Naval Research (ONR) through programs funded by the Naval Materials Division, Functional Polymer/Organic Materials and by the Bay Area Photovoltaic Consortium funded by the U.S. Department of Energy.

Acronyms

Amu	Atomic Mass Unit
EA	Electron Affinity
EQE	External Quantum Efficiency
F_n	Quasi Fermi level for electrons
F_p	Quasi Fermi level for holes
GWAXS	Grazing wide angle X-ray scattering
HOMO	Highest Occupied Molecular Orbital
ICBA	Indene-C 60 Bisadduct
IE	Ionization Energy
ITO	Indium-Tin-Oxide
LCOE	Levelized Cost of Electricity
LUMO	Lowest Unoccupied Molecular Orbital
P3HT	Poly(3-hexylthiophene)
$PC_{60}BM$	6,6-Phenyl-C 61-butyric acid methyl ester
PEDOT:PSS	Poly(3,4-ethylenedioxythiophene) polystyrene sulfonate
$W_{H,L}$	Work function

References

Ameri, T., Dennler, G., Lungenschmied, C., and Brabec, C.J. (2009) Organic tandem solar cells: a review. *Energy & Environmental Science*, **2**, 347–363.

Becquerel, E. (1839) Mémoire sur les effets électriques produits sous l'influence des rayons solaires. *Comptes Rendus de l'Académie des Sciences*, **9**, 561–567.

Brabec, C.J., Gowrisanker, S., Halls, J.J.M., *et al.* (2010) Polymer-fullerene bulk-heterojunction solar cells. *Advanced Materials*, **22**, 3839–3856.

Bredas, J. L., Beljonne, D., Coropceanu, V., and Cornil, J. (2004) Charge-transfer and energy-transfer processes in pi-conjugated oligomers and polymers: a molecular picture. *Chemical Reviews*, **104**, 4971–5003.

Chapin, D.M., Fuller, C.S., and Pearson, G.L. (1954) A new silicon p-n jucntion photocell for converting solar radiation into electrical power. *Journal of Applied Physics*, **25**, 676–677.

Chu, T.-Y., Chen, J.-F., Chen, S.-Y. *et al.* (2006) Highly efficient and stable inverted bottom-emission organic light emitting devices. *Applied Physics Letters*, **89**, 053503.

Ghosh, A.K. and Feng, T. (1978) Merocyanine organic solar-cells. *Journal of Applied Physics*, **49**, 5982–5989.

Green, M.A. (1982) *Solar Cells*. Prentice-Hall, Englewood Cliffs, NJ.

Green, M.A., Emery, K., Hishikawa, Y. *et al.* (2015) Solar cell efficiency tables (Version 45). *Progress in Photovoltaics*, **23**, 1–9.

Halls, J.J.M., Walsh, C.A., Greenham, N.C. *et al.* (1995) Efficient photodiodes from interpenetrating polymer networks. *Nature*, **376**, 498–500.

Kallmann, H. and Pope, M. (1959) Photovoltaic effect in organic crystals. *Journal of Chemical Physics*, **30**, 585–586.

Kearns, D. and Calvin, M. (1958) Photovoltaic effect and photoconductivity in laminated organic systems. *Journal of Chemical Physics*, **29**, 950–951.

Krebs, F.C. (2009) Polymer solar cell modules prepared using roll-to-roll methods: Knife-over-edge coating, slot-die coating and screen printing. *Solar Energy Materials and Solar Cells*, **93**, 465–475.

Li, G., Zhu, R., and Yang, Y. (2012) Polymer solar cells. *Nature Photonics*, **6**, 153–161.

Liu, Y.H., Zhao, J.B., Li, Z.K., *et al.* (2014) Aggregation and morphology control enables multiple cases of high-efficiency polymer solar cells. *Nature Communications*, **5**, 8.

Loutfy, R.O., Sharp, J.H., Hsiao, C.K., and Ho, R. (1981) Phthalocyanine organic solar cells- Indium-X-metal free phthalocyanine Schottky barriers. *Journal of Applied Physics*, **52**, 5218–5230.

Mishra, A. and Bauerle, P. (2012) Small molecule organic semiconductors on the move: promises for future solar energy technology. *Angewandte Chemie-International Edition*, **51**, 2020–2067.

Morel, D.L., Ghosh, A.K., Feng, T., *et al.* (1978) High-efficiency organic solar-cells *Applied Physics Letters*, **32**, 495–497.

Mu, C., Liu, P., Ma, W. *et al.* (2014) High-efficiency all-polymer solar cells based on a pair of crystalline low-bandgap polymers. *Advanced Materials*, **26**, 7224–7230.

Perez, M.D., Borek, C., Forrest, S.R., and Thompson, M. E. (2009) Molecular and morphological influences on the open circuit voltages of organic photovoltaic devices. *Journal of the American Chemical Society*, **131**, 9281–9286.

Potscavage, W.J., Yoo, S., and Kippelen, B. (2008) Origin of the open-circuit voltage in multilayer heterojunction organic solar cells. *Applied Physics Letters*, **93**, 193308.

Rivnay, J., Mannsfeld, S.C.B., Miller, C.E. *et al.* (2012) Quantitative determination of organic semiconductor microstructure from the molecular to device scale. *Chemical Reviews*, **112**, 5488–5519.

Tang, C.W. (1986) 2-layer organic photovoltaic cell. *Applied Physics Letters*, **48**, 183–185.

Thompson, B.C. and Frechet, J.M.J. (2008) Organic photovoltaics: polymer-fullerene composite solar cells. *Angewandte Chemie-International Edition*, **47**, 58–77.

Vandewal, K., Tvingstedt, K., Gadisa, A. *et al.* (2009) On the origin of the open-circuit voltage of polymer-fullerene solar cells. *Nature Materials*, **8**, 904–909.

Vandewal, K., Tvingstedt, K., Gadisa, A. *et al.* (2010) Relating the open-circuit voltage to interface molecular properties of donor:acceptor bulk heterojunction solar cells. *Physical Review B*, **81**, 8.

Wurfel, P. (2005) *Physics of Solar Cells*. Wiley-VCH, Weinheim.

Xue, J.G., Uchida, S., Rand, B.P., and Forrest, S.R. (2004) 4.2% efficient organic photovoltaic cells with low series resistances. *Applied Physics Letters*, **84**, 3013–3015.

You, J., Dou, L., Yoshimura, K. *et al.* (2013) A polymer tandem solar cell with 10.6% power conversion efficiency. *Nature Communications*, **4**, 1446.

Yu, G., Gao, J., Hummelen, J.C. *et al.* (1995) Polymer photovoltaic cells – enhanced efficiencies via a network of internal donor-acceptor heterojunctions. *Science*, **270**, 1789–1791.

Zhou, Y.H., Cheun, H., Choi, S. *et al.* (2010) Indium tin oxide-free and metal-free semitransparent organic solar cells. *Applied Physics Letters*, **97**, 153304.

Zhou, Y.H., Fuentes-Hernandez, C., Khan, T.M., *et al.* (2013) Recyclable organic solar cells on cellulose nanocrystal substrates. *Scientific Reports*, **3**.

Zhou, Y H., Fuentes-Hernandez, C., Shim, J. *et al.* (2012) A universal method to produce low-work function electrodes for organic electronics. *Science*, **336**, 327–332.

Zhou, Y.H., Khan, T.M., Shim, J.W. *et al.* (2014) All-plastic solar cells with a high photovoltaic dynamic range. *Journal of Materials Chemistry A*, **2**, 3492–3497.

6.2

Hybrid and Dye-Sensitized Solar Cells

Woojun Yoon
U.S. Naval Research Laboratory, Washington, DC, USA

6.2.1 Introduction

Photovoltaic (PV) technology based on hybrid and dye-sensitized solar cells presents a promising opportunity for point-of-use energy harvesting and possible large-scale installation due to its potential for low-cost production. Solar cells based on organic-inorganic hybrid materials that are of interest for PV applications include:

- *Colloidal quantum dot solar cells*: Solution-synthesized QDs possess an inorganic core (e.g., PbS, PbSe, CdTe, CdSe) surrounded by an encapsulating layer of organic material. The inorganic core provides a semiconducting function and the organic shell provides solvent compatibility and enables these materials to be manipulated in solution. In these devices, the absorber consists of layer-by-layer assembly of QDs via solution processes, such as spin-, spray- and dip-coating.
- *Silicon-organic hybrid solar cells*: Highly conducting *p*-type polymers, such as poly (3, 4-ethylenedioxythiophene) poly (styrenesulfonate) (PEDOT:PSS), is spin-coated onto a crystalline Si (c-Si) base as an emitter instead of conventional diffused junctions, creating a p^+-PEDOT:PSS/*n*-Si heterojunction.
- *Dye-sensitized solar cell*: The dye-sensitized solar cell (DSC) consists of nanostructured metal oxide electrodes (e.g. mesoporous TiO_2) covered with sensitizing dyes and liquid electrolytes. For all solid-state dye-sensitized solar cells, a liquid electrolyte has been replaced by a solid hole-transporting material, such as 2,2′,7,7′-tetrakis(N,N-di-p-methoxyphenylamine)-9,9′-spirobifluorene (spiro-OMeTAD).

Photovoltaic Solar Energy: From Fundamentals to Applications, First Edition.
Edited by Angèle Reinders, Pierre Verlinden, Wilfried van Sark, and Alexandre Freundlich.

Companion website: www.wiley.com/go/reinders/photovoltaic_solar_energy

- *Organic-inorganic hybrid perovskite*: Perovskite thin-film solar cells evolved from DSCs consist of metal-halide perovskite thin-film absorbers sandwiched between *p*-type and *n*-type transport layer, operating in a similar manner to a *p-i-n* heterojunction solar cell. More details of perovskite solar cell will be discussed in Chapter 6.3.

So far, the most successful hybrid PV technology is dye-sensitized solar cells, and several commercial providers are continuing their commercialization efforts, such as Dyesol, G24 Power, Exeger Sweden AB, Solaronix, Sony, and Sharp. Despite slow growth in DSC development, perovskite PV technology evolved from DSCs has become the subject of intensive research in the last five years due in particular to their potential for a high-efficiency tandem cell with *c*-Si cell. As an emerging PV technology, QD-based solar cells via solution process are also of great interest due to the use of quantum confinement to tune absorbance across the solar spectrum, making them very attractive for solution-processed multi-junction cells. Si-organic hybrid solar cells, often not classified as one of the emerging PV technologies, are promising for potential cost reduction in c-Si processes by reducing the complexity and number of processing steps, compared to conventional Si cell processing for high-volume manufacturing of Si solar cells.

In this chapter, the current status of hybrid and dye-sensitized solar cell performance will briefly be reviewed. With a brief summary of a notable prior work and background for each solar cell technology, we will focus on discussing the upper limit of the achievable open-circuit voltage (V_{oc}).

6.2.2 Current Status of Hybrid and Dye-Sensitized Solar Cell Performance

Most of the efficiencies of hybrid solar cells in Table 6.2.1 are adopted from the latest "Solar cell efficiency tables" by Green *et al.* (2016). While the highest confirmed cell efficiency for DSCs based on dyes and a liquid electrolyte is 11.9%, it is remarkable to find such a significant development beyond 22% (Green *et al.*, 2016) for small-area perovskite solar cells in 2016. Recently, solution-processed QD solar cells based on PbS QD/ZnO heterojuction show a significant improvement of cell efficiencies greater than 10% but module development has not been reported yet. For silicon-organic hybrid cells, the efficiency of 12.3% has been reported for 4 cm^2 area cells with the front PEDOT:PSS emitter (Zielke *et al.*, 2014), but this efficiency measurement has not yet been confirmed. Much higher efficiency of Si-organic hybrid cells has been reported for cells with PEDOT:PSS as back-junction and phosphorus diffused front junction (Zielke *et al.*, 2014). One of the many challenges of hybrid materials-based PV technologies is the short-term as well as the long-term stability. QD-based solar cells with a certified efficiency of 8.55% have demonstrated a decent short-term stability with no or minimal degradation after 150 days of storage in air without device encapsulation (Chuang *et al.*, 2014). For perovskite solar cells, the latest certified record efficiency of 22.1% announced by NREL in November 2014 has been classified as "*not stablized, initial efficiency*" (http://www.nrel.gov/ncpv/images/efficiency_chart.jpg) (Green *et al.*, 2016).

Table 6.2.1 Current status of hybrid and dye-sensitized PV technologies: Most of confirmed cell, minimodule, and submodule efficiency are adopted from (Green, 2011, 2014, 2015a, 2015b) and measured under the AM 1.5 global spectrum (100 mW/cm^2) at 25 °C. Note that efficiencies of Si-organic hybrid cells are reported to be measured under the one-sun condition (25 °C) but not confirmed yet

Classification	Efficiency (%)	Area[a] (cm^2)	V_{oc} (mV)	J_{sc} (mA/cm^2)	FF (%)	Test center /date	Ref.
Quantum dot							
Cell	8.55 ± 0.18	0.0137 (ap)	554.6	24.1	63.8	Newport (10/13)	(Chuang *et al.*, 2014)
Cell	11.28 ± 0.25	0.0498 (ap)	611.3	27.2	67.8	Newport (03/16)	Personal communication
Silicon-organic hybrid							
Cell	12.3[b]	4 (ap)	603	29.0	70.6	Not confirmed	(Zielke *et al.*, 2014)
Cell	17.4[b]	4 (ap)	653	39.7	67.2	Not confirmed	(Zielke *et al.*, 2014)
Dye sensitized							
Cell	11.9 ± 0.4	1.005 (da)	744	22.47	71.2	AIST (9/12)	(Green *et al.*, 2016)
Minimodule	10.7 ± 0.4	26.5 (da)	754	20.19	69.9	AIST (2/15)	(Green *et al.*, 2016)
Submodule	8.8 ± 0.3	398.8 (da)	697	18.42	68.7	AIST (9/12)	(Green *et al.*, 2016)
Inorganic-organic hybrid perovskite							
Cell	22.1 ± 0.7	0.046 (ap)	1105	24.97	80.3	Newport (3/16)	(Green *et al.*, 2016)
Cell	19.7 ± 0.6	0.9917 (da)	1104	24.67	72.3	Newport (3/16)	(Green *et al.*, 2016)

[a] ap = aperture area, da = designated illumination area.
[b] The cell efficiency is not confirmed.

6.2.3 Hybrid Quantum Dot Solar Cells

Many review articles have been published elsewhere regarding QD device physics, operation, materials consideration, and applications (Kamat, 2008; Nozik, 2002; Sargent, 2012). In addition, Bozyigit *et al.* (2015) published their outstanding work on the details of carrier transport based on PbS QD diodes. In QD solar cells, sequential layer-by-layer assemblies of QDs via solution process have been demonstrated as an excellent photoactive absorber layer. For example, PbS QD films with a bandgap of ~1.3 eV have yielded a high short-circuit current density (J_{sc}) of ~24 mA/cm^2 (Chuang *et al.*, 2014). which is comparable to a J_{sc} of 24.4 mA/cm^2 for a nanocrystalline Si absorber. (Benagli, 2009) However, the overall power conversion efficiency of these devices has been limited largely due to low V_{oc}, regardless of the type of junctions used. In QD solar cells, the V_{oc} values increase linearly with the QD bandgap energy (E_g). For example, a simple diode based on a Schottky junction between PbS QDs and metal (glass/ITO/PbS QDs/LiF/Al) showed V_{oc}(mV) = 553E_g/q-59, where q is the elementary electric charge (e) (Yoon *et al.*, 2013). In order to make useful V_{oc} assessment, the experimental bandgap voltage offsets $W_{oc} = (E_g/q) - V_{oc}$ under open-circuit condition and the ideal bandgap voltage offset limited by radiative recombination were plotted in Figure 6.2.1 using literature V_{oc} values measured under simulated one-sun condition. These V_{oc} values were obtained from the literatures regarding PbS QD and sintered CdTe QD solar cells, employing three different types of junctions; (1) a metal-QDs Schottky junction (Yoon *et al.*, 2013); (2) a p-n heterojunction using n-type wide bandgap semiconductors such as TiO_2, ZnO or CdS (Luther *et al.*, 2010; Brown *et al.*, 2011; Gao *et al.*, 2011; Jasieniak *et al.*, 2011; MacDonald *et al.*, 2012;

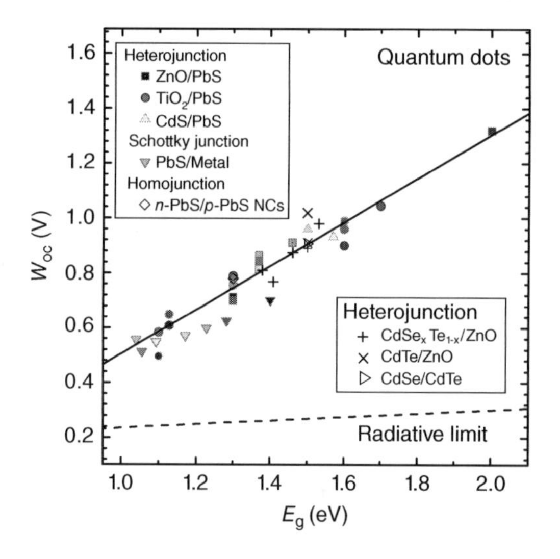

Figure 6.2.1 The bandgap-voltage offset (W_{oc}) under the open-circuit condition as a function of QD bandgap (E_g) in different configuration of QD solar cells; (1) PbS QDs/metal Schottky junction (∇) (Yoon *et al.*, 2013); (2) QD heterojunction devices based on *n*-type wide bandgap semiconductors for (□: ZnO/PbS, ○: TiO_2/PbS, △: CdS/PbS, +: $CdSe_xTe_{1-x}$/ZnO, ×: CdTe/ZnO and ▷: CdSe/CdTe) (Luther *et al.*, 2010; Brown *et al.*, 2011; Gao *et al.*, 2011; Jasieniak *et al.*, 2011; MacDonald *et al.*, 2012; Bhandari *et al.*, 2013; Chang *et al.*, 2013; Hyun *et al.*, 2013; Chuang *et al.*, 2014; Townsend *et al.*, 2014; Yoon *et al.*, 2014); and (3) *p-n* homojunction using *n*-type PbS NCs (◇) (Tang *et al.*, 2012)

Bhandari *et al.*, 2013; Chang *et al.*, 2013; Hyun *et al.*, 2013; Chuang *et al.*, 2014; Townsend *et al.*, 2014; Yoon *et al.*, 2014); and (3) a p-n homojunction using *n*-doped PbS QDs. (Tang *et al.*, 2012) Figure 6.2.1 shows a large deviation of the experimental W_{oc} from that calculated in the radiative limit for a broad bandgap range from 1 eV to 2 eV for all three different types of junctions. Such a large deviation of the experimental W_{oc} regardless of the types of QD-based junction and QD materials indicates that the non-radiative recombination contributions to the dark saturation current are dominant due to a large number of traps associated with QD surface defects. The possible origin of QD surface defects can be attributed to insufficient surface passivation during post-deposition ligand exchange required for appropriate charge carrier transport, as larger surface ligands that encourage organic solubility are often detrimental to charge conduction in the solid state. Further enhancement in V_{oc} could be achieved through improved passivation of QD surfaces through the use of core/shell structure of QDs (Neo *et al.*, 2014) or hybrid passivation schemes (Ip *et al.*, 2012).

6.2.4 Silicon-Organic Hybrid Solar Cells

Recently PV devices based on Si/organic heterojunction have been actively studied as potential candidates for high-efficiency, low-cost Si solar cell technology (Avasthi *et al.*, 2011; Jeong *et al.*, 2012; Avasthi *et al.*, 2014; Erickson *et al.*, 2014; Liu *et al.*, 2014; Nagamatsu *et al.*, 2014; Zielke *et al.*, 2014). In Si/organic hybrid configuration, the device typically consists of organic-based emitter and crystalline Si absorber base. The key advantage of this structure is that the emitter formation is based on a low-temperature and solution-based process, potentially leading to a cost reduction in the process by avoiding a diffusion process at high temperature or ion implantation. In addition, the ability to use the existing mature Si wafer technology coupled with recent large reductions in wafer price provides an opportunity for cost-effective and large-area cell production.

In these devices, the organic emitter layers are usually formed via a solution process, followed by low temperature sintering. Among intrinsically conductive polymers, PEDOT:PSS has been the most popular choice for emitter materials due to: (1) its tunability in electrical conductivity; (2) its good chemical stability; (3) high optical transparency in the visible ranges,; (4) good thermal/UV stability; and (5) ease of use. Commercially available PEDOT:PSS is a water-based dispersion of a wide bandgap organic semiconductor (~1.6 eV) doped with acid PSS. For the fabrication of heterojunction solar cells, PEDOT:PSS is chemically modified to modulate the conductivity, while improving the wettability of PEDOT:PSS to H-terminated crystalline Si surfaces. For example, dimethyl sulfoxide (DMSO) has been widely used to enhance the conductivity. Regardless of the chemical modification, PEDOT:PSS films show a carrier concentration on the order of $\sim 10^{21}$ cm^{-3}, acting as a heavily doped *p*-type semiconductor. Prior to PEDOT:PSS spin-coating, surface texturing was performed throughout the use of random pyramid through KOH etching or Si nanowires using $AgNO_3$/HF etching to improve the light trapping on the surface, while planar heterojunction devices have also been reported without any surface texturing. Most of the organic-Si heterojunction devices have been fabricated without back/front passivation, back-surface field (BSF) or anti-reflection (AR) coating until Schmidt *et al.* (2013) and Zielke *et al.* (2014) fully implemented a BSF and SiO_x front/back surface passivation (Figures 6.2.2 (a) and (b)).

Overall, this technology has shown above 10% power conversion efficiency for small area research-scale devices. Recently Schmidt *et al.* (2013) reported a 12.3% efficient

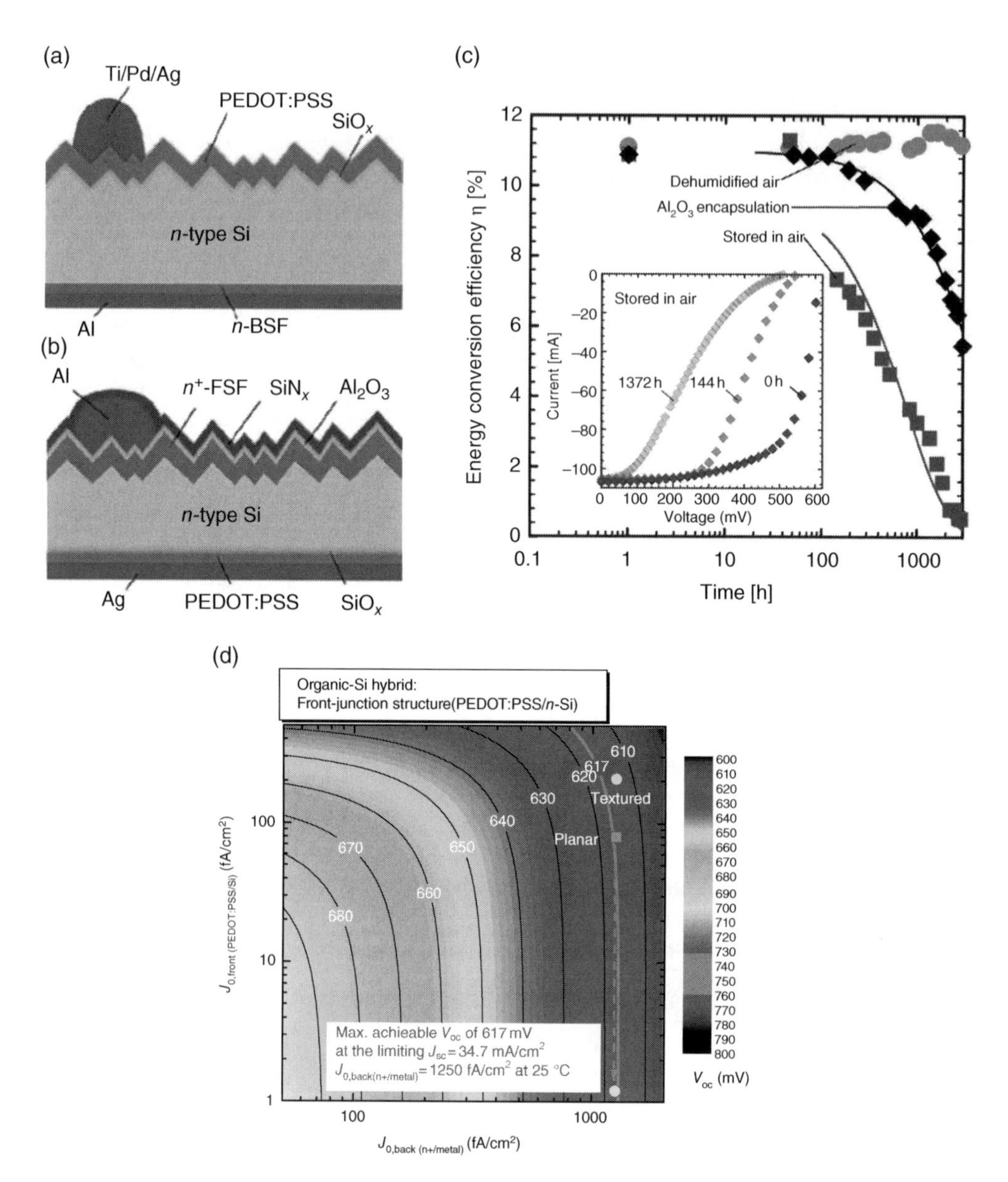

Figure 6.2.2 Schematic of (a) a front junction PEDOT:PSS/*n*-Si heterojunction and (b) a back junction *n*-Si/PEDOT:PSS heterojunction solar cell. Source: (Zielke *et al.*, 2014: 110). Copyright (2014), with permission from Elsevier; (c) Stability of a front junction PEDTO:PSS/*n*-Si heterojunction solar cell as a function of storage time in darkness. Source: Reprinted with permission from *Applied Physics Letters*, vol. 103, Schmidt *et al.*, 2013. Copyright 2013, AIP Publishing LLC; and (d) Achievable V_{oc} in the front-junction organic-Si heterojunction cells with phosphorus-doped BSF for the limiting J_{sc} of 34.7 mA/cm^2 and the saturation current density of P-diffused BSF ($J_{0,b}$) of 1250 fA/cm^2 (Zielke *et al.*, 2014). (*See insert for color representation of the figure*)

PEDOT:PSS/*n*-Si heterojunction cell with random pyramid texture and Al full BSF, showing $J_{sc} = 29$ mA/cm^2, $V_{oc} = 603$ mV and FF = 70.6%. Detailed analysis of the 12.3% cell showed a very low emitter saturation current density ($J_{0,front}$) of 80 fA/cm^2 in the high-injection condition, demonstrating a high level of front surface passivation of planar crystalline Si can be achieved by the use of organic PEDOT:PSS. However, Schmidt *et al.* reported a significant degradation of the PEDOT:PSS/*n*-Si heterojunction device efficiency in air even with Al_2O_3 encapsulation under of PEDOT:PSS as shown in Figure 6.2.2 (c).

For the planar PEDOT/Si heterojunction solar cells, the main contributions to the total saturation current density ($J_{0,total} = J_{0,front} + J_{0,back}$) are recombination at the unpassivated, full Al back metal contacts to phosphorus diffused BSF ($J_{0,back} = 1250$ fA/cm^2) and this resulted in a low V_{oc} of 603 mV (Zielke *et al.*, 2014). Assuming the limiting J_{sc} of 34.7 mA/cm^2 and ignoring the recombination contribution from the base and the front PEDOT:PSS/Si junction, the achievable V_{oc} of heterojunction cell is still limited only to ~617 mV (Figure 6.2.2 (d)).

A front-junction structure of PEDOT:PSS-Si hybrid cells also suffers from a significant loss in J_{sc} (>10 mA/cm^2) due to a parasitic absorption by the front PEDOT:PSS layer. This could be minimized by placing PEDOT:PSS on the rear side (Figure 6.2.2 (b)) without degrading high level of passivation of PEDOT:PSS, leading to an increase in V_{oc} to 663 mV (Zielke *et al.*, 2014). Further enhancement in V_{oc} for diffused junction free organic-Si hybrid cells can be achieved with an additional reduction in the saturation current by organic means for both front and back junctions.

6.2.5 Dye-Sensitized Solar Cells

Compared to other emerging PV technologies, DSC can be considered a more mature PV technology. Outstanding review articles have already been published elsewhere regarding DSC operation principles, limiting efficiency consideration, material development, characterization methods and module development (Goncalves *et al.*, 2008; Hagfeldt *et al.*, 2010; Hardin *et al.*, 2012). Therefore, the main focus here will be the upper limit of the attainable V_{oc} in DSC. The important energy levels of the component governing the overall efficiency of a DSC are the Fermi level of the nanostructured metal oxide electrode, the lowest unoccupied molecular orbital (LUMO) level and the highest occupied molecular orbital (HOMO) level of the sensitizing dye, and the potential of redox couples in the electrolyte (Figures 6.2.3 (a) and 6.2.3 (b)). The V_{oc} in a DSC depends on loss-in-potentials defined as the over-potentials required for an electron injection from dye to the Fermi level of oxide electrode and hole transport from the HOMO level of dye to the potential of redox components. Figure 6.2.3 (c) shows the contour map of the V_{oc} as a function of the absorption onset and the loss-in-potential with radiative limits. Due to the significant loss during the dye generation process, iodide/ trioiodide-based electrolyte limits an achievable V_{oc} to ~800 mV with the exception of N719 dye coupled with iodide complexes due to the higher position of the TiO_2 conduction band with an efficiency electron injection from excited dye molecules (Nazeeruddin *et al.*, 2005; Wang *et al.*, 2006).

Higher V_{oc} can be achieved by reducing the loss-in-potential through the use of new redox couples. For example, the use of cobalt-based complexes resulted in a very high V_{oc} of >900 mV in DSCs (Mathew *et al.*, 2014; Yella *et al.*, 2011). A more encouraging improvement in V_{oc} has been achieved using a solid-state small-molecule hole conductor instead of a liquid

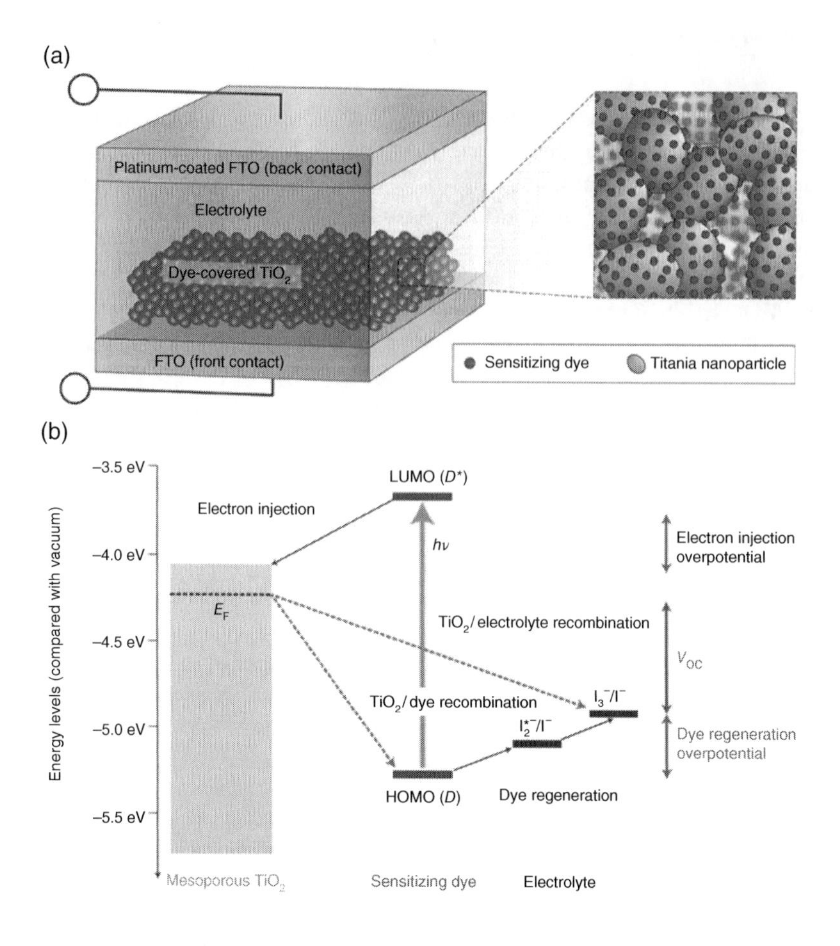

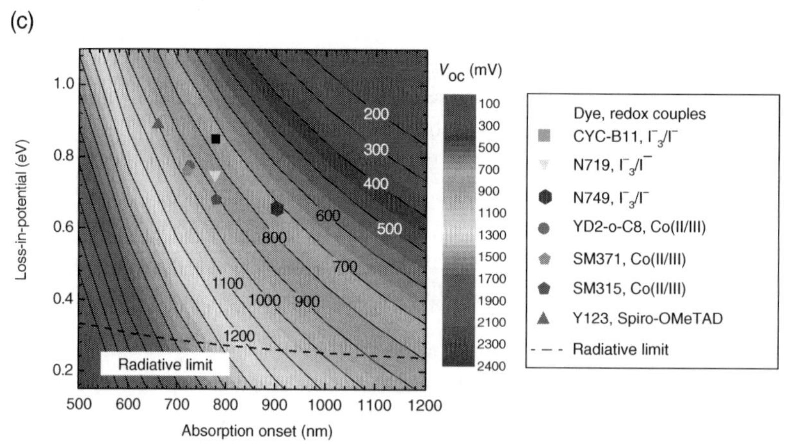

Figure 6.2.3 (a) Schematic and (b) energy level and operation of dye-sensitized solar cells Source: Reprinted by permission from Macmillan Publishers Ltd: Hardin *et al.*, 2012: 162, copyright (2012)); (c) Open-circuit voltage as a function of absorption onset and loss-in-potential with radiative limits. The literature values in the absorption onsets and the loss-in-potentials are shown for the ruthenium dye (CYC-B11)/iodide redox couple (C.-Y. Chen *et al.*, 2009), the N719 dye/iodide redox couple (Nazeeruddin *et al.*, 2005), the black dye N749/iodide redox couple (Nazeeruddin *et al.*, 2001), the co-sensitized donor–pi–acceptor dye (YD2-o-C8 and Y123)/cobalt redox couple (Yella *et al.*, 2011), the SM371/Co(II/III) (Mathew *et al.*, 2014), the SM315/Co(II/III) (Mathew *et al.*, 2014), and the Y123 dye/hole conductor spiro-OMeTAD (Burschka *et al.*, 2011). (*See insert for color representation of the figure*)

electrolyte, demonstrating a V_{oc} beyond 1 V (P. Chen *et al.*, 2009). The use of solid-state hole conductors enabled several orders of magnitude faster hole-transport (in tens to hundreds of picoseconds) from the oxidized dye to the HOMO level of the hole-conductors than the dye regeneration process, significantly minimizing the loss-in-potential.

6.2.6 Conclusion

In summary, we have reviewed the latest scientific and technical progress of hybrid and dye-sensitized solar cells, focusing on the upper limit of the achievable V_{oc} in these PV devices. These emerging materials-based PV technologies have shown an incredible progress in the past years. Based on abundant materials and scalable coating technologies, these emerging PV technologies show the potential for low cost, lightweight, and flexible solar power generation on a large scale. Despite this remarkable progress, much of the underlying physical processes and their limitations have yet to be better understood. Similarly, stability and scale-up in manufacturing volume have proven challenging in making fast progress towards commercialization. Addressing these issues, ranging from fundamental science to technological advances will lead to an opportunity close to full commercialization of PV technologies based on hybrid and dye-sensitized solar cells.

References

Avasthi, S., Lee, S., Loo, Y.-L., and Sturm, J.C. (2011) Role of majority and minority carrier barriers silicon/organic hybrid heterojunction solar cells. *Advanced Materials*, **23** (48), 5762–5766.

Avasthi, S., Nagamatsu, K.A., Jhaveri, J. *et al.* (2014) Double-heterojunction crystalline silicon solar cell fabricated at 250 oC with 12.9% efficiency. *Proceedings of IEEE 40th Photovoltaic Specialist Conference (PVSC)*, 2014, pp. 0949–0952.

Benagli, S., Borrello, D., Vallat-Sauvain, E. *et al.* (2009) High-efficiency amorphous silicon devices on LPCVD-ZNO TCO prepared in industrial KAI-M R&D reactor. *Proceedings of 24th European Photovoltaic Solar Energy Conference*, pp. 2293–2298.

Bhandari, K.P., Roland, P. J., Mahabaduge, H. *et al.* (2013) Thin film solar cells based on the heterojunction of colloidal PbS quantum dots with CdS. *Solar Energy Materials and Solar Cells*, **117** (0), 476–482.

Bozyigit, D., Lin, W.M.M., Yazdani, N. *et al.* (2015) A quantitative model for charge carrier transport, trapping and recombination in nanocrystal-based solar cells. *Nature Communications*, **6**.

Brown, P.R., Lunt, R.R., Zhao, N. *et al.* (2011) Improved current extraction from ZnO/PbS quantum dot heterojunction photovoltaics using a MoO3 Interfacial layer. *Nano Letters*, **11** (7), 2955–2961.

Burschka, J., Dualeh, A., Kessler, F. *et al.* (2011) Tris(2-(1H-pyrazol-1-yl)pyridine)cobalt(III) as p-type dopant for organic semiconductors and its application in highly efficient solid-state dye-sensitized solar cells. *Journal of the American Chemical Society*, **133** (45), 18042–18045.

Burschka, J., Pellet, N., Moon, S.-J. *et al.* (2013) Sequential deposition as a route to high-performance perovskite-sensitized solar cells. *Nature*, **499** (7458), 316–319.

Chang, L.-Y., Lunt, R.R., Brown, P.R. *et al.* (2013) Low-temperature solution-processed solar cells based on PbS colloidal quantum dot/CdS heterojunctions. *Nano Letters*, **13** (3), 994–999.

Chen, C.-Y., Wang, M., Li, J.-Y. *et al.* (2009a) Highly efficient light-harvesting ruthenium sensitizer for thin-film dye-sensitized solar cells. *ACS Nano*, **3** (10), 3103–3109.

Chen, P., Yum, J.H., Angelis, F.D. *et al.* (2009b) High open-circuit voltage solid-state dye-sensitized solar cells with organic dye. *Nano Letters*, **9** (6), 2487–2492.

Chuang, C.-H.M., Brown, P.R., Bulović, V., and Bawendi, M.G. (2014) Improved performance and stability in quantum dot solar cells through band alignment engineering. *Nature Materials*, **13** (8), 796–801.

Erickson, A.S., Zohar, A., and Cahen, D. (2014) n-Si–organic inversion layer interfaces: a Low temperature deposition method for forming a p–n homojunction in n-Si. *Advanced Energy Materials*, **4** (9).

Gao, J., Perkins, C.L., Luther, J.M. *et al.* (2011) n-Type transition metal oxide as a hole extraction layer in PbS quantum dot solar cells. *Nano Letters*, **11** (8), 3263–3266.

Goncalves, L.M., de Zea Bermudez, V., Ribeiro, H.A., and Mendes, A.M. (2008) Dye-sensitized solar cells: A safe bet for the future. [10.1039/B807236A]. *Energy and Environmental Science*, **1** (6), 655–667.

Green, M.A., Emery, K., Hishikawa, Y. *et al.* (2016) Solar cell efficiency tables (version 48) *Progress in Photovoltaics: Research and Applications*, **24** (7), 905–913.

Hagfeldt, A., Boschloo, G., Sun, L. *et al.* (2010) Dye-sensitized solar cells. *Chemical Reviews*, **110** (11), 6595–6663.

Hardin, B.E., Snaith, H.J., and McGehee, M.D. (2012) The renaissance of dye-sensitized solar cells. [10.1038/nphoton.2012.22]. *Nature Photonics*, **6** (3), 162–169.

Hyun, B.-R., Choi, J.J., Seyler, K.L. *et al.* (2013) Heterojunction PbS nanocrystal solar cells with oxide charge-transport layers. *ACS Nano*, **7** (12), 10938–10947.

Ip, A.H., Thon, S.M., Hoogland, S. *et al.* (2012) Hybrid passivated colloidal quantum dot solids. [10.1038/nnano.2012.127]. *Nature Nano*, **7** (9), 577–582.

Jasieniak, J., MacDonald, B.I., Watkins, S.E., and Mulvaney, P. (2011) Solution-processed sintered nanocrystal solar cells via layer-by-layer assembly. *Nano Letters*, **11** (7), 2856–2864.

Jeong, S., Garnett, E.C., Wang, S. *et al.* (2012) Hybrid silicon nanocone–polymer solar cells. *Nano Letters*, **12** (6), 2971–2976.

Kamat, P.V. (2008) Quantum dot solar cells. semiconductor nanocrystals as light harvesters. *The Journal of Physical Chemistry C*, **112** (48), 18737–18753.

Liu, R., Lee, S.-T., and Sun, B. (2014) 13.8% Efficiency hybrid si/organic heterojunction solar cells with MoO_3 film as antireflection and inversion induced layer. *Advanced Materials*, **26** (34), 6007–6012.

Luther, J.M., Gao, J., Lloyd, M. T. *et al.* (2010) Stability assessment on a 3% bilayer PbS/ZnO quantum dot heterojunction solar cell. *Advanced Materials*, **22** (33), 3704–3707.

MacDonald, B.I., Martucci, A., Rubanov, S. *et al.* (2012) Layer-by-layer assembly of sintered CdSexTe1–x nanocrystal solar cells. *ACS Nano*, **6** (7), 5995-6004.

Mathew, S., Yella, A., Gao, P. *et al.* (2014) Dye-sensitized solar cells with 13% efficiency achieved through the molecular engineering of porphyrin sensitizers. *Nature Chemistry*, **6** (3), 242–247.

Nagamatsu, K.A., Avasthi, S., Jhaveri, J., and Sturm, J.C. (2014) A 12% Efficient Silicon/PEDOT:PSS heterojunction solar cell fabricated at 100 oC. *IEEE Journal of Photovoltaics*, **4** (1), 260–264.

Nazeeruddin, M.K., De Angelis, F. *et al.* (2005) Combined experimental and DFT-TDDFT computational study of photoelectrochemical cell ruthenium sensitizers. *Journal of the American Chemical Society*, **127** (48), 16835–16847.

Nazeeruddin, M.K., Péchy, P., Renouard, T. *et al.* (2001) Engineering of efficient panchromatic sensitizers for nanocrystalline TiO2-based solar cells. *Journal of the American Chemical Society*, **123** (8), 1613–1624.

Neo, D.C.J., Cheng, C., Stranks, S.D. *et al.* (2014) Influence of shell thickness and surface passivation on PbS/CdS core/shell colloidal quantum dot solar cells. *Chemistry of Materials*, **26** (13), 4004–4013.

Nozik, A.J. (2002) Quantum dot solar cells. *Physica E: Low-dimensional Systems and Nanostructures*, **14** (1–2), 115–120.

Sargent, E.H. (2012) Colloidal quantum dot solar cells. [10.1038/nphoton.2012.33]. *Nature Photonics*, **6** (3), 133–135.

Schmidt, J., Titova, V., and Zielke, D. (2013) Organic-silicon heterojunction solar cells: Open-circuit voltage potential and stability. *Applied Physics Letters*, **103** (18), 183901.

Tang, J., Liu, H., Zhitomirsky, D. *et al.* (2012) Quantum junction solar cells. *Nano Letters*, **12** (9), 4889–4894.

Townsend, T.K., Yoon, W., Foos, E.E., and Tischler, J.G. (2014) Impact of nanocrystal spray deposition on inorganic solar cells. *ACS Applied Materials and Interfaces*, **6** (10), 7902–7909.

Wang, Q., Ito, S., Grätzel, M. *et al.* (2006) Characteristics of high efficiency dye-sensitized solar cells. *The Journal of Physical Chemistry B*, **110** (50), 25210–25221.

Yella, A., Lee, H.-W., Tsao, H.N. *et al.* (2011) Porphyrin-sensitized solar cells with cobalt (II/III)-based redox electrolyte exceed 12 percent efficiency. *Science*, **334** (6056), 629–634.

Yoon, W., Boercker, J.E., Lumb, M.P. *et al.* (2013) Enhanced open-circuit voltage of PbS nanocrystal quantum dot solar cells. *Science Reports*, **3**.

Yoon, W., Townsend, T.K., Lumb, M.P. *et al.* (2014) Sintered CdTe nanocrystal thin films: determination of optical constants and application in novel inverted heterojunction solar cells. *IEEE Transactions on Nanotechnology*, **13** (3), 551–556.

Zielke, D., Pazidis, A., Werner, F., and Schmidt, J. (2014) Organic-silicon heterojunction solar cells on n-type silicon wafers: The BackPEDOT concept. *Solar Energy Materials and Solar Cells*, **131** (0), 110–116.

6.3

Perovskite Solar Cells

Samuel D. Stranks[1] and Henry J. Snaith[2]
[1] *Research Laboratory of Electronics, Massachusetts Institute of Technology, Cambridge, MA, USA; Cavendish Laboratory, University of Cambridge, Cambridge, UK*
[2] *Clarendon Laboratory, University of Oxford, Oxford, UK*

6.3.1 Introduction

Metal-halide perovskites are crystalline materials with intriguing properties. Recently, perovskite solar cells have rapidly emerged as serious contenders to rival the leading photovoltaic (PV) technologies. Power conversion efficiencies have increased from 3% to over 22% in just a few years of academic research. In this chapter, we review the rapid progress in perovskite solar cells. In particular, we describe the broad tunability in bandgap and fabrication methods of these materials, and we highlight their enabling properties for high performance devices. We discuss key ongoing challenges facing perovskites, and give an outlook on future research avenues that might bring perovskite technology to commercialization.

6.3.2 Organic-Inorganic Perovskites for Photovoltaics

6.3.2.1 Tunability of the Crystal Structure

Organic-inorganic perovskites are crystallized from organic halide and metal halide salts to form crystals in the perovskite ABX_3 structure, where A is the organic cation, B is the metal cation and X is the halide anion (Figure 6.3.1). The predominant three-dimensional hybrid perovskites used in solar cells to date have short-chain organic cations such as methylammonium (MA) ($A = CH_3NH_3^+$), metal cations such as lead ($B = Pb^{2+}$) and halides ($X = I^-$, Br^-, Cl^- or mixtures) (Stranks and Snaith, 2015).

A remarkable versatility of the organic-inorganic metal trihalide perovskite system is the tunability of the crystal size and the ensuing bandgap of the absorber through both cation and anion substitution, for example, by the use of formamidinium cations (FA) ($A = HC(NH_2)_2^+$) or

Photovoltaic Solar Energy: From Fundamentals to Applications, First Edition.
Edited by Angèle Reinders, Pierre Verlinden, Wilfried van Sark, and Alexandre Freundlich.

Companion website: www.wiley.com/go/reinders/photovoltaic_solar_energy

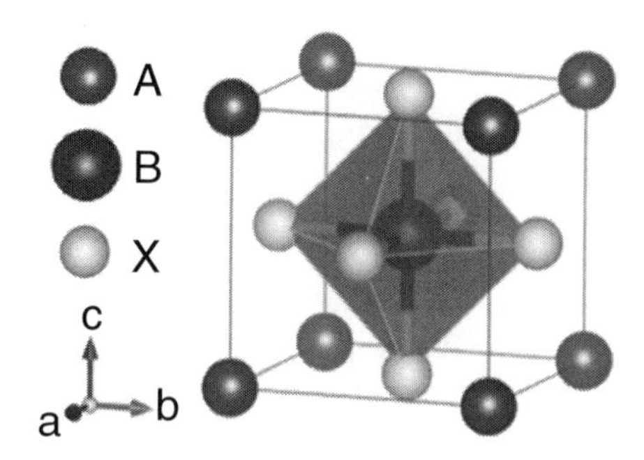

Figure 6.3.1 Perovskite ABX_3 crystal structure where typically $A = CH_3NH_3^+$, $B = Pb^{2+}$ and $X = I^-$, Br^-, Cl^-, or mixtures thereof. Source: (Stranks and Snaith, 2015). (*See insert for color representation of the figure*)

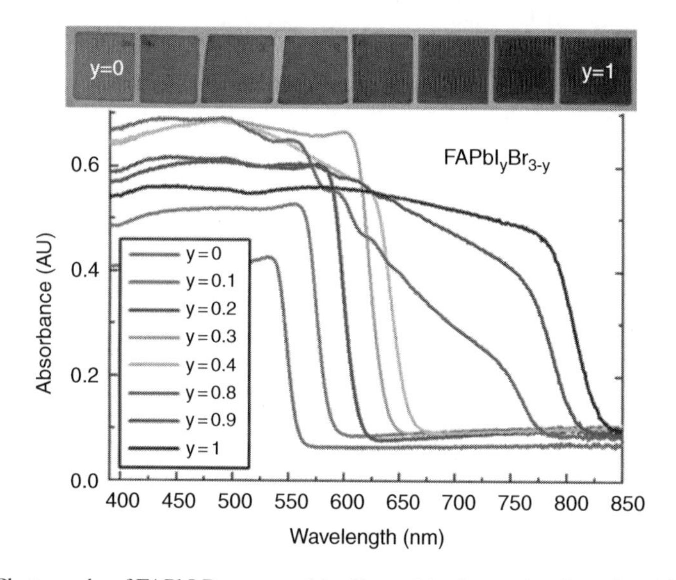

Figure 6.3.2 Photographs of $FAPbI_yBr_{3-y}$ perovskite films with y increasing from 0 to 1 from left to right, and corresponding absorption spectra. Source: (Eperon *et al.*, 2014b). (*See insert for color representation of the figure*)

alternative halides ($X=I^-$, Br^-, Cl^- or mixtures) (Stranks and Snaith, 2015). Figure 6.3.2 shows the range of absorption spectra and photographs of the different materials possible from $FAPbI_yBr_{3-y}$ mixed halide perovskites. This tunability will be useful for multijunction perovskite solar cells, or the creation of hybrid perovskite tandem cells with c-Si or CIGS (Snaith, 2013; Bailie *et al.*, 2015). Figure 6.3.3 shows the J-V curves of the highest performing device to date.

6.3.2.2 *Evolution from Dye-Sensitized Solar Cells*

Researchers studied the properties of hybrid perovskites and their use in transistors and other electronic applications in the 1990s (Mitzi *et al.*, 2001), but their potential for use in

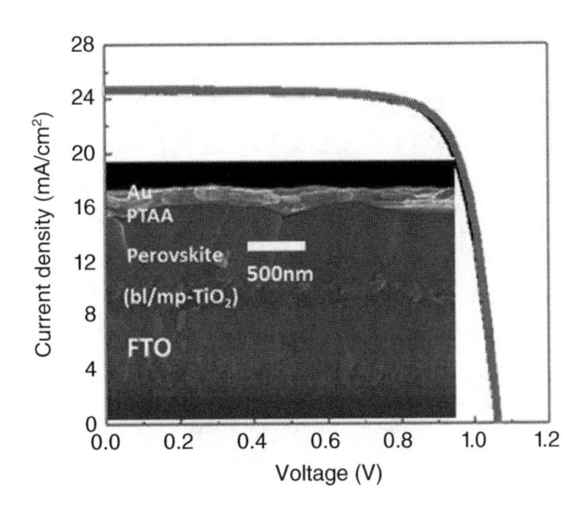

Figure 6.3.3 J-V curves of the highest performing device to date consisting of FTO-Glass/bl-TiO_2/mp-TiO_2/perovskite/PTAA/Au measured under standard AM 1.5G illumination, giving a power conversion efficiency of 20.2%. Source: (Yang *et al.*, 2015). Inset: a scanning electron microscope cross-section image of the device

solar cells was overlooked. In the quest for new absorbing "dyes" for dye-sensitized solar cells (DSSCs), Miyasaka and co-workers published the first perovskite solar cells in 2009 (Kojima *et al.*, 2009). In their work, small $MAPbI_3$ or $MAPbBr_3$ perovskite nanoparticles (only a few nanometers in size) endowed the surface of mesoporous TiO_2, akin to a dye in a dye-sensitized solar cell (see Chapter 6.2), and this resulted in device efficiencies of up to 3.8%. The efficiency was improved to 6.5% in 2011 through the optimization of the titania surface and perovskite processing (Im *et al.*, 2011). However, these first embodiments were highly unstable owing to the liquid electrolyte hole transport material (HTM) decomposing the perovskite absorber.

An important breakthrough came in 2012, when our group, in collaboration with Miyasaka and co-workers, simultaneously with Park in collaboration with Grätzel and co-workers, replaced the problematic liquid electrolyte HTM with a solid state analogue, 2,2',7,7'-tetrakis-(*N,N*-di-p-methoxyphenylamine)-9,9'-spirobifluorene (spiro-OMeTAD), resulting in power conversion efficiencies approaching 10% when employing the $CH_3NH_3PbI_{3-x}Cl_x$ mixed halide perovskite or $CH_3NH_3PbI_3$, respectively (Kim *et al.*, 2012; Lee *et al.*, 2012). In our work, we also replaced the mesoporous electron transporting TiO_2 layer with an insulating mesoporous Al_2O_3 scaffold, yielding a very similar meso-morphology. Remarkably, we found that the charge transport was improved and the photocurrent was unaffected, giving a device efficiency of 10.9%. This suggested that the shell comprised of perovskite crystals endowing the surface of the metal oxide can efficiently transport electrons (in addition to holes) itself without the need for a discrete mesoscopic electron transporting material. In hindsight, we now realize that the key to enabling efficient operation on an insulating scaffold was the serendipitous achievement of highly crystalline perovskite films with crystallite domains on the hundreds of nanometers to microns length scale. Since these reports we have witnessed a meteoric rise in device efficiency (Figure 6.3.4) that has been achieved primarily through the optimization of the device architecture and perovskite deposition, crystallization and composition.

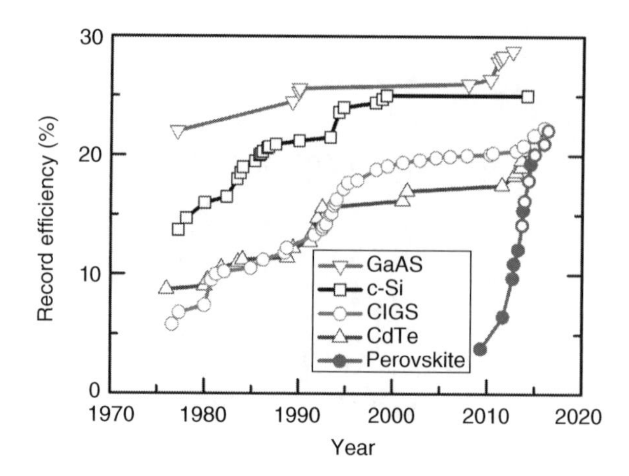

Figure 6.3.4 Record efficiencies of various established PV technologies over the years. Closed symbols are published laboratory results, open symbols represent certified efficiency values (NREL, 2016). Source: Figures adapted with permission from (1) Nature Publishing Group (NPG), (2) Royal Society of Chemistry (RSC), and (3) American Association for the Advancement of Science (AAAS). (*See insert for color representation of the figure*)

6.3.2.3 Emergence of a New Thin Film Technology

The first embodiments used a mesoporous TiO_2 scaffold to facilitate electron collection, but soon after it was established that the mesoporous scaffold could be thinned down and that a few hundred nanometer-thick solid perovskite film can even sustain charge generation and transport (Lee *et al.*, 2012;Ball *et al.*, 2013; M. Liu *et al.*, 2013). The device architectures are converging upon a simple planar heterojunction architecture, where a solid intrinsic perovskite layer is sandwiched between p- and n-type selective contacts. A typical device structure would be comprised of a thin compact n-type TiO_2 blocking layer (bl-TiO_2) on fluorinated tin oxide (FTO) and a p-type HTM such as Spiro-OMeTAD or polytriarylamine (PTAA); though high efficiencies have also been obtained with the "inverted" structure where poly(3,4-ethylenedioxythiophene):polys tyrene sulfonate (PEDOT:PSS) and phenyl-C61-butyric acid methyl ester (PCBM) are employed as the p- and n-type contacts, respectively (Docampo *et al.*, 2013; Shao *et al.*, 2014). The highest efficiencies to date (22.1% certified, details not yet published; 19.6% certified on a device active area of $1\,cm^2$) currently use a very thin (~50–100 nm) mesoporous TiO_2 layer located between the planar TiO_2 compact layer and the solid perovskite absorber layer (see Figure 6.3.3) (Yang *et al.*, 2015, Li *et al.*, 2016). Even for the researchers persisting with mesoporous TiO_2, the trend has been for the mesoporous layer to become thinner and thinner, and we expect the mesoporous layer eventually to be removed as the technology progresses.

6.3.3 Deposition Methods

A plethora of methods to deposit organic-inorganic perovskites have emerged, each resulting in varying degrees of surface coverage, crystal and film quality, and stability (Stranks *et al.*, 2015). Most methods are based on the same principle – the combination of an organic component such as MAI with an inorganic component such as PbI_2, $PbBr_2$ or $PbCl_2$ to form the perovskite ($MAPbI_3$, $MAPbI_{3-x}Br_x$ or $MAPbI_{3-x}Cl_x$, respectively).

6.3.3.1 One-Step Depositions

The simplest processing route is where the two precursor salts (metal halide and organic halide) are dissolved in an organic solvent, the mixture is spin-coated onto a substrate, and the sample is annealed at ~100 °C to form the perovskite. Smooth solid thin films with large crystalline domains can be manufactured by judiciously tuning the composition of the precursor solution and increasing the organic fraction. By increasing the temperature of the casting, and choosing the appropriate composition, crystal grains of a millimeter scale can even be realized (Nie *et al.*, 2015). To achieve semi-transparency, the coverage of the perovskite films can be varied from completely continuous to the creation of islands of perovskite surrounded by voids (Eperon *et al.*, 2014a). Many of the highest efficiencies reported recently have been achieved by employing solvent engineering techniques where the perovskite precursors are first cast from one solvent which is quickly followed by a quenching with a non-solvent, leading to extremely uniform and dense perovskite layers (Jeon *et al.*, 2014).

Understanding the role of chloride in employing $PbCl_2$ as opposed to PbI_2 as the starting salt has been a topic of intense study. Recent reports have suggested that the chloride ions play a crucial role in crystal growth and film morphology of the one-step solution-processed films (Williams *et al.*, 2014) but only trace levels of chloride remain in the final $MAPbI_{3-x}Cl_x$ films (Unger *et al.*, 2014). We note that the final chloride content and degree of incorporation into the lattice may depend on the film fabrication method. Moreover, the electronic impact of even small doping levels of chloride is not yet understood. Regardless, the resulting films are of better quality than the triiodide counterparts produced under the same single-step spin-coating conditions. The role the chloride plays in facilitating crystal formation but predominantly leaving the film ("spectator ions") is not unique, as it has recently been shown that the use of acetate or nitrate ions in the precursor solutions can lead to even higher quality films with exceptionally low levels of roughness (Moore *et al.*, 2015).

Dual source vapor deposition is also a feasible route to deposit extremely uniform polycrystalline films. Vapor phase deposition should be applicable to a wider range of salts, provided that the salts can sublime in vacuum at temperatures lower than their decomposition temperature (M. Liu *et al.*, 2013). Vapor phase deposition is not being followed as vehemently as solution-based techniques, as this is largely due to the requirement for specialist equipment. Vapor phase techniques may prove to be easier to scale up for manufacturing.

6.3.3.2 Two-Step Depositions

Burschka *et al.* (2013) employed a sequential solution-based deposition method to first deposit a thin coating of lead iodide within mesoporous TiO_2, and then converted this to the perovskite absorber by dipping it in a solution of MAI in isopropanol, achieving a device efficiency of 15.0%. Sequential deposition also works well on solid thin films of PbI_2 with both a solution phase (D. Liu and Kelly, 2014) and a vapor phase (Chen *et al.*, 2014) conversion. Il Seok and co-workers recently modified their solvent engineering method to first deposit PbI_2 and then induce crystallization by the direct intramolecular exchange of solvent molecules intercalated in PbI_2 with FAI to yield a certified efficiency of 20.1% when mixed with fractions of $MAPbBr_3$ (see Figure 6.3.3) (Yang *et al.*, 2015). Some of the smoothest films achieved to date are from spin-coating the precursor layers as a bilayer stack and the perovskite conversion is achieved by "inter-diffusion" of the layers (Xiao *et al.*, 2014).

6.3.4 Enabling Properties and Operation

6.3.4.1 Low-Loss Systems

One of the most promising aspects of the perovskite technology is the high open-circuit voltage V_{oc} that the cells can generate under full sun illumination with low loss-in-potential (Snaith, 2013). We show the open-circuit voltage versus the optical bandgap (lowest-energy absorbed photon that generates charge) for most established and emerging solar technologies in Figure 6.3.5, where the minimum energy loss is ~250–300 meV from thermodynamic considerations. The optical bandgap for $MAPbI_3$ of 1.55 eV and open-circuit voltages of >1.1 V give a loss of just ~450 meV, which places the technology only behind GaAs, c-Si and CIGS in terms of fundamental losses. Edri *et al.* (2014b) also demonstrated that $MAPbBr_{3-X}Cl_x$ solar cells (bandgap of 2.3 eV) could exhibit an open-circuit voltage of over 1.5 V, indicating that it is possible to raise the voltage by increasing the bandgap, which will be important for multi-junction solar cells.

The high open-circuit voltages in $MAPbI_3$ can be attributed in part to the low degree of energetic disorder in the materials. We show the effective absorption coefficient for a variety of typical PV materials in Figure 6.3.6, where the steepness of the absorption edge gives an indication of the quality of the material, as characterized by the Urbach energy. The perovskite films show a high absorption coefficient with a sharp absorption edge and an Urbach energy as low as 15 meV (De Wolf *et al.*, 2014). This value compares favorably

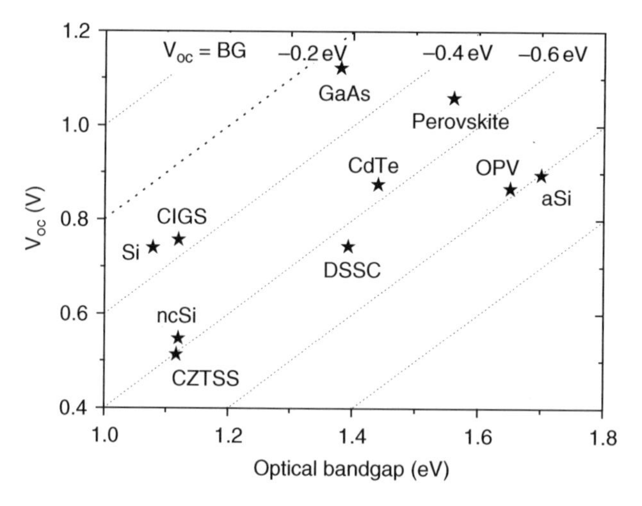

Figure 6.3.5 Open-circuit voltage (V_{oc}) versus optical band gap for the best-in-class solar cells for most current and emerging solar technologies as extracted from Green et al (Green et al., 2015). The optical band gaps have been estimated by taking the onset of the incident photon to converted electron (IPCE) for all technologies (Snaith, 2013). Source: adapted with permission from the American Chemical Society

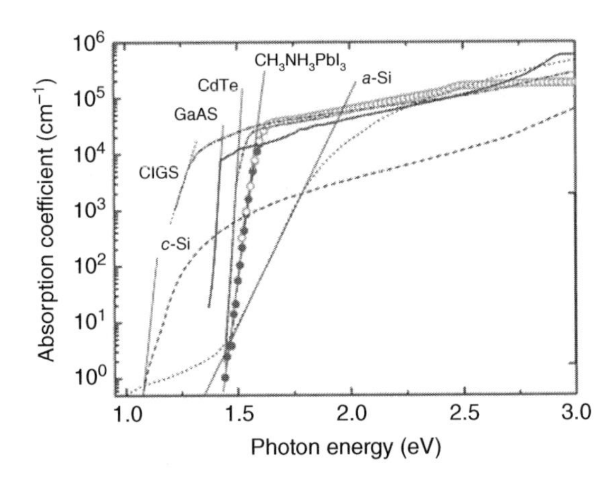

Figure 6.3.6 Effective absorption coefficient of a $CH_3NH_3PbI_3$ perovskite thin film compared with other typical PV technologies. The slopes of the Urbach tail are displayed on the plot (De Wolf *et al.*, 2014). Source: adapted with permission from the American Chemical Society

to other technologies, being lower than CIGS (~25 meV) and only slightly higher than GaAs (7 meV).

The low losses are also because the recombination in the materials is predominantly radiative, with room temperature photoluminescence quantum efficiencies demonstrated up to 70% (Deschler *et al.*, 2014). Such high luminescence efficiency is very encouraging for pushing perovskite solar cells towards the Shockley-Queisser efficiency limit in which all recombination is assumed to be radiative. These emissive properties will also make these perovskites attractive for use in light-emitting applications (Stranks and Snaith, 2015).

Early work showed that the inter-grain potential barrier is low compared to other thin film technologies (Edri *et al.*, 2014b). Nevertheless, recent work has shown that the grain boundaries act as significant non-radiative recombination sites, suggesting that the grain boundaries are far from benign (deQuilettes *et al.*, 2015). The empirical observations of enhanced device performance and charge carrier diffusion length with increasing grain size, along with recent work on single crystals, are consistent with this conjecture (Nie *et al.*, 2015). These findings are also consistent with reports of low densities of sub-gap electronic states (~10^{16} cm^{-3}) in perovskite samples fabricated using most standard processing methods (Stranks *et al.*, 2014). It remains a challenge to the community to find methods to switch off these non-radiative sub-gap pathways by either preventing their formation or by filling these states through selective doping or by chemically passivating the materials (Stranks *et al.*, 2014).

6.3.4.2 Electronic Properties

We recently used very high magnetic fields to make an accurate and direct spectroscopic measurement of the exciton binding energy, which we find to be only a few meV in the room temperature phase (Miyata *et al.*, 2015). This helps to rationalize their excellent device

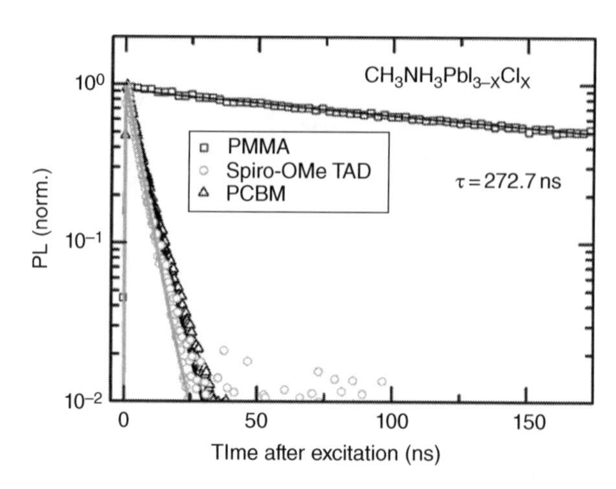

Figure 6.3.7 Time-resolved PL decays for $MAPbI_{3-x}Cl_x$ thin films without (PMMA) and with n-type (PCBM) or p-type (Spiro-OMeTAD) quenching layers (Stranks *et al.*, 2013). The mono-exponential lifetime τ is shown for the PMMA sample. Source; adapted with permission from AAAS

performance as being due to spontaneous free-carrier generation following light absorption, and that recombination is bimolecular (Manser and Kamat, 2014). The free electrons and holes have high mobilities in the range $\mu \sim 10\text{-}30\ cm^2V^{-1}s^{-1}$ (Wehrenfennig *et al.*, 2014), and their subsequent decay in polycrystalline films is very slow, up to tens of microseconds (Ponseca *et al.*, 2014).

The efficient operation of planar heterojunction cells requires that the electron and hole diffusion lengths are much longer than the film thickness required to obtain complete solar light absorption. The diffusion length has been reported using photoluminescence quenching measurements to be over one micron for both electrons and holes in films of $MAPbI_{3-x}Cl_x$ processed from mixed halide precursor solutions (Figure 6.3.7) (Stranks *et al.*, 2013; Xing *et al.*, 2013). Given the absorption depth is in the range ~100–400 nm, this diffusion length is suitably long for efficient operation in the planar heterojunction configuration. The long diffusion length arises from a very long electron-hole lifetime of the order of hundreds of nanoseconds to microseconds (Stranks *et al.*, 2013). We note that the carrier lifetime in the perovskite is also strongly dependent upon the polycrystalline nature of the film and upon the defect density or electronic trap density within the films. Recent work on single crystals of $MAPbI_3$, and $MAPbBr_3$ report defect densities as low as $\sim 10^{10}\ cm^{-3}$ and diffusion lengths on the order of tens or even hundreds of microns (Dong *et al.*, 2015; Shi *et al.*, 2015).

6.3.4.3 Device Operation

Cahen and co-workers performed electron beam induced current (EBIC) mapping of perovskite planar heterojunction solar cells, and showed that the solar cell is operating in an analogous manner to a thin film p-i-n heterojunction solar cell (Edri *et al.*, 2014b). The most accepted operating principle is that light is absorbed in the bulk of the film leading to generation of free charges (Manser and Kamat, 2014). These charges then diffuse throughout the film and

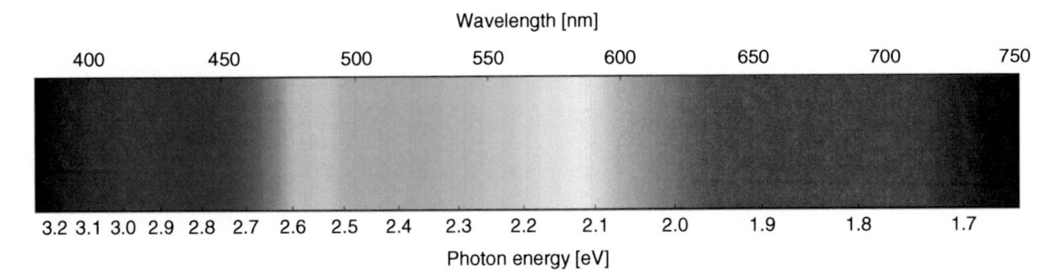

Figure 1.1.2 Relation between color of (visible) light, wavelength and its energy. Courtesy of A. Louwen, Utrecht University

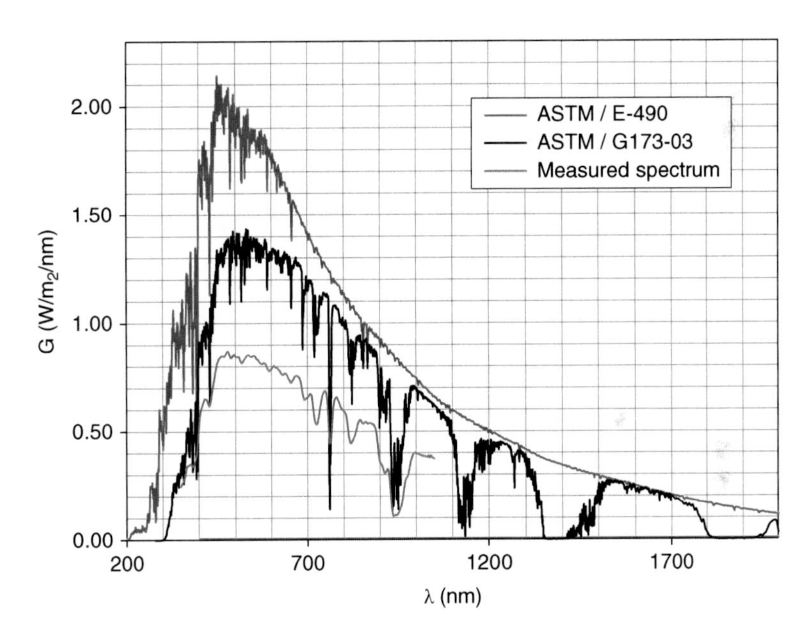

Figure 1.1.3 Solar spectra: ASTM E-490 representing AM0 (black line), ASTM G173-03 representing AM1.5 (red line), and a measured spectrum (green line) showing the differences that can occur in reality. Data from ASTM and University of Twente, The Netherlands. Courtesy of A. Reinders, University of Twente

Photovoltaic Solar Energy: From Fundamentals to Applications, First Edition.
Edited by Angèle Reinders, Pierre Verlinden, Wilfried van Sark, and Alexandre Freundlich.

Companion website: www.wiley.com/go/reinders/photovoltaic_solar_energy

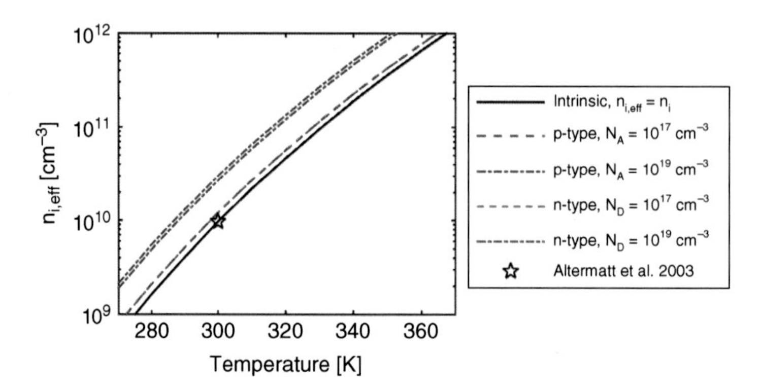

Figure 3.1.1 Effective intrinsic carrier density $n_{i,eff}$ as a function of temperature and doping level and type; lines are calculated using the models and assumptions discussed in the text; the symbol marks the measured value of n_i in. Source: (Altermatt *et al.*, 2003)

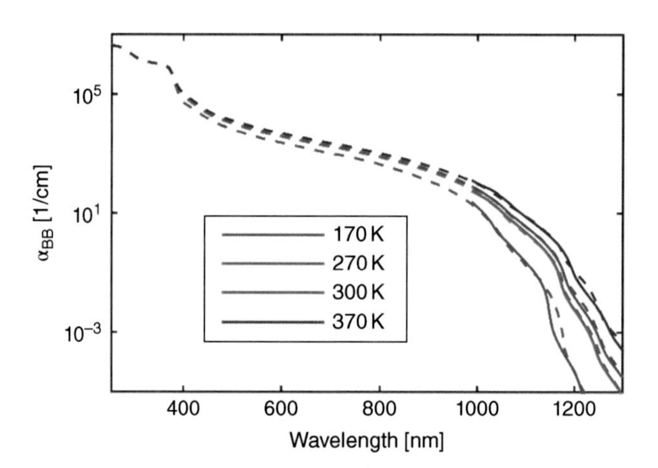

Figure 3.1.3 Absorption coefficient α_{BB} of silicon for several temperatures; *dashed lines*: calculated from (Green, 2008); *solid lines*: data from (Nguyen *et al.*, 2014b); though differences between the different data sets appear minor, they are significant for luminescence spectroscopy

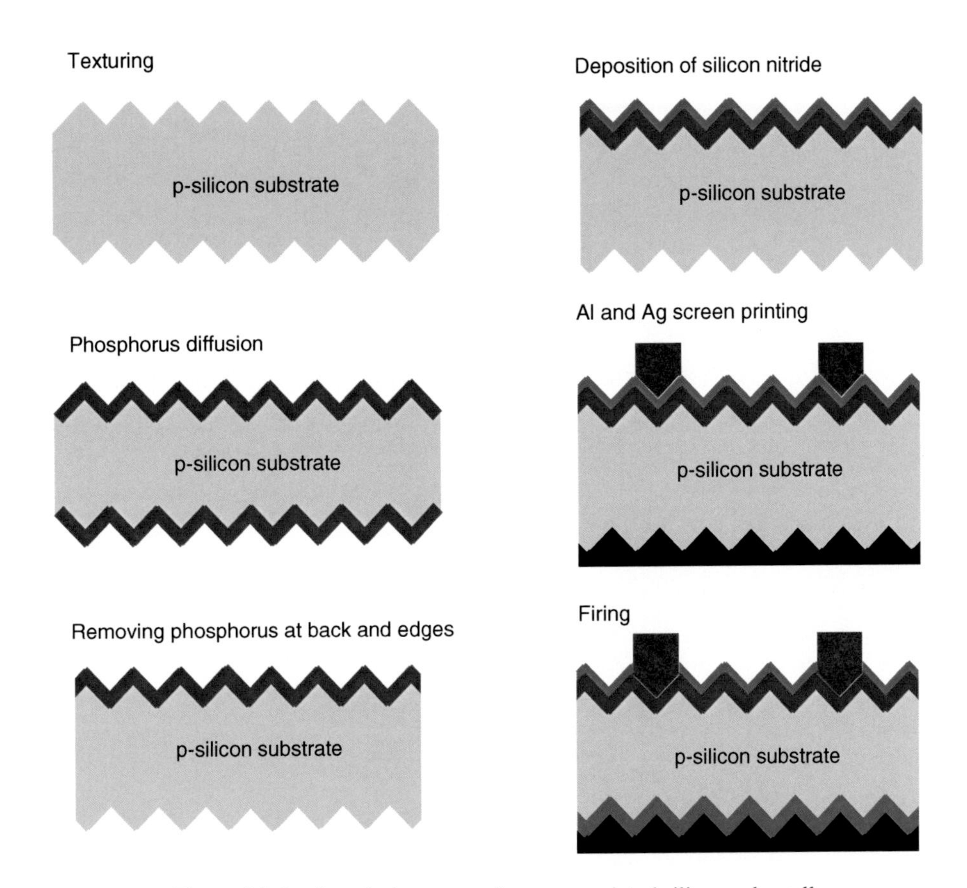

Figure 3.2.3 A typical process of a screen-printed silicon solar cell

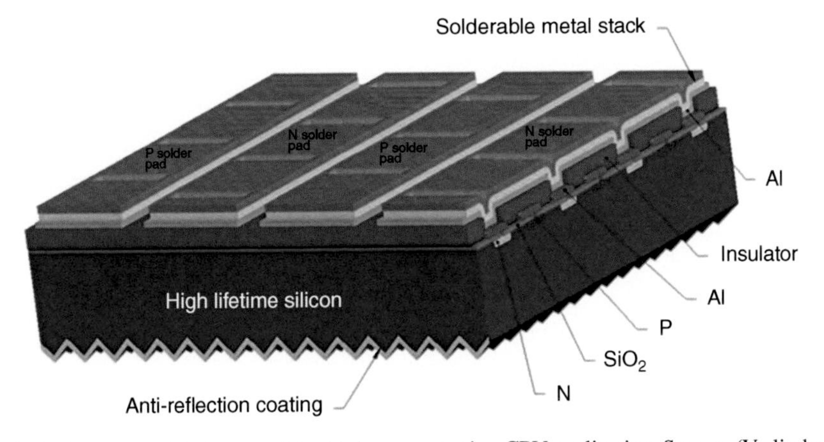

Figure 3.3.4 Structure of PC solar cell for high-concentration CPV application. Source: (Verlinden, 2005)

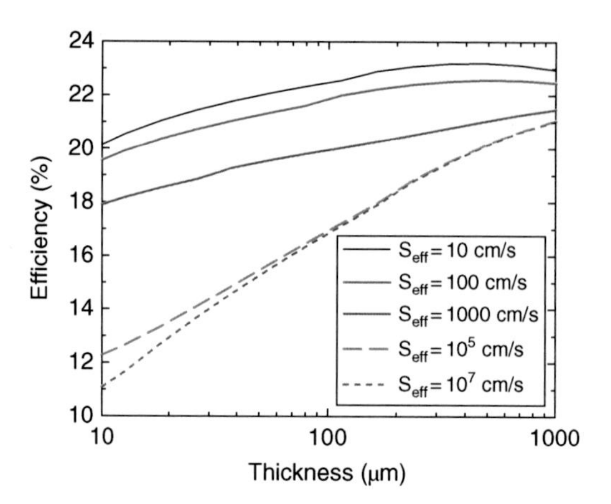

Figure 3.5.1 Simulated solar cell for a high-efficiency silicon wafer solar cell as a function of the solar cell thickness for various values of the effective surface recombination velocity at the rear side of the solar cell. The values that can be obtained by the standard aluminum back surface field are in the range of 200–600 cm/s while more advanced solar cell architectures can achieve values well below 100 cm/s. These simulations were conducted in the software package PC1D. Source: (Basore, 1990)

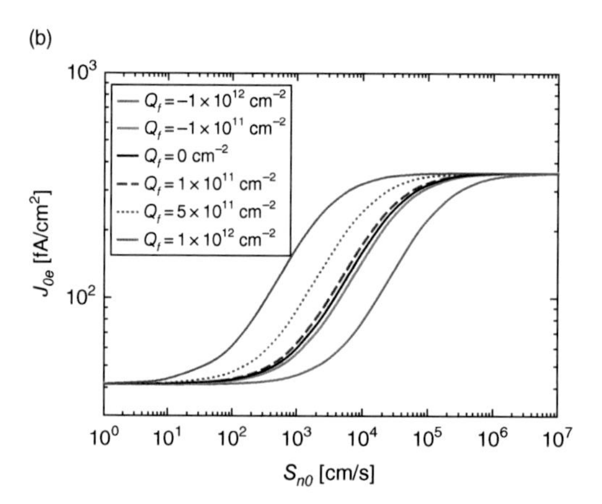

Figure 3.5.3b Resulting simulated J_{0e} value as a function of S_{n0} for various values of Q_f (in unit of elementary charges). The sheet resistance of the boron emitter was determined to be 30 Ω/□ by means of a four-point-probe measurement. The simulations were performed by Fajun Ma from the Solar Energy Institute of Singapore (SERIS) in the software package Sentaurus of Synopsys, assuming only intrinsic recombination in the emitter and an SRH recombination at the surface

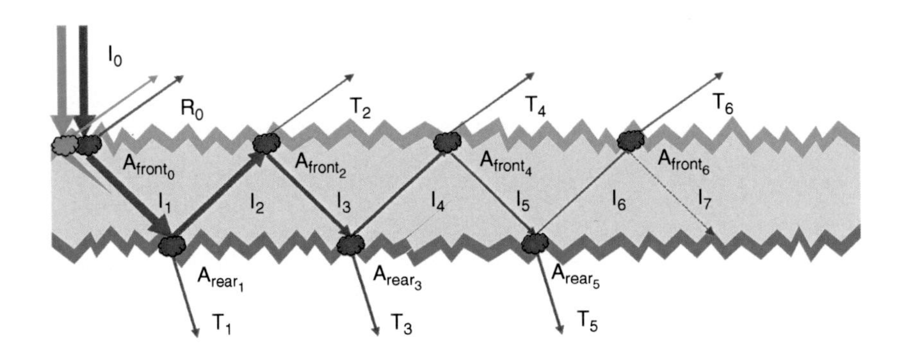

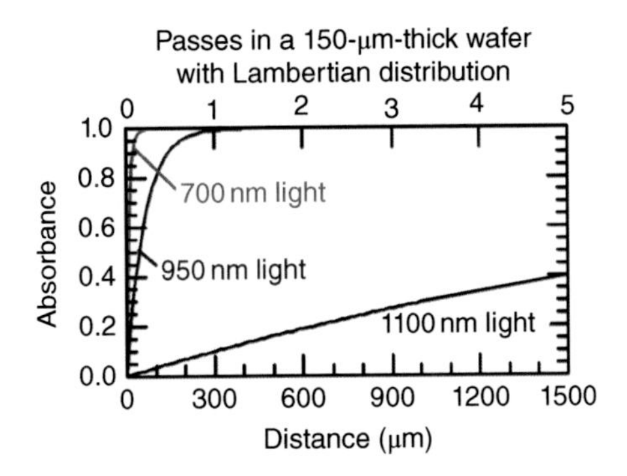

Figure 3.7.1 Schematic depiction of the path taken by visible and infrared light in a silicon solar cell and penetration depth of red and infrared light in a silicon wafer

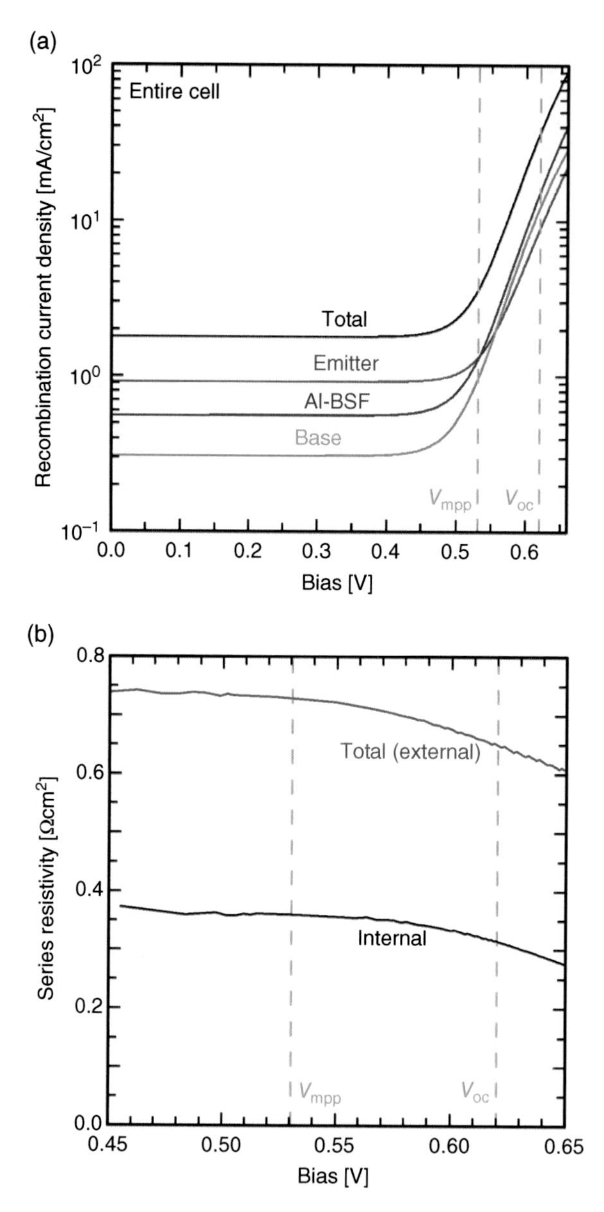

Figure 3.8.2 (a) Simulated recombination currents; and (b) simulated series resistivity of an mc-Si solar cell, under one-sun standard conditions

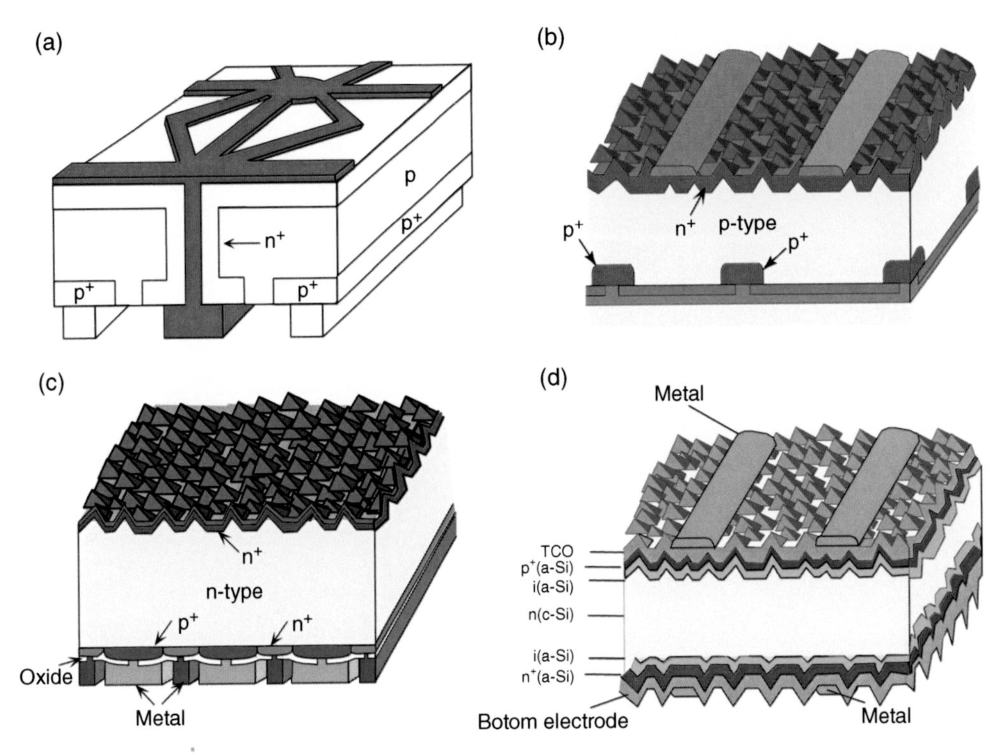

Figure 3.9.1 Four advanced silicon cell technologies: (a) Metal Wrap Through (MWT); (b) Passivated Emitter and Rear Cell (PERC); (c) Interdigitated Back Junction (IBJ); (d) Heterojunction Cell (HJT)

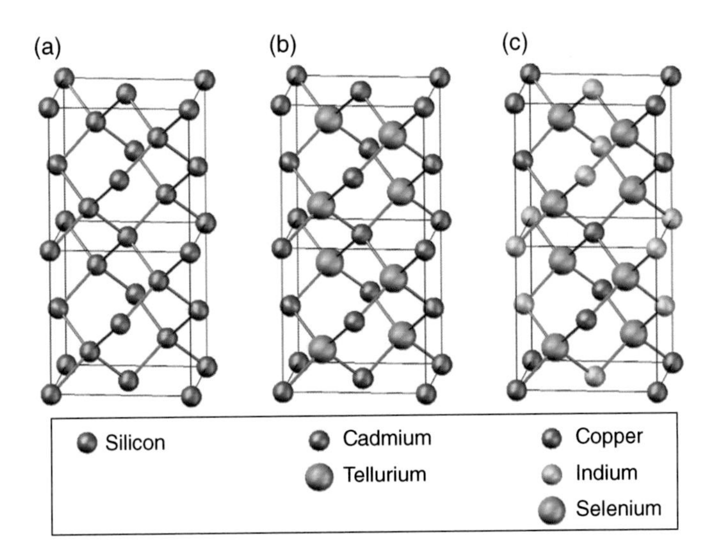

Figure 4.2.1 (a) FCC structure of a Si crystal with two identical atoms basis; (b) CdTe structure derived from the Si structure by replacing identical atoms basis by two different atoms; (c) $CuInSe_2$ structure derived from the CdTe structure by doubling the CdTe structure and replacing the cations by two different cations

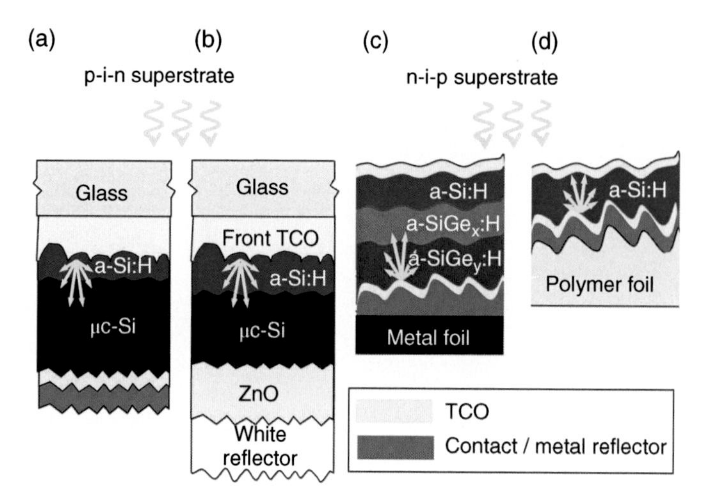

Figure 5.1.1 Schematics of four different device configurations

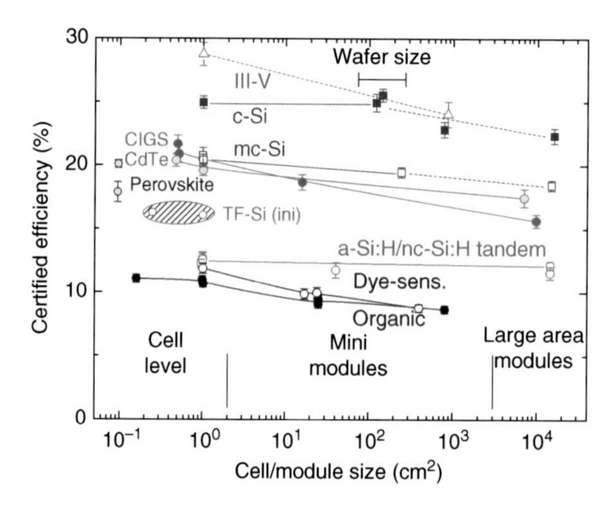

Figure 5.1.2 Independently certified efficiencies of different photovoltaic technologies, plotted with respect to the cell or module size. Source: adapted from Haug (2015) with permission of The Royal Society of Chemistry

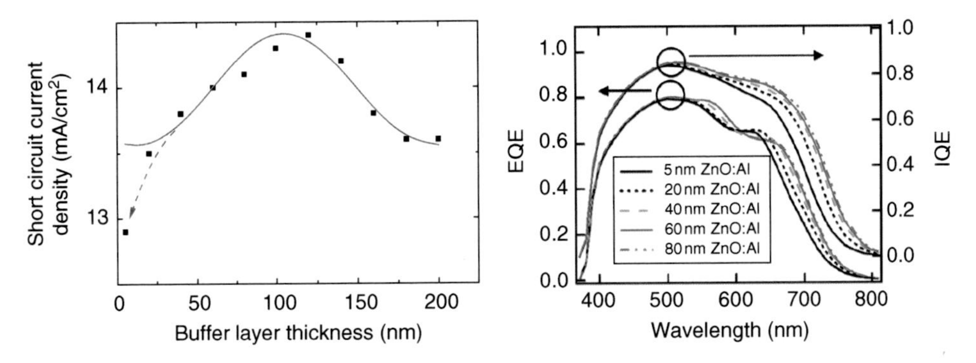

Figure 5.3.3 Photocurrent of n-i-p cells on textured silver electrode with varying buffer layer thickness (left panel). The right panel shows the underlying external quantum efficiency (EQE) and the EQE after correction for reflected light. Source: (Söderström, 2013)

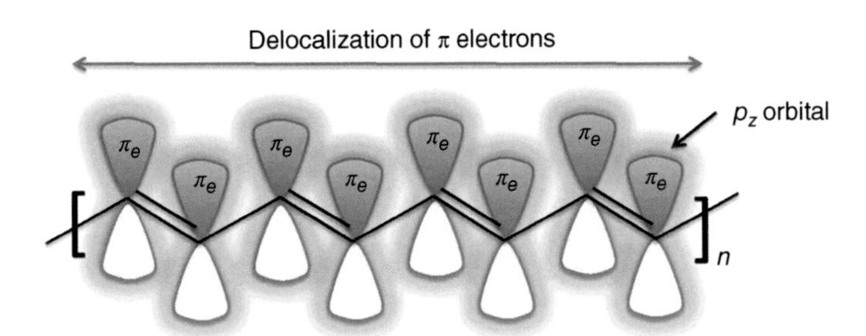

Figure 6.1.2 Schematics of the π electrons (π_e) in the p_z orbitals on each carbon atom and their delocalization over the conjugated polymer chain due to electronic coupling between neighboring carbon atoms

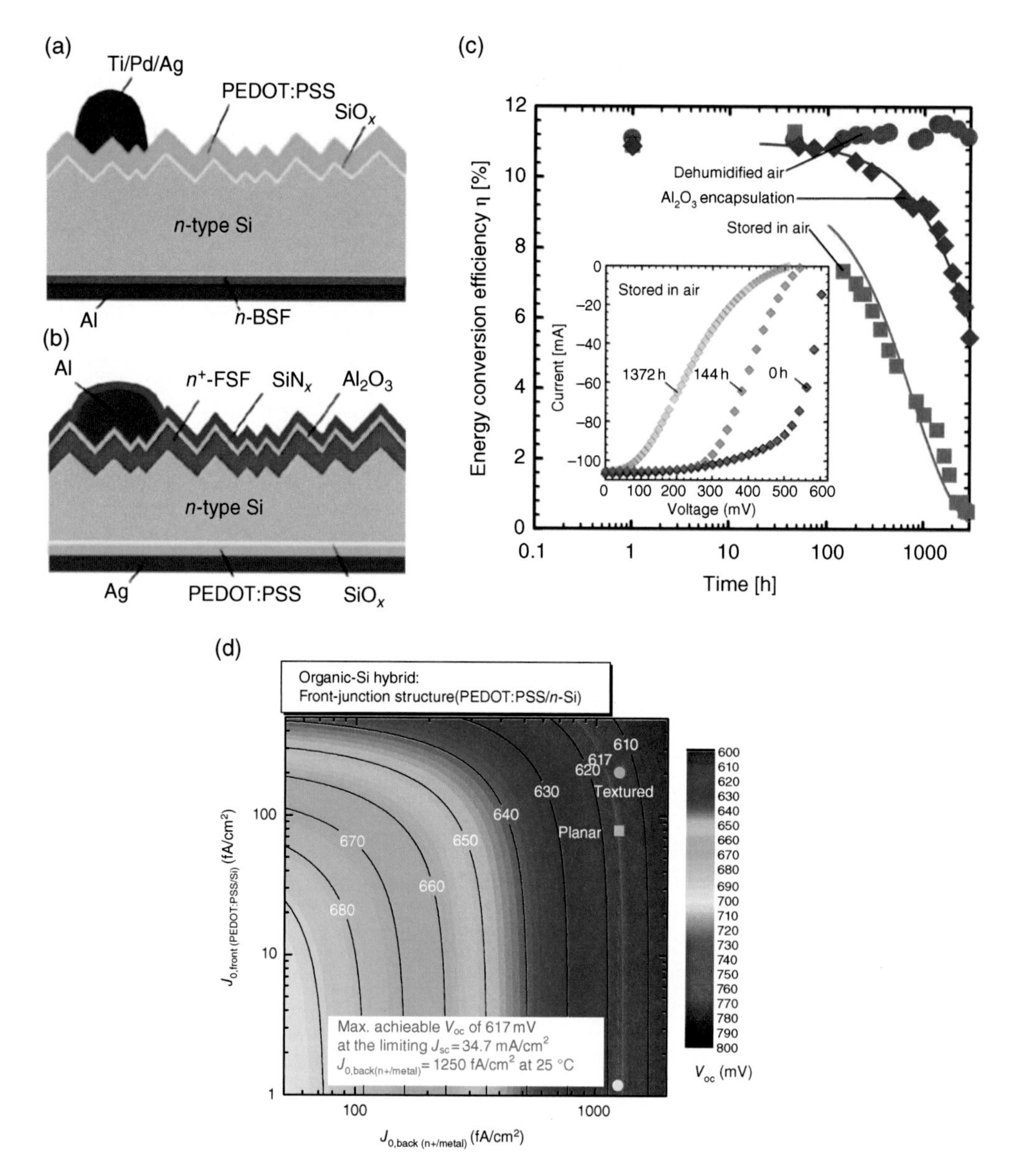

Figure 6.2.2 Schematic of (a) a front junction PEDOT:PSS/*n*-Si heterojunction and (b) a back junction *n*-Si/PEDOT:PSS heterojunction solar cell. Source: (Zielke *et al.*, 2014: 110). Copyright (2014), with permission from Elsevier; (c) Stability of a front junction PEDTO:PSS/*n*-Si heterojunction solar cell as a function of storage time in darkness. Source: Reprinted with permission from *Applied Physics Letters*, vol. 103, Schmidt *et al.*, 2013. Copyright 2013, AIP Publishing LLC; and (d) Achievable V_{oc} in the front-junction organic-Si heterojunction cells with phosphorus-doped BSF for the limiting J_{sc} of 34.7 mA/cm^2 and the saturation current density of P-diffused BSF ($J_{0,b}$) of 1250 fA/cm^2 (Zielke *et al.*, 2014)

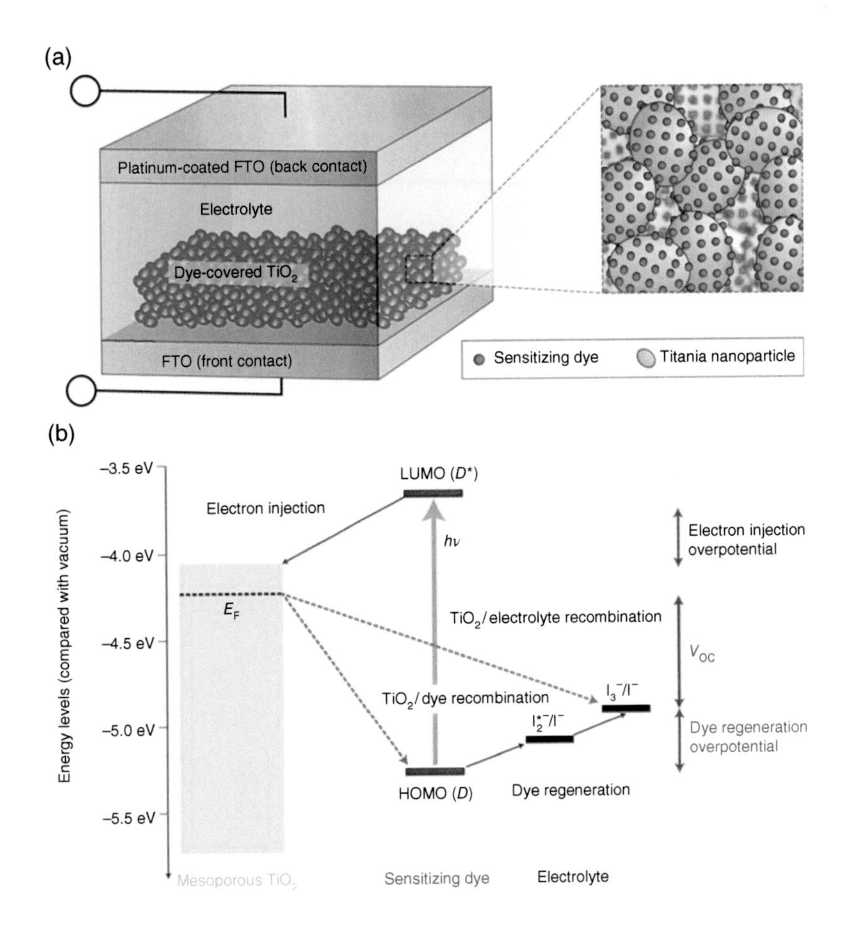

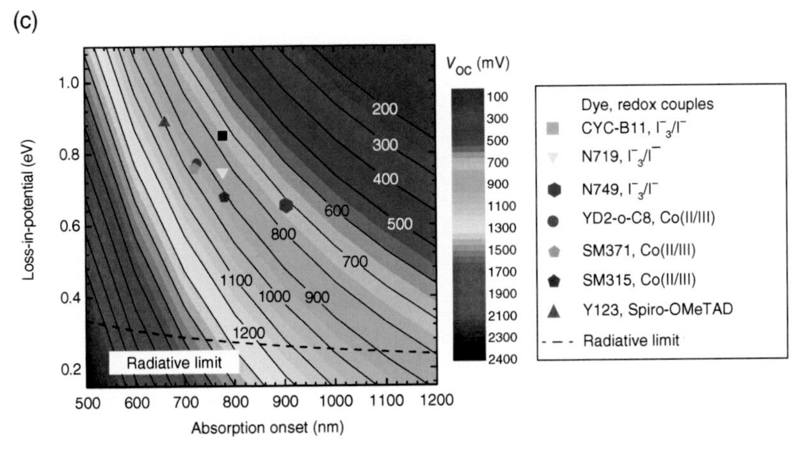

Figure 6.2.3 (a) Schematic and (b) energy level and operation of dye-sensitized solar cells Source: Reprinted by permission from Macmillan Publishers Ltd: Hardin *et al.*, 2012: 162, copyright (2012)); (c) Open-circuit voltage as a function of absorption onset and loss-in-potential with radiative limits. The literature values in the absorption onsets and the loss-in-potentials are shown for the ruthenium dye (CYC-B11)/iodide redox couple (C.-Y. Chen *et al.*, 2009), the N719 dye/iodide redox couple (Nazeeruddin *et al.*, 2005), the black dye N749/iodide redox couple (Nazeeruddin *et al.*, 2001), the co-sensitized donor–pi–acceptor dye (YD2-o-C8 and Y123)/cobalt redox couple (Yella *et al.*, 2011), the SM371/Co(II/III) (Mathew *et al.*, 2014), the SM315/Co(II/III) (Mathew *et al.*, 2014), and the Y123 dye/hole conductor spiro-OMeTAD (Burschka *et al.*, 2011)

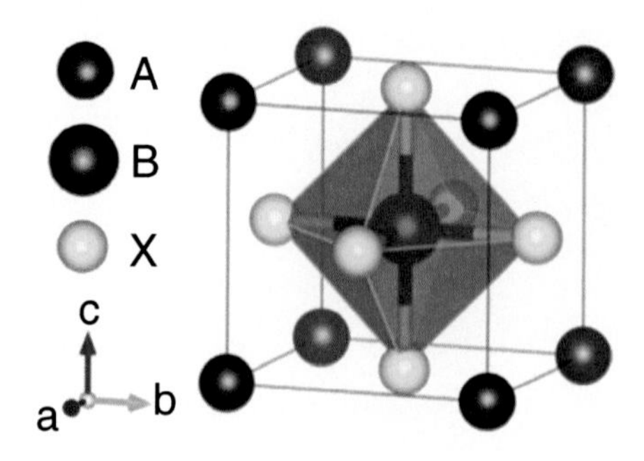

Figure 6.3.1 Perovskite ABX_3 crystal structure where typically $A = CH_3NH_3^+$, $B = Pb^{2+}$ and $X = I^-$, Br^-, Cl^-, or mixtures thereof. Source: (Stranks and Snaith, 2015)

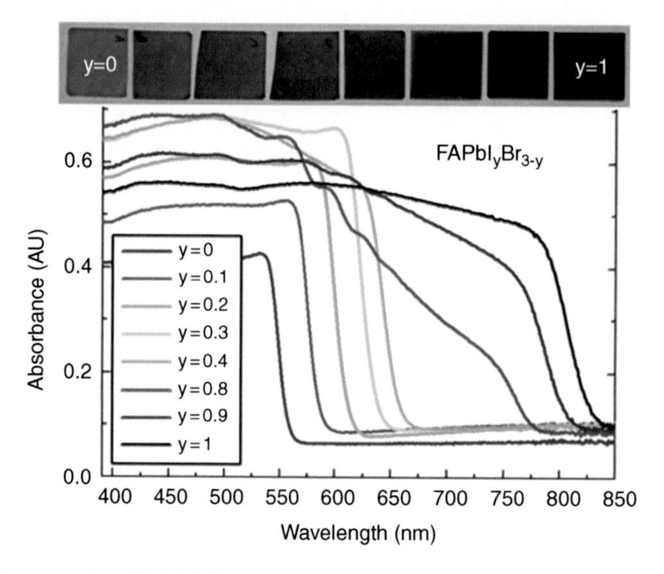

Figure 6.3.2 Photographs of $FAPbI_yBr_{3-y}$ perovskite films with y increasing from 0 to 1 from left to right, and corresponding absorption spectra. Source: (Eperon *et al.*, 2014b)

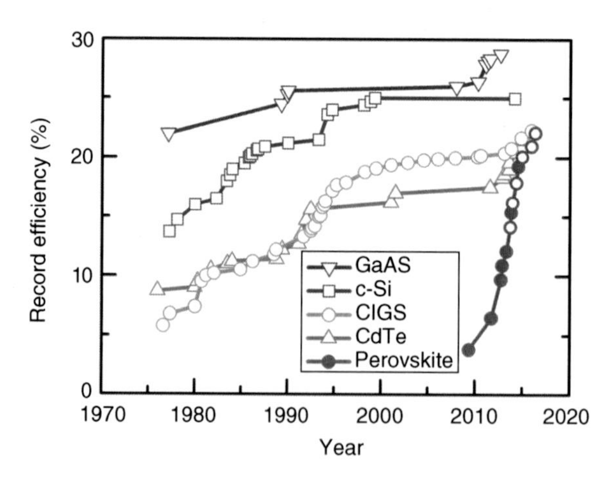

Figure 6.3.4 Record efficiencies of various established PV technologies over the years. Closed symbols are published laboratory results, open symbols represent certified efficiency values (NREL, 2016). Source: Figures adapted with permission from (1) Nature Publishing Group (NPG), (2) Royal Society of Chemistry (RSC), and (3) American Association for the Advancement of Science (AAAS)

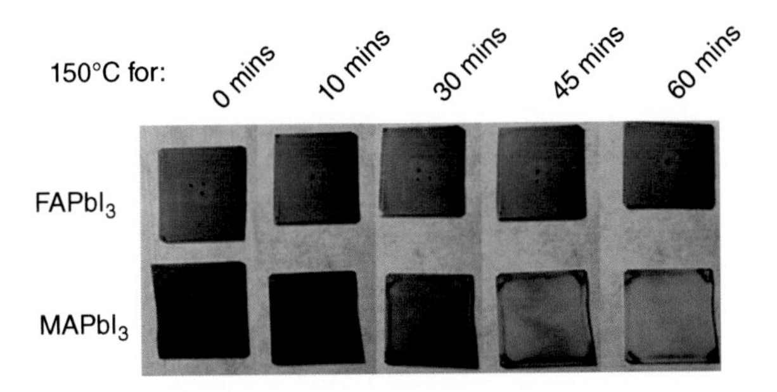

Figure 6.3.12 Photographs showing non-encapsulated $MAPbI_3$ and $FAPbI_3$ samples before and during heating on a hot plate at 150°C for 60 minutes in air (Stranks and Snaith, 2015). Source: Adapted with permission from Nature Publishing Group (NPG)

Figure 6.4.1 HeliaFilm™ encapsulation output and vision on low-cost solar cell in BIPV ©Heliatek

Figure 6.4.2 Two examples of free-form OPV modules; left, the Solarte garden lamp by Belectric OPV and, right, the world's first polychrome solar module by DisaSolar

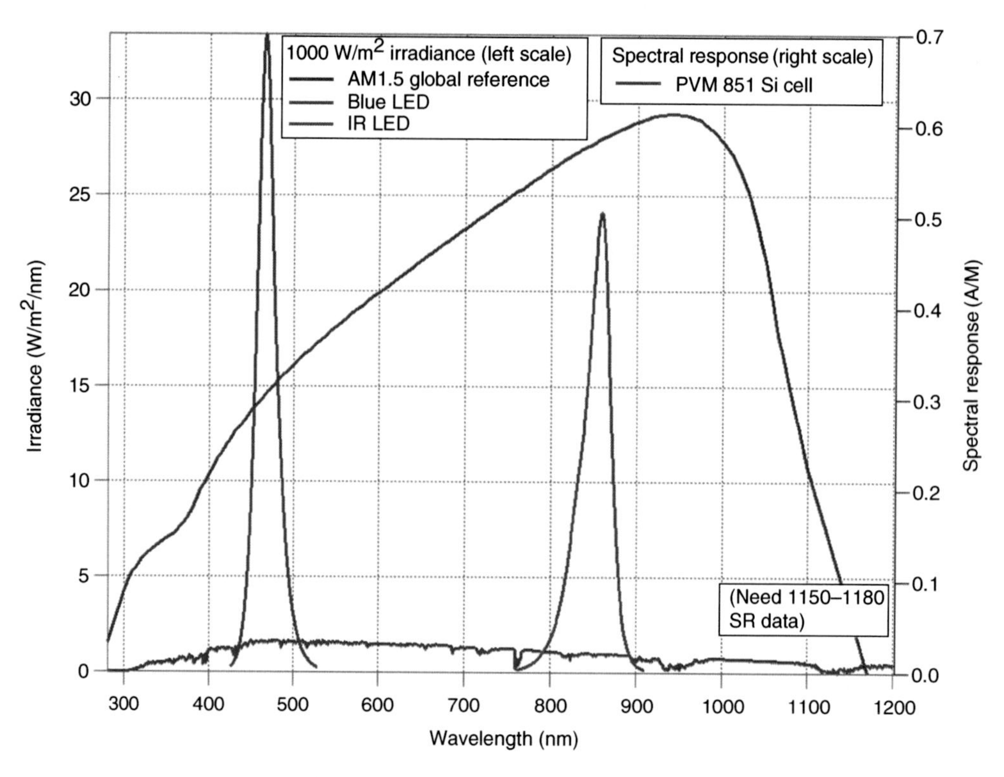

Figure 7.1.3 LED and sunlight irradiance spectra (equal energy content)

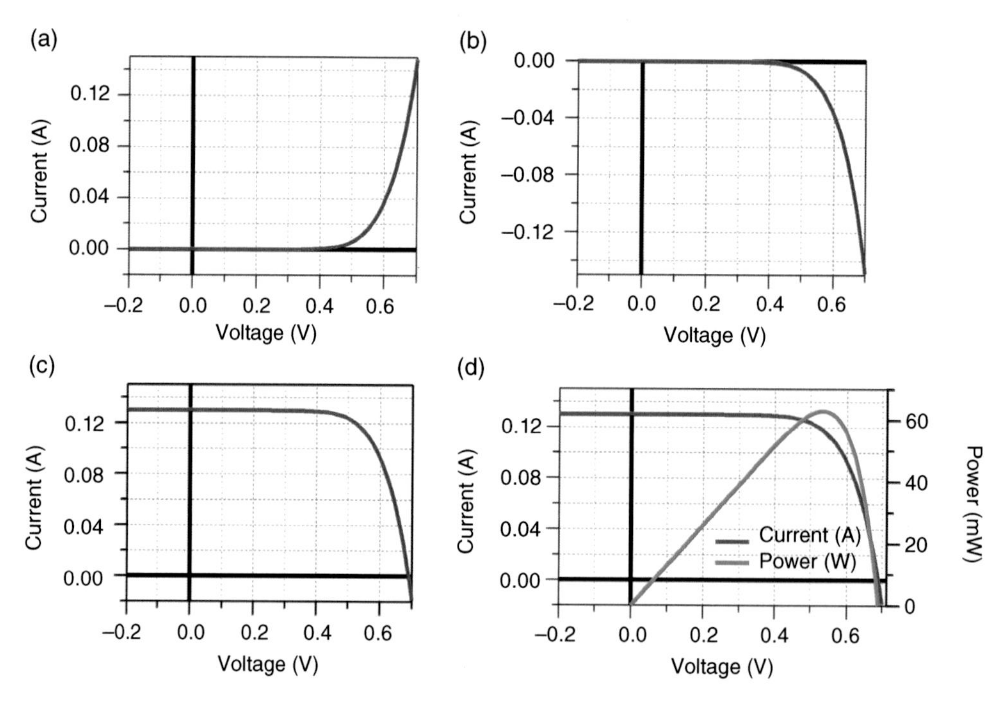

Figure 7.1.4 Solar cell I-V curve conventions: (a) Conventional diode I-V curve; (b) Diode I-V curve with current polarity inverted; (c) Inverted diode curve shifted for current generation; (d) Power curve added

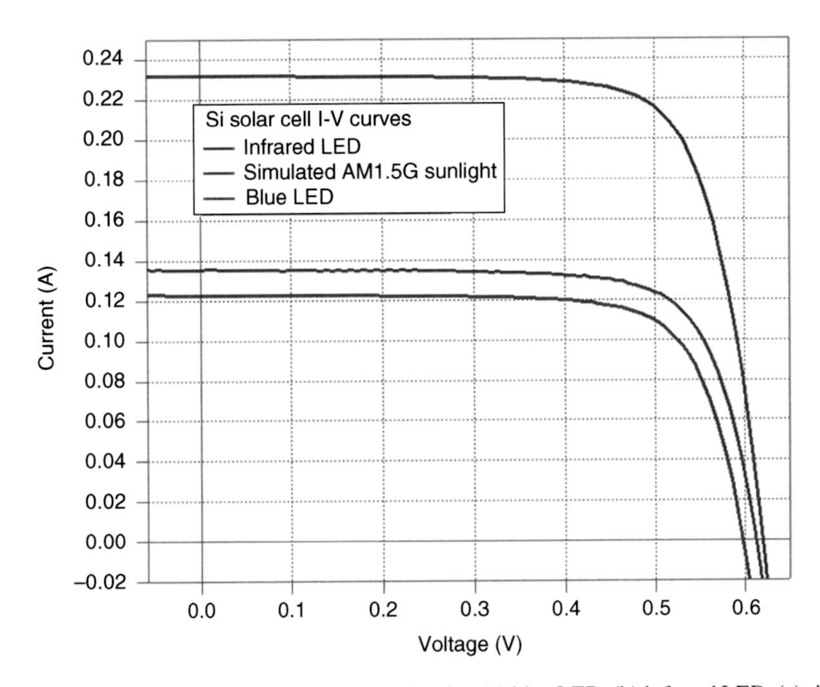

Figure 7.1.6 Solar cell I-V curves with LED illumination (a) blue LED, (b) infrared LED, (c) simulated AM1.5G sunlight

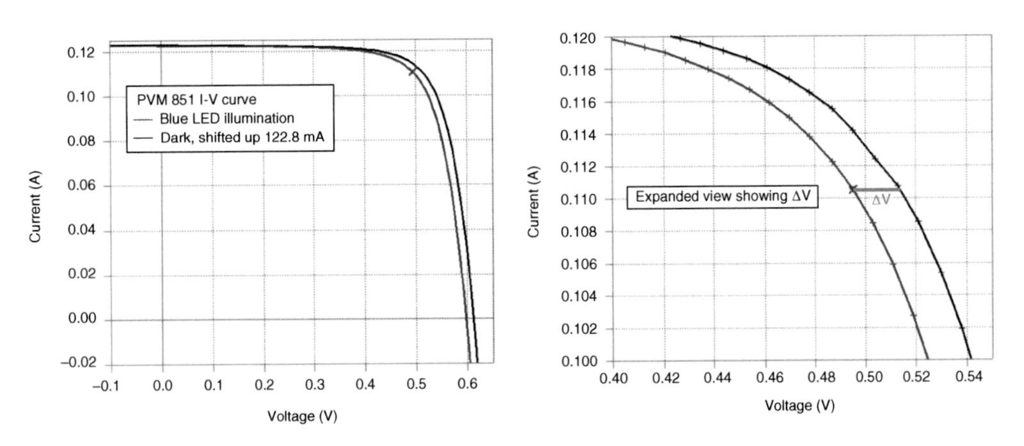

Figure 7.1.9 Solar cell I-V curve used to determine series resistance

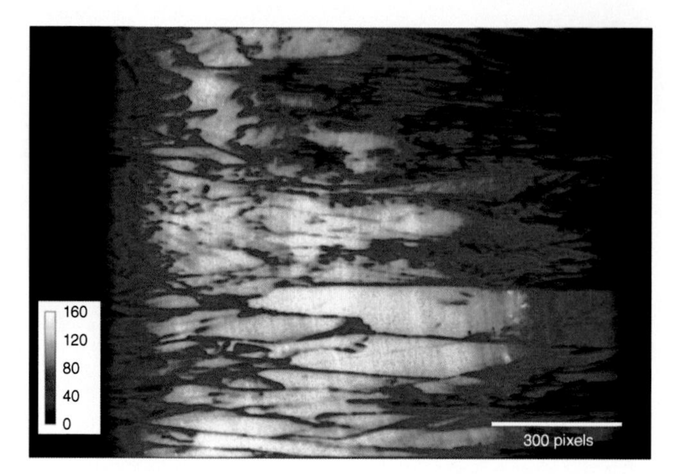

Figure 7.2.3 Bulk lifetime image of the side facet of a 6-inch, 23 cm high multicrystalline silicon brick prior to wafer slicing obtained using the PL intensity ratio method. The darkness scale represents the lifetime in microseconds, the right-hand side represents the top of the ingot. Source: (Mitchell *et al.*, 2011)

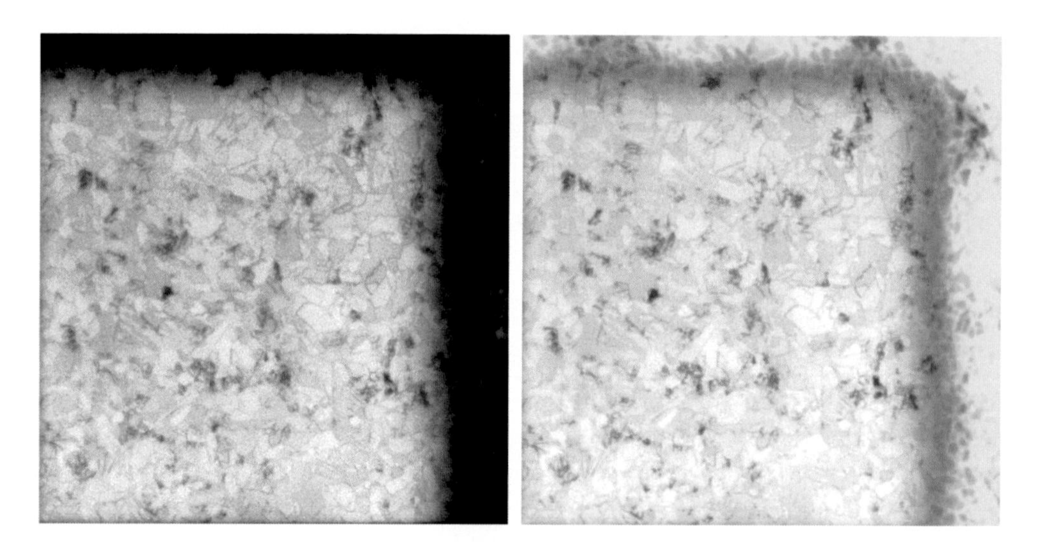

Figure 7.2.4 PL image of an as-cut high performance multicrystalline silicon wafer, measured on a BT Imaging iLS-W2 automated inline PL imaging system (left) and results of automated image processing (right) showing shaded overlays for areas of high impurity concentration (light) and for the location of structural defects (shaded)

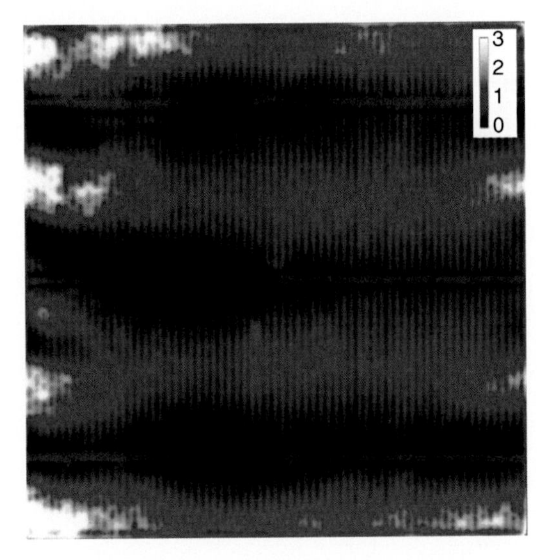

Figure 7.2.5 Series resistance image on a multicrystalline silicon solar cell measured using the method by Kampwerth (Kampwerth *et al.*, 2008). The darkness scale gives the local series resistance in absolute units of Ωcm^2

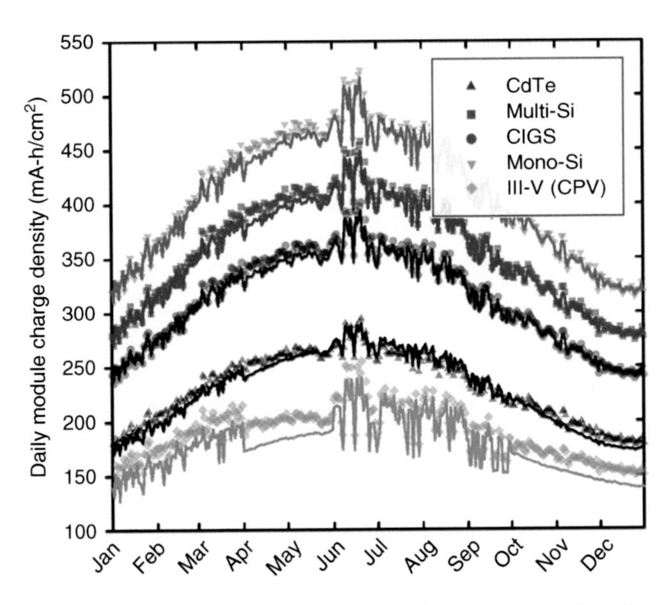

Figure 7.5.3 Predicted daily module short-circuit charge densities under clear sky conditions (solid lines) for various module designs. The symbols indicate values obtained by replacing the predicted spectrum with (normalized) reference spectra (AM1.5D for the III-V module, AM1.5G for the others). ("Module charge density" is the sum of the current per unit area over time for each day. This metric is used as an expedient parameter to increase the visual separation between the module types with similar current outputs.) Source: (Kinsey, 2015)

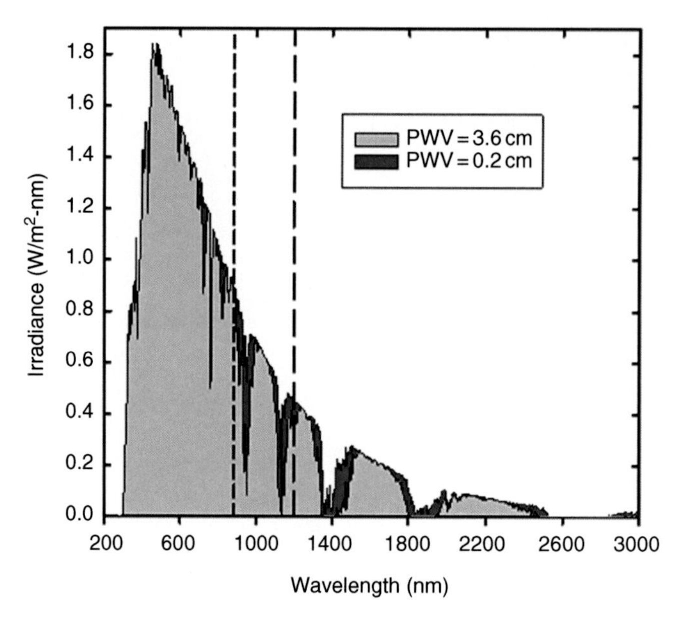

Figure 7.5.4 Modeled spectrum using TMY3 data (TMY3 database, available online) for Albuquerque, NM, USA and SMARTS Latitude-tilt global irradiance incident with the highest and lowest annual values for precipitable water vapor (PWV) in the atmosphere in Albuquerque are shown. Under the more arid conditions (PWV = 0.2 cm), irradiance in the range 280–4000 nm rises by 11%, whereas the irradiance below 1200 nm (near the Si band edge) rises by 7% and that below 880 nm (near the CdTe band edge) rises by only 2%. Source: (Gueymard, 2005)

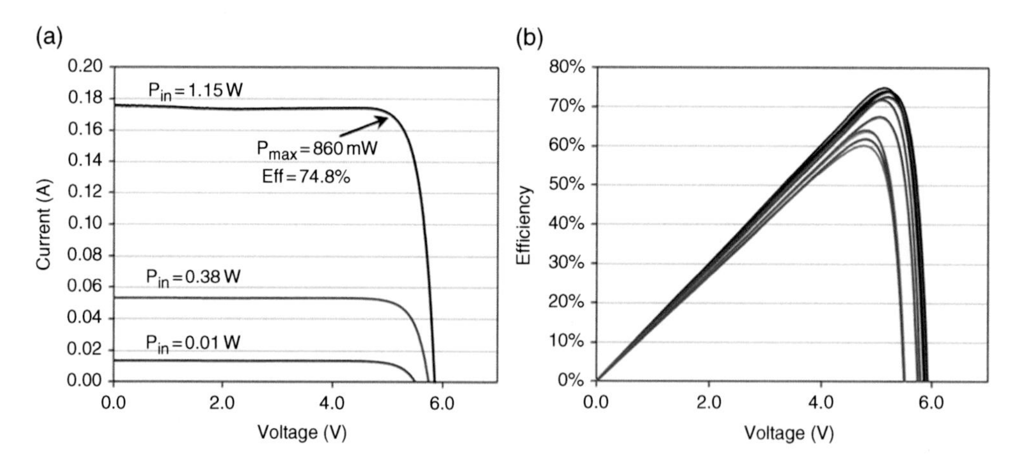

Figure 8.2.3 (a) I-V plots of a packaged phototransducer at 3 different input powers for an optical input at 835 nm. (b) Measured phototransducer efficiency for input powers of 0.01 W (lower curves), 0.38 W (middle) and 1.15 W (higher curves). The maximum conversion efficiency occurs near 5 V and increases with the input power from about 60% at 0.01 W to 74.8% at 1.15 W

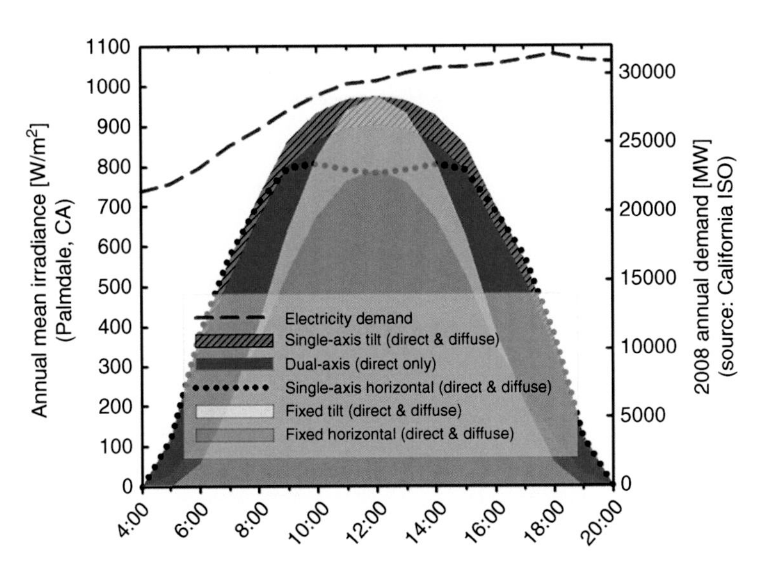

Figure 8.4.4 Available irradiance for various fixed rack and tracking configurations. The projected electrical load from the California Independent System Operator (CAISO) is shown for reference

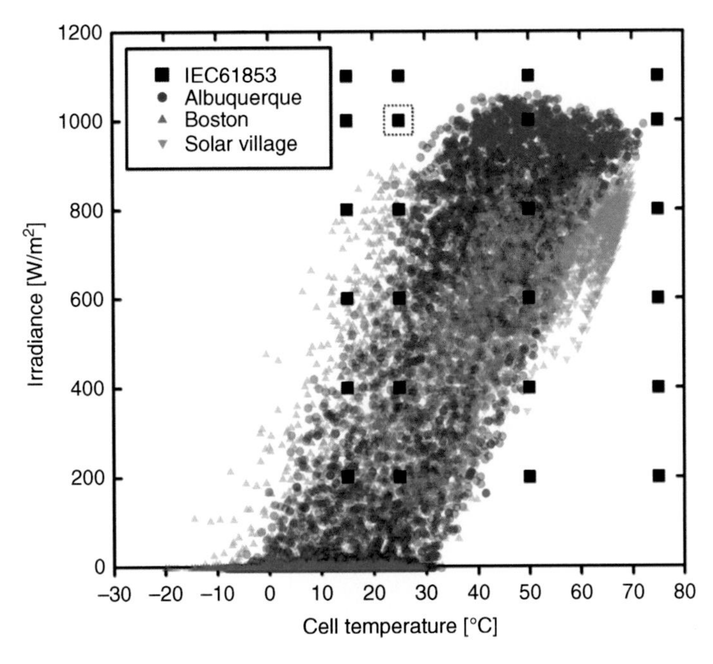

Figure 8.4.5 Comparison of test conditions specified in IEC 61853 and operating conditions derived for Albuquerque, USA, Boston, USA, and Solar Village, Saudi Arabia. The cell temperature was determined using the Sandia Photovoltaic Array Performance Model (King *et al.*, 2011) for DNI incident on a 22x linear concentrator. Note that STC conditions (1000 W/m^2, 25°C, outlined) are not encountered in operation at any of the three sites

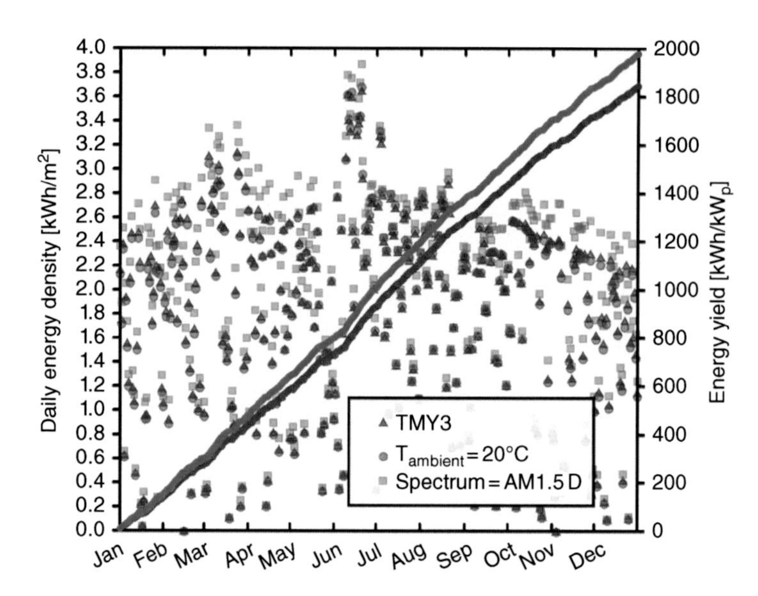

Figure 8.4.8 Sensitivity analysis of daily energy density (symbols) and energy yield (lines) for a 500x CPV system in Albuquerque, NM, USA. The result for TMY3 inputs is compared against cases where the temperature and spectrum are held constant. Cell temperature is assumed to be 40°C above the ambient. Energy yield is calculated based on peak power at CSTC (25°C, 1000 W/m^2)

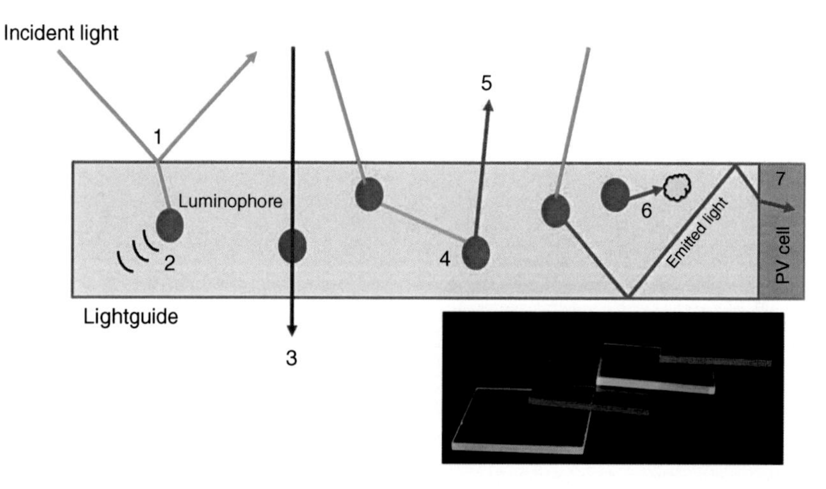

Figure 8.5.1 Main loss mechanisms of luminescent solar concentrators. (1) Reflected incident light; (2) non-unity quantum yield; (3) incomplete absorption; (4) re-absorption events; (5) emission outside the "capture cone"; (6) absorption by matrix; (7) spectral mismatch with PV cell. Inset photograph: Four LSCs exposed to UV light from above

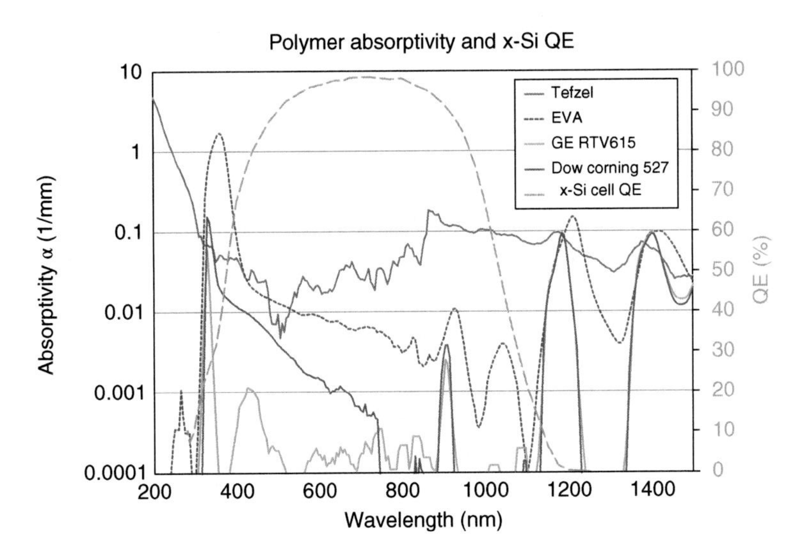

Figure 10.2.4 Plot of polymer absorptivity as a function of wavelength alongside the internal quantum efficiency of a typical crystalline silicon cell

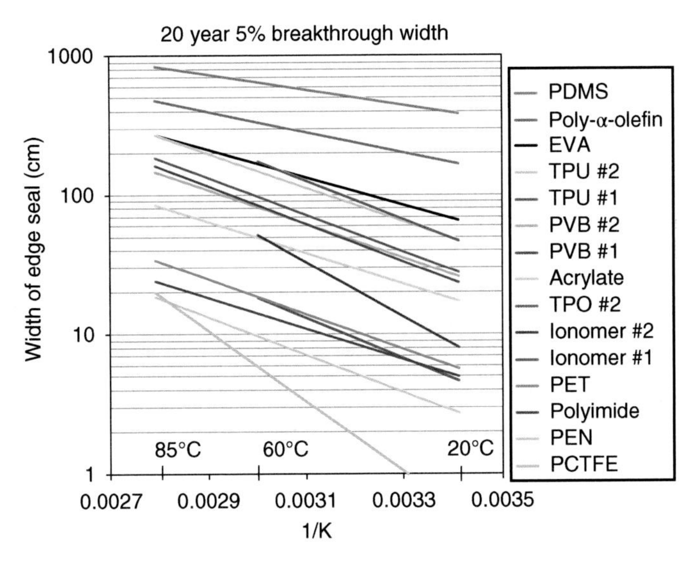

Figure 10.2.6 Width of edge seal made from different materials that would be necessary to keep moisture below 5% of equilibrium values at a given temperature. Source: (Kempe *et al.*, 2006)

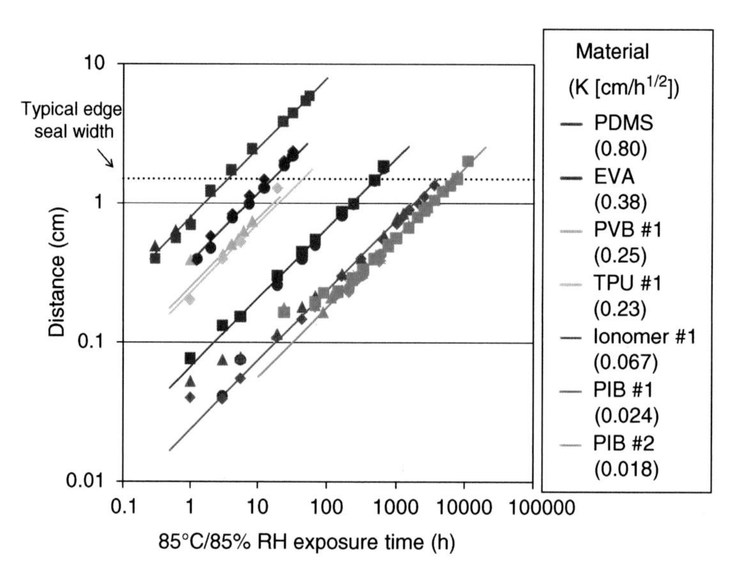

Figure 10.2.7 Penetration depth of moisture between glass plates laminated with different materials as measured by the oxidation of a 100 nm film of Ca. Source: (Kempe *et al.*, 2010)

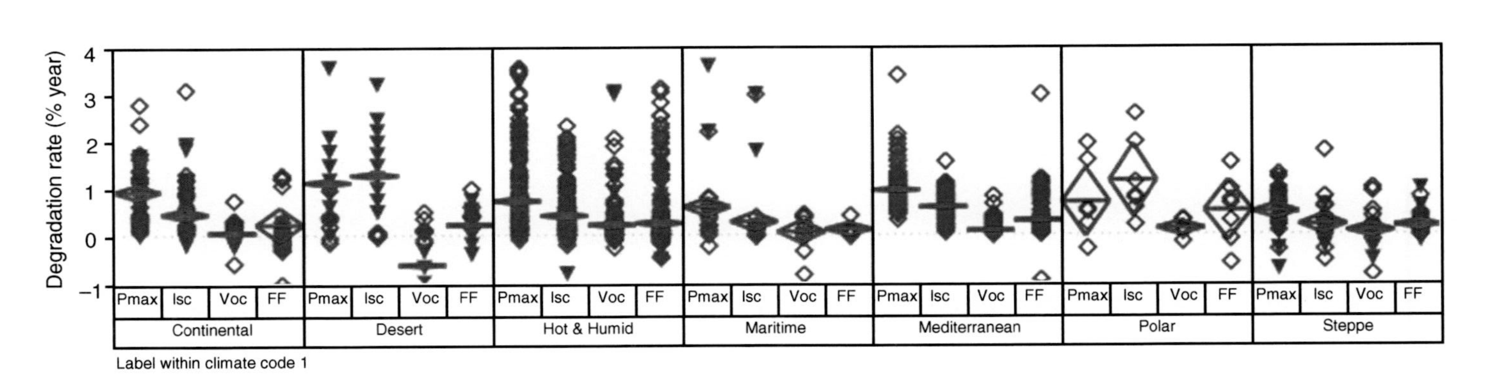

Figure 10.3.1 Summary of degradation rates reported in the literature for PV modules deployed in different climate zones. The degradation is described according to the changes (under standard test conditions) in short-circuit current (Isc), open-circuit voltage (Voc), fill factor (FF), and power (Pmax). Source: (Jordan *et al.*, 2012)

Figure 10.3.4 Silicon PV module after ~20 years in the field

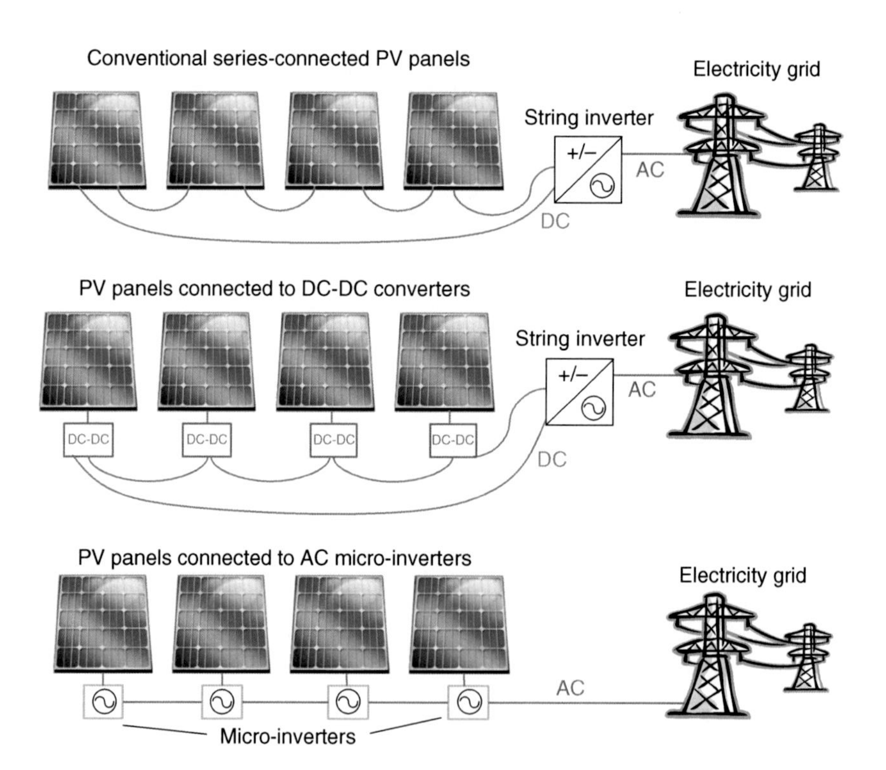

Figure 11.2.1 Schematic of conventional single-string PV system (top), DC-DC converter-equipped "Smart Modules" (middle), and AC micro-inverter-equipped PV system (bottom)

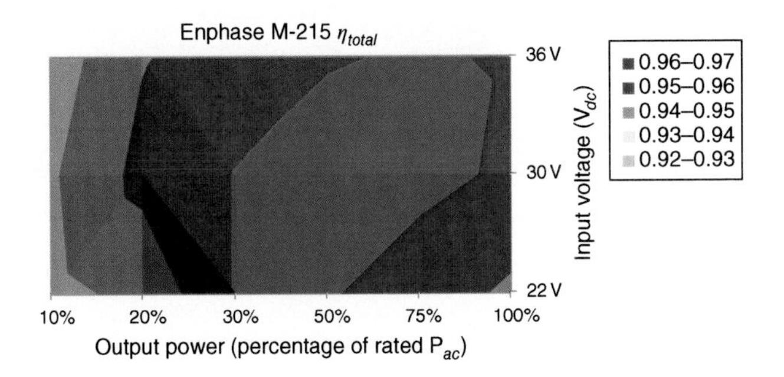

Figure 11.2.6 Total efficiency vs DC input voltage and AC output power for an Enphase M215 microinverter

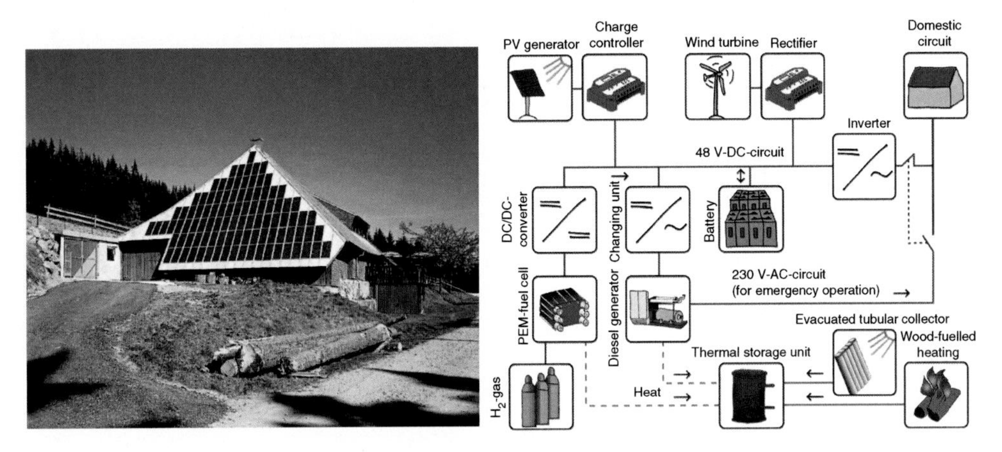

Figure 11.3.4 Hikers inn, Rappenecker Hütte near Freiburg has been operating since 1987 with a hybrid PV system, now equipped with a PV generator (3.8 kWp), a hydrogen fuel cell (4 kW), a diesel genset (12 kW), a wind generator (1.8 kW) and battery storage (45 kWh)

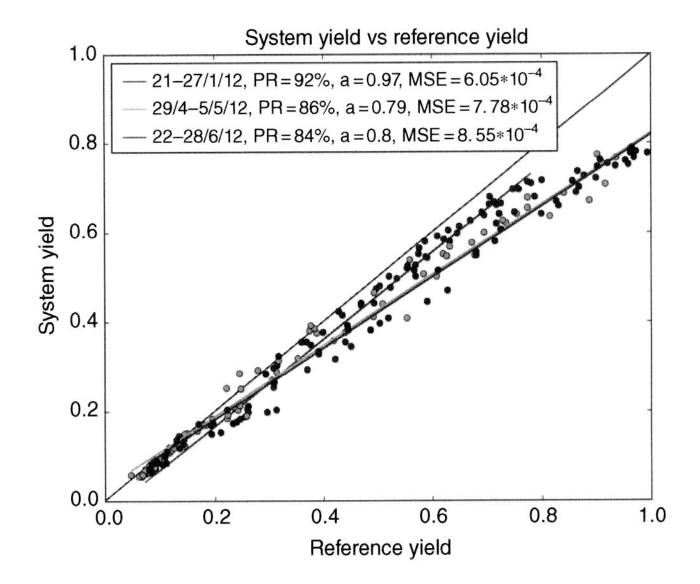

Figure 11.4.3 System yield versus reference yield for a 4.14 kWp PV system. Source: (Tsafarakis, 2014)

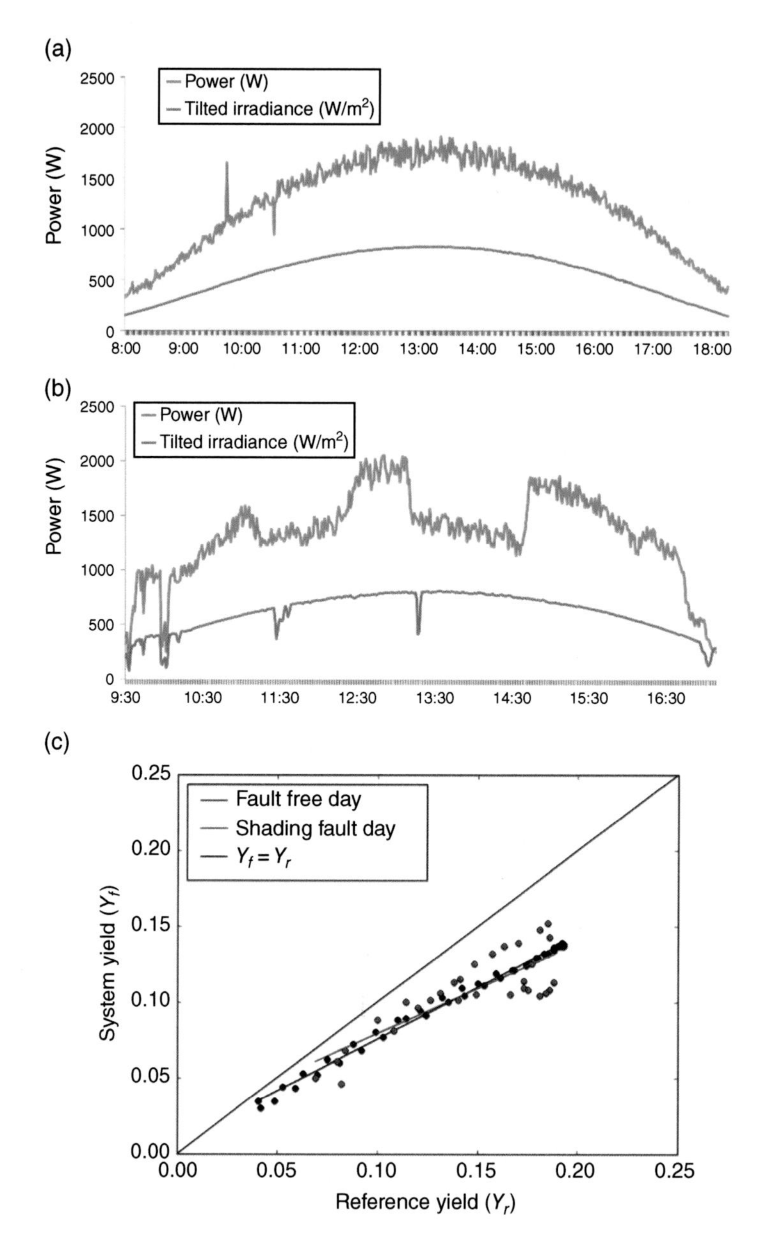

Figure 11.4.4 PV system power and irradiance for a clear day without shade (a), and with shade (b). Panel (c) shows system yield versus reference yield. Source: (Tsafarakis and Van Sark, 2014)

Table 11.6.1 BIPV categories, sub-categories and techniques

Category	Sub category	Technique	Example
Roof systems	Flat roof	Roofing material	
Roof systems	Pitched roofs	Opaque modules	
Roof systems	Pitched roofs	Colored cells/ modules	
Roof systems	Pitched roofs	Shingles	 Reproduced with permission by the Department of Geosciences, University of Wisconsin-Madison

(*Continued*)

Table 11.6.1 (*Continued*)

Category	Sub category	Technique	Example
Roof systems	Pitched roofs	Tiles	 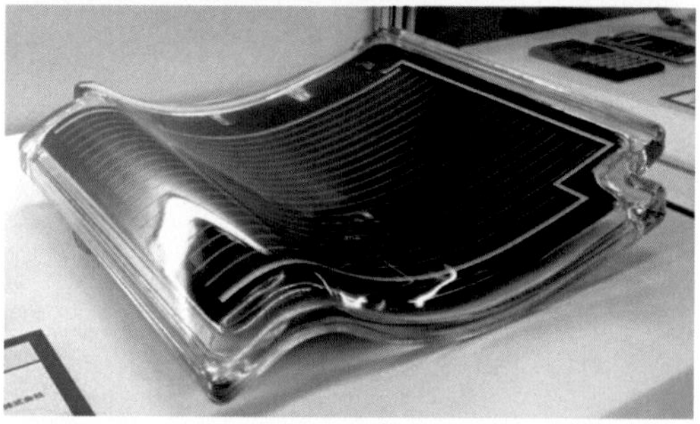
Roof systems	Pitched roofs	Skylights/ semitransparent modules	 

Table 11.6.1 (*Continued*)

Category	Sub category	Technique	Example
Façade systems		Curtain walls	 Reproduced with permission by Oskomera
Façade systems		Opaque panels	 Reproduced with permission by Brooks Scarpa Architects.
Façade systems		Semitransparent modules	
Façade systems		Shading devices	

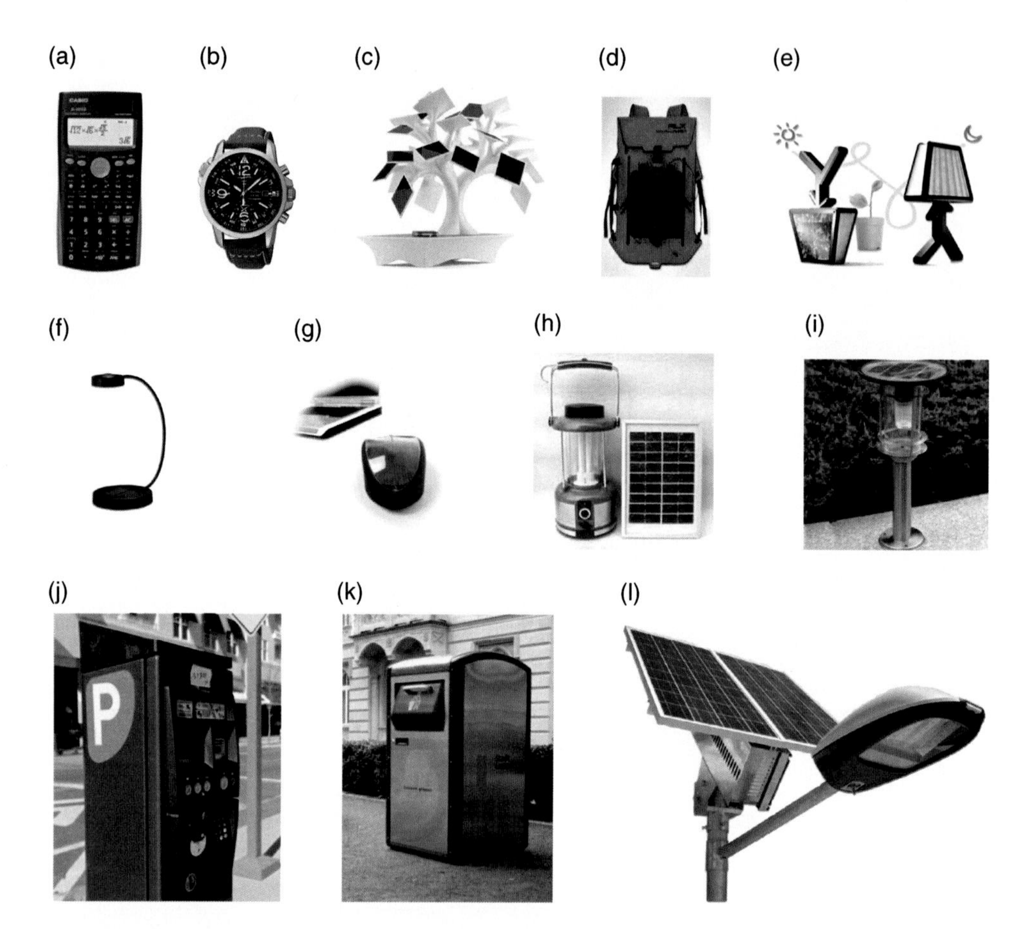

Figure 11.7.1 Photovoltaic products of various product-categories. (a) solar calculator (Sale Stores, 2015), (b) solar watch (Express Watches, 2015), (c) phone charger by Vivien Muller (Muller, 2015), (d) solar-powered bag (Ralph Lauren, 2015), (e) Spark lamp (Spark, 2015), (f) IKEA Sunnan lamp (IKEA, 2015), (g) PC computer mouse Sole-Mio (DDI, 2015), (h) solar lantern (Solar Lantern, 2015), (i) solar garden light (Solar Garden Light, 2015), (j) solar-powered parking meter in Virginia (Matray, 2015), (k) automated trash bin Big Belly (Big Belly, 2015), (l) solar traffic light, (m) Solar-powered car from University of Twente in 2011 (n) Planet Solar Catamaran (PlanetSolar, 2015), (o) Helios solar aircraft (NASA, 2015), (p) solar-powered tent (Harris, 2015), (q) PV-powered chandelier (Renewable Energy Magazine, 2015)

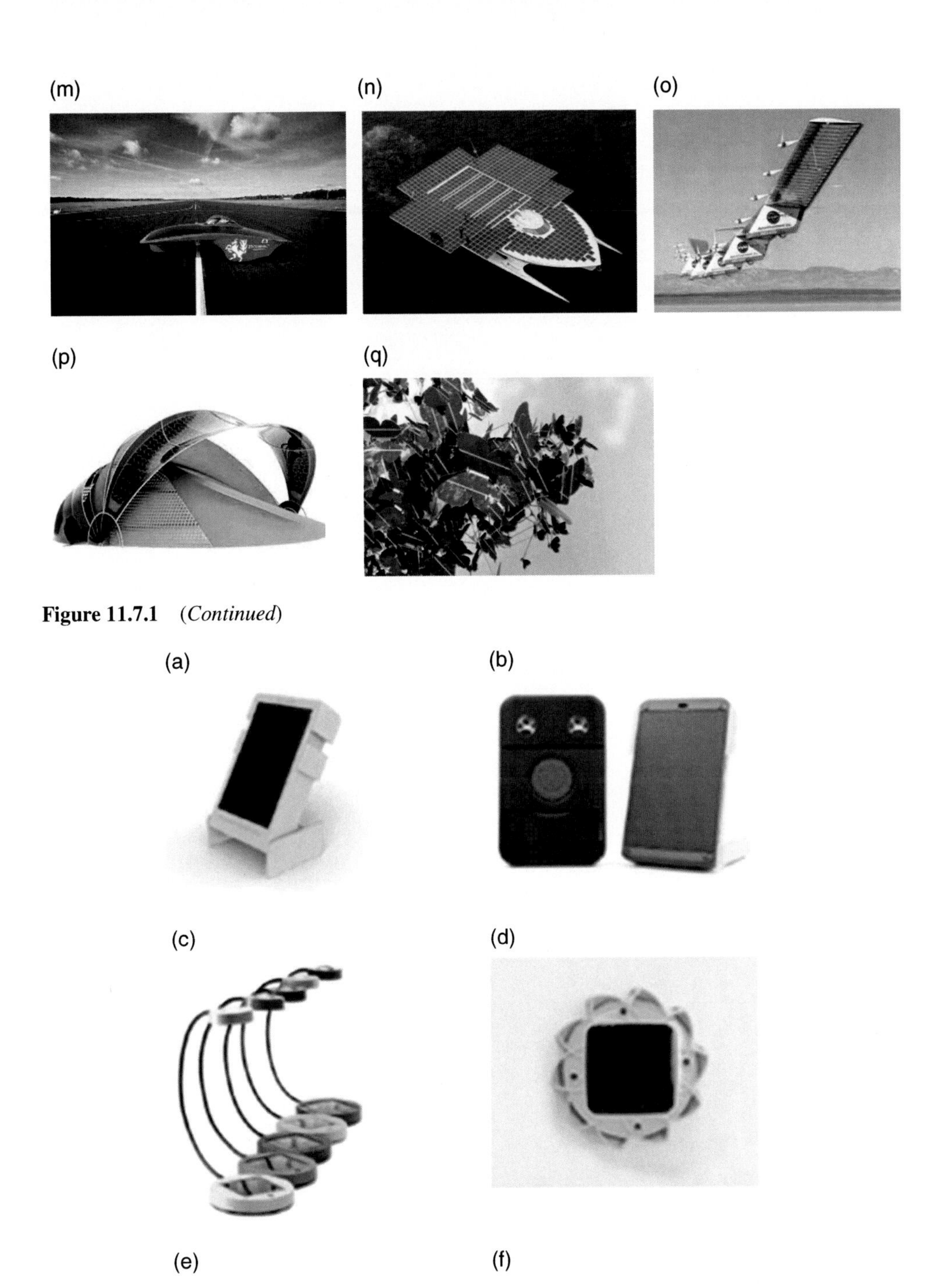

Figure 11.7.1 (*Continued*)

Figure 11.7.5 The tested PV products: (a) Waka Waka light; (b) Waka Waka Power light and charger; (c) Sunnan IKEA lamp; (d) Little Sun light; (e) Beurer kitchen weight scale; and (f) Logitech solar keyboard. Source: (Apostolou and Reinders, 2015, 2016b)

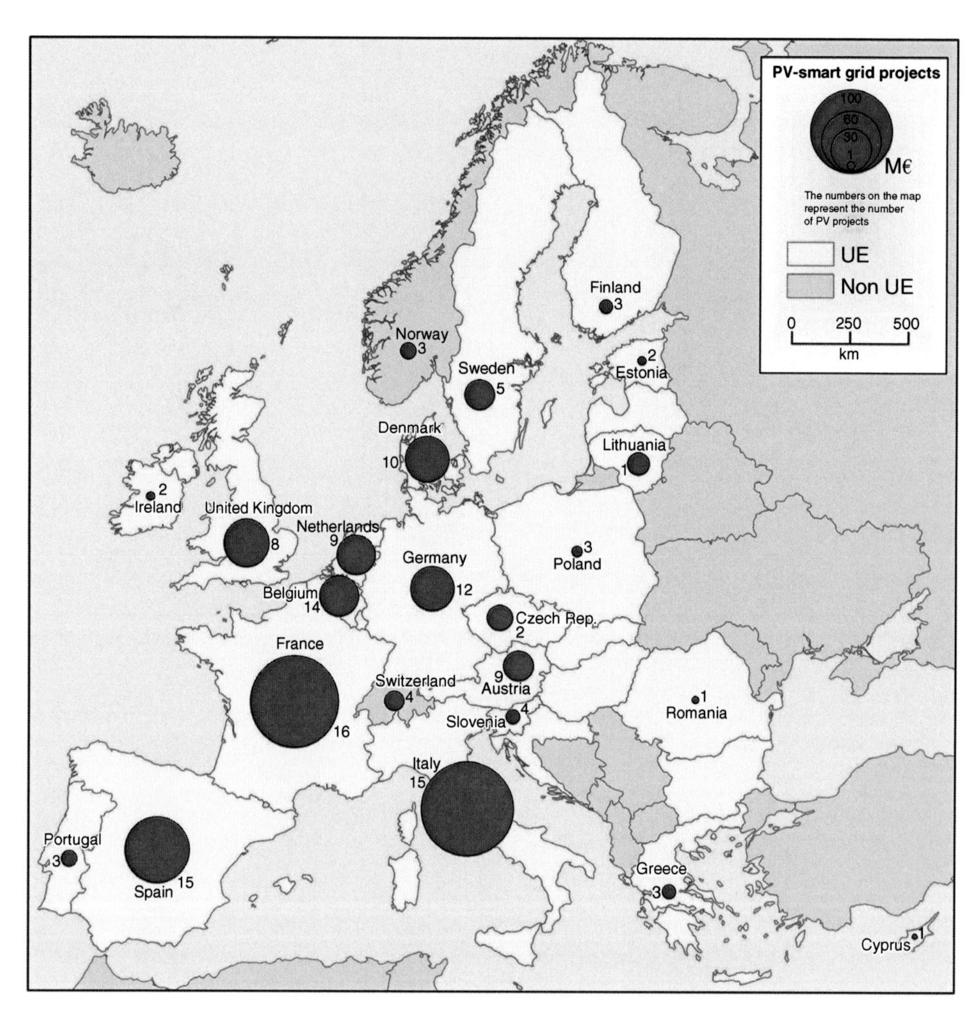

Figure 12.2.2 Investments in R&D and Demonstration PV smart grid projects in Europe (M€), 2002–2013

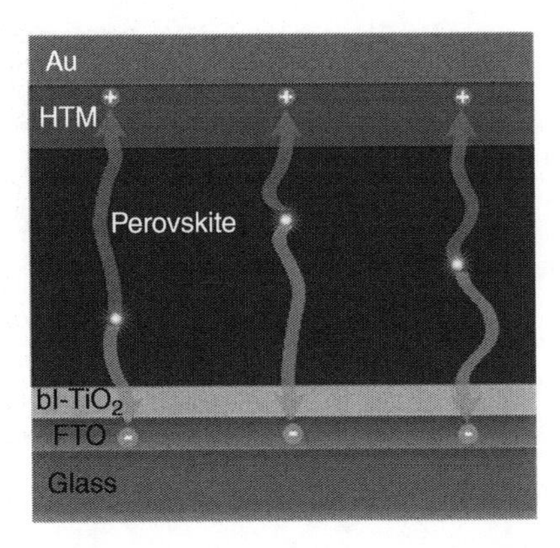

Figure 6.3.8 Illustration of the charge generation processes in a planar heterojunction perovskite solar cell (Stranks and Snaith, 2015). Source; adapted with permission from NPG

selectively transfer across their respective electrodes where they contribute to the current in the external circuit (Figure 6.3.8) (Edri *et al.*, 2014a).

6.3.5 Ongoing Challenges

6.3.5.1 Lead-free Alternatives

There is a perceived drive to replace lead in order to prevent potential toxicity issues. The use of cadmium in CdTe solar cells raised similar concerns, but life-cycle analyses and lobbying have led to the removal of regulations preventing their deployment in many countries. Full life-cycle analyses of perovskite solar cells are required, and a clear understanding of what is offset by employing lead-based perovskites needs to be realized.

Noel *et al.* (2014) and Hao *et al.* (2014) recently reported lead-free, $MASnI_3$-based perovskite solar cells, achieving power conversion efficiencies of 5–6% when processed on mesoporous TiO_2 scaffolds. The $MASnI_3$ perovskite has a bandgap of ~1.2–1.3 eV, allowing broader light harvesting (Figure 6.3.9) than its lead counterpart. However, the Sn-based perovskites are extremely unstable in air, and degrade rapidly even if processed in an inert atmosphere. It has been suggested that the key issue is the oxidation of Sn^{2+} to Sn^{4+}, which results in a high level of "self-doping." Other metal ions that can be stabilized in the 2+ state and could be promising alternatives to lead include, among others, Cu, Ge and Ni (Mitzi *et al.*, 2001).

6.3.5.2 Hysteresis

An anomalous hysteresis in the current-voltage curves of perovskite solar cells has been reported (Figure 6.3.10) (Snaith *et al.*, 2014). A number of possible causes have been proposed though the phenomenon is not yet fully understood. Although further understanding is required, most evidence points towards mobile ionic species being responsible: under an applied electric field, negative ionic species will migrate towards one

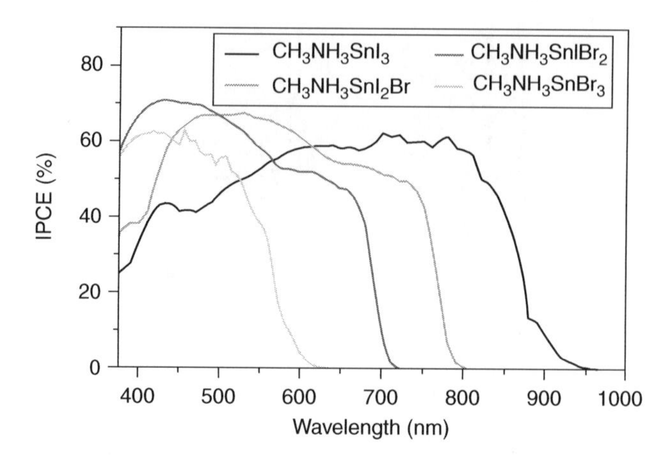

Figure 6.3.9 IPCE fraction for the Sn-based perovskite solar cells with a range of different halides to tune the bandgap (Hao et al., 2014). Source: Adapted with permission from NPG

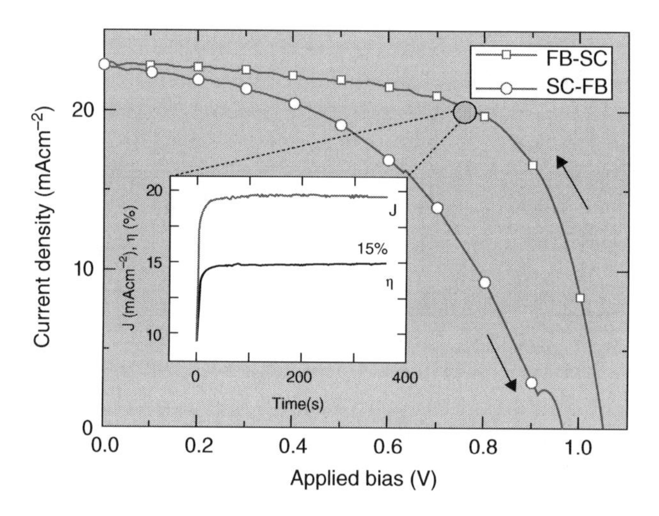

Figure 6.3.10 Forward bias to short circuit (FB-SC) and short circuit to forward bias (SC-FB) J-V curves measured under AM1.5 simulated sun light for a perovskite solar cell fabricated with a 400 nm thick mesoporous alumina film. Inset: Photocurrent density and power conversion efficiency as a function of time for the same cell held close to 0.75 V forward bias (Snaith *et al.*, 2014). Source: Adapted with permission from ACS

electrode and positive ionic species will migrate towards the other (Tress *et al.*, 2015). This results in a stabilization of positive electronic space charge near one electrode and negative electronic space charge near the other, effectively resulting in p- and n-doping of the perovskite near the charge collection layers. This leads to favorable electronic contact for charge extraction in forward bias but unfavorable charge extraction under short-circuit conditions.

Hysteresis is far less pronounced in the case of an inverted device structure, where the n-type fullerene layer is deposited on top of the perovskite (Shao *et al.*, 2014). It has been proposed that the fullerene infiltrates between perovskite grain boundaries, leading to the passivation of perovskite defect sites (Shao *et al.*, 2014). Hysteresis is also significantly reduced in devices employing mesoporous TiO_2 (Jeon *et al.*, 2014), which is consistent with a larger surface area, increasing the pathways for unimpeded electron injection. These results indicate that hysteresis is most apparent when there is poor electronic contact between the perovskite and the charge collection layer, and this contact is improved by the ionic displacement.

Understanding and mitigating hysteresis remain an ongoing challenge. However, the devices can stabilize within a short time at their maximum power point, with the efficiency at this point approaching that determined from the high efficiency scans (see Figure 6.3.10 inset) (Snaith *et al.*, 2014). We stress that maximum power point measurements are crucial to truly reflect the stabilized power output, as the absence of hysteresis between J-V scans is not sufficient to verify that the measurements are under quasi-steady state conditions.

6.3.5.3 Thermal and Operational Stability

The most pressing hurdles for the perovskite solar technology to reach commercialization are proving the ability to manufacture on a large scale with high yield, and proving perovskite modules are capable of surviving for at least 25 years in an outdoor environment without the need for overly expensive encapsulation methods.

$MAPbX_3$ perovskite solar cells aged under sunlight without encapsulation will rapidly irreversibly degrade, most likely due to moisture reacting with the perovskite. When encapsulated in an inert atmosphere, the degradation of the perovskite is largely resolved and the solar cells are capable of sustaining at least 1000 hours of exposure to simulated full spectrum sunlight with little drop in photocurrent (Leijtens *et al.*, 2013). However, if mesoporous TiO_2 is employed in the solar cell, then exposure of the sealed cells to the ultraviolet component of sunlight results in a rapid drop in photocurrent (Leijtens *et al.*, 2013). This is largely resolved by the removal of the mesoporous TiO_2.

Mei *et al.* (2014) recently demonstrated 1000 hours stability for an unsealed cell measured under "full AM 1.5 simulated sunlight in ambient air" by using a double layer of mesoporous TiO_2 and ZrO_2 with a thick hydrophobic carbon black electrode ("triple-layer" devices). The results are extremely encouraging and imply that the perovskite is not fundamentally unstable in ambient conditions, even under natural sunlight containing a UV component. It is likely, however, that the perovskite modules will still require sealing from environmental effects, especially considering the requirement for operation under elevated temperature and humidity, but the sealing methods developed for thin film materials or crystalline silicon could already be sufficient. Indoor heat stress tests of encapsulated triple-layer perovskite solar cells stored at 85 °C show very encouraging stability over several months (Figure 6.3.11) (Li *et al.*, 2015). Figure 6.3.12 shows photographs of perovskite films heated on a hotplate in air for 60 minutes at 150 °C. The $MAPbI_3$ film turns from brown to yellow as the material converts from the $MAPbI_3$ to PbI_2 (Stranks and Snaith, 2015). However, the loss of the organic cation can be largely mitigated by moving to the larger cation FA, suggesting that the $FAPbI_3$ perovskite may already be suitably thermally stable. Further thermal, chemical and operational stability improvements have been recently reported with the addition of low levels of Cs^+ cations and/or bromide (Saliba *et al.*, 2016; McMeekin *et al.*, 2016; Yang *et al.*, 2015).

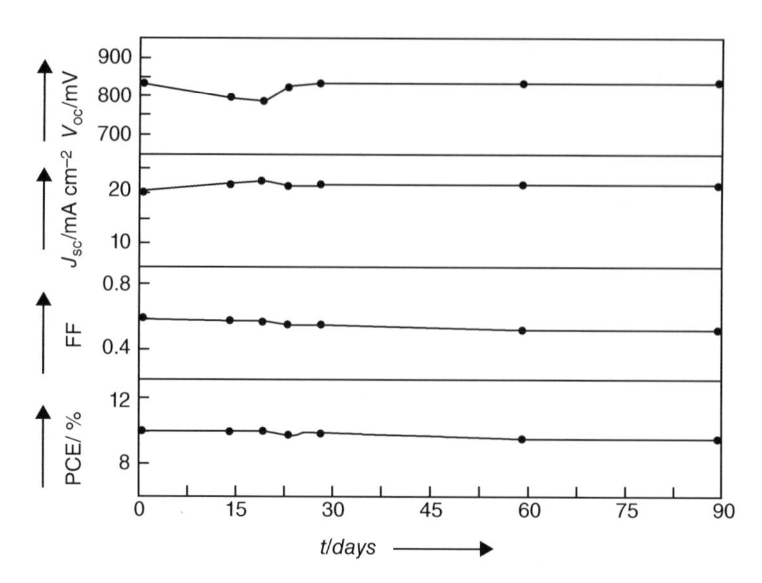

Figure 6.3.11 Indoor heat stress test of a triple-layer perovskite solar cell. The device was encapsulated and kept for 3 months in a normal oven filled with ambient air at ~85°C and removed periodically to measure J-V curves under full solar AM1.5 light at ambient temperature (Li *et al.*, 2015). Source: Adapted with permission from Wiley

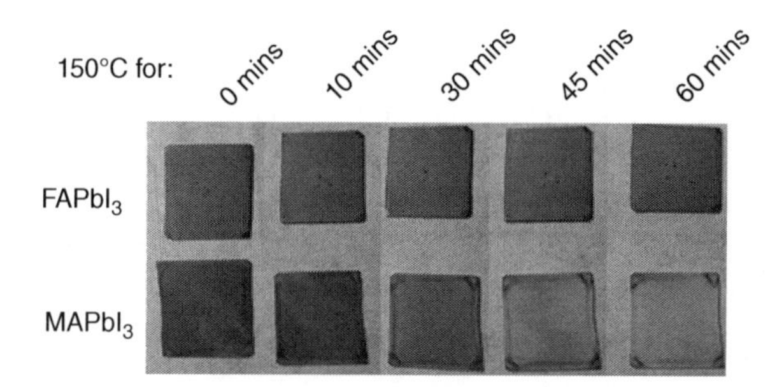

Figure 6.3.12 Photographs showing non-encapsulated $MAPbI_3$ and $FAPbI_3$ samples before and during heating on a hot plate at 150°C for 60 minutes in air (Stranks and Snaith, 2015). Source: Adapted with permission from Nature Publishing Group (NPG). (*See insert for color representation of the figure*)

6.3.6 Conclusion

We have witnessed unprecedented progress in the development of perovskite solar cells, though intense research has only just begun. Since their first use in a solar cell architecture akin to a dye-sensitized solar cell, there has been a rapid expansion in the diversity of architectures and fabrication routes. It appears that the community is converging on one general structure: the p-i-n planar heterojunction perovskite solar cell, with potential growth in efforts towards understanding the properties of and devices fabricated from single crystals. The challenge will then be to optimize the p- and n-type contact materials (which could be organic, inorganic or hybrid), the means of fabrication of the layers, and the substrates and choice of targeted deployment. While we cannot predict all of the potential areas for future research, the key areas which require pressing attention are: (1) understanding the nature of the heterojunctions; (2) understanding what governs and enhances the long-lived charge species in the perovskites; and (3) understanding and inhibiting material and device degradation mechanisms and hysteresis effects. Moreover, the search for new improved perovskite absorbers is very enticing. It is very possible that a commercially viable alternative to silicon photovoltaics, or a material capable of enhancing silicon photovoltaics (Snaith, 2013), will emerge in the next few years.

References

Bailie, C.D., Christoforo, M.G., Mailoa, J.P. *et al.* (2015) Semi-transparent perovskite solar cells for tandems with silicon and CIGS. *Energy and Environmental Science*, **8**, 956–963. DOI: 10.1039/C4EE03322A.

Ball, J.M., Lee, M.M., Hey, A., and Snaith, H.J. (2013) Low-temperature processed meso-superstructured to thin-film perovskite solar cells. *Energy and Environmental Science*, **6** (6), 1739–1743. DOI: 10.1039/C3ee40810h.

Burschka, J., Pellet, N., Moon, S.J. *et al.* (2013) Sequential deposition as a route to high-performance perovskite-sensitized solar cells. *Nature*, **499** (7458), 316–319. DOI: 10.1038/Nature12340.

Chen, Q., Zhou, H., Hong, Z. *et al.* (2014) Planar heterojunction perovskite solar cells via vapor-assisted solution process. *Journal of the American Chemical Society*, **136** (2), 622–625. DOI: 10.1021/ja411509g.

De Wolf, S., Holovsky, J., Moon, S.-J. *et al.* (2014) Organometallic halide perovskites: sharp optical absorption edge and its relation to photovoltaic performance. *Journal of Physical Chemistry Letters*, **5** (6), 1035–1039. DOI: 10.1021/jz500279b.

de Quilettes, D.W., Vorpahl, S.M., Stranks, S.D. *et al.* (2015) Impact of microstructure on local carrier lifetime in perovskite solar cells. *Science*, **348** (6235), 683–686. DOI: 10.1126/science.aaa5333.

Deschler, F., Price, M., Pathak, S. *et al.* (2014) High photoluminescence efficiency and optically pumped lasing in solution-processed mixed halide perovskite semiconductors. *Journal of Physical Chemistry Letters*, **5**, 1421–1426. DOI: 10.1021/jz5005285.

Docampo, P., Ball, J. M., Darwich, M. *et al.* (2013) Efficient organometal trihalide perovskite planar-heterojunction solar cells on flexible polymer substrates. *Nature Communications*, **4**, 2761. DOI: 10.1038/ncomms3761.

Dong, Q., Fang, Y., Shao, Y. *et al.* (2015) Electron-hole diffusion lengths > 175 mum in solution-grown CH3NH3PbI3 single crystals. *Science*, **347** (6225), 967–970. DOI: 10.1126/science.aaa5760.

Edri, E., Kirmayer, S., Henning, A. *et al.* (2014a) Why lead methylammonium tri-iodide perovskite-based solar cells require a mesoporous electron transporting scaffold (but not necessarily a hole conductor). *Nano Letters*, **14** (2), 1000–1004. DOI: 10.1021/nl404454h.

Edri, E., Kirmayer, S., Mukhopadhyay, S. *et al.* (2014b) Elucidating the charge carrier separation and working mechanism of CH3NH3PbI(3-x)Cl(x) perovskite solar cells. *Nature Communications*, **5**, 3461. DOI: 10.1038/ncomms4461.

Eperon, G.E., Burlakov, V.M., Goriely, A., and Snaith, H.J. (2014a) Neutral color semitransparent microstructured perovskite solar cells. *ACS Nano*, **8** (1), 591–598. DOI: 10.1021/nn4052309.

Eperon, G.E., Stranks, S.D., Menelaou, C. *et al.* (2014b) Formamidinium lead trihalide: a broadly tunable perovskite for efficient planar heterojunction solar cells. *Energy & Environmental Science*, **7** (3), 982–988. DOI: 10.1039/c3ee43822h.

Green, M.A., Emery, K., Hishikawa, Y. *et al.* (2015) Solar cell efficiency tables (version 46). *Progress in Photovoltaics: Research and Applications*, **23** (7), 805–812. DOI: 10.1002/pip.2637.

Hao, F., Stoumpos, C.C., Cao, D.H. *et al.* (2014) Lead-free solid-state organic-inorganic halide perovskite solar cells. *Nature Photonics*, **8** (6), 489–494. DOI: 10.1038/nphoton.2014.82.

Im, J.H., Lee, C.R., Lee, J.W. *et al.* (2011) 6.5% efficient perovskite quantum-dot-sensitized solar cell. *Nanoscale*, **3** (10), 4088–4093. DOI: 10.1039/c1nr10867k.

Jeon, N.J., Noh, J.H., Kim, Y.C. *et al.* (2014) Solvent engineering for high-performance inorganic-organic hybrid perovskite solar cells. *Nature Materials*, **13** (9), 897–903. DOI: 10.1038/nmat4014.

Kim, H.S., Lee, C.R., Im, J.H. *et al.* (2012) Lead iodide perovskite sensitized all-solid-state submicron thin film mesoscopic solar cell with efficiency exceeding 9%. *Scientific Reports*, **2**, 591. DOI: 10.1038/srep00591.

Kojima, A., Teshima, K., Shirai, Y., and Miyasaka, T. (2009) Organometal halide perovskites as visible-light sensitizers for photovoltaic cells. *Journal of the American Chemical Society*, **131**, 6050–6051.

Lee, M.M., Teuscher, J., Miyasaka, T., Murakami, T.N., and Snaith, H.J. (2012) Efficient hybrid solar cells based on meso-superstructured organometal halide perovskites. *Science*, **338** (6107), 643–647. DOI: 10.1126/science.122860.4

Leijtens, T., Eperon, G.E., Pathak, S. *et al.* (2013) Overcoming ultraviolet light instability of sensitized TiO(2) with meso-superstructured organometal tri-halide perovskite solar cells. *Nature Communications*, **4**, 2885. DOI: 10.1038/ncomms3885.

Li, X., Tschumi, M., Han, H. *et al.* (2015) Outdoor performance and stability under elevated temperatures and long-term light soaking of triple-layer mesoporous perovskite photovoltaics. *Energy Technology*, **3** (6), 551–555. DOI: 10.1002/ente.201500045.

Li, X. *et al.* (2016) A vacuum flash-assisted solution process for high-efficiency large-area perovskite solar cells. *Science*, **353**(6294): 58–62.

Liu, D. and Kelly, T.L. (2014) Perovskite solar cells with a planar heterojunction structure prepared using room-temperature solution processing techniques. *Naure Photonics*, **8** (2), 133–138. DOI: 10.1038/nphoton.2013.342.

Liu, M., Johnston, M.B., and Snaith, H.J. (2013) Efficient planar heterojunction perovskite solar cells by vapour deposition. *Nature*, **501** (7467), 395–398. DOI: 10.1038/nature12509.

Manser, J.S., and Kamat, P.V. (2014) Band filling with free charge carriers in organometal halide perovskites. *Nature Photonics*, **8** (9), 737–743. DOI: 10.1038/nphoton.2014.171.

McMeekin, D.P. *et al.* (2016) A mixed-cation lead mixed-halide perovskite absorber for tandem solar cells. *Science*, **351**(6269): 151–155.

Mei, A., Li, X., Liu, L. *et al.* (2014) A hole-conductor-free, fully printable mesoscopic perovskite solar cell with high stability. *Science*, **345** (6194), 295–298. DOI: 10.1126/science.1254763.

Mitzi, D.B., Chondroudis, K., and Kagan, C.R. (2001) Organic-inorganic electronics. *IBM Journal of Research and Development*, **45** (1), 29–45. DOI: 10.1147/rd.451.0029.

Miyata, A., Mitioglu, A., Plochocka, P. *et al.* (2015) Direct measurement of the exciton binding energy and effective masses for charge carriers in organic–inorganic tri-halide perovskites. *Nature Physics*, **11** (7), 582–587. DOI: 10.1038/nphys3357.

Moore, D.T., Sai, H., Tan, K.W. *et al.* (2015) Crystallization kinetics of organic-inorganic trihalide perovskites and the role of the lead anion in crystal growth. *Journal of the American Chemical Society*, **137** (6), 2350–2358. DOI: 10.1021/ja512117e.

Nie, W., Tsai, H., Asadpour, R., Blancon, J.C. *et al.* (2015) High-efficiency solution-processed perovskite solar cells with millimeter-scale grains. *Science*, **347** (6221), 522–525. DOI: 10.1126/science.aaa0472.

Noel, N.K., Stranks, S.D., Abate, A. *et al.* (2014) Lead-free organic-inorganic tin halide perovskites for photovoltaic applications. *Energy & Environmental Science*, **7** (9), 3061–3068. DOI: 10.1039/C4ee01076k.

NREL (2015) Renewable Energy Labs Efficiency Chart. http://www.nrel.gov/ncpv/images/efficiency_chart.jpg.

Ponseca, C.S., Jr., Savenije, T.J., Abdellah, M. *et al.* (2014) Organometal halide perovskite solar cell materials rationalized: ultrafast charge generation, high and microsecond-long balanced mobilities, and slow recombination. *Journal of the American Chemical Society*, **136** (14), 5189–5192. DOI: 10.1021/ja412583t.

Saliba, M. *et al.* (2016) Cesium-containing triple cation perovskite solar cells: improved stability, reproducibility and high efficiency. *Energy & Environmental Science*. DOI: 10.1039/c5ee03874j.

Shao, Y., Xiao, Z., Bi, C. *et al.* (2014) Origin and elimination of photocurrent hysteresis by fullerene passivation in CH3NH3PbI3 planar heterojunction solar cells. *Nature Communications*, **5**, 5784. DOI: 10.1038/ncomms6784.

Shi, D., Adinolfi, V., Comin, R. *et al.* (2015) Low trap-state density and long carrier diffusion in organolead trihalide perovskite single crystals. *Science*, **347** (6221), 519–522. DOI: 10.1126/science.aaa2725.

Snaith, H.J. (2013) Perovskites: the emergence of a new era for low-cost, high-efficiency solar cells. *Journal of Physical Chemistry Letters*, **4** (21), 3623–3630. DOI: 10.1021/Jz4020162.

Snaith, H.J., Abate, A., Ball, J.M. *et al.* (2014) Anomalous hysteresis in perovskite solar cells. *Journal of Physical Chemistry Letters*, **5** (9), 1511–1515. DOI: 10.1021/Jz500113x.

Stranks, S.D., Burlakov, V.M., Leijtens, T. *et al.* (2014) Recombination kinetics in organic-inorganic perovskites: excitons, free charge, and subgap states. *Physical Review Applied*, **2** (3), 034007. DOI: 10.1103/Physrevapplied.2.034007.

Stranks, S.D., Eperon, G.E., Grancini, G. *et al.* (2013) Electron-hole diffusion lengths exceeding 1 micrometer in an organometal trihalide perovskite absorber. *Science*, **342** (6156), 341–344. DOI: 10.1126/science.1243982.

Stranks, S.D., Nayak, P.K., Zhang, W. *et al.* (2015) Formation of thin films of organic-inorganic perovskites for high-efficiency solar cells. *Angewandte Chemie Int Ed*, **54** (11), 3240–3248. DOI: 10.1002/anie.201410214.

Stranks, S.D. and Snaith, H.J. (2015) Metal-halide perovskites for photovoltaic and light-emitting devices. *Nature Nanotechnology*, **10** (5), 391–402. DOI: 10.1038/nnano.2015.90.

Tress, W., Marinova, N., Moehl, T. *et al.* (2015) Understanding the rate-dependent J–V hysteresis, slow time component, and aging in CH3NH3PbI3perovskite solar cells: the role of a compensated electric field. *Energy & Environmental Science*, **8** (3), 995–1004. DOI: 10.1039/c4ee03664f.

Unger, E.L., Bowring, A.R., Tassone, C.J. *et al.* (2014) Chloride in lead chloride-derived organo-metal halides for perovskite-absorber solar cells. *Chemistry of Materials*, **26** (24), 7158–7165. DOI: 10.1021/cm503828b.

Wehrenfennig, C., Eperon, G.E., Johnston, M.B. *et al.* (2014) High charge carrier mobilities and lifetimes in organolead trihalide perovskites. *Advanced Materials*, **26** (10), 1584–1589. DOI: 10.1002/adma.201305172.

Williams, S. T., Zuo, F., Chueh, C.C. *et al.* (2014) Role of chloride in the morphological evolution of organo-lead halide perovskite thin films. *ACS Nano*, **8** (10), 10640–10654. DOI: 10.1021/nn5041922.

Xiao, Z. G., Bi, C., Shao, Y.C. *et al.* (2014) Efficient, high yield perovskite photovoltaic devices grown by interdiffusion of solution-processed precursor stacking layers. *Energy & Environmental Science*, **7** (8), 2619–2623. DOI: 10.1039/C4ee01138d

Xing, G., Mathews, N., Sun, S. *et al.* (2013) Long-range balanced electron- and hole-transport lengths in organic-inorganic CH3NH3PbI3. *Science*, **342** (6156), 344–347. DOI: 10.1126/science.1243167.

Yang, W.S., Noh, J.H., Jeon, N.J. *et al.* (2015) High-performance photovoltaic perovskite layers fabricated through intramolecular exchange. *Science*, **348**, 6240. http://science.sciencemag.org/content/348/6240/1234.

6.4

Organic PV Module Design and Manufacturing

Veronique S. Gevaerts
Energy Research Centre of the Netherlands – Solliance, Eindhoven, The Netherlands

6.4.1 Introduction

6.4.1.1 From Cells to Modules

World-record organic photovoltaic (OPV) devices have reached efficiencies up to 12% and up-scaling and industrialization of the technology are now in progress. In academia, cell efficiencies of up to 10.6% in the lab (Li *et al.*, 2012; You *et al.*, 2013) have been published and 11.5% has been the latest certified record in the NREL Best Research-Cell Efficiencies chart. In industry, the Heliatek company has announced a cell efficiency record of 12%, and Certified efficiencies of 11% at the cell level have been demonstrated by both Toshiba and Mitsubishi Chemical. In the first step towards modules, Toshiba also holds the record certified efficiency of 9.7% for an organic mini-module (Green *et al.*, 2015).

Overall, OPV holds great promise to become a cost-effective renewable energy platform, because it can be deposited on flexible substrates using low temperatures, enabling roll-to-roll production. Since the technology is still relatively new and there is no commercial production yet, the estimated cost prices for OPV modules vary enormously. It is good to note that OPV is expected to create new markets with its unique selling points such as flexibility, semitransparency and freedom of color and shape. As with all PV technologies, there is a balance between power conversion efficiency and the lifetime of the modules versus the cost of production and requirements of the application.

In this chapter, several methods and processes to realize and manufacture OPV modules are described in three stages. First, general considerations for large-scale production of OPV modules are described. Then, different processes for manufacturing are explained, with an emphasis on the different wet processing technologies that have been successfully

Photovoltaic Solar Energy: From Fundamentals to Applications, First Edition.
Edited by Angèle Reinders, Pierre Verlinden, Wilfried van Sark, and Alexandre Freundlich.

Companion website: www.wiley.com/go/reinders/photovoltaic_solar_energy

demonstrated for OPV layers. Finally, different processes of making interconnections for modules will be described.

6.4.1.2 Advantages of R2R Production

There are two methods of applying the layers of an organic solar cell: (1) small organic molecules can be applied using thermal evaporation under reduced pressure; and (2) alternatively, soluble organic molecules or polymers can be applied using solution processing. For solution processing, the active materials are dissolved in a solvent (mixture), creating an ink which is applied using a coating or printing technique (see Section 6.4.2). Both evaporated small molecule OPV as well as solution-processed OPV are compatible with roll-to-roll production, though there are differences in equipment and the respective cost for materials and processes. Roll-to-roll (R2R) production is a production process in which a roll of flexible material is used as a substrate, which is unrolled and fed into the production line and re-reeled after the process to create an output roll. The main advantages of R2R production of solution-based OPVs are given in Table 6.4.1.

In general, cheap production using R2R processes is possible owing to the high process efficiency through the sequential application of successive layers on a web. Since OPV is a relatively young PV technology, the development of stable processes and the equipment needed for manufacturing is still ongoing. Evaporation of small molecule OPV stacks in a R2R production facility is challenging, due to the need for low vacuum processing and the optimization of uniform deposition on large areas. Solution processing of OPVs on an industrial scale brings other additional challenges, such as finding alternative, non-chlorinated solvents and using roll-to-roll compatible processes. The latter will be further discussed in the second part of this chapter.

6.4.1.3 Module Design for Organic PV

Where module design for traditional silicon PV is limited due to the use of wafers, organic PV module design has more flexibility as a result of the application of the sequential OPV layers on a substrate. Also semi-transparency is one of the unique features of OPV modules that can be exploited for design purposes. Modules can literally become flexible through the use of substrates such as PET foils, but also the shape of the cells and modules can be adjusted to fit the purpose. R2R production of OPV is still in the pilot phase and typically is done on a substrate with a width of 30 cm, as shown in Figure 6.4.1. The length

Table 6.4.1 Benefits and challenges for roll-to-roll solution processing

Benefits	Challenges
Steady state processing; high throughput and high yield	Development of stable processing and defect repair
Flexible and low weight	Shorter drying times needed
Better scaling	New equipment needs
Energy and materials efficiency	Patterning
Low substrates costs	Use of benign solvents

of these modules can be adjusted to fit the purpose. As mentioned before, R2R production is scalable and also the width of the modules can be adjusted, depending on the equipment available. Nevertheless, R2R production of OPVs is mainly suitable for large-scale, high-volume market areas such as façade elements for building integrated photovoltaics (BIPV), as shown in Figure 6.4.1.

Next to roll-to-roll manufacturing, roll-to-sheet as well as sheet-to-sheet production methods are of interest for low-volume applications that require more flexibility in module design. Two examples of these free-form modules are shown in Figure 6.4.2. The fact that the size, shape, color and transparency of OPV modules are free for the designer's choice, makes

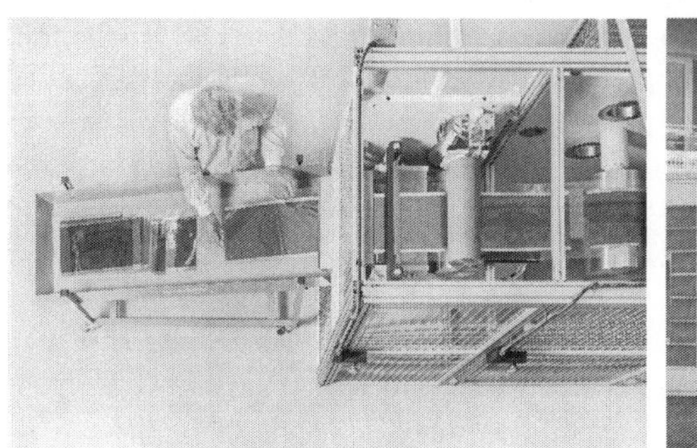

Figure 6.4.1 HeliaFilm™ encapsulation output and vision on low-cost solar cell in BIPV ©Heliatek. (*See insert for color representation of the figure*)

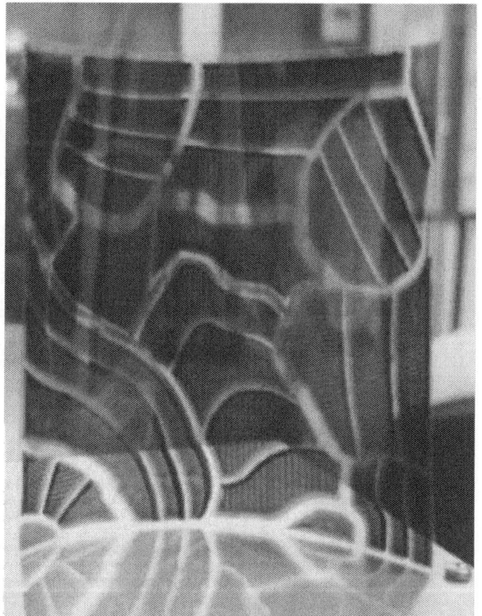

Figure 6.4.2 Two examples of free-form OPV modules; left, the Solarte garden lamp by Belectric OPV and, right, the world's first polychrome solar module by DisaSolar. (*See insert for color representation of the figure*)

custom-made OPV modules very attractive for architects and designers who want to integrate PV, for example, into the built environment.

6.4.2 Important Module Parameters

6.4.2.1 Cell Size

As a consequence of the low conductivity of the typical electrode and charge transport materials used in OPV, such as PEDOT:PSS, ITO and other metal oxides, the cell size that can be used efficiently in modules is limited to the centimeter scale, as Slooff *et al.* (2014) have shown when optimizing the silver grid lines that are used with PEDOT:PSS as a bottom electrode. These type of calculations are important when determining the cell dimensions that lead to the highest efficiencies of OPV modules.

6.4.2.2 Series Connection

By going from cells to modules, individual cells have to be connected in series and/or parallel. Serial interconnection is used most frequently in OPV modules and will be further discussed in Section 6.4.4 on interconnections. By using a serial interconnection, the voltages of the cell that are connected are added up (Sommer-Larsen *et al.*, 2013). At the Technical University of Denmark the *Infinity concept* has been developed that uses serial connections in large OPV modules that build up to a very high voltage. Krebs (Krebs *et al.*, 2014, p. 32) explains: "The rows that are 100 m long result in a voltage build-up along the row of >10 kV, achieved through serial connection of the 21 000 individual junctions along each lane of solar cell foil." These high voltages lead to increased need for safety precautions and in this case the need for a fence around the site.

6.4.2.3 Barrier Requirements

Another important parameter in the module production of OPV is the need for proper encapsulation to enhance the operational lifetime of the modules. OPVs are known for their degradation due to exposure to water and oxygen and therefore protection from the ingress of water and oxygen is essential. Where in crystalline silicon modules glass-glass encapsulation is used, which has great barrier properties, this is not the preferred solution for OPVs when the flexibility of these modules is considered such an important feature (Ahmad *et al.*, 2013). When OPV is processed on a flexible foil such as PET, typical bending angles of 60° are possible with ITO layers, and even higher when less rigid electrodes are used (Gomez de Arco *et al.*, 2010).

Flexible foils with oxygen and moisture barrier properties have been investigated to increase the lifetime of flexible organic electronics, such as the barrier made of alternating inorganic and organic layers on PET foil that was developed at the Holst Centre (van Assche *et al.*, 2008). The cost of these types of barrier foils is still relatively high compared to all the other materials used in the flexible OPV modules. The barrier and the adhesive actually make up the largest part of the weight and the volume. As a result, it has the major role in the energy balance, cost and environmental impact of OPV modules (Krebs *et al.*, 2014).

6.4.3 Wet Processing Technologies

In organic solar cells each layer has a specific function, such as the photoactive layer that converts light into charges and the electron or hole transporting layers as explained in Part 6. Each of these layers can be deposited in an ink using a variety of coating and printing processes. All functional layers have specific layer thicknesses and the specific processing of these different layers depends on the combination of the ink, the processing technology used, drying time, temperature, etc. All these parameters need to be optimized for each layer in the solar cell stack.

Going from cell processing to (roll-to-roll) module production, lab-scale technologies such as spin coating and blade coating need to be replaced by up-scalable coating and printing techniques such as slot-die coating, inkjet printing, and (rotary) screen printing. The ink formulations that can be used depend on the technique that is chosen and the layers that have been previously deposited. The commonest techniques for solution processing on a large scale are shown in Figure 6.4.3 and explained in the following sections.

Next to processing techniques, the ink formulations that are used for the application of the layers are a very important subject of research. The inks that are used should match the processing method and, for large-scale production, the solvents should be environmentally-friendly and nontoxic. The specific requirements for the ink vary with the technology that is chosen for the deposition, for example, inks for inkjet printing require viscosities that are in between 3 and 50 mPa·s, while for slot-die coating, the viscosity range is much wider and can be less than 1 mPa·s, but also several Pa·s. Sequential deposition of layers on top of each other, which is essential in OPVs, requires that the deposition of one ink does not destroy the functional layer underneath. This can be done by using combinations of so-called orthogonal solvents. A widely used combination is the use of organic hydrophobic solvents for the deposition of the photo-active layer on top of the water-soluble PEDOT:PSS layer. However, for other layers, it is less straightforward. Hence, careful optimization of the inks, the deposition method and the use of orthogonal solvents is key to the successful up-scaling of OPVs.

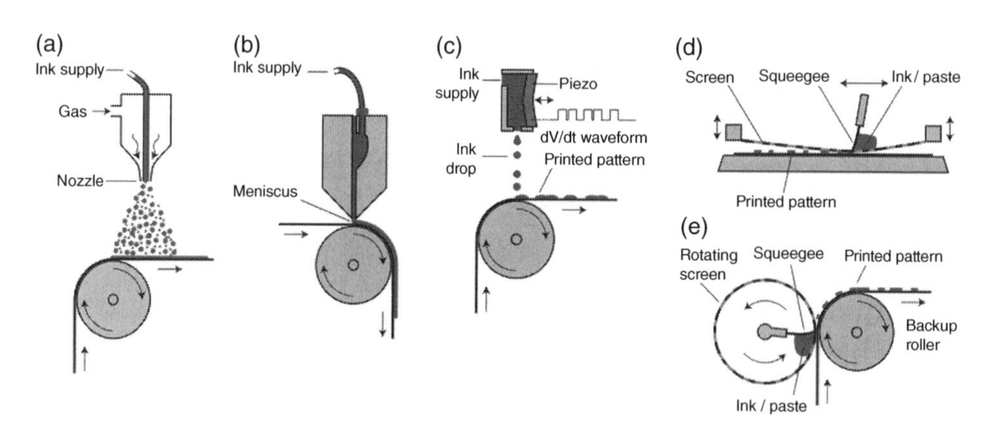

Figure 6.4.3 Schematic representations of (a) spray coating; (b) slot-die coating; (c) inkjet printing; (d) screen printing; and (e) rotary screen printing. Adapted from (Søndergaard *et al.*, 2013) © 2012 Wiley Periodicals, Inc

6.4.3.1 *Slot-Die Coating*

Homogeneous and uniform layers can be deposited on large areas using the slot-die coating process. Slot-die coating is a non-contact processing method that can use inks with a large variety in viscosities, including relatively volatile inks. The working principle is the deposition of a wet layer by supplying an ink through a die as shown in Figure 6.4.3. The film thickness obtained after drying depends on the flow rate of the ink, the coating width and the speed and the concentration of material in the ink. Stripe coating can be obtained by using slower speeds and an insert that splits the meniscus of the ink. In direct roll-to-roll OPV module production, this one-dimensional stripe coating is the preferred technique, since it allows for the alignment of the layers for series-connected cells in modules (Krebs, 2009a). Homogeneous coatings can also be used for module production when combined with (laser) scribing.

6.4.3.2 *Inkjet Printing*

Inkjet printing is probably the best-known wet processing technique, since nowadays it can be found in almost every household. The inkjet printing technique that is mostly used for OPVs is the drop-on-demand inkjet printing based on piezo-electrical elements as shown in Figure 6.4.3. The nozzle is filled with the ink that is maintained in the chamber due to surface tension and static pressure. When a voltage waveform is applied to the piezo element, a mechanical force is put on the chamber and droplets are ejected from the nozzle. Full OPV devices can be obtained by inkjet printing, as was shown by Eggenhuisen *et al.* (2015a).

One of the main challenges using inkjet printing for OPV is that the formation of the layers specific for the inkjet printing process (wetting and pinning) depends not only on the formulation of the ink that is used for the printing, but also on the surface energy of the layer/material that the layer is processed on. The formation of droplets and the merging of droplets on the surface are influenced by the print head, the printing parameters, and the ink formulation that is used. Also the interaction of the formed droplets with the surface and the environment plays an important role in the layers that are obtained.

Another challenge of inkjet printing on a large scale is the clogging of the nozzles, which needs careful optimization of the printing bar. In general, inkjet printing is a very complex technology that needs a lot of optimization of all the different parameters. On the other hand, inkjet printing is one of the few techniques that allows for waste-free printing and digital processing. Especially the latter makes inkjet very interesting for free form, smaller-scale, individualized production of OPV modules.

6.4.3.3 *Spray Coating*

Spray coating of OPV layers has been shown to be possible for lab-scale fabrication (Girotto *et al.*, 2011) and is compatible with high througput R2R processing. Similar to inkjet printing, there is low material waste for large area depositions. However, no patterning is possible without the use of shadow masking. In spray coating, as shown in Figure 6.4.3, the functional fluid or ink is forced out of the nozzle of the spray head using pressurized air or gas, which generates a continuous flow of droplets. The droplets hit the surface at high speed and dry quickly due to the small size of the droplets and the gas flow.

Spray coating allows for a broad range of solvent and material systems and the setup can be equipped with multiple spray heads. With this technique the surface tension, viscosity, fluid density, gas flow properties and nozzle design are important parameters for optimizing the spray production. To obtain uniform coatings with the sprayed inks, also the working distance, the coating speed, the droplet sizes, and all the interactions between the fluid and the surface are vital parameters to investigate. These are similar to the other techniques and include wetting behavior, surface properties, coating speed, spreading (of the droplets), and the surface temperature.

6.4.3.4 Other Techniques and Comparisons

Other techniques such as screen printing, gravure printing, or flexographic printing that are typically used in high volume traditional printing can also be used in the production of OPV modules. The most commonly used of those is screen printing.

Screen printing is commonly used in the manufacturing of printed circuit boards and the metallization of silicon solar cells and is illustrated in Figure 6.4.3 in the flatbed approach as well as the rotating application. First, the screen is covered with an ink by a floodbar, which fills the mesh. After that, a squeegee forces the ink through the open areas onto the substrate. In rotary screen printing, the cylindrical screen rotates with the same speed as the substrate, while the squeegee is kept at the same position, distributing the usually rather viscous ink. The features in the screen are transferred to the substrate, hence structured layers can be applied. This is also true for gravure and flexographic printing (Välimäki, 2015). However, since the structuring for all these techniques is based on the equipment, either a screen or a cylinder, changes to the structure involve a change in these elements.

When comparing the different techniques in Table 6.4.2, it is clear that there is no obvious choice for one single process for all processes and layers in an OPV layer stack. Hence, in the R2R solution processing of OPV, a combination of techniques is often used. All functional layers have specific layer thicknesses and a desired film morphology that need to be optimized for the chosen wet processing technique and ink. Also the accuracy of the positioning of the inks and the alignment of different layers depend on the combination of ink and technique used. In the end, all the parameters are optimized to lead to optimized solar module

Table 6.4.2 Comparing different major wet processing technologies

Technique	Webspeed	Coating options	Drawbacks
Slot-die coating	1-600 m/min	Uniform or stripe coating	Structuring only in stripes
Inkjet printing	>50 m/min	Full digital freedom	High requirements on inks
Spray coating	all speeds*	Uniform or patterned using shadow mask	Use of shadow masking in R2R is challenging
Screen printing	<180 m/min	Structured	New solar cell structure
Flexographic printing	>20 m/min	Structured	requires new equipment
Gravure printing	15-900 m/min	Structured	

*depending on the exact spray coating technology

efficiencies. An extensive review of all different coating and printing techniques was produced for solution-processed OPV specifically (Krebs, 2009b).

6.4.4 Interconnections

The sizes of efficient OPV cells are limited to about 1 cm in width. The resistance of the transparent conductive oxide (TCO) that is typically used is the limiting factor. For that reason, interconnections between solar cells are an important aspect of module fabrication. The interconnections that are used in modules are made by connecting the bottom electrode of one cell to the top electrode of the next cell, creating an electrical series connection. Through this series connection, the area of the module is split into cells. The area that is needed to create the interconnection does not contribute to the power output of the module and is referred to as the dead area. The active area or zone is the area that does contribute to the power and is the part where the layer stack is complete. The ratio of active area to total area of the modules is referred to as the geometric fill factor (GFF).

The series interconnection of cells results in lower current and a higher voltage output of the module, since for an electrical series connection, the voltages are added, while the currents should be identical. The total power output of the modules is the maximum power point current and voltage multiplied together. Increasing the number of cells in the module decreases the area per cell and hence lowers the current output of the module. Though this is compensated for by the increase in voltage, there is a trade-off due to the loss of active area.

There are several options to create interconnections between adjacent cells in an OPV module. Two of the most common are shown in Figure 6.4.4. In order to directly make interconnections using only (wet) processing techniques, structuring of the applied layers is needed as shown on the left side of Figure 6.4.4. This structuring is referred to as tiled. Alternatively, interconnections can be made by combining uniformly coated layers with the partial removal of the layers, also known as scribing, in between the (wet) processing steps. Both of these options can also be used in evaporated OPV modules, for the tiled structure, shadow masks are used.

The in-process scribing that is shown in Figure 6.4.4 (right) shows three labeled scribes that are used to create the interconnection. The P1, P2, P3 numbering of the scribes is related to the order in which they are made in the in-process procedure; however, they are now used as a convention for their function. The P1 scribe isolates the bottom electrode of the neighboring cells in the module, the P2 creates the interconnection between the top electrode of one cell to the bottom electrode of the other cell, and the P3 isolates the top electrode of the two neighbouring cells.

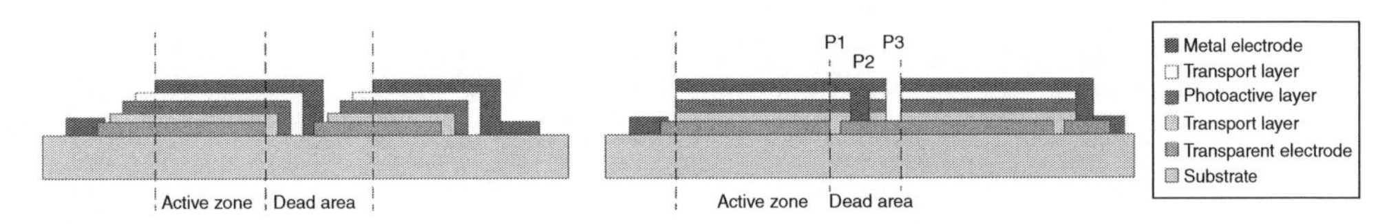

Figure 6.4.4 Schematic representations of interconnections using only wet processing techniques (left) and P1, P2, P3 in-process scribing (right)

6.4.4.1 Tiled Coating

Of all the wet processing techniques described before, the slot-die stripe coating is the most obvious choice for direct processing of interconnected modules. OPV modules have been shown that can use up to three sequential layers using inkjet printing (Eggenhuisen *et al.*, 2015b) and four layers using slot-die coating (Krebs, 2009a). In these modules the interconnection is made by the tiled application of stripes of active layers, creating a direct contact by application of the top metal electrode to a free section in the bottom electode. Next to slot-die stripe coating and inkjet printing, also other coating methods can be used (Välimäki *et al.*, 2015), as long as direct structuring of the applied layers is possible as indicated in Table 6.4.2.

The main drawback of this tiled coating for wet processing techniques is that a large area is needed for the interconnection, due to the low resolution of the coated line. The tiled coating means that wetting and pinning of the layers have to deal with interactions between the ink with the previously deposited layer and the substrate simultaneously. These tiled coatings typically have a 8 mm-wide interconnection zone (the dead area) and a similar cell width, which results in an active area which is only about 50% of the total module area. No significant improvements are expected on the width needed for interconnection in this tiled approach using wet processing techniques. However, development in the conductivity of the electrodes could increase the cell width and with that increase the GFF.

6.4.4.2 In-Process Scribing

The second most used option is combining full area homogeneous coatings with P1, P2 and P3 scribes, as indicated in Figure 6.4.4 (right). The scribing can be done by either mechanical scribing, by making a scratch in the layers with a stylus, or by laser ablation of part of the layer (Gebhardt *et al.*, 2013). With laser scribing, line widths below the mm scale can be obtained. Typical widths of a laser-scribed interconnection that have been published are between 100 and 500 μm, which results in >95% active area of the module (Kubis *et al.*, 2014). Although this can result in much higher power output of the module, it also makes the processing of the modules more complicated. Instead of processing layer after layer, a scribe has to be made inbetween the processing of layers. This invokes the need for cleaning inbetween steps, since the material that is removed will inevitably also end-up on the substrate. Also alignment becomes an issue, which is a costly and time-consuming step in high-throughput production.

6.4.4.3 Back-end (Laser) Processing

As an alternative to the inline fabrication of interconnections, a back-end approach can be used. In this approach, the large area production can be separated from the production of the modules. The module production can be digitally mastered, a specialized production which enables more freedom in the module design, especially when laser scribing and, for example, inkjet printing are combined.

A back-end interconnection can be created on full area deposited layers by first scribing a P1 through all layers, including the bottom contact, and scribing a P2 and a P3 by the selective removal of the layer stack without removing the bottom contact. Then P1 and P3 are filled

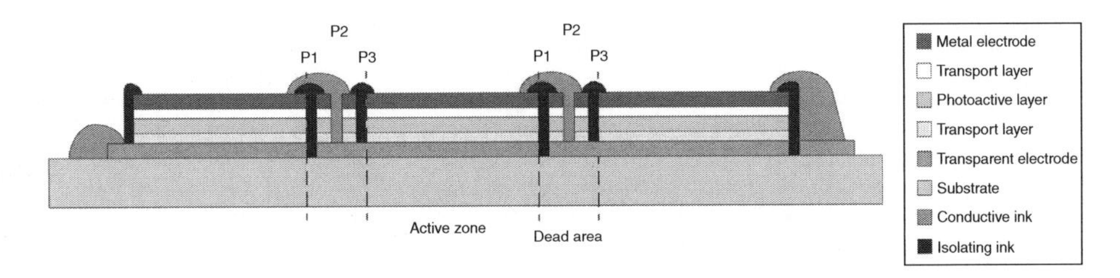

Figure 6.4.5 Schematic representations of back-end interconnections by combining P1,P2,P3 scribes and isolating and conductive inks

with isolating material and a conductive material is deposited onto the P2 and over the isolated P1 onto the top contact (Wipliez *et al.*, 2011). In this way, an interconnection is created between the top contact of the cell next to the P1 and the bottom contact of the cell adjacent to the P3, as indicated in Figure 6.4.5. This technology, however, is still relatively new for OPV and has not been demonstated in the literature yet.

6.4.5 Future Outlook

The progress in up-scaling production of OPV has been tremendous in the past five years. Though the market prospects for PV in general have not been too bright in the last few years, OPV is expected to create new markets owing to their unique selling points such as freedom of shape, color, and transparency. For the creation of OPV modules, a trend is seen towards the use of lasers in scribing for interconnections, which leads to increasing the power output and enables the creative design of modules. Though OPV modules are not commercially available yet, progress is being made in both evaporated as well as solution-processed OPV modules. The first pilot production plants have been built and products are being showcased and are expected to be commercialized soon.

List of Acronyms

Acronym	Description
BIPV	Building integrated photovoltaics
GFF	Geometric fill factor
ITO	Indium tin oxide
OPV	Organic photovoltaics
PEDOT:PSS	Poly(3,4-etheylendioxythiophene):poly(styrenesulfonate), a conductive polymer blend often dispersed in water
PET	Polyethylene terephthalate
R2R	Roll-to-roll
TCO	Transparent conductive oxide
V	Voltage

References

Ahmad, J., Bazaka, K., Anderson, L.J. *et al.* (2013) Materials and methods for encapsulation of OPV: A review. *Renewable and Sustainable Energy Reviews*, **27**, 104–117.

Assche, F. van, Rooms, H., Young, E. *et al.* (2008) Thin-film barrier on foil for organic LED lamps. Paper presented at AIMCAL Fall Technical Conference and 22nd International Vacuum Web Coating Conference, Myrtle Beach, NC.

Eggenhuisen, T.M., Galagan, Y., Biezemans, A.F.K.V. *et al.* (2015a) High efficiency, fully inkjet printed organic solar cells with freedom of design. *Journal of Materials Chemistry A*, **3**, 7255–7262.

Eggenhuisen,T.M., Galagan, Y., Coenen, E.W.C. *et al.* (2015b) Digital fabrication of organic solar cells by inkjet printing using non-halogenated solvents. *Solar Energy Material and Solar Cells*, **134**, 364–372.

Gebhardt, M., Hänel, J., Allenstein, F. *et al.* (2013) Laser structuring of flexible organic solar cells: manufacturing of polymer solar cells in a cost-effective roll-to-roll process. *Laser Technik Journal*, **10**, 25–28.

Girotto, C., Moia, D., Rand, B.P., and Heremans, P. (2011) High-performance organic solar cells with spray-coated hole-transport and active layers. *Advanced Functional Materials*, **21**, 64–71.

Gomez de Arco, L., Zhang, Y., Schlenker, C.W. *et al.* (2010) Continuous, highly flexible, and transparent graphene films by chemical vapor deposition for organic photovoltaics. *ACS Nano*, **4**, 2865–2873.

Green, M.A., Emery, K., Hishikawa, Y. *et al.* (2015) Solar cell efficiency tables (version 46) *Progress in Photovoltaics: Research and Applications*, **23**, 805–812.

Krebs, F.C. (2009a) All solution roll-to-roll processed polymer solar cells free from indium-tin-oxide and vacuum coating steps. *Organic Electronics*, **10**, 761–768.

Krebs, F.C. (2009b) Fabrication and processing of polymer solar cells: a review of printing and coating techniques. *Solar Energy Material and Solar Cells*, **93**, 394–412.

Krebs, F.C., Espinosa,N., Hösel, M. *et al.* (2014) 25th anniversary article: rise to power – OPV-based solar parks. *Advanced Materials*, **26**, 29–39.

Kubis, P., Lucera, L., Machui, F. *et al.* (2014) High precision processing of flexible P3HT/PCBM modules with geometric fill factor over 95%. *Organic Electronics*, **15**, 2256–2263.

Li, G., Zhu, R., and Yang, Y. (2012) Polymer solar cells. *Nature Photonics*, **6**, 153–161.

Slooff, L.H., Veenstra, S.C., Kroon, J.M. *et al.* (2014) Describing the light intensity dependence of polymer:fullerene solar cells using an adapted Shockley diode model. *Physical Chemistry Chemical Physics*, **16**, 5732–5738.

Sommer-Larsen, P., Jørgensen, M., Søndergaard, R.R. *et al.* (2013) It is all in the pattern: high-efficiency power extraction from polymer solar cells through high-voltage serial connection. *Energy Technology*, **1**, 15–19.

Søndergaard, R.R., Hösel, M., and Krebs, F.C. (2013) Roll-to-roll fabrication of large area functional organic materials. *Journal of Polymer Science B, Polymer Physics*, **51**, 16–34.

Välimäki, M., Apilo, P, Po, R. *et al.* (2015) R2R-printed inverted OPV modules – towards arbitrary patterned designs. *Nanoscale*, **7**, 9570–9579.

Wipliez, L., Löffler J., Heijna, M.C.R. *et al.* (2011) Monolithic series interconnection of flexible thin-film pv devices. Paper presented at 26th European Photovoltaic Solar Energy Conference and Exhibition, Hamburg, Germany.

You, J., Dou, L., Yoshimura, K. *et al.* (2013) A polymer tandem solar cell with 10.6% power conversion efficiency. *Nature Communications*, **4**, 1446.

Part Seven

Characterization and Measurements Methods

7.1

Methods and Instruments for the Characterization of Solar Cells

Halden Field
PV Measurements, Inc., Boulder, CO, USA

7.1.1 Introduction

Solar cell characterization instruments and techniques enable users to assess device performance, understand factors affecting performance, and characterize properties of device materials. This chapter introduces spectral response, current voltage (I-V) measurements, and associated techniques. It uses example data to illustrate central concepts and to portray common challenges in practical use.

7.1.2 External Quantum Efficiency

Also known as "Spectral Response," "Quantum Efficiency," or simply "QE," this measurement technique reveals the performance of the solar cell in the short-circuit condition as a function of wavelength of incident light. In a minimal system, broad spectrum light passes through a monochromator and focussing optics to illuminate a test device. Signal detection circuitry holds the device at a fixed voltage and measures the generated current. A computer compares the generated current to the incident photon flux to compute the conversion efficiency. The measurement repeats for all wavelengths of interest.

7.1.2.1 Apparatus

Most QE systems incorporate a bias light capable of one-sun intensity to enable the solar cell to operate under intended end-use conditions during the measurement. They also utilize a mechanical chopper to enable measurement circuitry to distinguish current generated by

Photovoltaic Solar Energy: From Fundamentals to Applications, First Edition.
Edited by Angèle Reinders, Pierre Verlinden, Wilfried van Sark, and Alexandre Freundlich.

Companion website: www.wiley.com/go/reinders/photovoltaic_solar_energy

monochromatic light from that generated by bias light, stray light, and noise. Most QE system monochromators employ a diffraction grating to separate the wavelengths of light. Rotation of the grating enables the light from selected wavelengths to reach a fixed exit slit. Most QE system monochromators contain multiple, interchangeable gratings to efficiently span the wavelength range to which solar cells respond. The additional stray light suppression offered by double monochromators is rarely needed. QE system monochromators use optical filters to reject light at harmonics of the selected wavelength. Finally, to enable compensation for light intensity variation between time of calibration and time of measurement, a fraction of the monochromatic beam diverges to a monitor detector that connects to a second channel of the current measurement circuitry. Current measurement circuitry commonly uses the lock-in technique with a synchronization signal from the mechanical chopper. Test device voltage is held at zero. Figure 7.1.1 illustrates the components and optical configuration of a QE system used for solar cell measurements.

7.1.2.2 Calibration

To calibrate an electrical current measurement apparatus, one can apply current from a known current source, compare instrument readings to true values, and adjust multiplicative calibration factors to ensure accurate readings. The number of photons reaching a test device is commonly determined by measuring the response of one or more photodiodes of known characteristics. This procedure has variants that depend on whether the system operates in power or irradiance mode.

A QE system operating in irradiance mode utilizes a beam homogenizer to apply a beam of light to the test device, irradiating the test device with light that is nearly uniform over its entire surface. The extent of its uniformity is described in terms of limits of non-uniformity, e.g., +/– 2%. In this configuration, the calibration photodiode typically contains an aperture that enables it to sample the irradiance of the incident beam. Irradiance mode gives more-accurate measurements for devices with substantial spatial variations in responsivity. However, it introduces an uncertainty proportional to the spatial non-uniformity of the monochromatic beam. In power

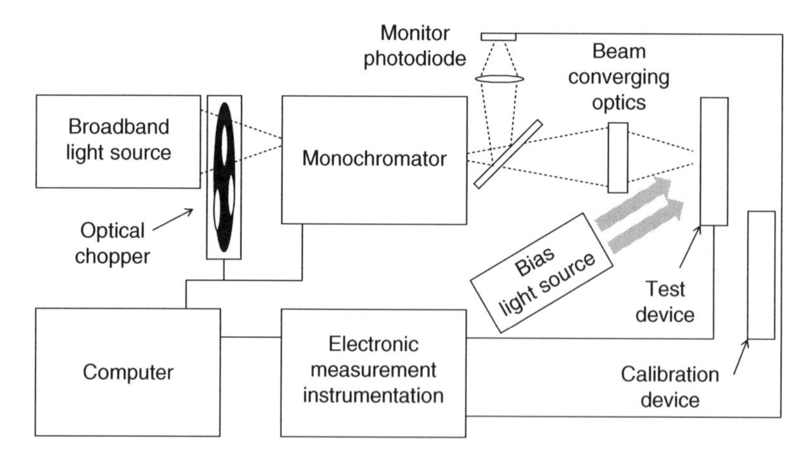

Figure 7.1.1 Schematic representation of a solar cell QE measurement configuration

mode, the entire monochromatic beam is smaller than the active region of the photodiode, enabling it to measure all the light applied to the solar cell. This mode gives the most accurate measurement, but only for the region of the device that the monochromatic beam illuminates.

When operating in power mode, a QE system computes the solar cell's QE in percent units according to

$$QE(\lambda) = 100 \times \frac{I_{TD}(\lambda)}{E_B(\lambda)} \times \frac{hc}{\lambda q} \tag{7.1.1}$$

where $I_{TD}(\lambda)$ is the current generated by the test device at each wavelength in amperes, $E_B(\lambda)$ is the monochromatic beam power in watts, λ is the wavelength in meters, q is the elementary charge ($1.602\ 10^{-19}$ C), h is Planck's constant ($6.626\ 10^{-34}$ J.sec), and c is the speed of light ($2.998\ 10^{8}$ m.sec^{-1}). This is also called External Quantum Efficiency (EQE) as it includes both reflectance losses and internal carrier recombination losses.

A QE system can also determine a test device's QE by simply comparing the current it generates to the current generated by a photodiode of known QE (the calibration photodiode) during the calibration scan performed prior to the test device measurement:

$$QE(\lambda) = \frac{I_{TD}(\lambda)}{I_{REF}(\lambda)} \times QE_{REF}(\lambda) \tag{7.1.2}$$

Here, I_{REF} is the calibration current measured during the calibration scan and QE_{REF} is the known QE of the calibration photodiode.

Light intensity in a QE system drifts due to: (1) changes in mechanical alignment as temperatures vary; (2) changes in the reflectance of optical coatings as they interact with the atmosphere; and (3) changes in lamp spectrum and efficiency. Therefore, system calibrations are often performed many times per day. Simply measuring the QE of a calibration photodiode and comparing its measurement results to its known characteristics enables the system calibration factors to accommodate light intensity changes as well as any changes in the electrical measurement circuitry.

7.1.2.3 Using Reflectance Data to Determine Internal Quantum Efficiency

By adding an integrating sphere with an appropriate irradiance sensor, one can enable a QE system to measure sample reflectance. The integrating sphere design must consider that solar cells typically have a combination of specular and diffuse reflectance characteristics. With reflectance information provided by a QE system accessory or a separate spectroradiometer, one can subtract the reflected light fraction from the incident light and determine the fraction of photons that enter the cell that are converted to device current. This quantity is known as IQE or Internal Quantum Efficiency:

$$IQE(\lambda) = \frac{EQE(\lambda)}{(1 - R(\lambda))} \tag{7.1.3}$$

where $R(\lambda)$ is the sample reflectance.

7.1.2.4 *Alternative Optical Configurations, Components, and Features*

Other optical configurations in use for solar cell and module QE measurements include:

- pulsed light instead of continuous;
- optical filters instead of a grating monochromator with harmonic suppression filters;
- a set of LEDs of various wavelengths instead of a light source with a monochromator.

Additional features common in QE systems used for solar cell measurements include:

- spectrally selective bias light using LEDs or optical filters;
- voltage bias capability in the current measurement instrumentation.

7.1.2.5 *QE Measurement Data*

Figure 7.1.2 shows the typical quantum efficiency of a mass-produced silicon solar cell. It shows that many UV photons do not result in device current, the vast majority of those in the visible and near-infrared do and those in the silicon bandgap region convert with rapidly decreasing efficiency as their wavelength increases. The highly energetic photons from the UV range are often absorbed by the dielectric material forming the anti-reflection coating of the cell, or are absorbed very close to the surface of the semiconductor creating electron–holes pairs that recombine easily due to the surface recombination velocity. The low-energy photons, with energy close to the bandgap, are poorly absorbed by the semiconductor and often escape before generating electron–hole pairs. In both cases, it results in low QE values.

Solar cells must be maintained at a constant temperature, typically at the standard temperature of 25 °C, during the entire QE measurement. At higher temperatures, the bandgap of the semiconductor will reduce and the solar cells will better absorb light of longer wavelengths. This results in an increased QE in the bandgap region of wavelengths, increasing the device's short-circuit current J_{sc} and decreasing the open-circuit voltage V_{oc}.

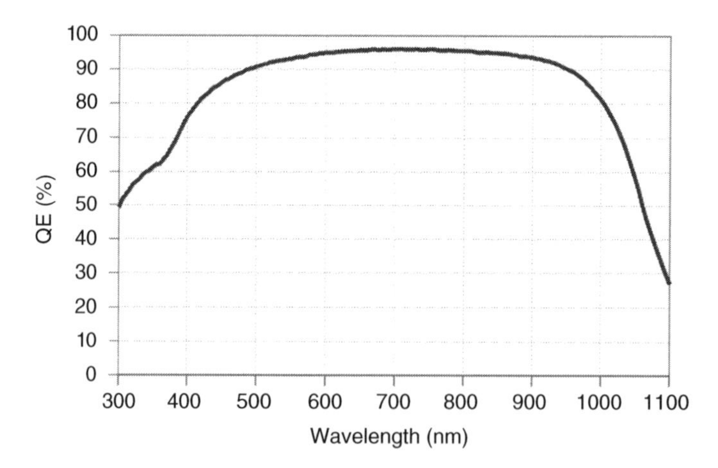

Figure 7.1.2 Example of silicon solar cell QE measurement result (measured at 25 °C)

QE information is normally graphed in percent or 0–1 units for quick portrayal of the performance of the device relative to ideal limits. To facilitate computations to determine spectral mismatch or theoretical current generation, one often converts it to Spectral Responsivity (*SR*) (in A/W) using

$$SR(\lambda) = \frac{QE}{100} \times \frac{\lambda q}{hc} = \frac{QE \times \lambda}{124{,}000} \tag{7.1.4}$$

where *QE* is as defined in Equation (7.1.1). Note, λ is wavelength in nm (Equation 7.1.1 used meters; nanometers is used here for convenience). Additional information regarding QE can be found in Honsberg and Bowden (2013a).

7.1.2.6 Determining Solar Cell Current for a Particular Light Source

LED illumination can show how light spectra and solar cell QE interact to cause solar cell current generation. Figure 7.1.3 shows the spectral content of light from a blue LED and an infrared LED plotted with the AM1.5G solar spectrum. The total optical energy of each of the three light sources is 100 mW/cm². The spectral response of the cell in Figure 7.1.2 is included to illustrate how well the solar cell converts the energy from the different sources.

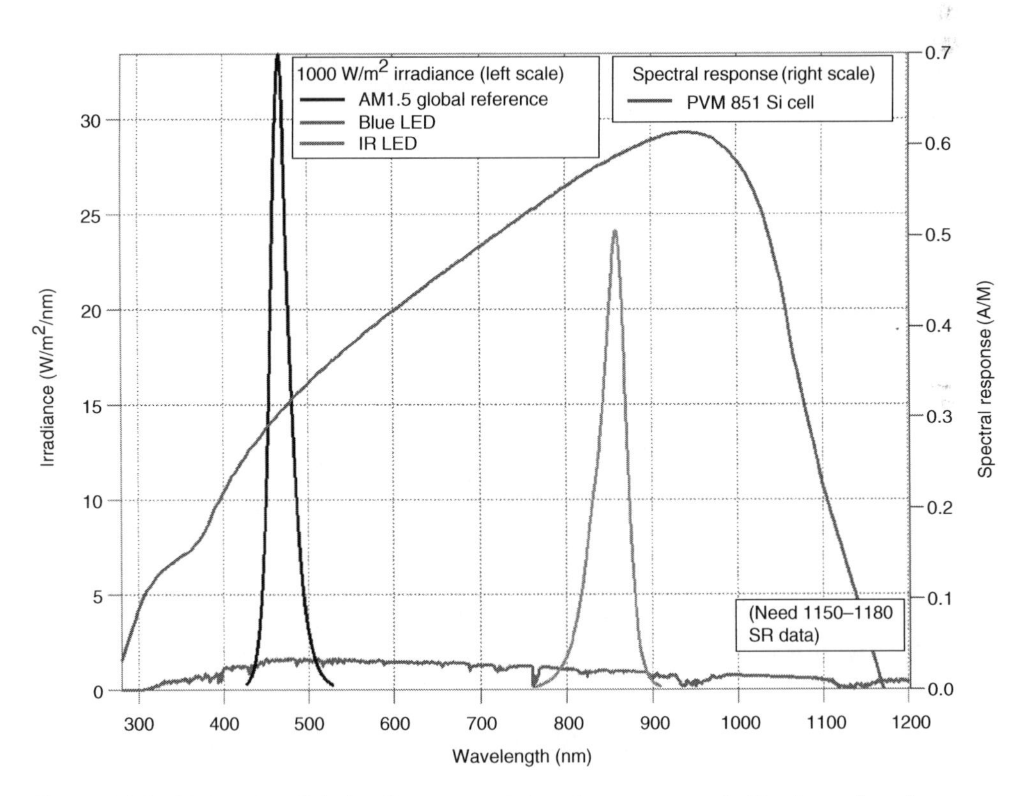

Figure 7.1.3 LED and sunlight irradiance spectra (equal energy content). (*See insert for color representation of the figure*)

The electrical current generated by the solar cell of Figure 7.1.1 when illuminated by the blue or infrared LED is the product of the incident light photon flux times the cell's quantum efficiency, integrated over the wavelengths in the LEDs' emission spectra.

$$I_{BL} = qA \int_{0.426}^{0.530} \Phi_{BL}(\lambda) \times QE(\lambda)\, d\lambda \tag{7.1.5}$$

or

$$I_{IRL} = qA \int_{0.761}^{0.910} \Phi_{IRL}(\lambda) \times QE(\lambda)\, d\lambda \tag{7.1.6}$$

where I is the current in mA, q is the charge of electron, A the illuminated area and Φ is the LED's photon flux (in $sec^{-1}cm^{-2}\mu m^{-1}$) (BL and IRL subscripts denote blue LED and infrared LED respectively). In this example, the solar cell generates I_{BL} mA when illuminated by the blue LED and I_{IRL} mA when illuminated by the infrared LED.

Equivalently, this current can be expressed as the integral of the light source total power times the solar cell's spectral response:

$$I_{BL} = A \int_{0.426}^{0.530} E_{BL}(\lambda) \times SR(\lambda)\, d\lambda \tag{7.1.7}$$

or

$$I_{IRL} = A \int_{0.761}^{0.910} E_{IRL}(\lambda) \times SR(\lambda)\, d\lambda \tag{7.1.8}$$

where E is the LED's power (in $Wm^{-2}\mu m^{-1}$).

By this principle, one can determine the short-circuit current that a solar cell generates when illuminated by any light source if one knows the spectral irradiance of the light source and the QE of the solar cell. In the case of a solar cell illuminated by a standard irradiance of AM1.5G (ASTM E927-10 (ASTM, 2010), the equation becomes

$$J_{sc} = \int_{\lambda_1}^{\lambda_2} E_0(\lambda) \times SR_{td}(\lambda)\, d\lambda \tag{7.1.9}$$

This equation resembles the equation used to illustrate the LED case and has the quantities of irradiance and response in different units. Irradiance E_0 is in W/m²/nm instead of flux units and spectral response SR_{td} is in A/W units instead of QE. The result, J_{sc}, is in A/m² units and is commonly converted to mA/cm². The PV device current is often presented and discussed using area-normalized units to enable easy comparison of different-sized devices.

7.1.3 Energy Conversion Efficiency

From many perspectives, the most important solar cell characterization parameter is its energy conversion efficiency. As with most energy conversion systems, a solar cell's efficiency is expressed as the ratio of energy in and energy out. A solar cell's energy input is the energy contained in the illumination light. Its output is in the form of electrical energy. The efficiency, η, is determined using the efficiency equation:

$$\eta = \frac{P_{max}}{E_{tot} \times A} \tag{7.1.10}$$

where P_{max} is the maximum electrical power (in W) generated by the solar cell, E_{tot} is the total irradiance (in Wm^{-2}) reaching it, and A is the device's surface area.

In the QE examples discussed above, the quantum efficiency of the solar cell exceeds 90% over a wide range of wavelengths, yet its overall solar energy conversion efficiency is less than 20%. This is because the voltage magnitude accompanying the current generated by the solar cell is governed by the solar cell's bandgap energy level which is lower than the energy of the incident photons.

7.1.3.1 I-V Curve Introduction

Solar cell I-V curves portray the relationship between output current and voltage of the test device at a particular set of illumination and temperature conditions. The I-V curve of an illuminated solar cell resembles that of a diode (Figure 7.1.4 (a), see Figure 7.1.5) which has been inverted to portray positive current as that generated by the device rather than applied to it (Figure 7.1.4 (b)) and then offset to represent that a current source exists electrically in parallel with the diode (Figure 7.1.4 (c)). Figure 7.1.4 (d) includes a power curve to illustrate the relationship between the device's power output and its operating voltage.

The conventional inversion of the I-V curve is often not performed for scientific presentations of devices and materials. It is generally included in discussions of device performance, especially that of PV modules and systems.

To be useful for performance optimization, a solar cell model includes not only a diode and current source as implied by the I-V curve convention above, but also series and shunt resistances that are minimized and maximized respectively in device design, Figure 7.1.5 exaggerates the size of the diode circuit symbol to emphasize that the entire solar cell is the diode.

Figure 7.1.6 shows I-V curves that represent the performance of a silicon solar cell under LED illumination. The solar cell's efficiencies are 13.7% when illuminated by the blue LED, 14.5% with AM1.5G simulated sunlight illumination, and 27.0% with illumination from the IR LED. In the QE system, the conversion efficiencies were zero because the device voltage was held at zero.

7.1.3.2 Illumination for I-V Curves

Most solar cells are intended for use with natural sunlight illumination which varies widely in intensity and spectrum. As illustrated in the example above using LEDs, a solar cell's performance depends on its spectral content. It also depends on the illumination intensity. With such widely varying end-use conditions, it is difficult to obtain a single number that represents a solar cell's performance. But since having such a performance number is valuable for documenting technological progress, the industry has adopted a set of measurement conditions which is, despite its shortcomings, widely used in the industry. An irradiance level of

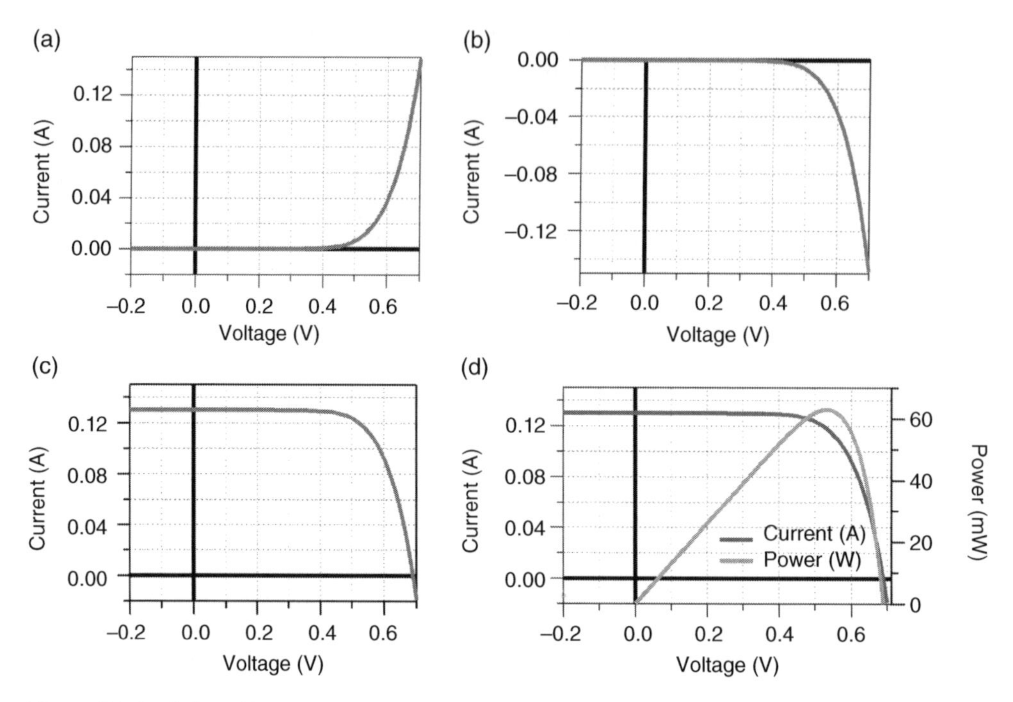

Figure 7.1.4 Solar cell I-V curve conventions: (a) Conventional diode I-V curve; (b) Diode I-V curve with current polarity inverted; (c) Inverted diode curve shifted for current generation; (d) Power curve added. (*See insert for color representation of the figure*)

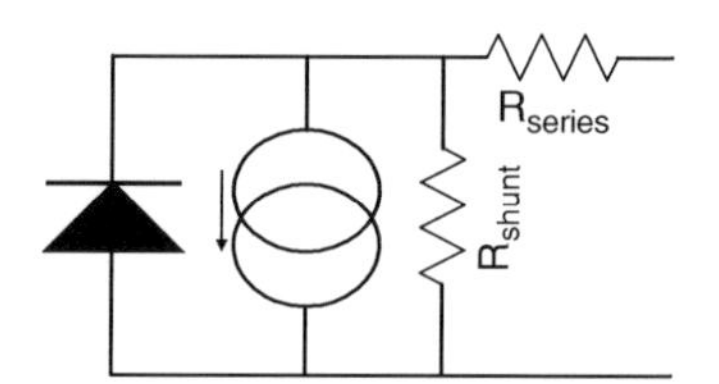

Figure 7.1.5 Solar cell diode model

1000 W/m^2 is used to represent a full-sunlight condition. A standard sunlight spectrum called AM1.5G, derived from measurements in space and an atmospheric transmittance model, represents full-sky illumination, i.e. without clouds, with a 37 degree path through a standard atmosphere to the solar cell located at sea level (ASTM E927-10, ASTM, 2010). An operating temperature of 25 °C represents conditions that are easy to obtain in a laboratory (although rarely encountered in conjunction with the other conditions in typical PV module operation). Figure 7.1.3 shows the spectral irradiance of the LEDs considered in the example, standard sunlight, and the QE of a silicon solar cell.

Recognizing the shortcomings of laboratory standard test conditions, those working with PV module performance use a different set of more-realistic standard conditions when characterizing and predicting the performance of fields of PV modules. PV modules typically operate at temperatures in excess of 45 °C which is significantly hotter than ambient laboratory conditions. Also, PV modules produce power from sunrise to sunset, encountering irradiance

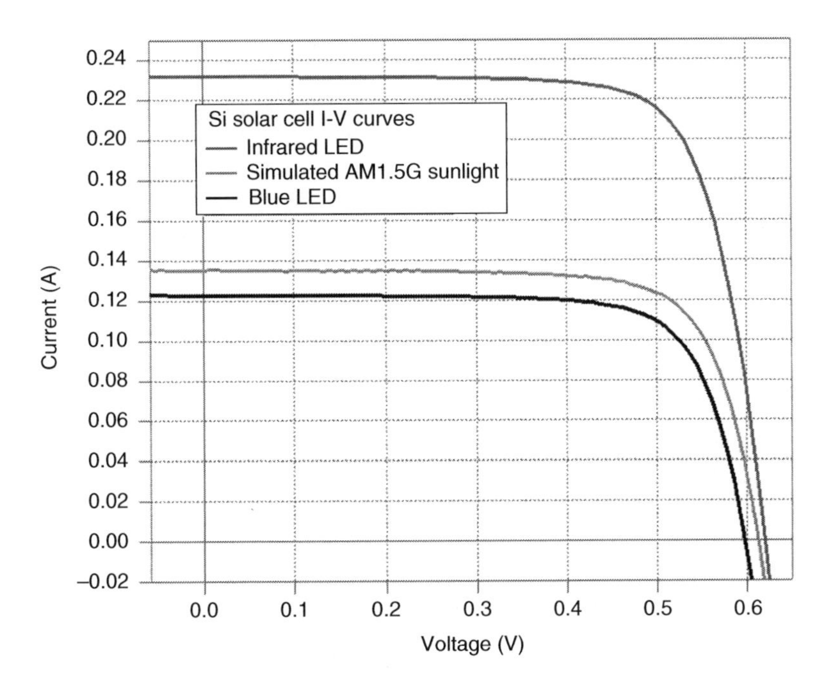

Figure 7.1.6 Solar cell I-V curves with LED illumination (a) blue LED, (b) infrared LED, (c) simulated AM1.5G sunlight. (*See insert for color representation of the figure*)

levels between zero and often exceeding the one-sun laboratory condition. Furthermore, the solar spectrum illuminating PV modules varies with time, location, and atmospheric conditions. Since PV field investors require precise forecasts of the energy output of PV fields, the planning process utilizes a more complete model of the PV panels characterizing their performance at different intensities and temperatures, as well as site-specific historical irradiance and weather data to calculate via software the predictable annual electrical energy generated by the PV system as described in Chapter 7.5 and Chapter 11.5.

7.1.3.3 I-V Curve Measurement Apparatus: Light Sources

I-V curve measurement systems incorporate a light source to simulate sunlight, a fixture to hold the solar cell while maintaining a constant cell temperature, and electronic apparatus to vary the load and measure current and voltage. Using standard reporting conditions for I-V measurements enables results in one laboratory to be meaningfully compared to those obtained in another laboratory. One aspect of the standard conditions is the light source, representing E_0 in the efficiency equation.

The irradiance spectrum component of the standard reporting conditions used for solar cell efficiency measurements never occurs in nature and cannot be synthesized in the laboratory. However, approximations are achieved using lamps and optical filters in special light sources called "Solar Simulators." In addition to meeting the challenges of approximating the spectral distribution of standard sunlight, these devices must use optical homogenizers to cause the light to have approximately the same intensity over the entire surface of the solar cell. Furthermore,

Table 7.1.1 Classification parameters of solar simulators

Classification	Spectral match	Spatial non-uniformity (%)	Temporal stability (%)
A	0.75–1.25	2	2 (1)
B	0.6–1.4	5 (3)	5 (3)
C	0.4–2.0	10	10

the light must have minimal variations in intensity over time. Standards ASTM E 927-10 (ASTM, 2010), IEC 60904-9 (IEC, 2007), JIS C 8912 (JIS, 2011a), and JIS C 8933 (JIS, 2011b) describe levels of solar simulator performance that correspond to solar simulator classifications. While the requirements associated with the classifications are nearly the same in all four standards, the techniques required to assess the performance vary. Classifications incorporate three letters, each ranging from A to C with A representing more-stringent criteria. The letters represent spectral match, spatial uniformity, and temporal stability respectively (Table 7.1.1).

The spectral match number represents the ratio of the amount of solar simulator irradiance in a spectral band relative to the amount of irradiance in the standard spectrum, usually AM1.5G (ASTM E927-10) (ASTM, 2010). To qualify for the classification, the solar simulator's ratios for all the bands must be within the range in Table 7.1.1. For the first three standards listed, five bands are 100 nm wide from 400–900 nm and the sixth spans from 900–1100 nm. The fourth standard is used when measuring amorphous silicon solar cells and has more bands and narrower bands covering a smaller spectral range.

Spatial non-uniformity expresses the maximum deviation of irradiance intensity relative to the average of the brightest and dimmest measurements in the illumination area. Similarly, temporal stability expresses the maximum excursion of intensity from the average intensity during the time period expected for a measurement.

When selecting a solar simulator, purchasers should consult and understand the latest version of the relevant standard, understand its applicability to their needs, and ensure that the seller's claims are based on the version consulted.

Typically, a xenon arc lamp with a current-regulated power supply provides the light while illumination optics and lens arrays convey the light to the test device. Optical filters reflect away or absorb portions of the spectrum, enabling the remainder to resemble sunlight.

Advancing LED technology enables solar simulators to use a set of LEDs in lieu of a xenon arc lamp. The LEDs are chosen to cover the range of wavelengths to which the solar cells to be measured are sensitive. Users of such solar simulators can adjust the illumination spectrum by controlling the current to subsets of the LEDs. Users should monitor the spectrum to prevent unknown spectral changes due to LED degradation and spectral shift that occurs with age and varying temperature. When evaluating spatial uniformity of illumination, users should incorporate a spectroradiometer to ensure that the light from multiple LEDs is sufficiently mixed at all locations in the illumination region.

7.1.3.4 I-V Curve Measurement Apparatus: Temperature Control

An I-V curve measurement system includes a fixture to hold the solar cells at a constant temperature, typically 25 °C, and provide an electrical connection to their terminals while minimizing the presence of components between the solar cell surface and the light source to prevent device shadowing. If possible, it also provides a facility to control and adjust the device temperature, commonly between 10 °C and 100 °C, to enable the study of variations in solar cell performance at the different operating temperatures likely to be encountered in common applications.

When measuring conventional silicon solar cells with a single rear terminal (usually the entire back surface), one can use an electrically-conductive vacuum chuck to achieve electrical contact and thermal control. Solar cells using a glass substrate can also use a vacuum chuck for temperature control. However, solar cells with both terminals on the side not facing the light, such as IBC or MWT cells, are not so easily controlled since the complex electrical contacts are difficult to incorporate into a vacuum plate. In such situations, users can employ temperature-controlled air directed at the sample to set the temperature and brief illumination periods to reduce temperature change during measurement.

7.1.3.5 I-V Curve Measurement Apparatus: Electrical Measurement

Since solar simulators with close spectral approximation to sunlight exhibit small variations in light intensity, an irradiance monitor is often provided adjacent to the test device. The electrical apparatus can synchronously measure the device current and irradiance intensity and adjust the current data to compensate for irradiance variations.

Providing an electrical connection to device terminals is often a challenge in solar cell testing. Since most solar cells are high-current, low-voltage devices, the current flowing through contact terminals and test leads can add substantial series resistance to the measurement circuit (see Figure 7.1.7). To avoid such series resistance from causing artificial fill factor reductions, test fixtures employ four-wire connections. At the device terminals, one set of probes connects to the voltmeter while another connects to the adjustable load. The voltage drops in the device connections occur only in the current circuit where they cannot interfere with the measurement results.

Such four-wire connections can be implemented easily on solar cells that are complete and do not rely on additional components when being assembled into a product. However, in the case of conventional silicon solar cells, the end product typically consists of a series-connected string of solar cells interconnected with conductive ribbons in contact with solar cell busbars. It is impractical to connect ribbons to every solar cell to be tested; yet without the ribbons, the solar cell's series resistance is far higher than it will be in the final product. As a compromise, typical silicon solar cell test fixtures use "ribbon simulators" consisting of narrow probing

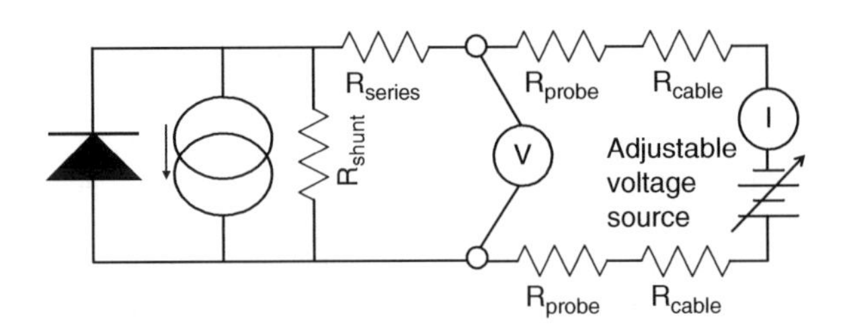

Figure 7.1.7 Solar cell measurement circuit

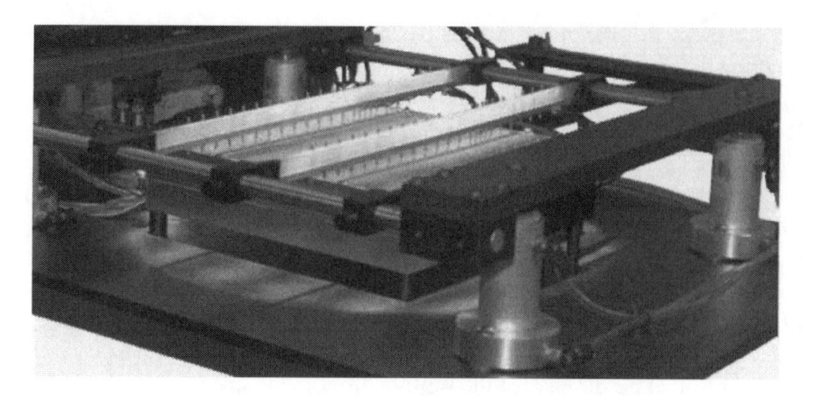

Figure 7.1.8 Solar cell "ribbon simulator" probing bars

bars with multiple current probes to collect current from multiple locations on the busbars (see Figure 7.1.8).

The configuration of the separate voltage and current probes varies among test fixture manufacturers. Some accompany each current probe with a voltage probe, with all the voltage probes wired in parallel. If such a configuration is used with a solar cell with spatial variations in performance, currents can flow between voltage probes, which introduces error in voltage measurement. Other configurations gather voltage probes near the end of each busbar. To reduce current flow between voltage probes, test fixtures can employ resistors in series with the probes.

As technology and device design narrow busbar widths, the shading from these probing bars causes larger dark regions on solar cells under test. Without compensation, this causes I_{sc} measurements to be lower than true values. The "low-shade probes" method is used in some laboratories to compensate for cell shadowing adjacent to probing bars. By removing the probing bars and substituting probes at the ends of the cell busbars, the I-V measurement apparatus measures an I_{sc} corresponding to a cell with ribbons of the same width as the busbars, albeit with a low fill factor due to the higher series resistance of a cell lacking ribbons. The user then reinstalls the probing bars, adjusts the light source intensity to obtain the I_{sc} just measured, and proceeds with the measurement.

7.1.3.6 Calibration

As in QE systems, calibration techniques for device current measurement have been developed for other applications and need little adaptation for solar cell measurements. Voltage calibration similarly requires little or no special attention. However, the numerator of the efficiency equation, power input, presents significant challenges. Almost all I-V curve measurement systems rely on a standard solar cell known as a "reference cell" to set the light source intensity. The user places the reference cell in the place where the test device will be mounted and adjusts the light intensity until the reference cell short circuit current (I_{sc}) measurement result is equal to the reference cell's I_{sc} provided by the calibration laboratory to represent its output in the one-sun condition. However, the reference cell's quantum efficiency characteristic seldom matches that of the device to be tested, which introduces spectral mismatch error (see Section 7.1.4 for details).

7.1.3.7 Comparing J_{sc} from QE and I-V Measurements

One can calculate J_{sc} from QE measurements using illumination spectrum data and from I-V measurements by dividing I_{sc} by the device area. The credibility of solar cell performance claims often follows this principle and discrepancies require explanation. Some common discrepancy sources are:

- *Errors in area measurement*: Often light incident outside the device boundaries either generates carriers collected by the cell (e.g., mesa structures) or reflects into the device (e.g., glass superstrate structures). This issue is often resolved by using an opaque mask to ensure that light outside the cell boundaries does not reach the cell.
- *Selection of area definition*: If an efficiency is reported on an *aperture area* basis, the area A in Equation (7.1.10) includes the smallest geometric region that includes all the photosensitive regions on the device. This typically excludes contact areas and busbars that are necessary for the device operation. Such performance claims represent the capabilities of the material and junction, but do not necessarily represent performance that can be achieved in devices.
- *QE scaling error*: The QE measurement may be performed in power mode with a beam incident on a region that does not represent the overall solar cell. In conventional silicon solar cells, this often happens when the beam does not intercept a share of gridlines and busbars. In others, it can represent non-uniformities in device performance or regions where carrier collection depends on series resistance. The issue is often resolved for conventional silicon solar cells by defocussing the QE system's monochromatic beam so that it covers a representational proportion of gridlines and busbars. For devices with TCO layers, one also has to set the bias light to one-sun so that the series resistance losses resemble those that would occur in actual operation.
- *Other errors*: When the above two error sources have been ruled out, a J_{sc} discrepancy may indicate a calibration, functional, or procedural failure in the QE or I-V measurement.

7.1.3.8 I-V Curves for Series and Shunt Resistance Measurements

In addition to portraying device performance, an I-V curve can portray the device's series and shunt resistance (see Figure 7.1.7) and their effects on performance. The slope of the I-V curve as it crosses zero voltage indicates the shunt resistance. Qualitatively, a flat I-V curve characteristic near I_{sc} indicates that the device does not have significant shunting. In practice, fitting a line to I-V points between V = −0.1 and V = 0.1 volts, for example, can indicate a high but negative shunt resistance in a device with minimal shunting. This is because there is a small amount of noise in the current data due to instability in the illumination intensity. When analyzing shunting data statistically, using conductance units can make the data easier to manage. Positive and negative conductances near zero represent devices with insignificant shunting, and the measurements that warrant attention will be positive.

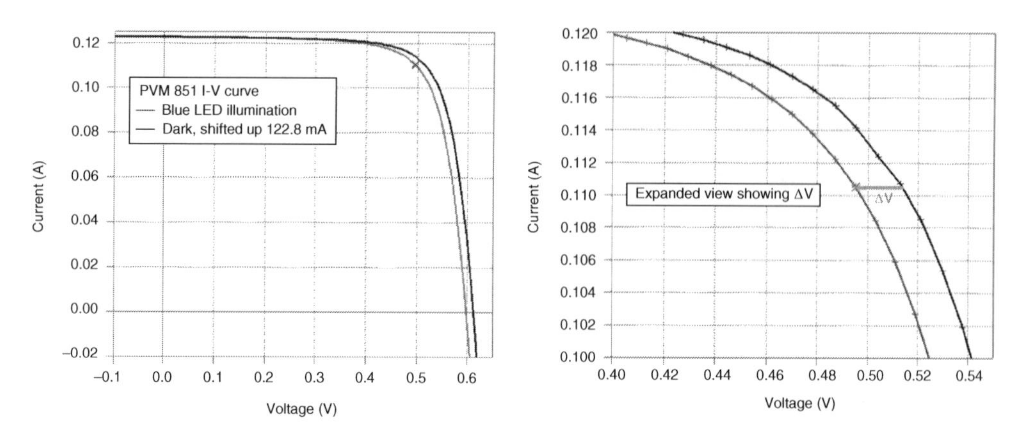

Figure 7.1.9 Solar cell I-V curve used to determine series resistance. (*See insert for color representation of the figure*)

By comparing the device's illuminated I-V curve to its dark I-V curve that has been shifted by I_{sc} or to its illuminated open circuit voltage (V_{oc}) curve, one can determine the extent to which the device's series resistance affects its performance (Figure 7.1.9).

When illuminated, the photogenerated current flows out of the solar cell, incurring a voltage drop equivalent to the current times the resistance. When measured without illumination, the current flows into the cell, causing the measured voltage to be higher than the cell's internal voltage. Selecting a point on the I-V curve as shown above allows the two voltage drops to be added, becoming $\triangle$V. The sum of the currents in the illuminated and dark curves is the amount of current by which the curve is shifted, i.e., I_{sc}.

The example in Figure 7.1.9 shows the dark I-V curve of cell "PVM 851" shifted by 122.8 mA to match the I_{sc} of the I-V curve measured with blue LED illumination. $\triangle V$ is 0.513 – 0.495 = 0.018 V. Dividing this by 0.1228A gives R_{series} of 0.15 Ω. Normalizing this 4 cm^2 cell's R_{series} measurement gives a specific series resistance value of 0.60 Ω/cm^2.

This technique relies on an assumption that R_{series} does not vary between dark and illuminated conditions. To avoid this assumption, one can use a partially-illuminated I-V curve or a Suns-Voc curve instead of the dark I-V curve. One can measure Suns-V_{oc} by collecting a set of V_{oc} points at various illumination levels. Associate each V_{oc} with a current that represents a fraction of I_{sc} that is the fraction of one-sun illumination that generated the V_{oc} measurement. This set of points creates an I-V curve that represents the performance of the solar cell without its series resistance, since all the voltages are measured with no current flowing in the external circuit.

A variety of other R_{series} measurement techniques exists, some of which involve fitting the solar cell model to I-V data. Additional information on series resistance measurements can be found in Honsberg and Bowden (2013b), Psych (2007) and Wolf and Rauschenback (1963).

7.1.4 Spectral Mismatch

As illustrated previously, a solar cell's conversion efficiency depends on the spectral content of the illumination source. Although a solar simulator's illumination is similar to that of sunlight, it is sufficiently different to introduce significant error into performance measurements.

The relevant solar simulator classification standards permit the amount of light in each 100-nm wide band to deviate from that in the standard spectrum by up to 25% while still qualifying for the highest-level classification (see Section 7.1.3.3). The consequence is that the user must attend to this issue to achieve high-accuracy solar cell performance measurements using an I-V system. Users of I-V curve measurement systems typically either quantify the spectral mismatch by the method described in ASTM E973-10 (ASTM, 2015) and other standards and apply it as an adjustment to the light intensity while preparing for an I-V curve measurement or take steps to minimize the mismatch so that it becomes insignificant.

The dimension-less spectral mismatch factor, *M*, for a solar cell I-V measurement is given by:

$$M = \frac{\int_{\lambda_1}^{\lambda_2} E_t(\lambda) \times SR_{td}(\lambda) d\lambda}{\int_{\lambda_3}^{\lambda_4} E_t(\lambda) \times SR_{ref}(\lambda) d\lambda} \times \frac{\int_{\lambda_3}^{\lambda_4} E_0(\lambda) \times SR_{ref}(\lambda) d\lambda}{\int_{\lambda_1}^{\lambda_2} E_0(\lambda) \times SR_{td}(\lambda) d\lambda} \quad (7.1.11)$$

where $E_t(\lambda)$ is the spectral irradiance of the light source (in $\mathrm{Wcm^{-2}\mu m^{-1}}$), $E_0(\lambda)$ is the spectral irradiance of the reporting spectrum, SR_{td} is the spectral response of the device being tested (in A/W), and SR_{ref} is the spectral response of the reference cell. λ_1 and λ_2 are the endpoints of the test device's range of sensitivity while λ_3 and λ_4 are the endpoints of the reference cell's range of sensitivity. If $E_t(\lambda)$ equals $E_0(\lambda)$, or if SR_{td} equals SR_{ref}, the spectral mismatch factor, *M*, equals 1.

A QE system can measure SR_{td} and SR_{ref}. $E_0(\lambda)$ is the reference spectrum. $E_t(\lambda)$ is measured using a spectroradiometer. As a guideline, users of xenon arc lamp-based solar simulators should repeat this measurement upon installing a new lamp, again after 50 and 100 hours of lamp use, and approximately every 100 hours thereafter.

Since spectral mismatch measurements require extra time and the availability of a calibrated spectroradiometer, many users strive to minimize *M* instead of computing it and applying it. This can be done by using a reference cell that is closely matched in QE characteristics to the test devices expected to be encountered or by investing in a solar simulator with excellent match to the reference spectrum. Initial attention is warranted in the former because the latter can be expensive.

7.1.5 Conclusion

Most solar cell researchers and developers can perform accurate device performance measurements when equipped with high-quality QE and I-V curve measurement systems accompanied by calibration devices and test fixtures appropriate for the device types measured. Understanding measurement instrumentation operation principles and sources of uncertainty enables users to guide laboratory investments, derive meaningful results, and identify anomalous data.

List of Symbols

Symbol	Description	Unit
V_{oc}	Open-circuit voltage	V
J_{sc}	Short-circuit current density	mA/cm^2
FF	Fill factor	%
λ	Wavelength of light	m, nm
h	Planck's constant	Js
c	Speed of light	m/s
q	Elementary charge	C
I_{TD}	Current through the test device	A
I_{REF}	Current through the reference device	A
QE_{REF}	Quantum efficiency of reference device	%
R	Reflectance	%
A	Area	m^2, cm^2
E	Irradiance	W/m^2, mW/cm^2
I	Current	A, mA
I_{sc}	Short-circuit current	A, mA
Φ	Photon flux	$1/(s * cm^2)$

When presented as a function of wavelength, photon flux and irradiance units include an inverse wavelength unit such as nm^{-1} or um^{-1}.

List of Acronyms

Acronym	Description
EQE	External quantum efficiency
IBC	Interdigitated back contact
I-V	Current voltage
IQE	Internal quantum efficiency
LED	Light-emitting diode
MWT	Metal wrap-through
PV	Photovoltaics
SR	Spectral response
TCO	Transparent conductive oxide
QE	Quantum efficiency
UV	Ultra-violet

Note: subscripts used with multiple symbols:
REF or ref = reference device; TD or td = test device.

References

ASTM (2010) ASTM E927-10 Standard specification for solar simulation for terrestrial photovoltaic testing http://www.astm.org/Standards/E927.htm (accessed 22 September 2015).

ASTM (2015) ASTM E973-10 (2015) Standard test method for determination of the spectral mismatch parameter between a photovoltaic device and a photovoltaic reference cell, http://www.astm.org/Standards/E973.htm (accessed 22 September 2015).

Honsberg, C. and Bowden, S. (2013a) PVCDROM quantum efficiency. Available at: http://www.pveducation.org/pvcdrom/solar-cell-operation/quantum-efficiency (accessed 30 August 2015).

Honsberg, C. and Bowden, S. (2013b) PVCDROM series resistance. Available at: http://www.pveducation.org/pvcdrom/solar-cell-operation/series-resistance (accessed 30 August 2015).

IEC (2007) IEC 60904-9 Ed. 2 Photovoltaic devices – Part 9: Solar simulator performance requirements. Available at: https://webstore.iec.ch/publication/3880 (accessed 22 September 2015).

JIS (2011a) JIS C 8912:1998 Solar simulators for crystalline solar cells and modules. Available at: http://www.techstreet.com/products/1230810 contains links for amendments 1 (2005) and 2 (2011) (accessed 22 September 2015).

JIS (2011b) JIS C 8933:1995 Solar simulators for amorphous solar cells and modules, includes all amendments and changes through Amendment 2, March 22, 2011. Available at: https://global.ihs.com/doc_detail.cfm?document_name=JIS%20C%208933&item_s_key=00275804 (Japanese only; accessed 29 June 2016).Pvsyst, PV module – Model parameters page. Available at: http://files.pvsyst.com/help/pvmodule_model_parameters.htm

Pysch, D., Mette, A., and Glunz, S.W. (2007) A review and comparison of different methods to determine the series resistance of solar cells. *Solar Energy Materials and Solar Cells*, **91**, 1698–1706.

Wolf, M. and Rauschenback, H. (1963) Series resistance effects on solar cell measurements. *Advanced Energy Conversion*, **3**, 455–479.

7.2

Photoluminescence and Electroluminescence Characterization in Silicon Photovoltaics

Thorsten Trupke
ARC Centre of Excellence for Advanced Silicon Photovoltaics, University of New South Wales, Sydney, Australia

7.2.1 Introduction

Luminescence-based characterization methods are widely used today for material and device characterization and for inspection across the silicon Photovoltaic (PV) value chain, starting with the characterization of silicon ingots, bricks, wafers and cells, through to luminescence imaging of complete modules and even PV systems.

Luminescence is the emission of light from specific classes of materials, which is observed under external excitation, and which is not caused by an increase in the temperature. Luminescent materials are characterized by having at least two distinct energy ranges, which are separated by an energy gap, in which ideally no electronic states exist. Examples of such materials include a range of organic molecules, dyes and semiconductors. In semiconductors, luminescence occurs when electrons from the conduction band undergo a transition to a valence band state, providing some or all of the excess energy to an emitted photon. In indirect band gap materials, such as crystalline silicon (c-Si), a phonon needs to be involved in the above electronic transition, which increases or decreases the photon energy, depending on whether the phonon is absorbed or emitted during the transition.

Photovoltaic Solar Energy: From Fundamentals to Applications, First Edition.
Edited by Angèle Reinders, Pierre Verlinden, Wilfried van Sark, and Alexandre Freundlich.

Companion website: www.wiley.com/go/reinders/photovoltaic_solar_energy

7.2.1.1 Electroluminescence Versus Photoluminescence

Luminescence is observed upon an external excitation, which increases the concentrations of electrons n_e and holes n_h, both in (cm^{-3}),above their equilibrium values. In PV applications the most widely used methods are electroluminescence (EL), where an external forward bias is applied (i.e. the solar cell is used as a light emitting diode) and photoluminescence (PL), where the excitation is by external illumination.

EL is limited to applications on finished cells and modules, while PL can be used on a wider range of samples, which enables a number of applications at an early stage of the PV value chain. On the other hand, the experimental requirements, particularly the design of illumination and detection optics, are more demanding in PL compared to EL systems. For solar cell characterization, the combination and analysis of both EL and PL experiments can yield important additional information, as will be discussed in more detail below.

In the following theory section we review some of the models underlying luminescence data analysis, which form the basis for a number of quantitative analysis methods. Some of these methods will then be discussed in more detail in the applications section.

7.2.2 Theory

7.2.2.1 The Generalized Planck Equation

The so-called generalized Planck equation describes the spectral rate of spontaneous emission per photon energy interval $dr_{sp}(\hbar\omega)$ in (cm^{-3}s^{-1}eV^{-1}) at photon energy $\hbar\omega$ in (eV), in other words, the emitted luminescence intensity, within a semiconductor as a function of the absorption coefficient for band–band transitions $\alpha_{BB}(\hbar\omega)$ in (cm^{-3}). In most practical cases that relationship can be written:

$$dr_{sp}(\hbar\omega) = \alpha_{BB}(\hbar\omega) c_\gamma D_\gamma \Omega \, exp\left(-\frac{\hbar\omega}{kT}\right) exp\left(\frac{\Delta\varepsilon_F}{kT}\right) d(\hbar\omega) \tag{7.2.1}$$

where c_γ is speed of light inside the medium in (m/s), D_γ is the photon density of states in (cm^{-3}), Ω is the solid angle into which emission occurs, T is the sample temperature in (K), k is Boltzmann's constant in (eV/K) and $\Delta\varepsilon_F$ is the separation of the quasi Fermi energies within the semiconductor in (eV). Equation 7.2.1 was first derived in similar form by Lasher and Stern in the context of describing the luminescence from direct semiconductors (Lasher and Stern, 1964), but it can be shown that the same emission law is also valid for indirect semiconductors such as silicon (Würfel, Finkbeiner, and Daub, 1995). The key dependencies in Equation (7.2.1) are, first, the linear correlation of $dr_{sp}(\hbar\omega)$ with $\alpha_{BB}(\hbar\omega)$, which enables optical material and device parameters to be extracted from luminescence data, and, second, the exponential dependence on $\Delta\varepsilon_F$.

7.2.2.2 Implied Voltage and Excess Carrier Density from Luminescence

The voltage of a solar cell is determined by the electrochemical potential difference between its terminals. A local separation of the quasi Fermi energies anywhere inside the volume of a sample is equivalent to a local electrochemical potential difference and can therefore be interpreted as a local voltage. Voltages that are not physically measurable, but rather inferred from other measurements are referred to as *implied* voltages. According to Equation (7.2.1), $dr_{sp}(\hbar\omega)$ increases exponentially with $\Delta\varepsilon_F$ and thus with the local implied voltage, which is the basis of a number of quantitative analysis methods discussed in more detail below. It is important to note that $dr_{sp}(\hbar\omega)$ in Equation (7.2.1) represents the *local* rate of spontaneous emission from a volume element within the semiconductor, which is a microscopic quantity, *not* the measurable luminescence intensity $I_{PL}(\hbar\omega)$ emitted from the surface of the sample as an emitted photon flux per sample area in ($cm^{-2}s^{-1}$). The latter is calculated by integrating $dr_{sp}(\hbar\omega)$ over the volume of the sample, taking into account multiple reflections and the specific sample geometry, a calculation which can be done analytically only for a few special cases, see e.g. (Schinke *et al.*, 2013).

The above interpretation of luminescence intensities in terms of an implied voltage is particularly accurate (i.e. it matches the measurable terminal voltage) for well-passivated devices with long diffusion length and under open circuit conditions, in which case, $\Delta\varepsilon_F$ is nearly constant across the device thickness. For lower quality samples or under operating conditions with simultaneous current extraction, the analysis becomes more complex and requires a number of corrections, for example, to account for the presence of diffusion limited carriers, which remain in the bulk of a solar cell even under external short circuit conditions (Abbott *et al.*, 2007).

From basic semiconductor physics we know that a fundamental relationship exists between n_e and n_h and the separation of the quasi-Fermi energies (Würfel, 2005):

$$\begin{aligned} n_e n_h &= n_i^2 exp\left(\frac{\Delta\varepsilon_F}{kT}\right) = \Delta n\left(\Delta n + N_{D/A}\right) \\ &= \Delta n^2 \text{ for high injection } \left(\Delta \mathrm{n} >> \mathrm{N_{D/A}}\right), \\ &= \Delta n N_{D/A} \text{for low injection } \left(\Delta \mathrm{n} << \mathrm{N_{D/A}}\right), \end{aligned} \tag{7.2.2}$$

with $N_{D/A}$ the background doping density in (cm^{-3}), Δn the excess carrier density in (cm^{-3}) and n_i the intrinsic carrier concentration in (cm^{-3}). Inserting Equation (7.2.2) into Equation (7.2.1) shows that the rate of spontaneous emission (and thus the measured luminescence intensity) are proportional to the product of minority and majority carrier concentrations $n_e n_h$, allowing Δn to be determined from the luminescence intensity, which is the basis of PL-based minority carrier lifetime measurements.

7.2.3 Applications

7.2.3.1 Spectral Luminescence Measurements

Figure 7.2.1 shows a typical PL spectrum from a crystalline silicon wafer at room temperature. A broad luminescence band is observed with a peak at typically around 1140 nm. This broad emission band is caused by electronic transitions between conduction band and valence band and unless stated otherwise, the analytical methods discussed below are based on luminescence within this band. Several emission bands are also observed from c-Si at longer wavelength (>1300 nm), which are equivalent to photon energies substantially smaller than the bandgap of silicon. These emission bands typically merge into one or two broad bands at room

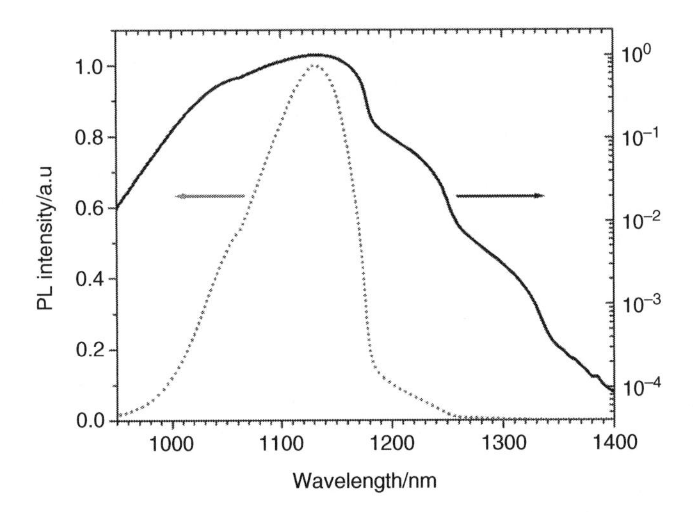

Figure 7.2.1 Room temperature band-to-band photoluminescence spectrum from a crystalline silicon wafer on linear (dotted line, left-hand scale) and on semi-logarithmic (solid line, right-hand scale). $T = 291$ K, data from (Trupke *et al.*, 2003)

temperature and are due to defect states within the bandgap of silicon, the latter commonly associated with the presence of decorated crystal defects (Ostapenko *et al.*, 2000; Tajima *et al.*, 2012).

7.2.3.2 Optical Material and Device Properties from PL

The linearity between $dr_{sp}(\hbar\omega)$ and the absorption coefficient from Equation (7.2.1) can be exploited to extract absorption coefficient data from measured luminescence spectra without significant impact of parasitic absorption processes such as free carrier absorption (Daub and Würfel, 1995).

Figure 7.2.2 shows the extremely low values for the absorption coefficient of crystalline silicon at various temperatures, obtained from PL spectra using Daub's method. The periodic pattern observed in the absorption coefficient at sub-band gap energies (also visible in the PL spectrum shown in Figure 7.2.1) represents photon absorption with simultaneous absorption of up to four phonons. The absorption coefficient data from PL spectra are fitted to conventional absorption coefficient data at short wavelengths and show excellent agreement with those data and also with the absorption coefficient data from spectral response measurements on silicon solar cells (Keevers and Green, 1995).

For a planar silicon slab with polished surfaces the analytical integration of the rate of spontaneous emission results in the following expression for the emitted energy current density per energy interval I_{PL} (Smestad, 1995):

$$I_{PL}(\hbar\omega) = CA_{BB}(\hbar\omega)\frac{(\hbar\omega)^3}{exp\left(\frac{\hbar\omega - \Delta\varepsilon_F}{kT}\right) - 1}d(\hbar\omega), \quad (7.2.3)$$

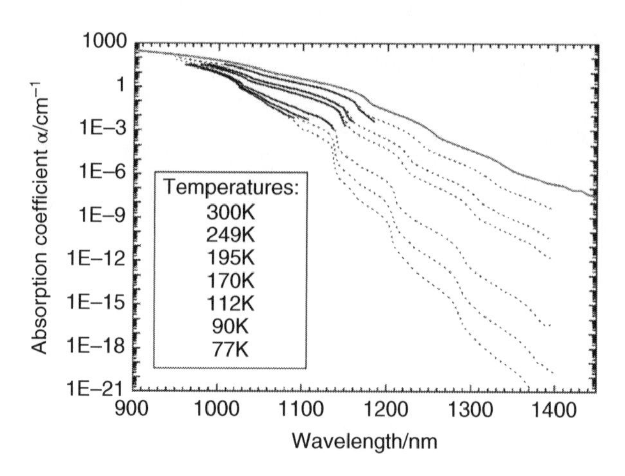

Figure 7.2.2 Spectral absorption coefficient of crystalline silicon for various temperatures. Data from PL (dotted lines) and from conventional optical transmission (solid lines) for various temperatures (given in the figure caption). The top curve represents absorption coefficient data from spectral response measurements at room temperature, which show an excellent match with data from PL

with C a proportionality constant and $A_{BB}(\hbar\omega)$ the absorptance for band–band transitions as an absolute dimensionless ratio. Equation (7.2.3) can be derived analytically only for simple device geometries, but holds generally also for non-planar sample geometries. It allows the absorptance of a specific sample (e.g. a silicon solar cell) to be extracted from the luminescence spectrum (Trupke, Daub, and Würfel, 1998). Comparison of different cells provides quantitative insight, e.g. into the impact of surface texture and antireflection coatings on light trapping properties.

7.2.3.3 Quantum Efficiency from EL

Equation (7.2.3) is derived with the explicit assumption of a uniform separation of the quasi Fermi energies across the device thickness, which is valid for well-passivated high bulk lifetime samples with a bulk diffusion length exceeding the sample thickness. The collection efficiency of photo-generated carriers is close to unity in that case, regardless of where the photogeneration occurs. The external quantum efficiency (EQE) of a solar cell is equivalent to the absorptance $A_{BB}(\hbar\omega)$ in that special case. As shown theoretically by Rau (2007) and experimentally by Kirchartz *et al.* (2009) the reciprocity between the two processes of photo generation and external current extraction, on the one hand, and the reverse process of electroluminescence (i.e. external carrier injection and photon emission) on the other, goes a lot further. The spectral electroluminescence emission $I_{EL}(\hbar\omega)$ (measured as an energy flux per photon energy interval) from a silicon solar cell is proportional to the spectral external quantum efficiency (EQE):

$$I_{EL}(\hbar\omega) \sim EQE(\hbar\omega)(\hbar\omega)^3 \, exp\left(\frac{qV - \hbar\omega}{kT}\right) d(\hbar\omega). \tag{7.2.4}$$

where qV in (eV) is defined as the separation of the quasi Fermi levels at the edge of the space charge region. The relationship between the spectral absorptance and the emitted energy flux from Equation (7.2.3), derived for a solar cell with uniform $\Delta\varepsilon_F$, is thus merely a special case of the more general relationship between the EL spectrum and EQE in Equation (7.2.4) (note that the -1 in the denominator of Equation (7.2.3) can be ignored in practical cases).

7.2.3.4 Quantitative Analysis of Luminescence Intensities

The analytical methods described above are based on relative spectral variations of the emitted luminescence intensity. In other applications the spectral variation is largely ignored and the absolute, spectrally integrated PL intensity is analyzed.

Quasi-steady state photoconductance (QSS-PC) measurements are widely used in PV research and development (R&D) for injection-dependent effective lifetime measurements of partially processed silicon wafers (Sinton and Cuevas, 1996). The incident light intensity is measured, calibrated and interpreted in terms of the average generation rate in ($cm^{-3}s^{-1}$), and the measured signal (excess conductance under illumination) is calibrated into Δn. Quasi steady state photoluminescence (QSS-PL) lifetime measurements are based on the same principle, except that Δn is extracted from the absolute PL intensity (Trupke and Bardos, 2005). Initially demonstrated with monochromatic LED illumination, QSS-PL measurements are unaffected by a number of physical effects that cause experimental artefacts in QSS-PC measurements at low to medium injection densities (Trupke and Bardos, 2005; Bardos *et al.*, 2006; Juhl *et al.*, 2013). QSS-PL is also extremely sensitive at low Δn, allowing τ_{eff} to be measured for $\Delta n < 10^9\,cm^{-3}$ (Trupke and Bardos, 2005; J.A. Giesecke *et al.*, 2012).

In QSS-PL lifetime measurements, the calibration of the measured PL signal into Δn requires accurate knowledge of a calibration constant that depends on the optical properties of the measurement system, the background doping of the sample and on the sample geometry. A separate calibration is therefore generally required for each specific type of sample, which can be performed, for example, by adjusting the calibration value so that the resulting lifetime data match the results of QSS-PC. The validity of that approach was investigated in detail by Herlufsen *et al.* (2013). In that case QSS-PL and QSS-PC complement each other by providing accurate data at low to medium injection (QSS-PL only) and at medium to high injection (QSS-PC and QSS-PL).

7.2.4 Luminescence Imaging

A recent breakthrough in the application of luminescence for PV applications is the development of camera-based luminescence imaging, a technique which was previously used only sporadically for the characterization of small area samples of direct bandgap semiconductor materials (Livescu *et al.*, 1990). Scanning luminescence measurements (mapping) have historically been used in PV-related research to obtain spatially resolved maps of the band-to-band luminescence (Daub *et al.*, 1994) and of the defect band luminescence (Ostapenko *et al.*, 2000) on silicon wafers. Performed in a point-by-point raster scanning mode, they result in rather long measurement times on 6" wafers. Scanning PL mapping is still used for particular applications requiring high spatial resolution on the micrometer scale (Gundel *et al.*, 2010), a resolution that is currently not achievable with luminescence imaging, where the spatial resolution is typically on the order of tens to hundreds of micrometres.

Scientific grade Silicon charge coupled device (CCD) cameras are commonly used in camera-based luminescence imaging to take megapixel images of the luminescence emission (Hinken *et al.*, 2011). InGaAs cameras are used in specific applications (Yan *et al.*, 2012), for example, to capture luminescence images of the defect band emission. In EL imaging of silicon solar cells, first demonstrated by Fuyuki *et al.* (2005), a forward bias is applied to a solar cell, the resulting forward current being equivalent to the total recombination rate within the cell. A small fraction of that recombination is radiative, i.e. associated with the emission of luminescence. In PL imaging of silicon wafers, introduced by Trupke *et al.* (Trupke *et al.*, 2006), a light source, often a high power fiber-coupled diode laser is expanded to illuminate the entire sample simultaneously uniformly. The luminescence emission is captured by the camera under continuous illumination, i.e. in true steady state conditions. PL imaging measurements therefore require sophisticated filtering, particularly for unpassivated samples, to separate the reflected laser light from the significantly weaker (up to 12 orders of magnitude) luminescence signal. PL imaging provides high resolution minority carrier lifetime measurements on passivated silicon wafers with measurement times of fractions of a second (Trupke *et al.*, 2006), where previously several hours were required, using, for example, microwave photoconductance decay scanning experiments.

Based on the rich information contained in luminescence images, both PL and EL imaging are now standard measurement techniques in PV R&D, and are widely used for a broad range of applications across the entire value chain, from the inspection of silicon ingots, bricks, wafers and cells through to EL and PL inspection of entire PV arrays. Some examples will be discussed below.

7.2.4.1 PL Imaging on Bricks and Ingots

Characterization of multicrystalline silicon bricks and monocrystalline ingots is an attractive proposition for immediate process feedback during crystallization and for quality control at an early processing stage. Minority carrier lifetime measurements on unpassivated silicon bricks are less strongly affected by surface recombination compared to unpassivated as-cut wafers due to the effective absence of a rear surface (Sinton *et al.*, 2004). This enables bulk lifetime information to be extracted over a significantly wider range, than is possible on wafers.

From Equation (7.2.2) we can see that under low injection conditions ($\Delta n << N_{D/A}$) the PL intensity is proportional to both Δn and to the net doping density $N_{D/A}$. As a result, the analysis of PL images on silicon bricks in terms of effective or bulk lifetime is complicated by the fact that the doping density changes with ingot height z. That variation therefore needs to be explicitly corrected for using either measured or empirical data for $N_{D/A}(z)$ (Trupke *et al.*, 2009). An elegant way to avoid this extra complexity is the so-called two-filter method or PL intensity ratio (PLIR) method. Initially demonstrated for the extraction of diffusion lengths on silicon solar cells (Würfel *et al.*, 2007), it is based on the ratio of two PL signals measured in the same sample location but with the PL detector sensitivity tuned to two different spectral ranges. In PL imaging experiments, this is achieved by mounting two different spectral filters in front of the camera lens. Combined with analytical models for the carrier distribution as a

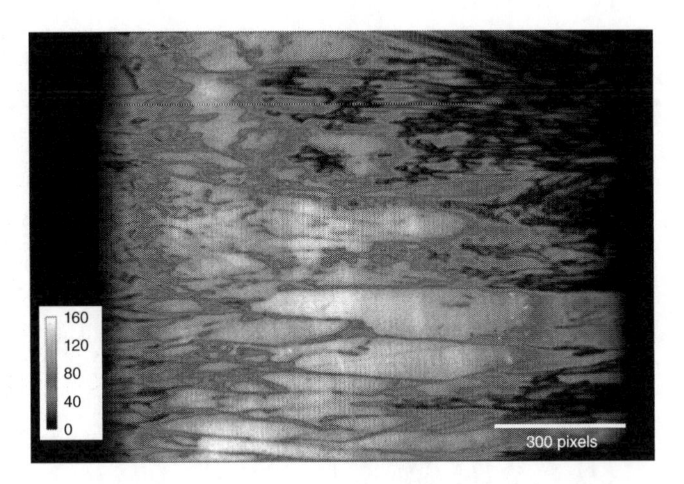

Figure 7.2.3 Bulk lifetime image of the side facet of a 6-inch, 23 cm high multicrystalline silicon brick prior to wafer slicing obtained using the PL intensity ratio method. The darkness scale represents the lifetime in microseconds, the right-hand side represents the top of the ingot. Source: (Mitchell *et al.*, 2011). (*See insert for color representation of the figure*)

function of bulk lifetime (or diffusion length), the PLIR measured on silicon bricks can be analyzed in terms of bulk lifetime (Mitchell *et al.*, 2011), with the advantage that the height-dependent background doping variation is cancelled out. Other approaches for absolute lifetime analysis of PL images on bricks are based on dynamic calibration approaches (Herlufsen *et al.*, 2012; Johannes *et al.*, 2013).

An example of a bulk lifetime image from the PLIR taken on multicrystalline silicon brick is shown in Figure 7.2.3. It shows very low bulk lifetime regions in the top and bottom sections, which is typical of cast multicrystalline ingots and a result of impurity (primarily iron (Fe)) diffusion from the crucible into the ingot from the bottom and segregation to and subsequent solid state impurity diffusion from the top. Dark lines in the image represent recombination active structural defects such as grain boundaries and dislocations, the density of the latter increasing strongly towards the top of the ingot (shown in Figure 7.2.3 on the right-hand side). With the adoption of more advanced higher efficiency cell structures (e.g., PERC), the bulk lifetime τ_b of the wafer will have a stronger impact on the efficiency distribution in mass manufacturing. Since τ_b is not measurable reliably on as-cut wafer, the brick inspection with PL imaging and the proven ability to obtain bulk lifetime data over a wide range is a promising candidate for early quantification of the material quality.

7.2.4.2 PL Imaging on Wafers

With unpassivated surfaces and several micrometre-thick surface damage layers on both sides, as-cut silicon wafers are limited by effectively infinite surface recombination velocities. The effective lifetime τ_{eff} in (s) is generally determined by the surface lifetime τ_s in (s) and the bulk lifetime τ_b in (s)

$$\frac{1}{\tau_{\text{eff}}} = \frac{1}{\tau_{\text{S}}} + \frac{1}{\tau_{\text{b}}} \tag{7.2.5}$$

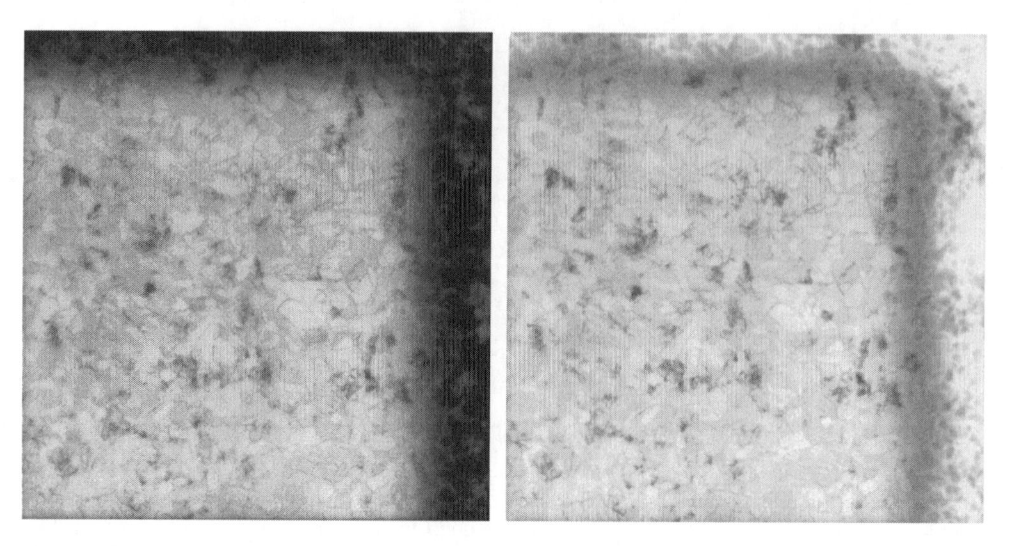

Figure 7.2.4 PL image of an as-cut high performance multicrystalline silicon wafer, measured on a BT Imaging iLS-W2 automated inline PL imaging system (left) and results of automated image processing (right) showing shaded overlays for areas of high impurity concentration (light) and for the location of structural defects (shaded). (*See insert for color representation of the figure*)

and is therefore dominated by the strongest recombination channel, i.e. the one with the lowest associated lifetime. In as-cut wafers τ_s is on the order of only 1–2 microseconds or less (depending on τ_b, illumination wavelength, carrier mobilities and wafer thickness). Any recombination mechanisms other than the surface recombination are therefore insignificant or only have a very small impact on τ_{eff} unless they have associated lifetimes that are comparable to τ_s. In the absence of strong doping variations within a wafer, PL images represent effective lifetime images. The latter only show any contrast in local areas with reduced τ_{eff}, i.e. in the presence of extremely highly recombination active areas, which can be expected to have a strong effect on solar cell performance.

Figure 7.2.4 shows a megapixel PL image taken on an industrial, so-called high performance multi-crystalline as-cut wafer from a corner brick. Line-shaped defects represent recombination active dislocation networks and grain boundaries within the wafer. The dark regions around the edges represent low lifetime regions caused by high Fe concentration, a result of the diffusion of Fe from the crucible walls into the crystal after solidification during casting.

The PL image shown in Figure 7.2.4 was measured using a line scanning PL system (www.btimaging.com) with a total measurement time of only 0.7 s, a capability that is enabled by the development of inline PL imaging systems, which can measure PL images on as-cut wafers at production speeds of 3,600 wafers per hour. Automated image processing algorithms, specifically developed for this application, generate a number of wafer quality metrics from the PL image for each wafer. The shaded overlays in Figure 7.2.4 represent the so-called defect and impurity metrics. A number of studies, some of them including large quantities of several 100,000 wafers, have revealed strong correlation between the efficiency of fully processed solar cells and predictive models based on the metrics obtained from PL images on unprocessed wafers. Only some of this data is publicly available (Giesecke *et al.*, 2009; Haunschild *et al.*, 2010; McMillan *et al.*, 2010; Trupke, Nyhus, and Haunschild, 2011; Haunschild *et al.*,

2012; Yan *et al.*, 2012). PL imaging can therefore be used not only for faster process feedback in wafer manufacturing, but also for effective quality control, with up to 100% inspection of production, enabling manufacturers to pre-sort or reject wafers.

PL imaging of partially processed wafers is used widely in PV R&D for process development, debugging and process optimization. That includes purely qualitative analysis of image features (Abbott *et al.*, 2006) and also a wide range of quantitative analysis methods, including imaging the concentration of important metallic impurities such as Fe (Macdonald, Tan, and Trupke, 2008), chromium (Habenicht, Schubert, and Warta, 2010), the Boron-oxygen complex (Lim, Rougieux, and Macdonald, 2013) of oxygen-induced stacking faults (Søndenå *et al.*, 2013), emitter saturation current density imaging (Bullock *et al.*, 2012; Müller *et al.*, 2012; Hameiri *et al.*, 2013), the correlation of PL intensities with crystal orientation (Sio *et al.*, 2012), or the quantification of the recombination activity of grain boundaries (Sio, Trupke, and Macdonald, 2014).

7.2.4.3 PL and EL Imaging on Finished Solar Cells

The fully processed silicon solar cell is the earliest production stage in the PV value chain at which EL imaging can be used. A large number of qualitative and quantitative luminescence imaging applications are used in solar cell R&D and in production today that are based on EL imaging, on PL imaging or on contacted PL imaging with simultaneous current extraction or current injection. These applications include recombination current density imaging (Ramspeck *et al.*, 2007; Glatthaar *et al.*, 2009;), shunt detection (Kasemann *et al.*, 2007; Breitenstein *et al.*, 2008; Augarten *et al.*, 2013), diffusion length imaging (Fuyuki *et al.*, 2005; Würfel *et al.*, 2007), minority carrier lifetime imaging on complete cells (Giesecke, *et al.*, 2011; Giesecke, Schubert, and Warta, 2013; Shen, Kampwerth, and Green, 2014), the quantification of surface recombination velocities (Haug *et al.*, 2012) and of the photocurrent collection efficiency (Rau *et al.*, 2014) and imaging of the reverse current density of highly doped wafer areas (Müller *et al.*, 2012). Recent work even showed that images of all relevant parameters from the one diode model can be extracted from a series or luminescence images taken under different operating conditions, the latter enabling, for example, the calculation of images of the local current density at the maximum power point (Shen *et al.*, 2013). Luminescence-based methods were also developed for short circuit current density J_{SC} imaging (Breitenstein, Höffler, and Haunschild, 2014). Michl *et al.* (2012) used PL data in combination with theoretical device modeling to predict the efficiency limits imposed on solar cells by the injection dependent lifetime. As an example of a widely used quantitative luminescence imaging based analytical method we will discuss series resistance imaging (Trupke *et al.*, 2006; Ramspeck *et al.*, 2007; Trupke *et al.*, 2007; Kampwerth *et al.*, 2008; Haunschild *et al.*, 2009; Breitenstein *et al.*, 2010; Glatthaar *et al.*, 2010) in some more detail.

The series resistance R_s has a strong impact on a solar cell's terminal characteristics and efficiency. The contributions to the global series resistance vary laterally across the cell area, a fast imaging technique that exposes such variations is therefore valuable for process monitoring and for R&D. Current flow between the two cell terminals via a specific cell location is associated with a voltage drop U_{Rs} in (V) across the local effective series resistance. In the simplified model of independent diodes, the local effective series resistance $R_{s,i}$ in ($\Omega\,\mathrm{cm}^2$) is given as

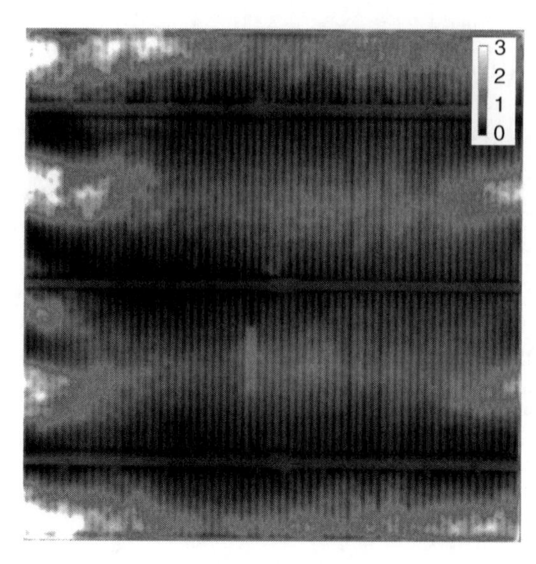

Figure 7.2.5 Series resistance image on a multicrystalline silicon solar cell measured using the method by Kampwerth (Kampwerth *et al.*, 2008). The darkness scale gives the local series resistance in absolute units of Ωcm^2. (*See insert for color representation of the figure*)

$$R_{S,i} = \frac{U_{Rs,i}}{J_i} \tag{7.2.6}$$

with J_i the locally injected or extracted current density in (Acm^{-2}). From a luminescence image on a finished cell an image of the local diode voltage can be obtained, based on the correlation of the luminescence signal with the local separation of the quasi-Fermi energies. This allows an image of $U_{Rs,i}$ to be obtained by comparison with the terminal voltage. A number of different R_S-imaging approaches based on PL with current extraction or on EL images have been proposed and demonstrated. These approaches rely on either finding approximate values for $J_{i,}$ on separate measurement of $J_{i,}$ or on circumventing the need for an actual quantification of J_i. An example of a series resistance image of an industrial multicrystalline silicon solar cell, measured using the method by Kampwerth (Kampwerth *et al.*, 2008) is shown in Figure 7.2.5. It demonstrates the ability of quantitative series resistance imaging to separate luminescence intensity variations caused by variations in the effective lifetime form series resistance variations.

7.2.4.4 EL Imaging on Modules and Systems

EL imaging has become a standard method for the inspection of silicon PV modules due to the rich information about electrical and mechanical defects contained in luminescence images and the simplicity of the measurement. Examples of EL images of industrial silicon PV modules are shown in Figure 7.2.6. Typical faults that are visible in EL images on modules include large cracks, lifetime-limiting recombination active defects, shunts, degraded cell areas,

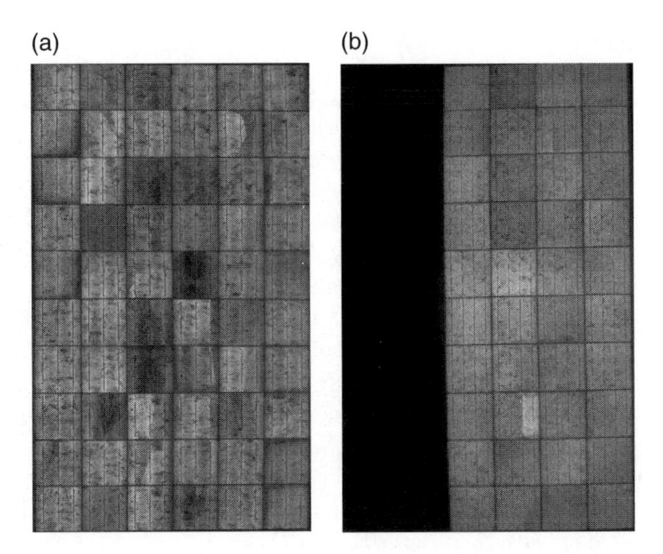

Figure 7.2.6 Electroluminescence images of industrial silicon modules showing (a) a large number of cracks as often induced by poor mechanical handling of fully assembled modules and (b) a module with one inactive string (i.e. two columns of inactive cells), as caused by a shunted by-pass diode or by one cell in the string being completely disconnected. Image data courtesy of Trina Solar

potential or light-induced degradation, electrically disconnected cells (or cell areas) and many more. A report on the EL inspection of modules, containing a catalogue of the types of defects visible in EL images on cells and modules and their typical fingerprint was recently provided by Köntges *et al.* (2014).

With the need for 25–30-year warranties and with the bankability of module suppliers becoming an increasingly important topic, the application of EL imaging for quality control of modules can be expected to increase and become a standard requirement in module manufacturing in the near future.

7.2.4.5 Inspection of PV Systems

EL imaging of one or several modules that are installed in an outdoor PV system can be achieved in similar fashion as described above by forward biasing an entire string, which can be advantageous, since no electrical wiring changes to individual modules (which can be awkward in practice) is then required. Such measurements must generally be performed at night, since the separation of the luminescence signal from ambient light cannot be achieved with conventional silicon camera-based luminescence systems. A lock-in based PL/EL imaging solution, in which the luminescence output is modulated electrically at low frequency by modulating the bias across a single module or across a string of modules, was also recently demonstrated and shown to enable luminescence imaging even in full daylight (Stoicescu, Reuter, and Werner, 2014). This is achieved by subtracting images with different bias conditions from each other, which results in any constant signals, such as reflected ambient light, to cancel out, leaving only the actual luminescence as the measured signal.

7.2.5 Conclusion

The use of luminescence for the characterization of silicon samples and devices has a long history from applications in microelectronics to applications in PV. Luminescence is an ideal probe for the electronic material and device quality, particularly for PV applications, due to its fundamental dependence on the separation of the quasi-Fermi energies and thus on the implied voltage. This has been recognized and exploited by a small number of research groups in highly specialized experimental work for many years. A major shift in the use of luminescence from niche applications to a standard measurement technique was caused by the development of high resolution camera-based luminescence imaging with measurement times on the order of seconds (EL and PL) in the years 2005–2006. The ability to gain information about a wide range of material- and device-related parameters and to detect processing problems with easy-to-use commercial luminescence systems makes luminescence measurements a valuable tool in R&D and in production across the entire PV value chain. While the emphasis of this chapter is on applications to wafer-based silicon solar cells, it is noted that luminescence characterization of other types of solar cells is also increasingly used, for example, for thin film solar cells (see e.g., Abou-Ras, Kirchartz, and Rau, 2011).

Acknowledgments

The author would like to thank Johannes Giesecke, Ziv Hameiri, Thomas Kirchhartz, and Sandra Herlufsen for very helpful discussions, comments and suggestions for improvement of this contribution.

List of Symbols

Symbol	Description	Unit
n_e	Electron concentration	cm^{-3}
n_h	Hole concentration	cm^{-3}
$dr_{sp}(\hbar\omega)$	Rate of spontaneous emission	$cm^{-3}s^{-1}$
$\alpha_{BB}(\hbar\omega)$	Absorption coefficient for band-band transitions	cm^{-1}
c_γ	Speed of light inside the medium	ms^{-1}
D_γ	Photon density of states	cm^{-3}
Ω	Solid angle	Steradian
T	Temperature	K
k	Boltzmann's constant	J/K
$\Delta\varepsilon_F$	Separation of quasi-Fermi Energies	J
$I_{PL}(\hbar\omega)$	Measured photoluminescence intensity	$cm^{-3}s^{-1}$
Δn	Excess carrier concentration	cm^{-3}
n_i	Intrinsic carrier concentration	cm^{-3}
G	Average generation rate per volume	$cm^{-3}s^{-1}$
$I(PL)$	Photoluminescence intensity	$Jcm^{-2}s^{-1}$
$N_{D/A}$	Background doping density	cm^{-3}
$A_{BB}(\hbar\omega)$	Absorptance for band-band transitions	%

Symbol	Description	Unit
$I_{EL}(\hbar\omega)$	Measurable electroluminescence intensity	$Jcm^{-2}s^{-1}$
τ_{eff}	Effective excess minority carrier lifetime	s
R_s	Series resistance	Ωcm^2
$R_{s,i}$	Local series resistance	Ωcm^2
J_i	Local current density	Acm^{-2}
q	Elementary charge	C
V	Voltage	V
z	Ingot height	cm
τ_b	Bulk lifetime	s
τ_s	Surface lifetime	s
U_{Rs}	Voltage drop over the series resistance	V

List of Acronyms

Symbol	Description
PV	Photovoltaic
c-Si	Crystalline Silicon
EL	Electroluminescence
PL	Photoluminescence
EQE	External quantum efficiency
CCD	Charge Coupled Device
InGaAs	Indium Gallium Arsenide
PLIR	Photoluminescence intensity ratio
QSS-PC	Quasi Steady State Photoconductance
QSS-PL	Quasi Steady State Photoluminescence
R&D	Research and development
LED	Light emitting diode
PERC	Passivated Emitter Rear Contact

References

Abbott, M.D., Bardos, R.A., Trupke, T. *et al.* (2007) The effect of diffusion-limited lifetime on implied current voltage curves based on photoluminescence data. *Journal of Applied Physics*, **102**(4), 44502.

Abbott, M.D., Cotter, J.E., Chen, F. W. *et al.* (2006) Application of photoluminescence characterisation to the development and manufacturing of high-efficiency silicon solar cells. *Journal of Applied Physics*, **100**, 114514.

Abou-Ras, D., Kirchartz, T., and Rau, U. (eds) (2011) Advanced Characterization Techniques for Thin Film Solar Cells. Wiley-VCH, Weinheim.

Augarten, Y., Trupke, T., Lenio, M. *et al.* (2013) Calculation of quantitative shunt values using photoluminescence imaging. *Progress in Photovoltaics: Research and Applications*, **21** (5), 933–941.

Bardos, R.A., Trupke, T., Schubert, M.C., and Roth, T. (2006) Trapping artifacts in quasi-steady-state photoluminescence and photoconductance lifetime measurements on silicon wafers. *Applied Physics Letters*, **88**, 53504.

Breitenstein, O., Bauer, J., Trupke, T., and Bardos, R.A. (2008) On the detection of shunts in silicon solar cells by photo- and electroluminescence imaging. *Progress in Photovoltaics: Research and Applications*, **16**, 325.

Breitenstein, O., Höffler, H., and Haunschild, J. (2014) Photoluminescence image evaluation of solar cells based on implied voltage distribution. *Solar Energy Materials and Solar Cells*, **128**, 296–299.

Breitenstein, O., Khanna, A., Augarten, Y. *et al.* (2010) Quantitative evaluation of electro- luminescence images of solar cells. *Physica Status Solidi RRL*, **4** (1), 7–9.

Bullock, J., Yan, D.,Thomson, A. and Cuevas, A. (2012) Imaging the recombination current pre-factor Jo of heavily doped surface regions: A comparison of low and high injection photoluminescence techniques. In *Proceedings of the 27th European Photovoltaic Solar Energy Conference*, pp. 1312–1318.

Daub, E., Klopp, P., Kugler, S., and Würfel, P. (1994) Diffusion length from spatial resolution of photoluminescence in silicon. In *Proceedings of the 12th European Photovoltaic Solar Energy Conference*, Amsterdam, *Netherlands*, pp. 1772–1774.

Daub, E. and Würfel, P. (1995) Ultralow values of the absorption coefficient of Si obtained from luminescence. *Physical Review Letters*, **74** (6), 1020–1023.

Fuyuki, T., Kondo, H., Yamazaki, T., Takahashi, Y., and Uraoka, Y. (2005) Photographic surveying of minority carrier diffusion length in polycrystalline silicon solar cells by electroluminescence. *Applied Physics Letters*, **86** (26), 262108.

Giesecke, J.A., Michl, B., Schindler, F. *et al.* (2011) Minority carrier lifetime of silicon solar cells from quasi-steady-state photoluminescence. *Solar Energy Materials and Solar Cells*, **95** (7), 1979–1982.

Giesecke, J.A., Niewelt, T., Rüdiger, M. *et al.* (2012) Broad range injection-dependent minority carrier lifetime from photoluminescence. *Solar Energy Materials and Solar Cells*, **102**, 220–224.

Giesecke, J.A., Schubert, M.C., and Warta, W. (2013) Carrier lifetime from dynamic electroluminescence. *IEEE Journal of Photovoltaics*, **3** (3), 1012–1015.

Giesecke, J.A., Sinton, R.A., Schubert, M.C. *et al.* (2013) Determination of bulk lifetime and surface recombination velocity of silicon ingots from dynamic photoluminescence. *IEEE Journal of Photovoltaics*, **3** (4), 1311–1318.

Giesecke, J.A., The, M., Kasemann, M., and Warta, W. (2009) Spatially resolved characterization of silicon as-cut wafers with photoluminescence imaging. *Progress in Photovoltaics: Research and Applications*, **17** (4), 217–225.

Glatthaar, M., Giesecke, J., Kasemann, M. *et al.* (2009) Spatially resolved determination of the dark saturation current of silicon solar cells from electroluminescence images. *Journal of Applied Physics*, **105**, 113110.

Glatthaar, M., Haunschild, J., Kasemann, M. *et al.* (2010) Spatially resolved determination of dark saturation current and series resistance of silicon solar cells. *Physica Status Solidi (RRL), Rapid Research Letters*, **4** (1–2), 13–15.

Gundel, P., Heinz, F.D., Schubert, M.C., and Giesecke, J.A. (2010) Quantitative carrier lifetime measurement with micron resolution. *Journal of Applied Physics*, **108**, 033705.

Habenicht, H., Schubert, M.C., and Warta, W. (2010) Imaging of chromium point defects in p-type silicon. *Journal of Applied Physics*, **108**, 034909.

Hameiri, Z., Chaturvedi, P., Juhl, M.K., and Trupke, T. (2013) Spatially resolved emitter saturation current by photoluminescence imaging. In *39 IEEE Photovoltaics Specialists Conference*, Tampa, USA.

Haug, H., Nordseth, Ø., Monakhov, E.V., and Marstein, E.S. (2012) Photoluminescence imaging under applied bias for characterization of Si surface passivation layers. *Solar Energy Materials and Solar Cells*, **106**, 60–65.

Haunschild, J., Glatthaar, M., Demant, M. *et al.* (2010) Quality control of as-cut multicrystalline silicon wafers using photoluminescene imaging for solar cell production. *Solar Energy Materials and Solar Cells*, **94** (12), 2007–2012.

Haunschild, J., Glatthaar, M., Kasemann, M. *et al.* (2009) Fast Series resistance imaging for silicon solar cells using electroluminescence. *Physica Status Solidi Rapid Research Letters*, **3** (7–8), 227–229.

Haunschild, J., Reis, I.E., Chipei, T. *et al.* (2012) Rating and sorting of mc-Si as-cut wafers in solar cell production using PL imaging. *Solar Energy Materials and Solar Cells*, **106**, 71–75.

Herlufsen, S., Bothe, K., Schmidt, J. *et al.* (2012) Dynamic photoluminescence lifetime imaging of multicrystalline silicon bricks. *Solar Energy Materials and Solar Cells*, **106**, 42–46.

Herlufsen, S., Hinken, D., Offer, M. *et al.* (2013) Validity of calibrated photoluminescence lifetime measurements of crystalline silicon wafers for arbitrary lifetime and injection ranges. *IEEE Journal of Photovoltaics*, **3** (1), 381–386.

Hinken, D., Schinke, C., Herlufsen, S. *et al.* (2011) Experimental setup for camera-based measurements of electrically and optically stimulated luminescence of silicon solar cells and wafers. *The Review of Scientific Instruments*, **82** (3), 033706.

Juhl, M., Chan, C., Abbott, M.D., and Trupke, T. (2013) Anomalously high lifetimes measured by quasi-steady-state photoconductance in advanced solar cell structures. *Applied Physics Letters*, **103** (24), 243902.

Kampwerth, H., Trupke, T., Weber, J., and Augarten, Y. (2008) Advanced luminescence based effective series resistance imaging of silicon solar cells. *Applied Physics Letters*, **93**, 202102.

Kasemann, M., Grote, D., Walter, B., *et al.* (2007) Shunt detection capabilities of luminescence imaging on silicon solar cells. In *22nd European Photovoltaic Solar Energy Conference*, Milan, Italy.

Keevers, M.J. and Green, M.A. (1995) Absorption edge of silicon from solar cell spectral response measurements. *Applied Physics Letters*, **66** (2), 174–176.

Kirchartz, T., Helbig, A., and Reetz, W. (2009) Reciprocity between electroluminescence and quantum efficiency used for the characterization of silicon solar cells. *Progress in Photovoltaics: Research and Applications*, **17** (6), 394–402.

Köntges, M., Kurtz, S., Packard, C. *et al.* (2014) Review of failures of photovoltaic modules. In *IEA PVPS Task 13, Final Report*.

Lasher, G. and Stern, F. (1964) Spontaneous and stimulated recombination radiation in semiconductors. *Physical Review*, **133**, A553–A563.

Lim, S.Y., Rougieux, F.E., and Macdonald, D. (2013) Boron-oxygen defect imaging in p-type Czochralski silicon. *Applied Physics Letters*, **103** (9), 092105.

Livescu, G., Angell, M., Filipe, J., and Knox, W.H. (1990) A real-time photoluminescence imaging system. *Journal of Electronic Materials*, **19** (9), 937–942.

Macdonald, D., Tan, J., and Trupke, T. (2008) Imaging interstitial iron concentration in boron-doped crystalline silicon using photoluminescence. *Journal of Applied Physics*, **103**, 73710.

McMillan, W., Trupke, T., Weber, J. *et al.* (2010) In-line monitoring of electrical wafer quality using photoluminescence imaging. In *25th European Photovoltaic Solar Energy Conference, Valencia*, Spain, pp. 1346–1351.

Michl, B., Giesecke, J.A., Warta, W., and Schubert, M.C. (2012) Separation of front and backside surface recombination by photoluminescence imaging on both wafer sides. *IEEE Journal of Photovoltaics*, **2** (3), 348–351.

Michl, B., Rüdiger, M., Giesecke, J.A. *et al.* (2012) Efficiency limiting bulk recombination in multicrystalline silicon solar cells. *Solar Energy Materials and Solar Cells*, **98**, 441–447.

Mitchell, B., Trupke, T., Nyhus, J., and Weber, J.W. (2011) Bulk minority carrier lifetimes and doping of silicon bricks from photoluminescence intensity ratios. *Journal of Applied Physics*, **109**, 083111.

Müller, J., Bothe, K., Herlufsen, S. *et al.* (2012) Reverse saturation current density imaging of highly doped regions in silicon: A photoluminescence approach. *Solar Energy Materials and Solar Cells*, **106**, 76–79.

Müller, M., Altermatt, P.P., Schlegel, K., and Fischer, G. (2012) A method for imaging the emitter saturation current with lateral resolution. *IEEE Journal of Photovoltaics*, **2** (4), 586–588.

Ostapenko, S., Tarasov, I., Kalejs, J.P. *et al.* (2000) Defect monitoring using scanning photoluminescence spectroscopy in multicrystalline silicon wafers. *Semiconductor Science Technology*, **15** (8), 840–848.

Ramspeck, K., Bothe, K., Hinken, D. *et al.* (2007) Recombination current and series resistance imaging of solar cells by combined luminescence and lock-in thermography. *Applied Physics Letters*, **90** (15), 153502.

Rau, U. (2007) Reciprocity relation between photovoltaic quantum efficiency and electroluminescent emission of solar cells. *Physical Review B*, **76** (8), 085303.

Rau, U., Huhn, V., Stoicescu, L. *et al.* (2014) Photocurrent collection efficiency mapping of a silicon solar cell by a differential luminescence imaging technique. *Applied Physics Letters*, **105** (16), 163507.

Schinke, C., Hinken, D., Schmidt, J. *et al.* (2013) Modeling the spectral luminescence emission of silicon solar cells and wafers. *IEEE Journal of Photovoltaics*, **3** (3), 1038–1052.

Shen, C., Kampwerth, H., and Green, M.A. (2014) Photoluminescence based open circuit voltage and effective lifetime images re-interpretation for solar cells: The influence of horizontal balancing currents. *Solar Energy Materials and Solar Cells*, **130**, 393–396.

Shen, C., Kampwerth, H., Green, M. *et al.* (2013) Spatially resolved photoluminescence imaging of essential silicon solar cell parameters and comparison with CELLO measurements. *Solar Energy Materials and Solar Cells*, **109**, 77–81.

Sinton, R.A., and Cuevas, A. (1996) Contactless determination of current-voltage characteristics and minority-carrier lifetimes in semiconductors from quasi-steady-state photoconductance data. *Applied Physics Letters*, **69** (17), 2510–2512.

Sinton, R.A., Mankad, T., Bowden, S., and Enjalbert, N. (2004) Evaluating silicon blocks and ingots with quasi-steady-state lifetime measurements. In *Proceedings of the 19th European Photovoltaic Solar Energy Conference, Paris, France* (eds W. Hoffman, J. L. Bal, H. Ossenbrink, W. Palz, and P. Helm), pp. 520–523.

Sio, H.C., Trupke, T., and Macdonald, D. (2014) Quantifying carrier recombination at grain boundaries in multicrystalline silicon wafers through photoluminescence imaging. *Journal of Applied Physics*, **116** (24), 244905.

Sio, H.C., Xiong, Z., Trupke, T., and Macdonald, D. (2012) Imaging crystal orientations in multicrystalline silicon wafers via photoluminescence. *Applied Physics Letters*, **101** (8), 082102.

Smestad, G.P. (1995) Absorptivity as a predictor of the photoluminescence spectra of silicon solar cells and photosynthesis. *Solar Energy Materials and Solar Cells*, **38**, 57–71.

Søndenå, R., Hu, Y., Juel, M. *et al.* (2013) Characterization of the OSF-band structure in n-type Cz-Si using photoluminescence-imaging and visual inspection. *Journal of Crystal Growth*, **367**, 68–72.

Stoicescu, L., Reuter, M., and Werner, J. (2014) DAYSY: Luminescence imaging of PV modules in daylihgt. In *29th European Photovoltaics Solar Energy Conference and Exhibition*, Amsterdam, the Netherlands, pp. 2553–2554.

Tajima, M., Iwata, Y., Okayama, F. *et al.* (2012) Deep-level photoluminescence due to dislocations and oxygen precipitates in multicrystalline Si. *Journal of Applied Physics*, **111** (11), 113523.

Trupke, T., and Bardos, R. (2005) Photoluminescence: a surprisingly sensitive lifetime technique. In 31st IEEE Photovoltaic Specialists Conference, *Orlando*, USA, pp. 903–906.

Trupke, T., Bardos, R.A., Abbott, M.D. *et al.* (2006) Fast photoluminescence imaging of silicon wafers. *4th World Conference on Photovoltaic Energy Conversion, WCPEC-4*, Waikoloa, USA.

Trupke, T., Bardos, R.A., Schubert, M.C., and Warta, W. (2006) Photoluminescence imaging of silicon wafers. *Applied Physics Letters*, **89**, 44107.

Trupke, T., Daub, E., and Würfel, P. (1998) Absorptivity of silicon solar cells obtained from luminescence. *Solar Energy Materials and Solar Cells*, **53** (1–2), 103–114.

Trupke, T., Green, M.A., and Würfel, P. (2003) Temperature dependence of the radiative recombination coefficient of intrinsic crystalline silicon. *Journal of Applied Physics*, **94** (8), 4930–4937.

Trupke, T., Nyhus, J., and Haunschild, J. (2011) Luminescence imaging for inline characterisation in silicon photovoltaics. *Physica Status Solidi RRL*, **5** (4), 131–137.

Trupke, T., Nyhus, J., Sinton, R.A., and Weber, J. (2009) Photoluminescence imaging on silicon bricks. In 24th European Photovoltaic Solar Energy Conference, *Hamburg*, *September*, pp. 1029–1033.

Trupke, T., Pink, E., Bardos, R.A., and Abbott, M.D. (2007) Spatially resolved series resistance of silicon solar cells obtained from luminescence imaging. *Applied Physics Letters*, **90**, 93506.

Würfel, P. (2005) Physics of Solar Cells. Wiley-VCH Verlag GmbH and Co. KGaA, Weinheim.

Würfel, P., Finkbeiner, S., and Daub, E. (1995) Generalized Planck's radiation law for luminescence via indirect transitions. *Applied Physics A: Materials Science and Processing*, **A60** (1), 67–70.

Würfel, P., Trupke, T., Puzzer, T. *et al.* (2007) Diffusion lengths of silicon solar cells obtained from luminescence images. *Journal of Applied Physics*, **101**, 123110.

Yan, F., Johnston, S., Zaunbrecher, K. *et al.* (2012) Defect-band photoluminescence imaging on multi-crystalline silicon wafers. *Physica Status Solidi (RRL) - Rapid Research Letters*, **6** (5), 190–192.

7.3

Measurement of Carrier Lifetime, Surface Recombination Velocity, and Emitter Recombination Parameters

Henner Kampwerth
University of New South Wales, Sydney, Australia

7.3.1 Introduction

Recombination losses have a major influence on the overall performance of a solar cell. Every minority carrier charge that is lost to recombination is a charge that is missing from the current or voltage that a cell could provide. This chapter will discuss the measurement of parameters that express this recombination activity: the effective minority carrier lifetime τ_{eff}, the surface recombination velocity S and the emitter recombination current J_{0e}. A brief introduction to τ_{eff} is given in Section 7.3.1. It also provides the motivation for the structure of the following sections, which are graphically displayed in Figure 7.3.1. Section 7.3.2 presents the two commonly used probing techniques to measure the excess carrier concentration Δn and generation rate G (in cm^{-3}s^{-1}), which are used to calculate the effective lifetime as demonstrated in Section 7.3.3. Section 7.3.4 discusses ways to separate τ_{eff} into its bulk τ_b and surface τ_S components. Section 7.3.5 breaks τ_S further down to the surface recombination velocity S and emitter recombination current J_{0e}. As this is an extensive research field, the different sections are relatively brief summaries to provide an overview.

Photovoltaic Solar Energy: From Fundamentals to Applications, First Edition.
Edited by Angèle Reinders, Pierre Verlinden, Wilfried van Sark, and Alexandre Freundlich.

Companion website: www.wiley.com/go/reinders/photovoltaic_solar_energy

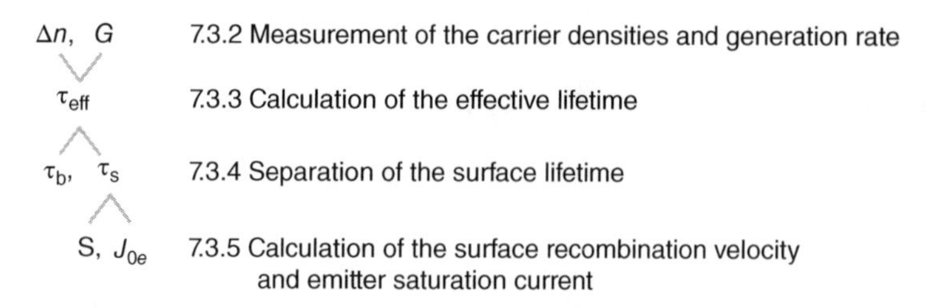

Figure 7.3.1 Breakdown of the lifetime values and associated sections

7.3.2 Carrier Lifetime

The excess minority carrier lifetime τ (in s) is the average time that photogenerated charge carriers, henceforth the excess minority carriers Δn (in cm^{-3}), exist before they recombine with a rate R (in $cm^{-3}s^{-1}$):

$$\tau(\Delta n) = \frac{\Delta n(t)}{R(t)} \tag{7.3.1}$$

Here the dependence on t indicates that these values vary in time. The "effective" carrier lifetime τ_{eff} is often used to express the total recombination in a material or device. It is the average lifetime of the entire structure (the wafer or finished solar cell). For its calculation an additional relationship between $G(t)$ and $R(t)$ is also needed:

$$\frac{\partial}{\partial t}\Delta n(t) = G(t) - R(\Delta n) \tag{7.3.2}$$

Together with Equation (7.3.1), this results in the generally valid equation:

$$\tau_{eff}(\Delta n) = \frac{\Delta n(t)}{G(\mathrm{t}) - \frac{\partial}{\partial t}\Delta n(t)} \tag{7.3.3}$$

The measurement of $\Delta n(t)$ and $G(t)$ is therefore the key to calculate $\tau_{eff}(\Delta n)$.

7.3.3 Measurement of the Carrier Densities and Generation Rate

Two time-resolved contact-less probing techniques are commonly used to measure minority carrier lifetimes in solar cells. Both techniques record the changes in time t of the minority carrier density $\Delta n(t)$ and Generation rate $G(t)$. The first technique calculates $\Delta n(t)$ by measuring the photoconductance (PC) of the semiconductor, whereas the second one calculates $\Delta n(t)$ by measuring the band-to-band photoluminescence (PL) emission. A third, more approximate method is the measurement of the device's open-circuit voltage V_{oc} under varying illumination.

All techniques typically rely upon the use of a reference photodetector which records the photon flux of an illumination source $\phi(t)$ (in $cm^{-2}s^{-1}$). It is installed in parallel to the detection of the main signal of the device or material under test to calculate $\Delta n(t)$. In order to scale $\phi(t)$ to $G(t)$, the average (or better weighted) optical absorption A (in cm^{-1}) of the wafer/cell to the particular spectrum of the light source must be known, that is

$$G(t) = A\phi(t) \tag{7.3.4}$$

An additional absorption measurement is therefore required.

7.3.3.1 Photoconductance (PC)

The photoconductance technique cannot be used on cells, as the metal contacts will interfere with the measurement. The PC method measures $\Delta n(t)$ via the electrical conductivity $\sigma(t)$ (in $ohm^{-1}cm^{-1}$) of the semiconductor, as shown in Figure 7.3.2. An excess of charge carriers in the conduction band of the material implies higher conductivity. Most commonly, $\sigma(t)$ is measured via the detection of an eddy current. The wafer is placed near an antenna that is driven by a radio-frequency oscillator circuit around 10 MHz; the electromagnetic waves from the antenna induce an eddy current within the wafer. As the induced current becomes a part of the overall oscillator circuit, values for $\sigma(t)$ can be extracted (Sinton, Cuevas, and Stuckings, 1996 and WCT-120). With appropriate conversion and calibration steps $\Delta n(t)$ can be calculated.

$$\Delta n(t) = \frac{\sigma(t)}{\left(\mu_n(\Delta n) + \mu_p(\Delta n)\right) q W} \tag{7.3.5}$$

The elementary charge q (in C) and wafer thickness W (in cm) are known values. The electron and hole mobilities μ_n and μ_p (in $cm^2V^{-1}s^{-1}$) are themselves functions of Δn (and doping) and can be found in the literature (see Chapter 3.1). Equation (7.3.5) is therefore an implicit function that has to be solved iteratively.

7.3.3.2 Photoluminescence (PL)

As shown in Figure 7.3.2, the PC and PL detection setups are nearly identical, except that the eddy-current antenna is replaced by a photodetector with sensitivity for wavelengths λ between 900 and 1200 nm. This means instead of conductivity, photons emitted by radiative

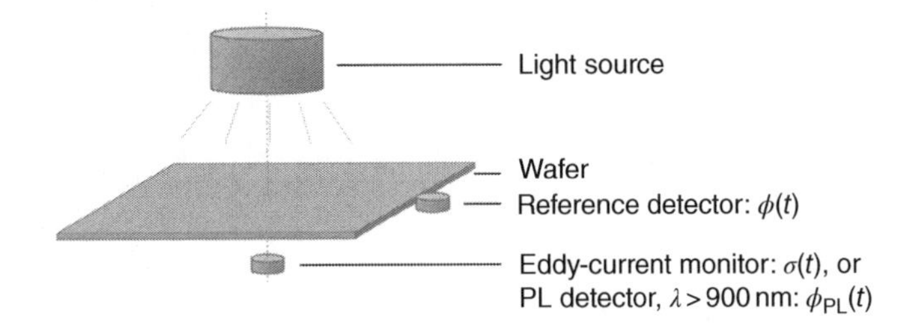

Figure 7.3.2 Schematic arrangement of a photoconductance or photoluminescence setup

recombination are detected; the rate of emission is expressed as the photoluminescence photon flux ϕ_{PL} (in cm^{-2}s^{-1}). The functional relationship between Δn and ϕ_{PL} can be converted to each other with appropriate calibration procedures.

$$\Delta n(t) = \frac{\phi_{PL}(t)}{C\,B\,N} \tag{7.3.6}$$

Here N (in cm^{-3}) is the doping density and B (in cm^{-3}s^{-1}) is a radiative recombination constant (a parameter containing various material constants) (Trupke *et al.*, 2003). C (in cm) is a optical scaling parameter that defines the fraction of generated photons actually being detected by the detector. This scaling parameter changes from sample to sample and its value is found by additional calibration procedures (Trupke, Bardos, and Abbott, 2005). Some commercial units that use PL for lifetime calculations are from Sinton (WCT-120PL) and BT Imaging (BTI).

7.3.3.3 *Open Circuit Voltage Analysis*

A lower limit for τ_{eff} may quite be easily extracted from the open circuit voltage V_{oc}[1]: Equation (7.3.7) shows the relationship between V_{oc} and Δn and doping density N, as well as its simplification for the common low injection conditions ($\Delta n << N$).

$$\Delta n(\Delta n + N) = n_i^2 \exp\left(\frac{qV_{oc}}{kT}\right) \xrightarrow{\Delta n \ll N} \Delta n = \frac{n_i^2}{N} \exp\left(\frac{qV_{oc}}{kT}\right) \tag{7.3.7}$$

Here n_i is the intrinsic carrier concentration (in cm^{-3}) and k is the Boltzmann constant (1.38 10^{-23} JK^{-1}). The temperature T (in K) and doping density N are usually sufficiently accurately known. Δn can be monitored in various ways. The generation rate G is here usually steady state, allowing a direct solving of Equation (7.3.8) in Section 7.3.4. A common commercial unit that make explicit use of this relationship is from Sinton Consulting (Suns-Voc).

7.3.4 Calculation of the Effective Lifetime

The time-resolved techniques of PC and PL provide a time-trace of the excess minority carrier density $\Delta n(t)$ and the generation rate $G(t)$. The conversion to injection-dependent effective lifetime $\tau_{eff}(\Delta n)$ is achieved by solving Equation (7.3.3). This would be a 'general' solution as provided by (Nagel *et al.*, 1999). A discussion of its solution would exceed the scope of this chapter. However, it is simple to solve the equation for two extreme conditions that are commonly achieved in practice: the quasi-steady-state (QSS) and transient conditions. The choice of an analytical approach depends on how fast $G(t)$ changes in relation to the effective lifetime τ_{eff}. If software is used where the general analysis is available, it would be the preferred default choice.

[1] To be more specific, here the implied voltage V_{oci} is meant. It is given by the split of the quasi-Fermi levels, which is directly related to the potential difference (or voltage) that characterizes the charge carrier distributions within the material. As V_{oc} is measured at the terminals, some voltage losses will have occurred and therefore $V_{oc} < V_{oci}$. As a result, τ_{eff} calculated from V_{oc} will form a lower limit of the wafer's actual lifetime. The difference is usually small.

Finally, a third, occasionally used and more approximate method will be discussed. It uses the open circuit voltage V_{oc} of an illuminated cell, such as that taken from an illuminated *I-V* curve, and calculates the minimum effective lifetime that must be present to achieve such a voltage.

7.3.4.1 Quasi-Steady-State (QSS) Analysis

Under quasi-steady-state (QSS) conditions, the changing generation rate $G(t)$ is so slow in the timeframe of τ_{eff} that it can be approximated as quasi-constant, or 'quasi-steady'. This results in a condition where $G(t)$ and $R(t)$ are in equilibrium with (and hence equal to) each other. This negates Equation (7.3.2) and simplifies Equation (7.3.3) to:

$$\tau_{eff}(\Delta n) = \frac{\Delta n(t)}{G(t)} \tag{7.3.8}$$

7.3.4.2 Transient Analysis

Under transient conditions, the generation rate $G(t)$ changes substantially within the timeframe of τ_{eff}. The equilibrium between $G(t)$ and $R(t)$ is interrupted and $\Delta n(t)$ will change in time until equilibrium is reached again. The transient technique is commonly used for signals *after* a light source is suddenly switched off, which is useful when the light source is a pulsed laser, see Figure 7.3.3.

For simplicity, this point shall be defined as $t=0$, and thus $G(t)=0$ for $t>0$. Under these assumptions, Equation (7.3.3) simplifies to:

$$\tau_{eff}(\Delta n) = \left. \frac{\Delta n(t)}{-\frac{\partial}{\partial t}\Delta n(t)} \right|_{t>0} \tag{7.3.9}$$

The lifetime $\tau_{eff}(t)$ can then be extracted by fitting the equation to the measurement data. Note that Equation (7.3.9) is a normalized equation, meaning that absolute values of Δn are not

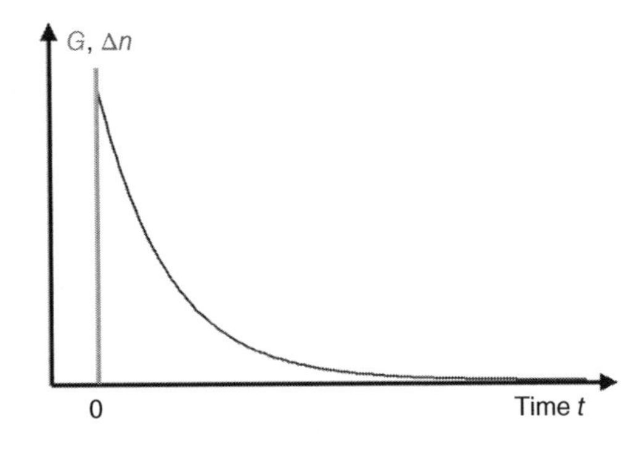

Figure 7.3.3 Typical exponential decay of the minority carrier decay $\Delta n(t)$ after a transient generation $G(t)$ at time t = 0

required. This particular case allows the optical constant C in Equation (7.3.6) to remain unknown without consequences; only the relative change of $\Delta n(t)$ must be correctly represented.

In the case where the absolute value of Δn is known, Equation (7.3.9) simplifies further to the general solutions for Δn and τ_{eff}:

$$\Delta n(t) = \exp\left(\frac{-t}{\tau(\Delta n)}\right)\Bigg|_{t>0} \tag{7.3.10a}$$

$$\tau(\Delta n) = \frac{-t}{\ln(\Delta n(t))}\Bigg|_{t>0} \tag{7.3.10b}$$

Equation (7.3.10b) is of all the presented techniques the most direct conversion of Δn to τ_{eff}.

7.3.5 Separation of the Surface Lifetime

The effective lifetime $\tau_{eff}(t)$, which was measured in Section 7.3.3, can be separated into two location-specific components: the surface lifetime $\tau_S(t)$ and the bulk lifetime $\tau_b(t)$. These terms are related to effective lifetime τ_{eff} by a sum of their inverse values.

$$\frac{1}{\tau_{eff}} = \frac{1}{\tau_b} + \frac{1}{\tau_s} \tag{7.3.11}$$

This separation can be carried out by various approaches, which are discussed below. The more popular ones will be given with basic equations. Note that often $\tau_b(t)$ is of interest for research. The separation works for both the surface and bulk. The following methods use the wavelength-dependent absorption depth; the second uses wafers of different thicknesses, the third uses a reference wafer and the fourth uses high-injection scenarios.

7.3.5.1 Absorption Variation

The absorption coefficient is wavelength-dependent. Longer-wavelength photons penetrate deeper into the material than shorter ones. The optically generated Δn has a similar depth profile to the optical absorption profile. With this the emphasis of recombination activity can be shifted to or away from the front surface. The selective different impacts that τ_S has on the overall τ_{eff} can be used to separate both. The first use of this technique was by Buczkowski, Radzimski, Rozgonyi, and Shimura (1991) and Gaubas and Vanhellemont (1996).

7.3.5.2 Thickness Variation

This method requires several identical wafers of different thickness W, but with otherwise identical parameters. Measuring the effective lifetime τ_{eff} of each wafer at the same level of Δn

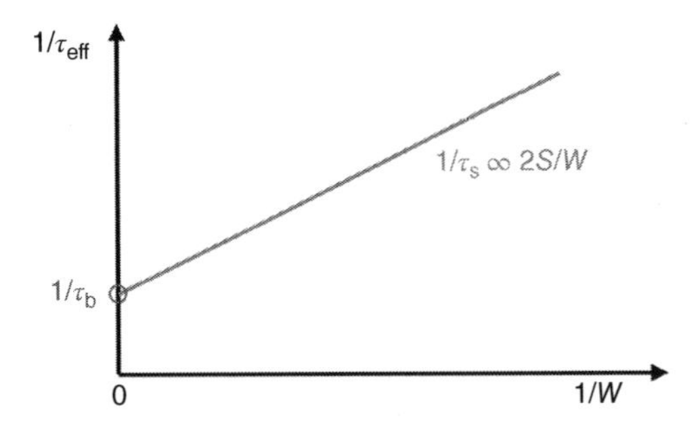

Figure 7.3.4 Graphic presentation of the inverse effective lifetime $1/\tau_{eff}$ vs the inverse wafer thickness W. The inverse bulk lifetime $1/\tau_b$ can be identified by the interception with the ordinate, leaving the inverse surface lifetime $1/\tau_S$

data are plotted in a graph with axis $1/\tau_{eff}$ vs. $1/W$, see Figure 7.3.4. The extension to $1/W=0$ provides the offset $1/\tau_b$. The $1/\tau_S(W)$ data follows approximately $2S/W$ where S is the surface recombination velocity (in cms^{-1}), but can be solved accurately via Equation (7.3.14) in Section 7.3.6.1.

7.3.5.3 *Reference Wafer*

It is common that a reference wafer with a high bulk lifetime τ_b is processed in the same batch as other wafers for reference purposes. This allows the monitoring of the impact of subsequent processes on the passivation of the wafer surfaces. Ideally the doping levels of the surfaces of the reference wafer and batch wafers should be similar as they impact the surface passivation characteristics.

The lifetime components of τ_{eff} are the bulk's radiative lifetime τ_{rad}, the bulk's Shockley-Read-Hall lifetime τ_{SRH} (Hall, 1952; Shockley and Read, 1952) and the bulk's Auger lifetime τ_{aug} (Auger, 1923). The surface component remains τ_S. They all add up with their inverse values.

$$\frac{1}{\tau_{eff}} = \left[\frac{1}{\tau_{rad}} + \frac{1}{\tau_{SRH}} + \frac{1}{\tau_{aug}}\right] + \frac{1}{\tau_S} \tag{7.3.12}$$

The first two terms $1/\tau_{rad}$ and $1/\tau_{SRH}$ can be ignored for high quality wafers. The remaining Auger lifetime term can be calculated (Richter *et al.*, 2012). With this, Equation (7.3.12) concludes to an Auger corrected measured effective lifetime τ_{eff} that equals τ_S.

$$\tau_S = \left(\frac{1}{\tau_{eff}} - \frac{1}{\tau_{aug}}\right)^{-1} \tag{7.3.13}$$

In cases where the surface passivation quality limits τ_{eff} to the extent that $\tau_b > 10\,\tau_{eff}$, then term $1/\tau_{aug}$ in Equation (7.3.13) can be ignored and $\tau_S = \tau_{eff}$.

7.3.5.4 *High Injection Scenario*

The extraction of τ_S via high injection scenarios is the main part of Kane and Swanson's work (Kane and Swanson, 1985). The technique uses a measured effective lifetime $\tau_{eff}(\Delta n)$ at high injection levels ($\Delta n >> N$ and numerically extracts the surface lifetime $\tau_S(\Delta n)$. The approach is similar to the one described in the Section 7.3.5.3 above. Here the τ_{rad} term is ignored and τ_{aug} is calculated via the method presented in (Richter *et al.*, 2012). However, τ_{SRH} is approximated to be constant.

7.3.6 Calculation of the Surface Recombination Velocity and Emitter Recombination Current

The surface lifetime τ_S from Section 7.3.5 can either be expressed as a function of surface recombination velocity S or emitter recombination current density J_{0e} (in A.cm^{-2}). The definition of the latter is often unclear in the literature: in some instances J_{0e} is used as a metric to describe recombination at the surface with the protection of a potential barrier (emitter); it is therefore similar to S. However, in other cases, it is used in a manner more in keeping with its name, which is the recombination current of the emitter itself, without the contribution of the surface.

7.3.6.1 *Surface Recombination Velocity S*

An often used approximation to convert τ_S to the surface recombination velocity S is from (Grivickas, Noreika, and Tellefsen, 1989; Sproul, 1994), as shown in Equation (7.3.14). Its error of <5% is sufficiently low for most calculations. Here W is the wafer thickness and D the diffusion constant of the minority carriers (in cm^{2}s^{-1}).

$$S = \frac{W}{2\left(\tau_s - \frac{1}{D}\left(\frac{W}{\pi}\right)^2\right)} \tag{7.3.14}$$

7.3.6.2 *Emitter Recombination Current J_{0e}*

The emitter recombination current density J_{0e} describes the recombination current density of the doped surfaces or emitter according to an ideal one-diode model, including surface recombination, SRH or Auger recombination, band gap narrowing within the emitter. It does not include the bulk properties. The relationship between the surface lifetime τ_S and J_{0e} in high-injection was provided by Kane and Swanson (1985).

$$J_{0e} = \frac{q W n_i^2}{\tau_S \Delta n} \tag{7.3.15}$$

The elementary charge q, thickness W and intrinsic carrier concentration n_i are known. The recombination current density J_{0e} of the surface can then be extracted from high-injection measurement (i.e. with $\Delta n >> N_A$ or N_D).

If it is necessary to know the emitter recombination current density without the surfaces, one can measure the emitter doping profile (for example, via secondary ion mass spectrometry (SIMS), electrochemical capacitance-voltage (ECV) profiling, or a spreading-resistance measurement) and couple it with a numerical simulation, such as EDNA (McIntosh and Altermatt, 2010), to separate surface recombination velocity *S* from the isolated emitter recombination current J_{0e*} and available as an online calculator (PVlighthouse). However, the need for this fine separation is not very common and surface and emitter are usually taken as one.

7.3.7 Conclusion

This chapter discussed various popular techniques to measure the time-dependent minority carrier density $\Delta n(t)$ as well as basic mathematical models to further convert them to effective lifetimes $\tau_{eff}(\Delta n)$. With $\tau_{eff}(\Delta n)$ known, additional techniques are discussed to separate $\tau_{eff}(\Delta n)$ into its bulk and surface components $\tau_b(\Delta n)$ and $\tau_S(\Delta n)$. Finally, the extraction of the emitter saturation current density J_{0e} was indicated.

List of Symbols

Symbol	Definition	Unit
B	Radiative recombination constant	$cm^{-3}s^{-1}$
C	Optical scaling parameter	cm
G	Generation rate	$cm^{-3}s^{-1}$
J_{0e}	Emitter recombination current density	$A\ cm^{-2}$
k	Boltzmann constant	$1.38 \cdot 10^{-23}\ JK^{-1}$
N	Doping density	cm^{-3}
N_A	Acceptor concentration	cm^{-3}
N_D	Donor concentration	cm^{-3}
n_i	Intrinsic carrier concentration	cm^{-3}
q	Charge of the electron	$1.6 \cdot 10^{-19}\ C$
R	Recombination rate	$cm^{-3}s^{-1}$
S	Surface recombination velocity	$cm\ s^{-1}$
t	Time	s
T	Temperature	K
V_{oc}	Open circuit voltage	V
V_{oci}	Implied voltage	V
W	wafer thickness	cm
Δn	excess carrier concentration	cm^{-3}
λ	wavelength	nm
μ_n	Electron mobility	$cm^2V^{-1}s^{-1}$
μ_p	Hole mobility	$cm^2V^{-1}s^{-1}$
Φ	Photon flux	$cm^{-2}s^{-1}$
ϕ_{PL}	Photoluminescence photon flux	$cm^{-2}s^{-1}$

(Continued)

(*Continued*)

Symbol	Definition	Unit
σ	electrical conductivity	$\Omega^{-1}cm^{-1}$
τ	Excess minority carrier lifetime	s
τ_{aug}	Auger lifetime	s
τ_b	Bulk minority carrier lifetime	s
τ_{eff}	effective minority carrier lifetime	s
τ_{rad}	radiative lifetime	s
τ_S	Surface minority carrier lifetime	s
τ_{SRH}	Shockley-Read-Hall lifetime	s

List of Acronyms

Acronym	Definition
I-V	Current *vs.* voltage
ECV	Electrochemical capacitance-voltage
QSS	Quasi-steady state
PC	Photoconductance
PL	Photoluminescence
SIMS	secondary ion mass spectrometry

References

Auger, P. (1923) Sur les rayons andbeta; secondaires produits dans un gaz par des rayons X. *C.R.A.S.*, 177.

BTI, http://www.btimaging.com/

Buczkowski, A., Radzimski, Z.J., Rozgonyi, G., and Shimura, F. (1991) Bulk and surface components of recombination lifetime based on a two-laser microwave reflection technique. *Journal of Applied Physics*, **69** (9), 6495. Available at: http://link.aip.org/link/JAPIAU/v69/i9/p6495/s1andAgg=doi

Gaubas, E. and Vanhellemont, J. (1996) A simple technique for the separation of bulk and surface recombination parameters in silicon. *Journal of Applied Physics*, **80** (11), 6293–6297. AIP. Available at: http://link.aip.org/link/?JAP/80/6293/1

Grivickas, V., Noreika, D., and Tellefsen, J.A. (1989) Surface and auger recombinations in silicon wafers of high carrier density. *Soviet Physics Collection*, **29** (5), 591–597.

Hall, R.N. (1952) Electron-hole recombination in germanium. *Physical Review*, **87** (2), 387. American Physical Society. Available at: http://link.aps.org/abstract/PR/v87/p387

Kane, D.E. and Swanson, R. (1985) Measurement of the emitter saturation current by a contactless photoconductivity decay method. *18th IEEE Photovoltaic Specialists Conference (PSC)* (p. 578). IEEE, New York.

McIntosh, K.R. and Altermatt, P.P. (2010) A freeware 1D emitter model for silicon solar cells. *Conference Record of the IEEE Photovoltaic Specialists Conference*, pp. 2188–2193.

Nagel, H., Berge, C., Aberle, A.G. *et al.* (1999) Generalized analysis of quasi-steady-state and quasi-transient measurements of carrier lifetimes in semiconductors. *Journal of Applied Physics*, **86** (11), 6218. Available at: http://link.aip.org/link/?JAP/86/6218/1

PVlighthouse, available at: http://www.pvlighthouse.com.au/calculators/EDNA%202/EDNA%202.aspx

Richter, A., Glunz, S.W., Werner, F., Schmidt, J., and Cuevas, A. (2012) Improved quantitative description of Auger recombination in crystalline silicon. *Physical Review B – Condensed Matter and Materials Physics*, **86**.

Shockley, W. and Read, W.T. (1952) Statistics of the recombinations of holes and electrons. *Physical Review*, **87** (5), 835. Available at: http://link.aps.org/abstract/PR/v87/p835

Sinton, R.A., Cuevas, A., and Stuckings, M. (1996) Quasi-steady-state photoconductance, a new method for solar cell material and device characterization. *Photovoltaic Specialists Conference, 1996, Conference Record of the Twenty Fifth IEEE*, pp. 457–460.

Sproul, A.B. (1994) Dimensionless solution of the equation describing the effect of surface recombination on carrier decay in semiconductors. *Journal of Applied Physics*, **76** (5), 2851–2854. AIP. Available at: http://link.aip.org/link/?JAP/76/2851/1 (accessed February 15, 2015).

Trupke, T., Bardos, R.A., and Abbott, M.D. (2005) Self-consistent calibration of photoluminescence and photoconductance lifetime measurements. *Applied Physics Letters*, **87** (18), 184102. AIP. Available at: http://link.aip.org/link/?APL/87/184102/1

Trupke, T., Green, M.A., Würfel, P. *et al.* (2003) Temperature dependence of the radiative recombination coefficient of intrinsic crystalline silicon. *Journal of Applied Physics*, **94** (8), 4930–4937.

WCT-120, Available at: http://www.sintoninstruments.com/Sinton-Instruments-WCT-120.html

WCT-120PL, Available at: http://www.sintoninstruments.com/Sinton-Instruments-WCT-120PL.html

WCT-Suns-Voc, Available at: http://www.sintoninstruments.com/Sinton-Instruments-Suns-Voc.html

7.4

In-situ Measurements, Process Control, and Defect Monitoring

Angus Rockett
University of Illinois, Urbana, IL, USA

7.4.1 Introduction

Manufacturing of photovoltaics requires the careful control of process conditions to enhance yield, improve performance, and maintain product quality. Process control requires a measurement of the properties of the grown material, which directly and accurately reflects both the status of the process and the performance of the material. Preferably these measurements are made continuously within the process environment so that they can be used for immediate feedback on the process and the quality of the material.

The majority of the techniques described here relate to vapor-deposited thin films because those are the commonest methods used to produce materials for microelectronics, and photovoltaics in particular. Bulk crystal growth methods such as the Czochralski process have their own set of unique challenges for in-situ monitoring and control. For the most part these will not be discussed here as they are specific to those methods rather than widely applicable. Most materials produced by these means are studied post-growth by methods such as microwave reflectivity, discussed below. Materials produced by techniques such as screen-printing are widely used in photovoltaics but are amenable to more macroscopic analysis by techniques, such as optical microscopy or even simple video imaging and image analysis. The more difficult problem is to analyse a defect-sensitive semiconductor continuously as the material is produced.

The fundamental requirements for a continuous process-monitoring technique providing immediate feedback control options are that it should be non-destructive (unless it monitors a sacrificial portion of the grown material), compatible with the process environment, and that it be fast enough to be performed as the material passes by the test station. Typical measurements may assess fluxes of atoms depositing on the sample surface during a thin film growth

Photovoltaic Solar Energy: From Fundamentals to Applications, First Edition.
Edited by Angèle Reinders, Pierre Verlinden, Wilfried van Sark, and Alexandre Freundlich.

Companion website: www.wiley.com/go/reinders/photovoltaic_solar_energy

process, the optical behavior of a semiconductor or dielectric, the electronic properties of a material, or the performance of a device shortly after it is produced.

This chapter covers a selection of some of the more useful techniques and methods that can be applied to a broad range of materials. There are many other methods to study materials during or shortly after growth. The approaches described here primarily concern characterization of bulk solids or surfaces, while liquids and liquid-solid interfaces are not considered, being much harder to study and of less broad applicability (but of high technological importance in some cases such as Si growth).

The techniques that are discussed below include thin film vapor phase flux monitoring approaches such as mass spectrometry, glow-discharge optical spectroscopy, and others. Post-growth characterization methods such as x-ray fluorescence spectroscopy are also considered. Ellipsometry can provide a similar measurement but is grouped with the optical methods. These also include optical beam-induced current and photoluminescence. Non-contact electrical measurements are more difficult to implement but one of the more important is the minority carrier lifetime measurements obtained from microwave reflectivity methods. In some cases, it is practical to produce contacts to the material as it is fabricated and so measurements are used, such as flash-lamp current/voltage and electroluminescence measurements, which can often be obtained reasonably shortly after completing the devices.

While the methods below provide very good control of many processes, one should not overlook simple optical techniques. Anyone who has operated a solar cell deposition tool for some time is likely to be able to tell simply by looking at a thin film whether it is acceptable because in many cases the film will be visibly different in color or reflectivity. This may be possible to monitor by a simple camera and image analysis software. More sophisticated techniques include light scattering, in which a laser or other light source reflects off the sample surface and the reflected intensity is detected. This can provide, for example, surface roughness information, which is directly related to deposition conditions in many cases and is often correlated with device performance.

7.4.2 Monitoring of Vapor Phase Film Growth

The defects in semiconductors, and particularly in compound semiconductors, are highly sensitive to the flux of material arriving at the sample surface as growth proceeds. The ratio of elements arriving at the surface often controls the stoichiometry of a compound, which determines the density of point defects on each sub-lattice in the compound structure. For example, the conductivity, defect density, formation of second phases, etc. are all determined by the Cu/In ratio in the growth of $CuInSe_2$ (CIS). At the same time, the overall flux of material determines the growth rate, which can have an impact on defect density. For example, in the epitaxial growth of Si, an excessive growth rate can result in the formation of dislocations and other extended defects.

7.4.2.1 Temperature

At the same time, the sample temperature is critical to the control of defects. Therefore, a few words on temperature control are worthwhile. One can measure temperatures very easily and inexpensively with thermocouples if they are in intimate contact with the sample

Figure 7.4.1 A compact portable optical pyrometer

(e.g., by soldering to the sample or insertion into a hole in the sample). This is often impractical. Furthermore, the sample is often not in direct contact with a heater or monitored solid. A generally highly reliable method for temperature measurement is a multi-wavelength optical pyrometer (Figure 7.4.1). This requires a clear and clean pathway to view the sample but measures the temperature very accurately. Such a pyrometer measures the optical emission intensity from a sample at two wavelengths. It then fits a black-body spectrum to the two intensities and uses that spectrum to determine the sample temperature. A two-wavelength pyrometer does not require knowledge of the emissivity of the sample surface and is less sensitive to the optical transmittance of things like windows in the optical path (as long as the transmittance is not wavelength-dependent in the range of the pyrometer). An important consideration in the selection of an optical pyrometer is that the sample be opaque to the wavelengths it uses. A sample that is transparent to these wavelengths can mean that the pyrometer may measure a surface behind the sample rather than the intended surface. For some materials, pyrometry is difficult to apply because of wavelength-dependent emissivity, particularly in the infrared spectrum where most pyrometers operate. This is a known problem for CIGS and should be kept in mind for other photovoltaic materials (Repins *et al.*, 2005).

An interesting opportunity that arises from temperature monitoring is the detection of changes in surface emissivity of the deposited material. In this case the temperature is monitored by a non-optical method, such as a thermocouple, which is used to control the sample heater power. When the surface emissivity changes during film deposition, for example, as a result of a change in the material surface chemistry, the heater power needed to maintain a fixed temperature changes. Monitoring the heater power while controlling measured temperature provides a measure of when this transformation occurs. This is the case when group I rich CIGS is transformed into group III rich material by the addition of group III element during growth. Details of this method may be found in Kohara *et al.* (1995), and Negami *et al.* (1996).

7.4.2.2 *Flux*

Flux monitors can be divided roughly into two categories: those that monitor what is in the vapor phase and those that monitor what is stuck to the sample surface. Examples of these include mass spectrometers or glow discharge optical spectrometers that measure species in the vapor phase, and x-ray emission spectrometers and ellipsometers that measure what is deposited on the sample surface. The major problem with vapor phase flux measurements is that they do not reflect what actually sticks to and stays on the sample surface. This is critical in cases such as the deposition of Cu_2ZnSnS_4 (CZTS) where the concentration of S may affect the desorption of Sn from the sample surface. In cases of high vapor pressure species, such as Group VI elements, the majority of the incident flux desorbs from the sample surface later. Therefore, the vapor phase gas pressure is only loosely connected to the concentration of that species on the surface. For example, in GaAs vapor phase growth, the As is supplied in massive excess, sometimes 40x the Ga flux, and the remainder desorbs from the sample surface. This excess flux may be required to maintain the surface in an As-stabilized state where the final layer of atoms on the surface is primarily As.

To apply a mass spectrometer to in-situ process control, the instrument is typically installed in a differentially-pumped chamber with a small opening for sampling the flux without contaminating the whole spectrometer (Figure 7.4.2). This method is well suited to evaporation-based processes but could be used for almost any deposition process. Implementation of a sampling mass spectrometer at atmospheric pressure is possible but requires a gas sampling system with a multistage differential pumping approach (Hastie *et al.*, 1973). The spectrometer must be located close to the surface on which deposition is occurring or the position must be chosen to sample the fluxes in the same ratios as occur at the sample surface. This may be problematic for large substrates such as full-scale solar panels if the sampling point cannot be located close to the substrate edge. Mass spectrometer control has been successfully

Figure 7.4.2 A simple quadrupole mass spectrometer such as this could be used to measure fluxes of atoms in a thin film deposition instrument

demonstrated by Stolt *et al.* for deposition of controlled compositions of CIGS (Stolt *et al.*, 1985; Hedstrom *et al.*, 1993). In this implementation a cluster of evaporation point sources deposited the CIGS onto a heated substrate over a significant distance. The mass spectrometer sampled individual Cu, In, Ga, and Se masses sequentially and the resulting signal is used to dynamically control the source temperatures to achieve a programmed flux profile that changed during the course of the deposition.

Glow discharge optical emission spectroscopy (GD-OES) (Greene and Whelan, 1973) is a method in which the light emission from a glow discharge in a gas phase is used to assess the composition of that gas. When a glow discharge is established in a gas, a few percent of the species in the gas are typically ionized. De-ionization of these species produces characteristic light emissions at specific wavelengths. These emissions can be analysed by wavelength to determine quantitatively how many atoms of a given type are present in the gas. Similar methods analyse the ions by mass spectrometry in the glow discharge mass spectrometry method (GD-MS). A recent review of various glow discharge-based methods for post-growth characterization of photovoltaic materials was given by Schmitt *et al.* (2014). However, GD-OES and GD-MS can also be effective for monitoring thin film deposition conditions. Deposition of a-Si:H by plasma-enhanced chemical vapor deposition is particularly well suited to GD-OES, as it occurs at a relatively high pressure where operation of a mass spectrometer is difficult and where a glow discharge is intrinsic to the deposition process. Examples of the application of GD-OES to a-Si:H process control were given by Paesler *et al.* (1980) and Bauer and Bilger (1981). This method can also be used in any other deposition process in which a glow discharge is present, for example, in any sputter deposition approach. It is important to note that the probability of ionizing individual elements in the gas phase depends on the gas composition. In particular, changes in the density of metal species in the gas can radically change the electron temperature in the discharge, which alters the probability of ionization dramatically. In a case where one is controlling a relatively stable process that exhibits only minor changes in gas chemistry, the method is quite reliable.

Several simpler methods related to GD-OES include electron impact emission spectroscopy (EIES) in which an electron beam is used to ionize the gas rather than a glow discharge; atomic absorption spectroscopy (AAS), where absorption of light by the gas is detected at wavelengths characteristic of specific atoms; and other methods. EIES requires that the process environment be compatible with an electron gun or other electron beam source. This works best in high vacuum or better and preferably in the absence of reactive gases. AAS works very well and produces quantitative results as long as the light source intensity at the detector in the absence of the detected species is well determined.

X-ray fluorescence (XRF) is a non-contact material analysis technique that is vacuum compatible and works well with thin film deposition tools. In this method an x-ray source illuminates the sample, ionizing the atoms by excitation of core level electrons to high energies (Figure 7.4.3). Deionization of the resulting ions leads to emission of x-rays with characteristic energies. The energy and flux of these may be detected with a Li doped Si detector as in a conventional energy dispersive x-ray analysis instrument or may be dispersed by diffraction from a reference crystal and different wavelengths detected by individual detectors or a movable detector, as in wavelength-dispersive spectroscopy. The general approach has been described in detail in Beckhoff (2006). A specific application to thin film photovoltaic *in-situ* process control for deposition of CIGS was given by Eisgruber *et al.* (2002) The latter contains a detailed description of the setup of the apparatus and data analysis. An example of more general

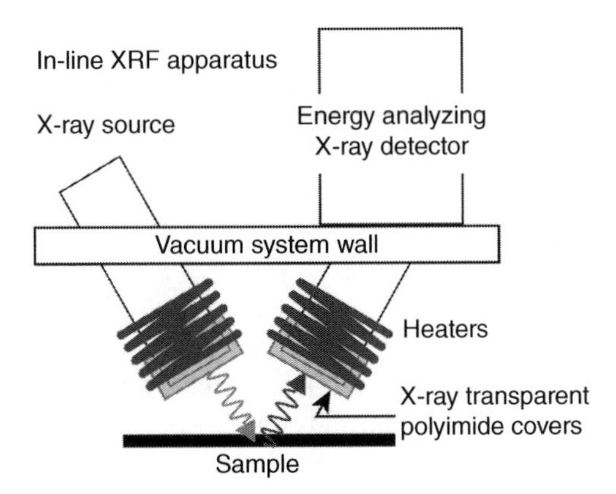

Figure 7.4.3 A schematic of the apparatus for an X-ray fluorescence system for monitoring film growth. Heating units are incorporated into the two units to prevent adsorption of species from the vapour phase. Source: (Eisgruber *et al.*, 2002)

application of X-ray reflectivity and X-ray fluorescence to materials for microelectronics, including both high and low dielectric constant materials, may be found in Wyon *et al.* (2004). A related method is X-ray reflectivity which simply detects the reflection of illuminating X-rays from the source. This method is sensitive, for example, to surface morphology and works well in characterizing multilayer stacks of films. One of the advantages of the use of XRF for process control is that typically the energies of the characteristic X-rays are well known, so a detection system optimized for only those elements can be used, reducing the system cost and complexity. In Eisgruber *et al.* (2002), it is pointed out that in some processes, species from the gas phase may condense on the detector. Therefore, they describe an approach to protect the X-ray source and detector. The source and detector can also be configured to translate across a large substrate area and therefore produce a map of the composition of the sample as a function of position. Because of the long mean-free-path of X-rays in solids, this method probes the entire thickness of relatively thick films (microns). This is an advantage because it can detect elements throughout typical photovoltaic device materials but is a disadvantage when compositions include gradients, to which the technique is not sensitive. Another concern is that fluorescent X-rays may be reabsorbed on the way out of the sample. This, in turn, can enhance the fluorescent emission from overlying species. Therefore, corrections are required for quantitative analysis for the atomic number of the species being studied, the absorption of fluorescent X-rays and re-emission fluorescence of additional X-rays resulting from that absorption. Quantitative analysis techniques are described in Beckhoff (2006) and various sources on the internet and are included in most XRF analysis software.

7.4.3 Optical Properties

Spectroscopic ellipsometry is a powerful technique for *in-situ* characterization of semiconductors and translucent metallic layers that provides information on thickness, composition, roughness, and chemical differences throughout the layers when appropriate modeling can be developed to understand the data. In this technique, a polarized light beam of a specific wavelength reflects off the sample and is analysed for polarization direction and measured for intensity. When the film contains multiple transparent or translucent layers with different dielectric constants, light is partially or fully reflected. The technique takes advantage of two factors: (1) interference of light reflected from an interior interface with light reflected from an interface lying closer to the surface; and (2) the difference in reflection of waves with their electric fields in or out of the interface plane. Reflection is strongest when the electric field of the light lies in the interface. For most orientations of the incident wave electric field, polarization of the outgoing wave is different from the incident wave polarization and the change in polarization depends on where and how the reflection occurs. Because the reflectivity of an interface depends on the difference of dielectric constant and because the interference of the reflected wave with waves reflecting farther out depends on the wavelength, a model can be constructed relating the observed reflection and outgoing polarization to the dielectric constant and thickness of all layers within the film. Measurements as a function of wavelength and incident polarization angle provide the greatest detail for the fit, although changing the incident polarization is not necessary for full simulation. If a film is not sufficiently transparent, then lower-lying layers may not be detectable. The complete theory and description of the apparatus have been described in detail by Fujiwara (2007).

As an example of its application, Collins and coworkers have implemented spectroscopic ellipsometry for *in-situ* process control (Figure 7.4.4). They used the technique to control the

Figure 7.4.4 Module-scale spectroscopic mapping tool under development as "In-line SE" by J.A. Woolam Co. Source: courtesy of Rob Collins, University of Toledo

deposition of CdTe (Jian *et al.*, 2005; Chen *et al.*, 2009); amorphous Si (Dahal *et al.*, 2010); Cu(In,Ga)Se_2 (Walker *et al.*, 2009); and other materials. They have also used the technique post-deposition to map large area films using the method (Aryal *et al.*, 2013).

A related technique is laser light scattering. This method measures the diffuse and specular reflection of a laser beam from a sample surface. When the surface roughness is of the order of half a wavelength or larger, then light is increasingly strongly scattered by the rough surface. By measuring the light intensity away from the specular reflection, the surface roughness can be deduced. A detailed discussion of the application of laser light scattering, including for measurement of polymer properties, may be found in Germer and Fasolka (2003).

It should be noted for completeness that a powerful but highly specialized method, reflection high energy electron diffraction (RHEED), may be used in thin film multilayers grown by molecular beam epitaxy or related methods and is only relevant to single crystal growth. In this technique, a moderate energy (20 keV) electron beam reflects off the semiconductor surface at a low angle. Because of the low angle, it only interacts strongly with the first atomic layer on the surface. By choosing the incident angle to produce diffraction in the off-Bragg condition, alternate layers of atoms interfere destructively. This results in a diffraction pattern, which can be analysed to yield the surface atomic periodicity. The intensity of the diffracted beams is found, for the off-Bragg geometry, to oscillate with sample surface roughness with monolayer periodicity. This can yield very precise measures of the film growth process. A discussion of this method may be found in Cho (1970), Howie (1970), Harris *et al.* (1981), and Van Hove *et al.* (1982).

Optical luminescence is a non-contact method similar to x-ray fluorescence but where light emission is characteristic of the energy gap and defects in the semiconductor rather than characteristic of the atoms therein (Perkowitz, 1993; Ozawa, 2007). Therefore, it is quite sensitive to the properties of the material being deposited. The technique relies on excitation of the sample, placing a number of electrons in conduction band states and holes in the valence band (Figure 7.4.5). This can be accomplished by short wavelength excitation, as in photoluminescence, or by forward bias if a diode device exists, as in electroluminescence. Because one is concerned here with monitoring the sample during growth, we will consider only photoluminescence (PL). In many implementations PL uses a short wavelength laser relative to the energy gap (e.g., in the green portion of the visible spectrum) to excite carriers. This wavelength can be varied to select the depth in the sample where excitation occurs based on the optical absorption coefficient of the sample. This method can be extended to produce photoluminescence excitation spectroscopy in which the detection wavelength is fixed and the source wavelength is swept. The optical absorption and energy gap of the semiconductor are then determined when the sample begins to luminesce. In general, a good sample will luminesce at wavelengths somewhat below the energy gap because at a minimum an electron and hole in the material would be electrostatically bound, forming an exciton, prior to recombination. Additional energy losses due to energy transfer to phonons can also occur. These are forms of "free to bound" transition. Similarly the sample may luminesce from defect states. For example an electron trapped in a donor or other high energy state may recombine with a hole trapped in an acceptor or other low energy state. The result is a "donor-acceptor pair transition". Because luminescence can occur in the second phases of lower energy gap, the material can detect these emissions as well. The technique is not generally well suited to indirect gap materials such as Si, which do not luminesce well. For most direct-gap materials, it can provide an effective method to monitor the electronic states in the material. An interesting option is to carry out PL in a spatially-resolved imaging mode (cf. Breitenstein *et al.*, 2011 or Nesswetter *et al.*, 2013).

Figure 7.4.5 An optical bench and cryostat set up for measuring photoluminescence. Source: courtesy of J. Soares, University of Illinois

This can provide local information on the behavior of the semiconductors in a photovoltaic device. This is generally difficult to do on a full-scale module after lamination but can be used to diagnose local defects selectively. A photoluminescence detection system can also be rasterized across a large area sample as in x-ray fluorescence.

Time-resolved photoluminescence (TRPL) can measure minority carrier lifetime in semiconductors that luminesce well (Thomas and Colbow, 1965). The carrier lifetime has been shown to correlate well with the performance of solar cells and can be configured to provide rapid feedback on material quality, potentially in a mapping mode (Repins *et al.* 2015). In this measurement an optical fiber or other method illuminates the sample with a very short pulse of photons, resulting in the generation of a large number of carriers in the semiconductor. These recombine over time, typically resulting in a rapid decay as carriers recombine that do not need to move to find an opposite charged state, a more gradual decay related to the normal minority carrier diffusion and formation of excitons, and a slower decay associated with trapping and later release of minority carriers. In the process monitoring TRPL method, it is the middle of the three – normal radiative recombination processes such as are observed in steady-state PL – that is used. The apparatus can be relatively simple to implement, is non-destructive, and is fast.

An important note is that samples generally emit light best at low temperatures while room temperature emissions are generally weak and spectrally broad. Therefore, while PL is an attractive method for monitoring semiconductor growth, it is difficult to implement for some semiconductors due to the low level of light emission under convenient process conditions.

Several tests on completed modules can provide diagnosis of performance after completion. While not true *in-situ* process control methods, these can provide rapid feedback on the status of the process. Electroluminescence (EL), referred to above, is one such example

(Potthoff, *et al.*, 2010). In this method the device is forward-biased and the luminescence is detected with an imaging camera or other detector with wavelength sensitivity selected for the material under test. This shows up poorly performing regions of the sample, hot spots that can cause degradation, and other problems (Kim and Krem, 2015). When properly set up, the EL intensity can be correlated with open circuit voltage in the device (Potthoff *et al.*, 2010).

Optical beam induced conductivity (OBIC) is similar to a quantum efficiency measurement in which a light beam is scanned across a module under short circuit conditions (typically) and the photocurrent is detected as a function of position (Ostendorf' and Endros, 1998; Vorasayan *et al.*, 2009). This method provides a rapid way to detect local areas that are not collecting current well. Because it can be done with various exciting wavelengths, it can detect the response of the device at different depths if the current collection is different there.

7.4.4 Electrical Behavior

Because most methods require electrical connection, as in the cases of EL and OBIC, electrical characterizations are usually after-the-fact measurements of the performance of a product or intermediate step in the module production process. However, they can be very valuable, especially when modules are assembled from individual stand-alone cells as is the case in Si and III-V based devices, that use single crystals, and for polycrystalline thin film devices deposited on metal foils and interconnected with tab-and-string approaches. In these cases, the cells are tested at the exit to their process line and may be binned by performance to match them for current output when series connected or voltage output when in parallel (Field and Gabor, 2002). This allows weak devices to be discarded, which is not possible in monolithically interconnected thin film modules. Both the devices and the modules are typically analysed in a flash testing apparatus that measures performance very rapidly detecting key portions or all of the current voltage curve for each device in fractions of a second Green *et al.* (2009), and Ferretti *et al.* (2013) Likewise flash testing on full modules provides feedback on performance, although diagnosis of problems is not straightforward as the measurement lacks position-sensitivity.

A scanning probe method that is particularly useful is microwave reflectivity. In this case, a tuned microwave cavity is positioned above the sample under test and the microwave reflectivity and absorption changes due to the sample are detected. Reflected power results from a change in the tuning of the microwave transmission line due to the capacitive or inductive interaction of the coil field with the charge carriers in the sample. Absorption of the microwave is usually related to the conductivity of the sample. Thus, both the dielectric properties of the sample and the free carrier density can be detected. If the sample is excited with a laser beam or other light source through the center of the coil, the excess carrier concentration can be detected. In this case, excess carrier concentration due to the optical excitation indicates the minority carrier lifetime. Microwave lifetime measurements are one of the most common methods to study Si wafers and have a strong correlation with subsequent device performance.

7.4.5 Conclusion

A wide range of *in-situ* process control methods have been developed applicable to a wide range of processes and process conditions for control of photovoltaic device processing systems. These include methods of monitoring gas phase reactant fluxes, deposited material composition, film thickness, semiconductor material quality, and more. Many of these can be integrated into a process tool to measure the properties of the deposited material dynamically during growth or shortly thereafter to provide rapid feedback for control of the process tool. These types of control systems are very important to maintain high process yields and achieve optimized device performances day after day. They can also be useful in smaller research systems for better control of the materials under study.

List of Acronyms

Acronym	Description
AAS	Atomic absorption spectroscopy
CIGS	$Cu(In,Ga)Se_2$
CIS	$CuInSe_2$
CZTS	Cu_2ZnSnS_4
EIES	Electron impact emission spectroscopy
EL	Electroluminescence
GD-MS	Glow discharge mass spectrometry
GD-OES	Glow discharge optical emission spectroscopy
OBIC	Optical beam induced conductivity
PL	Photoluminescence
RHEED	Reflection high energy electron diffraction
TRPL	Time resolved photoluminescence
XRF	X-ray fluorescence

References

Aryal, P., Attygalle, D., Pradhan, P. *et al.* (2013) Large-area compositional mapping of $Cu(In_{1-x}Ga_x)Se_2$ materials and devices with spectroscopic ellipsometry. *IEEE Journal of Photovoltaics*, **3** (1), 359–363.

Bauer, G.H. and Bilger, G. (1981) Properties of plasma-produced amorphous silicon governed by parameters of the production, transport, and deposition of Si and SiH_x. *Thin Solid Films*, **83** (2), 223–229.

Beckhoff, B. (2006) *Handbook of Practical X-Ray Fluorescence Analysis*. Springer, Berlin.

Breitenstein, O., Bauer, J., Bothe, K. *et al.* (2011) Can luminescence imaging replace lock-in thermography on solar cells? *IEEE Journal of Photovoltaics*, **1** (2), 159–167.

Chen, J., Li, J., Thornberry, C. *et al.* (2009) Through-the-glass spectroscopic ellipsometry of CdTe solar cells, *IEEE Photovoltaic Specialists Conference*.

Cho, A.Y. (1970) Epitaxial growth of gallium phosphide on cleaved and polished (LLL) calcium flouride. *Journal of Applied Physics*, **41** (2), 782–786.

Dahal, L.R., Huang, Z., Attygalle, D. *et al.* (2010) Application of real time spectroscopic ellipsometry for analysis of roll-to-roll fabrication of Si:H solar cells on polymer substrates. *IEEE Photovoltaic Specialists Conference*, pp. 631–636.

Eisgruber, I.L., Joshi, B., Gomez, N. *et al.* (2002) In situ x-ray fluorescence used for real-time control of $CuIn_xGa_{1-x}Se_2$ thin film composition. *Thin Solid Films*, **408** (1–2), 64–72.

Ferretti, N., Pelet, Y., Berghold, J. *et al.* (2013) Performance testing of high efficient PV modules using single 10 ms flash pulses. *European Photovoltaic Solar Energy Conference and Exhibition*.

Field, H. and Gabor, A.M. (2002) Cell binning method analysis to minimize mismatch losses and performance variation in Si-based modules. *IEEE Photovoltaic Specialists Conference*.

Fujiwara, H. (2007) *Spectroscopic Ellipsometry: Principles and Applications*. John Wiley & Sons, Ltd., Chichester.

Germer, T.A. and Fasolka, M.J. (2003) Characterizing surface roughness of thin films by polarized light scattering. *Advanced Characterization Techniques for Optics, Semiconductors, and Nanotechnologies*.

Green, E., Taylor, S., Cowley, S. and Xin, L. (2009) Correlation between collimated flash test and in-sun measurements of high concentration photovoltaic modules. *Optical Modeling and Measurements for Solar Energy Systems III*.

Greene, J E. and Whelan, J.M. (1973) Glow-discharge optical spectroscopy for analysis of thin-films. *Journal of Applied Physics*, **44** (6), 2509–2513.

Harris, J.J., Joyce, B.A. and Dobson, P.J. (1981) Comments on RED intensity oscillations during MBE of GaAs. *Surface Science*, **108** (2), 444–446.

Hastie, J.W. (1973) Mass-spectrometric analysis of 1 atm flames – apparatus and Ch_4O_2 system. *Combustion and Flame*, **21** (2), 187–194.

Hedstrom, J., Ohlsen, H., Bodegard, M. *et al.* (1993) ZnO/CdS/Cu(In,Ga)Se_2 thin film solar cells with improved performance. *IEEE Photovoltaic Specialists Conference*,

Howie, A. (1970) The theory of high energy electron diffraction. *Modern Diffraction and Imaging Techniques in Material Science*.

Jian, L., Jie, C., Zapien, J.A. *et al.* (2005) Real time analysis of magnetron-sputtered thin-film CdTe by multichannel spectroscopic ellipsometry. *Thin-Film Compound Semiconductor Photovoltaics Symposium*.

Kim, K.A. and Krein, P.T. (2015) Reexamination of photovoltaic hot spotting to show inadequacy of the bypass diode. *IEEE Journal of Photovoltaics*, in press.

Kohara, N., Negami, T., Nishitani, M. and Wada, T. (1995) Preparation of device-quality Cu(In,Ga)Se_2 thin-films deposited by coevaporation with composition monitor. *Japanese Journal of Applied Physics Part 2-Letters*, **34**(9A), L1141-L1144.

Negami, T., Nishitani, M., Kohara, N. *et al.* (1996) Real time composition monitoring methods in physical vapor deposition of Cu(In,Ga)Se_2 thin films. *Thin Films for Photovoltaic and Related Device Applications. Symposium*.

Nesswetter, H., Dyck, W., Lugli, P. *et al.* (2013) Series resistance mapping of III-V multijunction solar cells based on luminescence imaging. *IEEE Photovoltaic Specialists Conference*, pp. 76–80.

Ostendorf', H.C. and Endros, A.L. (1998) Laser induced mapping for separation of bulk and surface recombination. *Solid State Phenomena*, **63–4**, 33–44.

Ozawa, L. (2007) Cathodoluminescence and photoluminescence theories and practical applications.In *Phosphor Science and Engineering 2*. CRC Press, Boca Raton, FL.

Paesler, M.A., Okumura.T. and Paul, W. (1980) Emission-spectroscopy of glow-discharge and sputtering plasmas used in amorphous Si-H film deposition. *Journal of Vacuum Science & Technology*, **17** (6), 1332–1335.

Perkowitz, S. (1993) Optical characterization of semiconductors infrared, Raman, and photoluminescence spectroscopy. In *Techniques of Physics 14*. Academic Press, London.

Potthoff, T., Bothe, K., Eitner, U. *et al.* (2010) Detection of the voltage distribution in photovoltaic modules by electroluminescence imaging. *Progress in Photovoltaics: Research and Applications*, **18** (2), 100–106.

Repins, I.L., Egaas, B., Mansfield, M. *et al.* (2015) Fiber-fed time-resolved photoluminescence for reduced process feedback time on thin-film photovoltaics. *Review of Scientific Instruments*, **86** (1).

Repins, I. L., Fisher, D., Batchelor, W.K. *et al.* (2005) A non-contact low-cost sensor for improved repeatability in co-evaporated CIGS. *Progress in Photovoltaics: Research and Applications*, **13** (4), 311–323.

Schmitt, S.W., Venzago, C., Hoffmann, B. *et al.* (2014) Glow discharge techniques in the chemical analysis of photovoltaic materials. *Progress in Photovoltaics*, **22** (3), 371–82.

Stolt, L., Hedstrom, J. and Sigurd, D. (1985) Coevaporation with a rate control-system based on a quadrupole mass-spectrometer. *Journal of Vacuum Science & Technology A*, **3** (2), 403–407.

Thomas, D. F. and Colbow, K. (1965) Time resolution of photoluminescent spectra by phase controlled chopping. *Review of Scientific Instruments*, **36** (12), 1853–1856.

Van Hove, J. M., Pukite, P. and Cohen, P.I. (1982) RHEED streaks and instrument response. *Journal of Vacuum Science and Technology A*, **1** (2 pt 1), 609–613.

Vorasayan, P., Betts, T.R., Tiwari, A.N. and Gottschag, R. (2009) Multi-laser LBIC system for thin film PV module characterisation. *Solar Energy Materials and Solar Cells*, **93** (6–7), 917–921.

Walker, J.D., Khatri, H., Ranjan, V. *et al.* (2009) Dielectric functions and growth dynamics of $CuIn_{1-x}Ga_xSe_2$ absorber layers via in situ real time spectroscopic ellipsometry. *IEEE Photovoltaic Specialists Conference*, Vols **1–3**, 580–582.

Wyon, C., Delille, D., Gonchond, J.P. *et al.* (2004) In-line monitoring of advanced microelectronic processes using combined X-ray techniques. *Thin Solid Films*, **450** (1), 84–89.

7.5

PV Module Performance Testing and Standards

Geoffrey S. Kinsey
U.S. Department of Energy, Washington, DC, USA

7.5.1 Introduction

As a two-terminal device that is powered by the sun, a PV module is one of the easier pieces of hardware to measure for performance. With no more than a handheld multimeter, the open-circuit voltage and short-circuit current can be determined simply by pointing the module at the sun. However, at the ever-increasing scale of PV deployment worldwide, small improvements in performance testing are rewarded with significant avoided-cost savings. With the PV industry exceeding 40 GW per year, assuming an installed price of \$3/W, even a 1% error in performance measurements would translate to \$1 billion in misallocated investments. Both over-estimation and under-estimation carry a price. Over-estimation results not only in selection of an inferior module and disappointed customers, but also in excess investment in balance-of-system hardware such as inverters and cabling that are built to handle electrical power that seldom (or never) arrives. Under-estimation results in over-production of power in operation, which saturates inverters and leads to the inverter "clipping" the peak power to maintain its designated AC output. Within an array, mismatch in the output between modules causes the modules to operate from their maximum power point and to dissipate excess power as heat, causing decreased performance and accelerating degradation.

Accurate determination of PV module performance requires precise measurement of a module's current voltage (I-V) curve under controlled conditions. Modules are most commonly measured under Standard Test Conditions (STC): 25 °C module temperature and 1000 W/m^2 irradiance under the standard (AM1.5G or AM1.5D) spectra. The irradiance and spectrum values originated in the 1970s and are meant to represent "typical" radiation incident

Photovoltaic Solar Energy: From Fundamentals to Applications, First Edition.
Edited by Angèle Reinders, Pierre Verlinden, Wilfried van Sark, and Alexandre Freundlich.

Companion website: www.wiley.com/go/reinders/photovoltaic_solar_energy

on a surface tilted 37° with respect to a horizontal plane at mid-latitudes in the contiguous USA (ASTM G173, 2012). The choice of 25 °C is a matter of convenience: it enables indoor testing of modules near room temperature. As shown in Figure 7.5.1, module temperatures in operation ranges are considerably higher.

Small deviations from the standard conditions have a measurable impact on measured performance, which may be amplified in extrapolations to operating performance. International standards have been developed to dictate the acceptable tolerances for measurements conducted under STC. The International Electrotechnical Commission (IEC) has become the *de facto* source of international standards related to photovoltaics (Table 7.5.1). The IEC maintains Technical Committee 82 to oversee development of these standards. Volunteers drawn from academia, government, and industry meet on an ongoing basis to establish, codify, and revise standards as needed. IEC 60904 is the primary standard that specifies how performance testing is conducted. It specifies how measurements are to be conducted in both natural and simulated sunlight, methods for correcting to an equivalent cell temperature of 25 °C, and corrections for spectral mismatch. Measurements under IEC 60904 are made with the test plane of the module normal to the incident radiation.

Since modules operate under a wide range of irradiances, temperatures, spectra, and angles of incidence, multiple ancillary standards build on IEC 60904 (2007) to provide performance data under a wider range of conditions. Standards that complement IEC 60904 include: how to make temperature and irradiance corrections (IEC 60891, 2009), extension of the measurements to a range of temperature, irradiance, and angle-of-incidence conditions (IEC 61853,

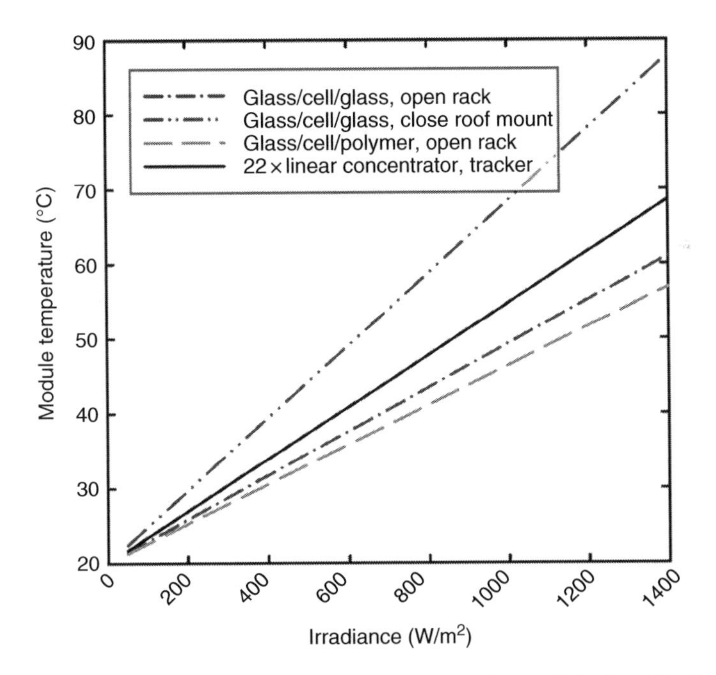

Figure 7.5.1 Module temperature calculated using the Sandia Photovoltaic Array Performance Model for four array configurations at an ambient temperature of 20°C and a wind speed of 1 m/s. Source: (King *et al.*, 2004)

Table 7.5.1 International standards related to performance testing of PV modules

IEC standard	Description
60891	Photovoltaic devices: procedures for temperature and irradiance corrections to measured I-V characteristics
60904	Photovoltaic devices: Measurement principles: current-voltage characteristics, reference devices, calibration traceability, equivalent cell temperature, spectral mismatch correction, solar simulator requirements, linearity
61724	Photovoltaic system performance: monitoring, capacity, and energy evaluation
61829	Crystalline silicon photovoltaic (PV) array: On-site measurement of I-V characteristics
61853	Photovoltaic (PV) modules performance testing and energy rating
62670	Photovoltaic concentrators (CPV): Performance testing
62892	Comparative testing of PV modules to differentiate performance in multiple climates and applications

2011) and methods specific to concentrator photovoltaic modules (IEC 62670, 2013). These standards are, in turn, integrated with the standards developed for qualifying commercial PV modules (IEC 61215, 2005; IEC 61646, 2008; IEC 62108, 2007, for silicon, thin-film, and concentrator PV modules, respectively).

7.5.2 Indoor Testing

Indoor testing on solar simulators provides the most stable and repeatable method for testing PV modules. The solar simulator must provide illumination that is uniform in spectrum and intensity across the test plane and is stable over time. Solar simulators are rated for their stability in these three categories per IEC 60904-9. A simulator that meets the highest standard in all three categories is known as an "AAA" simulator. (Simulators that achieve tighter specifications in a given category are informally referred to as "A+" simulators.) A xenon arc lamp is frequently used as the light source, as it can provide high-intensity illumination for hundreds of hours with a spectrum that approximates the target solar spectrum. As a lamp ages, its spectrum and intensity change, but these effects can be partially corrected with filters and by re-calibrating the voltage input to the arc. During calibration, the best match between the simulator's spectral irradiance and the target spectrum is obtained using reference cells. A reference cell (defined by IEC 60904-2) must be characterized in detail, including its spectral response. Reference cells are measured under a precisely calibrated standard spectrum at one of five designated test centers (Green *et al.*, 2012). Knowing the reference cell's current output under the standard spectrum allows for a spectrum mismatch correction to be made when the reference cell is measured during calibration of the solar simulator (IEC 60904-7, 2011). It is assumed that, when the reference cell produces the same current, it is being illuminated by a similar spectral irradiance. As illustrated by Figure 7.5.2, the validity of this assumption is limited, even when the bandgap of the reference cell and the test modules is the same.

Spatial uniformity over a typical module area of ~1 m^2 or more is a challenge. Steady-state simulators (such as those used to calibrate the reference cells) are able to achieve tight irradiance and spectrum distributions over small areas (a few cm^2) only. A xenon arc is an extended point source,

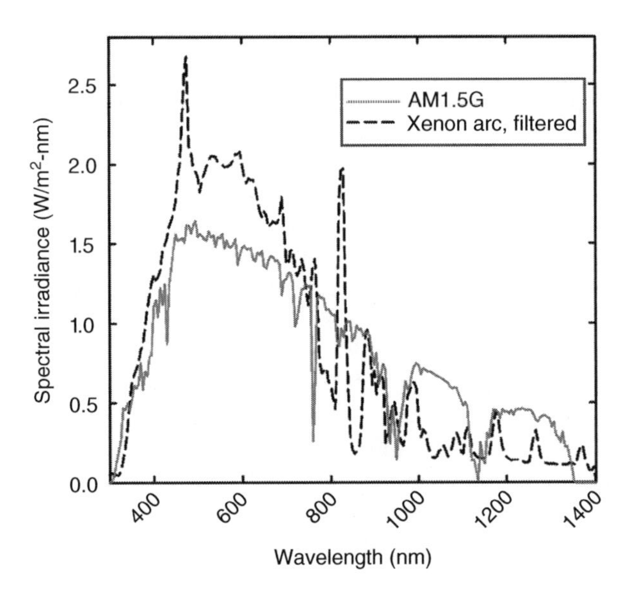

Figure 7.5.2 Spectral irradiance of a xenon arc filtered to approximate the AM1.5G spectrum (IEC 60904-7, 2011). The magnitude of the xenon arc has been scaled to match the photon flux of AM1.5G from 280 nm to 1200 nm. Source: IEC 60904-7 (2011). Photovoltaic devices – Part 7: Computation of the spectral mismatch correction for measurements of photovoltaic devices

Table 7.5.2 Standard ratings for solar simulators per IEC 60904-9

Classifications	Spectral match (over specified intervals)	Non-uniformity of irradiance (%)	Temporal instability	
			Short-term instability of irradiance (%)	Long-term instability of irradiance (%)
A	0.75–1.25	2	0.5	2
B	0.6–1.4	5	2	5
C	0.4–2.0	10	10	10

so improved uniformity can, in principle, be achieved by increasing the distance from the source to the test plane, but this intensity then falls off rapidly (as the square of the distance). The highest-grade ("A+A+A+") simulators use a working distance between source and module of several meters or more. Table 7.5.2 shows the standard ratings for solar simulators per IEC 60904-9.

In order to deliver the required uniformity and intensity, a steady-state solar simulator requires kilowatts to illuminate an area smaller than a typical module. The steady-state illumination also heats the test plane, requiring active cooling of the device under test. Pulsed solar simulators are therefore more practical for most applications. The temporal stability must be sufficient to enable a current-voltage (I-V) sweep to be completed; this usually takes at least a few milliseconds. If the solar cell parameters have long time constants (e.g., with thin-film modules and back-contact modules), longer pulse durations are required or the I-V curve may be reconstructed using data acquired from multiple pulses. In order to reach the temperature ranges of IEC 61853, the module is placed inside a closed, temperature-controlled chamber.

7.5.3 Outdoor Testing

By definition, outdoor testing provides performance data that captures operating conditions. Challenges familiar for indoor testing, such as maintaining spatial uniformity and short-term temporal stability, are effectively eliminated. Since outdoor conditions are variable, however, much of the challenge lies in obtaining the desired conditions that are broadly representative of the operating conditions for PV arrays at installation sites around the world. Tests conducted outside a factory in Germany may have limited application to operation of modules to be deployed in Morocco. Some of the variable conditions present in outdoor testing (such as soiling, wind speed and direction, and angle of incidence) may be eliminated as variables by applying standardized test procedures. The two conditions that are most often allowed to vary are the irradiance and temperature. IEC 60891 provides methods for translating measured performance to alternate irradiance and temperature conditions (such as STC). The translation for irradiance assumes that the PV device has a linear dependence on irradiance. This is a reasonable assumption (Emery *et al.*, 2006) as long as the sensitivity to spectrum variation is minimal; corrections for spectrum variation may also be conducted separately (Kinsey, 2015). An analysis of the impact of spectrum variation on module current is illustrated in Figure 7.5.3.

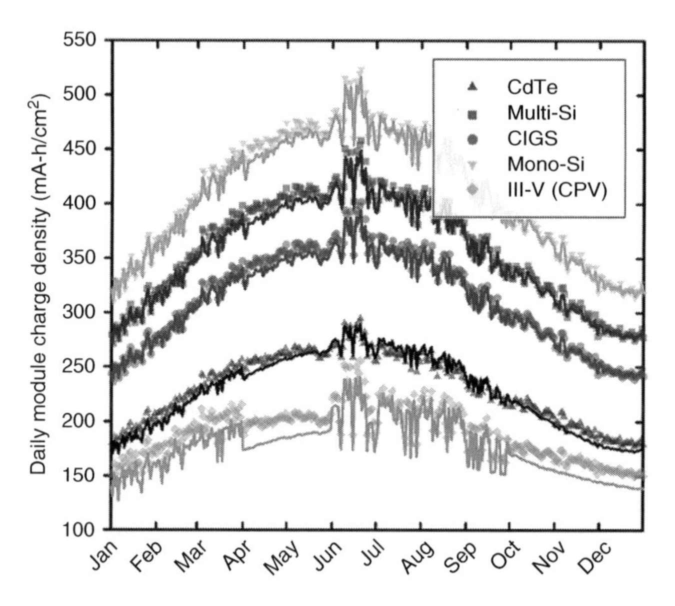

Figure 7.5.3 Predicted daily module short-circuit charge densities under clear sky conditions (solid lines) for various module designs. The symbols indicate values obtained by replacing the predicted spectrum with (normalized) reference spectra (AM1.5D for the III-V module, AM1.5G for the others). ("Module charge density" is the sum of the current per unit area over time for each day. This metric is used as an expedient parameter to increase the visual separation between the module types with similar current outputs.) Source: (Kinsey, 2015). (*See insert for color representation of the figure*)

7.5.4 Sandia Photovoltaic Array Performance Model

Over years of outdoor testing of PV arrays, combined with indoor testing and analysis of PV module optoelectronic characteristics, Sandia National Laboratories have developed the Sandia Photovoltaic Array Performance Model (King *et al.*, 2004). This model is frequently applied, either in part or as a whole, in predicting the field performance of PV arrays. Current voltage measurements are collected over several days to form a basis data set. Regression analysis is then applied to extract values for a series of coefficients used to predict PV module performance under a given set of target conditions. To choose one of the more complex examples, the value for voltage at maximum power, V_{mp}, is defined as:

$$V_{mp} = V_{mpo} + C_2 \cdot N_s \cdot \delta(T_c) \cdot \ln(E_e) + C_3 \cdot N_s \cdot \{\delta(T_c) \cdot \ln(E_e)\}^2 + \beta_{Vmp}(E_e) \cdot (T_c - T_o) \quad (7.5.1)$$

Where V_{mpo} (in V) is the reference V_{mp}, $\delta(T_c)$ (in V) is the thermal voltage, T_c (in K) and T_o (in K) are the cell temperature and reference cell temperatures, respectively, N_s is the number of cells in the string, and $\beta_{Vmp}(E_e)$ (in V/K) is the (irradiance dependent) temperature coefficient of V_{mp}. C_2 and C_3 are empirical coefficients that, together, are used to fit the linear and non-linear components of the voltage dependence on irradiance. To obtain C_2 and C_3, the basis data set is first corrected to a single temperature and then the coefficients are extracted using regression analysis.

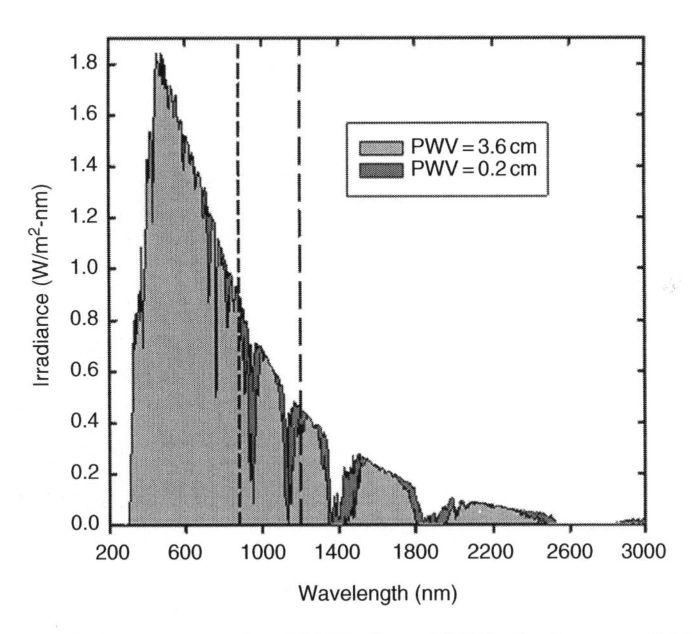

Figure 7.5.4 Modeled spectrum using TMY3 data (TMY3 database, available online) for Albuquerque, NM, USA and SMARTS Latitude-tilt global irradiance incident with the highest and lowest annual values for precipitable water vapor (PWV) in the atmosphere in Albuquerque are shown. Under the more arid conditions (PWV = 0.2 cm), irradiance in the range 280–4000 nm rises by 11%, whereas the irradiance below 1200 nm (near the Si band edge) rises by 7% and that below 880 nm (near the CdTe band edge) rises by only 2%. Source: (Gueymard, 2005). (*See insert for color representation of the figure*)

E_e is the "effective irradiance" that reaches the cells. To obtain this parameter, effects such as soiling, angle of incidence, and variations in spectrum must be corrected for. The acceptance angle of a module often differs from that of the instruments (such as pyranometers and pyrheliometers) used to measure irradiance, even if the instruments are oriented in the plane of the array. For example, a typical high-concentration PV module might have an acceptance angle of one degree or less (Kurtz *et al.*, 2014), but the pyrheliometers used to measure direct normal irradiance have a field of view that is a factor of five larger (Eppley Laboratory, online). Pyranometers and pyrheliometers are sensitive to a broader spectrum of sunlight than solar cells; this can result in non-linearities in the cell response that must be corrected for in the effective irradiance. For example, under arid conditions, the decrease in water absorption causes an increase in irradiance at wavelengths beyond the sensitivity of most solar cells (Figure 7.5.4). As a result, a PV module's response will fail to rise with the increase in irradiance; though output power will rise slightly; efficiency will fall as a result (Muller *et al.*, 2010).

7.5.5 Conclusion

Standards for performance testing provide the basis for comparison of photovoltaic technologies under a wide range of conditions. Under the controlled conditions of indoor testing, uncertainty is minimized by specifying a narrow set of test conditions. Under the inherent variability of the outdoor environment, a module's performance characteristics are measured under a wider range of conditions so that the measurements better reflect module performance in true operating conditions.

List of Symbols

Symbol	Definition	Unit
C_2	Empirical coefficient	–
C_3	Empirical coefficient	–
E_e	Effective irradiance	–
N_s	Number of cells in the string	–
T_o	Temperature of reference cell	K
T_c	Temperature of solar cell	K
V_{mp}	Voltage at the maximum power	V
V_{mpo}	Reference cell voltage at the maximum power	V
$\delta(T_c)$	Thermal voltage	V
$\beta V_{mp}(E_e)$	Temperature coefficient for voltage at the maximum power	VK^{-1}

List of Acronyms

Acronym	Definition
AC	Alternative Current
AM1.5D	Air mass 1.5 direct spectrum
AM1.5G	Air mass 1.5 global spectrum

Acronym	Definition
I-V	Current *vs.* voltage
IEC	International Electro-technical Commission
PV	Photovoltaic
PWV	Precipitable water vapor
STC	Standard test conditions
TMY	Typical meteorological year

References

ASTM G173 – 03. (2012) Standard Tables for Reference Solar Spectral Irradiances: Direct Normal and Hemispherical on 37° Tilted Surface. Available at: www.astm.org/DATABASE.CART/STD_REFERENCE/G173.htm

Emery, K., Winter, S., Pinegar, S., and Nalley, D. (2006) Linearity testing of photovoltaic cells. *Proceedings of the IEEE 4th World Conference on Photovoltaic Energy Conversion (WCPEC-4).*

Epply Laboratory, available at: eppleylab.com/instrumentation/normal_incidence_pyrheliometer.htm (accessed Jan. 2015).

Green, M.A., Emery, K., Hishikawa, Y. *et al.* (2012) Solar cell efficiency tables (version 39). *Progress in Photovoltaics: Research Applications*, **20**, 12–20.

Gueymard, C. (2005) SMARTS, a simple model of the atmospheric radiative transfer of sunshine: Algorithms and performance assessment. *Florida Solar Energy Center Tech. Rep. FSEC-PF-270-95.*

IEC 60891 (2009) Photovoltaic devices – procedures for temperature and irradiance corrections to measured I-V characteristics.

IEC 60904 (2007) Photovoltaic devices.

IEC 60904-2 (2007) Photovoltaic devices – Part 2: requirements for reference solar devices.

IEC 60904-7 (2011) Photovoltaic devices – Part 7: Computation of the spectral mismatch correction for measurements of photovoltaic devices.

IEC 60904-9 (2006) Photovoltaic devices – Part 9: solar simulator performance requirements.

IEC 61215 (2005) Crystalline silicon photovoltaic (PV) modules – design qualification and type approval.

IEC 61646 (2008) Thin-film terrestrial photovoltaic (PV) modules - design qualification and type approval.

IEC 61853-1 (2011) Photovoltaic (PV) module performance testing and energy rating – Part 1: Irradiance and temperature performance measurements and power rating.

IEC 62108 (2007) Concentrator photovoltaic (CPV) modules and assemblies - design qualification and type approval, IEC 62108.

IEC 62670 (2013) Photovoltaic concentrators (CPV) – performance testing.

King, D.L., Boyson, W.E., and Kratochvil, J.A. (2004) Sandia Photovoltaic Array Performance Model. Report, No. SAND2004-3535.

Kinsey, G.S. (2015) Spectrum sensitivity, energy yield, and revenue prediction of PV modules. *IEEE Journal of Photovoltaics*, **5** (1), 258–262, DOI: 10.1109/JPHOTOV.2014.2370256.

Kurtz, S., Muller, M., Jordan, D. *et al.* (2014) Key parameters in determining energy generated by CPV modules. *Progress in Photovoltaics: Research and Applications*, DOI: 10.1002/pip.2544.

Muller, M., Marion, B., Kurtz, S., and Rodriguez, J. (2010) An investigation into spectral parameters as they impact CPV module performance. *Proceedings of 6th International Conference Concerning Photovoltaic Systems: CPV-6, AIP Conference Proceedings*, vol. **1277**, pp. 307–311. DOI: 10.1063/1.3509218.

Newport-Oriel Technical Reference: Solar Simulation, Newport-Oriel. Available at: http://assets.newport.com/web Documents-EN/images/12298.pdf

Typical Meteorological Year 3 (TMY3) Database. Available at: http://rredc.nrel.gov/solar/old_data/nsrdb/1991-2005/tmy3/

Part Eight

III-Vs and PV Concentrator Technologies

8.1

III-V Solar Cells – Materials, Multi-Junction Cells – Cell Design and Performance

Frank Dimroth
Fraunhofer Institute for Solar Energy Systems, ISE, Freiburg, Germany

8.1.1 Historical Overview and Background of III-V Solar Cells

III-V-compound semiconductors have been used in photovoltaics since the 1970s. Nobel Prize laureate Zoran Alferov and his co-workers at the Ioffe Institute in St. Petersburg developed the first AlGaAs/GaAs heterojunction solar cells in 1970 (Alferov *et al.*, 1970). Liquid phase epitaxy was the method used to create these early devices which contained a p-$Al_xGa_{1-x}As$ emitter layer on n-GaAs, serving as the base layer and substrate. The use of $Al_xGa_{1-x}As$ hetero-structures instead of GaAs homo-junctions was the first attempt to prevent minority carriers from reaching the front surface with a high recombination velocity. Figure 8.1.1 shows another version of a similar device consisting of a GaAs absorber and a high bandgap p-$Al_xGa_{1-x}As$ front-surface field, often referred to as a window layer due to its transparency for incoming photons. Such hetero-structures are the basis of all modern III-V solar cells. PN-junctions are formed in an absorbing semiconductor material, surrounded by hetero-structures which are higher bandgap barrier layers, to prevent photo-generated minority charge carriers from leaving the active region of the device.

The technology of GaAs solar cells using liquid phase epitaxy was continuously improved over the 1970s and 1980s (Woodall and Hovel, 1972; Hovel, 1975; Kamath, Ewan, and Knechtli, 1977; Yoshida *et al.*, 1980) and the first GaAs hetero-structure solar cells were applied to power satellites and the Russian space station Mir. The advantage of GaAs compared to silicon was its high efficiency and low degradation of the IV-characteristics under high energy particle irradiation. This characteristic is referred to as radiation hardness and it is

Photovoltaic Solar Energy: From Fundamentals to Applications, First Edition.
Edited by Angèle Reinders, Pierre Verlinden, Wilfried van Sark, and Alexandre Freundlich.

Companion website: www.wiley.com/go/reinders/photovoltaic_solar_energy

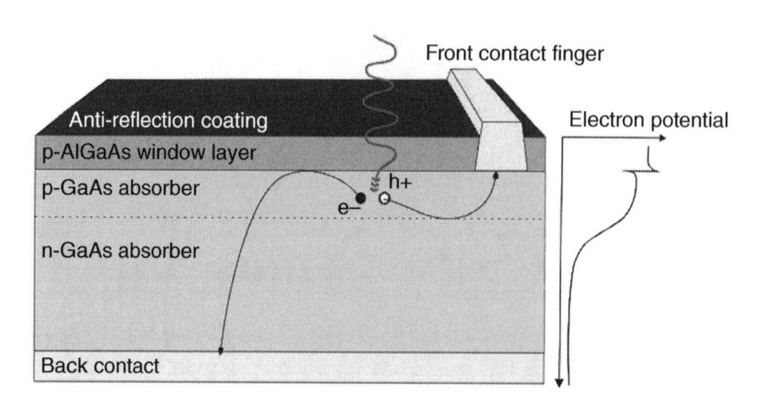

Figure 8.1.1 Schematic illustration of a GaAs heterostructure solar cell with n-GaAs base, p-GaAs emitter and p-AlGaAs front surface passivation layer also called the window layer

directly correlated to the property of GaAs being a direct bandgap semiconductor with a strong light absorption. A thickness of only 4 μm GaAs is sufficient to absorb 99% of all photons in the AM0 spectrum between 300–880 nm. The diffusion length of minority carriers consequently needs to be only of the same order of magnitude to prevent losses of minority carriers while moving through the material and before reaching the pn-junction. In comparison, this number needs to be about 100 times larger in the case of silicon. It is this material property of strong absorption, combined with a sufficiently large diffusion length and an ideal match of the GaAs bandgap to the solar spectrum, which made the material successful in the early days of photovoltaics. Single-crystalline GaAs was pulled by the liquid encapsulated Czochralski method (Weiner, Lassota, and Schwartz, 1971) and later Bridgman or vertical gradient freeze techniques (Rudolph and Jurisch, 1999), resulting in substrates with low dislocation density $< 5000\,cm^{-2}$. The liquid phase epitaxy of $Al_xGa_{1-x}As/GaAs$ hetero-structures extended the superb quality by using high purity metals and by growing close to thermal equilibrium. $Al_xGa_{1-x}As$ compounds could be realized with bandgap energies between 1.4 and 2.2 eV without substantially changing the lattice-constant (see also Figure 8.1.2 later), serving as transparent passivation layers. One-sun AM0 voltages of 996 mV, corresponding to an efficiency of 18.8% (AM0), were obtained in 1979 (Sahai, Edwall, and Harris, 1979) and improved to 20.3% (AM0) by 1987 (Tobin *et al.*, 1987). Also lift-off GaAs solar cells with a thickness of 20 μm and an efficiency of 13% were demonstrated as early as 1976 (Konagai and Takahashi, 1976). These technologies are the origin of today's most efficient GaAs solar cells, as will be discussed later in this chapter.

GaSb was another important III-V compound in the early days as it allowed extending the absorption of sunlight up to 1800 nm. Diffusion of Zn vapor was used to form a shallow pn-junction in n-GaSb. In 1990, workers at Boeing, Sandia and the Solar Energy Research Institute stacked a GaAs solar cell on top of a GaSb pn-junction and reached an efficiency of 32.6% under 100-fold concentration of AM1.5d (Yang *et al.*, 2015). This value was listed in the solar cell efficiency tables until 2009 (Green *et al.*, 2009) as one of the highest efficiencies.

Most of the building blocks for today's high efficiency III-V photovoltaics were developed 40–50 years ago. Since then, development has focused on realizing better material combinations,

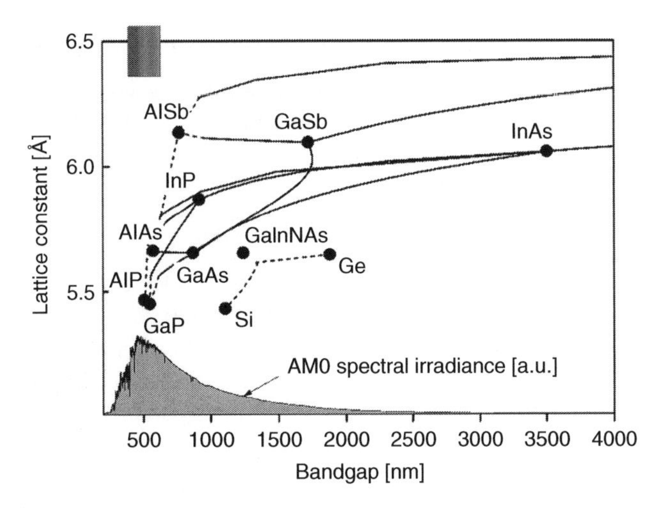

Figure 8.1.2 Lattice constant for arsenides, phosphides and antimonides versus the bandgap energy. The spectral irradiance of the AM0 solar spectrum is displayed as a reference and shows the important wavelength range which is covered by the III-V compounds

cell architectures, and growth conditions. In this context it is important to keep in mind, that a high crystalline quality is usually more important than the detailed solar cell design (Dimroth and Kurtz, 2007). It has taken more than two decades for metal organic vapor phase epitaxy (MOVPE) to achieve a similar material quality compared to liquid phase epitaxy but today MOVPE is the growth method of choice for the manufacturing of all advanced multi-junction cells. The transition from liquid phase epitaxy to MOVPE was necessary as the solar cell architectures incorporated more and more layers with up to six junctions. Such modern designs would not be feasible using liquid phase epitaxy.

Electronic and optoelectronic devices such as transistors, lasers and photodetectors have developed in parallel to III-V multi-junction solar cells, and over time new material compositions and hetero-structures became available. Figure 8.1.2 shows the landscape of the III-V compounds as a function of bandgap energy and atomic distance in the crystal (lattice constant). All compounds/alloys on a given horizontal line are referred to as being lattice matched, allowing growth without significant strain and consequently circumventing dislocation genesis in thick epitaxial layers. Materials which are lattice matched to one of the common substrate materials (GaAs, Ge, InP, InAs, GaSb, GaP) are more easily accessible with high material quality. In recent years, lattice mismatched growth has, however, seen serious developments (Lewis *et al.*, 1985; Sinharoy *et al.*, 2000; Dimroth *et al.*, 2000; Dimroth *et al.*, 2001; Fetzer *et al.*, 2004; France *et al.*, 2016). Today, a wide range of lattice matched and metamorphic compounds from arsenides, phosphides and antimonides are used for photovoltaic applications. Transition layers, referred to as metamorphic buffer structures, were developed to smoothly grade the lattice constant and relax the crystal while keeping low threading dislocation densities (typically $< 10^6$ dislocations/cm^2). Today, many of the highest efficiency solar cells incorporate metamorphic growth and a wide variety of III-V compounds such as $(Al_xGa_{1-x})_yIn_{1-y}P$, $Al_xGa_{1-x}As$, $Ga_yIn_{1-y}As$, $Ga_yIn_{1-y}As_zP_{1-z}$ or $GaAs_zSb_{1-z}$.

8.1.2 Minimizing Optical and Electrical Losses in III-V Solar Cells

Figure 8.1.2 displays the AM0 extra-terrestrial sun spectrum which spans a wide range of wavelengths from 300 nm to several microns. It can be seen that III-V compounds are available in the full desired wavelength range, even though high bandgap absorbers are more predominantly found for smaller lattice constants than low bandgap absorbers. The challenge of high efficiency photovoltaics is to absorb all wavelengths of the sun spectrum and convert them into electric current at the highest possible voltage. Unfortunately, this is hardly possible with a single semiconductor as high energy photons create excited charge carriers which thermalize to the band edge in picoseconds. On the other hand, low energy photons are not absorbed if the energy is below the bandgap. Multi-junction solar cells are the solution to this problem and have become the highest efficiency photovoltaic technology. In a multi-junction cell architecture, the shortest wavelengths of the sun spectrum are absorbed in a high bandgap semiconductor such as $Al_xGa_{1-x}As$ (1.4-2.2 eV) or $Ga_{0.50}In_{0.50}P$ (1.8-1.9 eV) and longer wavelengths are transmitted. Photons absorbed in the high bandgap semiconductors are able to generate high open-circuit voltages which is, for high quality materials, 350–400 mV below the bandgap of the semiconductor. As an example, a GaInP solar cell grown on GaAs reaches open-circuit voltages up to 1.46 Volt (Geisz *et al.*, 2013). Longer wavelength radiation is transmitted and can be used in a second junction made from a material like GaAs (1.4 eV). In this case voltages up to 1.12 Volt have been reported (Green *et al.*, 2015). This stacking of materials with declining bandgap energy can be continued until the dark current of the pn-junction does not support any meaningful voltage. Typical low bandgap absorbers are made of Ge, GaInAs or GaSb with bandgap energies around 0.7 eV.

The overall advantage of multi-junction solar cells is the simultaneous reduction of transmission and thermalization losses. This leads to an increase in efficiency which can be predicted based on thermodynamic considerations of detailed balance between the sun and the solar cell absorber. Figure 8.1.3 shows the result of such a calculation for the AM1.5g spectrum and for multi-junction solar cells with up to 6 pn-junctions. For each number of junctions, an optimum set of bandgap energies is assumed. One can clearly see how the theoretical efficiency limit increases from 33.8% for a single absorber to 59.2% for a six-junction cell. The efficiency is even higher if the illumination intensity and therefore, the charge carrier density in the semiconductor are increased. The open-circuit voltage rises proportional to the logarithm of the sunlight concentration and at 500-suns, the theoretical efficiency limit is 40.3% for a single-junction and 69.2% for a six-junction cell. This simplified calculation neglects specific material properties such as non-radiative recombination, reflection at interfaces, parasitic absorption, and resistance. Therefore, the highest reported efficiency values are well below the thermodynamic limit, as indicated by the white stars in Figure 8.1.3. But, the detailed balance calculation correctly predicts the trends of increasing efficiency with the number of junctions and under concentration. The highest reported efficiency of 46% was so far measured for a four-junction cell at 508-suns concentration (508 x AM1.5d) (Green e*t al.*, 2015). Five and six junction cells are still falling behind this value but future developments will certainly push record efficiencies towards more junctions.

Multi-junction solar cells can be realized in different designs. Three architectures have been successful. The first one is mechanical stacking of separate solar cell devices on top of each other. Usually this is accomplished in a way which allows the formation of separate contacts to the subcells. Parasitic absorption in the substrates and reflection at interfaces between

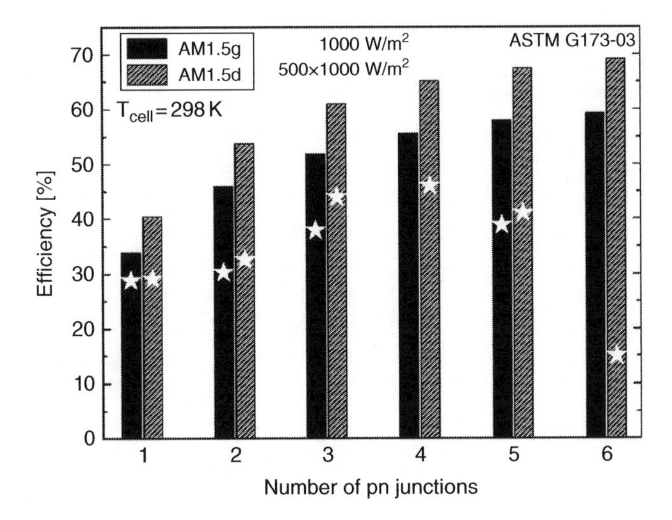

Figure 8.1.3 Increase of the thermodynamic efficiency limit for multi-junction solar cells, assuming that ideal bandgap combinations are used. Theoretical limits are given for both, one-sun AM1.5g and 500 suns AM1.5d conditions. Experimental values (stars) refer to the highest published 1-sun and concentrator cell performances (Chiu *et al.*, 2013; Essig *et al.*, 2011; Green, *et al.*, 2015). Concentration ratios may vary between 300–600 suns. Theoretical calculations have been performed with the program EtaOpt (Létay and Bett, 2001)

semiconductor and air limit the practical use to two or three junctions (Bett *et al.*, 2001). The second approach is obtained by spectral splitting. Dichroic mirrors are used to divide the solar spectrum into several components which are then directed towards single- or tandem cell devices, optimized for the specific spectral range (Mitchell *et al.*, 2011; Atwater *et al.*, 2013). This architecture is rather complex and requires very efficient mirrors but it can lead to very high performance as current matching of the junctions is not a requirement (Escarra *et al.*, 2013). The most successful commercial technology is monolithic integration of pn-junctions connected by Esaki tunnel diodes. The solar cells use only one front and back metal contact and allow a simple integration into space and terrestrial concentrator PV products. All subcells in the monolithic multi-junction stack are series connected which has the consequence that the smallest current limits the overall device. It is important to design such monolithic tandem cells carefully to achieve equal currents for all subcells (referred to as current matching). Up to six-junction devices have already been demonstrated successfully using the monolithic approach.

After understanding the principles of III-V multi-junction cells, we will discuss the most important prerequisites for reaching highest performance devices. Let's assume a simple GaAs solar cell with only one pn-junction. The device optimization goes through the steps of: minimizing reflection by an anti-reflection coating, minimizing grid shadowing, minimizing resistance losses in the emitter and at contacts, selecting GaAs pn-layer thickness to approach 100% absorption, realizing excellent crystalline quality to avoid non-radiative recombination in the bulk and at interfaces. After this optimization, the cell should have an internal quantum efficiency close to 100%. It will absorb all wavelengths up to the bandgap of GaAs (870 nm) and the lifetime of photogenerated minority carriers will be dominated only by radiative recombination. This means that photogenerated minority carriers are either separated by the junction and flowing through the external circuit where they are delivering electric power, or they are recombining internally by sending out a photon with a wavelength close to the bandgap of GaAs. Unfortunately, such photons may undergo free carrier absorption in the substrate, either be absorbed by the back side contact or leave the device through the front surface. It has been found in recent years that maximizing the re-absorption of photons emitted by the device (photon recycling) is a key element in reaching record performances. The highest efficiency GaAs solar cells today use a lift-off process like the one developed by (Konagai and Takahashi, 1976) to remove the GaAs pn-junction layers from the substrate and apply a highly reflective back mirror onto the rear side of the GaAs film. In this way, parasitic absorption processes are minimized. Randomly emitted photons hit the back mirror under all angles and total internal reflection at the front surface confines most of these photons within the GaAs absorber until they are reabsorbed. The photon recycling leads to an increase of the minority carrier density in the GaAs absorber and results in an increased voltage. GaAs solar cells with 28.8% efficiency under AM1.5g have been reported by Alta Devices (Green *et al.*, 2015), and these outstanding results are only possible due to the excellent material quality combined with an optimized photonic design. Unfortunately, the high photon recycling efficiencies which are necessary to boost the voltage are so far only obtained for the lowest bandgap absorber in a multi-junction cell. This is due to the fact that light emitted by higher bandgap materials will be re-absorbed in lower bandgap materials. Such photon coupling may shift current generation from high bandgap junctions to low bandgap junctions but it does not significantly increase photon and minority carrier densities, except for the lowest bandgap absorber. Alta Devices has presented GaInP/GaAs thin-film tandem solar cells with efficiency up to 30.8% under AM1.5g (Kayes *et al.*, 2014).

8.1.3 Monolithic III-V Multi-junction Cell Architectures

Monolithic multi-junction solar cells with a GaInP top cell absorber were first developed at the National Renewable Energy laboratory (Olson, Gessert, and Al-Jassim, 1985) and later manufactured by several companies, adding Germanium instead of GaAs as the substrate and lowest bandgap absorber (Stan *et al.*, 2005; King *et al.*, 2012; Strobl *et al.*, 2012). Metamorphic materials and strain-balanced quantum wells have been used to improve current matching between the junctions. Up to 41.6% conversion efficiency has been achieved for a metamorphic GaInP/GaInAs/Ge triple-junction cell under concentration (King *et al.*, 2012). Only recently, four-junction solar cells have overcome the best triple-junction devices. Detailed balance calculations predict the optimum set of bandgap energies for converting the AM1.5d spectrum to be 1.9 eV, 1.4 eV, 1.0 eV and 0.5 eV. Four different cell architectures with absorber materials close to this ideal bandgap combination are displayed, for example, in Figure 8.1.4.

Only the first cell uses lattice-matched materials. Approximately 2% of nitrogen is incorporated into GaInAs to reduce both the lattice-constant and the bandgap energy. It is a unique property of the dilute nitride alloys that small quantities of nitrogen lead to a lowering of the absorption threshold, even though GaN has a much wider bandgap of 3.4 eV compared to GaAs with 1.4 eV (Friedman *et al.*, 1998). Unfortunately, the nitrogen has to be forced into the crystal by growing at very low temperatures. Defects and impurities like carbon have hindered the success of this cell structure for many years (Essig *et al.*, 2011). But more recently, significant improvements in material quality became possible by a different growth method of molecular beam epitaxy (Sabnis, Yuen, and Wiemer, 2012) and have resulted in solar cells with higher than 44% efficiency.

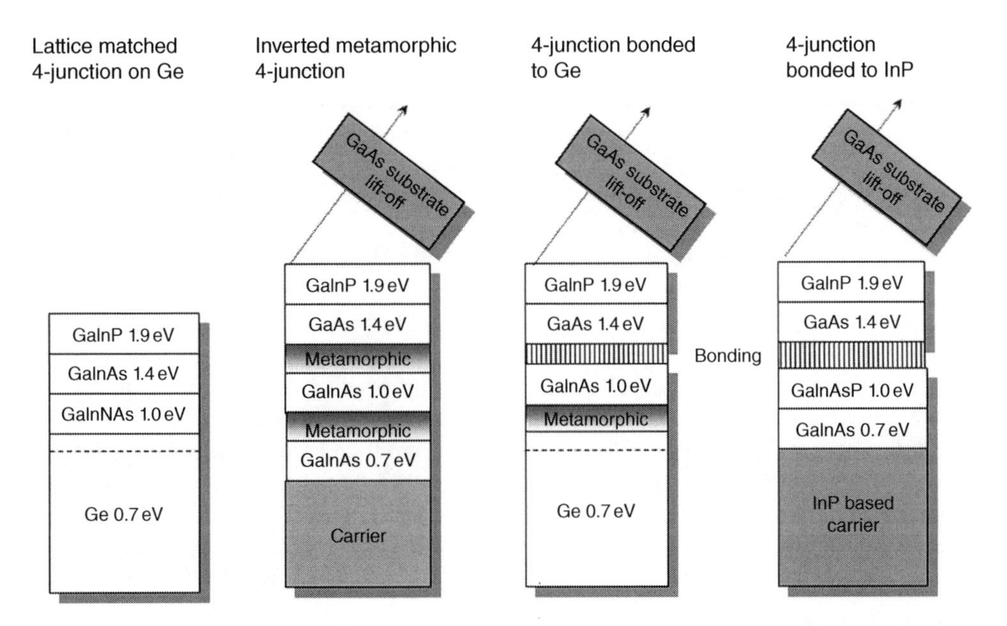

Figure 8.1.4 Four-junction solar cell architectures close to the optimum bandgap combination of 1.9 eV, 1.4 eV, 1.0 eV and 0.5 eV. Metamorphic growth or wafer bonding is used to overcome differences in the lattice-constant of the III-V compounds

All other concepts in Figure 8.1.4 use either metamorphic growth (Bett *et al.* 2005) or wafer bonding (Gösele *et al.*, 1996) to combine materials with different lattice constants. In the case of inverted metamorphic four-junction solar cells, a large lattice mismatch of 4% has to be overcome in two buffer layers. The cell is grown in an inverted manner with the top two GaInP/GaAs junctions being lattice-matched to GaAs. Afterwards the lattice is graded to the first and second GaInAs cell. The challenge is to relax the crystal by forming misfit dislocations, but keeping threading dislocations below $10^6\,cm^{-2}$. After the epitaxy growth, the solar cell layers have to be transferred to a mechanical carrier like silicon, before the GaAs substrate can be removed and processing of the front contacts can be completed. Solar cell efficiencies up to 45.7% have been announced at 234 suns concentration of the AM1.5d spectrum (NREL, 2014; France *et al.*, 2015).

Finally, the technology of wafer bonding allows the combination of the most suitable semiconductors without any constraints to the lattice constant and without compromising material quality. The most successful cell architecture today combines two separately grown tandem cells on GaAs and InP. The junctions of GaInP, GaAs, GaInAsP and GaInAs are close to the ideal bandgaps for a four-junction device and efficiencies up to 46.0% have been reported under 508-times AM1.5d (Dimroth *et al.*, 2014; Tibbits *et al.*, 2014; Green *et al.*, 2015). This solar cell exhibits excellent material quality for all junctions and an extended version with five-junctions has already been successfully realized, reaching 38.8% efficiency under one-sun AM1.5g (Chiu *et al.*, 2014). The technology of wafer bonding allows the combination of the best absorber materials in a multi-junction cell and this will certainly lead to new performance records in the future.

8.1.4 Conclusion

III-V compound semiconductors have proven to be excellent materials for high efficiency solar cells. The absorption properties of arsenide, phosphide, and antimonide compounds match perfectly to the wavelengths in the sun spectrum. Multi-junction devices with up to six III-V absorbers have been demonstrated. Such devices may contain up to 50 individual semiconductor layers, serving functions such as surface passivation, electrical series connection, lattice-constant transition or low resistance contact formation. But all of these individual layers are hidden inside the cell structure and externally, only the front and back metal contact as well as the anti-reflection coating are visible. Such solar cells are incorporated into power systems for satellites or concentrator photovoltaic modules. While triple-junction devices are the standard for these applications today, four-junction cells have already demonstrated superior performance with up to 46% efficiency at 508-times the AM1.5d spectrum. Optimization of each layer in the device structure will allow further improvements and hero devices may incorporate even more junctions. But unfortunately, until the cost is substantially reduced, the application of highest efficiency III-V solar cells will be restricted to Space and concentrator photovoltaics markets. Currently III-V wafer-based planar devices are at least 100 times more expensive per area compared to silicon solar cells. This challenge remains to be solved.

Acknowledgments

The author would like to thank all members of the "III-V epitaxy and solar cells" department at Fraunhofer ISE for their support. The preparation of this chapter was supported by the German Ministry for Economic Affairs and Energy through the contract HekMod4 (0325750).

List of Acronyms

Acronym	Description
AlGaAs	Aluminum gallium arsenide
AM0	Extraterrestrial air mass 0 spectrum
AM1.5d	Solar air mass 1.5 direct spectrum
AM1.5g	Solar air mass 1.5 global spectrum
GaAs	Gallium arsenide
GaInP	Gallium indium phosphide
GaSb	Gallium antimonide
MOVPE	Metal organic vapor phase epitaxy

References

Alferov, Z. I., Andreev, V.M., Kagan, M.B. *et al.* (1970). Solar energy converters based on p-n $Al_xGa_{1-x}As$-GaAs heterojunctions. *Soviet Physics Semiconductors*, **4** (12), 2378–2379.

Atwater, H.A., Escarra, M.D., Eisler, C.N. *et al.* (2013) Full spectrum ultrahigh efficiency photovoltaics. Paper presented at the 39th IEEE Photovoltaic Specialists Conference, Tampa, FL, USA, 16–21 June.

Bett, A.W., Baur, C., Beckert, R. *et al.* (2001) Development of high-efficiency mechanically stacked GaInP/GaInAs-GaSb triple-junction concentrator solar cells. Paper presented at the Proceedings of the 17th European Photovoltaic Solar Energy Conference, Munich, Germany, 22–26 October.

Bett, A.W., Baur, C., Dimroth, F., and Schöne, J. (2005) Metamorphic GaInP-GaInAs layers for photovoltaic applications. Paper presented at the Materials for Photovoltaics Symposium, Boston, MA, USA, 29 Nov.–2 Dec.

Chiu, P.T., Law, D.C., Woo, R.L. *et al.* (2013) Direct semiconductor bonded 5 J cell for space and terrestrial applications. *IEEE Journal of Photovoltaics*, **4** (1), 493–497.

Chiu, P.T., Law, D. C., Woo, R.L. *et al.* (2014) 35.8% space and 38.8% terrestrial 5 J direct bonded cells. *IEEE Journal of Photovoltaics*, **4**.

Dimroth, F., Beckert, R., Meusel, M. *et al* (2001) Metamorphic $Ga_yIn_{1-y}P/Ga_{1-x}In_xAs$ tandem solar cells for space and for terrestrial concentrator applications at C > 1000 suns. *Progress in Photovoltaics: Research and Applications*, **9** (3), 165–178.

Dimroth, F., Grave, M., Beutel, P. *et al.* (2014) Wafer bonded four-junction GaInP/GaAs//GaInAsP/GaInAs concentrator solar cells with 44.7% efficiency. *Progress in Photovoltaics: Research and Applications*, **22** (3), 277–282.

Dimroth, F. and Kurtz, S. (2007) High-efficiency multijunction solar cells. *MRS Bulletin*, **32**, 230–234.

Dimroth, F., Lanyi, P., Schubert, U., and Bett, A.W. (2000) MOVPE grown Ga1-xInxAs solar cells for GaInP/GaInAs tandem applications. *Journal of Electronic Materials*, **29** (1), 42–46.

Dimroth, F., Tibbits, T.N D., Niemeyer, F. *et al.* (2016) Four-Junction Wafer-Bonded Concentrator Solar Cells. *IEEE Journal of Photovoltaics*, **6** (1), 343–349.

Escarra, M.D., Darbe, S., Warmann, E.C., and Atwater, H.A. (2013) Spectrum-splitting photovoltaics: Holographic spectrum splitting in eight-junction, ultra-high efficiency module. Paper presented at the 39th IEEE Photovoltaic Specialists Conference, Tampa, FL, USA, 16–21 June.

Essig, S., Stämmler, E., Rönsch, S. *et al.* (2011) Dilute nitrides for 4- and 6- junction space solar cells. Paper presented at the 9th European Space Power Conference, St.-Raphael, France, 6–10 June.

Fetzer, C.M., King, R.R., Colter, P.C. *et al.* (2004) High-efficiency metamorphic GaInP/GaInAs/Ge solar cells grown by MOVPE. *Journal of Crystal Growth*, **261** (2–3), 341–348.

France, R.M., Geisz, J.F., Garcia, I. *et al.* (2015) Quadruple-junction inverted metamorphic concentrator devices. *IEEE Journal of Photovoltaics*, **5** (1), 432–437.

France, R.M., Dimroth, F., Grassman, T.J., King, R.R. (2016) Metamorphic epitaxy for multijunction solar cells. *MRS Bulletin*, **41**, 202–209.

Friedman, D.J., Geisz, J.F., Kurtz, S.R., and Olson, J.M. (1998) 1-eV GaInNAs solar cells for ultrahigh-efficiency multijunction devices. Paper presented at the Proceedings of the 2nd World Conference on Photovoltaic Solar Energy Conversion, Vienna, Austria, 6-10 July.

Geisz, J.F., Steiner, M.A., Garcia, I. *et al.* (2013) Enhanced external radiative efficiency for 20.8% efficient single-junction GaInP solar cells. *Applied Physics Letters*, **103** (4), 041115–041118.

Gösele, U., Stenzel, H., Reiche, M. *et al.* (1996) History and future of semiconductor wafer bonding. *Solid State Phenomena*, **47–48**, 33–44.

Green, M.A., Emery, K., Hishikawa, Y., and Warta, W. (2009) Solar cell efficiency tables (version 33), *Progress in Photovoltaics: Research and Applications*, **17** (1), 85–94.

Green, M.A., Emery, K., Hishikawa, Y. *et al.* (2015) Solar cell efficiency tables (version 45). *Progress in Photovoltaics: Research and Applications*, **23**, 1–9.

Hovel, H.J. (1975) *Solar Cells* (Vol. 11). Academic Press, New York.

Kamath, G.S., Ewan, J., and Knechtli, R.C. (1977) Large-area high-efficiency (AlGa)As-GaAs solar cells. *IEEE Transactions on Electron Devices, ED-24*(**4**), 473–475.

Kayes, B.M., Zhang, L., Twist, R. *et al.* (2014) Flexible thin-film tandem solar cells with >30% efficiency. *IEEE Journal of Photovoltaics*, **4** (2), 729–733.

King, R.R., Bhusari, D., Larrabee, D. *et al.* (2012) Solar cell generations over 40% efficiency. *Progress in Photovoltaics: Research and Applications*, **20** (6), 801–815.

Konagai, M. and Takahashi, K. (1976) Thin film GaAlAs-GaAs solar cells by peeled film technology. CA Conference Paper, 154–163.

Létay, G., and Bett, A. W. (2001) EtaOpt – a program for calculating limiting efficiency and optimum bandgap structure for multi-bandgap solar cells and TPV cells. Paper presented at the 17th European Photovoltaic Solar Energy Conference, Munich, Germany, 22–26 Oct.

Lewis, C.R., Ford, C.W., Virshup, G.F. *et al.* (1985) A two-terminal, two-junction monolithic cascade solar cell in a lattice-mismatched system. Paper presented at the Proceedings of the 18th IEEE Photovoltaic Specialists Conference, Las Vegas, Nevada, USA.

Mitchell, B., Peharz, G., Siefer, G. *et al.* (2011) Four-junction spectral beam-splitting photovoltaic receiver with high optical efficiency. *Progress in Photovoltaics: Research and Applications*, **19** (1), 61–72.

NREL (2014) News Release NR-4514: NREL demonstrates 45.7% efficiency for concentrator solar cell. Available at: http://www.nrel.gov/news/press/2014/15436.html (accessed 16 Dec. 2014).

Olson, J.M., Gessert, T., and Al-Jassim, M.M. (1985) GaInP/GaAs: a current- and lattice-matched tandem cell with a high theoretical efficiency. Paper presented at the Proceedings of the 18th IEEE Photovoltaic Specialists Conference, Las Vegas, Nevada, USA.

Rudolph, P., and Jurisch, M. (1999) Bulk growth of GaAs: An overview. *Journal of Crystal Growth*, **198–199**, 325–335.

Sabnis, V., Yuen, H., and Wiemer, M. (2012) High-efficiency multijunction solar cells employing dilute nitrides. Paper presented at the 8th International Conference on Concentrating Photovoltaic Systems, Toledo, Spain, 16-18 April.

Sahai, R., Edwall, D.D., and Harris, J.S., Jr. (1979) High-efficiency AlGaAs/GaAs concentrator solar cells. *Applied Physics Letters*, **34** (2), 147–149.

Sinharoy, S., Stan, M.A., Pal, A.M. *et al.* (2000) MOVPE growth of lattice-mismatched $Al_{0.88}In_{0.12}As$ on GaAs(100) for space solar cell applications. *Journal of Crystal Growth*, **221**, 683–687.

Stan, M.A., Aiken, D.J., Sharps, P.R., Hills, J., and Doman, J. (2005) InGaP/InGaAs/Ge high concentration solar cell development at Emcore. *Proceedings of the 31st IEEE Photovoltaic Specialists Conference, Orlando, Florida, USA*.

Strobl, G., Fuhrmann, D., Guter, W. *et al.* (2012) About Azur's "3G30-Advanced" space solar cell and next generation product with 35% efficiency. Paper presented at the 27th European Photovoltaic Solar Energy Conference and Exhibition Frankfurt, Germany.

Tibbits, T.N.D., Beutel, P., Grave, M. *et al.* (2014) New efficiency frontiers with wafer-bonded multi-junction solar cells. Paper presented at the 29th European Photovoltaic Solar Energy Conference and Exhibition, Amsterdam, the Netherlands.

Tobin, S.P., Bajgar, C., Vernon, S.M. *et al.* (1987) A 23.7% efficient one-sun GaAs solar cell. Paper presented at the 19th IEEE Photovoltaic Specialists Conference, New Orleans, LA, USA, 4–8 May.

Weiner, M.E., Lassota, D.T., and Schwartz, B. (1971) Liquid encapsulated Czochralski growth of GaAs. *Journal of the Electrochemical Society*, **118** (2), 301–306.

Woodall, J.M. and Hovel, H. J. (1972) High efficiency $Ga_{1-x}Al_xAs$–GaAs solar cells. *Applied Physics Letters*, **21** (8), 379–381.

Yang, Y., Wenzheng, Y., Weidong, T., and Chuandong, S. (2013) High-temperature solar cell for concentrated solar-power hybrid systems. *Applied Physics Letters*, **103** (8), 083902–083905.

Yoshida, S., Mitsui, K., Oda, T., Sogo, T., and Shirahata, K. (1980) High efficiency $Al_xGa_{1-x}As$-GaAs solar cells with high open-circuit voltage and high fill factor. [Conference Paper]. *Japanese Journal of Applied Physics*, **19** (Suppl. 19-1), 563–566.

8.2

New and Future III-V Cells and Concepts

Simon Fafard
Laboratoire Nanotechnologies Nanosystèmes, Institut Interdisciplinaire d'Innovation Technologique (3IT), Université de Sherbrooke, Canada

8.2.1 Introduction

In its simplest form, the optimization of an optoelectronic solar cell device consists in maximizing simultaneously the photovoltage and the photocurrent output of a photovoltaic heterostructure. The main steps of this process include the efficient absorption of the solar radiation in the first step, and then the efficient photocarrier extraction out of the device. A review of the underlying science and technology of interest can readily be found in the III-V compound semiconductor based solar cell literature[1] (Dominguez *et al.*, 2010; Luque and Hegedus, 2011; Philipps *et al.*, 2012), and by following the international CPV conferences, the IEEE Photovoltaic Specialists Conferences (PVSC), and the related proceedings.[2] High performance is typically obtained when the wide solar spectrum is subdivided in a number of smaller subset ranges, and that the corresponding "subcells" are designed to generate substantially equal amounts of photocurrent from those spectral slices. The subcells are typically photovoltaic p-n or p-i-n junctions. To minimize the carrier losses and maximize the conversion efficiency, the semiconductor materials comprised into the fabrication of such p-n junctions must feature long minority carrier diffusion lengths.

[1] See for example, Chapter 9, in J.M. Olson, D.J. Friedman and S. Kurtz, in *Handbook of Photovoltaic Science and Engineering* (eds: A. Luque and S. Hegedus) (John Wiley & Sons, Ltd.; Chichester, 2003) or A. Luque, and S. Hegedus (eds) *Handbook of Photovoltaic Science and Engineering* (2nd edition) (John Wiley & Sons, Ltd, Chichester, 2011).
[2] See www.cpv-11.org and www.ieee-pvsc.org.

Photovoltaic Solar Energy: From Fundamentals to Applications, First Edition.
Edited by Angèle Reinders, Pierre Verlinden, Wilfried van Sark, and Alexandre Freundlich.

Companion website: www.wiley.com/go/reinders/photovoltaic_solar_energy

8.2.2 Summary of Requirements

8.2.2.1 Design Considerations

The cell engineering must carefully consider the technical requirements and design rules for the application. Some of the fundamental cell design considerations include:

- In manufacturing and testing, the devices are evaluated under standardized conditions, whereas in the field the devices will be exposed to the operating conditions. For example, typically the standard test conditions (Ji and McConnell, 2006) specify an AM1.5D spectrum, a fixed input intensity such as 50 W/cm^2, and device temperature of 25 °C.
- The compound semiconductor design tools available include, for example, (a) changing the number of junctions; (b) the type of structure used such as lattice-matched (LM), pseudomorphic, metamorphic (MM), or inverted metamorphic (IMM); (c) the layer thicknesses used to optimize the absorption; (d) the types of dopants and the dopant levels; (e) the tunnel junction design; and (f) the metallization properties and layouts used to extract the photocurrents.
- There are a number of different optical bandgaps and lattice constants to choose from. However, the choices are limited and they are dictated by the availability of various III-V semiconductor alloys. Hybrid epitaxy, including Group IV or II-VI alloys, could widen the engineering options.

8.2.2.2 Challenges

The III-V cells are the engine powering the concentrator photovoltaic (CPV) systems. Because of the cost repartition, a performance gain at the cell level can typically be leveraged by a factor between 3 and 10 at the system level. There are, therefore, a number of opportunities in the optimization of new and future III-V cells. Most of those opportunities arise from the existing challenges which include:

- *Current matching*: multijunction cells are normally connected in series. Therefore, the overall cell current is limited by the junction with the lowest current (the weakest link). This drives the requirement for current matching between the subcells.
- *Lattice matching*: In order to optimize the minority carrier lifetime, as mentioned above, the crystal quality must typically be maximized. Each subcell is therefore preferably grown epitaxially using production-grade metalorganic chemical vapor deposition (MOCVD) on commonly available substrates. It must be possible to procure substrates with high quality-control, and in high volume, at price points that will be compatible with the targeted levelized cost of electricity (LCOE).
- *Electrical and thermal resistances*: Due to their capabilities of generating high current densities, CPV cells must have extremely low series resistances (R_s). The electrical resistance must typically be kept in the range of milliohms to minimize resistive losses ($P=R_sI^2$). Similarly, the thermal resistance must be minimized to permit efficient heat extraction away from the active areas.
- *Varying optimum design*: The field conditions are not always the same. Therefore, the cell optimization must consider various factors such as (a) the system's optical transfer function which can affect the incident solar spectrum; (b) the system's operating temperature which depends on the ambient conditions in the area of deployment and its thermal management design; and (c) the system's location which can also impact the incident solar spectrum.

8.2.2.3 *Design Rules*

The above considerations therefore give rise to various design rules. In some cases the research and development in new and future III-V cell concepts might focus on alleviating these design rules which typically include:

- The epitaxy on commonly available (and cheap) substrates is typically limited to Si, Ge, or GaAs. InP is also commonly available, but expensive.
- The absorption characteristics required for the performance optimization typically prescribe the following materials: AlInP, GaInP, AlGaAs, (In)GaAs, Ge. Other possible materials can include epitaxial quantum well or self-assembled quantum dot heterostructures, MM or IMM layers, or dilute nitrides alloys of III-Vs. The design should avoid inefficient leaky upper cell designs. The latter is because thin absorbing layers passing higher energy photons to a lower bandgap subcell result in significant carrier thermalization losses.
- The need for good crystal quality: Pseudomorphic or lattice-matched layers typically offer lower dopant diffusion. They enable well-controlled doping profiles which are necessary for CPV operations while they maintain good minority carrier lifetimes. It is also typically advantageous to avoid defects and the possible related reliability risks.
- The new designs must not compromise the LCOE: The high-volume affordable manufacturability must be maintained. Typically this means that the heterostructures must be achieved from a single epitaxial MOCVD growth run. It also implies that the new designs must avoid significantly increasing the thickness or growth time of the heterostructures. That is because in high volume, the thickness of the epitaxial materials and the equipment run-time will be the main cost drivers for the cell manufacturing. The device handling must also be compatible with high-throughput manufacturing tools. Therefore, it is typically preferable to avoid concepts that include designs involving the manipulations of thin cells, or designs with p-n junctions near the die-attach surface, or approaches that could compromise the thermal properties of the die-attach.

8.2.3 New and Future Cells and Concepts

In this section we will examine different III-V cell concepts and devices that are being researched, developed, or commercialized.

8.2.3.1 *Dilute-Nitride and Wafer-Bonded Cells*

The conventional triple junction (3 J) design is based on a GaInP top subcell, an (In)GaAs middle subcell, and Ge bottom subcell. One of the main limitation of this design is that the bottom germanium subcell generates excess photocurrents but yields a relatively low output photovoltage (Friedman *et al.*, 1998). As mentioned in Chapter 8.1, Solar Junction Inc has demonstrated that the dilute-nitride alloys such as GaInNAs or GaInNAsSb can be grown by molecular beam epitaxy (MBE) to provide the desired 1.0 eV material necessary to improve the power output of the bottom cell. This approach has been yielding record 3 J efficiencies in the range of ~44% (Wiemer *et al.*, 2011). A maximum gain of up to ~0.33 V in the V_{oc} of a 1 eV bottom subcell 3 J can be expected compared to the V_{oc} of a conventional 3 J having a 0.67 eV bottom subcell.

That is a 10% relative gain in V_{oc} which can translate into a 10% relative gain in efficiency, or roughly 4% absolute efficiency points with respect to a conventional 40% 3 J CPV cell. The future cell developments in this area include the addition of a 4th bottom Ge subcell to a heterostructure equivalent to the dilute nitride 3 J but where the lattice constant is changed to that of Ge instead of GaAs. Different R&D groups are pursuing the development of such new 4 J cells, including Solar Junction, according to their website, and the Sunlab in Ottawa and the 3IT in Sherbrooke, Canada (Wilins *et al.*, 2014). The 4 J configuration has also been successfully achieved using a wafer bonding technique. The wafer bonded cells achieved record efficiencies in excess of 45% (Dimroth *et al.*, 2014). They can combine the ideal tandem gaps grown on GaAs and on InP substrates for the high bandgaps and low bandgaps respectively.

8.2.3.2 Metamorphic (MM and IMM) Cells

As mentioned also in Chapter 8.1, the metamorphic epitaxial growth approach has permitted the demonstration of impressive device designs (Cornfeld *et al.*, 2012; Boisvert *et al.*, 2013; García *et al*, 2014; Strobl *et al.*, 2014). New cell developments will certainly continue in this area. For example, Cornfeld *et al.* (2012) obtained preliminary 6 J results that show the potential to exceed 50% efficiencies for CPV applications. The objective of the 3 J IMM design is to improve the bottom subcell V_{oc}, as discussed in Section 8.2.3.1. The developments of such 3 J IMM cell designs are therefore compatible with cell designs that seek to increase the current generated in the middle subcell by extending its absorption edge to longer wavelengths. For example, Section 8.2.3.3 covers quantum dot 3 J CPV cells where the middle cell produces more photocurrent by incorporating pseudomorphic quantum dot materials.

8.2.3.3 Self-Assembled Quantum Dot Cells

The QDEC® cells were developed and optimized in the 2003–2008 timeframe (Fafard, 2005). The strategy of this design is to improve the performance of the conventional 3 J design without significantly impacting the required total thickness or manufacturing time of the device process. This was achieved by incorporating well-controlled self-assembled quantum dots (Fafard *et al.*, 1999) in the middle subcell of a 3 J grown by MOCVD, as shown in Figure 8.2.1. The bandgap engineering thus obtained increased the middle cell photocurrent. Such QDEC® CPV cells with product efficiencies exceeding 40% have been routinely manufactured for the past few years on a 150 mm Ge wafer production platform. Other research groups have also worked on incorporating such QDs in multijunction cells. For example, recently 3 J heterostructures with similar quantum efficiency (QE) spectral response, including middle subcell QD contributions, have been reproduced by Ho *et al.* (2015). Future cell developments incorporating QDs would include designing a 3 J heterostructure with a higher performance bottom subcell. This can be achieved, for example, with the dilute nitrides or IMM approaches. It may be advantageous also to extend the QD photocurrent contribution within the middle subcell.

8.2.3.4 Thin Cells with Light Trapping and Photonic Confinement

Values of V_{oc} enhanced by 100 mV or more have been demonstrated for thin single-junction cells (Miller *et al.*, 2012; Massiot *et al.*, 2013; Steiner *et al.*, 2013; Rau *et al.*, 2014). However, the thin cell designs result in reduced short-circuit currents (J_{sc}) and efficiencies because the

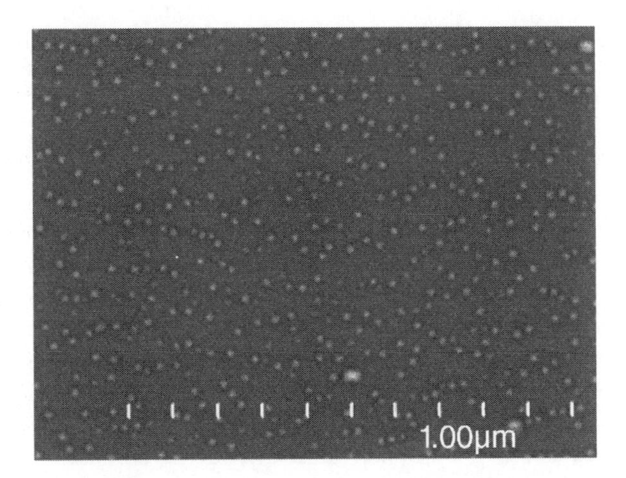

Figure 8.2.1 Nano-scale self-assembled InGaAs/GaAs QDs with a density of 400 μm^{-2}. The QDEC® heterostructure has many such layers incorporated in its middle subcell for bandgap engineering of its absorption

heterostructure does not absorb all the impinging photons. Light-trapping and photonic confinement schemes have been employed to gain back some of the escaping light. A related new development exploiting thin p-n junctions is demonstrated in Section 8.2.3.5. It is illustrated how simple thin heterostructures can be implemented in practical devices capable of unprecedented performance to convert optical power to useable electrical power. The devices in Section 8.2.3.5 therefore elegantly exploit such improved V_{oc} from thin junctions.

8.2.3.5 Photovoltaic Phototransducers with 75% Conversion Efficiencies

As mentioned in Section 8.2.3.4, the thin p-n junctions have the advantage that they can produce higher V_{oc}, but the disadvantage that they do not absorb all the photons. Azastra Opto Inc has recently demonstrated an unprecedented device performance exploiting the higher V_{oc} of thin p-n junctions with no need for any particular light management schemes for the case phototransducer applications (Masson and Fafard, 2015). This application typically uses a high-power narrow wavelength optical input beam to carry and deliver energy. The phototransducer needs to be capable of producing an output voltage greater than 5 V, as required for many electronic applications. For these purposes, single p-n junction PV devices have output voltages lower than desirable and multijunction PV devices cannot operate with a narrowband optical source due to current-matching constraints. Previous phototransducer developments used parquet geometries of single or tandem junction devices composed of planar configurations with multiple series connections (Virshup, 1994; Krut *et al.*, 2002; Schubert *et al.*, 2009; Werthen, 2011; Tao *et al.*, 2014; Khvostikov *et al.*, 2014). The segments were typically arranged in pie-shaped configurations, or in groups fabricated in circular patterns. The desired device output voltage can be obtained by selecting the number of segments connected in series. The maximum power conversion efficiency occurs when photocurrent matching is obtained between the series-connected segments. This prior approach therefore requires a segment geometry and alignment that is well matched to the intensity distribution of the impinging light. It makes it difficult to work with the non-uniform light distribution from multimode fibers. Furthermore, wafer processing in this approach requires first, separation of and then re-connection of the various segments, therefore reducing the active surface area and adding series resistance, which decreases the overall conversion efficiency.

Instead Azastra's phototransducer heterostructure has been designed to produce output voltages significantly higher than the corresponding photovoltage of the input light (Fafard and Masson, 2013; Valdivia *et al.*, 2015) by using a vertical stacking approach. The vertical heterostructure allows for higher overall power conversion efficiencies and operations with no particular restrictions on the impinging beam shape within the phototransducer active surface. The fabrication of such phototransducers is also significantly simplified by using a III-V heterostructure obtained in a single epitaxy run, having a total thickness comparable to a single junction solar cell. Azastra's new heterostructure includes tunnel junctions (TJs) joining together the multiple base segments stacked in the propagating direction of the incoming beam of light. The layers used for the phototransducer devices are shown in Figure 8.2.2, with light entering from the top. The III-V semiconductors were grown on standard 150 mm diameter (100) *p*-type GaAs substrates by MOCVD with an Aixtron 2600 multi-wafer reactor using standard GaAs growth conditions. The absorbing region of the phototransducer is made of five *np*-GaAs junctions, i.e. five *p*-type base segments separated by four TJs. The five thin

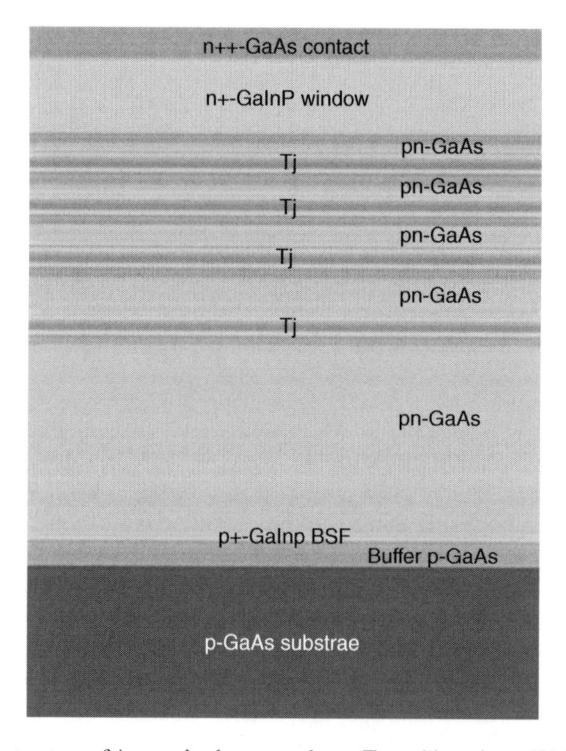

Figure 8.2.2 Heterostructure of Azastra's phototransducer. Tunnel junctions (TJs) are used to interconnect multiple GaAs base segments (*pn*-GaAs). The thickness of each *p*-GaAs base region is designed to obtain photocurrent matching of all the base segments. The total thickness of all the base segments allows the absorption of substantially all of the input light. Source: (Masson and Fafard, 2015)

np-GaAs base segments are the only absorbing layers of the device for the input light of interest for this application.

To optimize the power conversion, all GaAs base segments or sub-cells must generate the same currents, otherwise the overall current will be limited by the weakest subcell. In the structure shown in Figure 8.2.2, the top subcell must be designed to absorb 1/5th of the incoming light. The remaining light is transmitted to the second subcell also designed to absorb another 1/5th, and so on. Beer's law is applied to calculate the optimal thicknesses of each layer based on the material absorption coefficients. The amount of light absorbed in each base segment will be independent of both the beam non-uniformities and alignment. The photocurrents therefore remain matched despite such variations. This improved tolerance is a fundamental advantage over devices based on planar configurations.

The phototransducer devices were fabricated using a simple contact mask lithography. Measurements were done under illumination from a continuous-wave, fiber-coupled diode laser operating near 835 nm with a maximum output power of ~1.2 W. Typical results for the *I-V* response are shown in Figure 8.2.3 (a) for three different input powers. The short-circuit current (I_{sc}) increases linearly with input power while the open-circuit voltage shows a typical logarithmic increase with power. These high values for the open-circuit voltage, V_{oc} close to 6 V, arise from the five series connected GaAs p-n junctions. This yields an average of ~1.2 V per junction. Simulations of such devices suggest that the V_{oc} generated by the top subcells should be higher than the ones from the lower subcells because of the thinner bases (Valdivia *et al.*, 2015). The abrupt drop in current (large d*I*/d*V*) for voltages in excess of 5 V confirms minimal series resistance and demonstrates the high quality of the tunnel junctions connecting the subcells.

The device efficiency as a function of voltage for input powers ranging from 0.01 to 1.15 W is shown in Figure 8.2.3 (b). Maximum efficiencies occur near 5 V, reaching from ~60% to ~75% as the input power is increased. Other commercially available phototransducers typically show efficiencies between 30–40%. The results from Azastra's phototransducers approach

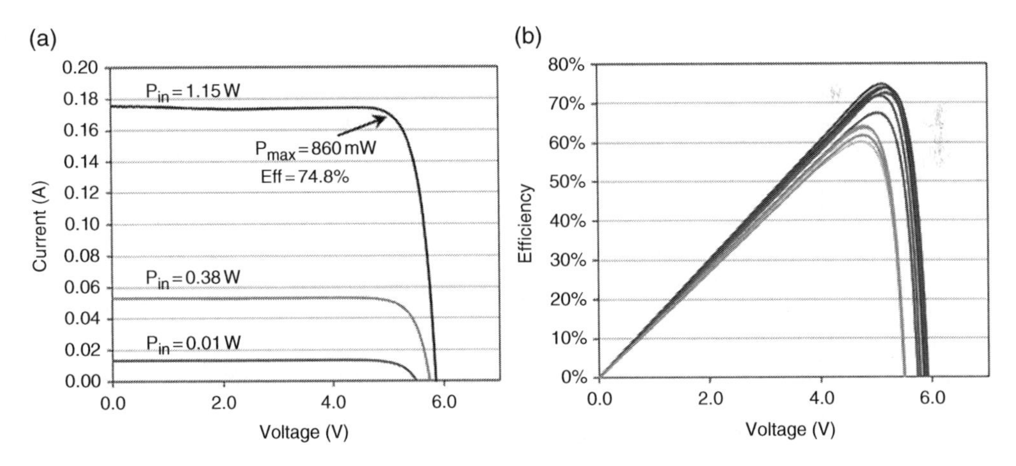

Figure 8.2.3 (a) I-V plots of a packaged phototransducer at 3 different input powers for an optical input at 835 nm. (b) Measured phototransducer efficiency for input powers of 0.01 W (lower curves), 0.38 W (middle) and 1.15 W (higher curves). The maximum conversion efficiency occurs near 5 V and increases with the input power from about 60% at 0.01 W to 74.8% at 1.15 W. (*See insert for color representation of the figure*)

theoretical limits. It is believed that these results represent the highest optical to electrical efficiency conversion ever reported. This impressive performance stems from the combination of near ideal quantum efficiencies, the enhanced V_{oc} of the thin individual partially-absorbing GaAs p-n junctions (Virshup, 1994), the very low series resistance of the interconnecting tunnel junctions, and the good current matching. This vertical heterostructure design is also very favorable for efficient radiative photon coupling and recycling. Future III-V solar cell developments could incorporate similar concepts to increase the voltage and/or improve the current matching in multijunction configurations, as explained in Section 8.2.3.6 for the duplicated junction approach.

8.2.3.6 Cells with Duplicated Junctions

Future CPV cell developments could also leverage the benefits that were clearly demonstrated in Section 8.2.3.5 for thinner junctions. For example, for solar cell applications, Boucherif *et al.* (2014) have recently proposed a CPV cell concept based on utilizing the well-proven III-V alloys currently used in conventional 3J devices, but by duplicating some or all of the subcells while making them thinner. As mentioned in Section 8.2.2.2, this would benefit the challenge of minimizing the electrical resistance. Indeed, duplicating the subcells will result in an increased operating voltage and a decreased operating currents. Everything else being equal, these changes in the cell operating conditions will therefore reduce the resistive losses and/or enable operations at higher concentrations. Furthermore, subcells that normally have an excess photocurrent, such as the bottom Ge subcell, could be separated in a higher number of partitions. The latter strategy could therefore also be beneficial for the current-matching optimizations. Development in this area would also be well aligned with the design rules of Section 8.2.2.3, namely, the growth can be done on commonly available substrates and the total thickness of the resulting novel heterostructures would not be impacted significantly.

8.2.3.7 Cells with Through Semiconductor Vias

The focus on the new III-V cell designs for CPV applications is typically predominantly on the optimization of the solar spectrum splitting with shared absorption bands between various subcells and the related current matching of the photo-generated carriers. However, as mentioned in Section 8.2.1, the efficient photocurrent extraction also needs to be a key step in the cell optimization. Minimizing the shadowing from metallized areas while minimizing the series resistance of the current extraction paths are essential research topics that could also benefit from further developments. For other electronic applications, the technology of through semiconductor vias has now been developed and optimized for mass deployment in consumer electronic devices, for example. It is therefore strategic to explore how new CPV cell designs could leverage the vast investments that have been made in the Through-Silicon-Via (TSV) technology. The TSV developments also enabled technology breakthroughs for flip-chip assembly and ball grid array contacts instead of wirebonding contacts. Aligned with this approach, new CPV cell R&D projects are currently being dedicated to developing similar processes in III-V CPV cells grown on Ge wafers (Zhao *et al.*, 2012; Fidaner and Wiemer, 2014; Richard *et al.*, 2015). As illustrated in Figure 8.2.4, the Through III-V Vias and the Through Germanium Vias will need to be etched through the thickness of the device going

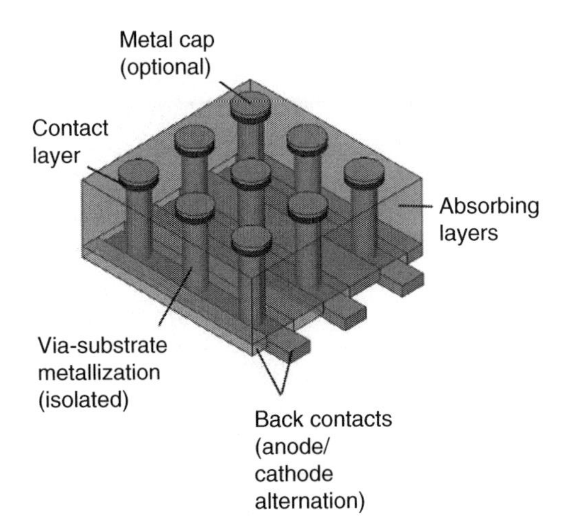

Figure 8.2.4 Illustration of a through-semiconductor-via CPV cell design where instead of top surface gridlines and busbars, vias are etched through the heterostructure and the substrate to extract the front surface photocurrent via the backside, according to Richard *et al.* (2015)

through several different layers. The holes will need to be passivated to minimize the performance degradation of the subcells. They will need to be electrically isolated to avoid shunting the semiconductor layers. The metallization of the holes will need to provide good electrical conductivity while minimizing stresses in the device.

8.2.3.8 Virtual Substrate Approach for New Cell Designs

Material developments focusing on engineering the semiconductor layer stacking of different III-V alloys with dissimilar lattice constants present a key opportunity for opening up new dimensions in the multijunction cell optimization. Similar to the metamorphic approach discussed in Section 8.2.3.2, the goal of this research is to address the perennial need to minimize the defect density of epilayers grown on a substrate to which they may be significantly lattice-mismatched. Recently, innovative methods have been explored based on porous materials to interface between the starting substrate and the modified lattice surface (Boucherif *et al.*, 2013). Such new cell developments have been exploring, for example, the growth of III-V junctions on a Si substrate using porous layer strategies (Boucherif *et al.*, 2015). The nucleation of epitaxial III-V layers on Ge surfaces modified by porosification is also being studied. The process development involves first depositing a thin strained layer with the desired lattice constant on a starting substrate. For example, by keeping the strained layer below the critical thickness, it can be kept coherent (pseudomorphic) with no interface dislocations. Then by using an electro-chemical process, a bi-layer of porous material is created. The porous material can be made to have very fine pores in the strained layer and larger pores in the underlying substrate region. The process can form small crystallites and favor a dislocation-free relaxation of the top surface. The top layer can then be homogenized with an epitaxial deposition to coalesce the islands, and/or with a heat treatment to redistribute the material. The optimization of this process could lead to a relaxed, dislocation-free, top surface. The surface can therefore be tailored to have a lattice constant determined by the material of the initial strained layer. The nano-voids created beneath the top layer can act as dislocation traps. With heat treatments, these voids can be made to form a mechanically-weak buried layer. The latter approach has been used successfully to separate a GaAs film from a Ge substrate (Boucherif *et al.*, 2013). Such material and cell developments could therefore lead to interesting methods for substrate recovery strategies.

8.2.3.9 Cell Designs Based on Mechanical Stacking or on Multi-Terminals

In Section 8.2.2.2, it was mentioned that some of the key challenges of multijunction solar cells are related to the fact that they normally need current-matching and lattice-matching between the various subcells. There are, therefore, potentially important opportunities in developing new cell concepts based on mechanical stacking or multi-terminal approaches, with the view to relieving these constraints. In its simplest form, the multi-terminal approach would still use a monolithic III-V multijunction heterostructure. However, the semiconductor stack might be slightly modified to allow photocurrent extraction in conducting layers between some of the subcells. This can be achieved, for example, by incorporating stop-etch layers and by using microfabrication techniques to create accessible electrical connections and to enable sheet conductivity in-between the epitaxially stacked subcells (Gray *et al.*, 2008; Braun *et al.*, 2012). The ability to connect the subcells individually would then release the current matching constraints. It would better exploit the individual subcells' photocarrier generation capabilities. Similar results can be obtained if the subcells are mechanically stacked (Zhao *et al.*, 2010; Makita *et al.*, 2013; Mathews *et al.*, 2014; Sheng *et al.*, 2014). A significant advantage of the mechanically stacked cells is that ultimately each subcell can be individually designed-for-manufacturing using simplified epitaxial processes and very high production yields. A large number of arrangements and numerous strategies are possible, given the relaxed lattice-matching requirements. The different scenarios of mechanical stacking for the development of new III-V cells must be evaluated in view of their potential to improve the efficiency/cost ratio. One challenge, which is commonly prevalent for the mechanical stacking approach, is typically related to the requirement for additional substrates for the growth of the various subcells. The other challenges are related to the additional fabrication and assembly steps necessary to efficiently and reliably join the subcells. The assembly strategy must yield simultaneously high-performance electrical, optical, and thermal interfaces between the mechanically assembled subcells.

8.2.4 Conclusion

III-V multijunction cells have already reached a very high level of perfection. The research cell record efficiency values have maintained a yearly growth of almost 0.5% absolute efficiency points for the past few years, reaching 46% conversion efficiency in December 2014. Such a high-efficiency has been obtained using a four-junction cell with each of its subcells efficiently converting precisely one quarter of the incoming photons in the solar spectral range between ~300 and ~1750 nm into electricity (Dimroth *et al.*, 2014). However, the 4 J cells have not yet gained commercial traction for CPV deployments. For now, the 3 J cells remain the workhorse for commercial applications. Several interesting strategies are being explored to further improve the III-V cell performance and some of them have been reviewed in this chapter. The appetite to further improve the efficiencies remains, but it is important to keep in focus the design-for-manufacturing considerations in the development of new III-V cells. Indeed, the efficiency of III-V cells can be highly leveraged, however, their performance is not only defined by their conversion efficiency, but rather by their efficiency/cost ratio. As the level of perfection of III-V cells improves, at some point the law of diminishing returns will put increasing pressure on which designs can become commercially viable. It is therefore strategic to favor the III-V heterostructure developments that also have built-in aspects of manufacturability in their new cell designs.

List of Acronyms and Symbols

Acronym	Description
3J	Triple-junctions
4J	Four-junctions
6J	Six-junctions
AM1.5D	Air mass 1.5 direct
BSF	Back surface field
CPV	Concentrator photovoltaic
DNI	Direct normal irradiance
IMM	Inverted metamorphic
I_{sc}	Short circuit current
J_{sc}	Short circuit current density
I–V	Current-voltage
LCOE	Levelized cost of energy
LM	Lattice-matched
MBE	Molecular beam epitaxy
MJ	Multi-junction
MM	Metamorphic
MOCVD	Metalorganic chemical vapor deposition
PV	Photovoltaic
QD	Quantum Dot
QDEC	Quantum dot energy converter
QE	Quantum efficiency
R_s	Series resistance
TJ	Tunnel junctions
TSV	Through-silicon-via
V_{oc}	Open circuit voltage

References

Boisvert, J., Law, D., King, R. *et al.* (2013) Conference Record of the IEEE Photovoltaic Specialists Conference, 6745051, pp. 2790–2792.

Boucherif, A., Beaudin, G., Aimez, V., and Ares, R. (2013) Mesoporous germanium morphology transformation for lift-off process and substrate re-use. *Applied Physics Letters*, **102** (1), 011915.

Boucherif, A. *et al.* (2014) US Patent Application.

Boucherif, A. *et al.* (2015) GaAs solar cells on mesoporous Silicon templates. Proceedings of the 11th International Conference on Concentrator Photovoltaic Systems: CPV-11. AIP Proceedings.

Braun, A., Vossier, A., Katz, E.A. *et al.* (2012) Multiple-bandgap vertical-junction architectures for ultra-efficient concentrator solar cells. *Energy & Environmental Science*, **5** (9), 8523–8527.

Cornfeld, A.B., Patel, P., Spann, J. *et al.*(2012) Evolution of a 2.05 eV AlGaInP top sub-cell for 5 and 6J-IMM applications. In *IEEE 38th Photovoltaic Specialists Conference (PVSC), 2012*, pp. 2788–2791.

Dimroth, F., Tibbits, T. N., Beutel, P. *et al.* (2014) Development of high efficiency wafer bonded 4-junction solar cells for concentrator photovoltaic applications. In IEEE 40th Photovoltaic Specialist Conference (PVSC), pp. 6–10.

Domínguez, C., Antón, I., and Sala, G. (2010) Multijunction solar cell model for translating I–V characteristics as a function of irradiance, spectrum, and cell temperature. *Progress in Photovoltaics: Research and Applications*, **18** (4), 272–284.

Fafard, S. (2011) Solar cell with epitaxially grown quantum dot material, US Patent Appl. No. 11/038,230 (Jan. 2005); US Patent 7,863,516 (Jan. 2011).

Fafard, S. and Masson, D.P. (2013) Harmonic photovoltaic up-converter for high-efficiency photo to direct-current (DC) phototransducer power conversion applications, US Patent application.

Fafard, S., Wasilewski, Z.R., Allen, C.N. *et al.* (1999) Manipulating the energy levels of semiconductor quantum dots. *Physical Review B*, **59** (23), 15368.

Fafard, S., York, M.C.A., Proix, F. *et al.* (2016) Ultrahigh efficiencies in vertical epitaxial heterostructure architectures. *Applied Physics Letters*, **108** (7), 071101.

Fidaner, O. and Wiemer, M.W. (2014) US Patent Appl. No. 2014019677.

Friedman, D.J., Geisz, J.F., Kurtz, S.R., and Olson, J.M. (1998) 1-eV solar cells with GaInNAs active layer. *Journal of Crystal Growth*, **195** (1), 409–415.

García, I., Geisz, J.F., France, R.M. *et al.* (2014) Metamorphic Ga0.76In0.24As/GaAs0.75Sb0.25 tunnel junctions grown on GaAs substrates. *Journal of Applied Physics*, **116** (7), 074508.

Gray, A.L., Stan, M., Varghese, T. *et al.* (2008) Multi-terminal dual junction InGaP/GaAs solar cells for hybrid system. In *IEEE 33rd Photovoltaic Specialists Conference*, 2008. PVSC'08, pp. 1–4.

Ho, W.J., Lee, Y.Y., Yang, G.C. and Chang, C.M. (2015) Optical and electrical characteristics of high-efficiency InGaP/InGaAs/Ge triple-junction solar cell incorporated with InGaAs/GaAs QD layers in the middle cell. *Progress in Photovoltaics: Research and Applications*, **24**, 554–559.

Ji, L., and McConnell, R. (2006) New qualification test procedures for concentrator photovoltaic modules and assemblies. In Conference Record of the 2006 IEEE 4th World Conference on Photovoltaic Energy Conversion (Vol. 1, pp. 721–724).

Khvostikov, V.P., Kalyuzhnyy, N.A. Mintairov, S.A. *et al.* (2014) AlGaAs/GaAs photovoltaic converters for high power narrowband radiation. *Proceedings of the 10th International Conference on Concentrator Photovoltaic Systems*, **1616**, 21.

Krut, D., Sudharsanan, R., Nishikawa, W. *et al.* (2002) Monolithic multi-cell GaAs laser power converter with very high current density. In *Conference Record of the IEEE Photovoltaic Specialists Conference, 29th IEEE PVSC*, pp. 908–911.

Luque, A. and Hegedus, S. (eds) (2011) *Handbook of Photovoltaic Science and Engineering* (2nd edition). John Wiley & Sons, Ltd, Chichester.

Makita, K., Mizuno, H., Komaki, H. *et al.* (2013) Over 20% efficiency mechanically stacked multi-junction solar cells fabricated by advanced bonding using conductive nanoparticle alignments. In *MRS Proceedings* (Vol. 1538, pp. 167–171). Cambridge University Press, Cambridge.

Massiot, C.I., Colin, N., Vandamme, N. *et al.* (2013) in 39th IEEE Photovoltaic Specialists Conference, PVSC, article no 6744089, pp. 17–21, Tampa, FL.

Masson, D., Prouix, F. and Fafard, S. (2015) Pushing the limits of concentrated photovoltaic solar cell tunnel junctions in novel high efficiency GaAs photodtransducers based on a vertical epitaxial heterostructure architecture. *Progress in Photovoltaics: Research and Applications*, **23** (12), 1687–1698.

Mathews, I., O'Mahony, D., Thomas, K. *et al.* (2014) Adhesive bonding for mechanically stacked solar cells. *Progress in Photovoltaics: Research and Applications*, **23** (9), 1080–1090.

Miller, O.D., Yablonovitch, E. and Kurtz, S.R. (2012) Strong internal and external luminescence as solar cells approach the Shockley-Queisser limit. *IEEE Journal of Photovoltaics*, **2**, 303.

Philipps, S.P., Dimroth, F., and Bett, A.W. (2012) High-efficiency III-V multijunction solar cells. In *Practical Handbook of Photovoltaics: Fundamentals and Applications* (eds A. McEvoy, T. Markvart, and L. Castaner). Elsevier, Oxford, pp. 417–448.

Rau, U., Paetzold, W. and Kirchartz, T. (2014) Thermodynamics of light management in photovoltaic devices. *Physics Review B*, **90**, 035211.
Richard, O., Jaouad, A., Bouzazi, B. et al, (2015) Simulation of a through-substrate vias contact structure for multi-junction solar cells under concentrated illumination. Proceedings of the 11th International conference on concentrator photovoltaic systems: CPV-11. AIP Proceedings.
Schubert, J., Oliva, E., and Dimroth, F. (2009) High voltage GaAs photovoltaic laser power converters. *IEEE Transactions on Electron Devices*, **56**, 170.
Sheng, X., Bower, C.A., Bonafede, S. *et al.* (2014) Printing-based assembly of quadruple-junction four-terminal microscale solar cells and their use in high-efficiency modules. *Nature Materials*, **13** (6), 593–598.
Steiner, M.A., Geisz, J. F., García, I. *et al.* (2103) Optical enhancement of the open circuit voltage in high quality GaAs solar cells. *Journal of Applied Physics*, **113**, 123109.
Strobl, G.F.X., Ebel, L., Fuhrmann, D. *et al.* (2014) Development of lightweight space solar cells with 30% efficiency at end-of-life. In *IEEE 40th Photovoltaic Specialist Conference (PVSC), 2014*, pp. 3595–3600.
Tao, He *et al.* (2014) High-power high efficiency laser power transmission at 100 m using optimized multi-cell GaAs converter. *Chinese Physics Letters*, **31**, 104203.
Valdivia, C.E., Wilkins, M.M., Bouzazi, B. *et al.* (2015) Five-volt vertically-stacked, single-cell GaAs photonic power converter. in *SPIE OPTO*, pp. 93580E. International Society for Optics and Photonics.
Virshup, G. (1994) Patent US Patent 5,342,451.
Werthen, J. (2011) Patent US Patent application no 2011/0108082.
Wiemer, M., Sabnis, V., and Yuen, H. (2011) 43.5% efficient lattice matched solar cells. In *SPIE Solar Energy Technology*. International Society for Optics and Photonics.
Wilkins, M.M., Gabor, A. M., Trojnar, A.H. *et al.* (2014) Effects of luminescent coupling in single and 4-junction dilute nitride solar cells. In IEEE 40th Photovoltaic Specialist Conference (PVSC), 2014, pp. 3601–3604.
Zhao, L., Flamand, G., and Poortmans, J. (2010) Recent progress and spectral robustness study for mechanically stacked multi-junction solar cells. In *6th International Conference on Concentrating Photovoltaic Systems: CPV-6* (Vol. 1277, No. 1, pp. 284–289).
Zhao, Y., Fay, P., Wibowo, A., Liu, J., and Youtsey, C. (2012) Via-hole fabrication for III-V triple-junction solar cells. *Journal of Vacuum Science and Technology B*, **30** (6), 06F401.

8.3

High Concentration PV Systems

Karin Hinzer[1], Christopher E. Valdivia[2], and John P.D. Cook[2]
[1] *School of Electrical Engineering and Computer Science, University of Ottawa, Canada*
[2] *SUNLAB University of Ottawa, Canada*

8.3.1 Introduction

The III-V semiconductor multi-junction solar cells (MJSC) described in the previous chapters have thus far demonstrated the highest achievable efficiencies of any solar power technology, both in the laboratory and in commercial products. However, this performance is achieved using relatively expensive high-purity materials with limited worldwide production, and manufacturing requires epitaxial processes to form high-quality single crystals. These necessities make this class of solar cell expensive in comparison to flat-panel silicon solar cells. While this cost can be accepted for some applications (e.g. satellites), terrestrial application must enable a levelized cost of electricity (LCOE) comparable to or better than incumbent technologies (e.g., fossil fuel power generation).

To realize cost competitiveness, high-concentration photovoltaic (HCPV) systems employ optics to gather a large area of sunlight and deliver it to a small-area solar device, concentrating the intensity of sunlight by a factor of 500–1000 or more. This reduces the requirement for solar cells by the same factor, replacing it with less expensive and less energy-intensive materials such as metals and plastics.

Employing this optical system imposes several limitations and requirements on the system design. First, the optics concentrate only the normally-incident sunlight, referred to as the direct normal irradiance (DNI). This leaves the diffuse component of the global irradiance unused, reducing the available solar resource by ~10% (AM1.5G). Second, a two-dimensional tracker must be employed to ensure the system continuously faces the sun, which, by comparison to fixed flat panel systems, has the added benefit of gathering ~30-40% more sunlight (see Figure 8.3.1).

Photovoltaic Solar Energy: From Fundamentals to Applications, First Edition.
Edited by Angèle Reinders, Pierre Verlinden, Wilfried van Sark, and Alexandre Freundlich.

Companion website: www.wiley.com/go/reinders/photovoltaic_solar_energy

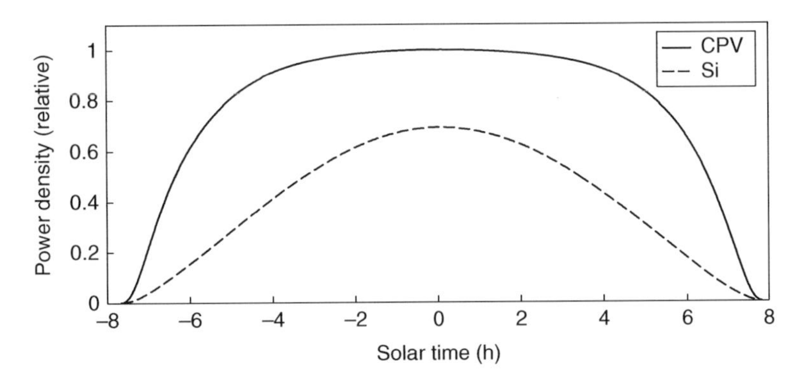

Figure 8.3.1 Relative power density produced by a two-dimensional tracking CPV module (30% efficiency) and a horizontal Si module (20% efficiency), on the summer solstice in Ottawa, Canada. The CPV module is capable of harvesting nearly twice the energy of the Si module

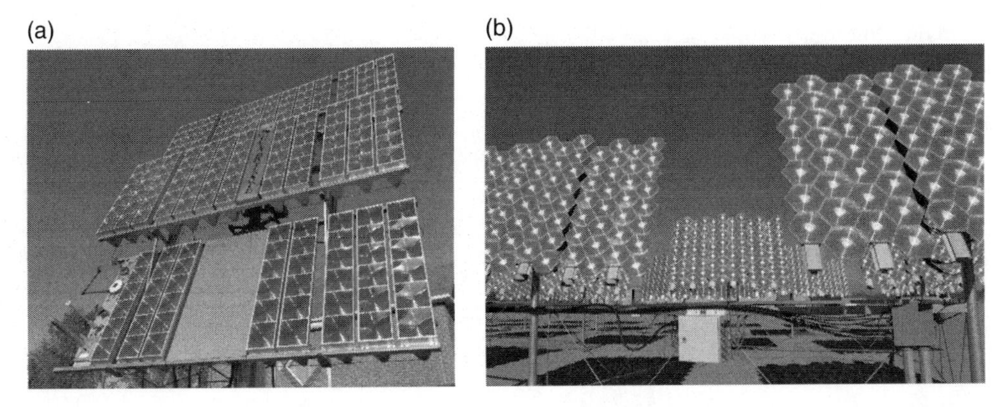

Figure 8.3.2 Two-axis tracker systems using (a) Fresnel-type modules by OPEL Solar Inc. in Ottawa, Canada, and (b) waveguiding optics by Morgan Solar Inc. in California, USA

Among the first concentrator photovoltaic (CPV) systems was the 1kW two-axis tracker by the USA National Sandia Laboratories in the 1970s, employing a Fresnel lens to concentrate ~60× on an actively-cooled crystalline Si solar cell (Burgess and Edenburn, 1976; Burgess and Pritchard, 1978). Since that time, a wide variety of designs have been proposed and commercially produced. Two examples are shown in Figure 8.3.2. As of this writing, CPV systems have achieved module efficiencies exceeding 35% (Green *et al.*, 2015; Keevers *et al.*, 2014; Steiner *et al.*, 2014, 2015).

8.3.2 Optics

CPV optical systems can be divided into two main categories based on their design strategies which employ either reflective or refractive optics, each typically consisting of 1–3 optical elements. Three examples of such CPV designs are illustrated in Figure 8.3.3. A lower number

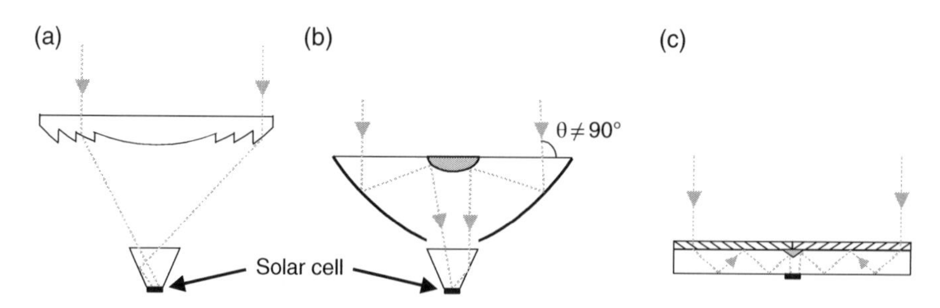

Figure 8.3.3 Concentrator designs with different optical trains: (a) Fresnel lens, (b) Cassegrain primary reflectors, and (c) waveguiding designs

of optical elements are employed to minimize costs, while a higher number enables the potential for improved irradiance uniformity, wider acceptance angles, and higher performance. Generally, an optical loss through the entire optical train of 10–30% can be expected with the present technologies.

Concentrator optics also introduce an optical transfer function defined as the transmissivity as a function of wavelength. This may substantially impact the spectrum of sunlight reaching the MJSC. These devices are more sensitive to sunlight spectral changes than conventional single-junction devices such as Si, but most CPV developers presently offer one module design, and the choice of the spectrum for which the device is optimized can be informed by detailed modeling (McDonald and Barnes, 2008; Muller *et al.*, 2010) and local solar resource metrology (Tatsiankou *et al.*, 2013). In future, it may be that developers will offer a suite of module designs, optimized for various regional markets.

8.3.2.1 Primary Optical Elements

Reflective (mirror-based) CPV optical systems are often considered for their broadband high-reflectivity and achromatic focus. These systems can be divided into two main categories. The first is the trough concentrator system, often used with single-axis trackers. The second is the spherical- or parabolic-type mirror system, which has been produced in a variety of sizes ranging from small sub-meter Cassegrain reflectors (Ludowise and Fraas, 2010) (Figure 8.3.3 (b)) to very large >10 m segmented dishes (Verlinden *et al.*, 2006), with either a single solar cell or a dense array of multiple cells per optical train, respectively. Dense arrays are also used in CPV systems composed of a field of heliostats and an actively-cooled central receiver (Lasich *et al.*, 2009).

Refractive CPV systems today are dominated by Fresnel lens designs (Figure 8.3.3 (a)). The Fresnel lens advantages are its flat profile, low weight, low manufacturing costs, and use of low-cost materials such as polymethylmethacrylate (PMMA) or silicone-on-glass (SoG). Some of the challenges associated with this type of optic are realizing high transparency in the full wavelength range of 300–1800 nm, limiting the performance degradation over service life, and chromatic aberrations. Fresnel lenses have been widely used since the advent of CPV systems in the 1970s (Burgess and Edenburn, 1976).

An innovative departure from these designs is the waveguiding concentrator, where light is directed into a planar waveguide toward a central photoconverter (Figure 8.3.3 (c)). Another design also employing waveguides is the luminescent concentrator (see Chapter 8.5).

8.3.2.2 Secondary Optical Elements

Following the primary optical element (POE), a secondary optical element (SOE) can optionally be included. These are often non-imaging optics, such as the compound parabolic concentrator (CPC), cone, dome, or inverted truncated pyramid, each with varying performance benefits (Victoria *et al.*, 2009). The SOE can also be constructed using reflective or refractive designs and need not be the same type as the POE.

A very important benefit of the SOE is the widening of the optical acceptance angle, which has several effects on the system (Victoria *et al.*, 2009; Benítez *et al.*, 2010). First, the SOE can reduce the alignment requirements between the optics and the PV device. Second, the SOE can reduce the need for pointing precision in the sun tracker, reducing its cost. This also reduces the requirement for rigidity, since sub-modules across the entire tracker area must be within the pointing accuracy, impacting its weight and shipping costs. The Fresnel-Köhler SOE design (Benítez *et al.*, 2010) provides useful further improvement on angular acceptance. Third, the SOE can improve the irradiance uniformity across the MJSC active area, helping to prevent tunnel junction current limitations and potential reliability issues caused by non-uniform thermal loads.

The introduction of an SOE to the optical train must be justified against several potentially significant disadvantages, including additional optical losses, alignment, and expense. Index-matching silicone or other optical adhesive is often used between the SOE and PV device to minimize reflections, for example. Presently, not all companies include an SOE in their optical train, so this element remains an optimization consideration specific to individual CPV system designs.

8.3.2.3 Optical Material Considerations

Optical materials are selected for cost, manufacturability, performance, and reliability. Refractive materials are typically chosen among glass, PMMA and silicones, due to their good ultra-violet (UV) transmission and relatively few absorption bands in the range of multi-junction cells. Fabrication is by injection or compression molding, using diamond-turned molds for optical finish. Polymers are sometimes preferred over glass for easier fabrication and lower mass, but can warp under residual internal stress and creep, and abrade under air-borne particulates, especially PMMA Fresnel lenses (Wineman and Rajagopal, 2000; Osswald *et al.*, 2006; Miller and Kurtz, 2011). Hybrid silicone Fresnel lenses have the mechanical stability of glass and the moldability of silicone, but must be checked for delamination and optical figure stability with temperature (Annen *et al.*, 2011). These refractive elements will have a ~4% reflection loss at each air/dielectric interface unless an anti-reflection coating is applied. Such coatings may be cost-effective on smaller downstream optical elements.

Refractors are subject to solarization (permanent changes in materials due to long-term exposure to sunlight), and the elements at highest optical concentration are under extreme stress. The literature on glass solarization under intense UV irradiation is sparse (Vaughnn and Poczulp, 2004), except for deep UV lithography optics (Matsumoto and Mori, 1998). Polymers are generally more susceptible to solarization than inorganic materials, and the polymer structure and additives must be engineered specifically for CPV applications. The literature on polymer Fresnel lens durability has been comprehensively surveyed (Miller *et al.*, 2010; Miller and Kurtz, 2011) for PMMA and silicone-on-glass technologies. Photokinetic mechanisms for degradation of silicones have also been studied (Zimmermann, 2008).

Metal reflectors reach high reflectivities with no absorption bands within the multi-junction absorption range, but may fall off in the UV. Aluminum mirrors trade slightly lower visible light reflectance for comparably high short-wavelength reflectivity. Aluminum also benefits from a transparent native oxide and lower cost compared to other metals, such as silver. Mirrors are either coated onto molded polymer forms, or fabricated by electroforming (Schmidt, 1966) on diamond-turned mandrels. Reflectors are also subject to degradation, through intrinsic internal reactions, and through extrinsic pathways such as reaction with ambient gases and airborne particulates. Reliable designs will protect reflective surfaces by internalization, passivation, encapsulation, or enclosure. Study of coatings on large astronomy telescope mirrors, a large investment, may offer useful insights.

8.3.2.4 Angular Response

The acceptance angle of the optical system is defined as the maximum angle for which it produces 90% of its maximum power, as shown in Figure 8.3.4. Light outside this angle – such as much of the diffuse component of solar radiation – will be attenuated or lost. Therefore, it is desirable to maximize the acceptance angle of a concentrator system to admit as much circumsolar light as possible and relax the tolerance on the tracker pointing accuracy. However, in general, the acceptance half-angle, θ_a, decreases as the maximum possible concentration, C_{max}, increases, as described by the equation $sin^2\theta_a = n_{ref}^2/C_{max}$ (Chaves, 2008; Chaves and Hernández, 2013) for a 2-dimensional concentrator. Therefore, one metric of a concentrator system is the concentration-acceptance angle product (CAP), defined as CAP = $\sqrt{C}\cdot sin\theta_a \leq n_{ref}$, where C is the geometrical concentration and n_{ref} is the refractive index of the medium surrounding the solar cell.

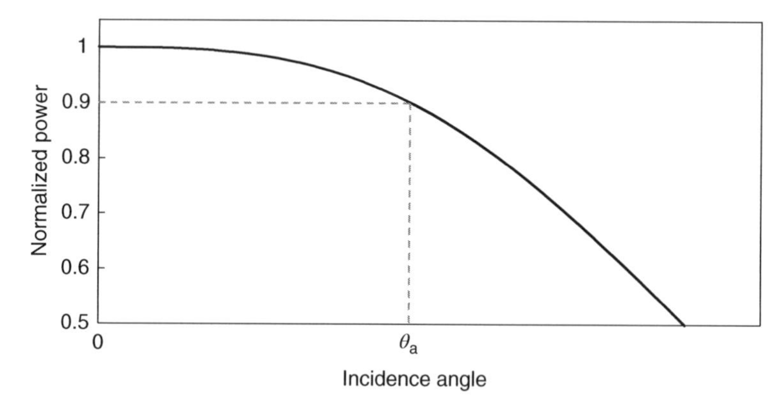

Figure 8.3.4 Illustration of normalized power as a function of incidence angle, depicting the acceptance half-angle, θ_a, as defined at 90% of maximum power

8.3.3 Trackers

While one- and two-axis sun trackers are optional performance enhancements in flat-panel PV fields, two-axis trackers with much tighter tracking precision are essential for HCPV. The most popular tracker architecture is the "azimuth-elevation"-style: a vertical column is planted in a buried concrete pedestal, and supports a horizontal axle through the center of mass of a single flat scaffold frame carrying an array of coplanar CPV modules, and driven through azimuth-elevation motion by two motors and gearing (see Figure 8.3.5 (a)). The "roll-tilt" architecture establishes a long roll bar, oriented north–south and driven by one motor, carrying many small module scaffolds on tilt mounts, controlled by a linkage to a second motor running the length of the roll bar (see Figure 8.3.5 (b)).

The relative merits of architectures are complex and difficult to quantify. Simple azimuth-elevation trackers must achieve large, stiff, yet lightweight scaffold frames, and wind loading increases with height above grade, requiring substantial concrete pedestals and columns, and the installation of larger devices requires heavy equipment. The roll-tilt system lies closer to the ground and can be assembled with little heavy equipment, but aligning the extended array may require a trained installation crew.

Emerging intermediate hybrid architectures include the Savanna™ design (Figure 8.3.2 (b)) by Morgan Solar Inc., which replaces costly excavated concrete pedestals with an above-ground extended-truss space-frame in which bending loads are transformed into pure tension and compression loads on cables and tubular spars, with substantial weight and cost saving. At each space-frame node a two-axis tracker head is attached, carrying a module scaffold of moderate size and wind load profile.

The aspiration for utility-grade solar power plant is zero maintenance, but field experience shows that trackers fail most frequently (Rubio *et al.*, 2011) compared to protection systems, inverters, modules, software, and communications.

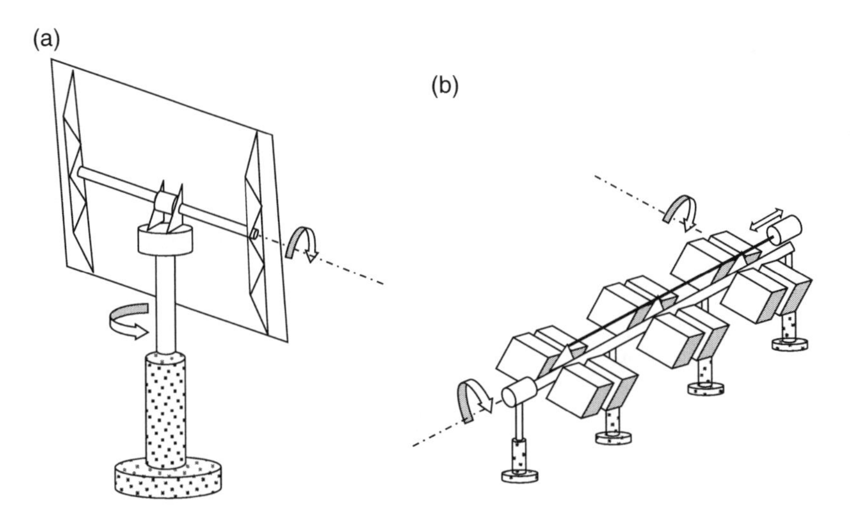

Figure 8.3.5 Schematic diagrams of popular CPV module and tracker architectures: (a) azimuth-elevation tracker (b) tilt and roll tracker

All tracked solar field designs must consider shading losses when the sun is close to the horizon (Kim *et al.*, 2013). This problem becomes greater with increasing latitude. The designer can control the aspect ratio of the receiving arrays, the field layout pattern, and the spacing between trackers, and is constrained by costs of real estate and inter-tracker copper power cable and trenching.

8.3.3.1 Pointing Accuracy

Most concentrator module optics have an angular acceptance in the range 0.5–1°, half-width 90%-maximum (HW90% M), which requires the combined contributions of tracking error and scaffold flatness to be a small fraction of a degree. Both PV and HCPV sectors now design for a service life of 25–35 years in outdoor plant conditions, and usually further design for one extreme weather event per service life. Wind is the most common degrader of pointing performance, followed by soiling.

Sun position algorithms are available to any precision (Lee *et al.*, 2009), but controller microprocessor resources limit the sophistication of real-time tracking calculations. Formula for tracking motion and mispointing, general to any orientation of two rotation axes, and related axis installation misalignment (Chong and Wong, 2009), show that a combined installation misalignment of as little as 0.4° in both axes gives rise to a substantial 0.7° worst-case error. It is costly to install and maintain tracker systems to precise axis alignments, so usually a sun sensor feedback control is added.

Sun sensors are photoelectric devices permanently mounted and aligned on the tracker scaffold, which develop analog signals proportional to tracking error in one or two dimensions, which can be developed into motor control feeds to minimize tracking error. Sun sensor enclosures must be specified to certification standards, and internal condensation mitigation similar to CPV modules should be established.

8.3.3.2 Re-acquisition Time

When the sun is briefly obscured, by intermittent cloud or sun sensor occlusion, the tracker switches from sun sensor tracking to calculated sun position tracking, and when direct sun is restored, the tracker switches back. However, the energy yield can be degraded if the discrepancy between the calculated and actual sun position is large, or if the tracking motion is slow, or if there is "hunting" as the feedback control point is re-established. On patchy cloud days the cumulative re-acquisition deficit may be substantial. This discrepancy arises when the tracker algorithm is not informed of tracker axis misalignment.

8.3.3.3 Control Systems

All trackers require a control unit (with battery backup sized to local average overcast) to support the following functions: integration of sensor signals; generation of motor drive signals; sun position calculation; and command and data communications. Usually control units support several modes of operation, including: approximate sun tracking in cloud, directed by calculated sun position; precision sun tracking, directed by a photoelectric sun sensor; and special modes for safety during installation, in adverse weather, and for night stowing and servicing.

8.3.4 Performance Evaluation and Fault Detection

CPV cells and systems are tested under two standard conditions, as defined by the IEC 62670 standard. The first, concentrator standard *test* conditions (CSTC), utilizes an AM1.5D spectrum with 1000 W/m^2 irradiance, and a cell temperature of 25 °C. The second, concentrator standard *operating* conditions (CSOC), uses an AM1.5D spectrum in conjunction with 900 W/m^2 irradiance, an ambient temperature of 20 °C, and a wind speed of 2 m/s. Part 2 of this standard establishes a testing methodology for energy measurements, which includes monitoring of gross, parasitic, and net powers, as well as the DNI solar resource. Part 3 of this standard defines how to filter measured data and to translate the conditions of measurement to both CSTC and CSOC.

CPV modules can be tested for their overall performance, while also non-invasively extracting performance from individual sub-modules (i.e. single solar cells) and identifying faults, for example, using arrays of shutters over individual sub-modules and real-time *IV* measurement (Yandt *et al.*, 2015). Component variability in PV and CPV is more important than in other electronic systems, because the performance of a module string connected electrically in series is limited by the single worst element in the string. For example, recently reported work (Steiner *et al.*, 2014) showed 36.7% module efficiency, despite some submodules at 38.9%; the shutter technique can be used to identify such an outlier submodule without breaking into the module.

Where local solar spectrum must be verified, spectroradiometers or spectroradiometer emulators (Tatsiankou *et al.*, 2013) can be used to determine both the spectral and DNI resource.

Metrology for precision long-term tracker pointing verification is not yet widely commercial: bolt-on turn-key inclinometers are available, but not long-term azimuth metrology to 0.1° precision. Custom sight tube/graticle alignment jigs can be machined locally.

8.3.5 Cost Optimization

As of this writing, most CPV equipment developers adopt the same nominal cost target as PV vendors: 1 $/W installed (utility-grade) by 2020, as articulated by the USA Dept. of Energy SunShot Initiative (US DoE, 2012), across ~40% CPV module cost, ~40% balance of system hardware cost, and ~20% soft costs (permitting, local resource and market assessments, forecasting, operating, etc.). Additional cost reduction pressure will arise where local grid integration costs are unusually high, for example, in older grids incapable of distributed power in-flows or two-way power flows. All components in a CPV field are in principle candidates for cost-reduction, but it is more effective first to improve the efficiency of the few photoconversion elements (such as the concentrating optics, HCPV solar cell and, less important, cell cooling), which is equivalent to cost-reducing the entire balance of system.

Tracker cost optimization, including costs of tracker, scaffolding, and trenching (Gombert, 2012), suggests that the optimal CPV module area per tracker is approximately 100 m^2.

A general empirical prediction of PV and CPV cost reduction trends has been made recently (Haysom *et al.*, 2014) using a formal "learning curve" analysis (Figure 8.3.6), and concluded that to achieve a fully installed CPV system price of 1 $/W in 2020, the cumulative deployment must increase at a cumulative annual growth rate (CAGR) of 67% (which is lower than the historical CAGR for CPV systems) to 7,900 MW.

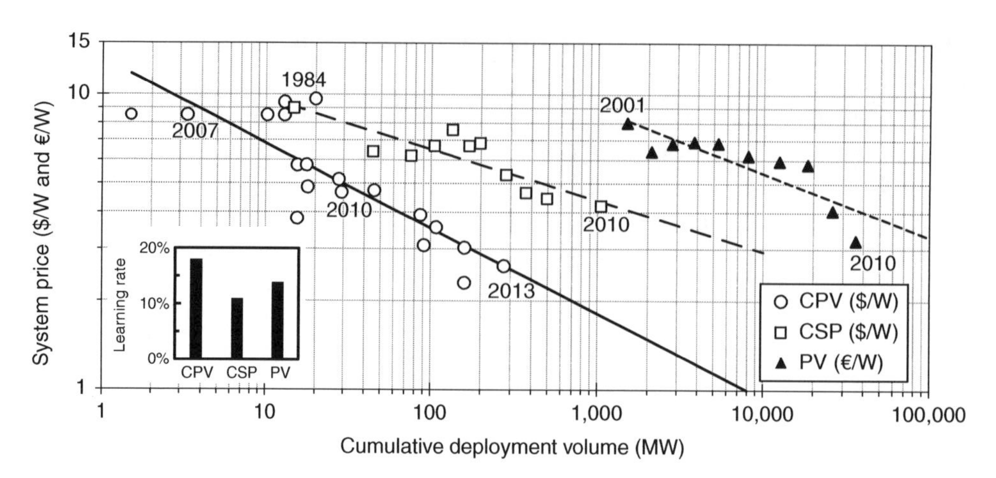

Figure 8.3.6 Global learning curve (log/log relationship between system price and cumulative deployment each year) for concentrated photovoltaic (CPV), concentrated solar (thermal) power (CSP), and flat-panel photovoltaic (PV) systems. Inset: The learning rate (proportional price reduction with doubling of volume) for each technology. Source: adapted from (Haysom, 2014)

8.3.6 Assembly and Reliability

We discuss these topics together, since early module fails are likely to come from assembly failures, not the MJSC die or the concentrator optical elements. The III-V compound semiconductor MJSC device is generally very reliable (the technology was developed for 20-year telecommunications optical networks). A few particular semiconductor solid-state failure mechanisms may affect CPV modules, such as hot spots, filamentary conduction, and thermal runaway (Bosco *et al.*, 2011).

The CPV module is a precision optoelectronic and thermomechanical device which must be inexpensive yet reliable over a span of 25–35 years in harsh shipping and service conditions. CPV module reliability must be achieved across all components and materials, in the presence of UV, high voltage, heat, cold, and moisture extremes and cycling, and with due regard for the possible presence of impurities, contaminants, pollutants, particulates, and other foreign elements, during both fabrication and operation. This is done mostly through the control of dimensional stability, chemical stability, and adhesion. Dimensional instability and mechanical stress arise from: temperature and humidity coefficient of expansion mismatch among bonded materials; residual internal stress; plastic deformation; and creep. Chemical instability arises from: UV photolysis of chemical bonds, especially in polymers; electrochemical corrosion between materials with appropriate galvanic index; diffusion of atoms followed by chemical reaction; or potential-induced diffusion of ions. Adhesion is essential for all assemblies joined with metallurgical solder or polymer adhesives, or by advanced deposition techniques: joint failure arises from adhesive and cohesive fracture; cyclic thermomechanical fatigue; and electrochemical corrosion.

Although assembly techniques may seem less sophisticated than the fabrication of the MJSC die, successful industrial designers appreciate and master the large body of advanced applied science necessary for high-yield high-reliability module assembly processes, in 3–6 years technology development cycles. Credible claims of CPV module long-term reliability must show documented due diligence in costly, difficult, and time-consuming reliability science studies, including accelerated studies where possible. The supply chain may already hold some results of such studies, but it is unlikely that comprehensive results are available across all processes and materials. Usually any service life warranty, and the necessary studies, are the burden of the final equipment manufacturer.

Assembly automation is essential for certain key receiver and module assembly steps, and must be regarded as enterprise-critical. All assembly processes should be selected in the context of eventual commercial throughput; a process which is adequate for pilot production may not be scalable to volume manufacturing. Some of these design-for-reliability elements already have commercial solutions for service life of 10 to 15 years in the automotive or power electronics industries.

8.3.6.1 Reliability of Optics Attaches and Optical Encapsulants

Most attachments of optical elements to the module housing can be handled with conventional industrial engineering design. Often, however, between the MJSC die and the final element in a refractive optical train there is an optical attach exposed to concentrated sunlight. Or, if the final optical element is reflective, there must be an optical encapsulant on the input face of the MJSC die to protect the high-voltage electrical grid lines. The fabrication process for such optical joints must be very carefully and thoroughly developed, taking into account adherend surface chemistry, topography, and pre-cleaning; adherend wetting; adhesion promoters; viable cure times and conditions (compatible with other components and materials present in the assembly); and cured material properties and uniformity. Process inputs, process tools and related equipment used for this fabrication process should be tightly controlled and completely automated, ensuring the highest yield and eliminating all operator dependencies. Clean-room conditions are advisable, since particulate contamination in a concentrated beam will cause excessive local heating. High-concentration optical joints should be thoroughly studied for performance and reliability, which is difficult since accelerated reliability studies must establish multiple simulated sources at ~10,000 suns, running for weeks or months.

Most designers have turned to optical silicones for this application, due to their excellent thermal stability and low UV absorption. However, their very high coefficient of thermal expansion (CTE), adjacent to very low CTE materials (Ge-based III-V cells, glass optics) may present a design challenge. Some designers have deliberately separated the optics attach problem into a mechanical element and an optical element, for independent optimization.

8.3.7 Receiver Assembly and Thermal Management

8.3.7.1 Cell-Carrier Attach

The attachment of the MJSC die to the carrier (which may also hold the bypass diode and two cable-attach lugs) can usually be done with conventional solder or epoxy techniques widely available in the microelectronics industry. Special needs include: very long service life with wide temperature excursions and about 10^4 lifetime thermomechanical fatigue cycles; possible high voltage gradients from long module strings; high-current capacity for the MJSC top contact wirebonds or ribbon bonds; thermal resistance compatible with 30 to 100 W/cm^2

heat extraction; and maximum die temperature of about 100 °C to satisfy long-term reliability and temperature coefficient of power of the MJSC die technology. Some jurisdictions may require lead-free solder. Where metal-loaded epoxy is contemplated, chemical corrosion should be assessed.

8.3.7.2 Carrier-Chassis Attach

The cell-carrier assembly (CCA) attachment to the module must meet diverse mechanical, thermal, chemical, and electrical requirements. Heat dissipation and cyclic thermomechanical fatigue in some CPV architectures favor smaller MJSC die (0.5 mm) using precision automated assembly tools. Thermal resistance of CCA materials and attach must keep the MJSC below 100 °C. Electrical resistance must be high if large string voltages appear between the chassis and CCA element. Thermal expansion mismatch between carrier materials and chassis may also be large, stressing the joint. Filled epoxy adhesives with spacer microbeads can be effective for carrier-chassis attach if carefully designed and verified. For example, boron nitride is a thermally conductive, electrically insulating fill, with no electrochemical corrosion issues.

8.3.7.3 Verification

Potential attach defects affecting thermal and electrical conductivity include internal voids, incomplete wetting of adherends, and incomplete cure. Attach integrity can be verified with several characterization techniques, including lock-in infra-red thermography, C-mode scanning acoustic microscopy, X-ray imaging, and electrical testing.

8.3.8 Housing

Most CPV module designs do not avoid an internal cavity, arising from the optical path among the concentrating optics and the solar cell, which must be housed, for safety and to minimize ambient degradation of optical and high-voltage elements, and be compatible with all requirements of assembled components at a viable cost. Where an internal cavity can be eliminated, such as in waveguide concentrator technologies, many of the following mitigations are unnecessary.

Internal cavity condensation events when ambient temperature and humidity drop sharply are undesirable: optical surfaces are degraded by temporary condensation droplets and by permanent drying marks, and condensation on any surface may facilitate movement of otherwise stationary contaminants. Prevention of internal condensation and internal pressurization is usually done by internal/ambient air exchange through a gas-permeable membrane which excludes liquids (rain), particulates, and biota, and with a pressure equilibration time constant fast enough to survive air shipping. Since even with a fast equilibration port there is risk of condensation, all high-voltage surfaces must be encapsulated. Some developers have studied replacing permeable ports with closed modules and permanent drying units, where filtered, dried air is actively pumped through chains of modules interconnected with air ducts (Gombert, 2012).

Selection of housing materials is a compromise among cost, weight, manufacturability, strength, stiffness, thermal and humidity expansion mismatch, and creep. Widely used materials are glass, aluminum, and fiber-reinforced plastics (FRP). Materials for a FRP housing should be selected for low particulation and outgassing throughout service life. All housing joints should be sealed to certification requirements against ingress from precipitation, particulates, and biota.

8.3.8.1 Mechanical Adhesives

Module designers often resort to adhesives for assembly, yet adhesives can be the first to fail in poorly-developed module assembly processes. Adhesives science is extremely broad (Pocius, 2002), but there are three main classes of adhesion mechanism: micromechanical interfacing, van der Waals forces, and covalent chemical bonds, of which the last is usually reserved for rare advanced specialty adhesive formulations. All adhesion mechanisms can be degraded by surface contamination during joint fabrication, by microvoids, or by mechanical, thermal, photolytic, or chemical stress during service life. Adhesives should be assessed for outgassing and particulation risk. Adhesion promoters are sometimes recommended and provided by adhesives vendors, and it is usually inadvisable to set aside such recommendations. Adhesion promoters function by a variety of mechanisms, including displacing surface contamination (cleaners), modifying adherend surface energy (surfactants), and improving chemical compatibility. Factory ambient temperature and humidity control are likely essential to successful use of adhesives. Adhesives contain a variety of proprietary additives, and it is advisable to consult closely with vendors, to assess technical detail.

8.3.8.2 Off-Axis Beam Damage

When the tracker is misaligned, there is the possibility that the intense focused beam spot from the primary optical element can be directed onto materials incapable of withstanding this energy density. In order to receive certification, CPV modules must be designed and demonstrated to be capable of stable safe operation through any misalignment.

8.3.9 Certification and Test Method Standards

The IEC Technical Committee 82 maintains an on-line record of standards established and in development, relevant to CPV technology. Certification must not be confused with long-term reliability. Certification testing ensures that the module type and design meet nameplate performance rating and safety requirements and conform to type definition. It does *not* demonstrate that the specified performance will be maintained throughout a 25- to 35-year service life. Existing CPV standards include UL 8703 and IEC 60904, 62108, 62670. New standards in preparation are: IEC 60904-8 (Measurement of Spectral Responsivity of Multi-junction PV Devices); 60904-9 (Solar Simulator Performance Requirements); 62108-9 (Retest Guidelines); 62670-2 (Energy Measurement); 62670-3 (Performance Measurements and Power Rating); 62688 (CPV Module and Assembly Safety Qualification); 62727 (PV systems-Specification for solar trackers); 62787 (CPV solar cells and cell-on-carrier assemblies – Reliability qualification); 62789 (PV concentrator cell documentation – specification for concentrator cell description); 62925 (Thermal cycling test for CPV modules to differentiate increased thermal fatigue durability); 62989 (Primary Optics for CPV Systems).

8.3.10 Future Directions

Commercial uptake of CPV systems may depend on the recognition of the present high-yield value proposition, the mitigation of perceived risk, and the progress to higher yields (Tomosk *et al.*, 2015). Cost and reliability mitigation has been achieved in materials count reduction (Gombert, 2012). Energy yield progress may include a higher junction count in the multi-junction cell (see Chapter 8.2), harvesting both direct and diffuse insolation using hybrid CPV plus PV in the same housing, CPV-T where thermal energy, presently lost to ambient, is captured in a circulating fluid system; or ultra-high concentration (UHCPV) to 2000 × or more.

List of Acronyms

Acronym	Description
AFM	Atomic force microscopy
AM	Airmass
AM1.5D	Air mass 1.5 direct spectrum, defined by ASTM G173 standard
CAP	Concentration-acceptance angle product
CCA	Cell-carrier assembly
CPV	Concentrator photovoltaic
CPV-T	concentrator photovoltaic thermal
CAGR	Cumulative annual growth rate
CTE	Coefficient of (linear) thermal expansion
CSOC	Concentrator standard operating conditions
CSTC	Concentrator standard test conditions
DNI	Direct normal irradiance
FRP	Fiber-reinforced plastic
FW90%M	Full width at 90% maximum
HCPV	High-concentration photovoltaic
HW90%M	Half width at 90% maximum
IEC	International Electrotechnical Commission
LCOE	Levelized cost of energy
MJSC	Multi-junction solar cell
PMMA	Polymethylmethacrylate
POE	Primary optical element
PV	Photovoltaic
SOE	Secondary optical element
UL	Underwriters Laboratories
UV	Ultra-violet

References

Annen, H.P., Fu, L., Leutz, R. *et al.* (2011) Direct comparison of polymethylmetacrylate (PMMA) and silicone on-glass (SOG) for Fresnel lenses in concentrating photovoltaics (CPV). *Reliability of Photovoltaic Cells, Modules, Components, and Systems IV*, **8112**, 811204. SPIE, San Diego.

Benítez, P., Miñano, J.C., Zamora, P. *et al.* (2010) High performance Fresnel-based photovoltaic concentrator. *Optical Express*, **18** (S1), A25–A40.

Bosco, N., Sweet, C., Silverman, T., and Kurtz, S. (2011) CPV cell infant mortality study. *7th International Conference on Concentrating Photovoltaic Systems*, **1407**, 323–326. AIP.

Burgess, E.L., and Edenburn, M.W. (1976) One kilowatt photovoltaic subsystem using Fresnel lens concentrators. *12th Photovoltaic Specialists Conference*, pp. 776–780. IEEE.

Burgess, E.L., and Pritchard, D.A. (1978) Performance of a one kilowatt concentrator photovoltaic array utilizing active cooling. *13th Photovoltaic Specialists Conference*, pp. 1121–1124. IEEE.

Chaves, J. (2008) *Introduction to Nonimaging Optics*. CRC Press, Boca Raton, FL, p. 21.

Chaves, J. and Hernández, M. (2013) Solar concentrators. In *Illumination Engineering: Design with Nonimaging Optics* (ed. R.J. Koshel). Wiley-IEEE Press, New York, pp. 148–153.

Chong, K.K. and Wong, C.W. (2009) General formula for on-axis sun-tracking system and its application in improving tracking accuracy of solar collector. *Solar Energy*, **83**, 298–305.

Gombert, A. (2012) Low cost reliable highly concentrating photovoltaics – a reality. *38th Photovoltaic Specialists Conference*, pp. 1651–1656. IEEE.

Green, M.A., Emery, K., Hishikawa, Y., *et al.* (2015) Solar cell efficiency tables (version 45). *Progress in Photovoltaics: Research Applications*, **23** (1), 1–9.

Haysom, J.E., Jafarieh, O., Anis, H. *et al.* (2014) Learning curve analysis of concentrated photovoltaic systems. *Progress in Photovoltaics: Research Applications*. DOI: 10.1002/pip.2567.

Keevers, M., Lau, J., Green, M., *et al.* (2014) High efficiency spectrum splitting prototype submodule using commercial CPV cells. Paper presented at the 5WeO. 4.4, 6th World Conference on Photovoltaic Energy Conversion, Kyoto.

Kim, Y. S., Kang, S.-M., and Winston, R. (2013) Modeling of a concentrating photovoltaic system for optimum land use. *Progress in Photovoltaics: Research Applications*, **21** (2), 240–249.

Lasich, J.B., Verlinden, P.J., Lewandowski, A. *et al.* (2009) World's first demonstration of a 140kWp heliostat concentrator PV (HCPV) system. 34th Photovoltaic Specialists Conference (pp. 002275–002280).IEEE.

Lee, C.-Y., Chou, P.-C., Chiang, C.-M., and Lin, C.-F. (2009) Sun tracking systems: a review. *Sensors*, **9**, 3875–3890.

Ludowise, M., and Fraas, L. (2010) High-concentration cassegrainian solar cell modules and arrays. In *Solar Cells and Their Applications* (eds L.M. Fraas, and L.D. Partain), (2nd ed.). John Wiley & Sons, Ltd., Chichester.

Matsumoto, K. and Mori, T. (1998) Lithography optics: its present and future. *International Optical Design Conference*, **3482**, pp. 362–368. SPIE.

McDonald, M. and Barnes, C. (2008) Spectral optimization of CPV for integrated energy output. In *Solar Energy+ Applications* (pp. 704604–704604). International Society for Optics and Photonics.

Miller, D.C., Gedvilas, L.M., To, B., Kennedy, C.E., and Kurtz, S.R. (2010) Durability of poly(methyl methacrylate) lenses used in concentrating photovoltaic modules. *Reliability of Photovoltaic Cells, Modules, Components, and Systems III. 7773*, p. 777303. SPIE, San Diego.

Miller, D.C. and Kurtz, S R. (2011) Durability of Fresnel lenses: A review specific to the concentrating photovoltaic application. *Solar Energy Materials and Solar Cells*, **95** (8), 2037–2068.

Muller, M., Marion, B., Kurtz, S., *et al.* (2010) An investigation into spectral parameters as they impact CPV module performance. In *AIP Conference Proceedings*, 1277 1, 307.

Osswald, T.A., Baur, E., Brinkmann, S., *et al.* (2006) *International Plastics Handbook. Carl Hanser Verlag GmbH and* Co. KG, Munich.

Pocius, A.V. (2002) *Adhesion and Adhesives Technology: An Introduction* (2nd edn). pp. 78–92, Carl Hanser Verlag GmbH and Co. KG, Munich.

Rubio, F., Martínez, M., Sánchez, D., and Aranda, R. (2011) Results of three years CPV demonstration plants in ISFOC. *37th Photovoltaic Specialists Conference*, pp. 3543–3546. IEEE.

Schmidt, F.J. (1966) Electroforming of large mirrors. *Applied Optics*, **5** (5), 719–725.

Steiner, M., Bösch, A., Dilger, A., *et al.* (2014) FLATCON CPV module with 36.7% efficiency equipped with four-junction solar cells. *Progress in Photovoltaics: Research Applications*. DOI: 10.1002/pip.2568.

Tatsiankou, V., Hinzer, K., Muron, A. *et al.* (2013) Reconstruction of solar spectral resource using limited spectral sampling for concentrating photovoltaic systems. *15th Photonics North Conference.* **8915**, p. 891506. SPIE.

Tomosk, S., Wright, D., Hinzer, K., and Haysom, J.E. (2015) Analysis of present and future financial viability of high concentrating photovoltaic projects. In *High Concentrator Photovoltaics* (eds P. Pérez-Higueras and E.. Fernandez). Springer, in press.

US Dept. of Energy, *SunShot Vision Study*, February 2012, Chapter. 4, p. 77. http://energy.gov/sites/prod/files/SunShot%20Vision%20Study.pdf (accessed 16 April, 2015).

Vaughnn, D., and Poczulp, G. (2004) Filter solarization, understanding and mitigating degradation. In *Nonimaging Optics and Efficient Illumination Systems* (eds R. Winston, and J. Koshel), **5529**, pp. 240–254. SPIE.

Verlinden, P.J., Lewandowski, A., and Lasich, J. B. (2006) Performance and reliability of a 30-kW triple-junction photovoltaic receiver for 500X concentrator dish or central receiver applications. *High and Low Concentration for Solar Electric Applications.* **6339**, p. 633907. SPIE, San Diego.

Victoria, M., Domínguez, C., Antón, I., and Sala, G. (2009) Comparative analysis of different secondary optical elements for aspheric primary lenses. *Optics Express*, **17** (8), 6487–6492.

Wineman, A.S. and Rajagopal, K.R. (2000) *Mechanical Response of Polymers.* Cambridge University Press, Cambridge.

Yandt, M.D., Cook, J.P., Hinzer, K., and Schriemer, H. (2015) Shutter technique for noninvasive individual cell characterization in sealed CPV modules. *IEEE Journal of Photovoltaics*, **5** (2), 691–696.

Zimmermann, C.G. (2008) On the kinetics of photodegradation in transparent silicones. *Journal of Applied Physics*, **103**, 083547.

8.4

Operation of CPV Power Plants: Energy Prediction

Geoffrey S. Kinsey
U.S. Department of Energy, Washington, DC, USA

8.4.1 Introduction

For the solar technology enthusiast, a large-scale CPV power plant is a gratifying sight. Hundreds of generating units stretch to the horizon, tracking the sun with graceful precision (Figure 8.4.1). From sunrise to sunset, the arrays generate megawatt-hours of emissions-free energy. Behind this pleasing image lies all the complexity of an operating solar power plant (Figure 8.4.2). A host of loss mechanisms must be managed to enable long-term outdoor operation. At the module level, output is subject to losses due to variation in spectral irradiance and temperature. Over time, performance degradation can occur due to multiple factors, including soiling, oxidation, moisture and particulate ingress, photo-degradation, freeze/thaw cycles, bird strikes, and hail. Relative to "conventional" PV installations with modules mounted on fixed racks, the trackers required for CPV plants provide higher irradiance to the modules, but introduce tracking errors due to design limitations, wind loading, and control issues. In particular, the mechanical actuators are subject to degradation and require periodic maintenance.

To be economically viable, a power plant must offset these limitations and perform as a reliable source of energy for the electrical grid for decades, under a wide range of environmental conditions and problems (Figure 8.4.3). A power purchase agreement (PPA) between an electrical utility and the power plant operator includes a "performance guarantee" that defines the contractual requirements for power to be delivered under agreed-upon conditions. The ability to accurately and precisely predict the energy output of a power plant will determine whether the plant makes or loses money. If the guaranteed performance is too low, the plant will not attract buyers; if the guaranteed performance is too high, the operators will have to make up for any shortfall and will lose money. Profit margins are often measured in

Photovoltaic Solar Energy: From Fundamentals to Applications, First Edition.
Edited by Angèle Reinders, Pierre Verlinden, Wilfried van Sark, and Alexandre Freundlich.

Companion website: www.wiley.com/go/reinders/photovoltaic_solar_energy

Figure 8.4.1 A Suncore CPV installation in western China

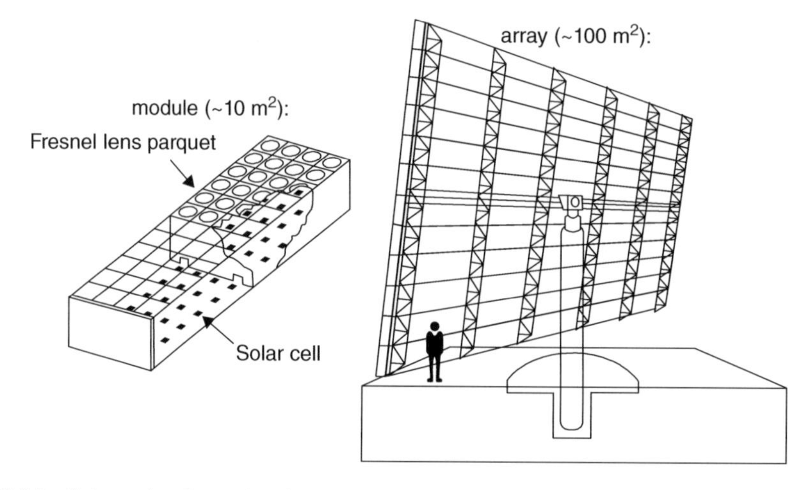

Figure 8.4.2 Schematic of a notional point-focus CPV module and array. Design specifics vary, but larger modules and an array area of around 100 m^2 are optimized for rapid, mechanized field installation. Source:adapted from IEC 62108 (2007)

single-digit percentages, so accurate and precise prediction is essential. Figure 8.4.4 shows the difference between the available irradiance and a projected target.

8.4.2 Performance Models

A number of tools for predicting solar energy output have been developed over the years. Open-source tools, such as PVWatts (Dobos, available online) and System Advisor Model (SAM, available online) (both developed by NREL) and the Sandia Performance Model (developed by Sandia National Laboratories) are designed for ease of use and accessibility for a wide range of users, but are somewhat limited in terms of the parameters that are included. Proprietary tools, which are generally more sophisticated, are, by design, less accessible and therefore make understanding of the underlying assumptions and comparison between

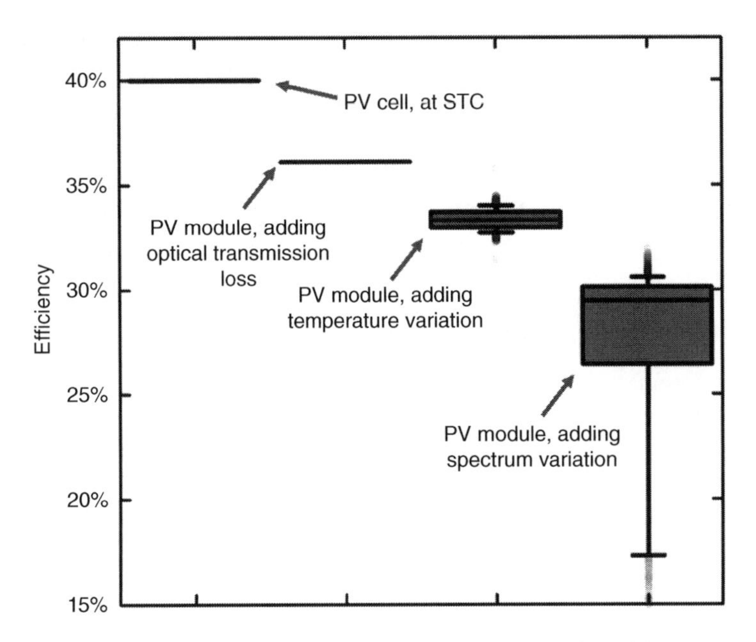

Figure 8.4.3 Some of the efficiency changes involved in moving from laboratory conditions to outdoor operation, as predicted for Albuquerque, USA, using the model described below. Source: (Kinsey *et al.*, 2015)

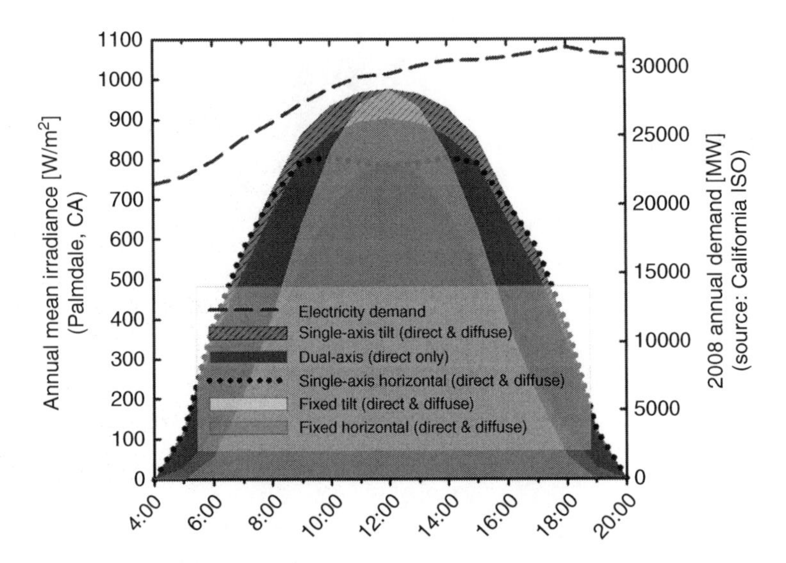

Figure 8.4.4 Available irradiance for various fixed rack and tracking configurations. The projected electrical load from the California Independent System Operator (CAISO) is shown for reference. (*See insert for color representation of the figure*)

technologies more difficult. The choice as to how sophisticated the prediction method should be depends on multiple factors, including the intended use of the results and the quality of the input variables available. When comparing different technologies, it is often preferable to employ one of the simpler methods, as this makes it more likely a true "apples-to-apples" comparison has been obtained. For practitioners involved in predicting performance for a given technology at an existing power plant, more sophisticated modeling (e.g., Kinsey *et al.*, 2013) is possible, making use of all the environmental input variables available at that site.

8.4.3 Performance Standards

The starting point for measuring power plant performance is IEC 61853-1, the international standard for "Photovoltaic (PV) module performance testing and energy rating." Additional methods specific to CPV are under development in the draft standard IEC 62670-2, "Photovoltaic concentrators (CPV)-performance testing – Part 2: energy measurement." IEC 61853 sets out test conditions that bracket the likely temperature and irradiance conditions of PV modules in operation (Figure 8.4.5). It is evident from Figure 8.4.5 that the 24 test conditions set out in the standard must be used with caution in predicting operating performance. The ranges of temperature and irradiance established in IEC 61853 do a good job of bracketing outdoor conditions, but, relative to the real-world examples, are skewed towards lower

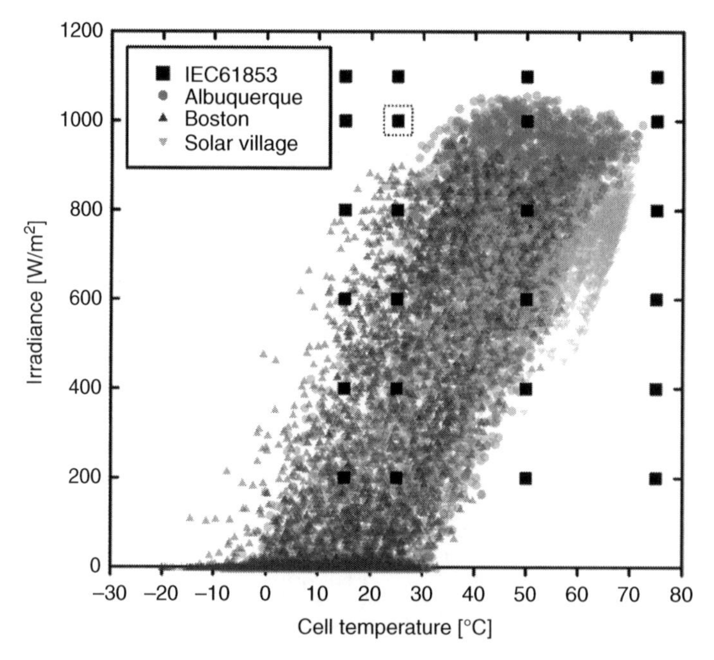

Figure 8.4.5 Comparison of test conditions specified in IEC 61853 and operating conditions derived for Albuquerque, USA, Boston, USA, and Solar Village, Saudi Arabia. The cell temperature was determined using the Sandia Photovoltaic Array Performance Model (King *et al.*, 2011) for DNI incident on a 22x linear concentrator. Note that STC conditions (1000 W/m², 25°C, outlined) are not encountered in operation at any of the three sites. (*See insert for color representation of the figure*)

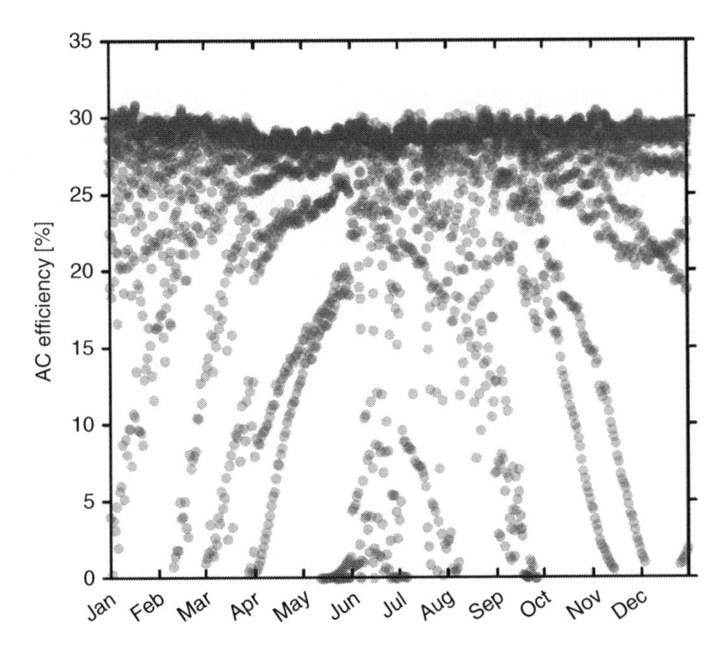

Figure 8.4.6 Hour-by-hour AC efficiency of a simulated CPV array (η_{STC}=33%) in Albuquerque, USA. Corrections have been made for: temperature, irradiance, spectrum, soiling, and inverter efficiency. The prediction near sunrise and sunset can be distinguished as the lower-efficiency points that fall off on either side of the summer solstice

temperatures and, relative to arid locations like Albuquerque, lower irradiances (Figure 8.4.6). The lowest two temperatures have been included mainly for reasons unrelated to outdoor operation. Standard Test Conditions (STC) stipulate a 25°C cell temperature, as that is a convenient temperature for indoor testing using a pulsed solar simulator. The lowest temperature (15°C) is useful in expanding the range so as to produce more accurate temperature coefficients. In order to provide an accurate prediction of operating performance, the data points must be weighted with respect to the anticipated operating conditions. This can be achieved by various combinations of predicted, expected, and measured performance in the field.

8.4.4 Prediction vs. Measurement

For an operational power plant with an on-site weather station and solar irradiance monitoring, three tiers of power plant performance may be obtained:

1. *Predicted*: based on performance specifications of the hardware and *projections* of the environmental conditions at the site (temperature, irradiance, etc.). For example, predicted power output of a module might be determined from the module's performance under standard test conditions, corrected for temperature and irradiance conditions from archived meteorological data. Predicted performance is most useful for technology comparisons and performance forecasting, but is subject to the greatest uncertainty in the input variables.

2. *Expected*: based on performance specifications of the hardware and *measured* environmental conditions. For example, expected power output of a module might be determined from the module's performance under standard test conditions, corrected for the temperature and irradiance conditions being measured during operation. Expected performance can only be determined after the outdoor operating conditions have been measured, but this remains a useful method for verifying that a plant is operating to specifications. It can also be useful for predicting performance in future years, particularly in regions where the meteorological conditions are stable from year to year. For example, the expected *target yield* of a PV array can be calculated based on its specified efficiency and the amount of energy measured incident on the plane of the array.
3. *Measured*: the performance recorded during operation.

Along with temperature, the spectral irradiance is an essential input parameter needed to obtain useful CPV performance prediction (e.g., Faine *et al.*, 1991; Betts *et al.*, 2004; Kinsey and Edmondson, 2009). LCPV systems using single-junction cells have a small (but non-negligible) spectrum sensitivity (Kinsey, 2015). For a CPV power plant employing III-V multijunction solar cells, the partitioning of the solar spectrum that enables the high efficiency leads to greater spectrum sensitivity than for systems using single-junction cells. One way to predict the incident spectrum is by modeling the radiative transfer that occurs as sunlight travels through the atmosphere. One such model, the Simple Model of the Atmospheric Radiative Transfer of Sunshine (SMARTS), was developed for NREL to predict spectral irradiance by taking the extraterrestrial (AM0) spectrum and modifying it for effects such as Rayleigh scattering and absorption by ozone, carbon dioxide, and aerosols (Gueymard, 2003). For a given location, the temporal variation in parameters such as "site pressure," "precipitable water," and "aerosol optical depth" must be considered in order to obtain a representative spectrum. Using sun photometers, these variables have been measured for years at numerous sites around the globe. In the USA, these parameters are part of the data set archived in the Typical Meteorological Year 3 database at 1020 sites (TMY3 database, available online). Aerosol data for sites around the world is also made available by the AERONET federation (AERONET, available online). At sites where a spectroradiometer is available (such as at Sandia National Laboratories in Albuquerque, NM, USA), the predicted values from SMARTS may be compared against direct measurement of the spectral irradiance (Gueymard, 2003).

Convolution of the spectral irradiance with the spectral response of a solar module gives the predicted short-circuit current. The open-circuit voltage and fill factor are corrected for the magnitude of the irradiance and all three parameters are corrected for temperature (Kinsey *et al.*, 2008). When used to predict the performance of CPV arrays operating in Las Vegas, NV, USA, the prediction was within 1% of the measured energy generated (Kinsey *et al.*, 2013). As an example, the approach is applied to a notional 500 x CPV system deployed in Albuquerque, NM, USA. Assuming a module with an efficiency of 33% under standard test conditions (STC), the predicted AC efficiency is shown in Figure 8.4.6.

The corresponding predictions for cumulative energy production, daily energy production and energy yield are shown in Figures 8.4.7 and 8.4.8. The sensitivity to temperature and spectrum may be illustrated by holding each of these variables constant. The ambient temperature has been fixed at 20°C (the temperature for Concentrator Standard Operating Conditions, CSOC). During daylight hours in Albuquerque, the mean annual ambient temperature from TMY3 data is 17°C, so the difference in annual energy yield with respect to a fixed, 20°C ambient condition

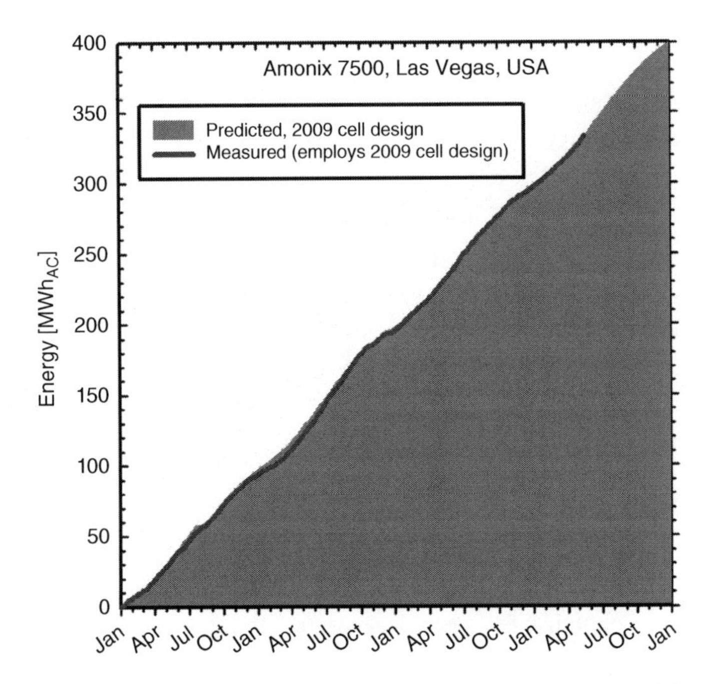

Figure 8.4.7 Predicted vs. measured cumulative energy for a 38-$kW_{AC\text{-}PTC}$ Amonix 7500 array deployed in Las Vegas, USA in 2009. After 3.5 years of operation, the cumulative energy differed from the prediction by less than 1%. Source: (Kinsey *et al.*, 2013)

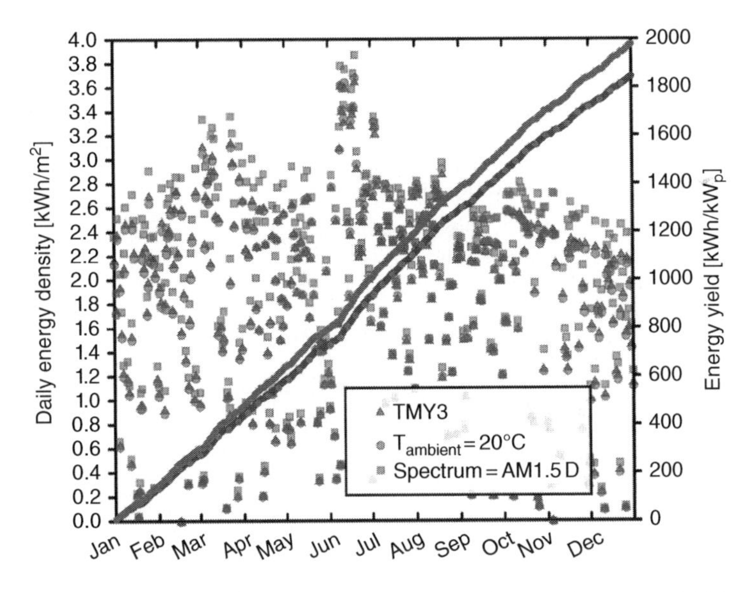

Figure 8.4.8 Sensitivity analysis of daily energy density (symbols) and energy yield (lines) for a 500x CPV system in Albuquerque, NM, USA. The result for TMY3 inputs is compared against cases where the temperature and spectrum are held constant. Cell temperature is assumed to be 40°C above the ambient. Energy yield is calculated based on peak power at CSTC (25°C, 1000 W/m^2). (*See insert for color representation of the figure*)

is slight. However, it can be seen from the daily energy values that, under the variable temperature conditions of TMY3, performance is higher in winter and lower in summer.

When the variable spectrum from TMY3 is replaced with a fixed, AM1.5D reference spectrum, the predicted energy yield rises by 7%. During the period of July to September, energy generation from the variable TMY3 spectrum closely matches that from the fixed spectrum, indicating that the average spectrum is close to AM1.5D during this period. It is unsurprising that energy output under the AM1.5D spectrum is higher, as most III-V multijunctions are designed for peak performance under this spectrum. Reference spectra such as AM1.5D were developed to represent average values of solar irradiance in the contiguous USA (ASTM G173-03, available online), which is not a particularly good match for the more arid regions where utility-scale CPV is deployed. Additional energy can be recovered by re-tuning the cell structure for operation under a spectrum that is more representative of the spectral irradiance at the narrower set of target locations for CPV plants (Kinsey *et al.*, 2013).

8.4.5 Conclusion

Relative to conventional silicon-based modules, high-concentration PV using multijunction cells has lower sensitivity to temperature, but a higher sensitivity to spectrum variation that must be addressed in optimization of practical designs. Accurate and precise calculation of the energy output of CPV systems is essential in determining a design's economic viability in a given location. The potential for lower cost of energy using high-efficiency CPV arrays can be realized via rigorous engineering in design, manufacturing, and operation.

List of Acronyms

Acronym	Description
CAISO	California independent system operator
CPV	Concentrator photovoltaic (s)
CSOC	Concentrator standard operating conditions
CSTC	Concentrator standard testing conditions
LCPV	Low concentration photovoltaic (s)
PPA	Power purchase agreement
SAM	System advisor model
SMARTS	Simple model of the atmospheric radiative transfer of sunshine
STC	Standard test conditions
SMARTS	Simple model of the atmospheric radiative transfer of sunshine
TMY3	Typical meteorological year 3 (database)

References

Araki, K. and Yamaguchi, M. (2009) Influences of spectrum change to 3-junction concentrator cells. *Solar Energy Materials and Solar Cells*, **75**, 707–714.

ASTM. ASTM G173 - 03(2012) Standard Tables for Reference Solar Spectral Irradiances: Direct Normal and Hemispherical on 37° Tilted Surface. Available: www.astm.org/DATABASE.CART/STD_REFERENCE/G173.htm

Betts, T.R., Infield, D.G., and Gottschalg, R. (2004) Spectral irradiance correction for PV system yield calculations. *Proceedings of the 19th European Photovoltaics Solar Energy Conference (EU PVSEC)*, pp. 2533–2536.

Chan, N.L.A., Brindley, H.E., and Ekins-Daukes, N.J. (2012) Quantifying the impact of individual atmospheric parameters on CPV system power and energy yield. *38th IEEE Photovoltaic Specialists Conference (PVSC)*, pp. 922–927.

Dobbin, A., Georghiou, G., Lumb, M. *et al.* (2011) Energy harvest predictions for a spectrally tuned multiple quantum well. *Proceedings of the 7th International Conference Conc. Photovoltaics.*

Dobos, A. (n.d.) PVWatts version 1 technical reference, National Renewable Energy Laboratory (U.S.) Golden, CO. Available at: pvwatts.nrel.gov/.

Faine, P., Kurtz, S.R., Riordan, C., and Olson, J.M. (1991) The influence of spectral solar irradiance variations on the performance of selected single junction and multijunction solar cells. *Solar Cells*, **31**, 259–278.

Gueymard, C.A. (2003) Direct solar transmittance and irradiance predictions with broadband models. Part I: detailed theoretical performance assessment. *Solar Energy*, **74** (5), 355–379.

Gueymard, C.A. (2005) SMARTS, a simple model of the atmospheric radiative transfer of sunshine: Algorithms and performance assessment. Florida Solar Energy Center, Cocoa, FL. *Technical Report FSEC-PF-270-95.*

King, D.L., Boyson, W.E., and Kratochvil, J.A. (2011) Report, No. SAND2004-3535, 2004. Photovoltaic (PV) module performance testing and energy rating – Part 1: Irradiance and temperature performance measurements and power rating. *IEC 61853-1.*

King, D.L., Kratochvil, J.A., and Boyson, W.E. (1997) Measuring solar spectral and angle-of-incidence effects on photovoltaic modules and solar irradiance sensors. *26th IEEE Photovoltaics Specialists Conference*, pp. 1113–1116.

Kinsey, G.S. (2015) Spectrum sensitivity, energy yield, and revenue prediction of PV modules. *IEEE Journal of Photovoltaics*, **5** (1), 258–262.

Kinsey, G.S., Bagienski, W., Nayak, A. *et al.* (2013) Advancing efficiency and scale in CPV arrays. *IEEE Journal of Photovoltaics*, **3** (2), 873–878, DOI: 10.1109/JPHOTOV.2012.2227992.

Kinsey, G.S. and Edmondson, K.M. (2009) Spectral response and energy output of concentrator multijunction solar cells. *Progress in Photovoltaics: Research Applications*, **17**, 279–288.

Kinsey, G.S., Hebert, P., Barbour, K.E., *et al.* (2008) Concentrator multijunction solar cell characteristics under variable intensity and temperature. *Progress in Photovoltaics: Research Applications*, **16**, 503–508, DOI: 10.1002/pip.834.

Kinsey, G.S., Nayak, A., Liu, M., and Garboushian, V. (2011) Increasing power and energy in Amonix CPV solar power plants. *IEEE Journal of Photovoltaics*, **1** (2), 213–218.

Letay, G., Baur, C., and Bett, A.W. (2004) Theoretical investigations of III–V multi-junction concentrator cells under realistic spectral conditions. *Proceedings of 19th European Photovoltaics Solar Energy Conference (EU PVSEC)*, pp. 187–190.

Nann, S. and Riordan, C. (1990) Solar spectral irradiance under overcast skies (solar cell performance effects). *Proceedings of the 21st IEEE Photovoltaics Specialists Conference (PVSC)*, 2, pp. 1110–1115, DOI: 10.1109/PVSC.1990.111789.

NASA. Aerosol Robotic Network. Available at: http://aeronet.gsfc.nasa.gov/new_web/data_description.html

Phillips, S.P., Peharz, G., Hoheisel, R., *et al.* (2010) Energy harvesting efficiency of III–V triple junction concentrator solar cells under realistic spectral conditions. *Solar Energy Materials and Solar Cells*, **94**, 869–877.

Pooltananan, N., Sripadungtham, P., Limmanee, A. and Hattha, E. (2010) Effect of spectral irradiance distribution on the outdoor performance of photovoltaic modules. *International Conference of Electrical Engineers/Electron. Comp. Telecom. Information Technology (ECTI-CON)*, pp. 71–73.

Sandia National Laboratories, Albuquerque Solar Resource (Online) Available at: energy.sandia.gov/wp/wp-content/gallery/uploads/24YearSummary.pdf

System Advisor Model Version 2014.1.14 (SAM 2014.1.14) National Renewable Energy Laboratory. Golden, CO (Online) Available at: sam.nrel.gov/content/downloads.

Typical Meteorological Year 3 (TMY3) Database. Available at: http://rredc.nrel.gov/solar/old_data/nsrdb/1991-2005/tmy3/

Ye, J.Y., Reindl, T., Aberle, A.G., and Walsh, T.M. (2014) Effect of solar spectrum on the performance of various thin-film PV module technologies in tropical Singapore. *IEEE Journal of Photovoltaics*, **4** (5), 1268–1274.

8.5

The Luminescent Solar Concentrator (LSC)

Michael Debije
Technical University, Eindhoven, the Netherlands

8.5.1 Introduction

The luminescent solar concentrator (LSC) was first suggested in an article back in 1976 (Weber and Lambe, 1976). Over the past almost forty years, there has been little change in the basic concept. The LSC is usually depicted as a plastic plate, which serves as a lightguide, see Figure 8.5.1, which is either filled with or topped by a thin layer of luminescent materials, either organic or inorganic. The luminophores absorb incident sunlight and re-emit this light at a longer wavelength. A fraction of this re-emitted light is trapped within the lightguide by total internal reflection due to the higher refractive index of the plastic plate. Eventually, some of the trapped light will exit from the thin edges of the lightguide, where one can place long, thin photovoltaic (PV) cells to convert the light into electrical current. Due to the loss of energy of the light during the emission event, the LSC is able to overcome the traditional concentration limits imposed by the *étendue*, and could potentially attain quite high concentration factors at the lightguide edges. In addition to this, the LSC exhibits a number of other advantages over other solar energy devices such as the traditional silicon-based solar panel. For instance, they could be produced in a wide variety of colors and in almost any shape, allowing for an easy and aesthetic integration of the device in the building environment. The LSC also functions well under both normal and off-normal sunlight illuminations, this characteristics being particularly important in an urban setting where shading from trees, pedestrians, dirt, and graffiti are common occurences. Also being made of plastic, LSCs would be lighter in weight than glass-covered PV panels, making easier attachments to façades. In summary, the only real limitation on application of the LSC devices would be the imagination of the designer. So, why has the LSC not had more of an impact in the solar energy arena?

Photovoltaic Solar Energy: From Fundamentals to Applications, First Edition.
Edited by Angèle Reinders, Pierre Verlinden, Wilfried van Sark, and Alexandre Freundlich.

Companion website: www.wiley.com/go/reinders/photovoltaic_solar_energy

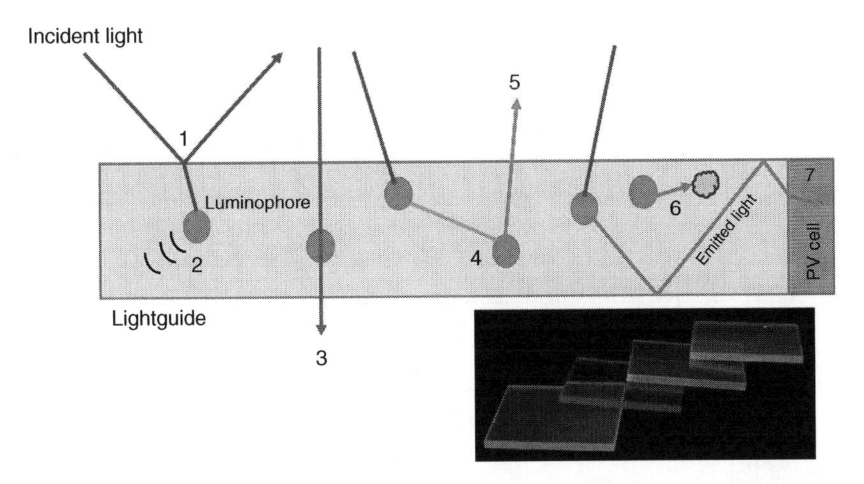

Figure 8.5.1 Main loss mechanisms of luminescent solar concentrators. (1) Reflected incident light; (2) non-unity quantum yield; (3) incomplete absorption; (4) re-absorption events; (5) emission outside the "capture cone"; (6) absorption by matrix; (7) spectral mismatch with PV cell. Inset photograph: Four LSCs exposed to UV light from above. (*See insert for color representation of the figure*)

8.5.2 Challenges for the Deployment of the LSC

The LSC suffers from a number of loss mechanisms which collectively have limited their overall performance. A summary of the main loss mechanisms identified is shown in Figure 8.51. These include: (1) reflections from the lightguide surface, around 8% for glass or PMMA lightguides; (2) non-unity fluorescent quantum yields of the embedded dyes; (3) transmission of incident light outside the absorption spectra of the dyes; (4) re-absorption by subsequent dye molecules, which may either result in the loss described in (2); or (5) emission outside the 'capture cone' of the lightguide; (6) matrix (scattering or absorption) losses; and (7) spectral mismatch of the edge emission from the lightguide and the response spectrum of the attached photovoltaic cell. The collective actions of these losses have limited the overall light-to-electron efficiency of the devices to quite modest levels. Thus far, the reported efficiencies, which depend on the nature of the attached PV cell, range from 4.2% for an LSC system using silicon cells (Desmet *et al.*, 2012) up to 7.1% for LSCs using edge-mounted III-V cells (Slooff *et al.*, 2008) and 6.8% for rear-mounted silicon cells (Corrado *et al.*, 2013). These efficiencies are only for LSCs of rather modest sizes: larger devices (i.e. on the square meter scale) have been reported with efficiencies up to the 3% range (Friedman and Parent, 1985). These reported efficiencies (η) describe how much electrical power is effectively obtained from the solar cells attached to the LSC relative to the incident optical power on the total LSC surface: the conversion efficiency of the attached photovoltaic cell(s) and the overall photon-in/photon-out system efficiency of the LSC are embedded in these numbers. If we describe the naked photovoltaic cell conversion efficiency (η_0) as describing the amount of electrical power obtained from the directly illuminated PV cell without an attached lightguide with respect to the incident optical power (integrated over the whole spectrum and surface), the optical collection probability (P) is the ratio of photons that reach the surfaces of the PV cells to the photons incident on the total LSC surface; that is: $P=\eta/\eta_0$. Geometric gain (G) describes the ratio of the total lightguide surface area to the total PV cell surface area present in the LSC. Finally, the concentration (C) is the ratio between incoming and outgoing optical irradiance of the solar concentrator, determined by the relation C = PG (Desmet *et al.*, 2012). Ideally, C should be greater than 1 for a true concentrator, which still is a challenge for this family of devices. For the LSC to be able to enter the marketplace, efficiencies and concentration ratios must be improved.

In the following sections, some strategies to reduce losses in the LSC are discussed.

8.5.2.1 Internal losses in LCS

8.5.2.1.1 Luminophores

The luminophore is the single most important material used in the LSC device, and is responsible for collecting and redistributing the light. In choosing the correct luminophore number of features must be considered, including the absorption spectrum (a broader spectrum absorbs more sunlight, but given the importance of the appearance of the LSC in many applications, more restricted spectra may be preferable), the quantum yield of the emission, the Stokes shift (that is, the overlap between the absorption and emission spectra, meaning that light emitted by one molecule can be re-absorbed by another molecule), the solubility of the material in the lightguide host/filler-media, and most importantly the stability of the dye to the exposure to the sunlight.

There is a large variety of luminophores available, including organic dyes, inorganic phosphors, and even quantum dots. Each material type has its own advantage and disadvantage (see the over-simplified Table 8.5.1). Organic dyes often are very efficient at absorption and can have very high quantum yields. They are often quite soluble in the organic polymer used as filler-media in lightguides, but their photostability is often in sufficient, their absorption/emission overlap is often too wide, and they are limited in the spectral bandwidth range they can process. Phosphors can be very stable to light exposure with large separation between absorption and emission, but often are poor absorbers with lower quantum yields and are sparingly soluble. Quantum dots and rods hold a great promise for future application, but at this time often suffer from poor stability, absorption, and quantum yield.

Table 8.5.1 Overview of typical luminophores and their advantages and disadvantages

Luminophore type	Examples	Advantages	Disadvantages
Organic	Perylene(bisimides) Coumarins Rhodamines	High absorption High quantum yield Good solubility in polymers	Low photostability Small Stokes shift
Inorganic	CdSe	Large Stokes shift	Poor solubility in polymers
	PbS CdSe/ZnS SrB4O7:5%Sm2+,5%Eu2+ Nd^{3+} UO_2^{2+} Cr^{3+}	Good photostability	Low absorption Quantum yield

At this time, the major directions in LSC research are the organic dyes. Years of research on laser dyes have resulted in promising candidates for use in the LSC. In particular, perylene-based dyes are widely used: some have quantum yields approaching the unity with extended photostability (Seybold and Wagenblast, 1989; Kinderman *et al.*, 2007; Wilson and Richards, 2009). The most common perylenes have absorption peaks below 600 nm, a disadvantage as a significant fraction of sunlight is beyond 600 nm. There are dyes, however, that show promise of extending the absorption spectra at reasonable efficiency beyond 600 nm (Debije *et al.*, 2011, Zhao *et al.*, 2014).

Phosphors have been also applied to LSCs, and recent work demonstrates novel phosphors with exceptionally large separation between absorption and emission, so that light transport is only limited by the characteristics of the lightguide itself (De Boer *et al.*, 2012, Erickson *et al.*, 2014). However, the absorption spectra of these phosphors are also limited primarily to wavelengths shorter than 600 nm. In addition, employing the inorganic phosphors can be challenging for their lack of solubility in standard polymeric materials, resulting in excessive scatter in the final devices. One suggestion for an alternative employment of the phosphor is to embed it in a separate rear scattering layer (see Section 2.2.1) to effectively downshift light that is not well absorbed by the lightguide dyes to a wavelength that can be absorbed (Debije and Dekkers 2012).

The use of quantum dots in LSCs continues to grow as they offer better absorption/emission characteristics (Krumer *et al.*, 2013, Meinardi *et al.*, 2014). In particular, there are several on-going research efforts focusing on these materials for use in colorless LSC systems, essentially for absorption of UV light for direct conversion to IR light, in order to allow the device to be used as a window (Chatten *et al.*, 2011).

A very interesting area of research is the use of the plasmonic enhancement generated by nearby metal particles to increase both absorption and emission from the dye molecules (Chandra *et al.*, 2010, Tummeltshammer *et al.*, 2013). By bringing small metallic particles in close proximity to luminophores, it is possible to effectively increase the performance of the dye molecules.

8.5.2.1.2 Lightguide

For the most part, polymer (plastic) plates are employed as lightguides, given their ease of manufacture, low cost, and reduced weight compared to glass. However, production of optically clear polymeric lightguides is challenging, as the materials making up the lightguide itself can have a parasitic absorption to the wavelengths emitted by the luminophores (Kastelijn *et al.*, 2009), especially as the materials approach infrared emission wavelengths. General solutions, like the employment of fluorinated polymers, are too expensive. Surface imperfections, whether abrasions or scratches, create locations for outcoupling of light, so care must be taken to keep the surfaces as optically smooth as possible: this means applying anti-scratch coatings at the minimum.

While rectangular is the most common shape for lightguides found in the literature, there are alternatives, including hexagonal and triangular (Sidrach de Cardona *et al.*, 1985). Cylindrical devices have been suggested to reduce surface reflection losses and enhance concentration (McIntosh *et al.*, 2007), and fiber lightguides have also been used with a variety of cross-sectional geometries that show promise (Wang *et al.*, 2010; Edelenbosch *et al.*, 2013). Lightguides do not have to remain rigid. In addition to the highly-flexible fiber designs, lightguides made from rubbery polysiloxanes (Buffa *et al.*, 2012) have also been proposed. In fact, if the challenge of reduced solubility in the lightguides can be overcome, such materials could prove quite useful as light transport through the devices was shown to be about as good as transport through (poly)carbonate plates, a common LSC lightguide material.

8.5.2.2 Losses from the Waveguide Surfaces

Losses through the top and bottom surfaces of the waveguide can account for more than 50% of the absorbed energy in LSCs (Debije *et al.*, 2008). The source of these losses is dye emission directed outside the capture cone of total internal reflection, a result of either a direct emission from an embedded luminophore, or from light reabsorbed and re-emitted by a subsequent luminophore. A rear mirror or scattering element is normally used to increase the total path-length of incident light through the lightguide to increase the absorption. In this case, the surface emission may also result from light being reflected by the backside light management element. In order to control these losses, number of strategies have been employed and are described hereafter.

8.5.2.2.1 Backside Reflectors

The simplest way to prevent losses through the bottom side of the LSC (the side not exposed to the incident sunlight) is to employ some type of reflector at the back surface. This back reflector, often silver-based, serves an additional purpose of reducing losses of incident light by allowing unabsorbed light a second chance to pass through the device and increasing the probability of absorption. Applying such mirrors allows for reducing the consumption relatively expensive dyes while maintaining a high absorption. However, care must be taken to keep a medium with a low refractive index between the lightguide and the mirror: depositing a silver reflector directly on the surface of the lightguide results in additional losses for light undergoing total internal reflection, as even the best metallic mirrors will absorb a few percent of the incident light at each encounter.

As an alternative to mirrors, white scattering layers are often employed for a threefold benefit. First, they return the initially non-absorbed light for a second pass through the device, just as the silver mirror. Second they also help prevent the loss of light emitted into the bottom escape cone by redistributing the angles of the escaping light, with a probability the scattered light will be directed towards the edge mounted cells. Finally, there is the possibility to collect additional light that cannot be absorbed by the dye but could be scattered via the white layer to the PV cell. This latter effect is quite short-range, effective to a distance on the order of 5 cm or so. It is important to separate the white scattering layer from the bulk lightguide by a low refractive index layer for reasons similar to those of the silvered mirrors (Debije *et al.*, 2009).

8.5.2.2.2 Wavelength-Selective Mirrors

Another way to limit surface losses is to apply a wavelength-selective reflector to the surface of the lightguide. In this way, incident light, that can be absorbed by the dye, is allowed to pass through the top surface, but wavelengths emitted by the dye are reflected, preventing it from escaping the waveguide. Such reflectors have been made from both inorganic (Goldschmidt *et al.*, 2009) and organic (Debije *et al.*, 2010; Rodarte *et al.*, 2014) materials. Both have proven capable of managing incoming and emitted light, and have been found to increase the amount of light reaching the lightguide edge. However, these materials are limited in effectiveness because they demonstrate angular dependence in their reflection. This is a consequence of the existence of a blue shift in the reflection band at increased incident angle, resulting in additional losses for the LSC: incident light at steeper angles that would normally enter the device is instead reflected away from the surface instead of entering the LSC to be absorbed. Recent work has focused on developing organic-based reflectors that show reduced angular dependence (De Boer *et al.*, 2012). If this can be achieved, the impact of the selective reflectors, especially in overcast conditions, will increase significantly.

8.5.2.2.3 Luminophore Alignment

Many luminophores, organic dye molecules and quantum rods in particular, are dichroic, which means they are not isotropic in absorption nor emission. Rather, they possess an internal absorption/emission dipole that is sensitive to the polarization state of the incident light: they absorb light parallel with their dipole, and emit light in a direction generally perpendicular to the dipole (see Figure 8.5.2). Researchers have taken advantage of this and work to reduce LSC losses by physical alignment of the luminescent materials, and concurrently alignment of the absorption/emission dipoles. By laying organic dye molecules flat on a lightguide surface and aligning the dipoles all in one direction, edge emissions from the edges parallel to the dye alignment direction have demonstrated to be 25% greater than any single edge of an isotropic LSC (Verbunt *et al.*, 2009). This alignment could reduce coverage of the PV cells to two edges rather than four. Even a higher order has demonstrated using extended, multichromophoric dyes (Kendhale *et al.*, 2013) and sterically engineered dyes (Benjamin *et al.*, 2014). However, planar alignment does not reduce losses through the surface of the lightguide.

To prevent light escaping from the top and/or bottom surfaces, homeotropic alignment has been applied, which means that the dye molecules are made to stand perpendicular to the

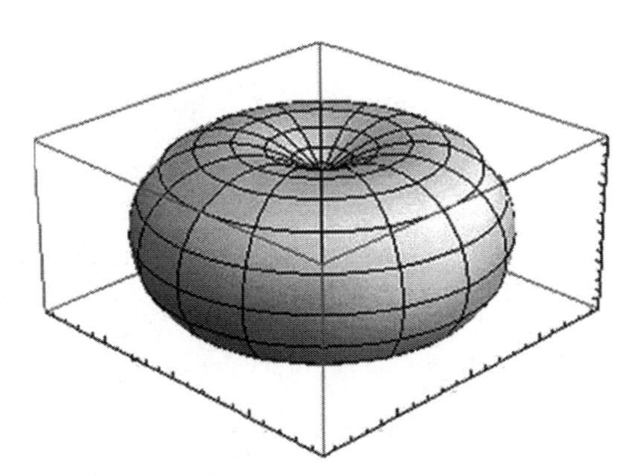

Figure 8.5.2 Calculated emission profile of dichroic dye ensemble oriented perpendicular to the plane of the lightguide (homeotropic) illuminated from above. Source: (Verbunt *et al.*, 2014)

lightguide surface, reducing surface losses to near-zero (Verbunt *et al.*, 2009; MacQueen *et al.*, 2010; Mulder *et al.*, 2010) with all the emission being directed into the lightguide. However, in this orientation, total absorption of incident light is reduced. Recent work has described the use of an absorber/emitter dyad, i.e. coupling a donor to a homeotropic emitter, to overcome these limitations (MacQueen and Schmidt, 2013).

Another option is to allow a continuous tuning of the dye orientation. This can be achieved by employing a liquid crystal host doped with guest dye molecules sandwiched in a conductive cell. By applying an electrical potential across the cell, the liquid crystalline material can be reoriented from its natural planar state to stand at any angle with respect to the lightguide surface, all the way to a homeotropic configuration. In this way, one can tune the amount of light absorbed by the embedded dye molecules and control the emission light that becomes trapped in the lightguide. This arrangement has been employed to produce so-called 'smart' windows that are capable of both controlling incident sunlight entering a room space as well as generating electrical power from the collected light (Debije, 2010).

8.5.3 The Future of the LSC

A considerable amount of work needs to be done before LSC PV systems are found everywhere in buildings. Hereafter we will tackle on several that appear as promising for the future of LSCs.

8.5.3.1 The Materials

The single most important advance that must be made is in the area of the luminophore. Without an effective light harvester with long-term stability, there is no chance for the LSC to be deployed to any significant extent. Essentially, this means advances in effective collection of light beyond 600 nm. Because of their nature, organic materials have a drop-off in quantum yield as the wavelength increases. In addition, improving knowledge of damage mechanisms for the organic luminophores and processes to extend their lifetimes will allow photostability to meet requirements for application on a much greater scale (Earp *et al.*, 2010; Griffini *et al.*, 2013). Perhaps some organic/inorganic hybrid materials will be the solution, taking advantage of the high absorbance and fluorescent yield of the organic materials and the photostability and emissive spectral shift of the inorganic ones. As the light-processing window is pushed further towards the infrared, care must be taken in selecting appropriate materials for lightguide manufacture, to avoid unnecessary parasitic losses from the host material itself.

Another area to consider for improving performance is correct matching of the PV cell to the spectral characteristics of the emitting luminophore. Sophisticated design are implemented to make multijunction PV cells sensitive over the entire solar spectral range. However, LSCs do not generate broadband light for the PV cells. The use of a single junction solar cell, made of a III-V compound semiconductor or silicon, is suitable for the narrow band of emission of the LSC. An intriguing possibility is the use of organic PV cells, which often present a peak sensitivity to precisely the wavelengths emitted by the dyes, and where the exposure to UV light, responsible for their performance degradation in conventional flat panel applications, can be avoided.

Preliminary demonstrator projects designed to test the outdoor performance of the devices are necessary, as simulation can only go so far in predicting the performance of the LSC in real conditions. This author is involved in a project to install a sound barrier panel in 's Hertogenbosch, the Netherlands in early 2015, and it is field tests such as this that will be required both to sell to the public the concept of the LSC as well as proving their functionality and determine their potential for longer-term performance.

8.5.3.2 Alternative Applications

While the vast majority of LSC-related research programs focus on the use of the device for the generation of electricity, there are other options. For example, rather than conversion of the edge emission via a photovoltaic cell, light from two or three stacked LSCs is collected,

Figure 8.5.3 Photographs of the Palais des Congrès, Montreal, Canada. (Images courtesy Palais des Congrès de Montréal, bottom image by Bob Fisher)

mixed, and used to generate white light for daylighting purposes (Earp *et al.*, 2004). Other options are to collect portions of the incident light for electricity generation, but allow the surface-emitted light to be used in a greenhouse (Lamnatou and Chemisana, 2013). This is particularly useful for devices effective at absorbing green light (which is not used by the plants) and emitting red light (which is used by the plants).

8.5.3.3 Change in Approach

It is the opinion of this author that if the LSC is to really emerge from the laboratory environment and impress itself in the marketplace, there needs to be a change in perception about the device. Originally, the LSC was introduced as a potential replacement for the roof-mounted silicon-based solar panel. Given the rapidly dropping prices of solar modules and the increased efficiency of standard PV devices, this objective appears as irrelevant. Rather, the introduction of the LSC needs to take advantage of the numerous attributes of the device that make it unique and different from the standard PV panel. There are many examples where designers, architects, and planners have included the use of non-functional colorful plastic or glass panels in their designs and structures (see Figure 8.5.3 for one example). These installations have employed panels mainly for their appearance (Kerrouche *et al.*, 2014). LSC offers an elegant alternative to these applications given its additional functionality of generating electrical power to go along with their regular structural, sound blocking or other functions. If the additional costs are modest, the add-on expenses can be paid off over a reasonable period of time and, given the sheer area potential available to a system that is relatively insensitive to changes in light quality and temperature, a significant amount of electricity can also be generated. The LSC is a very adaptable device, and this needs to be emphasized and exploited to its fullest extent if commercialization is ever to occur.

List of Symbols

Symbol	Description
η_0	Photovoltaic cell efficiency
η	Luminescent solar concentrator efficiency

List of Acronyms

Acronym	Description
C	Concentration
G	Geometric gain
LSC	Luminescent solar concentrator
P	Optical collection probability
PMMA	(Poly) methyl methacrylate
PV	Photovoltaic

References

Benjamin, W.E., Veit, D. R., Perkins, M.J. *et al.* (2014) Sterically engineered perylene dyes for high efficiency oriented fluorophore luminescent solar concentrators. *Chemistry of Materials*, **26** (3), 1291–1293.

Buffa, M., Carturan, S., Debije, M.G. *et al.* (2012) Dye-doped polysiloxane rubbers for luminescent solar concentrator systems. *Solar Energy Materials and Solar Cells*, **103**, 114–118.

Chandra, S., McCormack, S., Doran,, J. *et al.* (2010) New concept for luminescent solar concentrators. *Proceedings of the 25th European Photovoltaic Solar Energy Conference and Exhibition*, Valencia, Spain.

Chatten, A.J., Farrell, D.J., Bose, R. *et al.* (2011) Luminescent and geometric concentrators for building integrated photovoltaics. Paper presented at *37th IEEE Photovoltaic Specialists Conference* (PVSC).

Corrado, C., Leow, S.W., Osborn, M. et al. (2013) Optimization of gain and energy conversion efficiency using front-facing photovoltaic cell luminescent solar concentratordesign. *Solar Energy Materials and Solar Cells*, **111** (0), 74–81.

De Boer, D.K.G., Broer, D.J., Debije, M.G. *et al.* (2012) Progress in phosphors and filters for luminescent solar concentrators. *Optics Express*, **20** (S3), A395–A405.

Debije, M.G. (2010) Solar energy collectors with tunable transmission. *Advances in Functional Materials*, **20**, 1498–1502.

Debije, M.G. and Dekkers, W. (2012) Functionalizing the rear scatterer in a luminescent solar concentrator. *Journal of Renewable Sustainable Energy*, **4**, 013103.

Debije, M G. Teunissen, J.P., Kastelijn, M.J. et al. (2009) The effect of a scattering layer on the edge output of a luminescent solar concentrator. *Solar Energy Materials and Solar Cells*, **93**, 1345–1350.

Debije, M. G., Van, M.-P., Verbunt, P. P. C. *et al.* (2010) The effect on the output of a luminescent solar concentrator on application of organic wavelength-selective mirrors. *Applied Optics*, **49** (4), 745–751.

Debije, M.G., Verbunt,P.P.C., Nadkarni, P.J. *et al.* (2011) Promising fluorescent dye for solar energy conversion based on a perylene perinone. *Applied Optics*, **50** (2), 163–169.

Debije, M.G., Verbunt, P. P.C., Rowan, B.C. *et al.* (2008) Measured surface loss from luminescent solar concentrator waveguides. *Applied Optics*, **47** (36), 6763–6768.

Desmet, L., Ras, A.J.M. De Boer, D.K.G. and Debije, M.G. (2012) Monocrystalline silicon photovoltaic luminescent solar concentrator with 4.2% power conversion efficiency. *Optical Letters*, **37**, 3087–3089.

Earp, A.A., Rawling, T., Franklin, J.B. and Smith, G.B. (2010) Perylene dye photodegradation due to ketones and singlet oxygen. *Dyes and Pigments*, **84** (1), 59–61.

Earp, A.A., Smith, G.B., Franklin, J. and Swift, P.D. (2004) Optimization of a three-colour luminescent solar concentrator daylighting system. *Solar Energy Materials and Solar Cells*, **84**, 411–426.

Edelenbosch, O.Y., Fisher, M., Patrignani, L. *et al.* (2013) Luminescent solar concentrators with fiber geometry. *Optics Express*, **21** (S3), A503–A514.

Erickson, C.S., Bradshaw, L.R., McDowall, S. *et al.* (2014) Zero-reabsorption doped-nanocrystal luminescent solar concentrators. *ACS Nano*, **8** (4), 3461–3467.

Friedman, P.S. and Parent, C.R. (1985) Luminescent solar concentrator development, *SERI*, **204**.

Goldschmidt, J.C., Peters, M., Bösch, A. *et al.*(2009) Increasing the efficiency of fluorescent concentrator systems. *Solar Energy Materials and Solar Cells*, **93** (2), 176–182.

Griffini, G., Levi, M. and Turri, S. (2013) Novel crosslinked host matrices based on fluorinated polymers for long-term durability in thin-film luminescent solar concentrators. *Solar Energy Materials and Solar Cells*, **118**, 36–42.

Kastelijn, M J., Bastiaansen, C.W.M. and Debije, M.G. (2009) Influence of waveguide material on light emission in luminescent solar concentrators. *Optical Materials*, **31**: 1720–1722.

Kendhale, A.M., Schenning A.P.H.J. and Debije, M.G. (2013) Superior alignment of multi-chromophoric perylenebisimides in nematic liquid crystals and their application in switchable optical waveguides. *Journal of Material Chemistry, A*, **1**, 229–232.

Kerrouche, A., Hardy, D.A., Ross, D. and Richards, B.S. (2014) Luminescent solar concentrators: From experimental validation of 3D ray-tracing simulations to coloured stained-glass windows for BIPV. *Solar Energy Materials and Solar Cells*, **122** (0), 99–106.

Kinderman, R., Slooff, L.H., Burgers, A.R. *et al.* (2007) I-V performance and stability study of dyes for luminescent plate concentrators. *Journal of Solar Energy*, **129**, 277–282.

Krumer, Z., Pera, S.J., van Dijk-Moes, R.J.A., *et al.* (2013) Tackling self-absorption in luminescent solar concentrators with type-II colloidal quantum dots. *Solar Energy Materials and Solar Cells*, **111** (0), 57–65.

Lamnatou, C. and Chemisana, D. (2013) Solar radiation manipulations and their role in greenhouse claddings: Fluorescent solar concentrators, photoselective and other materials. *Renewable and Sustainable Energy Reviews*, **27** (0), 175–190.

MacQueen, R.W., Cheng, Y.Y., Clady C.R. and Schmidt T.W. (2010) Towards an aligned luminophore solar concentrator. *Optics Express*, **18** (102), A161–A166.

MacQueen, R.W. and Schmidt, T.W. (2013) Molecular polarization switching for improved light coupling in luminescent solar concentrators. *The Journal of Physical Chemistry Letters*, **4** (17), 2874–2879.

McIntosh, K.R., Yamada, N. and Richards, B.S. (2007) Theoretical comparison of cylindrical and square-planar luminescent solar concentrators. *Applied Physics B*, **88**, 285–290.

Meinardi, F., Colombo, K.A., Velizhanin, R. *et al.* (2014) Large-area luminescent solar concentrators based on 'Stokes-shift-engineered' nanocrystals in a mass-polymerized PMMA matrix. *Nature Photonics*, **8** (5), 392–399.

Mulder, C.L., Reusswig, P.D. Velázquez, A.M. *et al.* (2010) Dye alignment in luminescent solar concentrators: I. Vertical alignment for improved waveguide coupling. *Optics Express*, **18**(S1), A79–A90.

Rodarte, A.L., Cisneros, F., Hirst, L.S. and Ghosh, S. (2014) Dye-integrated cholesteric photonic luminescent solar concentrator. *Liquid Crystals*, **41** (10), 1442–1447.

Seybold, G. and Wagenblast, G. (1989) New perylene and violanthrone dyestuffs for fluorescent collectors. *Dyes and Pigments*, **11**, 303–317.

Shen, Y.Y., Jia, X., Sheng, L. *et al.* (2014) Nonimaging optical gain in luminescent concentration through photonic control of emission étendue. *ACS Photonics*, **1** (8), 746–753.

Sidrach de Cardona, Carrasscosa, M.M., Meseguer, F. *et al.* (1985) Edge effect on luminescent solar concentrators. *Solar Cells*, **15**, 225–230.

Slooff, L.H., Bende, E.E., Burgers, A.R. *et al.* (2008) A luminescent solar concentrator with 7.1% power conversion efficiency. *physica status solidi R*, **2** (6), 257–259.

Tummeltshammer, C., Brown, M.S., Taylor, A. *et al.* (2013) Efficiency and loss mechanisms of plasmonic luminescent solar concentrators. *Optics Express*, **21** (S5), A735–A779.

Verbunt, P.P.C., Bastiaansen, C., Broer, D.J. and Debije, M.G. (2009) The effect of dye aligned by liquid crystals on luminescent solar concentrator performance. Paper presented at 24th European Photovoltaic Solar Energy Conference, Hamburg, Germany.

Verbunt, P.P.C., de Jong, T.M., de Boer, D.K.G. *et al.* (2014) Anisotropic light emission from aligned luminophores. *European Physics Journal – Applied Physics*, **67** (01), 10201.

Verbunt, P.P.C., Kaiser, A., Hermans, K., *et al.* (2009) Controlling light emission in luminescent solar concentrators through use of dye molecules planarly aligned by liquid crystals. *Advanced Functional Materials*, **19**, 2714–2719.

Wang, C., Abdul-Rahman, H. and Rao, S.P. (2010) Daylighting can be fluorescent: Development of a fiber solar concentrator and test for its indoor illumination. *Energy and Buildings*, **42**, 717–727.

Weber, W.H. and Lambe, J. (1976) Luminescent greenhouse collector for solar radiation. *Applied Optics*, **15** (10), 2299–2300.

Wilson, L.R. and Richards, B.S. (2009) Measurement method for photoluminescent quantum yields of fluorescent organic dyes in polymethyl methacrylate for luminescent solar concentrators. *Applied Optics*, **48** (2), 212–220.

Zhao, Y., Meek, G.A., Lebvine, B.G. and Lunt, R.R. (2014) Near-infrared harvesting transparent luminescent solar concentrators. *Advanced Optical Materials*, **2**, 606–611.

Part Nine

Space Technologies

9.1

Materials, Cell Structures, and Radiation Effects

Rob Walters
US Naval Research Lab, Washington, DC, USA

9.1.1 Introduction

Photovoltaics are, by far, the primary space power source for Earth orbiting satellites and many interplanetary spacecrafts. The space environment is often characterized by a harsh radiation environment; therefore, to be used in space, the radiation response of a solar cell must be well understood. In this chapter, the typical materials used for space solar cells and the solar cell structures are presented. The mechanisms for radiation effects in space solar cells are then investigated to establish an understanding of how radiation-induced degradation is addressed in designing a satellite power system.

The first semiconductor material used for a space solar cell was silicon (Si). Indeed, it was the use of Si solar cells to power Vanguard, the first US Earth orbiting satellite, which was built and launched by the US Naval Research Laboratory,[1] that ushered in the modern era of photovoltaics. It is fascinating to note that the launch of Vanguard was only a few years after Bell Laboratories had demonstrated the first silicon p-n junction solar cell (Chapin *et al.*, 1954). From the time of Vanguard up to the early 1990s, all satellites, except the very few nuclear-powered deep space probes, were powered by silicon solar cells.

In contrast to silicon, compound semiconductors formed from elements of the III and V columns of the Periodic Table (the so-called *III-V semiconductors*), like GaAs, typically

[1] nssdc.gsfc.nasa.gov. Vanguard was the first man-made solar-powered satellite, and was launched in March of 1958. The grapefruit-sized satellite incorporated several silicon solar cells developed by the Bell Laboratories and was the first to demonstrate the suitability of solar cells as a power source for long-term operation of spacecraft in space.

Photovoltaic Solar Energy: From Fundamentals to Applications, First Edition.
Edited by Angèle Reinders, Pierre Verlinden, Wilfried van Sark, and Alexandre Freundlich.

Companion website: www.wiley.com/go/reinders/photovoltaic_solar_energy

display a direct bandgap, and thus have much higher potential solar cell efficiencies than Si (Madelung, 1964). One year after Bell Laboratories demonstrated the first Si solar cell, RCA Laboratories reported the first GaAs solar cell, which was 6% efficient (Jenny *et al.* 1956). With the emergence of growth methods like metalorganic chemical vapor deposition (MOCVD) during the 1980s, high quality, compound semiconductor materials could be achieved (Yeh. 1984), enabling GaAs solar cells with efficiencies far exceeding that of Si cells to be produced (Iles, 1990; Kaminar *et al.*, 1988). GaAs solar cells were also shown to be more radiation-resistant than their Si counterparts (Anspaugh, 1991), and as a result, as the new millennium approached, GaAs quickly replaced Si as the space solar cell industry standard (Brown *et al.*, 1996).

The next step in space PV was the adoption of multijunction (MJ) solar cells. In 1980, Henry showed theoretically (Henry, 1980) that higher efficiencies than possible with a single-junction can be attained if multiple p-n junctions of different semiconductors could be grown in a stack, with the highest band gap material topmost, as had been described elsewhere in this book. The first dual-junction solar cell to achieve an efficiency greater than 30% had an InGaP top cell and a GaAs bottom cell (Bertness *et al.*, 1994a; Bertness *et al.*, 1994b), which quickly evolved into a triple junction device when the Ge substrate was made active (Chiang *et al.*, 1996). This cell was designed specifically for space applications, and the first space flight of MJ solar cells occurred around 1997. The continued MJ cell development, in particular, the inverted metamorphic (IMM) technology, has brought an even wider pallet of materials to space photovoltaics, with these newer technologies making their way into initial space experiments over the past decade, as will be discussed in the next section.

9.1.2 Radiation Response Mechanisms

Turning attention to the radiation response of a space solar cell, the wide range of materials used in space PV may appear to significantly complicate the analysis. Fortunately, nature has allowed us to address radiation effects in a coherent way, independent of the specific material under study.

The primary effect of particle irradiation of a solar cell is displacement damage where collisions between the irradiating particle and atoms in the crystal lattice of the target semiconductor cause the atoms to be ejected from their equilibrium position to form point defects. These defects can form energy levels within the forbidden gap of the semiconductor forming charge-trapping centers. The existence of these defect centers affects charge transport in essentially five basic ways as shown schematically in Figure 9.1.1 (Srour, 1988).

Carrier generation (labeled #1 in Figure 9.1.1) occurs when the existence of the defect energy level makes it statistically favorable for a charge carrier to move from a bound to a free energy level. The liberated charges are then swept away by the junction field, which produces a current. This causes the dark IV characteristic to increase, which degrades the photovoltage (i.e., Voc). Recombination (labeled #2 in Figure 9.1.1) occurs when it is statistically favorable for an electron-hole pair to recombine at the defect site. When this occurs, free charge carriers are lost, resulting in degradation of the photocurrent (i.e. Isc). Trapping (labeled #3 in Figure 9.1.1) occurs when a defect level is able to capture and temporarily localize free charge carriers, which are then thermally reemitted. The fourth mechanism illustrated in Figure 9.1.1 is referred to as compensation. Compensation occurs when a defect level permanently

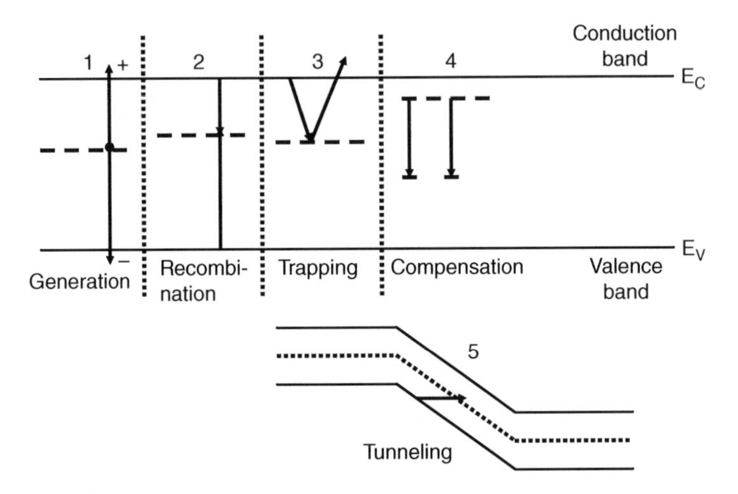

Figure 9.1.1 A schematic representation of the effects that radiation-induced defect levels can have on current transport in a solar cell. Source: (Tada *et al.*, 1982)

localizes a free charge carrier supplied by the dopant atoms. This reduces the majority charge carrier density and is referred to as carrier removal. The fifth mechanism is trap-assisted tunneling where the position of the defect level effectively lowers the tunneling potential.

9.1.3 Effect of Radiation on Space Solar Cells

The effects of irradiation by 1 MeV electrons (commonly used in the space radiation effects community for characterizing solar cell response) on the maximum power output of Si (Tada *et al.*, 1982), GaAs/Ge (Anspaugh, 1996), and InP (Walters *et al.*, 1996a; Keavney *et al.*, 1993) solar cells are shown in Figure 9.1.2, and the general shape of the degradation curves is seen to be similar. This is because the primary degradation mechanism in these cells has been shown to be a decrease in minority carrier diffusion length (L (μm) in unit) due to radiation-induced recombination centers, and the implication is that this mechanism applies generally to all crystalline semiconductor-based photovoltaic devices (Bertness *et al.*, 1991; Hovel, 1975; Yamaguchi *et al.*, 1996). This is quite important, since most space PV technology in use or under development is formed from a combination of these materials.

Degradation of L reduces the carrier collection efficiency since those photogenerated carriers created in the cell base are less likely to reach the junction. This effect is seen in the QE data where most of the radiation-induced degradation appears in the response to longer wavelengths of light, which are absorbed deeper in the cell (Figure 9.1.3). Degradation of the solar cell QE results in degradation of Isc.

Using the formalism of Hovel (1975), an estimate of L can be extracted from analysis of the QE data, which has been done for the data of Figure 9.1.3. (Keavney *et al.*, 1993). The decrease in L with the introduction of defects is given by (Walters *et al.*, 2013).

$$\frac{1}{L^2(D_d)} = \frac{1}{L_0^2} + \sum \frac{\sigma_i \upsilon I_{ti}}{D} \phi = \frac{1}{L_0^2} + K_L \phi \tag{9.1.1}$$

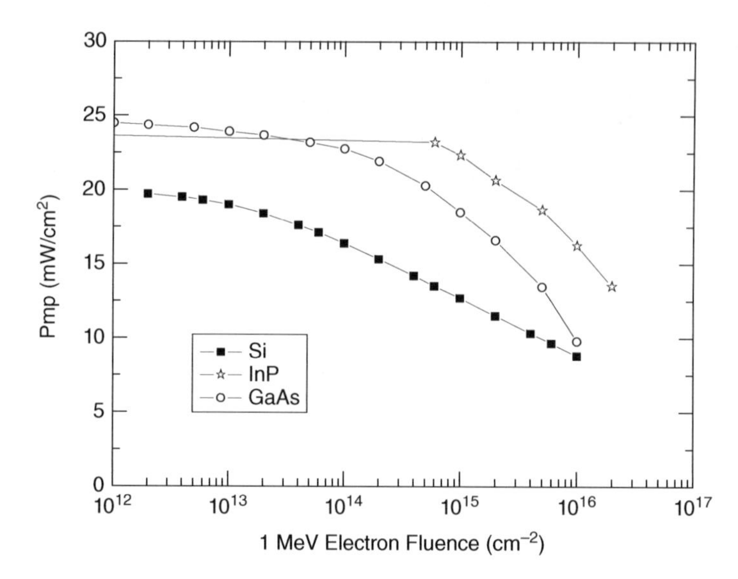

Figure 9.1.2 The radiation response of single-junction crystalline semiconductor solar cell technologies that have been developed for space use. Source: (Keavney *et al.*, 1993)

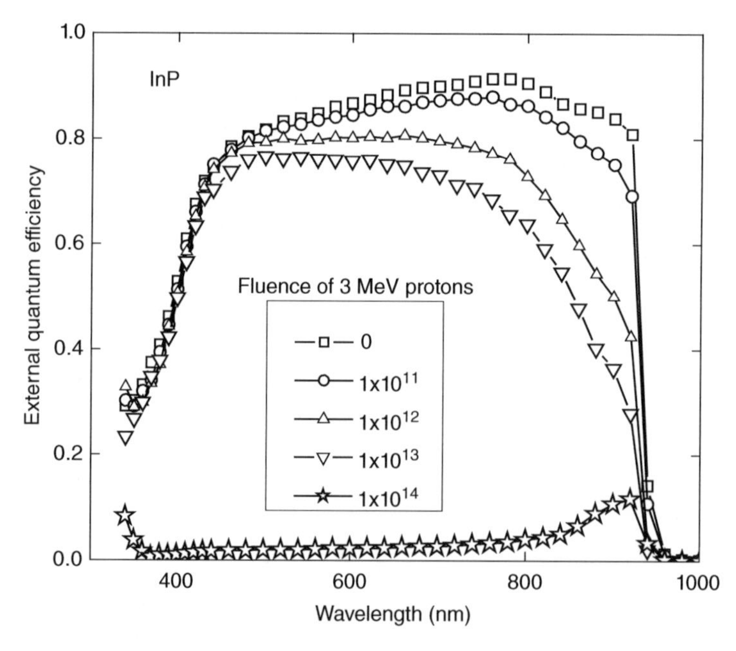

Figure 9.1.3 The degradation in QE of an InP solar cell due to 3 MeV proton irradiation. The particle fluence is given in the legend in units of cm^{-2}. The long wavelength response degrades due to diffusion length degradation. Source: (Keavney *et al.*, 1993)

where L_o is the pre-irradiation value of L in um, σ_i is the minority carrier capture cross section of the ith recombination center in cm^{-2}, I_{ti} is the introduction rate of the ith recombination center in cm^{-1}, υ is the thermal velocity of the minority carriers (cm/s), D is the diffusion coefficient ($cm^2 s^{-1}$), and ϕ is the particle fluence (cm^{-2}). As shown in Equation (9.1.1), the specific parameters for each defect are typically lumped into a constant, K_L (dimensionless), referred to as the damage coefficient for L. By determining L at various fluences, K_L may be determined, as shown for the example of an InP solar cell grown on a Si substrate in Figure 9.1.4 (Shockley, 1950).

The introduction of recombination/generation centers also causes an increase in the dark IV characteristic of the solar cell, which degrades Voc. An analysis of dark IV data measured in an irradiated InP solar cell is shown in Figure 9.1.5 (Keavney *et al*, 1993). The irradiation is seen to cause a two orders of magnitude increase in the dark IV characteristic. The solid lines in Figure 9.1.5 represent fits of the measured data to a theoretical expression for diode dark current. It is the current at the higher voltages (>~0.65 V in this case) that most strongly affects the photovoltaic output. This portion of the dark IV curve is the diffusion current as given by Shockley (Sah *et al.*, 1957), in which the magnitude of the diffusion current is shown to be inversely proportional to L. Thus, the introduction of recombination/generation centers by irradiation causes L to degrade, which degrades Isc and the dark IV characteristic to increase, which degrades Voc. These effects combine to reduce the solar cell maximum power (Pmp).

The dark IV curve at lower voltages also shows an increase due to the irradiation. This is due to an increase in the recombination/generation current as described in Figure 9.1.1, which is the dark current produced by defects within the depleted region of the diode junction (Choo, 1968; Walters *et al.*, 1996b). The magnitude of this current is directly proportional to the defect concentration and exponentially dependent on the difference between the trap energy

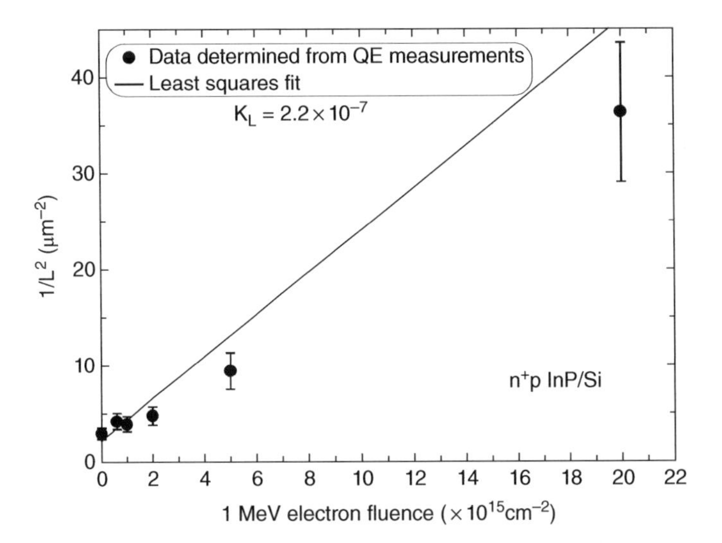

Figure 9.1.4 Minority carrier diffusion length data determined from analysis of QE data as a function of particle fluence (Figure 9.1.3). The line represents a linear regression of the data from which the diffusion length damage coefficient, K_L, can be determined according to Equation (9.1.1). Source: (Shockley, 1950)

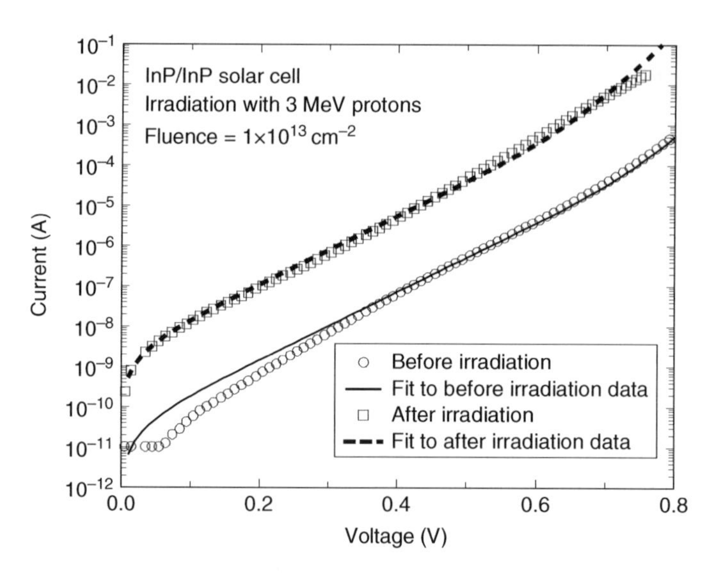

Figure 9.1.5 Dark current data measured in the InP solar cell from the preceding figures. The proton irradiation causes an increase in the dark current. Analysis of these data shows that this increase is due to an increased diffusion current brought on by radiation-induced recombination/generation centers. Source: (Keavney *et al.*, 1993)

level and the intrinsic Fermi level. The closer the trap level lies to mid-bandgap, the more efficient it will be as a recombination/generation center and the larger the recombination/generation current will be. This portion of the dark current is typically several orders of magnitude less than the diffusion dark current, so it has a proportionally smaller effect on the illuminated IV curve. However, at large enough defect concentration levels, the recombination/generation current can become significant, which results in a degradation of the fill factor (FF) of the current versus voltage curve.

It can now be seen that the radiation response of space solar cells is controlled by the diffusion length degradation caused by the radiation-induced displacement damage and that the variation in degradation rate among various solar cell technologies is controlled by the sensitivity of each technology to the radiation induced defects. The superior radiation tolerance of InP solar cells (Figure 9.1.2) has been shown to be due to the relatively low activation energy for annealing of the radiation-induced defects, such that equivalent amounts of irradiation results in a lower, permanent defect concentration in these cells compared to GaAs and Si (Yamaguchi, 1995; Walters *et al.*, 1998). Si, on the hand, as an indirect bandgap material with a correspondingly weak absorption coefficient, requires an extremely long diffusion length for an efficient PV operation, and is thus very sensitive to radiation induced defects.

9.1.4 Effect of Radiation on Multijunction Space Solar Cells

Extending this analysis to MJ solar cells is straightforward since a MJ cell can be considered to be a series connection of multiple single junction solar cells, each of which exhibit the standard degradation mechanisms described thus far. Considering the InGaP/GaAs/Ge

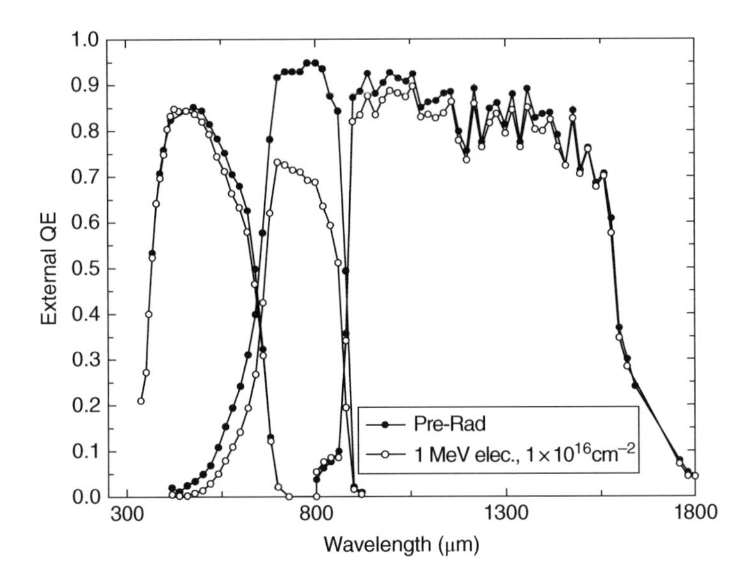

Figure 9.1.6 Quantum efficiency measurements made on a triple-junction $InGaP_2$/GaAs/Ge solar cell before and after irradiation with 1 MeV electrons to a fluence of 1×10^{15} cm^{-2}. The GaAs subcell (middle cell) is seen to degrade most rapidly of the three. Source: (Summers *et al.*, 1995)

three-junctions (3J) cell, which is presently the industry standard for space PV, studies have shown the radiation response of the InGaP top cell to be similar to that of InP (Yamaguchi, 1995; Walters *et al.*, 2000), leaving the 3J device to be controlled by the GaAs middle cell (Summers *et al.*, 1995). This can be seen by the QE data shown in Figure 9.1.6, where the GaAs middle cell shows the largest amount of degradation.

9.1.5 Correlating Radiation Damage

Throughout this analysis, different types of irradiation have been considered, e.g. 1 MeV electrons or 3 MeV protons. Knowledge of the solar cell response to irradiation by different particles at different energies is essential because the space radiation environment consists of spectra of various particles. Extensive research at the US Naval Research Laboratory (NRL) has developed a methodology for this analysis based on the quantity referred to as the Non-ionizing energy loss (NIEL) (given in units of MeV/cm^2g) (Summers *et al.*, 1993). NIEL is the density-normalized rate at which an incident particle loses energy to create atomic displacements, and provides the energy dependence of displacement damage effects in a variety of semiconductors. A plot of NIEL values for the materials of the MJ cell in Figure 9.1.6 are given in Figure 9.1.7 (Walters *et al.*, 2000).

It has been shown that there is a direct proportionality between damage coefficients and NIEL for a wide variety of solar cells (Messenger *et al.*, 2001), and an example for Si solar cells is shown in Figure 9.1.8. Building on this result, the NRL methodology establishes the quantity referred to as Displacement Damage Dose (Dd) (given in units of

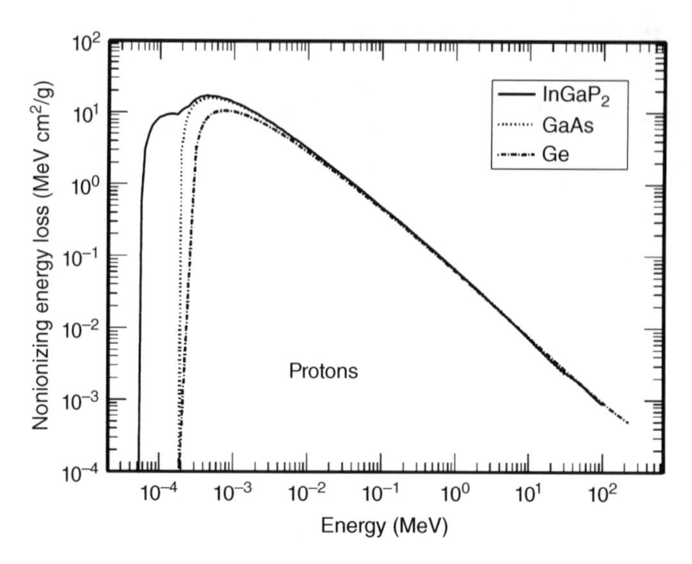

Figure 9.1.7 Calculated NIEL values for protons in the materials comprising the triple-junction $InGaP_2$/GaAs/Ge solar cell. Source: (Walters *et al.*, 2000)

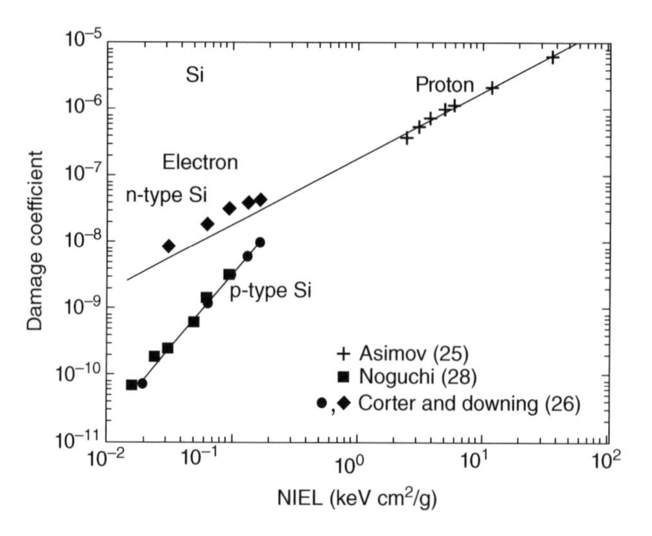

Figure 9.1.8 A plot showing the relationship between the NIEL and diffusion damage coefficients for Si. Source: (Messenger *et al.*, 2001)

MeV/g), which is given as the product of the particle fluence and NIEL (Summers *et al.*, 1993) as follows:

$$D_{e,eff}(1.0) = D_e(E)\left[\frac{NIEL_e(E)}{NIEL_e(1MeV)}\right]^{(n-1)} \tag{9.1.2}$$

where the quantity in square brackets is included to account for any nonlinear behavior as observed with p-type Si in Figure 9.1.8 and p/n diffused junction GaAs solar cells and the parameter n is determined empirically (Walters *et al.* 2005). Plotting solar cell degradation data as a function of Dd results in a collapse of the data to a single curve that is characteristic for the given solar cell technology. This is shown in Figure 9.1.9 for a GaAs solar cell, and as

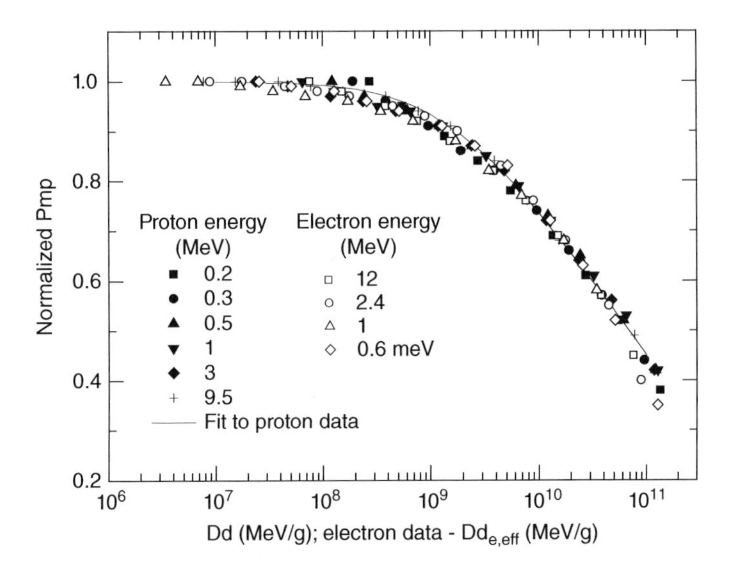

Figure 9.1.9 A plot showing the degradation of GaAs solar cells under electron and proton irradiation at a wide range of energies. When plotted as a function of Dd, the data collapse to a single, characteristic curve. Source: (Messenger *et al.*, 2001)

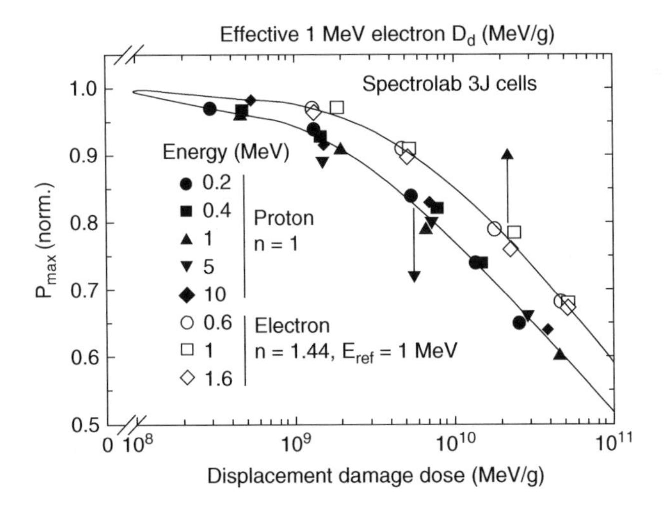

Figure 9.1.10 A plot showing the degradation of 3J InGaP/GaAs/Ge solar cells under electron and proton irradiation at a wide range of energies. As with the single-junction GaAs data in Figure 9.1.9, when plotted as a function of Dd, the data collapse to a single, characteristic curve. Source: (Walters *et al.*, 2005)

shown in detail in (Messenger *et al.* 2001), this curve can now be used to predict the solar cell response in a complex radiation environment.

The final aspect to address is the extension of the Dd analysis to MJ solar cells. Given that the individual junctions of the MJ devices are well described within the NRL methodology, the same is expected for the MJ devices as well. This is shown explicitly for a 3J InGaP/GaAs/Ge solar cell in Figure 9.1.10 (Walters *et al.* 2005). With this result, the performance of this 3J solar can be predicted in a variety of space environments.

9.1.6 Conclusion

This section has reviewed space solar cell technologies. Focusing on the primary technologies presently used on spacecraft, namely, Si and single-junction and 3J GaAs, the radiation-induced degradation mechanisms have been identified to be diffusion length degradation and carrier removal. The effects of these degradation mechanisms on the solar cell response have been explained and detailed comparison of solar cell before and after radiation exposure was presented. A method of correlating solar cell degradation due to irradiation by different particles at various energies was described, which is required since the space radiation consists of a complex spectrum of particles. With the analysis presented here, the performance of a solar cell in a complex space radiation environment can be accurately predicted.

List of Symbols and Units

Symbol	Description
D	Diffusion coefficient ($cm^2 s^{-1}$)
Dd	Displacement Damage Dose (MeV/g)
FF	Fill factor of the solar cell current vs. voltage curve (dimensionless)
Isc	Short circuit current (mA)
I_{ti}	Introduction rate of the ith recombination center (cm^{-1}),
K_L	Damage coefficient for L (dimensionless)
L	Minority carrier diffusion length (μm)
L_o	Pre-irradiation value of L (μm)
NIEL	Non-ionizing energy loss ($MeV/cm^2 g$)
Pmp	Solar cell maximum power (mW/cm^2)
QE	Quantum efficiency
Voc	Open circuit voltage (V)
υ	Thermal velocity of the minority carriers (cm/s)
φ	Particle fluence (cm^{-2})
σ_i	Minority carrier capture cross-section (cm^{-2})

References

Anspaugh, B.E. (1991) Proton and electron damage coefficients for GaAs/Ge solar cells. *Proceedings of 22nd IEEE Photovoltaic Specialists Conference*, pp. 1593–1598.

Anspaugh, B.E. (1996) *GaAs Solar Cell Radiation Handbook*. JPL Publication, New York.

Bertness, K.A., Cavicchi, B.T., Kurtz,. S R. *et al.* (1991) Effect of base doping or radiation damage in GaAs single-junction solar cells. *Proceedings of IEEE Photovoltaics Specialists Conference*, pp. 1582–1587.

Bertness, K.A., Kurtz, S.R., Friedman, D.J. *et al.* (1994a) High-efficiency GaInP/GaAs tandem solar cells for space and terrestrial applications. *Proceedings of the 1st World Conference on Photovoltaic Energy Conversion*, **2**, 1671–1678.

Bertness, K.A., Kurtz, S.R., Friedman, D.J., *et al.* (1994b) 29.5% Efficient GaInP/GaAs tandem solar cells. *Applied Physics Letters*, **65** (8), 989–991.

Brown, M.R., Garcia C.A., Goodelle G.S., *et al.* (1996) Characterization testing of measat GaAs/Ge solar cell assemblies. *Progress in Photovoltaics: Research and Applications*, **4**, 129–138.

Chapin, D.M., Fuller, C.S., and Pearson, G.L. (1954) A new silicon pn junction photocell for converting solar radiation into electrical power. *Journal of Applied Physics*, **25** (5), 676–677.

Chiang, P., Ermer, J.H., Nishikawa, W. *et al.* (1996) Experimental results of $GaInP_2$/GaAs/Ge triple junction cell development for space power systems. *Proceedings of the 25th IEEE Photovoltaic Specialists Conference*, 183–186.

Choo, S.C. (1968) Carrier generation: recombination in the space charge region of an asymmetrical p-n junction. *Solid State Electronics*, **11**, 1069.

Henry C.H. (1980) Limiting efficiencies of ideal single and multiple energy gap terrestrial solar cells. *Journal of Applied Physics*, **51** (8), 4494–4500.

Hovel, H.J. (1975) *Solar Cells, Semiconductors and Semimetals* (eds R.K. Willardson and A.C. Beer). Academic Press, New York, pp. 17–20.

Iles, P.A., Yeh Y.C.M., Ho, F.H. *et al.* (1990) High efficiency (>20% AM0) GaAs solar Cells grown on inactive Ge substrates. *IEEE Electron Device Letters*, **11** (4), 140.

Jenny, D.A., Loferski, J. J. and Rappaport, P. (1956) Photovoltaic effect in GaAs p-n junctions and solar energy conversion. *Physical Review*, **101** (3), 1208–1209.

Kaminar, N., Liu, D., MacMillan, H. *et al.* (1988) Concentrator efficiencies of 29.2% for a GaAs cell and 24.8% for a mounted cell lens assembly. *Proceedings of the 20th IEEE Photovoltaic Specialists Conference*, **1**, 766–768.

Keavney, C.J., Walters R.J., and Drevinsky, P.J. (1993) Optimizing the radiation resistance of InP solar cells: effect of dopant level and cell thickness. *Journal of Applied Physics*, **73**, 60.

Madelung, O. (1964) *Physics of III-V Compounds* (trans. D. Meyerhofer). John Wiley & Sons, Inc., New York.

Messenger S.R., Summers, G. P., Burke E.A. *et al.* (2001) Modeling solar cell degradation in space: A comparison of the NRL displacement damage dose and the JPL equivalent fluence approaches, *Progress in Photovoltaics: Research and Applications*, **9** (2), 103–121.

Sah, C., Noyce, R.N. and Shockley, W. (1957) Carrier generation and recombination in P-N junctions and P-N junction characteristics. *Proceedings of IRE*, **45**, 1228.

Shockley, W. (1950) *Electrons and Holes in Semiconductors*. Van Nostrand, New York.

Srour, J.R. (1988) Displacement damage effects in electronic materials, devices, and integrated circuits, Tutorial Short Course Notes, IEEE Nuclear and Space Radiation Effects Conference, Portland, OR, July.

Summers, G.P., Burke, E.A., Shapiro, P. *et al.* (1993) Damage correlations in semiconductors exposed to gamma, electron, and proton radiations. *Transactions on Nuclear Science*, **40**(6), 1372.

Summers, G.P., Burke, E.A. and Xapsos, M.A. (1995) Displacement damage analogs to ionizing radiation effects. *Radiation Measurements*, **24** (1), 1–8.

Tada, H.Y., Carter, J.R. Jr., Anspaugh, B.E. and Downing, R.G. (1982) *Solar Cell Radiation Handbook*, 3rd edition, JPL Publication, pp. 82–109.

Walters, R.J., Messenger, S.R., Cotal, H.L. *et al.* (1996a) Electron and proton irradiation-induced degradation of epitaxial InP solar cells. *Solid-State Electronics*, **39** (6), 797.

Walters, R.J., Messenger, S.R., Cotal, H.L., *et al.* (2013) Radiation response of heteroepitaxial n(+)p InP/Si solar cells. *Journal of Applied Physics*, **82** (5), 2164–2175.

Walters, R.J., Messenger, S.R., Cotal, H. L. and Summers, G.P. (1996b) Annealing of irradiated epitaxial InP solar cells, *Journal of Applied Physics*, **80** (8), 15 October.

Walters, R.J., Summers, G.P. and Messenger, S.R. (2000) Analysis and modeling of the radiation response of multijunction space solar cells. *Proceedings of the 28th IEEE Photovoltaics Specialists Conference*.

Walters, R.J., Warner, J.H., Summers, G.P. *et al.* (2005) Radiation response mechanisms in multijunction III-V space solar cells. *Proceedings of 31st IEEE Photovoltaic Specialists Conference*, pp. 542–547.

Walters, R.J., Xapsos, M.A., Cotal, H.L. *et al.* (1998) Radiation response and injection annealing of p+n InGaP solar cells, *Solid State Electronics*, **42** (9), 1747–1756.

Yamaguchi, M. (1995) Radiation resistance of compound semiconductor solar cells. *Journal of Applied Physics*, **78** (3), 1476.

Yamaguchi, M., Taylor, S J., Yangm M., *et al.* (1996) High-energy and high-fluence proton irradiation effects in silicon solar cells. *Journal of Applied Physics*, **80**, 4916.

Yeh, Y.C.M. (1984) *Proceedings of the 17th IEEE Photovoltaic Specialists Conference*. pp. 37–40.

9.2

Space PV Systems and Flight Demonstrations

Phillip Jenkins
U.S. Naval Research Laboratory, Washington, DC, USA

9.2.1 Introduction

Solar arrays intended for use in space have very little in common with their terrestrial counterparts other than the economic considerations of minimizing cost and maximizing reliability. It is the cost and reliability metrics that push the design of space solar arrays toward ever more radiation hard and high efficiency cells rather than low-cost silicon cells. How is it that space solar cells, that currently cost more than 100 times that of a terrestrial solar cell, make for the most cost-effective solar array? Simply stated, the operating conditions in space are so much harsher than on Earth, combined with the limited mass and volume that can be devoted to the solar panels in order to achieve the satellite mission, and drive the design to use higher-performing materials and cells (Fatemi *et al.*, 2000). The radiation damage is so important that array designers work to what is the end-of life (EOL) power requirement, and less often the beginning-of-life (BOL) power, which can be significantly higher. In Chapter 9.1, it was shown that the III-V materials' (e.g., GaAs, InP) radiation resistance is superior to silicon. At the system level, a silicon solar cell array would have to have so many more cells to meet the EOL power requirement, that it either would not fit under the rocket fairing, or would add substantially to the launch mass. Launch mass is a substantial cost driver in space systems. Although specific launch costs are often considered proprietary, and not generally published, the nominal cost to Low Earth Orbit (LEO) is roughly \$12,500/kg and to Geosynchronous Transfer Orbit (GTO) is \$22,000/kg. In this chapter we will briefly describe the typical solar array module construction, what type of materials are used, and how each component supports the function of the solar cell in the space environment. Then we touch on some of the next generation solar array designs.

Photovoltaic Solar Energy: From Fundamentals to Applications, First Edition.
Edited by Angèle Reinders, Pierre Verlinden, Wilfried van Sark, and Alexandre Freundlich.

Companion website: www.wiley.com/go/reinders/photovoltaic_solar_energy

9.2.2 The Building Block of the Solar Array: The Cell with Interconnect and Cover Glass

The building block for space solar arrays, since their first use until now, is a solar cell with attached interconnects and cover glass referred to as a CIC (may also be referred to as a "SCA," solar cell assembly). These individual assemblies are connected in series and parallel combinations to meet the power requirements of the solar array. The purpose of the cover glass is to protect the solar cell from particle radiation. The cover glass is attached to overhang the solar cell to protect the edges of the cell from the radiation field, which is considered to be omnidirectional. Low energy particles (<50keV) can be stopped with just a few micrometers of glass, but to provide shielding for higher energy particles, thicker glass, up to 500 micrometers, is available. Performance trade-offs between the weight of the glass and the shielding provided for the solar cell are part of the engineering trade study.

The cover glass is bonded to the cell with an optically clear silicone adhesive. Such adhesives are susceptible to UV degradation, so the cover glass must either absorb UV in its bulk, or reflect it using a thin film optical coating. The CIC is adhesively bonded to the array substrate, and must remain pliable over a wide temperature range in order to accommodate the thermal expansion mismatch between solar cells and the array substrate. These adhesives must also have low outgassing properties. The volatile by-products of the adhesives must be low enough that they do not contaminate the solar array. These by-products are known to darken and fix to surfaces in the presence of UV and proton radiation (Liu *et al.*, 2010; Liu, 2011). This is a known mechanism of damaging solar arrays (Fodor *et al.*, 2003).

Cell to substrate bonding serves two main purposes; to secure the CICs on the substrate during launch phases with have high acoustic and dynamic loadings, and to thermally conduct waste heat to the array substrate. The adhesive selection is important to ensure adequate thermal conduction, yet also an adequate bond line thickness to provide a mechanical buffer layer for the coefficient of thermal expansion (CTE) mismatch between the array substrate and the CICs.

Accounting for CTE mismatch is critical in space solar arrays, because the temperature excursions are wide. In Earth orbit, satellites experience the solar intensity, $1360.8 \pm 0.5\,\mathrm{W\,m^{-2}}$ with corrections for Earth–sun distance and the solar 11-year cycle (Jopp and Lean, 2011). While in the Earth's shadow, a solar array, which by design is lightweight and consequently has a low thermal mass, will radiate its heat to the Earth and to deep space. Thermal cycles of −100 °C to +80 °C are not uncommon. LEO satellites (orbital height above the Earth $\lesssim$ 2000 km) may endure 15 orbits (cycles) or more per day. For Earth geosynchronous orbit (known as GEO, orbital height of ~36,000 km above the earth), the solar array temperatures can be −180 °C to +80 °C. Here the array temperatures are generally cooler and the thermal cycles deeper due to smaller angular view of the Earth, which is warm compared to deep space, and the longer eclipse period, up to 72 minutes compared with 30–40 minutes for LEO (Maini and Agrawal, 2007). Transition from cold to hot can occur in less than a minute, as the spacecraft comes out of the eclipse into the full sun. These thermal transitions exacerbate thermal stress, particularly on the electrical interconnect to the cell or the cell-substrate bond interface. Interconnects are typically made of a super alloy, such as Kovar® to match the thermal expansion of the cell. Since the common solar array substrates have graphite composite facesheets, they typically have 0 to slightly negative CTE. The CTE mismatch to the substrate can result in the solar cell and interconnector expanding and contracting relative to the substrate. For that reason, a stress relieve feature, either in-plane (like a S-curve), or out-of-plane (like a loop) is formed in the interconnect to accommodate the substrate thermal expansion.

9.2.3 System Considerations

Ideally, the solar array contributes power to the spacecraft and demands no resources itself. This is rarely the case, as high power solar arrays require some degree of pointing to maximize power production and minimize mass. This may at times interfere with the other operations of the spacecraft. Conversely, the satellite, as part of its normal operation may shadow parts of the solar array. If a solar cell string becomes partially shaded, the shaded cells may experience a reverse bias voltage from cells in series that are illuminated. When this voltage is high enough, it may cause reverse breakdown in the solar cell diode and permanently damage the cell. For that reason, with modern III-V solar cells, each cell is connected in parallel to a bypass diode of opposite polarity. These diodes are discrete, or monolithically integrated (Sharps *et al.*, 2003). The bypass diode is forward-biased if the solar cell is reversed-bias, thus effectively bypassing the solar cell and appearing as if the shadowed cell is a simple forward-biased diode.

9.2.4 Solar Array Interactions with the Space Environment

A solar array may potentially have a large current loop area creating a magnetic moment that will create a torque on the satellite as it passes through the Earth's magnetic field. The satellite will also experience a Lorentz Force as the satellite electrical current passes through the Earth's magnetic field. Solar arrays are designed to minimize the effective loop area of the solar cell circuits. While it may appear the resulting torque is small, in the absence of any other forces, it will change the attitude of the spacecraft, forcing the spacecraft to supply a restoring torque, expending energy or fuel. This feature drives the array design to lay out symmetric current loops with opposing current flow to minimize the net magnetic dipole moment. A typical value for solar arrays magnetic moment is <3 A/m^2. Fuel for attitude control and station keeping is often the life-limiting resource for satellites. For GEO satellites, the magnetic moment, the moment of inertia and the natural frequency drive the design to smaller and stiffer solar arrays, which drives the design to use higher efficiency solar cells. For more information on spacecraft magnetic torques and environmental interactions, see (NASA, 1969; Garrett, 1990).

The ability to maintain an orbit in very low Earth orbit is strongly influenced by the cross-section the satellite presents in the orbital velocity vector. The Earth's ionosphere creates a drag force on the satellite, reducing its orbital energy. Solar arrays potentially have a large area and act to increase the drag of the satellite. Again, the array design is steered to higher efficiency solar cells to reduce the size of the array.

The ionosphere is dynamic and its structure is influenced by the 11-year solar cycle and the day/night cycle. The varying energy from the sun changes the height and energetics of the ionosphere constituents. Atomic oxygen (AO), a constituent of the ionosphere, can erode surfaces. Kapton®, a high performance polymer material often used in spacecraft solar arrays, is particularly susceptible to AO and must be protected with a coating to prevent erosion (Degroh *et al.*, 1995).

Space solar arrays must operate in the local plasma environment and the high energy portion of the solar spectrum. In GEO, for example, plasma particles have a very high energy >10 keV (Davis *et al.*, 2008). These particles embed themselves in solar array surfaces and can charge surfaces up to many thousands of volts, creating the potential for an electric arc. The UV and the x-ray portion of the sun's energy will create photoemission of electrons, and secondary electrons are created by initial electron impacts, and these can help to de-ionize surfaces. However, different surfaces (materials) will have different photoemission and secondary electron emission characteristics, and if not electrically connected, surfaces will experience differential charging, creating a potential for an arc between surfaces.

Solar cell cover glass is an effective insulator, and will charge up with respect to the solar cell or substrate, creating a risk of an arc. This discharge arc is known as a "primary" arc. The primary arc creates more plasma in the vicinity of the edge of the cells. The layout of cells on the solar array may place adjacent cells at different voltages, potentially as high as the string voltage. If the plasma density created by the primary arc is conductive enough, and the electric field between two adjacent cells is high enough, a "secondary" arc between two adjacent cells may be created. These "secondary arcs" have at their disposal the power of the solar cell string, and thus can deliver a current equal to the string current. Secondary arcs may not extinguish themselves until significant damage to the solar array has occurred. Figure 9.2.1 shows damage to a solar array due to a secondary arc. Spacecraft thrusters or space debris impacts may also contribute to an elevated plasma density near the solar array, potentially triggering secondary arcs.

Making surface materials electrically "leaky" to bleed off accumulated charge and controlling photoemission and secondary electron emission rates by material or coating choice are design features to limit differential charging. The solar cell cover glass may be made

Figure 9.2.1 Photograph of secondary arc damage (Ferguson and Hillard, 2003). Source: NASA

electrically conductive using a transparent conductive coating. This coating is then electrically bonded to the spacecraft. These are all attempts to suppress the formation of primary arcs.

To prevent secondary arcs, other methods are used; limiting the voltage between adjacent cells, limiting the string current, increasing the spacing between cells, and filling voids between cells with a potting compound (known as "grouting" the array). Thruster plume-related plasma may be mitigated operationally, or by placing the thrusters on the spacecraft to minimize exposure to the array. Further reading of interest on the topic can be found in (NASA, 2007).

All of the mitigation methods mentioned above for thermal cycling, particle and high energy solar radiation, atomic oxygen and plasma interactions all take their toll on array efficiency, mass, and spacecraft effectiveness. In the next section we will see what the current research and development trends are to reduce cost, mitigate environmental effects, and meet the expanding needs of space missions.

9.2.5 Space Solar Array Research and Development Trends

Research and development in new solar array technology at its core can be distilled down to reducing costs. In space systems, reducing cost means reducing mass, volume, and touch labor. Below are just a few examples of current research efforts to lower each of these metrics.

9.2.5.1 Solar Array Structures

The common construction of solar arrays consists of thick aluminum honeycomb panels (1–3 cm) with thin composite facesheets to serve as the substrate for laying down the CICs. These panels support cells and are mechanically strong enough to withstand launch acoustic and vibration environments. The panels serve as a primary structural component of the array, and have a high natural frequency, which is important to minimize attitude control resources. An example of the ridge panel design is shown in Figure 9.2.2.

The honeycomb panel represents a significant fraction of the array mass, so eliminating the honeycomb panel (while maintaining structural integrity) with something lighter and thinner, can significantly lower the cost by lowering mass and stowed volume. While eliminating the honeycomb substrate is not a new idea (Ray, 1967), the latest developments brings this concept to more mission platforms, for both commercial and government projects.

9.2.5.2 Orbital ATK, UltraFlex Array

The Orbital ATK UltraFlex™ and MegaFlex™ array design provides low stowed volume (>40 kW/m^3), low mass (up to 200 W/kg), and high deployed strength and stiffness (Murphy, 2012; Murphy *et al.*, 2015). The MegaFlex™ version of this technology can be scaled up to 300 kW using current rocket fairing volumes. This family of arrays is a "fanfold" design and deploys to a round shape. Structural efficiency comes from the dual use of a pair of graphite composite panels that serve both as launch stowage containment and support structure of the deployed array. The panels deploy to a round, shallow umbrella-like shape. The deployed

Figure 9.2.2 Deployment of one of two solar arrays after it has been integrated into the Global Precipitation Mission (GPM) Core spacecraft, launched in 2014. Source: NASA/Goddard

shape of these systems allows the blanket to be self-tensioning. The blanket tension does not need to be reacted by relatively heavy deployment booms such as those typically required by rectangular blanket systems. Examples of three different wings that have been manufactured, ranging from 2 m to 10 m in diameter, are shown in Figure 9.2.3.

9.2.5.3 Next Generation Solar Array (NGSA)

The Airbus DS Eurostar 3000 solar array designed for high power commercial communication satellites produces up to 22 kW BOL on orbit with a specific power of 70 W/Kg and a stowed volume specific power of 4.2 kW/m^3. The Airbus Next Generation Solar Array (NGSA) concept is a hybrid solar array, first proposed by V. G. Baghdasarian (Baghdasarian, 1998). Here, a conventional in-line deploying rigid panel solar array is combined with lightweight, semi-rigid lateral panels.

The NGSA has a configuration of two rigid and five lateral panels (Figure 9.2.4). The lateral panels are manufactured from a sheet of high strength material, which is reinforced by corrugations around the perimeter. Two additional corrugations support the inner area of the panel. These reinforcements provide the required stiffness during deployment of the lateral panels as well as in the fully deployed wing.

An attractive cell to be used for this configuration for missions in which beginning of life power is important is the inverted metamorphic (IMM) solar cell. In a four-junction configuration, efficiencies of 35% are expected (Patel *et al.*, 2012). Combining this new cell type with

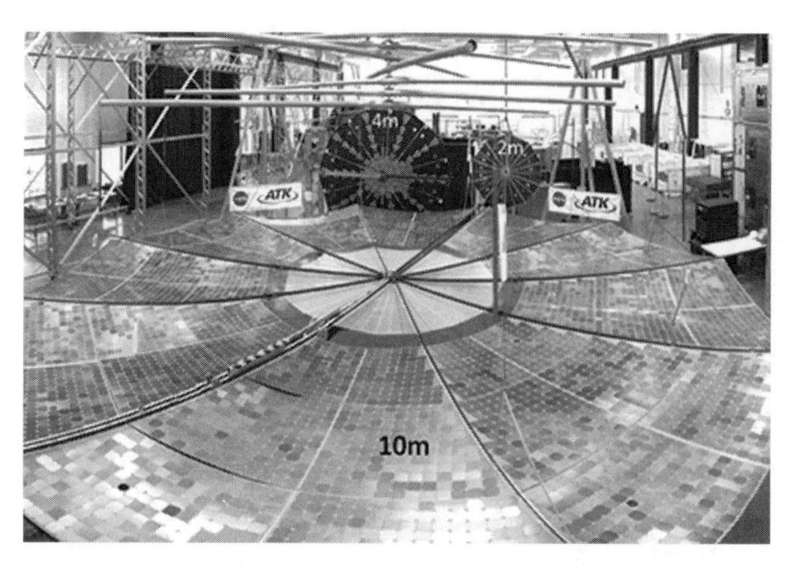

Figure 9.2.3 Manufactured examples of 2- and 4-meter diameter UltraFlex™ arrays (background) and a 10-meter MegaFlex™ array (foreground). Source: courtesy of ATK Orbital, 2014

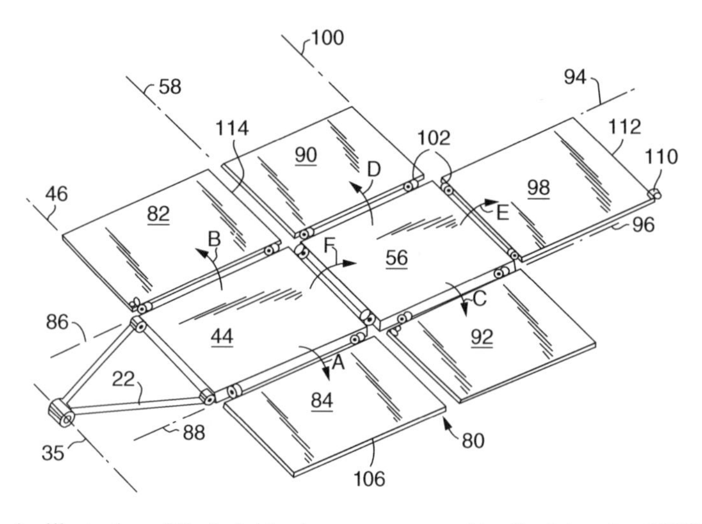

Figure 9.2.4 An illustration of the hybrid solar array, proposed by Baghdasarian (1998) showing two ridge panels (44 and 56) and five lateral panels

the hybrid array concept, a specific power of 200 W/Kg and a stowed volume specific power of 8.3 kW/m^3 can be achieved (pers. comm., Claus Zimmerman, Airbus DS, 2015).

9.2.5.4 Roll-Out Solar Array (ROSA)

Deployable Space Systems Inc. (DSS) has developed a tensioned membrane solar array technology called ROSA (Roll-Out Solar Array) with improved stowed volume efficiency and specific power over conventional ridge panel solar arrays. The ROSA is composed of two longitudinally-oriented thin-walled composite reinforced elastically-deployable slit-tube booms that are attached laterally at the tip and root ends with a mandrel and yoke/offset structure, respectively. A modular integrated lightweight flexible PV blanket assembly (called IMBA – Integrated Modular Blanket Assembly) is attached to and tensioned between the mandrel and the yoke/offset structure and reacted against the compressively preloaded booms. IMBA consists of a tensioned lightweight very-openweave glass-fiber mesh (or other insulating mesh material) that serves as a backplane/interface to the array structure on which CICs are mounted. A picture of the ROSA solar array (with the IMBA blanket assembly) in stowed and deployed configurations, and the deployment sequence is shown in Figure 9.2.5. The slit-tube and thin-walled elastic nature of the booms allow them to be flattened and then rolled up into an extremely compact stowage volume, and they provide the motive force for deployment. For ZTJ/XTJ solar cells, the ROSA is projected to provide 150–175 W/kg BOL specific power and 45–60 kW/m^3 specific stowed volume (pers. comm. Brian Spence, Deployable Space Systems Inc., 2015).

The above examples attack the mass, stowage, and array costs by eliminating the ridge honeycomb panel substrate while incorporating conventional CICs. Replacing the CIC is also a strategy for reducing cost. What is current practice to reduce CIC costs is to use larger cells (Boisvert *et al.*, 2011; Cho *et al.*, 2011). A larger cell translates as fewer cells for the array, reducing the touch labor in constructing the array. Another trend is to investigate how to eliminate the individual CICs altogether and use a continuous film to replace the cover glass as well as to automate the array laydown process (Wrosch *et al.*, 2013; Walmsley *et al.*, 2014). Still others have attempted to use other thin-film solar cells, such as amorphous silicon or CIGS. These materials have not shown high enough efficiency, or sufficient robustness to currently warrant consideration in space PV arrays.

The IMM technology mentioned earlier, which is still in the development stage at the time of writing, holds the promise of higher efficiency solar cells than the current GaInP/GaInAs/Ge technology. IMM cells are separated from the growth substrate leaving just the active solar cell 10–20 micrometers thick (Boisvert *et al.*, 2010; Cornfield *et al.*, 2010). These cells have a very high specific power approaching 3650 W/kg and are *flexible* (Cornfield and Dias, 2009). These attributes open up the design space to have even higher specific power and compact volume than with conventional CICs. The Sharp Corporation has developed a solar blanket technology referred to as a "Space Solar Sheet" using flexible IMM cells (Takamoto *et al.*, 2014). The Japanese Aerospace Exploration Agency (JAXA) successfully conducted a flight demonstrated using a small module of Space Solar Sheet (Shimazaki *et al.*, 2014). Other organizations are also pursuing options to take advantage of thin IMM technology as well (Breen *et al.*, 2010).

9.2.6 Conclusion

This chapter has attempted to make the reader aware of the issues facing designers of space PV systems. The demanding environment and required high reliability make incorporating new solar array technology difficult. Flight demonstrations and ground testing in a space-like environment are expensive and the required reliability of the power system is so high, it is difficult to accept the risk of introducing new technology. New technology makes its way into spacecraft because it is either mission-enabling or so substantially reduces cost as to encourage accepting the risk.

Figure 9.2.5 ROSA Solar Array and Deployment Sequence. Source: Courtesy Deployable Space Systems, 2012

Acknowledgments

The author wishes to acknowledge helpful conversations with, Dr. Dale Ferguson, USAF, Kirtland Air Base, Mr. Bao Hoang, Space Systems Loral, Mr. Michael Eskenazi, ATK Orbital, Dr. Claus Zimmerman, Airbus DS, Mr. Brian Spence, Deployable Space Systems, Dr. Marc Breene, Boeing Corporation, and Dr. Mitsuru Imaizumi, Japan Aerospace Exploration Agency in the writing of this chapter.

List of Acronyms

Acronym	Description
AO	Atomic oxygen
BOL	Beginning-of-life
CIC	Solar cell with attached interconnects and cover glass
CTE	Coefficient of thermal expansion
DSS	Deployable Space Systems Inc.
EOL	End-of life
GaAs	Gallium arsinide
GaInAs	Gallium indium arsenide
GaInP	Gallium indium phosphide
Ge	Germanium
GEO	Geosynchronous Earth orbit
GPM	Global Precipitation Mission
GTO	Geosynchronous transfer orbit
HEO	Highly elliptical Earth orbit
III-V	Materials made from the group III and group V materials as they appear in the periodic table of elements
IMBA	Integrated Modular Blanket Assembly
IMM	Inverted metamorphic
InP	Indium phosphide
JAXA	Japanese Aerospace Exploration Agency
KeV	Thousands of electron volts
LEO	Low Earth orbit
NGSA	Next generation solar array
ROSA	Roll-Out Solar Array
SCA	Solar cell assembly
UV	Ultra-violet radiation
XTJ	Spectrolab, a Boeing Company, product designator, NeXt Triple Junction solar cell
ZTJ	Product name for SolAero Technologies™ currently available, highest efficiency solar cell

References

Baghdasarian, V.G. (1998) Hybrid solar panel array, patent US5785280 A.

Boisvert, J., Law, D., King, R. *et al.* (2010) Development of advanced space solar cells at Spectrolab. *35th IEEE Photovoltaic Specialists Conference Record*, pp. 000123–000127.

Boisvert, J., Law, D., King, R. *et al.* (2011) Development of space solar cells at Spectrolab. *37th IEEE Photovoltaic Specialists Conference Record*, pp. 001528–001531.

Breen, M.L., Streett, A., Cokin, D. *et al.* (2010) IBIS (Integrated Blanket/Interconnect System), Boeing's solution for implementing IMM (Inverted Metamorphic) solar cells on a light-weight flexible solar panel. *35th IEEE Photovoltaic Specialists Conference Record*, pp. 000723–000724.

Cho, B., Stan, M.A., Patel, P. *et al.* (2011) The large-area One-per-wafer ZTJ and ZTJM solar cells from Emcore. *37th IEEE Photovoltaic Specialists Conference Record*, pp. 001945–001948.

Cornfeld, A.B., Aiken, D., Cho, B. *et al.* (2010) Development of a four sub-cell inverted metamorphic multi-junction (IMM) highly efficient AM0 solar cell. *35th IEEE Photovoltaic Specialists Conference Record*, pp. 000105–000109.

Cornfeld, A.B. and Diaz, J. (2009) The 3J-IMM solar cell: Pathways for insertion into space power systems. *34th IEEE Photovoltaic Specialists Conference Record*, pp. 000954–000959.

Davis, V.A.M., Mandell, M.J., and Thomsen, M.F. (2008) Representation of the measured geosynchronous plasma environment in spacecraft charging calculations. *Journal of Geophysical Research*, **113**, A10204.

Degroh, K.K., Banks, B.A., and Smith, D. C. (1995) Environmental durability issues for solar power systems in low Earth orbit, NASA Report, TM-106775, E-9226, NAS 1.15:106775.

Fatemi, N.S., Pollard, H.E., Hou, H.Q., and Sharps, P.R. (2000) Solar array trades between very high-efficiency multi-junction and Si space solar cells. *28th IEEE Photovoltaic Specialists Conference Record*, pp. 1083,1086.

Fodor, J.S., Frey, M.A., Gelb, S.W. *et al.* (2003) In-orbit performance of space solar arrays, *Proceedings of 3rd World Conference on Photovoltaic Energy Conversion*, **1**, 638, 641.

Garrett, H.B. (1990) Space environments and their effects on space automation and robotics. *Proceedings of the Third Annual Workshop on Space Operations Automation and Robotics (SOAR)*, pp. 361–380.

Kopp. G. and Lean, J.L. (2011) A new, lower value of total solar irradiance: evidence and climate significance. *Geophysical Research Letters*, **38**, L01706.

Liu, D.L., Liu, S.H., Panetta, C.J., *et al.* (2010) Effects of contamination on solar cell coverglass. 35th *IEEE Photovoltaic Specialists Conference Record*, pp.002563–002568.

Liu, D.L., Liu, S.H., Panetta, C.J. *et al.* (2011) Synergistic effects of contamination and low energy space protons on solar cell current output. *37th IEEE Photovoltaic Specialists Conference Record*, pp. 001595–001600.

Maini, A.K. and Agrawal, V. (2007) *Satellite Technology: Principles and Applications*. John Wiley & Sons, Inc., New York.

NASA (1969) *Technical Publication: Spacecraft Magnetic Torques*, SP-8018. NASA.

NASA (2007) Technical Standard NASA-HDBK-4006, *Low Earth Orbit Spacecraft Charging Design Handbook*. NASA.

Patel, P., Aiken, D,. Boca, A. *et al.* (2012) Experimental results from performance improvement and radiation hardening of inverted metamorphic multijunction solar cells. *IEEE Journal of Photovoltaics*, **2**, 377.

Ray, K.A. (1967) Flexible solar cell arrays for increased space power, *IEEE Transactions on Aerospace and Electronic Systems*, **AES-3** (1), 107–115.

Sharps, P.R., Stan, M.A., Aiken, D.J. *et al.* (2003) Multi-junction cells with monolithic bypass diodes, *Proceedings of 3rd World Conference on Photovoltaic Energy Conversion*, **1**, 626–629.

Shimazaki, K., Kobayashi, Y., Takahashi, M. *et al.* (2014) First flight demonstration of glass-type space solar sheet. *40th IEEE Photovoltaic Specialists Conference Record*, pp. 2149–2154.

Takamoto, T., Washio, H., and Juso, H. (2014) Application of InGaP/GaAs/InGaAs triple junction solar cells to space use and concentrator photovoltaic. *40th IEEE Photovoltaic Specialists Conference Record*, pp. 0001–0005.

Walmsley, N., Wrosch, M., and Stern, T. (2014) Low cost automated manufacture of high specific power photovoltaic solar arrays for space. *40th IEEE Photovoltaic Specialists Conference Record*, pp. 2166–2169.

Wrosch, M., Walmsley, N., Stern, T. *et al.* (2013) Laminated solar panels for space using multi-cell transparent covers. *39th IEEE Photovoltaic Specialists Conference Record*, pp. 2809–2811.

9.3

A Vision on Future Developments in Space Photovoltaics

David Wilt
Air Force Research Laboratory, Ohio, USA

9.3.1 Current Status and Near-Term Challenges/Opportunities

Before launching into a starry-eyed set of prognostications regarding the future of space power, it may be prudent to quickly assess the past predictions as well as the current state of spacecraft power system technology, as exemplified by the Dawn solar array (Figure 9.3.1). The age of solar-powered spacecraft began in 1958 with the launch of Vanguard I, which used six small silicon cells to power a 5 mW radio transmitter. Early predictions (Wood *et al.*, 1965) suggested that most spacecraft would require tens to thousands of kW of electrical power. In that period, photovoltaics was only seen as suitable for very small power systems and the future of space power was certain to be based on nuclear-fueled sources (i.e. SNAP-8). Given the difficulty in scaling reactor-based technologies to low power levels and the low efficiency of the electronics of that period, the grandiose power requirements predicted might not be surprising. Safety concerns raised by a well-publicized launch failure with a nuclear source on board (TRANSIT 5BN-3, 1964) limited the future of nuclear-fueled systems to all but deep space applications. In the 1990s, spacecraft electrical power requirement predictions ranged from a few hundred watts to megawatts of power (Massie, 1991). As before, the high power missions in this study would utilize nuclear reactor-based systems as solar was envisioned to be suitable only for missions up to 100 kW.

Looking at these past predictions, it is clear that the range of mission power requirements has been and continues to be large. In fact, with the advent of CubeSats (Woellert *et al.*, 2011) and the ever-increasing capability and efficiency of microelectronics, the current range of power requirements spans from a few watts to nearly 30 kW, not including the International Space Station (110 kW). It is also clear that the amount of electrical power able to be supplied by solar continues to expand (>100 kW), pushing the opportunity space for nuclear to even

Photovoltaic Solar Energy: From Fundamentals to Applications, First Edition.
Edited by Angèle Reinders, Pierre Verlinden, Wilfried van Sark, and Alexandre Freundlich.

Companion website: www.wiley.com/go/reinders/photovoltaic_solar_energy

Figure 9.3.1 Dawn solar array. Source: courtesy of NASA

higher power levels (>MW). Finally, it is clear that the actual sizes of power systems typically flown are nearly universally on the lower end of the predicted power spectrum. Currently there is much interest and activity surrounding large (~300 kW) solar power systems for electric-powered vehicles. Given the costs of current solar array technologies, these high power systems could remain concepts unless dramatic reductions in power system costs are developed. Hence we define as *Challenge 1: Power System Cost Reduction.*

Although slower than predicted by some, actual spacecraft electrical power requirements have grown and continue to do so. For example, commercial Geosynchronous Earth Orbit (GEO) communication satellites had traditionally been in the 5–10 kW range, but have recently grown to over 20 kW with plans for even higher power levels. The United States Global Positioning System (GPS) has experienced a power growth of over 400% in recent modifications. All of these power increases are driven by the advantage gained through increased capability within a single spacecraft as well as the demands that improved sensors, high data volumes and near real time communications bring. For GEO communications satellites, the conventional rigid solar array designs are limiting the total array power that can fit inside the launch fairing to approximately 30 kW. Increasing the volumetric specific power (W/m^3) of solar arrays is a near-term challenge to address that limitation. Hence we define as *Challenge 2: Volumetric Specific Power.*

A recent study of power system reliability demonstrated that nearly $9 billion in losses can be attributed to spacecraft electrical power system failures from 1990 thru 2013 (Landis, 2013). The largest number of major failures can be grouped into two general categories: solar array mechanical failures and array plasma interactions. Space plasma is the charged particle environment (electrons and ions) that surround the spacecraft. As shown in a related study (Frost & Sullivan, 2004), solar array-related failures account for 50% of the value of all spacecraft insurance claims. A principal reason for the high cost of array failures is that they tend to occur early in the spacecraft mission life, thus the economic impact (revenue loss) of such a failure is much higher than failures which occur near the end of the spacecraft mission life. Given the frequent failures associated with spacecraft solar arrays and their high negative economic impact, there is a need to improve solar array reliability. Hence we define as *Challenge 3: Solar Array Reliability.*

Current spacecraft solar arrays are manufactured very much as custom components, with very high labor content and very long development schedules. Frequently, solar array development is the long pole in the spacecraft development timeline, placing it on the critical path for mission success. This approach to array development not only creates long development schedules, it raises serious cost and reliability challenges. Hence we define as *Challenge 4: Solar Array Development Schedule*.

Now that we have identified several pressing challenges, let us take a quick look at current growth opportunities for space photovoltaics. Recently, GEO Communications satellite providers have begun to offer all-electric vehicles that replace the second stage chemical motors with electric thrusters. This switch provides a ~50% mass savings at the spacecraft level and can thereby offer the potential to either step down to a smaller launch vehicle or the ability to launch two spacecraft on a single rocket. Either of these options offers significant costs advantage (~$50M). Electric propulsion technology has also been used to save a satellite whose second stage chemical rocket motor failed to function (Penn *et al*., 2014) and left it days away from re-entering the Earth's atmosphere. The reliability of electric thruster technology has resulted in a reduction in the insurance rates for these vehicles as compared to conventional spacecraft, which has the potential also to provide significant savings.

Finally, the range of missions for space photovoltaics continues to expand with the latest example being the NASA Juno spacecraft (Lewis, 2014) powered by photovoltaics on its way to Jupiter. This solar array provides 12–14 kW of electrical power at Earth, but only ~480 W once the spacecraft reaches Jupiter. Fortunately, the scientific mission is able to be accomplished with this limited power. And while it may seem expensive to fly a 14 kW array to provide 480 W of power, given the scarcity and cost of Pu238 (radio-isotope fuel for all deep space missions) and the cost of obtaining approval to fly a nuclear power system, using solar arrays for this mission undoubtedly provided a significant overall cost savings.

9.3.2 Near-Term Technologies to Address Challenges

In this section, a variety of power system technologies will be reviewed, specifically considering their ability to impact the challenges mentioned previously.

9.3.2.1 Advanced Solar Cell Technologies

Currently, ~30% efficient triple-junction solar cells based on germanium are the state of the art (SOA) for space-qualified products. Several new solar cell technologies are being investigated as potential next generation cells to provide >30% efficiency, including four-junction Inverted Metamorphic Multijunction (IMM) solar cells, dilute nitride-based multijunction cells and mechanically bonded multijunction cells. These technologies are likely to provide cells with beginning of life (BOL) efficiencies of ~35% (~15% relative increase over SOA). This increase in cell efficiency has a direct impact on array cost (Challenge 1) and array volume (Challenge 2). The array volume benefit is most obvious as these higher performance cells have the potential to reduce array area requirements by ~15% for the same electrical power requirement, provided they have similar environmental durability. This reduction in array area translates into reduced stowed volume, although the impact is likely smaller than the reduction in array area.

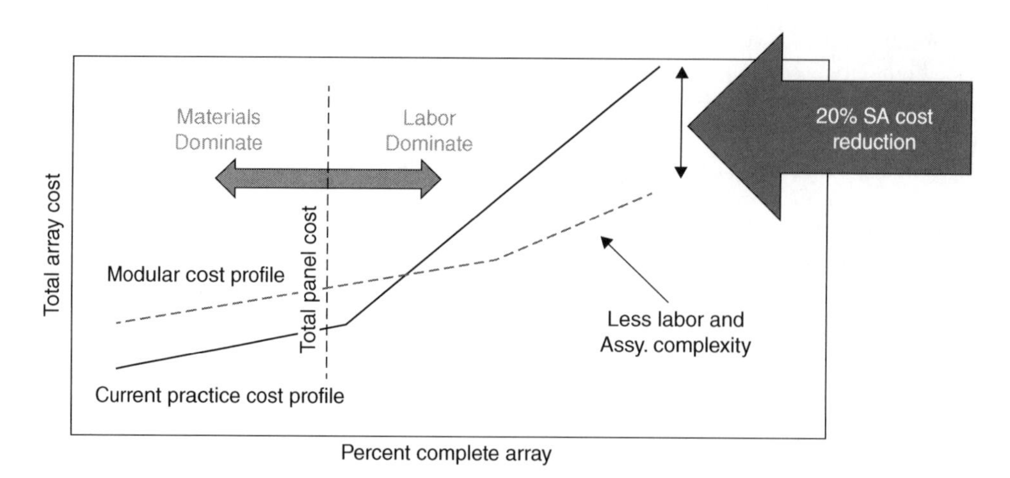

Figure 9.3.2 Cumulative cost for space solar panel assembly, starting with cells on the left and ending with full panels on the right

The cost benefit of higher performance and higher cost cells is related to the current process for manufacturing solar panels. As shown in Figure 9.3.2, the costs of space solar panel fabrication are dominated by the cell laydown/panel-integration processes, shown in the Labor Dominate portion of the graph. Panel assembly is heavily dependent upon skilled touch labor, welding interconnects on the cells, applying cover glass, stringing the cells, and finally bonding them to the panel. Higher performance cells offer a cost benefit as fewer cells need to be integrated in order to achieve the same power level. This fact has helped new, more efficient and more expensive cell technologies justify inclusion on new missions as they provide cost savings at the panel and array level. In the future, this cost benefit may be diminished significantly based on the development of new fabrication technologies, as described below. However, smaller solar arrays have other spacecraft benefits (lower drag, reduced torques, etc.) that may warrant inclusion in future missions.

9.3.2.2 Modularity

Currently, solar arrays are custom-fabricated, with each array optimized for a specific spacecraft. This level of customization has technical advantages but also introduces significant costs, long development schedules, and reliability concerns. Cost pressures in the aerospace industry, particularly for smaller spacecraft, are opening up opportunities for the introduction of modularity features in solar array technologies. Even using high touch labor panel assembly methods, modularity can offer significant cost benefits. The dashed line in Figure 9.3.2 represents a notional cost profile for a modular array system. Modularity and standardization offer the potential to dramatically reduce the non-recurring engineering (NRE) as this design development work is spread across a large number of panels. In addition, custom tooling can easily be developed to facilitate module fabrication, thereby reducing assembly costs. Using modular systems on multiple missions allows the design to be refined, potentially leading to higher reliability systems. Finally, modular solar arrays can be fabricated and assembled much more quickly as module manufactures can spool up their production capability and potentially develop a warehouse of common solar array module products. Based on this discussion, it is readily apparent that modularity constructs within space power technology have the potential to address three of the aforementioned challenges: cost, schedule, and reliability.

9.3.2.3 Automated Panel and Blanket Manufacturing

Leveraging the development of advanced manufacturing technologies in other industries, there is a significant potential for automating the integration of advanced solar cells onto rigid and flexible assemblies. This approach has been investigated in the past, however, the equipment lacked the flexibility and capability to run a variety of different-sized solar cells and incorporate them in a wide variety of configurations. Given the extremely high cost of cell integration (Figure 9.3.2), there is strong financial justification for development of automated assembly technology. In addition, bringing in automation is likely to provide reliability benefits.

9.3.2.4 Advanced Solar Arrays

Current SOA solar arrays are based on rigid aluminum honeycomb/composite panels interconnected with a variety of hinges and actuators. These technologies are problematic from a cost, reliability, and volumetric stowage efficiency perspective. As indicated in the aforementioned power system reliability studies (Frost & Sullivan, 2004; Landis, 2013), a significant fraction of the array failures are due to mechanical problems early in the mission life. These problems can largely be attributed to the complexity of conventional array stowage and synchronized deployment as well as the difficulties and costs associated with validating the deployment mechanism in a 1-g environment. To address these cost, complexity, and reliability challenges, new solar arrays are being developed that have considerably fewer components, offer improved stowage volume, and allow for repeated deployment testing in a laboratory environment (Mikulas *et al.*, 2015).

9.3.3 Far-Term Power Needs and Technology Options

As stated previously, the range of spacecraft power system's projected needs is extremely large, however, virtually all current spacecraft are in the <30 kW class. It would be reasonable to project that this will remain the case with spacecraft and that over the next 20+ years the primarily power requirement will be <50 kW per spacecraft. This prediction is in line with the current trends of spacecraft systems' efficiencies increasing, thereby reducing the amount of power needed to accomplish a given mission. In addition, numerous smaller spacecraft are being considered in lieu of fewer larger spacecraft, as this provides several potential benefits.

The trends of electric spacecraft propulsion are likely to continue as well as solar-powered satellites in harsh and ever more remote locations (Van Allen radiation belts, lunar surface/poles, outer planets, etc.). Solar power satellites (collecting power on orbit and beaming it to the Earth's surface) never seem to be completely forgotten and may one day be feasible from a technical and financial perspective (Landis, 2012). There is also a growing trend to consider in-situ manufacturing of spacecraft-related technologies and while the capability to directly print solar arrays is many years off, having such a capability offers some interesting opportunities for systems optimized for space operation without having to be concerned with fitting inside a launch fairing and surviving the launch.

The photovoltaic technologies likely to be needed to meet future needs could include ever more efficient solar cells, low-cost solar cells, novel solar array technologies and novel power system solutions. As mentioned previously, ~30% efficient three-junction solar cells are SOA technology with four-junction cells (IMM and potentially mechanically stacked configurations) being developed as the next generation. The trend of achieving ever-increasing cell performance via increasing the number of sub-cells may have challenges continuing into the future. As demonstrated with the four-junction IMM cell, the radiation tolerance of advanced cells may have difficulties achieving the same performance as older generation technologies (Boisvert *et al.*, 2013). This effect, when combined with the increase in voltage temperature coefficient as more subcells are added, could result in a 4+ junction cell that has minimal performance advantage over older generation cells at end of life (EOL) and at operational temperature.

In order to address this challenge to more advanced multijunction cell technologies (>4 junction), it may be appropriate to consider alternative operational incarnations, such as concentrator solar arrays, which may be beneficial to both reducing the radiation degradation sensitivity as well as potentially providing a performance boost at higher optical concentration levels. As shown in a recent TacSat4 experiment (Jenkins *et al.*, 2013), even low-level optical concentrators (<10×) can offer mass efficient options for increasing radiation tolerance. In that flight experiment, the conventional solar arrays demonstrated a ~25% efficiency reduction in a single year of operation. During that same period, the stretched lens concentrator experiment showed a ~10% efficiency reduction. This improved radiation tolerance is due to the thicker cover glass that can be added to concentrator cells without having a significant array mass penalty.

Concentrator solar arrays are also important for alternative solar cell concepts, such as intermediate band or hot carrier cells. The physics of these advanced devices require high photon flux in order to operate properly. For example, the two-photon absorption process for carrier generation in intermediate band solar cells is proportional to the square of the photon density. At one-sun operation, the probability of carrier generation is very small, thus high concentration (>100×) will likely be required for these advanced devices to demonstrate their potential.

Another approach worth considering for improving the radiation tolerance of multijunction devices is the inclusion of advanced photon management strategies. As demonstrated by Alta (Kayes, 2011), thin III-V devices with appropriate photon management strategies have demonstrated record performance. The inclusion of photon management features enables the use of thin absorbers, which reduce the overall dark current and provide the voltage boost to produce record efficiencies. Thin absorbers with photon management also reduce the minority carrier diffusion length requirement for full carrier collection. This reduced diffusion length requirement suggests that such a device should be capable of withstanding additional radiation impingement before showing loss of carrier collection. To take full advantage of this approach, the photon management should be incorporated subcell by subcell in a multijunction device. This is a significant technical challenge and will significantly complicate the device fabrication process, but does offer interesting opportunities.

An area of ongoing interest and possibilities is very low cost photovoltaics for space applications. Leveraging the encouraging developments in the terrestrial photovoltaic market, there may be future opportunities for cadmium telluride (CdTe), copper indium gallium diselenide (CIGS), peroskovites, silicon or other new materials. These technologies have demonstrated encouraging efficiency gains (~20% AM1.5) and several have the potential for good radiation tolerance. Prior flight testing of terrestrial thin film photovoltaics has demonstrated that additional efforts are likely required to make them fully compatible with the environmental challenges of space operation (radiation, UV, vacuum, etc.). One might suspect that these technologies may find their initial space application in surface power applications (lunar, Mars, etc.), where landed mass is critically important and the lower efficiency of these technologies does not penalize them in array structural metrics.

In conclusion, photovoltaics have been critically important for space virtually from the beginning of spaceflight. It is quite likely that spacecraft photovoltaic power requirements will continue to expand, both at the very low power level and the very high power level in future years. And photovoltaics will continue to offer compelling power solutions for virtually every space mission.

List of Symbols

Symbol	Description
kW	kilowatt
m^3	cubic meter
MW	megawatt
mW	milliwatt
W	watt

List of Acronyms

Acronym	Description
AEHF	Advanced extremely high frequency
BOL	Beginning of life
CdTe	Cadmium telluride
CIGS	Copper indium gallium diselenide
GEO	Geosynchronous Earth orbit
GPS	Global positioning system
IMM	Inverted Metamorphic Multijunction
NASA	National Aeronautics and Space Administration
NRE	Non-recurring engineering
SNAP-8	Systems nuclear auxiliary power number 8
SOA	State of the art
UV	Ultra violet
III-V	Semiconductor materials from columns III and V of the Periodic Table of Elements

References

Boisvert, J. *et al.* (2013) High efficiency inverted metamorphic (IMM) solar cells. *Proceedings of the 39th IEEE Photovoltaic Specialists Conference.*

Frost & Sullivan (2004) Commercial Communications Satellite Bus Reliability Analysis, http://www.lr.tudelft.nl/index.php?id=29218&L=1

Jenkins, P.P. *et al.* (2013) Initial Results from the TacSat-4 Solar Cell Experiment. *Proceedings of the 39th IEEE Photovoltaic Specialists Conference.*

Kayes, B.M. (2011) 27.6% Conversion efficiency, a new record for single-junction solar cells under 1 sun illumination. *Proceedings of the 37th IEEE Photovoltaic Specialists Conference (PVSC)*, 2011. DOI: 10.1109/PVSC.2011.6185831.

Landis, G.A. (2013) Tabulation of power-related satellite failure causes. *Proceedings of the 11th International Energy Conversion Engineering Conference* (AIAA 2013-3736).

Landis, G. (2012) Solar power satellites. In *Photovoltaic Technology* (ed. W.G.J.H.M. van Sark) Vol. 1 in *Comprehensive Renewable Energy* (ed. A. Sayigh). Elsevier, Oxford, pp. 767–774.

Lewis, J. (2014) Juno spacecraft operations lessons learned for early Cruise mission phases. *Proceedings of the 2014 IEEE Aerospace Conference.*

Massie, L.D. (1991) Future trends in space power technology. *IEEE Aerospace and Electronic Systems Magazine*, **6** (11), 8–13.

Mikulas, M., Pappa, R., Warren, J., and Rose, G. (2015) Telescoping solar array concept for achieving high packaging efficiency, (AIAA 2015-1398), 2nd AIAA Spacecraft Structures Conference, 10.2514/6.2015-1398.

Penn, J.P., Mayberry, J., Ranieri, C., and O'Brian, R. (2014) Re-Imagining SMC's fleet with high-power solar arrays and solar electric propulsion (AIAA 2014-4328). AIAA SPACE 2014 Conference and Exposition, 10.2514/6.2014-4328.

Woellert, K., Ehrenfreund, P., Ricco, A.J., and Hertzfeld, H. (2011) Cubesats: cost-effective science and technology platforms for emerging and developing nations, *Advances in Space Research*, **47** (4), 663–684.

Wood, L.H., Vachon, R.I., and Seitz, R.N. (1965) Electrical power requirements for future space exploration, *IEEE Transactions on Nuclear Science*, **12** (1), 189–196.

Part Ten

PV Modules and Manufacturing

10.1

Manufacturing of Various PV Technologies

Alison Lennon and Rhett Evans
University of New South Wales, Sydney, Australia

10.1.1 Introduction

Silicon wafer-based technologies comprise more than 90% of annual PV production. Most silicon PV modules are fabricated using p-type multi-crystalline silicon wafers with screen-printed metal electrodes. This standard technology follows a sequence of steps that represents a continued refinement from the cell process developed at Spectrolab in the 1970s (Ralph, 1975). Improvements in cell and module efficiency, wafer production, process engineering, and other benefits associated with large-scale manufacturing over the past 5 to 10 years have resulted in the cost of silicon PV modules being below US$1 per Watt now for a number of years (ITRPV, 2014). Module costs continue to decrease and remain the primary driving force for future innovation.

More advanced cell structures which use new screen-printing approaches, ion-implantation, selective emitters and rear-surface passivation will continue to increase in future years as manufacturers seek to differentiate themselves in a competitive and diversifying market. Many of these higher efficiency cell structures are more beneficially applied to Czochralski (Cz) mono-crystalline wafers (Annis, 2014), which although more expensive, can often maximize the performance benefits of advanced cell structures.

10.1.2 Manufacturing of Screen-Printed p-Type Silicon Cells

The sequence of steps typically followed in the manufacture of screen-printed cells is shown in Figure 10.1.1. High-throughput equipment and modern large production facilities (see Figure 10.1.2) can now annually produce cells and modules in very large volume, in the range of several GWs per year. This section describes in more detail the individual steps depicted in Figure 10.1.1, in terms of current practice, recent advances and possible future developments.

Photovoltaic Solar Energy: From Fundamentals to Applications, First Edition.
Edited by Angèle Reinders, Pierre Verlinden, Wilfried van Sark, and Alexandre Freundlich.

Companion website: www.wiley.com/go/reinders/photovoltaic_solar_energy

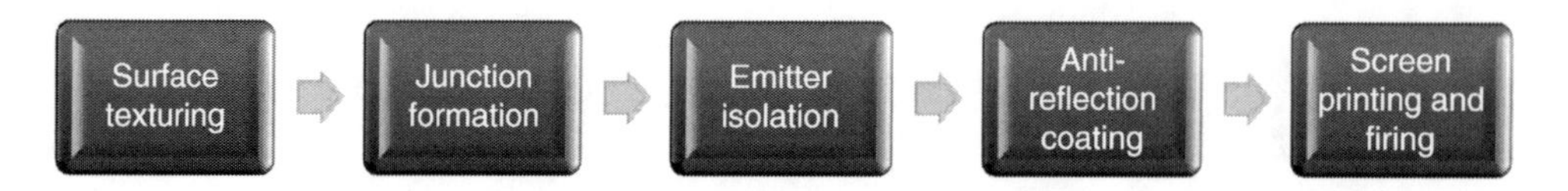

Figure 10.1.1 Manufacturing sequence for screen-printed p-type silicon solar cells

Figure 10.1.2 A screen-printed silicon solar cell manufacturing facility showing multiple production lines. Source: Courtesy of Trina Solar

10.1.2.1 Surface Texturing

Surface texturing, to reduce front surface reflection and increase light trapping, is typically performed using chemical etching processes that remove silicon non-uniformly across the surface. Mono-crystalline wafers are textured in chemical baths containing dilute alkaline solutions with additives (e.g., isopropanol (IPA)) to produce a pyramidal silicon surface. This process, used by Spectrolab in their early development of p-type cells (Stella and Scott-Monk, 1976), can reduce the surface front surface reflection from approximately 36% to minimum values of ~11%. Historically, ~10 μm of silicon was etched from each wafer surface before texturing, to remove the damage that was created by the wire saws, however, many manufacturers now rely on removing this damage during the texturing process. This can reduce process cost, complexity, and time as one less etching bath is required. Improvements have also been made to the additives (Ximello-Quiebras *et al.*, 2012)) and there are likely to be many proprietary developments in this area.

The alkaline texturing process is not typically used for multi-crystalline wafers because multiple crystal orientations are exposed at the surface and these etch at different rates. Consequently, multi-crystalline wafers are usually textured in cooled inline chemical baths, containing acidic solutions of nitric (HNO_3) and hydrofluoric (HF) acids. Etching occurs preferentially at crystal defects that remain from the sawing process used to cut wafers from a cast block (Kaiser *et al.*, 1991; Einhaus *et al.*, 1997). The slope of the surface features formed by acidic texturing is less than achieved with alkaline texturing, consequently the reflectivity of acidic-textured wafer surfaces in the range of 23 to 25% (Einhaus *et al.*, 1997). However, when cells are coated with an antireflection coating (ARC) and encapsulated, total internal reflection at the glass-laminate interface can reduce or nearly eliminate the effect of this higher post-texturing reflectance (Gee *et al.*, 1994). The optimum etch loss is typically 4–5 μm per side (Hauser *et al.*, 2003) with the etch loss being monitored by wafer weight loss and surface reflectivity measurements. The etched silicon loss can be tuned by altering the speed at which the wafers are transported through the inline bath.

The increased recent adoption of diamond-tipped wire saws has presented some challenges for texturing processes that rely on the presence of defects on the surface, because these saws leave a distinctively different pattern of saw marks on the surface which can result in less than optimal texturing (Bidiville *et al.*, 2010). This has motivated interest in alternative texturing, such as plasma texturing, especially for multi-crystalline cells. Plasma texturing is a "dry" texturing process in which fluorine radicals, generated from plasmas typically comprising sulphur hexafluoride (SF_6) and O_2, selectively etch the wafer surface (Ruby *et al.*, 2000; Chan *et al.*, 2013), reducing front surface reflection on multi-crystalline wafers to values similar to that achieved with alkaline texturing (Seiffe *et al.*, 2013). Plasma texturing can result in reduced water consumption and allows for single-sided texturing which can be advantageous for advanced cell designs (e.g., rear-passivated cells). However, the equipment costs are significantly higher than for chemical etching processes, the global warming potential impact of the gases like SF_6 needs to be carefully evaluated and gas abatement must be tightly controlled.

10.1.2.2 Junction Formation

Screen-printed cells use a p-n junction formed in a p-type silicon wafer to separate the electrical charge carriers that are generated by the absorption of light. The n-type region, which is commonly referred to as an emitter, is formed by solid state thermal diffusion of phosphorous atoms into the surfaces of the wafer. The diffusion can be performed using a tube furnace in a batch process where approximately 300 wafers are loaded in a quartz boat as shown in Figure 10.1.3. A phosphorus-containing gas, phosphoryl chloride ($POCl_3$), is introduced into the quartz tube furnace, where it reacts with O_2 to form solid phosphorus pentoxide (P_2O_5) which deposits on the oxidized wafer surface, providing the source of phosphorus atoms. In the subsequent drive-in step, the $POCl_3$ flow is stopped and the P_2O_5 is reduced by the silicon, producing SiO_2 and elemental phosphorus which diffuses into the wafer to form an n-type layer on the wafer's surface. A typical tube furnace diffusion process takes about 50 min. and is performed at temperatures in the range of 800–900°C.

Emitters can also be formed by depositing a phosphorus-containing layer on the surface (e.g., by spray coating the surface with the relatively inexpensive phosphoric acid (H_3PO_4)) and then placing the wafers on a belt or rollers, which transports them through an inline diffusion

Figure 10.1.3 Tube furnace showing a quartz boat loaded with silicon wafers in preparation for phosphorus diffusion

furnace. A typical inline diffusion process takes 15–20 min. and is performed at a temperature of 850–980°C. Although promising reduced process costs due to a faster process and lower chemical costs, inline diffusion can result in higher phosphorus concentrations at the surface which increase the recombination that occurs in the doped region, thereby reducing cell efficiency (Kim *et al.*, 2006). Furthermore, inline diffusion of multi-crystalline wafers from spray-on sources can result in localized areas of light diffusion which can lead to intensification of the electric fields in reverse bias, resulting in failure in hot spot tests (Bauer *et al.*, 2009).

Although earlier emitters were simply characterized in terms of junction depth and sheet resistance, greater attention is now being given to the engineering of emitter profiles. Historically, silver pastes required a high surface concentration of phosphorus for ohmic contact. In recent times newer pastes have been developed which can make low resistance ohmic contact to silicon surfaces with phosphorus concentrations less than $2\times10^{20}\,cm^{-3}$ (Shanmugam *et al.*, 2014). Reducing the surface concentration of phosphorus reduces recombination in the emitter region while still enabling low contact resistance (Scardera *et al.*, 2012; Shetty *et al.*, 2013). These improvements, coupled with improvements in the phosphorus diffusion process, have enabled cell manufacturers to continually reduce emitter recombination currents to values as low as 175 fA/cm^2 with values of <100 fA/cm^2 being predicted in 10 years (ITRPV, 2014).

10.1.2.3 Emitter Isolation

Most phosphorus diffusion processes result in phosphorus diffusing into all surfaces of the wafer, including the rear surface and the edges. It is necessary to isolate the front-surface emitter to prevent the photo-generated electrons, collected by the front junction, from flowing

Figure 10.1.4 Wafers being transported through a single-side etch tool to remove the phosphorus-doped silicon on the rear surface. Source: Courtesy of RENA Technologies GmbH

down the shunting pathway created between the front junction and the rear surface by the phosphorus-diffused edges. Cell shunt resistance is a critical parameter (higher is better) because it impacts a module's low light performance and thereby significantly affects the kWh/kWp energy yield (Grunow *et al.*, 2004). The latter has become an important figure of merit for ranking cell technologies and manufacturers' products.

The generally preferred approach to isolating the emitter on the front surface of the cell is to etch the rear surface junction in a bath of cooled HF and HNO_3 in which the wafers are transported through the etchant in a way that allows only the rear surface to be etched (e.g., see Figure 10.1.4). These multi-lane single-side etch tools can etch the rear side of 7,500 wafers per hour and result in shunt resistances exceeding 10 kΩ. The same tool can be used to remove the phosphorus glass remaining on the wafer after diffusion and slightly etch-back the emitter so as to remove the surface layer, which may contain a very high concentration of phosphorus atoms.

10.1.2.4 Antireflection Coating

Screen-printed solar cells use silicon nitride (SiN_x) deposited by plasma-enhanced chemical vapour deposition (PECVD) as the ARC. In this process, a plasma is created between two radiofrequency or microwave electrodes, where the space between the electrodes is filled with sub-atmospheric mixtures of reactive gases, including silane (SH_4) and ammonia (NH_3). The excitation of the plasma can be achieved by direct or remote methods. In direct PECVD reactors the wafers are located directly within the plasma (see Figure 10.1.5), whereas in remote plasma reactors the excitation of the plasma is spatially separated from the wafer. Plasma-enhanced CVD reactors are typically large and complex tools which require strict safety systems for gas management and abatement procedures due to the pyrophoric nature of the SH_4 precursor gas.

Silicon nitride films with refractive indices in the range from 1.9 to 2.3 can be deposited by varying the PECVD deposition parameters. Optimally, the refractive index should be about 2.0 when the cell is in air and about 2.3 when encapsulated. However, more light is absorbed in an ARC with a higher refractive index (Duerinckx and Szlufcik, 2002). Consequently a moderate value of ~2.1 is typically used (Karunagaran *et al.*, 2006) to minimize both parasitic absorption and reflection losses. An important attribute of SiN_x layers deposited by PECVD is

Figure 10.1.5 Graphite wafer carrier inside a direct plasma PECVD tube reactor. Source: courtesy of Amtech Tempress

their ability to incorporate hydrogen in a way that it can be released into the silicon during subsequent thermal processes. Hydrogen can reduce recombination associated with defects or impurities within the silicon wafer or at the surfaces, thereby effectively improving the minority carrier effective lifetime and increasing cell voltages (Hallam *et al.*, 2013a; Liu *et al.*, 2014). Of particular interest is the role of hydrogen in the quasi-permanent deactivation of boron-oxygen defects (Fischer and Pschunder, 1973) by thermal annealing under carrier injection (Herguth *et al.*, 2006, 2008; Münzer, 2009; Hallam *et al.*, 2013b; Wilking *et al.*, 2013) which can result in stabilized improved efficiencies of 1.0% absolute for screen-printed cells (Schmidt *et al.*, 2013).

10.1.2.5 Screen Printing and Firing of Metal Electrodes

Screen printing is a low-cost way of depositing a material on a surface according to a pattern formed in a screen. Screens are typically made from woven steel fabric that is stretched and glued to a metal frame and coated with an emulsion. Openings are formed in this emulsion using UV exposure, and paste can be pushed through the openings with a squeegee to form a pattern of paste on the wafer surface under the screen. Screens can withstand up to 10,000 passes of a high-speed high-pressure print squeegee without distortion or breakage.

Most of the rear surface of the cell is screen-printed with aluminium paste to form the rear electrode, however, busbars, or short busbar electrodes, are also printed with a predominantly silver paste to enable the cells to be interconnected by soldering. On the front side, a silver grid

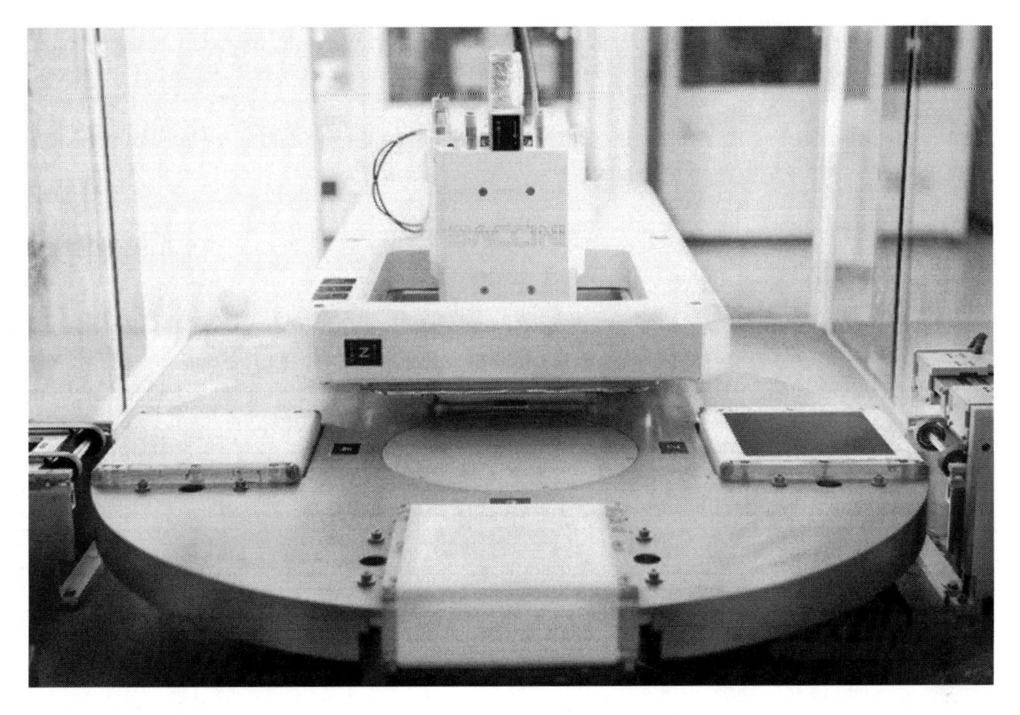

Figure 10.1.6 Screen printer in operation. Source: courtesy of Trina Solar

pattern, comprising fingers and a number of busbars, is printed directly over the SiN_x ARC. Cells are then "fired" in an inline infrared firing furnace where they experience a peak firing temperature in the range of 750 to 870 °C for about 5 s. The rear-surface aluminum alloys with the silicon at the interface to form an aluminum-doped (p-type) silicon layer that acts as a back surface field (BSF). This reduces the minority carrier concentration at the aluminum electrode and in doing so reduces the recombination rate at that surface, resulting in increased voltages. During the firing process, front surface contacts are also formed. Glass frits in the silver paste melt and etch the underlying SiN_x and silver-containing crystallites grow into the silicon surface from the glass, enabling electrical contact (Hilali, 2005).

Screen printing improvements have seen finger widths decrease to below 50 µm (ITRPV, 2014; Osborne, 2014), reducing the front surface shading. Higher aspect ratios metal fingers, made possible by print-on-print screen printing (Hannebauer *et al.*, 2013) and stencil printing (Hoornstra and Heurtault, 2009; Hannebauer *et al.*, 2013), can reduce metal grid resistive losses (Figure 10.1.6). The need to reduce cost is driving innovations to decrease the amount of silver used, as it accounts for a significant proportion of the cell cost (ITRPV, 2014). Silver consumption is predicted to reduce to about 30 mg per cell or even lower, this reduction being made possible by new low-silver content pastes and process innovations such as dual printing where thinner busbars are printed in a separate printing step using less expensive low etching-activity pastes (Osborne, 2014). Interconnection regimes that increase the number of busbars are also gaining popularity (Edwards *et al.*, 2013) and this approach will also reduce the amount of silver required in the front grid.

One of the challenges facing screen-printed metallization is the increasing fragility of wafers as they become thinner. As wafers approach 160 μm in thickness they cannot withstand edge contacting, so edge grippers and the practice of driving a cell into a hard stop for alignment cannot be used. This means that sensors and vision-assisted alignment techniques must be used for double printing processes.

10.1.3 Advanced p-Type Cell Technologies

For most manufacturers, the path to higher p-type cell efficiencies involves adopting rear-surface passivation with localized metal contact regions. This so-called passivated emitter and rear cell (PERC) (Blakers *et al.*, 1989) involves coating the rear surface with a dielectric and then making openings in the dielectric though which the screen-printed aluminum can alloy with the exposed silicon. The approach can increase average efficiencies of mono-crystalline cells by up to 1% absolute (Tjahjono *et al.*, 2013) and reduce the $ per Watt of modules by ~ 1% (Basore, 2014). A rear dielectric comprising a thin layer of aluminum oxide capped with a thicker layer of SiN_x is typically used to reduce the rear recombination and therefore enable higher operating cell voltages. Higher short circuit current densities also result, due to the reflectivity of the rear surface being increased to 90–95% at 1000 nm compared with about 65% for fully alloy aluminum surfaces (Lauermann *et al.*, 2012). In 2014, the crystalline silicon market share for PERC cells was only 10% but with increased experience in production, this is expected to increase to greater than 45% by 2024 (ITRPV, 2014).

Improvements in the rear surface have been paralleled with improved emitter engineering so as to reduce emitter recombination losses. One approach is to reduce the surface phosphorus (as described in Section 10.1.2.2 on junction formation). Alternatively, a selective emitter, which has heavily doped regions where the metal contacts are located while the remaining surface is lightly doped, can be used to shield the minority carriers from the metallized areas. Selective emitters can be formed using a masking process (Lauermann *et al.*, 2009; Raabe *et al.* 2010), ion implantation (Dubé *et al.*, 2011; Low *et al.*, 2011; Rohatgi *et al.*, 2012) or laser-doping (Jäger *et al.*, 2009; Hallam *et al.*, 2011; Wang *et al.*, 2012; Lee *et al.*, 2014). The metallization process must ensure that the electrode metal is deposited in the heavily doped regions (e.g., by aligned screen printing or self-aligned metal plating (Wang *et al.*, 2012)).

The need to reduce silver usage is also driving increased interest in replacing silver screen printing with copper plating. Unlike silver, which is an investment metal, copper is a commodity metal with a relatively stable lower cost and so from a manufacturing perspective a transition from silver to copper is attractive. Copper-plated laser-doped cells were manufactured under the brand name of Pluto (Shi *et al.*, 2009), however, concerns regarding the adhesion of the plated metal fingers, the ability to use soldered interconnection, and the possible ingress of copper into cells have plagued the adoption of plated copper metallization (Lee *et al.*, 2014; Lennon *et al.*, 2014). Improved copper plating chemistry (Letize *et al.*, 2013) and processes (Horzel *et al.*, 2014) have recently led to the demonstration of laser-ablated nickel-copper plated metal contacts where cell efficiencies of 21.3% have been achieved with high sheet resistance emitters (Metz *et al.*, 2014).

10.1.4 Higher Efficiency n-Type Technologies

The market for higher-efficiency crystalline silicon PV modules is being driven by space-constrained solar PV deployment, typified by the growing Japanese consumer market. These higher efficiency technologies predominantly use n-type Cz wafers due to their longer minority carrier lifetimes and greater tolerance to common metal impurities. A number of n-type cell technologies have been manufactured (e.g., Yingli Green's PANDA and metal-wrap-through (MWT) cells) and have been reviewed by Rehman (Rehman and Lee, 2013), however, market share remains low largely due to the higher cost of n-type wafers.

One strategy to achieve higher efficiencies is to form metal contacts to both carrier polarities on the rear surface of the cell, leaving the front surface unshaded. SunPower have pioneered the development of interdigitated back contact (IBC) cells, having developed three industrial solar cell technology generations with cell efficiencies going from 20% in 2005 to 23% in 2012, and recently reporting a 25% efficiency for an optimized Gen III cell with an area of 121 cm^2 (Smith *et al.*, 2014). Although attractive from a cell design perspective, back contact cells involve more complex processing and so their cost of production per Watt of power produced is greater than the more commonly-produced screen-printed cells. Higher module efficiency provides system-level benefits due to reduced balance of system costs and so consequently higher efficiency cells are more competitive against utility screen-printed modules when their levelized costs of electricity (LCOE) are compared. The electron and hole collection on the rear surface of IBC cells can be achieved using diffused or ion implanted regions, a key design aspect being to maintain the metal contact regions as small as possible to minimize both voltage losses while not incurring high resistive losses.

Another promising direction for higher efficiency is to use heterojunctions to extract the carriers from a crystalline silicon absorber. The best-known example of this approach is the heterojunction with intrinsic thin layer (HIT) technology pioneered by Sanyo and now being manufactured by Panasonic with energy conversion efficiencies exceeding 20% in production (De Wolf *et al.*, 2012). Research in this area has advanced in other companies with at least one equipment manufacturer now offering turnkey lines based on this approach (Zhao *et al.*, 2014). Sharp and Panasonic have achieved cell efficiencies exceeding 25% (Masuko *et al.*, 2014; Nakamura *et al.*, 2014) by moving the heterojunction contacts to the rear, however, the fabrication of these cells is complex and the path to high-volume manufacturing is not immediately so clear. A further challenge for heterojunction cells, in general, is the need to carefully prepare and clean the silicon wafer interface (Schmidt *et al.*, 2009). This can make high electrical yields difficult to achieve when scaling to high-volume manufacturing.

10.1.5 Conclusion

The photovoltaic market remains dominated by the production of screen-printed p-type crystalline silicon modules with most manufacturing currently focused on production of the cheaper multi-crystalline modules. However, there is an increasing demand for higher efficiency PV products and so it is predicted that many manufacturers will look to more advanced cell structures as they seek to differentiate themselves in a competitive and diversifying market. The production of screen-printed PERC modules is expected to increase in coming years, with higher efficiencies arising from reduced emitter recombination and improvements in screen printing. However, more substantial increases in module performance are likely to arise from alternative technologies such as back contact and heterojunction cells. Although challenges remain for these technologies in terms of cost-effective manufacturing and yield, they represent an opportunity as the return on investment for research and development is potentially higher.

Acknowledgements

Stuart Wenham is thanked for his advice and help with proof-reading.

List of Abbreviations

ARC	Anti-reflection coating
BSF	Back surface field
Cz	Czochralski
HIT	Heterojunction with intrinsic thin layer
IBC	Interdigitated back contact
ITRPV	International Technology Roadmap for Photovoltaics
LCOE	Levelised cost of electricity
MWT	Metal-wrap-through
PECVD	Plasma-enhanced chemical vapour deposition
PERC	Passivated emitter and rear cell
PV	Photovoltaics
UV	Ultraviolet
Chemical Formulae	
HF	Hydrofluoric acid
HNO_3	Hydrofluoric acid
IPA	Isopropyl alcohol (isopropanol)
NH_3	Ammonia
$POCl_3$	Phosphoryl chloride
P_2O_5	Phosphorus pentoxide
H_3PO_4	Phosphoric acid
SH_4	Silane
SiN_x	Silicon nitride
SiO_2	Silicon dioxide

References

Annis, C. (2014) Mono wafers to recapture share from multi in solar PV. Available at: http://www.displaysearchblog.com/2014/07/mono-wafers-to-recapture-share-from-multi-in-solar-pv/(accessed 22 February 2014).

Basore, P.A. (2014) Understanding manufacturing cost influence on future trends in silicon photovoltaics. *IEEE Journal of Photovoltaics*, **4** (6), 1477–1482. DOI: 10.1109/JPHOTOV.2014.2358081.

Bauer, J., Wagner, J.M., Lotnyk, A. *et al.* (2009) Hot spots in multicrystalline silicon solar cells: avalanche breakdown due to etch pits. *physica status solidi (RRL) – Rapid Research Letters*, **3** (2–3), 40–42. DOI: 10.1002/pssr.200802250.

Bidiville, A., Wasmer, K., Michler, J. *et al.* (2010) Mechanisms of wafer sawing and impact on wafer properties. *Progress in Photovoltaics: Research and Applications*, **18** (8), 563–572. DOI: 10.1002/pip.972.

Blakers, A., Wang, A., Milne, A. *et al.* (1989) 22.8% efficient silicon solar cell. *Applied Physics Letters*, **55** (13), 1363–1365. DOI: 10.1063/1.101596.

Chan, B.T., Kunnen, E., Kaidong, X. *et al.* (2013) Two-step plasma-texturing process for multicrystalline silicon solar cells with linear microwave plasma sources. *IEEE Journal of Photovoltaics*, **3** (1), 152–158. DOI: 10.1109/JPHOTOV.2012.2216510.

De Wolf, S, Descoeudres, A., Holman, Z.C., and Ballif, C. (2012) High-efficiency silicon heterojunction solar cells: A review, *Green*, **2**, 7.

Dubé, C.E., Tsefrekas, B., Buzby, D. *et al.* (2011) High efficiency selective emitter cells using patterned ion implantation. *Energy Procedia*, **8** (0), 706–711. DOI: 10.1016/j.egypro.2011.06.205.

Duerinckx, F. and Szlufcik, J. (2002) Defect passivation of industrial multicrystalline solar cells based on PECVD silicon nitride. *Solar Energy Materials and Solar Cells*, **72** (1–4), 231–246. DOI: 10.1016/S0927-0248(01)00170-2.

Edwards, M., Ji, J., Sugianto, A. *et al.* (2013) High efficiency at module level with almost no cell metallisation: Multiple wire interconnection of reduced metal solar cells. Paper presented at the 39th IEEE Photovoltaic Specialists Conference, Tampa, FL, 16 June– 21 June.

Einhaus, R., Vazsonyi, E., Szlufcik, J. *et al.* (1997) Isotropic texturing of multicrystalline silicon wafers with acidic texturing solutions [solar cell manufacture]. Paper presented at the 26th IEEE Photovoltaic Specialists Conference, 29 September–3 October.

Fischer, H. and Pschunder, W. (1973) Investigation of photon and thermal induced change in silicon solar cells. Paper presented at the 10th IEEE Photovolataic Conference, Palo Alto, CA.

Gee, J.M., Schubert, W.K., Tardy, H.L *et al.* (1994,) The effect of encapsulation on the reflectance of photovoltaic modules using textured multicrystalline-silicon solar cells. Paper presented at the 24th IEEE Photovoltaic Specialists Conference, 5–9 December.

Grunow, P., Lust, S., Sauter, D. *et al.* (2004) Weak light performance and annual yields of pv modules and systems as a result of the basic parameter set of industrial solar cel*ls* Paper presented at the 19th European Photovoltaic Solar Energy Conference, Paris, France, 7–11 June.

Hallam, B., Hamer, P., Wenham, S. *et al.* (2013a) Advanced bulk defect passivation for silicon solar cells. *IEEE Journal of Photovoltaics*, **4** (1), 1–8. DOI: 10.1109/JPHOTOV.2013.2281732.

Hallam, B., Wenham, S., Hamer, P. *et al.* (2013b) Hydrogen passivation of B-O defects in Czochralski silicon. *Energy Procedia*, **38** (0), 561–570. DOI: 10.1016/j.egypro.2013.07.317.

Hallam, B., Wenham, S., Sugianto, A. *et al.* (2011) Record large-area p-type CZ production cell efficiency of 19.3%; based on LDSE technology. *IEEE Journal of Photovoltaics*, **1** (1), 43–48.

Hannebauer, H., Dullweber, T., Falcon, T. *et al.* (2013) Record low Ag paste consumption of 67.7 mg with dual print. *Energy Procedia*, **43** (0), 66–71. DOI: 10.1016/j.egypro.2013.11.089.

Hauser, A., Melnyk, I., Fath, P. *et al.* (2003) A simplified process for isotropic texturing of mc-Si. Paper presented at the 3rd World Conference on Photovoltaic Energy Conversion, Osaka, Japan, 11–18 May.

Herguth, A., Schubert, G., Kaes, M., and Hahn, G. (2006) A new approach to prevent the negative impact of the metastable defect in boron doped CZ silicon solar cells. Paper presented at the IEEE 4th World Conference on Photovoltaic Energy Conversion, Conference Record.

Herguth, A., Schubert, G., Kaes, M., and Hahn, G. (2008) Investigations on the long time behavior of the metastable boron–oxygen complex in crystalline silicon. *Progress in Photovoltaics: Research and Applications*, **16** (2), 135–140. DOI: 10.1002/pip.779.

Hilali, M.M. (2005) Understanding and development of manufacturable screen-printed contacts on high sheet-resistance emitters for low-cost silicon solar cells. PhD, thesis. Georgia Institute of Technology, USA.

Hoornstra, J. and Heurtault, B. (2009) Stencil print applications and progress for crystalline silicon solar cells. Paper presented at the 24th European Photovoltaic Solar Energy Conference, Hamburg, Germany, 21–25 September.

Horzel, J., Bay, N., Passig, M. *et al.* (2014) Low cost metallisation based on Ni/Cu plating enabling high efficiency industrial solar cells. Paper presented at the 29th European Photovoltaic Solar Energy Conference, Amsterdam, 22–26 September.

ITRPV (2014) International Technology Roadmap for Photovoltaic: Results 2013. 5th edition.

Jäger, U., Okanovic, M., Hörteis, M. *et al.* (2009) Selective emitter by laser doping from phosphosilicate glass. Paper presented at the 24th European Photovoltaic Solar Energy Conferecne, Hamburg, Germany, 21–25 September.

Kaiser, U., Kaiser, M., and Schindler, R. (1991) Texture etching of multicrystalline silicon. In *Proceedings of Tenth E.C. Photovoltaic Solar Energy Conference* (eds A. Luque, G. Sala *et al.*), pp. 293–294. Springer, Dordrecht.

Karunagaran, B., Jeong, J.P., Nagarajan, S. *et al.* (2006) Low-temperature deposition of silicon-nitride layers by using PECVD for high efficiency Si solar cells. *Journal of the Korean Physical Society*, **48** (6), 1250–1254.

Kim, D.S., Hilali, M.M., Rohatgi, A. *et al.* (2006) Development of a phosphorus spray diffusion system for low-cost silicon solar cells. *Journal of the Electrochemical Society*, **153** (7), A1391–A1396. DOI: 10.1149/1.2202088.

Lauermann, T., Dastgheib-Shirazi, A., Book, F. *et al.* (2009) INSECT: An inline selective emitter concept with high efficiencies at competitive process costs improved with inkjet masking technology. Paper presented at the 24th European Photovoltaic Solar Energy Conference, Hamburg, Germany, 21–25 September.

Lauermann, T., Frohlich, B., Hahn, G., and Terheiden, B. (2012) Design considerations for industrial rear passivated solar cells. Paper presented at the 38th IEEE Photovoltaic Specialists Conference Austin, Texas, 3-8 June.

Lee, K., Kyeong, D., Kim, M. *et al.* (2014) Copper metallization of silicon PERL solar cells: 21% cell efficiency and module assembly using conductive film. Paper presented at the IEEE Photovoltaics Specialist Conference, Denver, 9–13 June.

Lennon, A., Flynn, S., Young, T. *et al.* (2014) Addressing perceived barriers to the adoption of plated metallisation for silicon photovoltaic manufacturing. Paper presented at the Solar 2014 Conference, Melbourne, Australia, 8–9 May.

Letize, A., Lee, B., Crouse, K., and Cullen, D. (2013) Light-induced plating of silicon solar cell conductors using a novel low acid, high speed copper electroplating process. Paper presented at the 28th European Photovoltaic Solar Energy Conference and Exhibition, Amsterdam, 22–26 September.

Liu, A., Sun, C., and Macdonald, D. (2014) Hydrogen passivation of interstitial iron in boron-doped multicrystalline silicon during annealing. *Journal of Applied Physics*, **116** (19), DOI: 10.1063/1.4901831.

Low, R., Gupta, A., Gossmann, H. *et al.* (2011) High efficiency selective emitter enabled through patterned ion implantation. Paper presented at the 37th IEEE Photovoltaic Specialists Conference, 19–24 June.

Masuko, K., Shigematsu, M., Hashiguchi, T. *et al.* (2014) Achievement of more than 25%; Conversion efficiency with crystalline silicon heterojunction solar cell. *IEEE Journal of Photovoltaics*, **4** (6), 1433–1435. DOI: 10.1109/JPHOTOV.2014.2352151.

Metz, A., Adler, D., Bagus, S. *et al.* (2014) Industrial high performance crystalline silicon solar cells and modules based on rear surface passivation technology. *Solar Energy Materials and Solar Cells*, **120**, Part A(0), 417–425. DOI: 10.1016/j.solmat.2013.06.025

Münzer, A. (2009) Hydrogenated silicon nitride for regeneration of light induced degradation. Paper presented at the 24th European Photovoltaic Solar Energy Conference, Hamburg, Germany, 21–25 September.

Nakamura, J., Asano, N., Hieda, T. *et al.* (2014) Development of heterojunction back contact Si solar cells. *IEEE Journal of Photovoltaics*, **4** (6), 1491–1495. DOI: 10.1109/JPHOTOV.2014.2358377.

Osborne, M. (2014) ISFH partnership pushes industrial PERC solar cell to record 21.2% efficiency, available at: http://www.pv-tech.org/news/isfh_partnership_pushes_industrial_perc_solar_cell_to_record_21.2_efficiency

Raabe, B., Book, F., Dastgheib-Shirazi, A., and Hahn, G. (2010) The development of etch-back processes for industrial silicon solar cells. Paper presented at the 25th European Photovoltaic Solar Energy Conference and Exhibition/5th World Conference on Photovoltaic Energy Conversion, Valencia, Spain, 6–10 September.

Ralph, E.L. (1975) Recent advances in low-cost solar cell processing. Paper presented at the 11th IEEE Photovoltaics Specialists Conference, Scottsdale, Arizona, USA.

Rehman, A., and Lee, S.H. (2013) Advancements in n-type base crystalline silicon solar cells and their emergence in the photovoltaic industry. *The Scientific World Journal*, **2013**, 13. DOI: 10.1155/2013/470347.

Rohatgi, A., Meier, D.L., McPherson, B. *et al.* (2012) High-throughput ion-implantation for low-cost high-efficiency silicon solar cells. *Energy Procedia*, **15** (0), 10–19. DOI: 10.1016/j.egypro.2012.02.002

Ruby, D.S., Zaidi, S.H., and Narayanan, S. (2000) Plasma-texturization for multicrystalline silicon solar cells. Paper presented at the 28th IEEE Photovoltaic Specialists Conference.

Scardera, G., Meisel, A., Mikeska, K.R. *et al.* (2012) High efficiency phosphorus emitters for industrial solar cells: Comparing advanced homogeneous emitter cells and selective emitters using silicon ink technology. Paper presented at the 27th European Photovoltaic and Solar Energy Conference, Frankfurt, Germany.

Schmidt, J., Lim, B., Walter, D. *et al.* (2013) Impurity-related limitations of next-generation industrial silicon solar cells. *IEEE Journal of Photovoltaics*, **3** (1), 114–118. DOI: 10.1109/JPHOTOV.2012.2210030.

Schmidt, J., Veith, B., and Brendel, R. (2009) Effective surface passivation of crystalline silicon using ultrathin Al2O3 films and Al2O3/SiNx stacks. *physica status solidi (RRL) – Rapid Research Letters*, **3** (9), 287–289. DOI: 10.1002/pssr.200903272.

Seiffe, J., Bremen, B., Rappl, S. *et al.* (2013) Two-step plasma texturing process for industrial solar cell manufacturing. Paper presented at the 28th European Photovoltaic Solar Energy Conference, Paris, 30 September–3 October.

Shanmugam, V., Cunnusamy, J., Khanna, A. *et al.* (2014) Electrical and microstructural analysis of contact formation on lightly doped phosphorus emitters using thick-film ag screen printing pastes. *IEEE Journal of Photovoltaics*, **4** (1), 168–174. DOI: 10.1109/JPHOTOV.2013.2291313.

Shetty, K.D., Boreland, M.B., Shanmugam, V. *et al.* (2013) Lightly doped emitters for high efficiency silicon wafer solar cells. *Energy Procedia*, **33** (0), 70–75. DOI: 10.1016/j.egypro.2013.05.041

Shi, Z., Wenham, S.R., and Ji, J. (2009) Mass production of the innovative PLUTO solar cell technology. Paper presented at the IEEE Photovoltaics Specialists Conference, Philadelphia.

Smith, D.D., Cousins, P., Westerberg, S. *et al.* (2014) Toward the practical limits of silicon solar cells. *IEEE Journal of,Photovoltaics*, **4** (6), 1465–1469. DOI: 10.1109/JPHOTOV.2014.2350695.

Stella, P., and Scott-Monk, J. (1976) Development of high efficiency, radiation tolerant, thin silicon solar cell. In Spectrolab (ed.), (pp. 1–9).

Tjahjono, B., Yang, M.J., Wu, V. *et al.* (2013) Optimizing celco cell technology in one year of mass production. Paper presented at the 28th European Photovoltaic Solar Energy Conference and Exhibition, Paris, 30 September–4 October.

Wang, Z., Han, P., Lu, H. *et al.* (2012) Advanced PERC and PERL production cells with 20.3% record efficiency for standard commercial p-type silicon wafers. *Progress in Photovoltaics: Research and Applications*, **20** (3), 260–268. DOI: 10.1002/pip.2178.

Wilking, S., Herguth, A., and Hahn, G. (2013) Influence of hydrogen on the regeneration of boron-oxygen related defects in crystalline silicon. *Journal of Applied Physics*, **113** (19), 194503.

Ximello-Quiebras, J.N., Junge, J., Seren, S., and Hahn, G. (2012) Solutions used in the texturization of monocrystalline silicon. *Photovoltaics International, 16th Edition*, 68–74.

Zhao, J., Wißen, A., Decker, D. *et al.* (2014) Silicon heterojunction solar cells in Roth and Rau's pilot line: Process performance improvement on mass production tools. Paper presented at the 29th European Photovoltaic Solar Energy Conference, Hamburg, Germany, 22–26 September.

10.2

Encapsulant Materials for PV Modules

Michael Kempe
National Renewable Energy Laboratory, Golden, CO, USA

10.2.1 Introduction

Encapsulant materials used in photovoltaic (PV) modules serve multiple purposes. They physically hold components in place, provide electrical insulation, reduce or halt moisture ingress, optically couple superstrate materials (e.g., glass) to PV cells, protect components from mechanical stress by mechanically de-coupling components via strain relief, and protect materials from corrosion. To provide these functions, encapsulant materials must adhere well to all surfaces, remain compliant, and transmit light after exposure to temperature, humidity, and UV radiation exposure. Here, a brief review of some of the polymeric materials under consideration for PV applications is provided, with an explanation of some of their advantages and disadvantages.

10.2.2 Types of Encapsulant Materials

Many types of encapsulant resins have been considered for use in PV modules. When PV panels were first developed in the 1960s and the 1970s, the dominant encapsulants were based on polydimethyl siloxane (PDMS) (Dow Corning Corporation, 1979; Green, 2005). This was chosen because of its exceptional intrinsic stability against thermal decomposition and its resistance to degradation caused by ultraviolet (UV) radiation (Kempe, 2008). However, in an effort to reduce module materials and manufacturing costs, alternative materials were investigated and developed, leading to the emergence of poly(ethylene-co-vinyl acetate) (EVA) as the dominant PV encapsulant.

Recently, there has been renewed interest in using alternative encapsulant materials with some significant manufacturers switching from EVA to polyolefin elastomer-based (POE)

Photovoltaic Solar Energy: From Fundamentals to Applications, First Edition.
Edited by Angèle Reinders, Pierre Verlinden, Wilfried van Sark, and Alexandre Freundlich.

Companion website: www.wiley.com/go/reinders/photovoltaic_solar_energy

alternatives (Strevel *et al.*, 2013). The reasons for this switch include concerns over potential induced degradation (PID) (Pingel *et al.*, 2010; Hacke *et al.*, 2012; Koch *et al.* 2012; Reid *et al.*, 2013), the desire for faster lamination processes through the use of different resins or cure systems (Cattaneo *et al.*, 2015), or concerns over acetic acid production (Kempe *et al.*, 2007). PID is a general term for voltage-induced degradation. One of the most common forms of this is a decrease in shunt resistance in silicon-based PV cells for the negatively charged string. Diffusion of Na^+ from the glass superstrate is often implicated in this degradation pathway. This PID can be mitigated by the use of encapsulants with a resistivity greater than around 10^{15} Ω·cm or 10^{16} Ω·cm (Koch *et al.*, 2012; Reid *et al.*, 2013).

Many of the alternatives to EVA (see Figure 10.2.1), ionomer, polyvinyl butyral (PVB), and polyolefin POEs, have a backbone consisting of only carbon-carbon (C—C) bonds. Alternatively, thermoplastic polyurethane (TPU) formulations have nitrogen and oxygen incorporated into the backbone in the form of a urethane bond (R—NH—OOR'). The ester bond (R—COOR') is susceptible to hydrolysis; however, the presence of hydrolytically unstable bonds in the backbone TPUs is of greater concern because depolymerization can significantly affect the rheological properties of these materials. If the side groups of PVB or EVA become cleaved, one would expect to see stronger hydrogen bonding between polymer chains and surfaces. This can lead to embrittlement of polymers, however, a substantially greater extent of hydrolysis (compared to breaking of the backbone bonds in TPUs) must occur for these effects to be significant.

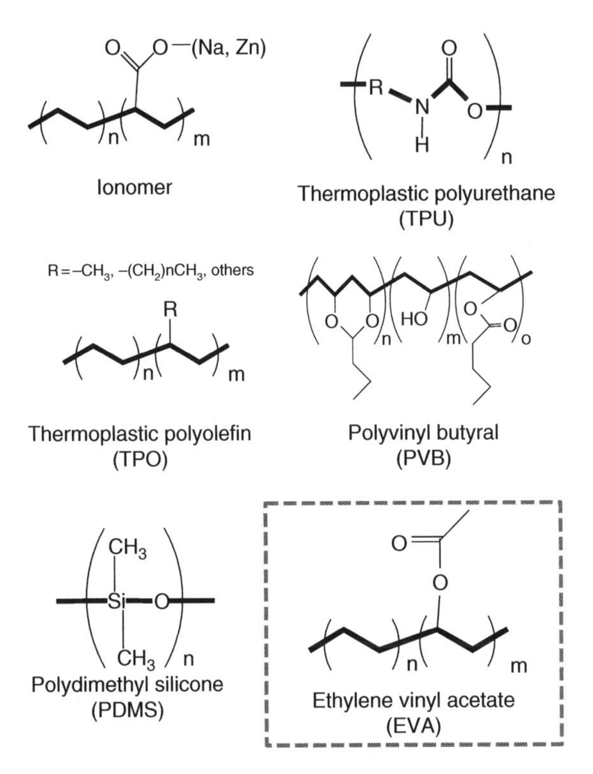

Figure 10.2.1 Structures of common PV encapsulant resins

In contrast, PDMS has a backbone consisting of alternating atoms of Si and O. Because the silicon atom is much larger than oxygen or carbon atoms, there is greater freedom of motion for rotation and bending of Si—C side-group and Si—O back-bone bonds for silicone-based polymers compared to hydrocarbon-based polymers. This enhanced mobility in PDMS results in polymers with extremely low glass transition temperatures and, so long as the cross-link density is low, they will have lower mechanical moduli. The glass transition is the temperature where polymer chain segments begin to have substantial mobility and can slide past one another. This further acts to lower the mechanical moduli, providing better protection against mechanical stress being applied to the cells. A lower mechanical modulus at all temperatures is needed to reduce the mechanical stress being applied to the cells. Additionally, the bond dissociation energy (the energy required to cleave a chemical bond) of Si—O is ~108 kcal/mol compared to 83 kcal/mol for C—C bonds. This contributes to the enhanced durability of silicone encapsulants relative to hydrocarbon-based materials.

Typically, ionomers, TPUs, TPOs, and PVBs are formulated as thermoplastic (non-cross-linked) materials, though there is no inherent reason why they could not be made to form cross-links and/or chemically bond to surfaces. For PVBs, plasticizers are also added to lower their mechanical moduli and to tailor their phase-transition temperatures. As is also summarized in (Miller *et al.*, 2010), TPUs and PVBs typically have a glass transition around or below room temperature and are therefore in a rubbery state during much of their use, and potentially are susceptible to shear-induced flow. TPUs and PVBs are typically designed to have a high viscosity at PV operating temperatures to prevent creep (Miller *et al.*, 2010). Ionomers and TPOs typically have a melt transition around 90° to 100 °C. Below the melt temperature, polyethylene segments in these polymers are aligned, forming physical cross-links whose formation is reversible upon heating.

To overcome concerns with polymer creep or flow at elevated temperatures, EVA and PDMS materials are typically formulated to form chemical cross-links. Additionally, for PDMS, a Pt-based catalyst combines vinyl groups (of vinyl-terminated PDMS) to silane groups of a polymethyl-co-dimethyl siloxane (see Figure 10.2.2). This chemistry will proceed at room temperature, but is significantly accelerated at elevated temperature. Chemical cross-linking restricts material flow to only occur when mechanical stresses are large enough to

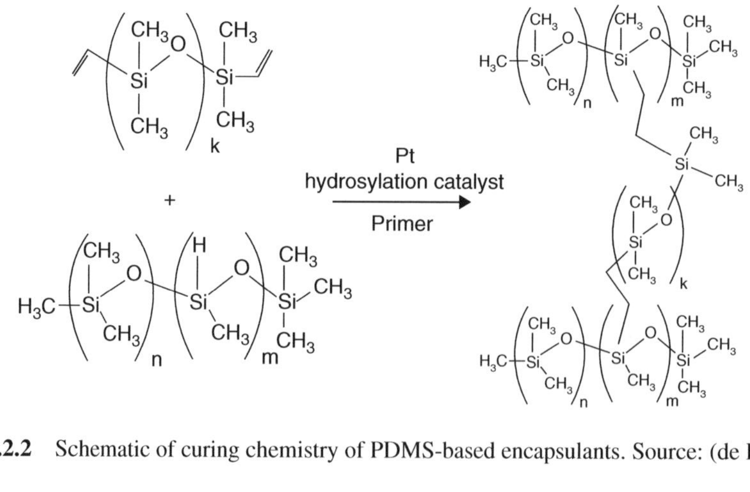

Figure 10.2.2 Schematic of curing chemistry of PDMS-based encapsulants. Source: (de Buyl, 2001)

break chemical bonds. Additionally, the use of chemical cross-links enables more effective use of primers to promote adhesion at surface interfaces.

Thus, a cross-linked system can be chemically bonded to surfaces, whereas a thermoplastic systems must rely on a combination of ionic, hydrogen, and/or Van der Waals forces for adhesion. When water reaches an interface between the polymer and an inorganic material, the polar water molecules will compete with the less polar polymer at adhesion sites. If the polymer is displaced by the water, delamination will occur. In contrast, with a chemically bonded encapsulant, chemical bonds must be broken in addition to the physical bonds, making it easier for chemically bonded, cross-linked encapsulants to be formulated for durable interfacial adhesion.

PDMS-based materials are inherently UV and thermally stable, but hydrocarbon-based materials (EVA, TPU, PVB, and ionomer) require stabilizers to be durable. An EVA formulation is not just simply EVA resin, but a complex mixture of components. A typical EVA formulation is shown in Figure 10.2.3 (Klemchuk *et al.*, 1997; Pern, 2000). The majority of the

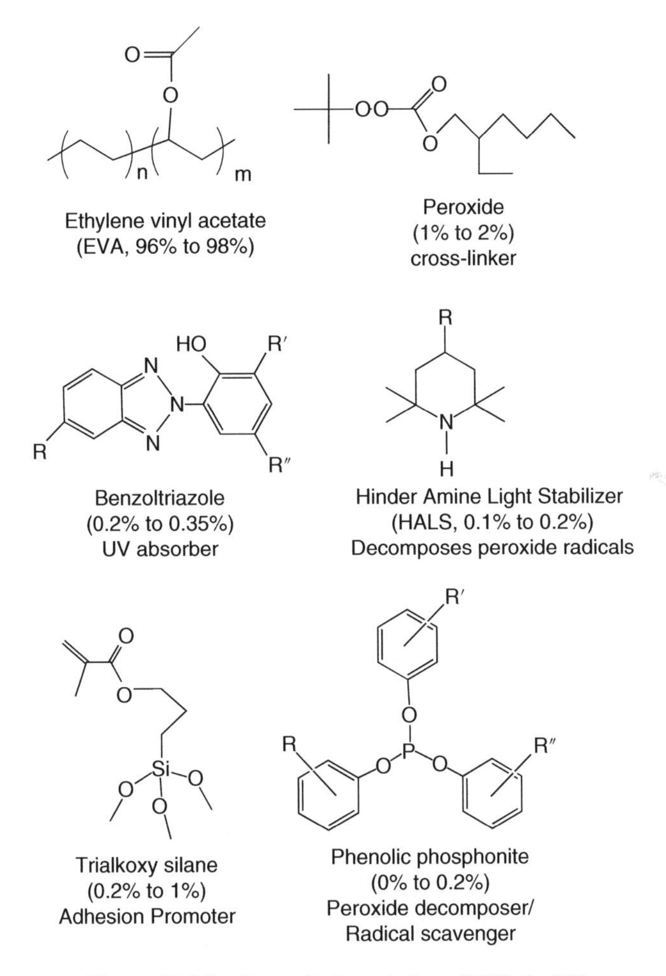

Figure 10.2.3 Example formulation of EVA for PV

material is the EVA resin. Typically a 27 wt% to 33 wt% vinyl acetate copolymer content EVA is used to provide a balance of properties, which include: a low glass transition, low modulus, low crystallinity/highly light transmittance resin, and a convenient melting temperature (45 °C to 65 °C), enabling easy melting for processing. EVA resins are also designed with molecular weight distributions and branching characteristics to facilitate extrusion into a film, which may minimize shrinkage in subsequent laminations.

Typically, about 1 to 2 wt% of an EVA film is a thermally activated peroxide used for cross-linking at elevated temperatures during lamination. The peroxide decomposes to produce radicals, which react with the polymer using non-specific chemical pathways to form cross-links. At temperatures above 140 °C, a typical peroxide such as tertbutyl-2-ethylhexyl-peroxycarbonate (TBEC) will decompose sufficiently to facilitate the cure within two minutes (Cuddihy *et al.*, 1983). The time required to heat the polymer in a module to this temperature range is therefore the most significant factor limiting the speed of lamination.

A trialkoxy silane is used to promote adhesion between EVA and inorganic surfaces. The silane end tends to be attracted to polar surface hydroxyl groups and is able to react with them, creating methanol as a leaving group and forming a covalent chemical bond in place of the hydroxyl group (Cuddihy *et al.*, 1986). The other two alkoxy groups may further react with other surface groups or with other trialkoxy silane groups, forming a three-dimensional network that ensures good adhesion. This interfacial structure also helps to passivate inorganic surfaces against corrosion by limiting the movement of corrosion by-products away from the interface.

The effects of UV radiation are mitigated by the inclusion of a UV absorber based on functional groups such as a benzotriazole or benzophenone. Early work on EVA formulations found an interaction between benzophenone, lupersol 101 (peroxide), and a phenyl phosphonite that had a significant tendency to form chromophores (Holley and Agro, 1998). These early formulations resulted in extreme degradation of early PV modules made before 1990, as demonstrated in the 7 MW PV installation at Carrizo Plains, California. Initially, the loss in power of the modules was attributed primarily to EVA discoloration (Gay and Berman, 1990), but subsequent analysis demonstrated that solder joint breakage was the more significant problem (Wohlgemuth and Petersen, 1993).

Finally, a hindered amine light stabilizer (HALS) and possibly a phenolic phosphonite may be added as antioxidants. The HALS acts to decompose peroxide radicals that may form due to thermal or UV exposure. In this process, the HALS is not consumed as opposed to the phenolic phosponite, which is oxidized to produce phosphate and phenols.

10.2.3 Polymer Light Transmittance

PV encapsulants optically couple PV cells to a transparent superstrate such as glass through the elimination of high refractive index change interfaces (e.g., air/glass and air/cell) where the reflection of light would be higher. Therefore, high transmittance encapsulants are desirable. Hemispherical transmittance of light through encapsulant samples laminated between two pieces of 3.2 mm thick glass was measured to enable comparison of different materials (see Table 10.2.1) (Kempe, 2010). From these and similar measurements of bare glass, the optical absorptivity of the polymers was determined (Figure 10.2.4). For EVA, one can see that most of the absorption occurs below 400 nm. With this, the photon transmission through a glass superstrate and 0.45 mm of encapsulant to a hypothetical cell interface was estimated.

Table 10.2.1 Solar photon (300–1100 nm) weighted average optical density determined from transmittance measurements through polymer samples of various thickness (1.5–5.5 mm) laminated between two pieces of 3.18-mm-thick, Ce-doped, low-Fe glass (Kempe, 2010)

Encapsulant	AM 1.5 Solar Photon and x-Si QE Weighted Absorptivity	Transmission to Cells through 3.18 mm glass and 0.45 mm Encapsulant	Comments
	(1/mm)	%	
GE RTV615	0.000 ± 0.003	94.8 ± 0.3	PDMS, Addition Cure
Dow Corning Sylgard 184	0.002 ± 0.004	94.7 ± 0.3	PDMS, Addition Cure
Dow Corning 527	0.004 ± 0.003	94.7 ± 0.3	PDMS Gel, Addition Cure
Polyvinyl Butyral	0.011 ± 0.005	94.3 ± 0.4	
EVA	0.012 ± 0.005	94.3 ± 0.4	
Thermoplastic Polyurethane	0.024 ± 0.004	93.8 ± 0.3	
NREL Experimental	0.027 ± 0.006	93.7 ± 0.4	Poly a-Olefin Copolymer
Thermoplastic Ionomer #1	0.049 ± 0.007	92.7 ± 0.4	Copolymer of Ethylene and Methacrylic acid
DC 1199 SSL	0.064 ± 0.004	92.1 ± 0.3	PDMS, One Part Neutral and Condensation Cure
DC 700	0.068 ± 0.004	92.0 ± 0.3	PDMS, Acetic Acid Condensation Cure
Thermoplastic Ionomer #2	0.149 ± 0.007	88.7 ± 0.4	Copolymer of Ethylene and Methacrylic acid

The PDMS samples had the best transmittance, about 0.6% better than the best hydrocarbon-based materials. Part of this difference is attributable to the absence of UV absorbers in PDMS. This analysis considered only normal transmittance. A more thorough analysis, by McIntosh *et al.* (2009), using ray tracing models and considering multiple reflections, non-normal incidence, and reflections off the backsheet between cells, estimated this difference to be as high as 1.5%.

10.2.4 UV Durability

Depending on its composition, glass may block much of the UV-B radiation, but typically blocks very little of the UV-A (Gay and Berman, 1990; Kempe *et al.*, 2009b). Therefore, the UV stability of the encapsulation material used in front of the cell is important. Figure 10.2.5 shows the results of a highly accelerated stress test designed to investigate the possible use of non-silicone-based encapsulants in medium-concentration PV applications (Kempe *et al.*, 2009a). Sample encapsulants were laminated between two pieces of low-Fe, UV-transmitting glass while monitoring the transmittance weighted to the photon flux of the solar spectrum and the quantum efficiency of a typical crystalline silicon cell. They were exposed to 42 UV-suns at a temperature between 80° and 95 °C. Here, a UV-sun is defined as the

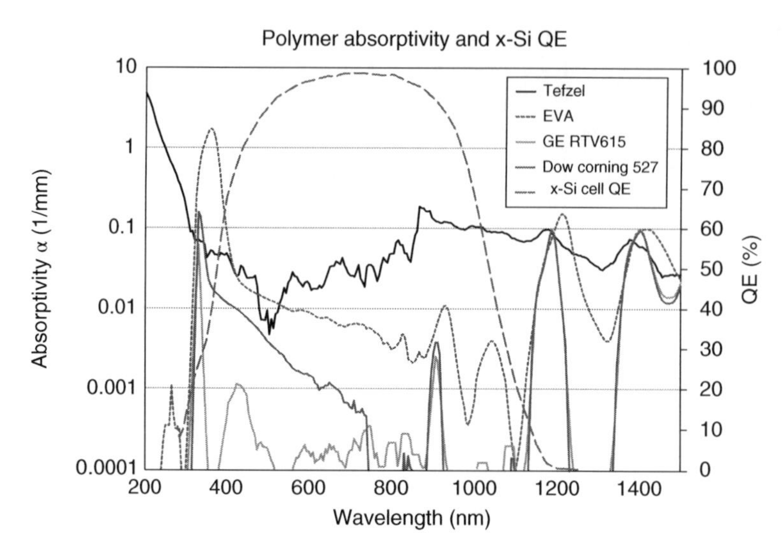

Figure 10.2.4 Plot of polymer absorptivity as a function of wavelength alongside the internal quantum efficiency of a typical crystalline silicon cell. (*See insert for color representation of the figure*)

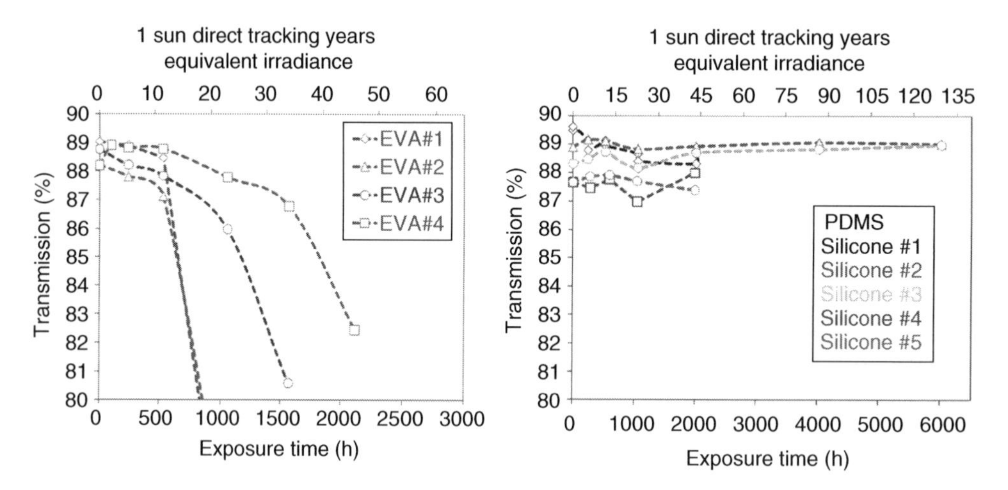

Figure 10.2.5 Solar and crystalline silicon quantum efficiency-weighted transmittance of test samples exposed to 42 global-UV suns in a Xenon arc Weather-Ometer. Samples consist of 0.5 mm encapsulant laminated between two 2.5-cm-square, 3.18-mm-thick, low-Fe, non-Ce glass samples (i.e., highly UV transmissive glass). The top axis corresponds to the amount of UV radition that would be seen with a system tracking the sun and utilizing only the direct spectrum Source: (Kempe *et al.*, 2009a)

ratio of the xenon arc lamp to the AM1.5 spectrum integrated between 300 nm and 400 nm. None of the five different PDMS silicone samples demonstrated any significant loss in transmittance after up to 6000 h of exposure. Under the same conditions, the four different EVA formulations showed very significant degradation after only 750 to 1700 h of exposure.

This demonstrates the inherently greater stability of PDMS relative to EVA as the PDMS samples do not contain stabilization chemistries such as antioxidants and UV absorbers.

Also important in Figure 10.2.5 is the great variation in performance of the EVA formulations provided by different manufacturers. This is attributable to differences in either the type or the amount of additives similar to those described in Figure 10.2.3. Considering the extreme conditions of this test, these formulations performed quite well. Similar experiments were also performed with PVB, TPU, and ionomer formulations (Kempe *et al.*, 2009a). Here PVB performed exceptionally poorly, TPU was comparable to EVA, and the ionomer was more durable than EVA. It must also be kept in mind that this test addressed only light transmittance, which is only one of several important characteristics such as adhesion, and that the results are more accurately applied to low concentration PV applications.

10.2.5 Resistivity

Typically the backsheet provides electrical insulation to a PV module, but the volume resistivity of encapsulants may be relevant to electrical insulation and to module durability Volume resistivity is an intrinsic property that quantifies how strongly a given material opposes the flow of electrical current. In practice, it is determined by measuring the electrical resistance (R) across a material, multiplying it by the cross-sectional area (A), and dividing by the length (l) of the electrical path as:

$$\rho = \frac{RA}{l} \tag{10.2.1}$$

Relatively low resistance in encapsulant materials has been linked to electrochemical corrosion and potential induced degradation (PID) (Hacke *et al.*, 2012; Koch *et al.*, 2012; Mon and Ross, 1985; Mon *et al.*,1987; Pingel *et al.*, 2010; Reid *et al.*, 2013). Mon and his colleagues (Mon and Ross, 1985; Mon *et al.*, 1987) found that, for PVB and EVA, temperature had a much greater effect on resistivity than absorbed water. Mon *et al.* also found good correlation between degradation induced by electrochemical corrosion and total leakage from cells to the frame in amorphous silicon-based PV modules.

There is a great range in the value of resistivity among polymers, which can be a significant determining factor for PID. Reid *et al.* (2013) investigated the resistivity of several EVAs and POEs from Specialized Technology Resource (STR) and from some competitors (Table 10.2.2). For EVAs, they found that they could vary the volume resistivity by almost two orders of magnitude, but the POEs still had resistivities about another two orders of magnitude higher. This indicates that, unless very drastic and expensive steps are made to increase the resistivity of EVA, there may be practical limitations to what is achievable. This is a good motivator for manufacturers to switch to POEs, which several large manufacturers have already done.

Reid *et al.* also investigated the ability of these different encapsulants to make cells more resistant to PID. While there is definitely a good correlation between volume resistivity and PID resistance, the type of impurity enabling conductivity in the poymer has a significant effect on the volume resistivity. To achieve good resistance to PID, both a good choice of encapsulant and improved cells is required.

Table 10.2.2 Data taken from Reid *et al.* (2013) showing the variability in resistivity of a number of PV encapsulant materials

Code	Encapsulant Grade Name	Volume Resistivity at 23 °C (Ω·cm)	WVTR at 25 °C ($g/m^2/day$)
EVA-1	STR 15295P	5×10^{13}	23
EVA-2	STR 15420P	2×10^{14}	23
EVA-3	STR 15455P	5×10^{14}	23
EVA-4	STR 15580P	3×10^{15}	18
EVA-5	Chinese Source	1×10^{14}	25
EVA-6	Japanese Source	1×10^{15}	17
POE-1	STR X-28-138	1×10^{17}	1.5
POE-2	STR X-44 Series	1×10^{16}	1.7
POE-3	STR X-31 Series	1×10^{16}	5.2
POE-4	Chinese Source	1×10^{13}	2.3

10.2.6 Moisture Ingress Prevention

Typical transparent encapsulant materials by themselves do not completely prevent water vapor ingress (Coyle, *et al.*, 2009; Kempe, 2006; Kempe et al., 2010), but if they are well adhered, they will prevent the accumulation of liquid water, providing protection against corrosion as well as electrical shock. For example, at 25 °C EVA has a diffusivity (D) for water of $1.0.10^{-6}$ cm^2/s. The characteristic time (τ) for the center of a square cell to reach half of its equilibrium value as a function of cell width (X) is given by:

$$\tau = \frac{X^2}{D} \tag{10.2.2}$$

Using half the cell width for $X=7.8$ cm, the characteristic time for moisture ingress in front of a PV cell is 1.93 years. Thus, PV cells in typical crystalline silicon modules will be exposed to significant concentrations of water vapor for the majority of their useful life. Therefore, good adhesion is more important than low permeation as most module constructions will equilibrate with moisture in a time frame on the order of less than a year (Kempe, 2006).

Even for a module construction with an impermeable backsheet (e.g., glass), water may still diffuse in from the edges. Figure 10.2.6 shows estimates of how far moisture can penetrate different PV polymeric materials as a function of temperature. Of the encapsulant materials evaluated here, Ionomer #1 has the lowest diffusivity, and even if it was installed in a cold climate with an average effective temperature of 20 °C (Kempe *et al.*, 2014), moisture would still penetrate to a depth of 4 or 5 cm over the course of 20 years. Depending on the sensitivity of a PV material, this may not be sufficient (Coyle *et al.*, 2009).

In another set of experiments, encapsulant materials and edge seal materials composed of polyisobutylene (PIB) filled with desiccants were laminated between two pieces of glass, one of which had a 100 nm film of Ca metal deposited on the surface (Kempe *et al.*, 2010). As moisture permeates the polymer, the Ca metal is oxidized to transparent CaO giving a simple visual indicator of the extent of moisture ingress. As shown in Figure 10.2.7, these PIB-based edge seal materials are an order of magnitude better than the ionomer with respect

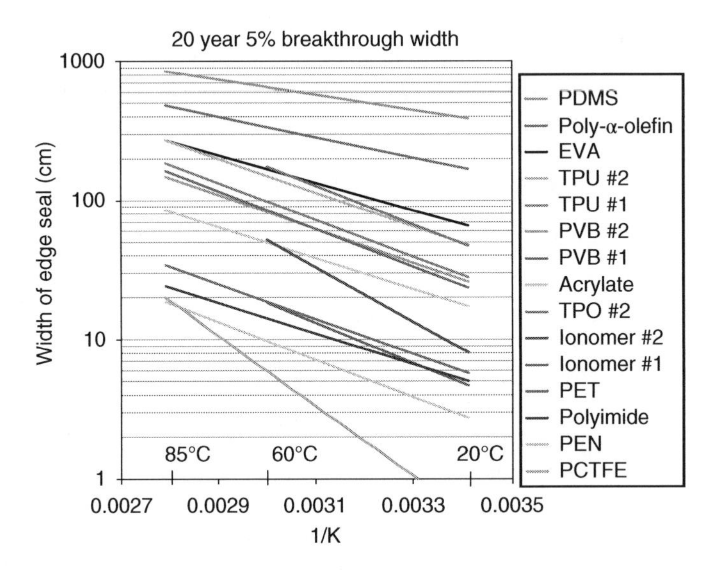

Figure 10.2.6 Width of edge seal made from different materials that would be necessary to keep moisture below 5% of equilibrium values at a given temperature. Source: (Kempe *et al.*, 2006). (*See insert for color representation of the figure*)

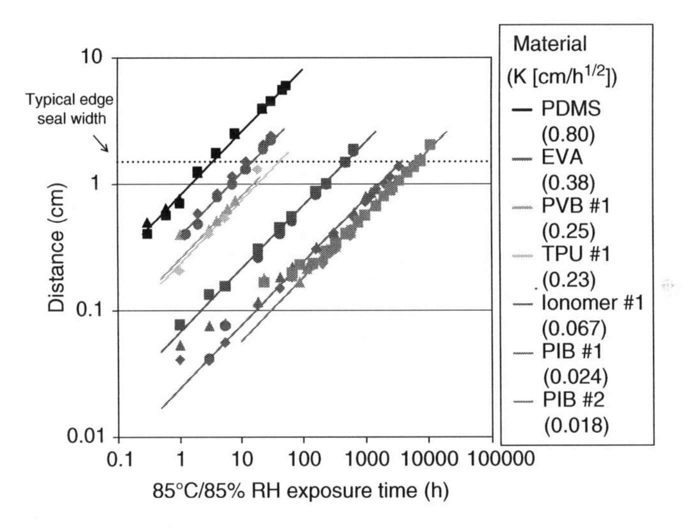

Figure 10.2.7 Penetration depth of moisture between glass plates laminated with different materials as measured by the oxidation of a 100 nm film of Ca. Source: (Kempe *et al.*, 2010). (*See insert for color representation of the figure*)

to their ability to prevent moisture ingress. This should enable them to restrict moisture ingress to less than a cm from the perimeter of a module for hot and humid environments over its expected lifetime if the material properties do not change with aging in the field. If a PV material is very sensitive to moisture (Coyle *et al.*, 2009), a desiccant-filled PIB edge seal material is needed to prevent moisture ingress.

10.2.7 Conclusion

An encapsulant provides optical coupling of PV cells and protection against environmental stress. Polymers must perform these functions under prolonged periods of high temperature, humidity, and UV radiation. The base polymer structure is the first thing to consider as it dominates subsequent properties. However, encapsulant films of the same base polymer have varying amounts and types of stabilization additives, resulting in different durabilities among manufacturers.

With the development of thin film-based PV technologies, concerns with moisture ingress are important. Some effort has been focused on flexible barriers with permeation rates less than 10^{-4} g/m^2/day (Coyle *et al.*, 2009; Kempe, 2006), but even with an impermeable frontsheet and backsheet, moisture vapor will penetrate a module unless extremely low permeability materials are used such as desiccant-filled PIBs.

EVA is currently the dominant encapsulant chosen for PV applications, not because it has the best combination of properties, but because it is an economical option with an established history of acceptable durability. Getting new products onto the market is challenging because there is no room for dramatic improvements (e.g., transmittance or price reductions), and one must balance the initial cost and performance with the unknowns of long-term service life.

Acknowledgments

This work was supported by the U.S. Department of Energy under Contract No. DE-AC36-08-GO28308 with the National Renewable Energy Laboratory.

List of Symbols

Symbol	Meaning	Unit
A	Sample test area	m^2
D	Diffusivity	cm^2/s
l	Sample thickness	m
τ	Characteristic diffusion time to reach half of the equilibrium value	s
R	Resistance	Ω
P	Volume resistivity	Ω cm
X	Characteristic distance for diffusion	m
$WVTR$	Water Vapor Transmission Rate	g/m^2/day

List of Acronyms

Acronym	Meaning
EVA	Poly(Ethylene-co-Vinyl Acetate)
HALS	Hindered Amine Light Stabilizer
PDMS	Polydimethylsiloxane
PIB	Polyisobutylene

Acronym	Meaning
PID	Potential Induced Degradation
POE	Polyolefin elastomer
PVB	Polyvinyl Butyral
TPU	Thermoplastic Polyurethane
UV	Ultraviolet
UV-A	The part of the solar spectrum between 320 nm and 400 nm
UV-Sun	The integrated light intensity between 300 and 400 nm relative to that in AM1.5

References

Cattaneo, G., Faes, A., Li, H.Y. *et al.* (2015) Lamination process and encapsulation materials for glass-glass PV module design. *Photovoltaics International*, **27**.

Coyle, D.J., Blaydes, H.A., Pickett, J.E. *et al.* (2009) Degradation kinetics of CIGS solar cells. *Proceedings of the 2009 34th IEEE Photovoltaic Specialists Conference (PVSC 2009)*, pp. 001943–001947. DOI: 10.1109/pvsc.2009.5411551

Cuddihy, E.F., Coulbert, C.D., Gupta, A., and Liang, R. (1986) Electricity from photovoltaic solar cells, flat-plate solar array project final report, Volume VII: Module encapsulation. *DOE/JPL-1012-125.*

Cuddihy, E.F., Gupta, A., Coulbert, C.D. *et al.* (1983) Applications of ethylene vinyl acetate as an encapsulation material for terrestrial photovoltaic modules. *DOE/JPL/1012-87 (DE83013509).*

de Buyl, F. (2001) Silicone sealants and structural adhesives. *International Journal of Adhesion and Adhesives*, **21** (5), 411–422.

Dow Corning Corporation (1979) Develop silicone encapsulation systems for terrestrial silicon solar arrays. *DOE/JPL/954995-80/6.*

Gay, C.F. and Berman, E. (1990) Performance of large photovoltaic systems. *Chemtech*, **20** (3), 182–186.

Green, M.A. (2005) Silicon photovoltaic modules: A brief history of the first 50 years. *Progress in Photovoltaics*, **13** (5), 447–455.

Hacke, P., Smith, R., Terwilliger, K. *et al.* (2012) Testing and analysis for lifetime prediction of crystalline silicon PV modules undergoing degradation by system voltage stress. Paper presented at the 38th IEEE Photovoltaic Specialists Conference (PVSC), 3–8 June.

Holley, W.W. and Agro, S.C. (1998) Advanced EVA-based encapsulants, Final Report January 1993–June 1997. *NREL/SR-520-25296.*

Kempe, M.D. (2006) Modeling of rates of moisture ingress into photovoltaic modules. *Solar Energy Materials and Solar Cells*, **90** (16), 2720–2738.

Kempe, M.D. (2008) Accelerated UV test methods for encapsulants of photovoltaic modules. *33rd IEEE Photovoltaic Specialist Conference Program.*

Kempe, M.D. (2010) Ultraviolet light test and evaluation methods for encapsulants of photovoltaic modules. *Solar Energy Materials and Solar Cells*, **94** (2), 246–253. DOI: 10.1016/j.solmat.2009.09.009

Kempe, M.D., Dameron, A.A., and Reese, M.O. (2014) Evaluation of moisture ingress from the perimeter of photovoltaic modules. *Progress in Photovoltaics: Research and Applications*, **22** (11), 1159–1171. DOI: 10.1002/pip.2374.

Kempe, M.D., Dameron, A.A., Moricone, T.J., and Reese, M.O. (2010) Evaluation and modeling of edge-seal materials for photovoltaic applications. Paper presented at 35th IEEE PVSC, Honolulu, HI, 20–25 June

Kempe, M.D., Jorgensen, G.J., Terwilliger, K.M. *et al.* (2007) Acetic acid production and glass transition concerns with ethylene-vinyl acetate used in photovoltaic devices. *Solar Energy Materials and Solar Cells*, **91** (4), 315–329.

Kempe, M.D., Kilkenny, M., Moricone, T.J., and Zhang, J.Z. (2009a) Accelerated stress testing of hydrocarbon-based encapsulants for medium-concentration CPV applications. Paper presented at Philadelphia, PA, 7–12 June.

Kempe, M.D., Moricone, T., and Kilkenny, M. (2009b) Effects of cerium removal from glass on photovoltaic module performance and stability. *Proceedings of the SPIE – The International Society for Optical Engineering, 7412*. DOI: 10.1117/12.825699

Klemchuk, P., Ezrin, M., Lavigne, G., *et al.* (1997) Investigation of the degradation and stabilization of EVA-based encapsulant in field-aged solar energy modules. *Polymer Degradation and Stability*, **55** (3), 347–365.

Koch, S., Berghold, J., Okoroafor, O. *et al.* (2012) Encapsulant influence on the potential induced degradation of crystalline silicon cells with selective emitter structures, 1991 to 1995. *27th European Photovoltaic Solar Energy Conference and Exhibition*. DOI: 10.4229/27thEUPVSEC2012-2CV.7.4

McIntosh, K.R., Cotsell, J.N., Cumpston, J.S. *et al.* (2009) An optical comparison of silicone and EVA encapsulants for conventional silicon PV modules: a ray-tracing study. Paper presented at Philadelphia, PA, 7–12 June.

Miller, D., Kempe, M.D., Glick, S.H., and Kurtz, S. (2010) Creep in photovoltaic modules: examining the stability of polymeric materials and components. Paper presented at 35th IEEE PVSC, Honolulu, HI, 20–25 June.

Mon, G. and Ross, R.G., Jr. (1985) Electrochemical degradation of amorphous-silicon photovoltaic modules. Paper presented at conference, Las Vegas, 21–25 October.

Mon, G., Wen, L., and Ross, R., Jr. (1987,) Encapsulant free-surfaces and interfaces: critical parameters in controlling cell corrosion [photovoltaic modules]. Paper presented at conference, New Orleans, LA, 4–8 May.

Pern, F.J. (2000) Composition and method for encapsulating photovoltaic devices. Patent# 6,093,757.

Pingel, S., Frank, O., Winkler, M. *et al.* (2010) Potential Induced Degradation of solar cells and panels. Paper presented at the 35th IEEE Photovoltaic Specialists Conference (PVSC), Honolulu, HI, 20–25 June.

Reid, C., Ferrigan, S., Fidalgo, I., and Woods, J.T. (2013) Contribution of PV encapsulant composition to reduction of potential induced degradation of crystalline silicon PV cells. Paper presented at 28th European Photovoltaic Solar Energy Conference and Exhibition.

Strevel, N., Trippel, L., Kotarba, C. and Khan, I. (2013) Improvements in CdTe module reliability and long-term degradation through advances in construction and device innovation. *Photovoltaics International*, 22nd edition.

Wohlgemuth, J.H. and Petersen, R.C. (1993) Reliability of EVA modules. Paper presented at Louisville, KY, 10–14 May.

10.3

Reliability and Durability of PV Modules

Sarah Kurtz
National Center for Photovoltaics, U.S. Department of Energy National Renewable Energy Laboratory, Golden, CO, USA

10.3.1 Introduction

10.3.1.1 Importance and Challenge of PV Module Reliability and Durability

Each year the world invests tens of billions of dollars or euros in PV systems with the expectation that these systems will last approximately 25 years. Solar energy has the potential to meet the world's energy needs if reliable products can be integrated into a stable grid in a cost-effective manner. PV module reliability has been positively described (Hasselbrink *et al.*, 2013) and continues to be important because:

- excellent PV module reliability is a prerequisite for reliability of a solar-powered grid;
- excellent PV module reliability supports continued and increased investment in PV;
- a thorough understanding of PV reliability enables continued reductions in the levelized cost of electricity (LCOE) for PV-generated electricity.

Standard tests, such as IEC 61215 ("Crystalline silicon terrestrial photovoltaic (PV) modules – Design qualification and type approval, "2005) and IEC 61730 ("Photovoltaic (PV) module safety qualification," 2004), see also Chapter 13.4, identify most problems that lead to early failures.

Photovoltaic Solar Energy: From Fundamentals to Applications, First Edition.
Edited by Angèle Reinders, Pierre Verlinden, Wilfried van Sark, and Alexandre Freundlich.

Companion website: www.wiley.com/go/reinders/photovoltaic_solar_energy

10.3.1.2 *Challenge of Determining Reliability and Service Lifetime*

Although the disciplines of reliability, quality, and service life prediction have been well established for numerous products, a full understanding of these is currently challenging for PV modules because:

- the desired service lifetimes are decades, preventing direct verification of lifetime predictions;
- the use environment of PV modules can be highly variable so that a service life prediction or reliability assessment for one location and mounting configuration is not universally applicable;
- the PV industry is rapidly changing product designs in order to reduce cost and boost performance. The product cycle is typically too short to allow for lengthy testing.
- the size and cost of PV modules prevent testing of large numbers of samples.

10.3.2 PV Module Durability, Quality, and Reliability Issues

A number of excellent reviews can be found in the literature summarizing the types of failures that are commonly observed for PV modules (Hasselbrink *et al.*, 2013; Kurtz *et al.*, 2013b; Köntges *et al.*, 2014; Makrides *et al.*, 2014). In the 1970s and 1980s, many PV modules failed quickly in the field. However, the failure rates dropped when accelerated stress testing was used to qualify module designs for adequate durability. Rosenthal reported how early failure rates of close to 50% were reduced to ~1% by using the "Block V" test procedure for qualifying a module design (Rosenthal *et al.*, 1993). Today's failure rates can be highly variable depending on the care taken by the manufacturer, but failure rates <0.1% can be common. For example, Hasselbrink's study of three million module-years of field data reported <0.5% failure rates overall, and <0.005% for the best module design (Hasselbrink *et al.*, 2013).

Although very low failure rates are commonly reported, higher failure rates are also reported, and there is emerging evidence that deployment in hotter climates often leads to degradation greater than expected based on the warranty. For example, a study (Yedidi *et al.*, 2014) of>3000 modules deployed for 16 years in a hot, dry climate quantified safety failure rates of 0.5% to 1.7%, which is just slightly higher than reported by Hasselbrink. However, the more striking effect of the climate appears to be in the degradation rates. Yedidi reported degradation that frequently topped 1%/y, leading to failure according to the warranty claims for about 75% of the modules (Yedidi *et al.*, 2014). Similarly, a study of modules deployed in different parts of India concluded that hot climates resulted in higher degradation rates (~1.4%/y) than cooler climates (~0.2%/y) (Dubey *et al.*, 2014b). There are reports of thousands of measurements of degradation rates in the literature (Chandel *et al.*, 2015; Jordan and Kurtz, 2013). Care must be taken in interpreting these reports since it is often difficult to detect the small changes in performance for modules that have been deployed a small number of years, and modules that were deployed 20 years ago may have started with power outputs that differed significantly from the power rating on the nameplates. Nevertheless, statistical analysis can identify trends that can be used as the basis of risk calculations or for setting research agendas.

Jordan analysed climate effects on the reported degradation rates for PV modules, as shown in Figure 10.3.1 (Jordan *et al.*, 2012). Modules deployed in the desert frequently show larger

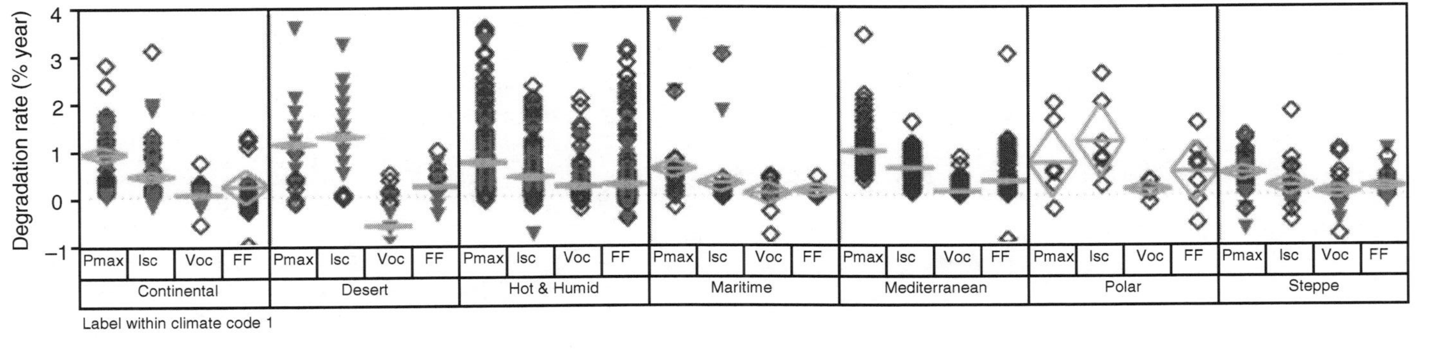

Figure 10.3.1 Summary of degradation rates reported in the literature for PV modules deployed in different climate zones. The degradation is described according to the changes (under standard test conditions) in short-circuit current (Isc), open-circuit voltage (Voc), fill factor (FF), and power (Pmax). Source: (Jordan *et al.*, 2012). (*See insert for color representation of the figure*)

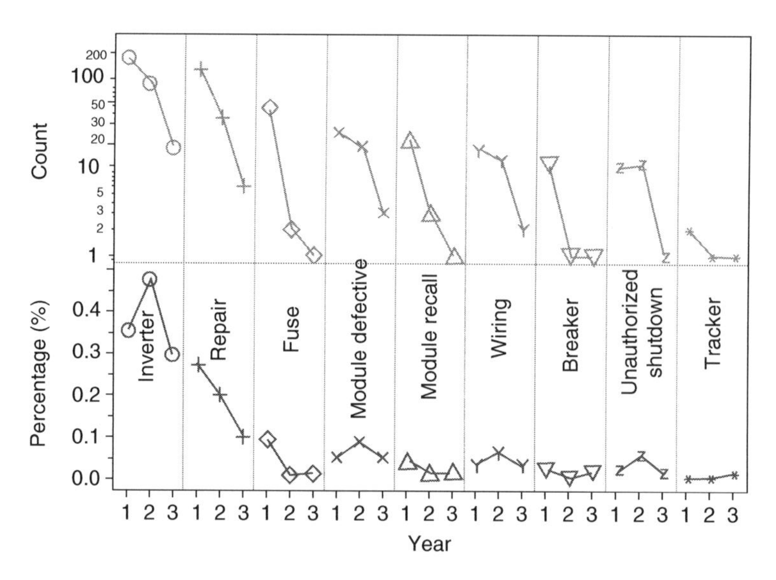

Figure 10.3.2 Statistics of PV system problems that were flagged in a study of ~50,000 systems. Source: (Jordan and Kurtz, 2014)

degradation of the Isc, which is commonly linked to discoloration of the encapsulant, motivating a study of the effects of temperature on encapsulant discoloration (see below). Statistical analyses such as described in Figure 10.3.1 guide the prioritization of research and standards development by identifying the types of problems that are observed most commonly in the field.

While module degradation or failure can have debilitating implications on the financial health of a PV project, it is important to understand that PV system degradation may not be identical to module degradation. In some cases, the PV system degradation may mirror the degradation of the worst modules rather than of the average module (Jordan *et al.*, 2015). In other cases, PV system performance may be degraded by causes unrelated to the modules (Figure 10.3.2). A study of ~50,000 PV systems deployed in the USA between 2009 and 2012 found that problems with the modules were reported in $<0.1\%$ of the systems each year; issues were reported with the inverters much more frequently (Figure 10.3.2).

Here, we discuss key failure/degradation mechanisms selected to highlight how the kinetics of failure rates can and cannot be confidently predicted. It is essential to be able to test the design to demonstrate the adequate robustness using accelerated stress tests that can be completed in a few months; it is also essential to test products coming from the production line to ensure that the robustness is duplicated using tests that identify deviations from the intended design in a timely way (Table 10.3.1).

10.3.2.1 Electrical Circuit Failures

A number of studies (Hasselbrink *et al.*, 2013; Dubey *et al.*, 2014a) have identified electrical discontinuities within the laminate, including ribbon, solder bond, and cell breakage (Table 10.3.1) to be common causes of PV module failure. Ribbon breakage results from

Table 10.3.1 Failure mechanisms, related test methods, and example of current research needs

Issue	Accelerated test	Acceleration factor*	Quality control	Research needs
Ribbon breakage	Mechanical cycling	Mechanical cycling can approach X100,000	Test incoming ribbons; destructive module test for process control	Non-destructive test
Solder bond failure	Thermal cycling	Maybe X100–X250	Solder and flux material qualification; test solder bond toughness	Quantify acceleration factor
Cell cracking followed by power loss	Mechanical stress, then thermal cycling and humidity freeze	X10–X20; seen in the field in<2 years	Electroluminescent imaging identifies cracks after accelerated stress	Need to identify which cracks are acceptable
Encapsulant discoloration	UV exposure at elevated temperature	~X10	Control encapsulant time on shelf and contamination/ formulation	Understand chemistry that causes discoloration so that it can be controlled
Corrosion	UV then humidity-freeze causes delamination, then damp heat causes corrosion	Maybe ~X100	Control of incoming materials, surface preparation and lamination process	Best accelerated test sequence

*Acceleration factor is estimated from the expected time in the field to failure divided by the time in the test lab to duplicate that failure.

mechanical fatigue (from thermal cycling, snow loading, and other mechanical deformation) of the ribbons that connect the cells and, in at least some situations, the robustness of the design can be tested in a highly accelerated fashion using mechanical cycling with a cycle time on the order of 1 second (Bosco *et al.*, 2013). This may be the only common failure mechanism for which the accelerated test can be completed in a very short time (1/100,000 of the associated time in the field, providing an acceleration factor of 100,000) without inducing irrelevant failures.

In contrast, mechanical cycling does not duplicate thermal fatigue of solder bonds, so accelerated testing of solder bonds benefits from thermal cycling, even though a thermal cycle typically requires hours rather than seconds, reducing acceleration factors to ~100. Thermal cycling to test for 25-year solder bond durability can be completed within the product development time frame of<6 months, but does not provide timely feedback for the Quality Management System (QMS). Thus, for the QMS to give timely feedback, faster measurements (such as checking solder composition and solder bond strength) are needed (Table 10.3,1). The tests and research needs shown in Table 10.3.1 are described and explained below.

Especially as silicon wafers have been thinned to reduce cost, cracked cells have become increasingly problematic. The cells may crack during tabbing and stringing, lamination, transportation, during installation, or deployment during times of high mechanical load (in wind or snow). In some cases, the function of the module is unaffected at the time the cells are cracked,

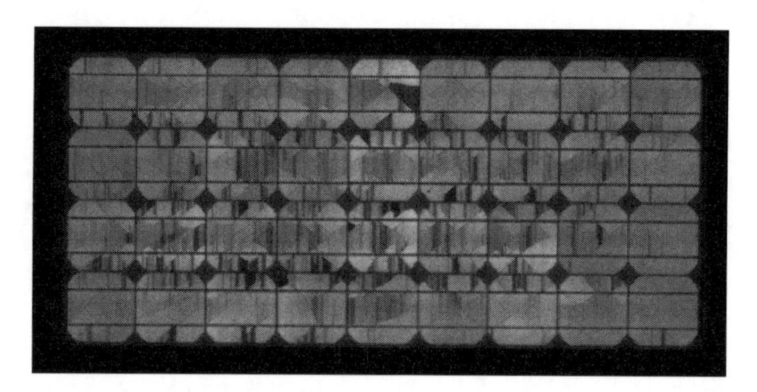

Figure 10.3.3 Silicon PV module after ~2 years in the field, showing about 9% degradation in power at standard test conditions. Source: courtesy of (Marion and Glick, 2013)

resulting in the module passing quality control at the module and system level. However, subsequently, probably because of thermal cycling, some gridlines may break, disconnecting the cracked piece from the rest of the cell, resulting in decreased power production (see Figure 10.3.3 for a module that was deployed for ~ 2 years).

These (and other electrical problems with connections in the junction box and diode failures) often dominate the statistics of PV module field failures, but some companies have demonstrated success in reducing failure rates to < 0.1%, reflecting the substantial knowledge of how to qualify a design and control the manufacturing process for these types of failures (DeGraaff, Lacerda, and Campeau, 2011; Hasselbrink *et al.*, 2013; Kurtz *et al.*, 2013b).

10.3.2.2 Encapsulant Discoloration and Delamination

While electrical issues are cited as the most common cause of outright failure, an even greater proportion of modules exhibit discoloration and/or delamination (Jordan and Kurtz, 2013). These do not usually cause module failure, but cause decreased output and may be a key contributor to the slow, ~0.5%/y degradation typically observed for PV modules. The degradation of crystalline silicon modules is most commonly observed to be a decrease in the short-circuit current, which may be explained by the discoloration and/or delamination (Figures 10.3.4 and 10.3.5).

Ultra-violet (UV) light-induced discoloration of ethylene-vinyl acetate (EVA, the most commonly used encapsulant material) is usually determined more by the additives (impurities) in the formulation than by the host polymeric material. Additionally, the detailed mechanism is highly dependent on in- and out-diffusion of reactants (such as oxygen) and reactant products, the temperature and the spectrum of the UV light. Figure 10.3.4 shows that a test result may depend on the size and geometry of the test sample since oxygen bleaching may reduce discoloration. The end result reflects competition between discoloration and bleaching processes and if these are not accelerated equally, the result obtained by accelerated testing may not correlate with field results. The conventional wisdom for UV-induced weathering is that acceleration factors higher than 10 are unlikely to correlate with field results, suggesting a stress test longer than 2 years to qualify new designs when a full lifetime dose is required to

Figure 10.3.4 Silicon PV module after ~20 years in the field. (*See insert for color representation of the figure*)

Figure 10.3.5 Silicon PV module after 22 years in Florida. Source: Courtesy of Wohlgemuth

detect a problem. However, in the case of EVA discoloration, which is commonly observed to be a linear change rather than a rapid change at the end of life, good success has been found with identifying susceptibility to discoloration after a shorter exposure (Reid, Bokria, and Woods, 2013).

New encapsulant materials are being introduced into PV modules. Each needs to be tested for potential issues with discoloration. If the kinetics of the discoloration are fairly similar, then the same stress test may be used for all encapsulant materials, but this is still a topic of research.

10.3.2.3 Corrosion

Corrosion is sometimes observed, especially for modules deployed in warm and humid climates or close to the ocean. For EVA-encapsulated modules, corrosion is observed to follow delamination, which then allows water droplets to directly contact the metallization. As PV is deployed more frequently in tropical climates, we anticipate that corrosion could become more common. The damp heat test (1000 hours at 85 °C and 85% relative humidity) has been applied as a test of the package, and prolonged exposure to the damp-heat condition leads to corrosion in many modules. However, the type of corrosion caused by prolonged exposure to damp heat differs from the corrosion that is observed in the field.

The exposure to UV followed by the humidity freeze test in IEC 61215 has been very effective at causing delamination, apparently because UV exposure weakens the interface, then the humidity freeze stresses the interface when the moisture in the EVA freezes. Based on the observation that corrosion in the field occurs following delamination, it may be effective to test for corrosion by using a similar sequence followed by a damp-heat exposure. When a failure is caused by a sequence of events, as described here, the acceleration factor quoted in Table 10.3.1 was estimated from the total time to duplicate the field failure, but the prediction of failure rates may be quite complex.

10.3.2.4 Potential-Induced Degradation

PV module durability is typically tested at the module level, but when deployed in a system, PV modules create their own electrical stress, especially for systems generating 600 V or 1000 V. Bias-related issues have been reported and addressed historically, but a new set of issues became apparent in recent years for systems using transformer-less inverters, for which the PV modules may operate at potentials either positive or negative to ground. Although this issue is uncommon, when it does occur, modules may lose more than half of their performance in less than a year. In conventional silicon modules, it appears that operating under negative bias with respect to ground causes sodium to move from the glass, through the encapsulant, and into the cells. The effect is sometimes reversible, but ultimately the degradation can be irreversible (Pingel *et al.*, 2010; Naumann *et al.*, 2013).

10.3.2.5 Thin Film Module Failure/Degradation Mechanisms

Thin film modules (especially copper indium gallium di selenide) can be very sensitive to moisture, so they are commonly packaged using two glass sheets with a seal around all edges. Using such packages, the most common cause of thin-film PV module failure has been broken glass. While silicon modules use a sheet of tempered glass that is quite robust and seldom breaks, most thin film products today deposit the thin film layers directly on a sheet of glass at a temperature that affects the tempering of the glass, decreasing the strength of the glass. An accelerated test for glass breakage is difficult to conceive since breakage is so often dependent on damage related to a rock thrown during a mowing operation or other inadvertent damage.

Degradation rates of thin film PV have decreased to be comparable to those of silicon (Jordan and Kurtz, 2013), though some thin film systems still suffer from somewhat higher degradation as quantified in a recent analysis of 12 systems (Makrides *et al.*, 2014).

The degradation is most commonly associated with a decreased fill factor (Jordan, Wohlgemuth, and Kurtz, 2012). The fill factor may decrease because of increased series resistance or shunting and a number of mechanisms are found, reflecting the diversity of design (Quintana *et al.*, 2002).

10.3.3 Strategy for Improving PV Reliability

10.3.3.1 Applying Today's Knowledge

As described above, much is known about how PV modules fail and how to quickly identify problematic designs. Standard tests, such as IEC 61215 ("Crystalline silicon terrestrial photovoltaic (PV) modules: Design qualification and type approval, " 2005) and IEC 61730 ("Photovoltaic (PV) module safety qualification, " 2004) identify most problems that lead to early failures. However, edition 2 of IEC 61215 does not identify susceptibility to potential-induced degradation, power loss from cracked cells, or slow discoloration of the encapsulant. Extended test protocols such as Qualification Plus (Kurtz *et al.*, 2013a) were created to apply today's knowledge to address these known problems while standards groups update standard tests through the consensus process. Compared with IEC 61215 and IEC 61730, Qualification Plus adds tests for potential-induced degradation, additional mechanical stress to crack cells before thermal cycling, a more stressful hot-spot test, more thermal cycles, and more stressful UV durability testing. In addition, it requires testing of five modules instead of two for each leg of the test, a thorough audit of the quality management system and an ongoing stress-testing program to ensure consistency of the implementation of the design. Each of these features is under consideration as a modification to an existing or added IEC standard and the IEC will begin publishing these revisions by the end of 2015.

10.3.3.2 Creating Qualification Standards that Differentiate Use Environments

Historically, most companies have manufactured a single PV module design for all applications. But, in some cases, it may be helpful to tailor the design to the use environment. For example, thicker glass and frame may be used for a module that will need to withstand high snow loads. PV deployment first grew to the multi-GW scale in Germany, providing extensive field data in this climate. As PV deployment grows in desert and tropical regions, designs may need to be modified to withstand higher temperature and humidity. The IEC is currently writing a standard that differentiates testing for three climate zones (desert, tropical, and temperate) and two mounting configurations (open rack and closed roof). This tool will enable manufacturers and customers to explore the design options that are best suited for each application. Although these tests are designed to quickly replicate known failure modes rather than to predict exact lifetimes, a key step in designing these tests is to model the expected equivalency of the applied stress relative to each of the use environments.

10.3.3.3 Addressing Quality Control and Manufacturability

While qualification tests are very useful for identifying designs that will be successful in the field, that product must be consistently manufactured. With the frantic pace of today's

PV market, manufacturers often switch material suppliers and may not adequately retest the product design, assuming that the new supplier is providing a material that is identical to the original supplier. Similarly, some companies do not carefully identify the root cause of problems, which is needed to refine the design specification to ensure consistent manufacturability.

10.3.3.4 Creating a Standard for Service Life Predictions

The procedure for making a service life prediction is relatively straightforward: (1) identify field failure/degradation mechanisms that determine end of life; (2) quantify kinetic rates for the rate-limiting steps using accelerated stress tests; (3) for each use environment, apply kinetic rates within a model to estimate expected lifetime; and (4) Verify the model by comparing field and accelerated test data.

However, depending on the primary failure modes that are identified in the first steps, the tests that are needed will vary. A test lab wanting accreditation to complete this sort of test would not even know what equipment to purchase. So, implementation of this into a standard is not straightforward.

Furthermore, creation of a service life prediction based on an initial design may not be meaningful if that design varies during the manufacturing process. To have confidence in a service life prediction, the QMS must control the design to be consistent with the modeled durability. Thus, a standard for a service life prediction may best be implemented as part of the QMS.

Although today most companies are modifying their products too frequently to be able to provide a quantitative service life prediction, a sign of maturity of the industry will be an increasing number of companies taking on the challenge of a quantitative service life prediction based on a solid understanding of failure/degradation modes that they have not been able to eliminate.

10.3.4 Conclusion

Today, the PV industry is fairly knowledgeable about the most common causes of failures in the field, but a complete understanding of the acceptable process and design windows is challenging, especially as companies may make changes to the bill of materials every few months. PV reliability standards are maturing by:

- addressing failure mechanisms that have recently become problematic;
- differentiating the durability of PV module designs as a function of the use environment;
- improving manufacturing consistency toward a consistent implementation of product design, moving toward inclusion of service life prediction in the most mature quality management systems.

Together, these improvements will provide increasing confidence in an industry that already has a strong track record, supporting further growth of the PV industry.

Acknowledgments

This work was completed under contract no. DE-AC36-99GO10337 with the U.S. Department of Energy.

References

Bosco, N., Silverman, T.J., Wohlgemuth, J. *et al.* (2013) Paper presented at the IEEE 39th.Photovoltaic Specialists Conference (PVSC), 2013.

Chandel, S., Naik, M. N., Sharma, V., and Chandel, R. (2015) Degradation analysis of 28 year field exposed mono-c-Si photovoltaic modules of a direct coupled solar water pumping system in western Himalayan region of India. *Renewable Energy*, **78**, 193–202.

De Graaff, D., Lacerda, R., and Campeau, Z. (2011) Degradation mechanisms in si module technologies observed in the field; their analysis and statistics. Paper presented at the PV Module Reliability Workshop.

Dubey, R., Chattopadhyay, S., Kuthanazhi, V. *et al.* (2014a) Performance degradation in field-aged crystalline silicon PV modules in different Indian climatic conditions. Paper presented at the IEEE 40th Photovoltaic Specialists Conference (PVSC), Denver, CO.

Dubey, R., Chattopadhyay, S., Kuthanazhi, K. *et al.* (2014b) All India survey of PV module degradation: 2013. National Centre of Photovoltaic Research Edu- cat., Mumbai, India. Available at: www.ncpre.iitb.ac.in/uploads/All_India_Survey_of_Photovoltaic_Module_Degradation_2013.pdf

Hasselbrink, E., Anderson, M., Defreitas, Z. *et al.* (2013) Validation of the PVLife Model using 3 million module-years of live site data. Paper presented at the 39th IEEE Photovoltaic Specialists Conference.

IEC Central Office (2004) Photovoltaic (PV) module safety qualification. IEC, Geneva, Switzerland.

IEC Central Office (2005) Crystalline silicon terrestrial photovoltaic (PV) modules: Design qualification and type approval. IEC, Geneva, Switzerland.

Jordan, D.C. and Kurtz, S.R. (2013) Photovoltaic degradation rates: an analytical review. *Progress in Photovoltaics: Research and Applications*, **21**, 12–29.

Jordan, D.C. and Kurtz, S R. (2014) Reliability and Geographic Trends of 50,000 Photovoltaic Systems in the USA. Paper presented at the 29th European Photovoltaic Solar Energy Conference, Amsterdam, the Netherlands.

Jordan, D.C., Sekulic, B., Marion, B. and Kurtz, S.R. (2015) Performance and aging of a 20-year-old silicon PV system. *IEEE Journal of Photovoltaics*, DOI: 10.1109/JPHOTOV.2015.2396360.

Jordan, D.C., Wohlgemuth, J.H., and Kurtz, S.R. (2012) Technology and climate trends in PV module degradation. Paper presented at the 27th European Photovoltaic Solar Energy Conference, Frankfurt, Germany.

Köntges, M., Kurtz, S., Jahn, U. *et al.* (2014) *Review of Failures of Photovoltaic Modules Final* (No. T13-01). IEA-PVPVS.

Kurtz, S., Wohlgemuth, J., Kempe, M. *et al.* (2013a) Photovoltaic module qualification plus testing. *Progress in Photovoltaics*, **21**.

Kurtz, S., Wohlgemuth, J., Yamamichi, M. *et al.* (2013b) A framework for a comparative accelerated testing standard for PV modules. Paper presented at the 39th IEEE PVSC.

Makrides, G., Zinsser, B., Schubert, M., and Georghiou, G.E. (2014) Performance loss rate of twelve photovoltaic technologies under field conditions using statistical techniques. *Solar Energy*, **103**, 28–42.

Naumann, V., Lausch, D., Graff, A. *et al.* (2013) The role of stacking faults for the formation of shunts during potential induced degradation of crystalline Si solar cells. *physica status solidi (RRL) Rapid Research Letters*, **7** (5), 315–318.

Pingel, S., Frank, O., Winkler, M. *et al.* (2010) Potential induced degradation of solar cells and panels. Paper presented at the 35th IEEE Photovoltaic Specialists Conference (PVSC), 2010.

Quintana, M.A., King, D.L., McMahon, T.J., and Osterwald, C. (2002) Commonly observed degradation in field-aged photovoltaic modules. Conference Record of the 29th IEEE Photovoltaic Specialists Conference.

Reid, C.G., Bokria, J.G., and Woods, J.T. (2013) UV aging and outdoor exposure correlation for EVA PV encapsulants. Paper presented at the SPIE, San Diego.

Rosenthal, A.L., Thomas, M.G., and Durand, S.J. (1993) A ten-year review of performance of photovoltaic systems. Paper presented at the 23rd IEEE Photovoltaic Specialists Conference.

Yedidi, K., Tatapudi, S., Mallineni, J. *et al.* (2014) Failure and degradation modes and rates of PV modules in a hot-dry climate: results after 16 years of field exposure. Paper presented at the 40th IEEE Photovoltaic Specialist Conference (PVSC), 8–13 June.

10.4

Advanced Module Concepts

Pierre Verlinden
Trina Solar, Changzhou, Jiangsu, China

10.4.1 Introduction

In several chapters of this book, in particular in Part 3, the research in improving the efficiency of silicon solar cells is discussed in detail. It is, however, also important to consider the efficiency of PV modules. For example, a standard 1.65 m × 0.992 m PV module, made of 60 cells (each 156 mm × 156 mm) with an average cell efficiency of 18%, would have a typical power output of 250 W and efficiency of 15.3%. The difference in efficiency between cells and module is due to several factors:

- On the negative side: inactive area of the frame and module border, gaps between cells, light reflection from the front glass, electrical losses due to the series resistance of cell interconnections, busbars, cables and connectors, and mismatch between cells connected in series.
- On the positive side: improvement (i.e. reduction) of the reflectivity of the cell due to total internal reflection at the front glass/air interface and capture of part of the light reflected by the backsheet, fingers and ribbons.

Following Haedrich *et al.* (2014), the biggest power losses are due to the front reflection at the glass/air interface (~4%) and the cell interconnection losses (~3%). The ratio between the power output of a module and the sum of the power outputs of the different cells composing the module is called the Cell-to-Module (CTM) ratio. A typical CTM value is in the range of 95 to 100%. As seen below, it is possible to achieve a CTM value greater than 100% by using several advanced technologies to reduce the interconnection losses and to redirect toward the cell the light that is reflected by the ribbons and the backsheet. The first section of this chapter is, however, concerned with reliability improvement. The following sections discuss the improvement of the optical efficiency of the modules, the reduction of interconnections losses, and finally the improvement of the energy efficiency.

Photovoltaic Solar Energy: From Fundamentals to Applications, First Edition.
Edited by Angèle Reinders, Pierre Verlinden, Wilfried van Sark, and Alexandre Freundlich.

Companion website: www.wiley.com/go/reinders/photovoltaic_solar_energy

10.4.2 Double-Glass Modules

It is very hard to produce the most reliable PV module at the lowest cost. Reliability comes at a cost. Current PV modules are designed to operate in any location, under any climatic conditions. This approach means that the lifetime of the PV modules may significantly vary from one type of climate to the other. A better approach is to design modules and select the best module material for a particular climate rather than for all climates. Although tropical, desert and mountain climates are quite different from each other, they are the harshest environments for PV modules, exhibiting a combination of at least four of the following conditions: high ambient temperatures in the day, high irradiation, strong UV irradiation, high relative humidity, stagnating water on modules at low tilt angles, high wind or snow load, or large number of thermal or deep cycles. For the best reliability, PV modules require high quality module components. For example, some recent studies of the reliability of modules have shown that the degradation rate of modules in a desert environment is significantly higher than the average degradation rate in temperate climates, up to as much as ~2.5%/year (Jordan *et al.*, 2012; Jordan and Kurtz, 2013; Singh *et al.*, 2013). In addition, tropical and mountain regions require PV modules with high resistance to mechanical stress due to snow load or wind load in the event of a hurricane. In a traditional PV module using backsheet, the cells are located very close to the back surface of the module, i.e., very close to the area with the highest tensile or compressive stress during positive or negative mechanical loading. Although good quality backsheets are perfectly fine for 25-year survival in most climates, the replacement of the backsheet by a second glass panel offers an additional protection against long-term degradation in harsh environments.

In a symmetrical double glass "sandwich" structure, i.e. with two glass panels of the same thickness and strength, the solar cells are maintained at the mechanical neutral fiber, i.e. the line of zero stress in the middle of the module (Figure 10.4.1). The dual-glass module design

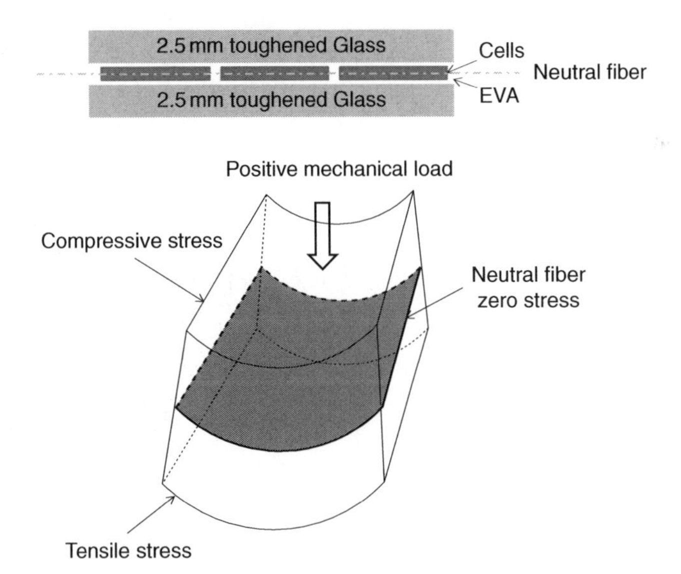

Figure 10.4.1 Top: Symmetrical double-glass module structure with cells being sandwiched between two 2.5 mm thick heat-strengthened glass panels. Bottom: The mechanical stress to the cells is negligible during bending because the cells are located at the "Neutral Fiber," i.e. the middle of the symmetrical structure

and properties have recently been reported on by Zhang *et al.* (2013). With this design, it is believed that the dual-glass structure virtually eliminates the generation of any micro-cracks in the silicon solar cells during handling, transportation, installation (even when installers walk on the modules), or operation.

A second major advantage of double-glass PV modules is the drastic reduction in water vapor permeability. This characteristic has a significant impact on the elimination of reliability issues related to moisture ingress into the module, for example, corrosion, hydrolysis, potential induced degradation (PID), "snail tracks" (the discoloration of the silver front fingers observed around the location of the cell cracks and in the case of backsheets with high water vapor permeability), and degradation due to humidity-freeze conditions (Verlinden *et al*, 2014). For example, it has been shown that double-glass modules easily pass the hardest PID test of 600 hours at 85 °C and 85% R.H., with a maximum system voltage of 1500 V (positive or negative) and with copper foils to simulate the presence of stagnating water. Finally, double-glass modules also offer excellent resistance to UV degradation.

10.4.3 Anti-Reflection Coated Glass

Anti-reflection coating (ARC) on the front surface of the glass of PV modules has been widely used in production since about 2010. It reduces the light reflection at the front glass/air interface from typically 4% for non-coated glass to about 1 to 3% for ARC glass at normal light incidence. The gain is even more significant for non-normal incidence. Therefore, the gain in energy efficiency for ARC glass compared to non-coated glass, typically in the order of 2 to 4%, is usually greater than the gain in power efficiency. Two main methods are uses to form an ARC on glass for PV applications: etching the surface of the glass or depositing a film of porous glass. In both cases, the objective is to create a glass-type film, if possible as hard as the glass, with a low refractive index, ideally 1.2 being the geometrical mean between the refractive index of glass and air, and a controlled thickness of about 100 nm, forming a quarter wavelength adaptation between the two media. Different techniques of deposition are commonly used: chemical vapor deposition (CVD), physical vapor deposition (PVD) and sol-gel deposition, including dip coating and roll, spray, and slot-die coating. Sol-gel coating is by far the most widely used technique to produce ARC glass.

The ARC for PV applications can be classified into three categories: open-pore, closed-pore and dense structure (Pan *et al.*, 2015). ARCs with a closed-pore structure have been shown to have the best optical performance, while the ARCs with a dense structure have a better resistance to abrasion and are more suitable for desert application.

Schneider *et al.* (2014) studied a combination of ARC technology, the use of thin 2 mm glass, polyvinyl butyral (PVB) with high UV transmittance as encapsulant and light harvesting strings (LHS) or light capturing ribbons (LCR). In this study, it is demonstrated that a total relative gain of 5% can be achieved compared to standard modules.

10.4.4 Half-cell Modules

Cell interconnection losses can represent 2–4% of the total power output of a PV module. Mitsubishi (2011) introduced modules made with cells cut in half to reduce the cell interconnection losses. By doing so, the current of each cell is reduced by a factor of two and the power loss due to the interconnection between cells is reduced by a factor of four. Bagdahn *et al.* (2014) demonstrated that modules made with half-cells have an improved CTM of 98.1%, compared to 93.6% for full-cell modules. Zhang *et al.* (2015) reported a module made of 60 mono-crystalline PERC cells cut in half (120 half-cells) with a world record power output of 335 W. This module also included special features like ARC glass, LCR and light reflective films (LRF) between cells. The CTM value of this particular module was 111%.

Cutting finished solar cells in smaller pieces to assemble a PV module with series and parallel interconnections has another advantage as demonstrated by Carr *et al.* (2015). It makes the module more tolerant to partial shading. It is well known that most PV modules do not respond linearly to partial shading. A typical PV module has three sub-strings, each one of them protected by one bypass diode. When one cell, representing 1.7% of the total module area, is shaded, the power output can decrease by as much as 33%. Carr demonstrated that a PV module made of a combination of series and parallel interconnection of small 39 mm x 39 mm cells (1/16 of a standard 156 mm cell) can have a shade linearity of up to 92%, whereas a standard module presents a shade linearity of 66%.

10.4.5 Light Capturing Ribbon

When measuring the efficiency of cells in air, about 6–7% of the light is reflected by the front fingers and busbars. After encapsulation of the cells into a PV module, a small part of this reflected light can be reflected back to the cell due to the surface roughness of the interconnecting ribbons and the rounded shape of the metal fingers, and due to the fact that some of the reflected rays reaching the glass/air interface at an angle greater than 44° are totally internally reflected inside the module. As a consequence, the effective shading factor of the cell metallization is usually smaller than the real metal coverage fraction by about 1% absolute.

The CTM factor can be improved by creating grooves at the top surface of the ribbons (Figure 10.4.2) with a facet angle greater than 22°, which would force the reflected light to be

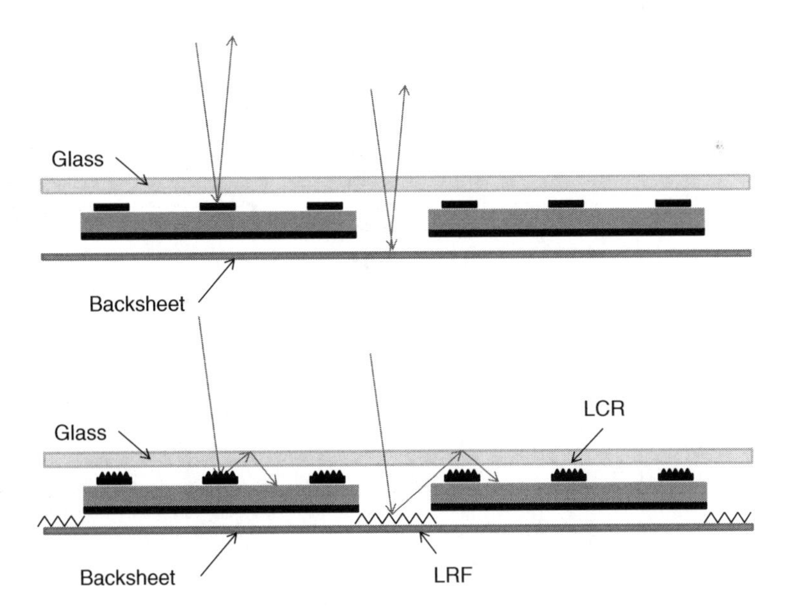

Figure 10.4.2 Examples of light trapping in advanced PV modules with light capturing ribbons (LCR) and light reflective film (LRF)

totally internally reflected at the glass/air interface. Examples of PV modules with LCR ribbons have been presented by Schneider *et al.* (2014) and Zhang *et al.* (2015).

10.4.6 Light Reflective Film

The typical gap between cells within a PV module is around 2–3 mm. To improve the efficiency of PV modules, the module may be designed with smaller gaps between cells. For example, the gap between cells can be reduced to 1–2 mm. On the other hand, if a large power output is desired regardless of efficiency, the gap between cells can be increased to benefit from the increased current generation due to the light reflected by the backsheet and totally internally reflected inside the module (Indeok *et al.*, 2012). This effect is even more pronounced when a structured light reflective film (LRF) is placed between the cells (Figure 10.4.3). The LRF structure is composed of grooves with a good specular reflection and a facet angle greater than 22° to force the reflected light to be totally internally reflected inside the module (Indeok *et al.*, 2014). As the gap between cells increases, the module efficiency decreases, while the power output, CTM and cost increase. As a consequence, the cost per Watt peak reaches a minimum, as indicated in Figure 10.4.3, for a particular cell gap that is determined by the quality of the reflector, and the cost of the module materials (Zhang *et al.*, 2015).

Indeok *et al.* (2014) showed that LRF can be applied on top of the ribbons or between cells. When applied on top of the ribbons, the gain in short-circuit current, J_{sc}, can be as high as +1.7% compared to conventional ribbons. When applied between cells, the gain in J_{sc} depends on the gap between cells. For normal spacing between cells, the gain due to LRF compared to conventional white backsheet is about 0.8% (Indeok *et al.*, 2014). Zhang *et al.* (2015) showed that, for a 10 mm gap, the gain can be as high as 6% compared to a normal 2 mm gap (Figure 10.4.3). However, even if the optimum cell gap for module cost in $/Watt is around 10 mm, the optimum

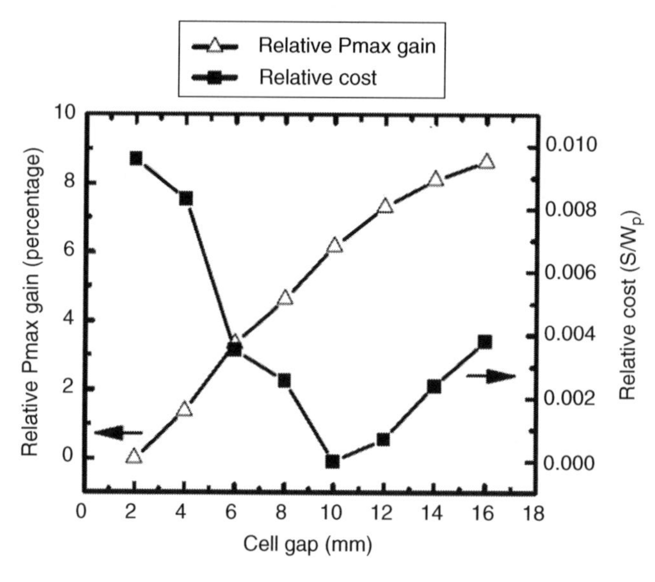

Figure 10.4.3 Relative power output increase and relative cost per Watt peak as a function of the gap between cells for a module made of 60 cells (156 mm × 156 mm) with LRF between cells

for the levelized cost of electricity (LCOE) is much lower, depending on other factors such as the cost of labor for installation, the cost of cable and racking, and the cost of land.

10.4.7 Smart Wire and Multi-Busbars

Another method to recapture the light reflected by the cell interconnections is to replace the conventional flat ribbons by small-diameter wires (Söderström *et al.*, 2015). In this case, about 36.8% of the light reaching the top of the wire is reflected back and escapes the module. The other 63.2% is reflected at an angle greater than 43.6 degrees and is totally internally reflected at the glass/air interface to be recaptured by the cell. The wires are about 300 to 400 μm in diameter (about the same thickness as the encapsulant) and are soldered directly onto the silver fingers of the cell. The Smart Wire technology currently being developed by Meyer Burger (Söderström *et al.*, 2015) uses Indium-coated copper wires that are soldered to the fingers of the solar cells at a low temperature during the module lamination process. The multi-busbar technology also uses round wires instead of conventional flat ribbons, but uses a standard soldering process with modified automatic stringers. Cost reduction and manufacturability improvement of these two techniques are still ongoing researches.

10.4.8 Smart PV Modules

Smart PV modules are modules that include electronic circuits inside the junction box to provide information, a safety function or power conditioning. This includes:

- monitoring functions (current, voltage, power, temperature, irradiance, status of the bypass diodes);
- safety switch disconnect or bypass, either with remote control or automatically in case of malfunction or ground fault;
- maximum power point tracking through DC-DC conversion, either at the module level or sub-string level (Doubleday *et al.*, 2015);
- reshaping of the I-V characteristics, for example, to fix a maximum module voltage between V_{mp} and V_{oc} to allow more modules per strings for a given maximum system voltage;
- DC to AC conversion via a micro-inverter;
- communication via Power Line Communication (PLC), ZigBee, WiFi or Bluetooth to a hub that is connected to a central control unit or to a Cloud server via the internet.

The different functions integrated in the junction box of the module have all the same objective: reducing the cost of the electrical energy generated by the PV system or LCOE, for example:

- monitoring and communication improve the operation and maintenance of a PV system, allowing the operator to react faster when a problem occurs;
- a safety switch protects the PV system against the consequence of a malfunction;
- maximum power point trackers, DC-DC or DC-AC converters improve the energy efficiency of PV systems in the case of partial shading, non-uniform soiling, mismatch, or non-uniform degradation;
- reshaping of the I-V characteristics reduces the cost of the Balance of Systems (BOS) by increasing the number of modules per strings (e.g., see Trina, 2015).

10.4.9 Conclusion

The technology of PV modules is rapidly developing to provide better reliability and durability with a cost-effective design for a particular climate. This is, for example, the case of double-glass modules, which, among other benefits, protect the solar cells against micro-cracks, water ingress, and PID. The excellent durability of double-glass modules comes at the price of higher weight.

In the category of power output improvements, recent developments have been presented to improve the optical efficiency of PV modules (ARC, LCR, LRF, Smart Wire, multi busbars). For example, ARC glass reduces the front reflection by 1% or 2% absolute, or even more for non-normal incidence, but a trade-off must be considered between optical efficiency and durability. The most durable ARCs for desert application, or other areas where sand abrasion is an issue, do not necessarily present the best optical performance. LCR and LRF are technologies that allow light rays that are reflected by the ribbons or that fall between cells to be totally internally reflected within the module and captured by the solar cells. The design optimization of LRF strongly depends on the objective, i.e. maximizing efficiency or power, or minimizing cost per Watt or LCOE. Smart Wire and multi-busbar technologies are other possible future technologies that use round wires instead of flat ribbons to reduce the reflection of light due to cell interconnections. The improvement of manufacturability of these two technologies is the topic of ongoing research.

Also in the category of power output improvement, PV modules using half-cell are being developed to reduce by a factor of four the power loss due to the series resistance in the cell interconnection. Although very promising for high efficiency modules with high CTM, this technique requires one additional step in the cell process and reduces by a factor of two the throughput of the cell tester and sorter at the end of the production line.

Smart modules are also being developed to improve the energy efficiency of PV systems, providing multiple functions like monitoring, communication, safety, power optimization and DC to AC conversion. This also comes with an additional cost (in $/Watt terms), but results in most cases in a reduction of LCOE, particularly for PV systems challenged with partial shading or with modules with different orientations.

List of Symbols

Symbol	Description	Unit
J_{sc}	Short-circuit current density	A/cm^2
V_{mp}	Maximum power point voltage	V
V_{oc}	Open-circuit voltage	V

List of Acronyms

Acronym	Description
ARC	Anti-Reflection Coating
CTM	Cell-To-Module power ratio
CVD	Chemical Vapor Deposition

Acronym	Description
LCOE	Levelized Cost of Electricity
LCR	Light Capturing Ribbon
LHS	Light Harvesting String
LRF	Light Reflective Film
PERC	Passivated Emitter and Rear Contact Cell
PID	Potential Induced Degradation
PLC	Power Line Communication
PV	Photovoltaic or photovoltaics
PVB	Polyvinyl Butyral
PVD	Physical Vapor Deposition
UV	Ultra-Violet

References

Bagdahn, J., Dassler, D., Ebert, M. *et al.* (2014) Opportunities and challenges of half-cells solar modules. Paper presented at SNEC, Shanghai, May 21.

Carr, A.J., de Groot, K., Jansen, M.J. *et al.* (2015) Tessera: scalable, shade robust module. *Proceedings of the 42th IEEE Photovoltaic Specialists Conference*, New Orleans.

Doubleday, K., Deline, C., Olalla, C., and Maksimovic, D. (2015) Performance of differential power-processing submodule DC-DC converters in recovering inter-row shading losses. *Proceedings of the 42th IEEE Photovoltaic Specialists Conference*, New Orleans.

Haedrich, I., Eitner, U., Wiese, M., and Wirth, H. (2014) Unified methodology for determining CTM ratios: Systematic prediction of module power. Paper presented at the 4th International Conference on Crystalline Silicon Photovoltaics (SiliconPV 2014), 's-Hertogenbosch, The Netherlands, March 25–27, 2014. *Solar Energy Materials and Solar Cells*, **131**, 14–23.

Indeok, C., Baek, U., Moon, I. *et al.* (2012) Analysis of current gain by varying the spacing between cells in a PV module with quantum efficiency measurement. *Proceedings of the 38th IEEE Photovoltaic Specialists Conference (PVSC)*, Austin, Texas, pp. 2388–2390, DOI: 10.1109/PVSC.2012.6318078.

Indeok C., Lee, W., Cho, E., Moon, I. *et al.* (2014) Light capturing film for power gain of silicon PV modules. *Proceedings of the 40th IEEE Photovoltaic Specialists Conference (PVSC)*, Denver, Colorado, pp. 2689–2692, DOI: 10.1109/PVSC.2014.6925484.

Jordan, D.C., Wohlgemuth, J.H., and Kurtz, S.R. (2012) Technology and climate trends in PV module degradation. *27th European Photovoltaic Solar Energy Conference*, Frankfurt, Germany, 24–28 Sept.

Jordan, D.C. and Kurtz, S.R. (2013) Photovoltaic degradation rates: an analytical review. *Progress in Photovoltaics: Research and Applications*, **21** (1), 12–29.

Maxim Integrated (2015) Available at: http://www.maximintegrated.com/content/dam/files/design/technical-documents/product-briefs/maxim-solar-cell-optimizer-product-brief.pdf (accessed 30 August 2015).

Mitsubishi (2011) Available at: http://www.mitsubishielectric.com/bu/solar/pv_modules/pdf/monocrystalline.pdf (accessed 30 August 2015).

Pan, X., Zhang, S., Xu, J. *et al.* (2015) Performance and reliability of different anti-reflective coated glass for PV modules. Paper presented at the 31st European Photovoltaic Solar Energy Conference (EU PVSEC), Hamburg.

Schneider, J., Turek, M., Dyrba, M. *et al.* (2014) Combined effect of light harvesting strings, anti-reflective coating, thin glass, and ultraviolet transmission encapsulant to reduce optical losses in solar modules. *Progress in Photovoltaics Research and Applications*, **22** (7), 830–837. DOI: 10.1002/pip.2470.

Singh, J., Belmont, J., and Tamizhmani, G. (2013) Degradation analysis of 1900 PV modules in a hot-dry climate: results after 12 to 18 years of field exposure. *Proceedings of the 39th IEEE Photovoltaic Specialists Conference (PVSC)*, Tampa, FL.

Söderström, T., Yao, Y., Gragert, M. *et al.* (2015) Low cost high energy yield solar module lines and its application. *Proceedings of the 42nd IEEE Photovoltaic Specialists Conference (PVSC)*, New Orleans.

Trina (2015) Available at: http://www.trinasolar.com/HtmlData/downloads/pdf/Trinasmart_SmartCurve_Note.pdf (accessed 2 September 2015).

Verlinden, P.J., Zhang, Y., Xu, J. *et al.* (2014) Optimized PV modules for tropical, mountain and desert climates. *Proceedings of the 29th European Photovoltaic Solar Energy Conference.* Amsterdam.

Zhang, S., Deng, W., Pan, X. *et al.* (2015) 335 Watt world record p-type mono-crystalline module with 20.6% efficiency PERC solar cells. *42nd IEEE Photovoltaic Specialists Conference*, New Orleans.

Zhang, Y., Xu, J., Shu, Y. *et al.* (2013) High-reliability and long-durability double-glass module with crystalline silicon solar cells with fire-safety class A certification. *Proceedings of the 28th European Photovoltaic Solar Energy Conference (EU PVSEC)*, Paris.

Part Eleven

PV Systems and Applications

11.1

Grid-Connected PV Systems

Greg J. Ball
Solar City, Oakland, CA, USA

11.1.1 Introduction

Simply defined, grid-connected PV systems include all PV systems that produce AC power while electrically connected and synchronized to an electric utility grid. Grid-connected PV systems are ubiquitous in the countries that have established programs and incentives for the use of PV and other renewable sources of energy. They are found on rooftops delivering as little as a few hundred watts of power, as well as in fields as large-scale PV power plants or "solar farms," ranging in power output from hundreds of kilowatts to hundreds of megawatts. The modular nature of PV technology and its lack of mechanical machinery lend itself to use where the electric loads are being consumed, including in residences, small commercial buildings, empty lots adjacent to industrial loads, and so on. This characteristic distinguishes PV from most all other types of electrical generation, including wind. It also impacts the characteristics and requirements for integrating PV systems with the local utility grid.

Grid-connected PV system capacity worldwide has grown at a tremendous rate over the past decades in response to demand for clean energy technology and as a result of dramatic system cost reductions. The European PV Industry Association (EPIA, 2014) reports that annual installations of PV systems have increased from less than 300 MW per year in the year 2000 to nearly 40 GW per year in 2013, and as of 2013, the cumulative installed capacity is nearly 140 GW. The vast majority of these systems are grid-connected (IEA, 2013).

This chapter discusses the fundamentals of grid-connected systems, the range of types and technologies, and the relevant areas of development for improved system engineering, performance, protection and safety.

Photovoltaic Solar Energy: From Fundamentals to Applications, First Edition.
Edited by Angèle Reinders, Pierre Verlinden, Wilfried van Sark, and Alexandre Freundlich.

Companion website: www.wiley.com/go/reinders/photovoltaic_solar_energy

11.1.1.1 System Design Basics

At a high level, a grid-connected PV system design is uncomplicated. Small systems (e.g., 2 to 10 kW residential) consist largely of PV modules and a DC-AC power converter (inverter), connected together with wires, switches, and fuses. Larger systems are scaled-up versions, with additional equipment to aggregate smaller subsystems together. PV array sizes are normally defined by their "Watt-peak" (or Wp) DC power rating, which corresponds to the summed rating of all PV modules at Standard Test Conditions (STC) of 1,000 W/m^2 solar irradiance, 25 °C module cell temperature, and a solar spectrum corresponding to an air mass of AM 1.5 (IEC 61215). The size of the grid-connected system in total is also defined by its AC power "Watt" (W) or "Watt-AC" (Wac) rating, which is most simply defined by the AC rating of the inverter. Figure 11.1.1 shows the electrical

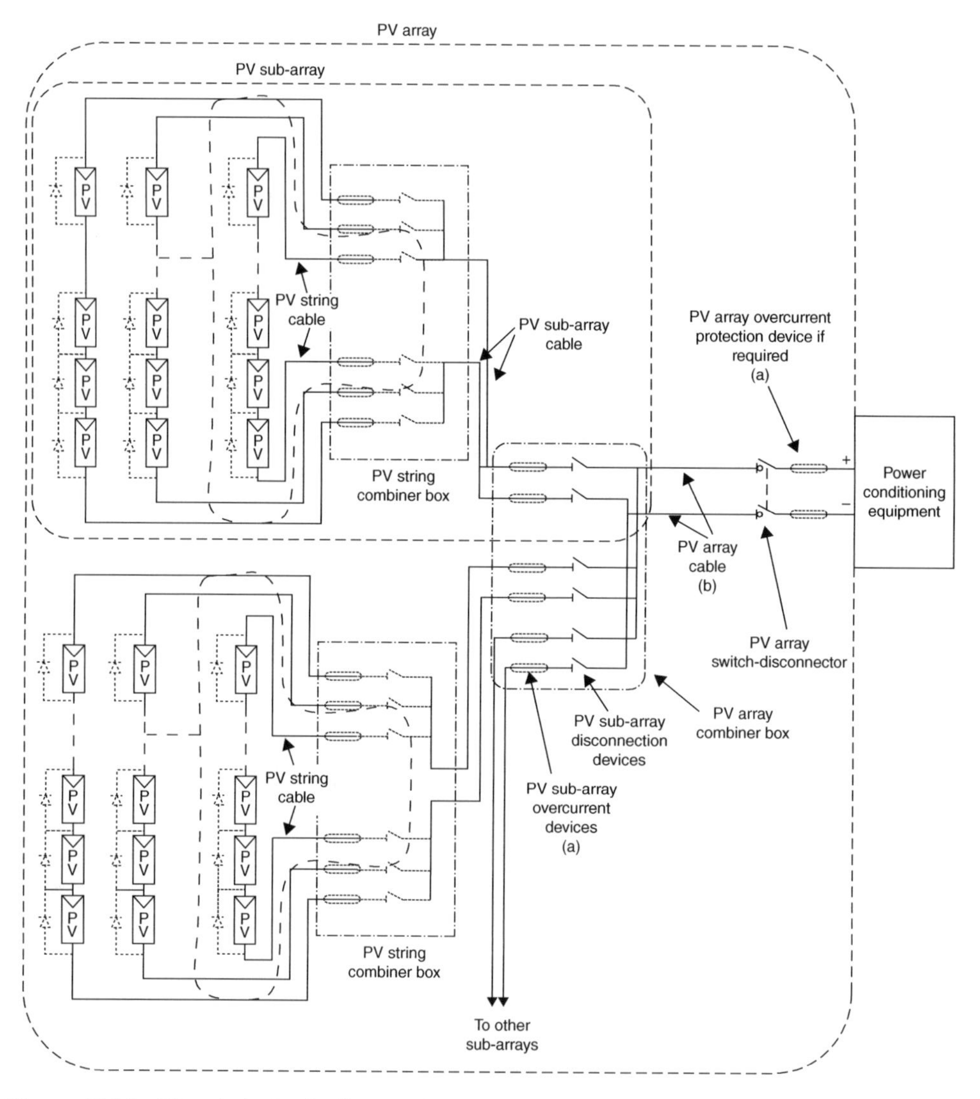

Figure 11.1.1 Electrical circuit diagram for large PV system. Source: (IEC TS 62548, ed. 1.0). Copyright © 2013 IEC Geneva, Switzerland

schematic for a large array and central inverter configuration, e.g., 500kW. PV modules are shown connected in series "strings" to reach a desired voltage suitable for input into the inverter. Each string, typically rated from 1 to 5 kWp, has its own pair of wires and is connected in parallel with other strings in combiner boxes. Here they are bussed together allowing a single larger pair of conductors to deliver power from the sub-array. The sub-array conductors are in turn connected in parallel in array combiners, where they are fused in the range of 100 to 400 Amps and connected to the input of the large inverter. Disconnecting devices are installed in the combiners as well, to allow for isolation and maintenance of individual circuits. There are multiple variations on this approach, such as paralleling AC circuits with string inverters rather than DC circuits, using module level "micro-inverters" or "AC modules" so that all interconnecting circuits are AC, and the use of DC/DC converters which regulate the DC voltage at the module, string, or combiner box level.

11.1.2 Grid-Connected System Types

By definition, grid-connected PV systems produce power in parallel with the utility on the grid network. Not all grid-connected systems actually produce or "export" power into the grid, however. Systems that do not export power are installed on the customer side of a utility meter and do not have the capacity to generate more power than is consumed by the connected load, or use batteries to store the excess generated energy. Non-export systems may have simpler interconnection requirements or may be financially incentivized by the utility relative to those that do export, but it is more common that PV systems export to the grid either continuously or during parts of the day when the system output exceeds the load.

This section covers the prominent application of grid-connected PV systems, such as residential and commercial rooftop systems, and ground-mounted systems. In each the basic engineering concepts, unique aspects, challenges in innovation are addressed.

11.1.2.1 PV on Buildings and Structures

Grid-connected PV systems on buildings and structures bring electric generation directly to the load and in many ways simplify its usage.

11.1.2.1.1 Residential Rooftop

The proportion of residential rooftop PV systems relative to total PV capacity varies widely among different countries but in many places it is a very significant portion. Residential rooftop systems usually are in the 1 to 10kW range, but most are well under 10kW. A system can produce enough energy to significantly offset the energy supplied by the utility – the amount depending on many factors such as system size, location, solar resource, and house end use load energy usage. The PV output occurs during sunlight hours of course and cannot offset night-time loads, but utility "net-metering" programs allow systems to sell power to the utility during the day (at the same rate that they purchase power), effectively banking the power for buyback during night-time hours. Alternatively, there is a growing trend to store excess PV power in energy storage systems such as batteries (Colthorpe, 2014). In this sense it is possible for residents to achieve net-zero energy usage on an annualized basis.

The most common approach for residential rooftop installations involves the use of low profile racking systems oriented in the same plane as the roof. Low profile installations are advantageous for both cost and aesthetic reasons. Structural requirements for wind-loading

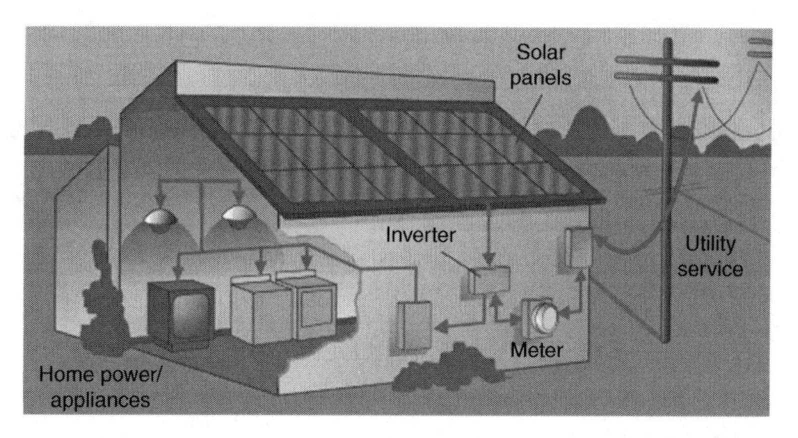

Figure 11.1.2 Typical residential rooftop system configuration. Source: (DOE EERE, 2010)

are minimized, which in turn minimize the weight and volume of structural components. This also reduces the reliance upon roof penetrations to achieve the structural strength, and therefore reduces opportunities for roof leaks. The most productive systems are those that are south-facing (in the northern hemisphere), but it is not uncommon to see arrays or portions of arrays oriented at various angles to the east and/or west, or even north.

Residential systems use either string inverters or module level "micro-inverters" to convert to the utility-supplied AC voltage, or utilization voltage. Figure 11.1.2 shows an example residential system configuration.

In Figure 11.1.2, a string inverter is mounted on the exterior wall and converts the DC power directly to the house utilization voltage. Module-level micro-inverters are often used instead of string inverters in small rooftop applications, and are experiencing rapid growth in the market. Micro-inverters generally do not achieve the same DC-AC conversion efficiency of string inverters, and do not yet have well-understood long-term reliability statistics. They can be advantageous particularly in residential and small commercial rooftop systems however, by reducing losses due to mismatch and shading. Inverter comparisons are discussed in more detail in Chapter 11.2.

Any grid-connected PV system may also have a battery back-up system to supply loads during utility outages, but it is most common to see these in residential systems. Relatively small battery banks can provide power during brief utility supply interruptions.

11.1.2.1.2 Commercial Rooftops and Carports

Larger rooftop PV systems are installed on commercial and industrial buildings, stores, warehouses and the like. Commercial (or industrial) rooftop systems range typically from the tens of kW to as much as several megawatts. The typical commercial building has a flat or low sloped roof allowing PV arrays to be laid out like a jigsaw puzzle conforming to the building shape and rooftop obstacles, such as air-handlers, skylights, parapets, vents, etc. Modules can be laid flat or installed at a low tilt – common is south-facing at 5–20° (in the northern hemisphere) to achieve greater year-round energy capture. Figures 11.1.3 (a) and 11.1.3 (b) show a typical commercial rooftop system and carport system, respectively.

(a)

(b)

Figure 11.1.3 (a) Typical commercial rooftop and (b) carport PV arrays. Source: (a) Topher Donahue (2009); (b) California State University, Fresno (2007)

As with residential systems, structural elements and wind loading significantly limit the ability to cost-effectively incorporate fixed tilt arrays at greater angles or to utilize rooftop trackers. Flat rooftop system economics drive mounting technologies that are lightweight to meet dead load requirements, are easy to install, and require minimal roof penetrations. The result is commonly ballasted arrays incorporating wind deflectors to minimize the ballast

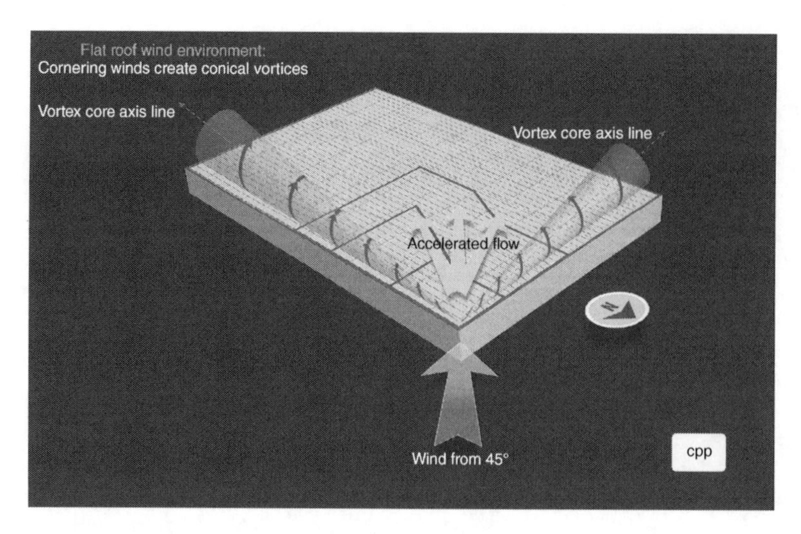

Figure 11.1.4 Illustration of exacerbated wind profiles impacting flat roof PV systems. Source: CPP Inc. (2012)

weight requirement. Understanding the dynamics of commercial rooftop wind flow is critical to identify structural weak points and achieve cost-effective designs that can safely withstand high wind conditions (O'Brien and Banks, 2012). Figure 11.1.4 illustrates how wind directed at the rooftop corner accelerates the wind across the roof by as much as 20% and creates conical vortices. These vortices lead to the highest wind loads by virtue of the way they hit the PV modules. Research in this area has demonstrated that building codes can under-predict the wind loads on PV arrays, and in some cases has led to code revisions to improve those predictions.

Electrically, commercial systems differ from residential by virtue of the scale of electrical components needed to deliver power to the grid. Systems are commonly configured with central, string or micro-inverters. The majority of large commercial rooftop systems installed to date incorporate central inverters – typically rated from 30 kW to 1 MW.

Central inverters were historically favored in these larger systems because smaller string and module inverters produced single phase low voltage AC (120 to 240 V), requiring substantially more conductor (copper or aluminum) per unit of power to carry the current from the arrays to the building electrical panels. Electrical losses are also proportionally larger per Watt at the lower voltages. The central inverters operating between 600 to 1000 V enabled lower-cost transmission of power from the arrays.

Newer string inverters in the 12 to 50 kW range and rated 1000 V_{dc} and 400 to 480 V_{ac} effectively challenge the central inverter cost model, however. These units are not much larger than combiner boxes and can be installed on the roof adjacent to arrays. Three-phase AC circuits carry more power per unit volume of conductor than similarly rated DC circuits. Moreover, DC circuits in PV systems have greater oversizing requirements than inverter AC output circuits, due to the need to account for short-circuit and high irradiance conditions. As an example, a 1000 V_{dc} array using three-phase, 400 V_{ac} string inverters requires less than 2/3 the volume of conductor of a system designed using DC combiner boxes and output cables.

Central inverters also produce three-phase AC power, but their size and weight generally make it prohibitive to install them on rooftops. As with residential systems, there is also the growth of microinverters or module level dc/dc converters in the commercial system market. Micro-inverters can now be configured to create three-phase ac circuits exceeding $400V_{ac}$ as well. These are discussed in more detail in Chapter 11.2.

11.1.2.1.3 Building-Integrated PV (BIPV)

PV systems in and on buildings can also be fully integrated into the structure itself, such as in exterior glass, shingles or other roofing materials. Although this can be more expensive than simple add-on systems, there are great opportunities to use innovative techniques and materials during the design and construction phases of a new building. There are numerous design and performance considerations owing to the wide variety of implementation methods. BIPV is discussed in detail in Chapter 11.6.

11.1.2.1.4 Ground-Mounted Systems

The category of ground-mounted systems covers everything from a pair of modules on a pole to power plants containing hundreds of thousands of modules. At a very high level, the basic configurations and components for ground-mounted systems are similar to those of large commercial rooftop systems. Key features of ground-mounted systems are the array orientation methods, which include fixed tilt racks similar to rooftop systems but with greater allowable tilt angles, and mechanical trackers to actively align the modules to the path of the sun.

The most common tracker types include:

- single axis trackers, horizontal N-S axis (modules track east to west over the course of the day – see Figure 11.1.5 (a));
- single axis trackers that have a tilted N-S axis (towards the south – see Figure 11.1.5 (b));
- two-axis trackers, which track north to south seasonally, and east to west daily.

Small ground-mounted systems may be installed in yards, property or lots adjacent to commercial or industrial buildings. These can interconnect to the grid similar to rooftop systems, by bringing the AC power conductors to the facility electrical panels and utility service meters.

11.1.2.1.5 Utility-Scale PV Power Plants

The largest growth sector for grid-connected PV is the utility-scale plant, sometimes referred to as solar farms (Figure 11.1.6).

These plants are essentially aggregates of multiple large array-inverter configurations presented in Figure 11.1.1. Equipment pads or containers house multiple large inverters (e.g., four 500 kW inverters for an aggregate 2 MW capacity) and AC transformers which boost the voltage to utility distribution levels (e.g., 10–40 kV). Alternatively, three phase string inverters may be distributed throughout the array and their AC outputs combined at the equipment pads. The outputs of the transformers are collected with parallel underground feeder circuits and interconnected to the utility via a dedicated plant substation. The evolution of these plants as they get larger has been towards higher DC and AC voltages. 1000 V_{dc} systems

(a)

(b)

Figure 11.1.5 PV systems with (a) a single-axis horizontal tracker, and (b) a single-axis tilted tracker. Sources: (a) NEXTracker™ (2015); (b) Nellis AFB (2007)

Figure 11.1.6 250 MWac ground-mounted system in California. Source: SunPower

are the most common but 1500 V_{dc} systems are making inroads, and the possibility for higher voltages remains. The AC voltage is mostly dictated by the total size of the plant because of the total currents and distances involved. An important development area for PV power plants is how they are addressed in codes and standards, which evolved from practices and policies meant for small systems. Some requirements fitting for residential or commercial systems are unnecessarily expensive or immaterial for power plants, where operations are similar to conventional utility generation and access is highly controlled to qualified personnel. National and international codes and standards such as IEC design and installation standards (IEC TS 62548, IEC TS 62738 [draft]), the U.S. NEC, and other country-specific codes are therefore being revised to address the unique aspects of utility-scale PV.

11.1.3 Performance

The PV industry expends a significant amount of resources and energies toward monitoring, understanding, and improving the performance of grid-connected PV systems. This starts at the module cell level efficiency and ends at financially driven utility dispatching. This section focuses on PV system-level performance considerations. PV system monitoring is discussed in detail in Chapters 11.5 and 11.6.

11.1.3.1 Maximizing Energy Capture

A major objective of PV system design is to maximize the energy capture from the modules. This was especially true prior to the last decade when module prices were orders of magnitude higher, and the cost of delivered energy justified spending more on the balance of system components such as two-axis trackers. The more the modules track the path of the sun, the more energy they deliver. The historically high cost of module silicon also incentivized concentration techniques as well. Apart from the solar resource at the site location, which clearly has an enormous impact on production, there are numerous design and operation factors that influence system efficiency. These are summarized in Table 11.1.1.

Table 11.1.1 Design and operation factors impacting performance

Factor	Feature or impact	Considerations
Array orientation	Two-axis tracking	Energy capture is maximized by keeping the plane of array normal to the direct beam of the sun throughout the day and seasonally. Energy yield is as high as 40% over fixed tilt systems. High structural, space, and cost requirements.
	Single-axis tracking (N-S) axis	Energy capture is enhanced as much as 25% over fixed tilt systems. Simpler and less expensive than two-axis tracking systems. Typical tracker rotation range is 45 degrees East and West. "Backtracking" arrays away from low sun angles in the morning and night limit shading from adjacent rows of trackers.
	Fixed tilt systems	Least cost balance of system approach. Ideal tilt for annual production is the site azimuth angle (determined by latitude) ± 15–20 degrees. Practical considerations such as row-to-row shading, available space, and wind loading tend to favor tilts less than azimuth angle.
Array shading	Row to row shading	Shading of one row of modules by another, influenced by the tilt angle and space between rows. Cost and performance trade-off involves consideration of available space, structure costs, annual energy lost from shading. 3–5% shading loss typical.
	Object shading	Shading of modules by trees, structures, roof objects or features, chimneys, etc. Shading a portion of the array causes disproportionally high losses on series and parallel connected DC strings. Designing to minimize shading by objects can have a significant impact on annual production.
	Horizon shading	Shade created by landscape features such as hills, distant treelines, terrain slopes. Accurate accounting for these losses is important for determining project economics.
Electrical losses	Wiring losses	Resistive losses in potentially very long DC conductors, create voltage drop. Array designs have lower loss targets (0.5–1.5% on DC side, 1% or less on AC side) than typical AC industrial wiring (2–5%). Lower losses achieved by increasing conductor sizes.
	Array voltage	Wiring and inverter losses are impacted by the array voltage as determined by the number of modules connected in series per string. Losses occur if the maximum array voltage is outside the inverter's maximum power tracking range. Higher voltages reduce wiring resistive losses on a per Watt basis, offset somewhat by larger switching losses in the inverter. Lower voltages create higher wiring resistive losses on a per Watt basis.
	Inverter efficiency	Inverter DC-AC conversion efficiency is a significant competitive factor for manufacturers. Typical range is 94–98% and expressed in terms of peak and weighted output efficiency. Inverter maximum power point tracking (MPPT) efficiency typically can result in additional 0.5–1% loss.
	Transformer losses	Additional external transformers are sometimes required to interconnect with utility distribution or transmission systems. Losses on an annual basis typically in range of 0.5–1.5%.
Soiling		Modules soiled by dirt, dust, and other particulates can cause annual energy production losses between 1–10% or more. Periodic cleaning in operations phase can be justified depending on cost, safe access, etc.
Mismatch		Mismatch losses occur as a result of variations in module nameplate current and voltage ratings when connected in series and parallel strings, variations in the array plane angles, and non-uniform shading, soiling, age-related degradation. Impact ranges from less than 0.5% in well-designed ground mount systems to 5% on systems with significant object shading.

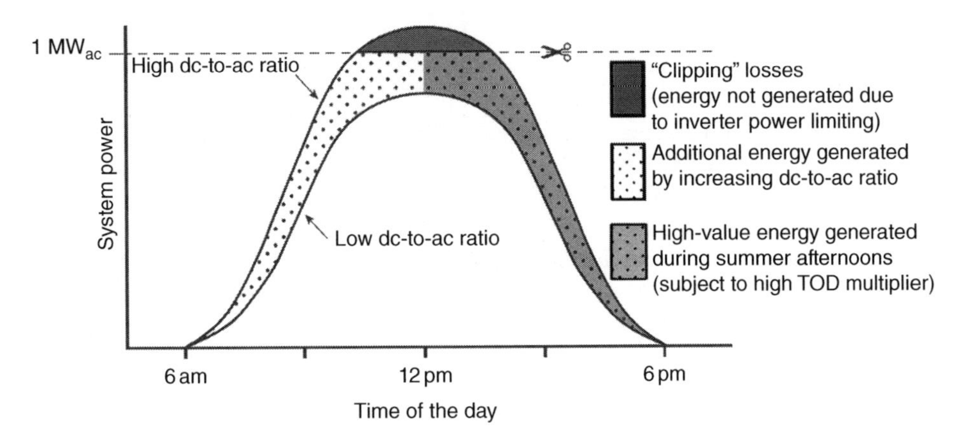

Figure 11.1.7 Impact of varying PV system DC/AC ratio. Source: *SolarPro Magazine* (2013)

11.1.3.2 Designing for Financial Return

Maximizing system efficiency by optimizing the design factors described above comes at a cost, and project economics such as the return on investment can drive very different design decisions. This is particularly true in more recent years (2005–2014), as module efficiency and manufacturing improvements have greatly reduced the cost of modules on a per-Watt basis. The result of the reduced module cost is a shift in strategy towards boosting system production rather than system efficiency. For example, a fixed-tilt system may be designed with tighter row-to-row spacing in order to fit more rows into the project boundaries. This lowers the array production efficiency by causing more row-to-row shading in the winter, but it also results in greater overall annual energy production relative to that of an array with fewer rows. Perhaps the best example of production-/cost-driven design is to increase the system's DC/AC ratio – that is, the array's DC rating relative to the inverter capacity. A design optimized for efficiency ensures that the array is not sized to produce more power than the inverter can convert. When the array is oversized, the inverter limits or "clips" power during high irradiance conditions by controlling away from the maximum power point of the array I-V curve. Figure 11.1.7 shows the energy production of a one megawatt AC inverter with a low and high DC/AC ratio array. The output of the high DC/AC ratio system is curtailed during peak sun hours as shown, resulting in a clipping loss that reduces its efficiency relative to the low DC/AC ratio design. However, the inverter produces significantly more energy from off-peak sun hours with the larger array, and the revenue from the additional energy can justify the marginal added cost of modules, wiring and mounting structures (Evarts and Leducq, 2013).

Additional details regarding the economics of PV can be found in Chapter 13.1.

11.1.4 Safety and Fire Protection

The importance of safety and fire protection in grid-connected PV systems is paramount. Due to the distributed nature of the technology as discussed earlier, PV is a general consumer product installed in residences, in yards, on carports, and other publicly accessible places. Yet it is still a source of generation, producing dangerous voltages and currents at the system level. Fortunately, PV systems can be designed to be very safe thanks to decades-long maturation of product safety standards installation codes and relevant safety procedures. Electrical fires in PV systems are uncommon but with the proliferation of systems, they have and will continue to occur as with other electrical products. Technology and standards are taking large strides at present to improve upon fault protection methods in order to reduce the likelihood and spread of fire in faulted cables or components. Fault protection – the basis for fire protection – is a set of measures to prevent and/or mitigate unintended currents and voltages on conductors and equipment, such as short-circuits (Reil *et al.*, 2011; Ball *et al.*, 2013). In PV systems, the primary fault concerns discussed here are short circuits between DC circuits, short circuits between DC circuits and ground, and arcing faults.

11.1.4.1 Overcurrent Protection

Overcurrent protection measures use fuses, circuit breakers, and other devices to limit currents to safe and appropriate levels and interrupt short circuits. Overcurrent protection in PV systems is unique relative to conventional generation and end use load circuitry due to the fact that PV is a limited current source. The short-circuit current from a PV module is only marginally higher (e.g. 10%) than its ideal operating current. Moreover, the short-current capacity of the module drops with low irradiance. On a cloudy day therefore, a string of modules generating into a hard short produces less fault current than its normal operating current at rated conditions (1000 W/m^2). This makes it impossible to size a fuse to interrupt the source fault current as one would for circuits supplied by the utility or conventional generator. For grid-connected systems with only a few strings, high fault current can only come from the utility through the inverter. In systems where there are multiple strings, a short circuit in one string causes current from all of the parallel connected strings to divert from the inverter and flow back into the shorted string. Thus fault current flows in the opposite direction of the normal string current, and is limited by the number of parallel strings. One might think that low fault current is a good thing, but it is actually better to have high available fault current to ensure that a fuse or circuit breaker opens quickly to interrupt the short.

A fuse should just be large enough to prevent unintended opening (or tripping) with normal currents in high irradiance and temperature conditions. The fuses are located in a combiner box where the individual strings are connected in parallel (see Figure 11.1.1). Combiner box output circuits (called array circuits), if combined again before or at the DC input terminals of the inverter, are protected similarly with larger fuses rated according to the total number of strings feeding the circuit. As in the case of strings, the fault currents in array circuits vary considerably with irradiance and the number of parallel connected circuits, making their effectiveness less than ideal. For array circuits, a more effective approach is the use of reverse current detection. Reverse current of any magnitude beyond noise and leakage levels is indicative of a problem in array circuits, and is a reason for interrupting the current flow (Albers and Ball, 2014).

11.1.4.2 Grounding and Ground Fault Protection

Ground faults are short circuits between a live conductor and a grounded conductive surface, such as metallic enclosures, cable conduits or trays, module frames, mounting structures, etc. They occur when a conductor's insulation is damaged or nicked, and the exposed wire makes contact with the metal object. Two potentially serious issues can occur as a result if adequate protection is not in place: (1) a potentially large fault current can flow because of the voltage difference between the conductor and ground (e.g. up to 1000 V_{dc} on standard PV systems) leading to dangerous arcs and potentially fire; and (2) the metal object, if not properly grounded, can have the conductor's voltage imposed on it, creating a potentially lethal hazard if touched. Ground faults are more common than faults between conductors since only one cable needs to be damaged to occur, and it is often the metal surface causing the damage to the insulation.

Ground fault impacts and protection techniques differ depending on the PV system's overall DC *system* grounding design. PV system grounding is the deliberate connection or referencing of one polarity of the DC circuit to ground. This is distinct from *equipment* grounding, which is the connection of conductive metal enclosures and parts to ground to ensure they are touch-safe. All systems employ equipment grounding, but system grounding is optional. The vast majority of grid-connected systems employ one of three system grounding methods (Wiles, 2012). The methods and the protection implications are summarized in Table 11.1.2.

Table 11.1.2 PV system grounding and protection methods

System grounding type	Definition and use	Ground fault protection considerations
Ungrounded	Neither the positive or negative DC circuit is grounded or referenced to ground. Used historically in most locations outside of North America, and is still the preferred method unless ground referencing is required to prevent module potential induced degradation (PID).	The first ground fault in the array does not create the fault current, since there is no path for the short circuit flow. The fault simply creates a grounded circuit. The fault is detected by insulation resistance monitoring, which measures the resistance between each polarity conductor and ground on a continuous basis. Ground faults are indicated if the measured resistance drops below a threshold value. The faulted circuits are isolated until they can be repaired. Short circuits occur only if a second fault in the opposite polarity occurs before the first fault is isolated or repaired.
Grounded	One of the DC circuits (either the positive or negative polarity conductor, but usually the negative) is connected to ground through a protective device.	The circuit connection to ground typically takes place in the inverter, which controls and/or monitors the current through the protective device. Ground faults in the opposite polarity circuit result in the array being shorted, and cause significant short circuit currents. The fault is quickly interrupted by the protective device. Ground faults in the same polarity circuit are more difficult to detect, but are done so with recent advanced methods, including residual current monitoring – which isolates a circuit if there is a difference between currents in the positive and negative polarity conductors.
AC referenced grounded	The DC circuits' voltage relative to ground varies during inverter operation.	This occurs in systems that do not use AC isolation transformers and is an artefact of the inverter control method. Ground faults in the DC system cause short circuit currents to be fed back through the inverter from the AC grid, and are readily mitigated by AC-side residual current detectors.

11.1.4.3 Arc Fault Protection

An arc is formed when current jumps a gap between conductive parts and can be sustained by continuous ionization of the air in the path. They are dangerous if sustained because of the intense flash and heat that can melt conductors and enclosures and lead to system fires. Arcs in DC systems can persist more readily than those of AC systems, since DC voltage is maintained, while AC current and voltage fall to zero twice per cycle. Electrical arcs take place, for example, in switches and circuit breakers, if the conductive parts do not separate quickly enough to interrupt the flowing current. In PV systems, arc faults typically occur in string connections – in the connectors between modules, at connections in the module junction box, at terminations in combiner boxes. They occur when loose connections build up heat from the high resistances, damaging the conductors or connector until the physical connection has deteriorated and current jumps an emerging air gap. Arcs faults in these connections are called series arcs, since the arcing current makes a series connection in the normal path of current. As such, they are not detected by short-circuit or ground fault protection devices. Series arc faults are detected by devices that measure circuit currents and look for a signature high frequency spectrum that is characteristic of the ionization and plasma discharge in the arc. If the series arc is detected, it can be extinguished by simply turning the inverter off and stopping the current flow. PV arc fault detection and mitigation are problematic with the current technology because of the difficulties of accurately or correctly detecting the arc current spectrum, and distinguishing it from conducted noise from different inverter topologies, from radiated noise in the array, or short-term arcs that naturally occur in switches or breakers (Johnson and Armigo, 2014). Despite the fact that several countries now require arc fault protection in PV systems, and that they are establishing some success at the string inverter level, significant research is still needed to improve effective detection approaches that do not lead to unacceptable nuisance tripping (Armijo *et al.*, 2014).

Common to each of these advanced fault protection developments is the use of active switching devices such as contactors, instead of passive fuses and circuit breakers. The contactors allow circuits to be opened and isolated by control circuits in response to multiple detected issues: reverse current in string or array conductors, residual ground current in the milliamp range, low conductor resistivity to ground, arc fault signatures, etc. The contactors or similar active switches can be located at any level of the PV DC circuit – at the string level, the output of combiner boxes, the input of array combiners, or inverters.

11.1.4.4 Safety of the Public

The advances in fault protection focus on prevention and mitigation of abnormal events that can lead to fire, and enhance what is already a very safe technology from the standpoint of day-to-day personnel interaction. Safety of the public is achieved most significantly by ensuring exposed equipment is touch-safe, and this is in turn achieved through product safety testing and effective grounding of the assembled components in a system. Modules, for example, are certified to product safety standards, e.g., IEC or UL standards, which dictate construction requirements and tests to reduce the likelihood of voltage on the exposed frames and parts. The effective grounding of mounting systems or structures is well understood, but there is significant innovation in this area to utilize mechanical connections and structural members as the ground path instead of relying on ground conductors to connect to the numerous parts.

11.1.4.5 Safety of the Emergency Personnel

PV systems involved in building fires, whether the cause of the fire or not, require special procedures and attention by firefighters and other emergency personnel. When a building fire occurs, one of the first steps taken by firefighters is to turn off the utility-supplied electricity to the building. Grid-connected PV systems shut off automatically when the grid voltage is removed, thereby reducing all system currents to zero. However, in daylight, the PV modules and connected conductors and live parts still do remain energized with voltage, and therefore remain a potential hazard if contact is made, for example, by a firefighter's axe cutting through roofing. This issue has received considerable attention in the industry and mainstream press, the latter of which has sometimes badly mischaracterized the threat and source of problems. Established studies, most notably from Germany, have demonstrated the actual impact of PV systems in these instances and document methods and procedures developed jointly by PV industry experts and firefighters to safely work around PV systems during firefighter operations (Laukamp *et al.*, 2013). The attention has led to real improvements as well in installation codes, such as enhanced pathways and access on roofs around arrays, and better clearances at the roof edges. Enhanced circuit control is also being required in some countries, most notably the USA, involving active switching (such as contactors) to ensure conductors leaving the arrays and connected to the inverters are disconnected and de-energized (Brooks, 2015). Many believe this will ultimately result in requirements to shut off systems at the module level in emergency conditions – thereby removing any presence of high DC voltage on the roof. Significant research and testing are needed to develop certification standards to ensure that such a capability can be provided safely and reliably over the long life of a PV system.

11.1.5 Conclusion

The dramatic growth of grid-connected PV in the past decades should continue for decades to come, given the continuous lowering of costs, the modular nature of the technology, and the increased demand for clean generating sources. This chapter has presented the basic design, application, performance, and safety aspects of grid-connected PV, as well as the most important technological developments and trends. Despite its apparent simplicity, there is considerable opportunity for advances in all areas of grid-connected system design. Beyond the topics discussed here, there will be continued advances in "smart" components: Smart modules, combiner boxes, and inverters that can communicate and respond to system or operator triggers, advanced diagnostic capabilities that reduce personnel maintenance requirements, racking, and methods that greatly reduce installation and maintenance labor costs, to name a few. Innovation to improve safety and fire prevention will also continue for some time, to keep up with the proliferation of systems in public spaces.

Acknowledgments

The author thanks the International Electrotechnical Commission (IEC) for permission to reproduce Information from its International Standards. All such extracts are copyright of IEC, Geneva, Switzerland, all rights reserved. Further information on the IEC is available from www.iec.ch. IEC has no responsibility for the placement and context in which the extracts and contents are reproduced by the author, nor is IEC in any way responsible for the other content or accuracy therein.

List of Acronyms

Acronym	Unit	Description
I	A, or Amp	Electrical current
AC		Alternating Current, used to describe current, voltage and power that periodically reverses direction, and which is used in electric utility grids.
AM		Air Mass, expressed in AM units
BIPV		Building Integrated PV
T	°C	Temperature in degrees Celsius
DC		Direct Current, used to describe current, voltage and power that is unidirectional, and which is produced by solar modules, batteries, etc.
EPIA		European PV Industry Association
GW		Gigawatt, or 1 billion Watts
IEC		International Electrotechnical Commission, a developer of international standards for products, performance, qualification and installation
IEEE		Institute of Electrical and Electronics Engineers, a U.S. standards organization with large focus in electrical power.
kW		KiloWatt or 1 thousand Watts
MW		Megawatt, or 1 million Watts
MPPT		Maximum Power Point Tracking
NEC		National Electrical Code, sets requirements for electrical installations in the USA
P	W	Electrical power
P_{STC}	Wp	Watt peak – the DC power rating of PV modules and arrays at STC condition
PID		Potential Induced Degradation
Single Phase AC		AC electricity supplied as a single alternating voltage source, for low power and used in most residences
STC		Standard Test Conditions, the test condition for rating PV cells and modules: 1,000 W/m^2 solar irradiance, 25 °C module cell temperature, and an air mass of AM 1.5
Three-Phase AC		AC electricity supplied as three alternating voltage sources offset from each other in time, efficient for utility power delivery and for create rotating magnetic fields for generators and motors
UL		Underwriters Laboratories, a developer of product safety and performance standards.
V	V	Electrical voltage potential
V_{ac}	V	AC Electrical voltage
V_{dc}	V	DC Electrical voltage
P	W	Electrical power
P_{STC}	Wp	Watt peak - the DC power rating of PV modules and arrays at STC condition

References

Albers, M. and Ball, G. (2014) Comparative evaluation of DC fault mitigation techniques in large PV systems, Conference Record of the 40th IEEE PVSC, Denver, Colorado.

Armijo, K. M., Johnson, J. *et al.* (2014) Characterizing fire danger from low power pv arc-faults, Conference Record of the 40th IEEE PVSC, Denver, Colorado.

Ball, G., Brooks, B. *et al.* (2013) *Inverter Ground Fault Detection Blind Spot and Mitigation Methods*. Public Report, Solar America Board for Codes and Standards, June.

Brooks, B. (2015) Rapid shutdown for PV systems: understanding NEC 690.12, *SolarPro Magazine*, **8** (1), 26–40.

California State University, Fresno, 2007; http://www.fresnostatenews.com/archive/2007/11/solarparking.htm.

Colthorpe, A. (2014) IHS: Residential PV storage market to leap tenfold by 2018. PV Tech Storage, Nov. 12, 2014, http://storage.pv-tech.org/.

CPP Inc., 2012; http://solarprofessional.com/articles/design-installation/wind-load-analysis-for-commercial-roof mounted-arrays.

DOE EERE, *Homeowners Guide to Financing a Grid-Connected Solar Electric System*, Solar Energy Technologies Program publication, October, 2010. http://www1.eere.energy.gov/solar/pdfs/48969.pdf.

EPIA (2014) *Global Market Outlook for Photovoltaics, 2014–2018*, European PV Industry Association.

Evarts, G. and Leducq, M. (2013). Designing for value in large-scale PV systems. *SolarPro Magazine*, **6** (4), 14–24.

IEA, (2013) *Trends 2013 in Photovoltaic Applications: Survey Report of Selected* IEA *Countries between 1992 and 2012*. International Energy Agency, Report IEA-PVPS T1-23:2013.

IEC 61215, Edition 2.0 (2005) Crystalline silicon terrestrial photovoltaic (PV) modules: design qualification and type approval. International Electrotechnical Commission.

IEC TS 62548, Edition 1 (2013) Photovoltaic (PV) arrays: design requirements. International Electrotechnical Commission.

Johnson, J. and Armijo, K.M. (2014) Parametric study of PV arc-fault generation methods and analysis of conducted DC spectrum. Conference Record of the 40th IEEE PVSC, Denver, Colorado.

Laukamp, H., Bopp, G., *et al.* (2013) PV fire hazard: analysis and assessment of fire incidents. Conference Record of the 28th EU PVSEC Conference, Paris.

Nellis AFB, Sr. Airman Larry Reid, Jr., 2007; http://www.nellis.af.mil/news/story_print.asp?id=123066656.

O'Brien, C. and Banks, D. (2012) Wind load analysis for commercial roof-mounted arrays. *SolarPro Magazine*, **5** (4), 72–92.

Reil, F., Vaaßen, W. *et al.* (2011) Determination of fire safety risks at PV systems and development of risk minimization measures, Conference Record of the 26th EU PVSEC, Hamburg, Germany.

SolarPro Magazine, (2013); http://solarprofessional.com/articles/design-installation/designing-for-value-in-large-scale-pv-systems.

SunPower; http://www.nrg.com/renew/projects/solar/california-valley-solar-ranch/.

Topher Donahue, (2009) http://solarprofessional.com/articles/project-profiles/commercial-grid-direct-photovoltaic system-denver-museum-of-nature-science.

Wiles, J. (2012) *Photovoltaic System Grounding*. Public Report, Solar America Board for Codes and Standards, October.

11.2

Inverters, Power Optimizers, and Microinverters

Chris Deline
National Renewable Energy Laboratory, Golden, CO, USA

11.2.1 Introduction

A grid-tied PV inverter's function is to convert direct current (DC) power generated by the solar array to alternating current (AC) power suitable for export on the local electrical grid. To perform this conversion effectively, three separate functions are required. The most important function is to generate a sinusoidal waveform by switching the DC input to the appropriate grid frequency and filtering higher-order harmonics. A second function of the device is to provide maximum-power-point tracking (MPPT) of the DC source. MPPT ensures that the maximum available DC power is processed by the inverter circuit. A third function is to monitor the grid for appropriate interconnection conditions (anti-islanding or voltage/frequency support) and to monitor the DC side for operation conditions free of ground and arc faults.

Inverters span a wide range of sizes, topologies, and connection voltages: from utility-scale megawatt inverters to string inverters. Systems can include module-level micro-inverters or DC-DC converters as well, to increase the granularity of MPPT which improves performance under conditions of partial shading or orientation mismatch (Figure 11.2.1).

11.2.2 Power Conversion

Switch-mode power conversion relies on high frequency (typically 10s to 100s of kHz) chopping of DC signal to periodically charge and discharge energy storage elements, such as inductors and capacitors. To match the AC voltage at any instant in time, the input voltage may need to be increased (boosted) and will need to decrease to zero (bucked). In a residential 240 V_{ac} system, for example, the peak of the sine wave is 330 V, so a solar array with an input

Photovoltaic Solar Energy: From Fundamentals to Applications, First Edition.
Edited by Angèle Reinders, Pierre Verlinden, Wilfried van Sark, and Alexandre Freundlich.

Companion website: www.wiley.com/go/reinders/photovoltaic_solar_energy

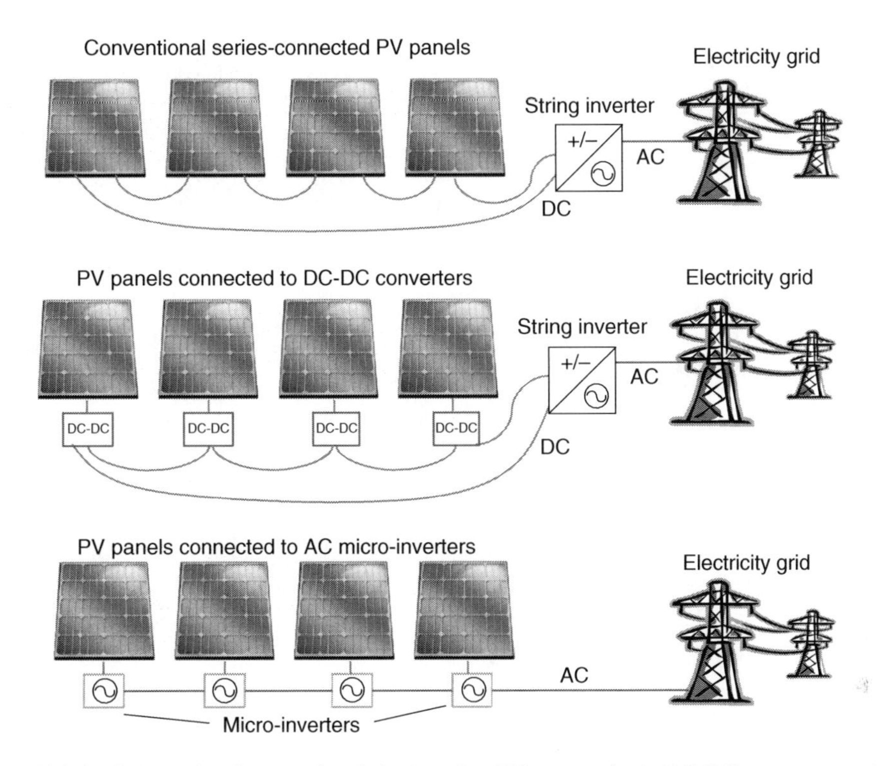

Figure 11.2.1 Schematic of conventional single-string PV system (top), DC-DC converter-equipped "Smart Modules" (middle), and AC micro-inverter-equipped PV system (bottom). (*See insert for color representation of the figure*)

less than that value must incorporate a boost stage followed by a DC-AC buck stage. A microinverter requires the greatest boost ratio because of the low DC input voltage, typically requiring a step-up transformer instead of a conventional inductor. Many inverter designs rely on transformers to enable galvanic isolation. However, for efficiency, weight and cost reasons, transformerless designs are becoming more common, as depicted in Figure 11.2.2 (Xue *et al.*, 2004).

The high-frequency transformer is an improvement on older designs using large, lossy line-frequency transformers. Additional DC-DC or DC-AC stages can be included to improve efficiency or reduce ripple, at the cost of greater circuit complexity. In very large (utility-scale) inverters, cost and efficiency can be improved by more closely matching the allowed DC input voltages to the AC line voltage. This enables voltage conversion stages to be removed because the DC input voltage is directly unfolded to AC through the line-frequency inverter stage, hence enabling high efficiency, single-stage inverters.

Transformerless inverters can be smaller, lighter, and more efficient because of the removal of the transformer and its associated conduction and magnetic losses. Unlike in a galvanically isolated inverter, the negative DC input of a transformerless inverter can remain floating with respect to ground. Transformerless inverters have recently been allowed by residential electrical codes in the United States, but have long been allowed in Europe, and they are shown to improve ground-fault safety margin relative to inverters with a solidly grounded negative DC conductor (Kjaer *et al.*, 2005).

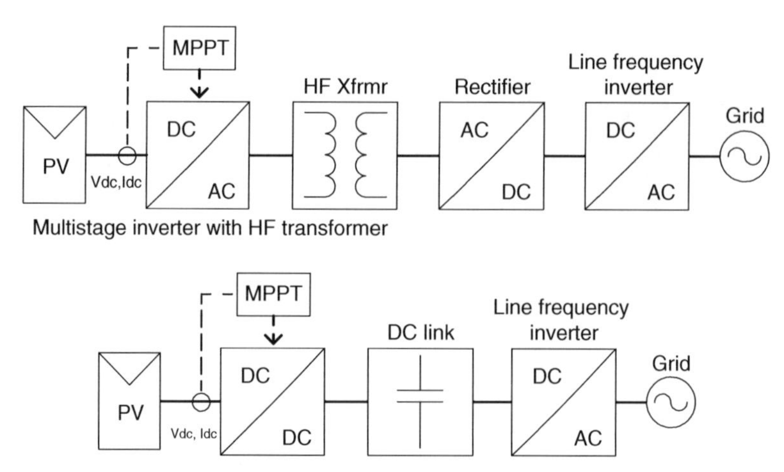

Figure 11.2.2 Block diagrams of PV inverter examples: (top) high-frequency isolated multistage; and (bottom) DC-DC transformerless designs

11.2.2.1 Practical Converter Considerations

Additional circuit components are required to address practical issues in inverters such as voltage ripple and harmonic distortion. The instantaneous AC power injected to the grid (assuming unity power factor) is:

$$p_{AC} = 2\overline{p}_{AC} \sin^2(\omega t), \tag{11.2.1}$$

where $\overline{p}_{AC}$ is the average injected power in (W), ω is the grid cycle frequency (Hz), and t is time in (s). The sine-squared term results in a double-frequency (100 Hz or 120 Hz) voltage ripple present on the DC input. Conceptually the sinusoidal power output has zero-crossings where no power is exported to the grid. Energy storage capacitors must buffer the full PV power production during these half-cycle null points. The input filter capacitors represented in Figure 11.2.3 can be quite substantial, and many inverter topology concepts are intended to reduce the size of these capacitors, which can be expensive, bulky, and present a reliability concern.

Additionally, unwanted high-frequency content can occur on the AC or DC side of the inverter because of the switched nature of the converter. Insulated-gate bipolar transistors (IGBTs) or Power MOSFETs are operated by square wave pulses, or pulse-width modulation (PWM), to reduce on-off transition losses. A simple PWM signal which determines the switches' on state, can be generated by comparing an AC voltage reference signal with a triangle-wave carrier signal. The carrier frequency ranges from 1 kHz to tens of kHz depending on inverter size, topology, and switch technology. In general, higher switching frequencies result in increased cost of components and losses, but allow smaller and lighter filter components to be used (Hong and Zuercher-Martinson, 2013). Figure 11.2.4 shows an example PWM signal that drives the inverting switches' on state. Higher PWM duty cycle results in a higher instantaneous current output.

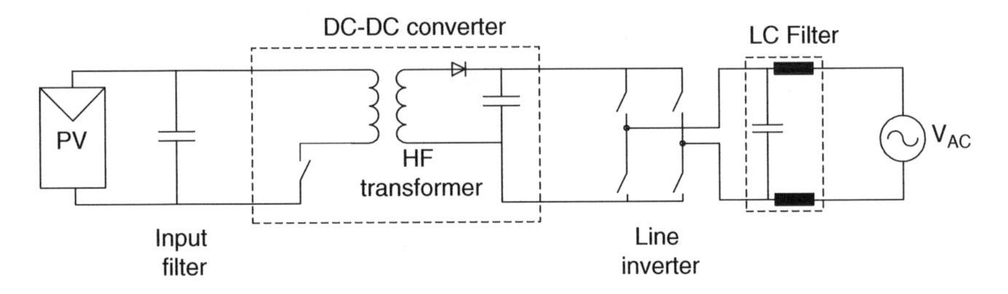

Figure 11.2.3 High-frequency isolated multi-stage inverter. Source: (Burger *et al.*, 2010)

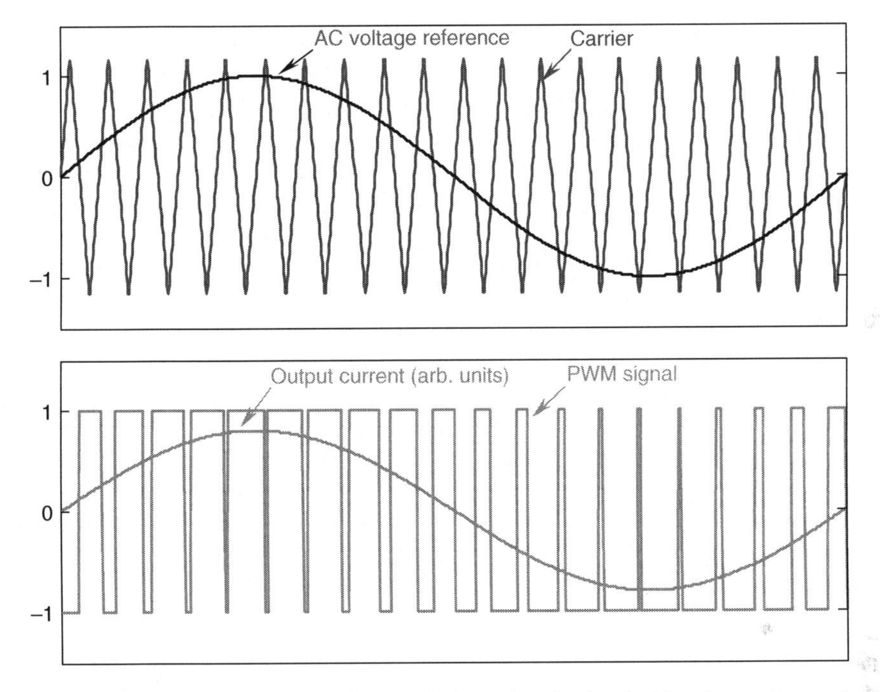

Figure 11.2.4 PWM signal (top) generated by comparison of carrier signal with voltage reference (bottom)

The output current signal in Figure 11.2.4 is idealized; additional filtering is required to remove the kHz switching frequency from the AC output, as represented by the LC filter in Figure 11.2.3. Additionally, hard switching of transistors generates increased harmonic content, which arises from the Fourier decomposition of a square wave, and from additional switching transients, such as ground bounce or inductive "ringing" in the current path. This harmonic content can typically be filtered using electromagnetic interference (EMI) filters to meet grid-quality and emitted radio-frequency interference (RFI) standards, as described in (IEEE, 2008).

11.2.2.2 Wide Band Gap Components

Inverters are beginning to incorporate components with a bandgap above should be 3 eV, such as SiC and GaN (Tolbert *et al.*, 2005). These lower-loss devices enable higher switching frequencies and/or higher operating voltages than conventional Si-based components. As a result, higher-efficiency designs may be possible, as well as taking advantage of the large reverse-breakdown voltage of SiC to convert directly to 1.5 kV distribution voltage levels without distribution transformers.

11.2.3 DC Maximum Power Point Tracking

PV modules respond dynamically to changing temperature and irradiation conditions. Thus, maximum DC power extraction requires periodic adjustment of the PV voltage and current operating point. This function is accomplished in the inverter's MPPT stage whereby the DC operating voltage (or current) is adjusted slightly as DC power is monitored for maximum production (Hohm and Ropp, 2003). The MPPT algorithm can be tuned based on the inverter's size and dynamic response, with the main algorithm category being "perturb and observe" (P&O). P&O is popular because of its simplicity, a small perturbation in voltage ΔV results in a measured change in power ΔP. If ΔP is positive, an additional ΔV is applied in the same direction. If ΔP is negative, ΔV is applied in the opposite direction. This method is described as a hill-climbing algorithm because the inverter's operating point climbs the P-V curve from the open circuit voltage V_{oc} to the voltage at maximum power V_{mp} to find the local maximum DC power. The tuning of the MPPT algorithm is important in balancing static response with dynamic irradiance response. In the static case, small ΔV steps are desired to limit power loss from operating far from the local V_{mp}. In the dynamic case, changing irradiance leads to dynamic changes in the local V_{mp}. Small ΔV steps mean that the inverter does not track the changed V_{mp} as quickly. Other MPPT algorithms (e.g., Esram *et al.*, 2006) have been shown to improve the dynamic response of inverters under rapidly changing irradiance conditions by using the inverter's inherent ripple for voltage perturbation of the array input.

An additional consideration for inverter MPPT is mismatch caused by partial shading. In the P&O strategy described above, the algorithm operates around a local maximum point,

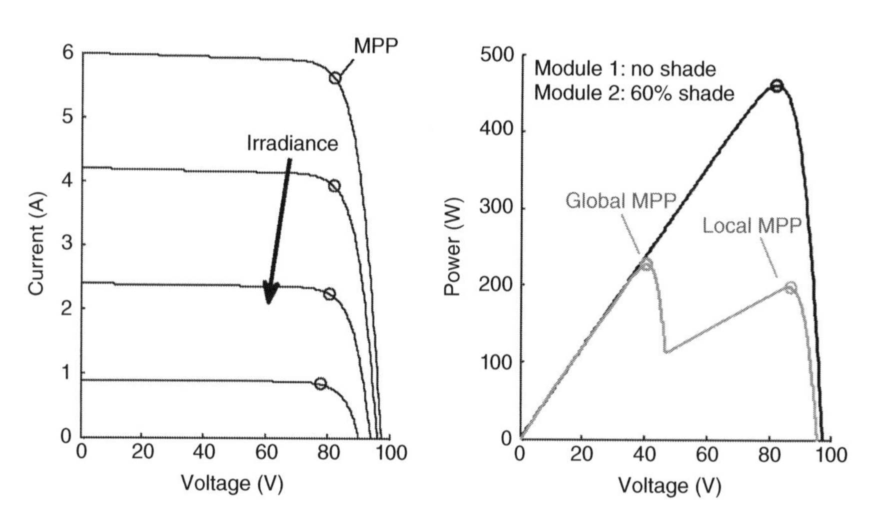

Figure 11.2.5 Comparison of MPP voltage and current with change in irradiance (left). Partial shading (right) can result in multiple local maxima, potentially affecting the MPPT operation

which may not necessarily coincide with the global maximum (Figure 11.2.5). Additional algorithms are designed to seek out a global maximum specifically for this purpose. This feature has also been implemented in some commercial inverters as a global optimization, or shadow mode. Because of the additional time spent away from MPP to check for additional local operating points, static MPPT efficiency is reduced slightly for these methods.

Additional options in overcoming partial-shade mismatch loss include distributing the MPPT function throughout the PV system through the use of microinverters or DC-DC-equipped "smart modules." Mismatch loss is reduced by moving from a single central inverter, to string inverters, to microinverters. The amount of performance improvement depends on the extent of shade and the system configuration, but module-level solutions have been shown to typically recover about a third of the loss from partial shading (Deline *et al.*, 2014).

11.2.4 Inverter Efficiency

An inverter's total efficiency is measured by the product of its conversion efficiency and the MPPT efficiency (how closely the MPPT circuit tracks the actual maximum P_{mp}):

$$\eta_{total} = \eta_{Conv}\eta_{MPPT} = \frac{V_{ac}I_{ac}\cos(\phi)}{V_{dc}I_{dc}}\frac{V_{dc}I_{dc}}{V_{mp}I_{mp}}, \tag{11.2.2}$$

where $V_{mp}\cdot I_{mp}$ is the PV module maximum power, $V_{dc}\cdot I_{dc}$ is the DC input power to the inverter, and $\cos(\phi)$ is the AC power factor. This static efficiency is based on steady-state operation. Additional efficiency losses can occur during dynamic changes or in the presence of mismatch, for example, from partial shading.

11.2.4.1 CEC vs European Efficiency

The total efficiency of an inverter can be measured at several DC input voltage and operating power conditions, as shown for the example microinverter in Figure 11.2.6. Measures of overall efficiency in a single value have been defined based on different use scenarios. The California Energy Commission (CEC) efficiency is a weighted average based on a

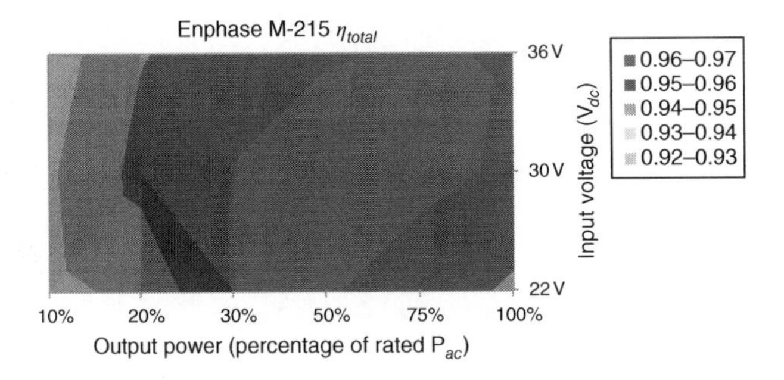

Figure 11.2.6 Total efficiency vs DC input voltage and AC output power for an Enphase M215 microinverter. (*See insert for color representation of the figure*)

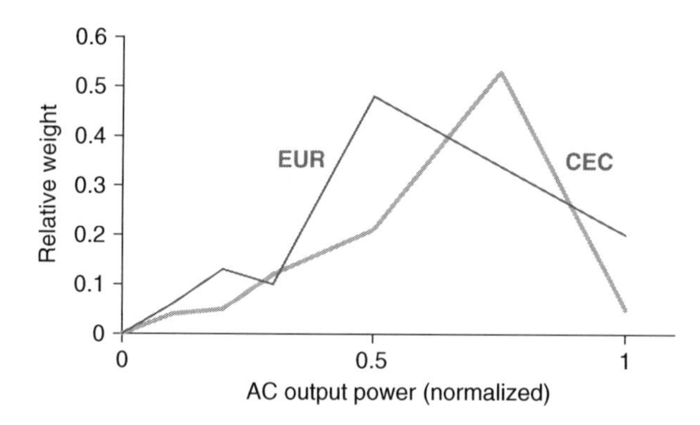

Figure 11.2.7 Relative weight of measured efficiency at given output power levels for CEC and European conversion efficiency

high-irradiance operating scenario, so that the inverter is operating close to its maximum power for a large portion of the year. The European efficiency is a similar weighted average based on a cloudier climate (Figure 11.2.7). The equations for calculating the overall efficiency are (CENELEC, 2010; IEC, 2014):

$$\eta_{CEC} = 0.04\eta_{10\%} + 0.05\eta_{20\%} + 0.12\eta_{30\%} + 0.21\eta_{50\%} + 0.53\eta_{75\%} + 0.05\eta_{100\%} \quad (11.2.3)$$

$$\eta_{EUR} = 0.03\eta_{5\%} + 0.06\eta_{10\%} + 0.13\eta_{20\%} + 0.10\eta_{30\%} + 0.48\eta_{50\%} + 0.20\eta_{100\%} \quad (11.2.4)$$

11.2.5 Auxiliary Functions

In addition to the primary functions listed above, inverters can include auxiliary capabilities, such as monitoring of DC and AC performance, and other error reporting. In the case of micro-inverters or DC power optimizers, this performance monitoring can be at the module level, which aids in troubleshooting system performance or reliability issues. Additional balance-of-system components such as disconnect switches, fuses, and distribution transformers can also be packaged within the inverter to reduce overall system cost. To support grid interactivity, inverters can be required to provide system safety benefits such as DC conductor ground-fault/arc fault detection, anti-islanding control, or rapid system shut-down capability in case of an emergency. Also, auxiliary grid support is increasingly required as PV energy becomes more common on the grid. These grid support features can include low-voltage ride-through, VAR support and frequency support.

11.2.6 Conclusion

Inverters play an important role as the interface between the PV modules and the AC grid. While inverters' cost and efficiency are of paramount importance, consideration must also be given to factors such as reliability, match to the system's DC operating voltage and power, and the presence of shade or mismatch in the system. Modern inverters are becoming ever more efficient, reliable, and cost-effective, speeding the adoption of PV from the smallest microinverter system to the largest multi-megawatt utility-scale system.

List of Symbols

Symbol	Description
$\cos(\phi)$	AC power factor (unitless)
$\eta_{10\%}$	Inverter total efficiency at e.g. 10% of rated AC power (unitless)
η_{CEC}	California Energy Commission weighted inverter total efficiency (unitless)
η_{Conv}	Inverter conversion efficiency (unitless)
η_{EUR}	European weighted inverter total efficiency (unitless)
η_{MPPT}	Inverter MPPT efficiency (unitless)
η_{total}	Total inverter efficiency (unitless)
I_{ac}	AC output current from the inverter (A*rms*)
I_{dc}	Input current to the inverter (A)
I_{mp}	PV module maximum-power current (A)
P_{mp}	PV module maximum DC power (W)
$\underline{p}_{AC}$	Instantaneous injected power (W)
$\overline{p}_{AC}$	Average injected power (W)
t	Time (s)
V_{ac}	AC output voltage from the inverter (Vrms)
V_{dc}	DC input voltage to the inverter (V)
V_{mp}	PV module maximum-power voltage (V)
V_{oc}	PV module open-circuit voltage (V)
ω	Grid cycle frequency (Hz)

List of Acronyms

Acronym	Description
AC	Alternating Current
DC	Direct Current
EMI	ElectroMagnetic Interference
GaN	Gallium Nitride
HF	High Frequency
IGBT	Insulated-Gate Bipolar Transistor
LC	Inductor-Capacitor
MOSFET	Metal-Oxide Field-Effect Transistor
MPPT	Maximum Power-Point Tracking
P and O	Perturb and Observe
PWM	Pulse-Width Modulation
RFI	RadioFrequency Interference
SiC	Silicon Carbide

References

Burger, B., Goeldi, B., Rogalla, S., and Schmidt, H. (2010) Module integrated electronics: an overview. In *Proceedings of 25th European Photovoltaic Solar Energy Conference and Exhibition*, pp. 3700–3707.

CENELEC (2010) Overall efficiency of grid connected photovoltaic inverters, EN 50530, Apr. 2010.

Deline, C., Meydbray, J., and Donovan, M. (2014) Photovoltaic Shading Testbed for Module-Level Power Electronics: 2014 Update. NREL Report No. TP-5J00-62471. Available at: http://www.nrel.gov/docs/fy14osti/62471.pdf

Esram, T., Kimball, J.W., Krein, P.T. *et al.* (2006) Dynamic maximum power point tracking of photovoltaic arrays using ripple correlation control. *IEEE Transactions on Power Electronics*, **21** (5), 1282–1291.

Hohm, D.P. and Ropp, M.E. (2003) Comparative study of maximum power point tracking algorithms. *Progress in Photovoltaics: Research and Applications*, **11** (1), 47–62.

Hong, S. and Zuercher-Martinson, M. (2013) Harmonics and noise in photovoltaic (pv) inverter and the mitigation strategies. Solectria Renewables White Paper. Available atL http://www.solectria.com//site/assets/files/1482/solectria_harmonics_noise_pv_inverters_white_paper.pdf

IEC (2014) Indoor testing, characterization and evaluation of the efficiency of photovoltaic grid-connected inverters, IEC 62891 Ed. 1.0 Committee Draft, Aug. 2014.

IEEE (2008) Application Guide for IEEE Std. 1547, IEEE Standard for Interconnecting Distributed Resources with Electric Power Systems, IEEE 1547.2-2008.

Kjaer, S.B., Pedersen, J.K., and Blaabjerg, F. (2005) A review of single-phase grid-connected inverters for photovoltaic modules. *IEEE Transactions on Industry Applications*, **41** (5), 1292–1306.

Tolbert, L.M. *et al.* (2005) Power electronics for distributed energy systems and transmission and distribution applications. *ORNL/TM-2005/230, UT-Battelle, LLC, Oak Ridge National Laboratory, 8*. Available at: http://web.ornl.gov/sci/decc/Reports/PE%20For%20DE%20and%20TandD%20Applications%20%28ORNL-TM-2005-230%29.pdf

Xue, Y., Chang, L., Baekhj Kjaer, S. *et al.* (2004) Topologies of single-phase inverters for small distributed power generators: an overview. *IEEE Transactions on Power Electronics*, **19** (5), 1305–1314, Sept. 2004. DOI: 10.1109/TPEL.2004.833460.

11.3

Stand-Alone and Hybrid PV Systems

Matthias Vetter and Georg Bopp
Fraunhofer Institute for Solar Energy Systems ISE, Freiburg, Germany

11.3.1 Introduction

In this chapter, various stand-alone PV systems and hybrid PV system solutions as well as their corresponding components will be described. Furthermore, typical applications ranging from technical devices to so-called rural electrification systems are considered. In the field of stand-alone PV, the power range extends from some Watts in the case of “Pico-PV” devices to isolated mini-grid applications with several MW. All these systems have one important boundary condition in common: The available power sources are limited as there is no available connection to a supervisory power layer, neither for small stand-alone systems nor for isolated mini-grids, as there is no transformer to any transmission grid. Therefore, a proper system design, starting with a detailed analysis of the load patterns, is crucial. Furthermore, an optimized energy management structure is needed, which integrates generation, storage, and demand-side management to fulfill all the requirements of a reliable and resilient state-of-the-art power supply system. Therefore, in this chapter these aspects of stand-alone and hybrid PV solutions are presented and discussed.

11.3.2 Solar Pico Systems

Solar pico systems, also called pico PV, have followed a strong development path in the last few years, because they combine the use of very efficient lighting (mostly LEDs) with sophisticated charge controllers and efficient types of batteries, particularly lithium-ion batteries. Already with a small PV panel of 0.3 Wp up to 10 Wp, essential lighting services can be provided and replace air-polluting kerosene-burning lights, candles and throw-away batteries. Figure 11.3.1 shows a variety of different solar lanterns available on the market.

Photovoltaic Solar Energy: From Fundamentals to Applications, First Edition.
Edited by Angèle Reinders, Pierre Verlinden, Wilfried van Sark, and Alexandre Freundlich.

Companion website: www.wiley.com/go/reinders/photovoltaic_solar_energy

Figure 11.3.1 Examples of different solar lanterns, Source: (Bopp *et al.*, 2012)

With these small PV systems, additional services can be covered: powering a small radio, a music player and furthermore charging a mobile phone. Recently expandable solar pico systems have entered the market. Households can start buying a small kit, serving small loads, such as two lights and a radio. Gradually they can add an extra kit, so additional lights and devices such as a small black-and-white TV can be connected. The equipment costs are much lower than the cost of solar home systems, so a much larger market can be reached, with simpler business models, like cash and carry or small microcredits.

On the other hand, it should be realized that solar pico systems can only provide an initial level of service which does not imply that the users should be considered "electrified" – as a matter of fact, in a number of regions, these systems are being sold as simple off-the-shelf consumer appliances.

11.3.3 Solar Home Systems

With so-called Solar Home Systems (SHS), the basic electrification of rural houses without a grid connection can easily be realized. This consists of a PV module, a battery and a charge controller, as well as directly connected 12 V DC appliances. Typically such systems do not deliver AC power and are only able to provide power for lighting, radio, a small television, small fans and the charging of mobile phones. They do not allow the connection of machines for productive use. The dimensioning of the PV module and the battery must be chosen so that it matches the calculated electricity demand. As an early example, an SHS in Indonesia, consisting of a PV module with 50 Wp and a 12 V/70 Ah lead acid battery, operates three compact fluorescent lights (each 8 W), a radio (8 W) and a television (20 W) (Hoeke *et al.*, 1993). The technological development during the last 20 years, especially in the field of lighting with highly efficient white LEDs, permits with such a system the operation of more lights, instead of a black-and-white TV, permits a color TV, and the charging of mobile devices like phones (Bopp *et al.*, 2015). To guarantee reliable operation the selection of well-tested components, a proper dimensioning, and regular maintenance are necessary (Nieuwenhout *et al.*, 2001).

Currently a few million solar home systems (SHS) have been installed in remote areas in developing countries. One of the most successful SHS programmes has been realized in Bangladesh with approximately 3 million SHS, which were installed until April 2014 (IDCOL, 2015) (Figure 11.3.2). These systems provide electricity for 13 million people, which corresponds to approximately 9% of the population. This could be achieved thanks to strong government support, via the Infrastructure Development Company (IDCOL), and by financial

Figure 11.3.2 A typical solar home system in Bangladesh

support from donor agencies, such as the World Bank, KfW, GTZ, and ADB. Micro-finance institutions, so-called Partner Organizations (PO), finance the loans, install and maintain the solar systems, and collect the fees.

11.3.4 Hybrid PV Systems for Stand-Alone Applications

Hybrid PV systems offer a cost-efficient and reliable energy supply in a lot of cases. Besides a PV generator, these systems consist of battery storage as well as additional components like small wind power generators, small hydro power generators and diesel generators (gensets).

11.3.4.1 Technical Applications

Stand-alone PV systems provide electricity for remote technical applications not connected to the electricity grid, such as repeater stations and especially for taking environmental measurements. There are many remote locations, which are suitable for environmental measuring campaigns; however, they cannot be used at present. Due to the existing environmental conditions and the short-term as well as the seasonal fluctuations in PV power production, additional power sources are needed. Among others, in winter, iced-up PV modules in the measurement stations in the mountains are an example of such power shortages of stand-alone PV systems. Furthermore, in such periods, additional power is often needed, e.g., to heat sensors during

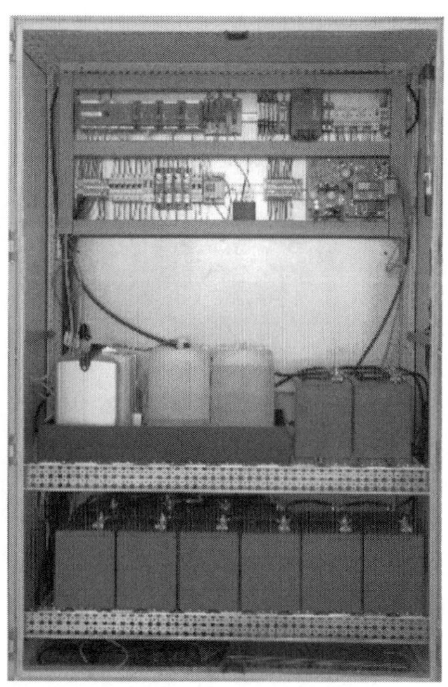

Figure 11.3.3 Hybrid PV power supply system for autonomous wind measurement station with a system voltage of 12 V, consisting of a PV generator (110 Wp), a direct methanol fuel cell (65 W), a battery storage (660 Ah, C10) and an energy management system. For monitoring purposes, a data logger and several measurement components are integrated into the electric control cabinet. Source: (Vetter *et al.*, 2006)

periods of bad weather. This results in even more critical situations, when the battery storage is discharged and no additional power generator is available.

With modular hybrid PV systems, this issue can be resolved and reliable operation of measurement stations in remote locations will be possible. Therefore, PV systems and batteries are supplemented with an auxiliary power supply, e.g., in the form of wind generators or fuel cells (Figure 11.3.3). With such an approach, measurement stations with a peak load up to 1000 W and more can be supplied with power.

The core of such a system is an intelligent energy management system, which ensures a highly efficient and reliable operation even under harsh environmental conditions and which requires less maintenance. For this purpose it is essential to operate the auxiliary power generator – in this case, the direct methanol fuel cell – as efficiently as possible to minimize the fuel consumption and therefore to extend the operating period without maintenance. The optimization criteria can therefore be described as follows with the boundary conditions, that in such technical applications the energy production and energy consumption during one day should be equalized and the state of charge (SOC) of the batteries should be higher than a defined minimum value of the state of charge SOC_{min}. In such applications, typically SOC_{min} is defined as a value of 80% to secure high autonomy times if none the power generators are unable to deliver electricity, e.g. due to bad weather conditions or system failure. For the implementation of optimized control strategies, a day is typically divided into several intervals with a defined time span $t_2 - t_1$.

11.3.4.1.1 Optimization Criteria 1

$$\int_{t1}^{t2} P_{residual}\,dt = \int_{t1}^{t2} \left(P_{load} - P_{PV} - P_{wind}\right)dt \rightarrow minimum \tag{11.3.1}$$

$P_{residual}$ (in W) represents the actual net load as a difference between the actual load P_{load} (in W) and the actual PV power P_{PV} (in W) and the actual wind power P_{wind} (in W).

11.3.4.1.2 Boundary Conditions

$$\int_{t=0h}^{t=24h} P_{el,generated}\,dt \approx \int_{t=0h}^{t=24h} P_{el,consumed}\,dt \tag{11.3.2}$$

$$SOC_{Battery} \geq SOC_{min} \tag{11.3.3}$$

$P_{el,generated}$ represents the power of all connected power generators and $P_{el,consumed}$ represents the power of all connected loads. $SOC_{Battery}$ is the state of charge of the battery and SOC_{min} the allowed minimal state of charge value of the battery.

If both boundary conditions have to be fulfilled in a single time interval (t_1 to t_2) or in certain critical applications, even at each single time step, the residual load is covered by the fuel cell and/or the battery:

$$\int_{t1}^{t2} P_{residual}\,dt = \int_{t1}^{t2} P_{fuel\,cell}\,dt + \int_{t1}^{t2} P_{battery}\,dt \tag{11.3.4}$$

$P_{residual}$ (in W) represents the actual net load, $P_{fuelcell}$ (in W) represents the actual power of the fuel cell system and $P_{battery}$ (in W) represents the actual power of the battery system.

If the fuel cell is switched off, only the battery covers the residual load as long as the state of charge is above its absolute limit:

$$P_{battery} = P_{residual},\ if\ P_{battery} \leq P_{battery,discharge,max}\ and\ SOC \geq SOC_{min} \tag{11.3.5}$$

$P_{battery}$ (in W) represents the actual power of the battery system, $P_{residual}$ (in W) represents the actual net load, $P_{battery,discharge,max}$ presents the maximum discharge power of the battery system, SOC is the actual state of charge of the battery system and SOC_{min} represents the minimum allowed state of charge value of the battery system.

If the residual load is negative, meaning that PV and wind generation exceed the actual load, the battery is charged:

$$P_{battery} = P_{residual},\ if \left|P_{battery}\right| \leq P_{battery,charge,max}\ and\ SOC \leq 1 \tag{11.3.6}$$

Whereas $P_{battery,charge,max}$ represents the maximum allowed charge power, which is also dependent on the actual state of charge (SOC) of the battery.

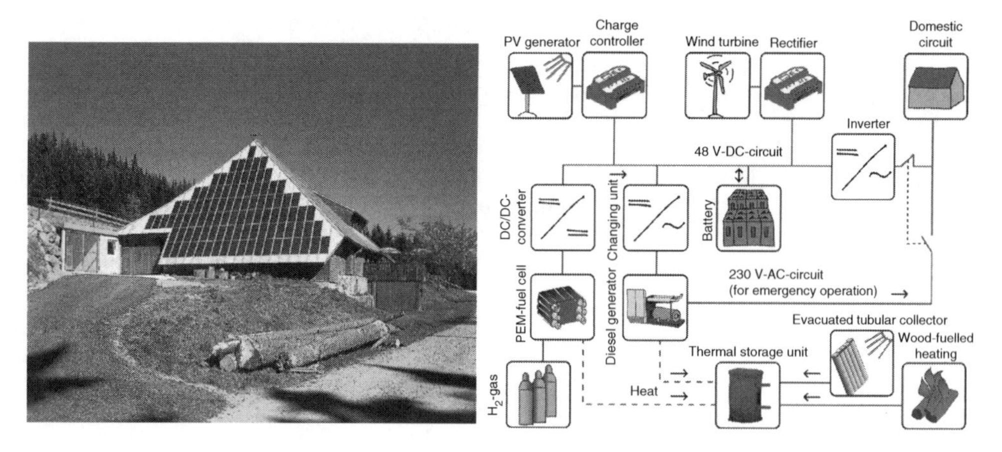

Figure 11.3.4 Hikers inn, Rappenecker Hütte near Freiburg has been operating since 1987 with a hybrid PV system, now equipped with a PV generator (3.8 kWp), a hydrogen fuel cell (4 kW), a diesel genset (12 kW), a wind generator (1.8 kW) and battery storage (45 kWh). (*See insert for color representation of the figure*)

11.3.4.2 Residential Applications

In Europe, several 100,000 remote houses, small farms, refuges, alpine huts have no grid connection. As an example. there are some 1500 mountain huts in the European Alps (Austrian Alpine Club, 1995). Around 200 of the 320 mountain huts without a grid connection, in the German and Austrian Alpine Association, are supplied with a PV stand-alone system. Most of these systems are designed as hybrid solutions to overcome the power and energy restrictions of a pure solar home system by adding an additional generator, which is operated by diesel or also vegetable oil. An inverter converts the DC power into common AC power to meet the requirements of typical residential appliances. In some cases additional auxiliary generators like wind turbines, small hydro turbines or fuel cells are integrated. As a very good example of such a system configuration, we refer to the well-known Rappenecker Hof (Figure 11.3.4).

The PV size is typically in the range of 0.5 and 15 kWp and depends on the local boundary conditions, mainly on the load profile and solar irradiation. Due to cost reasons and the lack of solar power during the winter season, especially in Central Europe, the solar fraction[1] of the annual consumption lies in the range between 30% (mainly in large systems) and 80% (mainly in small systems). To ensure a reliable power supply, well-tested components, a proper system design and regular maintenance (typically once per year) are necessary (Rappeneck, 2012).

11.3.5 PV Diesel Mini-Grids

With increasing diesel costs and decreasing PV module prices, high solar fractions up to 80% are economically viable in mini-grid applications, when project lifetimes of 15–20 years are considered. In such configurations the diesel generator is only used a few hours per day and storage is of great importance. Nevertheless such system configurations still lead to relatively

[1] Solar fraction is defined as the amount of annually used PV electricity divided by the annual consumption.

high investment costs, which are combined in certain applications and/or regions with a high financial risk due to external boundary conditions, e.g., political decisions or local demographic developments. Therefore, it might still be more suitable to install only a PV generator of a certain size without any storage, to keep investment costs reasonable, and accept the fact that only a certain solar fraction can be realized. Nevertheless remarkable diesel savings can be achieved with such a system configuration, and in fact these solutions have been introduced onto the mini-grid markets as so-called "fuel-savers." In the following two sections both system solutions with and without storage are described.

11.3.5.1 Systems Without Storage

Especially the integration of PV generators in already existing diesel mini-grids systems without any additional storage is a fast and cost-efficient opportunity to reduce diesel consumption. Such approaches work generally up to solar fractions of 10% without any additional control efforts, as shown in Vetter *et al.* (2006). Only the grid-tied inverters have to be configured with higher tolerances, as voltage and frequency fluctuate more in mini-grids than in robust distribution grids. In such a system configuration, at least one diesel generator is always operating and is the grid-forming unit. The PV inverters are working in grid parallel mode and are only feeding in.

To further increase the solar fraction, additional control measures have to be integrated into the power supply system. Such units are called "fuel savers" and fulfill the following tasks:

- optimized control of the diesel engine;
- reduction of PV inverter output to ensure minimum diesel generator power, which has to be typically above 30–50% of the nominal power.

To secure the stable operation of the mini-grid, it is necessary in some cases to reduce the current feed-in power of the PV system temporarily, which means the maximum power point tracker (MPPT) has to be switched off and the inverter control follows the set point of the supervisory control algorithm of the fuel saver unit. Otherwise the feed-in of too much PV power would lead to unstable operating conditions in the mini-grid. Therefore, the specifications of the diesel generator, which define its permitted operating states, have to be considered.

11.3.5.2 Systems with Storage

In (Vetter *et al.*, 2014) case studies for three different applications have been presented, showing the annual diesel savings potential depending on the PV generator size in relation to the peak load of the power system (Figure 11.3.5). These examples show that remarkable diesel savings in liters can be achieved by the integration of a PV system without any storage. A storage device in the mini-grid configuration is necessary. For this purpose, typically batteries are used as described in Section 11.3.4.

In principle, there are three possibilities to combine the PV generator, the battery storage and additional power generators. The corresponding principles as well as the specific advantages and disadvantages are discussed in the following sections.

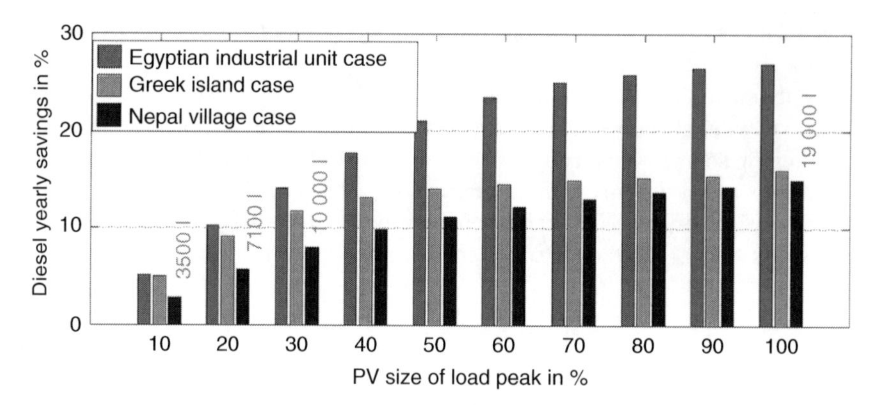

Figure 11.3.5 Three case studies showing the annual diesel saving potential depending on the PV generator size in relation to peak load of the power system. Source: (Vetter *et al.*, 2014)

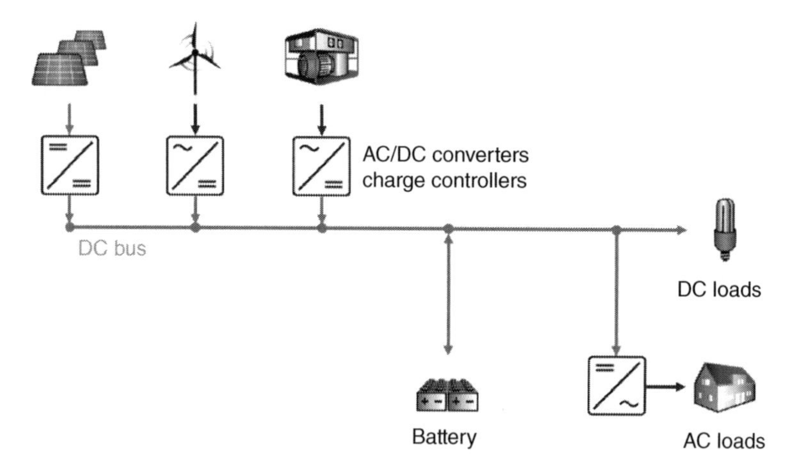

Figure 11.3.6 A DC coupled hybrid PV system with different power generators, DC appliances and an inverter supplying AC loads

11.3.5.3 DC Coupled Systems

In pure DC coupled systems, all the generators are coupled on a DC bus as shown in Figure 11.3.6. Single DC appliances are also connected directly to this DC bus. For AC loads, an inverter is connected to this DC bus as well. Depending on the specific application and also the size of the entire power supply, this inverter can be a one-phase, three-phase or a split-phase device.

The main advantages of such a pure DC coupled solution for hybrid PV systems can be summarized as follows:

- Only one conversion step for battery charging via PV generator is needed.
- Many different manufacturers and products are available.
- Comparably high overall system efficiency for PV-dominated systems and for load curves with high energy demand during the night.

The main disadvantages of these system configurations are:

- Two conversion steps needed for the direct supply of AC appliances with PV power.
- Power extension is combined with a comparably high effort.
- Comparably low overall system efficiency for load curves with high energy demand during sunshine hours in a PV-dominated system.

11.3.5.4 AC Coupled Systems

In pure AC coupled systems, all the generators as well as the battery storage are coupled on the AC side, as shown in Figure 11.3.7. In principle, it is possible to supply single DC loads by connecting them directly to the battery busbar.

The main advantages of the AC coupled system solution can be summarized as follows:

- Only one conversion step is needed for the AC supply in the case of direct use of PV power.
- Higher system efficiency for load profiles with peak demands during sunshine hours.
- Power extension is easy to handle.
- Standard AC installation.

Disadvantages of this pure AC coupled hybrid systems are:

- Two conversion steps needed in case of battery charging.
- Lower system efficiency for load profiles with high energy demands during the evening and night.
- Only a few manufacturers provide such configurations for hybrid PV systems.

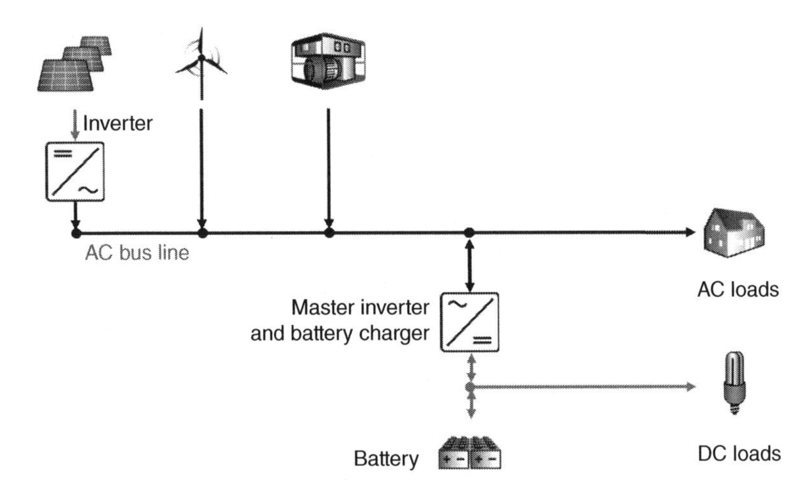

Figure 11.3.7 An AC coupled hybrid system with different generators and a battery storage supplying typical AC devices

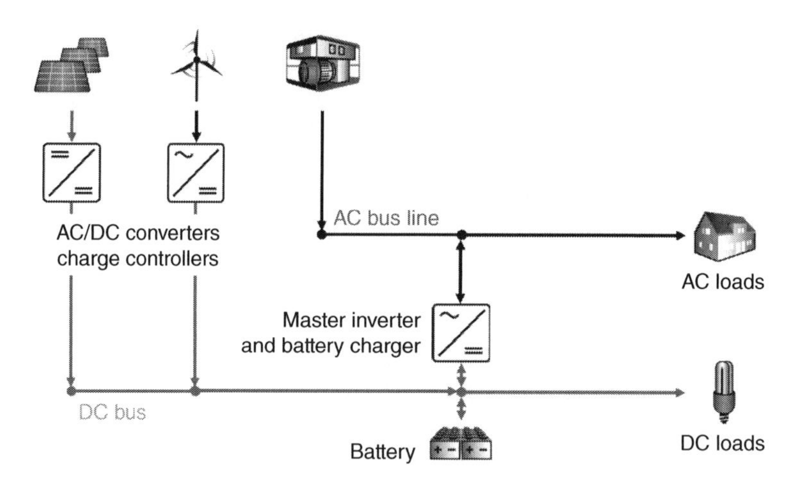

Figure 11.3.8 A DC/AC mixed configuration for hybrid PV systems

11.3.5.5 DC/AC Mixed Solutions

In DC/AC mixed configurations for hybrid systems, DC sources and DC storages are coupled directly to a DC bus, whereas certain additional sources such as diesel generators can be coupled directly to the AC side. Figure 11.3.8 shows an example of such a DC/AC mixed hybrid PV system.

The main advantages of a DC/AC mixed configuration can be summarized as follows:

- Only one conversion step needed for battery charging via PV generator.
- Comparably high overall system efficiency for PV-dominated systems and for load curves with energy demand during the night.
- The inverter and the diesel generator can be operated in parallel, therefore high peak power can be provided.

Disadvantages of this DC/AC mixed configuration are:

- Two conversion steps for the direct supply of AC appliances with PV power.
- Comparably low overall system efficiency for load curves with energy demands during sunshine hours in PV-dominated systems.
- Comparably high control and communication efforts.

Typically the PV generator is connected via a charge controller to the battery, whereas the diesel generator is coupled on the AC side. The battery inverter can be used as a master inverter, as a battery charger and furthermore also in a so-called power assistance mode in case additional power is needed to cover peak loads.

In all three system solutions described, it is in principle possible to shut off the diesel gensets as the battery inverter, in the case of an AC coupled solution or the combined battery/ PV inverter in the case of a DC coupled or DC/AC mixed solution, is able to build up the grid

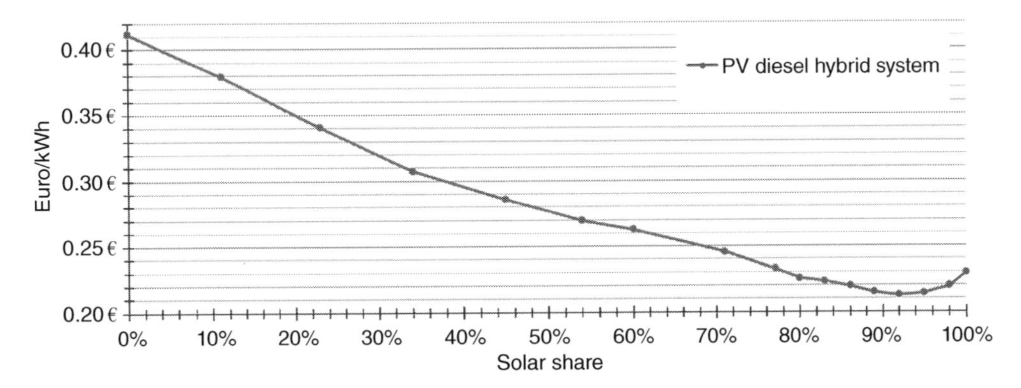

Figure 11.3.9 Levelized cost of electricity depending on the solar share for a PV Diesel battery hybrid system. For this study at a location in Uganda the following parameters were used: Peak load 200 kW, Annual consumption: 574 MWh, cost for PV system (incl. power electronics) 1.5 Euro/Wp, cost for battery system 220 Euro/kWh, cost for Diesel genset 273 $/kW invest, 1$/l fuel, 0.7 $/h maintenance. Source: (Vetter *et al.*, 2014)

and to secure a stable operation of the power supply system. In the AC coupled or in DC/AC mixed system solutions there exist in principle three operation modes:

- Diesel genset builds up the grid; battery inverter is operated in its power assistance mode to cover peak loads.
- Diesel genset builds up the grid; battery inverter is operated in its battery charging mode in case of low state of charge levels of the battery.
- Battery inverter builds up the grid; diesel genset is switched off or is operated in its power assistance mode.

With such PV diesel battery hybrid systems, solar fractions of 80% and more are economically viable, when project lifetimes of 20 years can be considered as shown in Figure 11.3.9, from a case study for a site in Uganda.

11.3.6 Battery Storage

Battery storage is a very crucial component in stand-alone PV systems. The purpose of batteries is to store the electrical energy generated by PV modules on sunny days to be consumed at night or on cloudy days. The most common types of batteries currently used in PV systems are lead acid, nickel-metal-hybrid, nickel cadmium and lithium-ion batteries (Table 11.3.1). The lead acid technology is the cheapest; all other technologies are more expensive. Lithium-ion and nickel-metal-hybrid batteries are used mainly in small portable PV applications like Pico PV appliances, because they are about two or three times lighter in comparison to lead acid batteries. Nickel cadmium batteries are mainly used for very cold climates. From a technical point of view the lithium-ion battery technology is the best choice for stand-alone applications, because of its high efficiency and the robustness for long periods at low states of charge levels, but it is still more expensive in terms of its investment cost in comparison to

Table 11.3.1 Comparison of different selected battery technologies (Vetter, 2014)

	Lead acid	NiMH	Li NMC/ graphite	LiFePO4/ graphite	Vanadium-redox-flow
Energy density (Wh/kg)	40	75	160	110	45
Power density (W/kg)	350	600	1300	4000	120
Cycle lifetime	600	900	2500	5000	12000
Calendar lifetime (years)	7	5	7	14	15
Efficiency (%)	85	75	93	94	80
Self-discharge (%) per month	8	20	3	3	5
Cost per kWh (€)	60–300	400–600	200–2000	200–2000	150–800

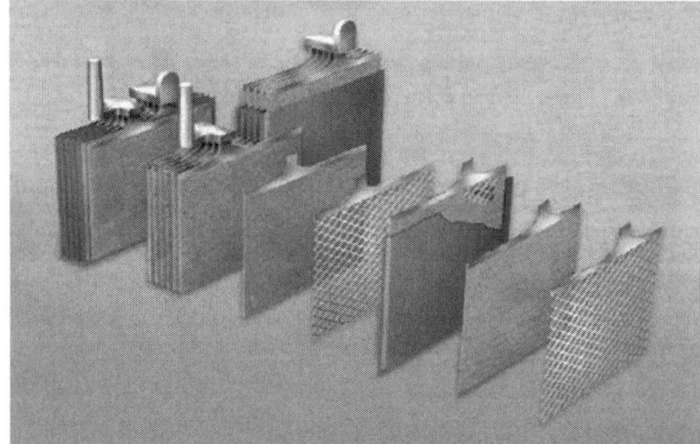

Figure 11.3.10 Typical construction of tubular (left) and flat plates (right) lead acid-battery. Source: (DETA, 1998)

lead acid batteries. It is expected that prices will drop significantly for certain lithium-ion batteries within the next few years. Such predictions are related to the developments of battery systems for electric vehicles.

Due to current prices in most stand-alone applications such as solar homes and hybrid systems, lead acid batteries are still the favorites. It is recommended to use special stationary batteries, which reach a higher number of cycles in comparison to starting lighting ignition (SLI) batteries. Stationary batteries are produced with two different types of positive plates, see Figure 11.3.10. The so-called tubular plate achieves a longer lifetime in comparison to the flat (pasted) plate, and it is especially more robust against deep discharge.

In many cases the so-called solar batteries are modified SLI batteries with thicker plates that achieve more cycles in comparison to SLI batteries, but do not reach the cycle numbers of stationary batteries, see Figure 11.3.11.

For long lifetimes of lead-acid batteries it is recommended to keep the depth of discharge (DOD) very low. The DOD of a battery is defined as the percentage of capacity that has been withdrawn from the battery, compared to the total full charge capacity. For deep discharge batteries, the DOD can reach up to 80%. However, a lower DOD results in an increased

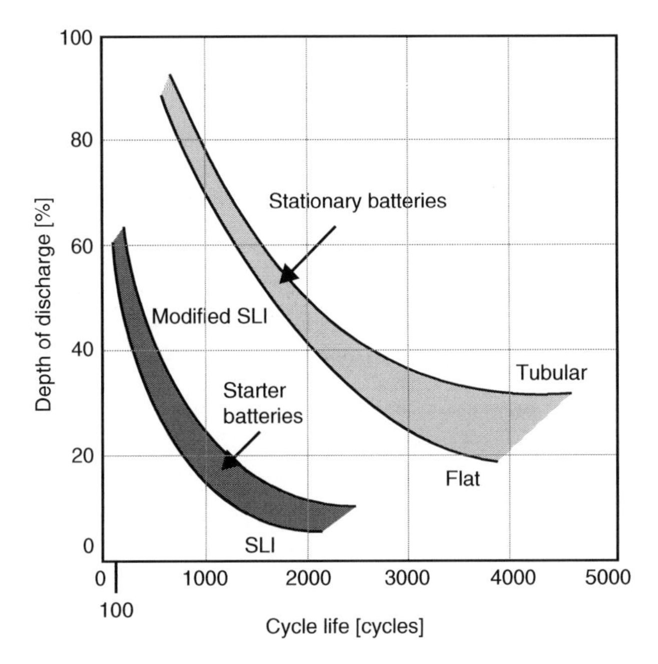

Figure 11.3.11 Life cycles vs DOD of different types of lead-acid batteries. Source: (Varta, 2014)

achievable number of cycles and extended battery lifetime. Often, a charge controller controls the DOD of the battery bank. This means that, at a certain DOD, it disconnects the load or starts the backup generator. If a DOD>70% is reached, it is important to fully charge the battery within a few days, otherwise the battery lifetime will be reduced. Because in most stand-alone systems this regular full charging is not possible, the batteries never reach the cycle numbers which are shown in Figure 11.3.11 as other ageing effects limit the real lifetime to 4 or a maximum of 10 years.

Lead acid batteries with liquid electrolyte need to be maintained regularly, such as by keeping the electrolyte at the level assigned by the manufacturer, by adding distilled water. When using maintenance-free batteries with a fixed electrolyte, it is strongly recommended to use gel types. When AGM (absorbent glass mat) types are used, it is recommended to install them in such a way that the internal lead plates are orientated horizontally.

11.3.7 Conclusion

There are various PV stand-alone and hybrid system solutions available on the market today, covering a wide range of different applications and enabling access to energy even in remote areas far from the national electricity grid. This chapter has shown the range from the smallest pico PV devices up to the hybridization of diesel mini-grids with PV systems enabling high solar fractions, which are in most cases today the cost-efficient solution compared to fossil power generators and also compared to grid extension. Therefore, this market segment can be considered a very promising multi-GW market, worldwide and annually.

Nevertheless these markets are more challenging than on-grid markets as a lot of "special" boundary conditions, risks and uncertainties have to be taken into account. Besides technical issues, financial schemes, locally adapted operating models and a proper risk management are needed to address investments in such projects.

References

Austrian Alpine Club (1995) Messages of the Austrian Alpine Club, **1**, 1995.

Bopp, G., Dresch, E.A., Pfanner, N. *et al.* (2012) *Assuring the Quality of Solar Lanterns*. Fraunhofer ISE annual report, Freiburg.

Bopp, G. and Gabler, H. (2015) Technical trends and innovations. *Proceedings of 4th Symposium Small PV-Applications*, Munich.

DETA (1998) Product flyer of the former lead acid battery company DETA.

Hoeke, P. (1993) Solar electricity in Lebak, Indonesia, Perpetuum mobile of the 21st century?, *Sun World*, **17** (1).

IDCOL (2015) website, www.idcol.org

Nieuwenhout, F.D.J., van Dijk, A., Lasschuit, P. E. *et al.* (2001) Experience with solar home systems in developing countries: a review. *Progress in Photovoltaics*, **9**, 455–474.

Rappeneck (2012) Flyer, "Rappenecker Hütte im 21. Jahrhundert."

Rüther, R., Schmid, A., Beyer, H. *et al.* (2003) Cutting on diesel, boosting PV: The potential of hybrid Diesel/PV systems in existing mini-grids in the Brazilian Amazon. *Proceedings of the 3rd World Conference on Photovoltaic Energy Conversion (30th IEEE Photovoltaic Specialists Conference*), Osaka, Japan.

Varta (2014) Fact sheet.

Vetter, M. (2014) Energy storage: renewable energy's key "blade" for grid integration. Canada Energy Storage Summit, Toronto, 12 November.

Vetter, M. and Schies, S. (2014) Simulation of solar diesel hybrid systems. *Proceedings of Intersolar Conference of North America*, San Francisco, 9 July.

Vetter, M., Schwunk, S., Thomas, R. *et al.* (2006) Reliable power supply for measurement stations on the basis of hybrid PV systems. *Proceedings of PVSEC Conference, Milan.*

11.4

PV System Monitoring and Characterization

Wilfried van Sark, Atse Louwen, Odysseas Tsafarakis, and Panos Moraitis
Copernicus Institute, Utrecht University, The Netherlands

11.4.1 Introduction

In the past few years the development of photovoltaic (PV) solar technology has led to enormous price reductions that, in combination with low installation and maintenance costs, have made PV modules a popular form of renewable electricity generation (IEA-PVPS, 2014). Besides impressive developments in very large-scale (VLS) PV power plants of >100 MWp, this especially pertains to small and medium-sized residential systems whose owners have embraced PV as a profitable means to reduce their electricity bill as a result of grid parity in many countries around the globe. Economic benefits can be guaranteed only if the performance of the PV system is as expected. Hence proper monitoring of power and energy generation is a prerequisite. Implementation of monitoring depends on the size of the PV system and its purpose, such as the control of the commercial systems for which any malfunction directly may lead to financial loss. This pertains to VLS PV systems as well as much smaller grid-connected residential systems. Monitoring of stand-alone PV systems usually has a different purpose in that not the energy but rather the function that is provided should be guaranteed, such as water pumping or communication.

Two approaches to monitoring can be discerned, i.e., analytical and global monitoring (Blaesser and Munro, 1995a, 1995b). Analytical monitoring requires measurement of many parameters, such as array output voltage and current, irradiance, ambient temperature, inverter DC input and AC output power, where all these are measured at a certain time interval, such as hourly, every 15 minutes, every minute, or even at the level of seconds. In view of the cost and complexity of analytical monitoring, global monitoring of energy (instead of power) can provide sufficient data to ascertain the expected performance, and requires fewer measured parameters, such as array output energy, energy to grid and irradiation (kWh/m^2) on a monthly, weekly, or daily time resolution.

Photovoltaic Solar Energy: From Fundamentals to Applications, First Edition.
Edited by Angèle Reinders, Pierre Verlinden, Wilfried van Sark, and Alexandre Freundlich.

Companion website: www.wiley.com/go/reinders/photovoltaic_solar_energy

The majority of residential PV electricity generation is due to a large amount of small (1 to 5 kWp), geographically scattered systems. Many of these systems have insufficient monitoring so that system owners cannot assess by themselves whether their systems are generating the expected amount of energy. Consequently, failures and energy losses may remain undetected for a long time. For example, Jahn and Nasse (2004) showed in a study of German grid-connected residential PV systems of 1 to 5 kWp size that once every 4.5 years a statistical failure occurred per system. In 63% of the cases, inverters caused the failure; in 15% of the cases modules were to blame, and in 22% of the cases other system components. Today, fee-based services are offered to PV system owners that allow also small PV systems to assess the performance of their system in an automatic way. Large systems (>10 kWp to multi-MW size) usually are equipped with proper extensive monitoring equipment, as its cost is small compared to the possible incurred financial losses due to energy losses. Requirements for this analytical monitoring are laid down in the IEC 61724 standard (IEC, 1998): an automatic data acquisition system with a minimum set of parameters that are to be monitored.

The performance ratio (*PR*) is usually employed as a proxy for system performance; it is a measure of the degree of utilization of an entire PV system (Blaesser and Munro, 1995a, 1995b; IEC, 1998) with respect to its nameplate nominal power. It indicates the overall effect of losses on the overall performance of the PV system, and thus includes effects of PV module temperature, intensity dependence of the module efficiency, partial shading, soiling, incomplete utilization of irradiation, system component limited efficiencies, module mismatch and any malfunctions (Reich *et al.*, 2012; Van Sark *et al.*, 2012). Systems that perform well reach $PR=0.85$ (note that *PR* is also commonly expressed in percentages, i.e., 85%). PV systems installed in the late 1980s showed *PR* values between 0.5 and 0.75 (Decker and Jahn, 1997). Nordmann (2007) analyzed the performance of systems installed since 1991: a trend in *PR* was observed from $PR=0.65$ to $PR=0.72$ for the systems installed in 2005. Reich *et al.* (2012) reported maximum $PR=0.90$ for systems installed in 2010, and argued that the upper bound may be $PR=0.92$.

Although a large PR value is a good indicator of system performance, the customer usually is interested in maximized energy yields and low levelized cost of electricity. Therefore, other figure of merits, such as energy efficiency (energy out divided by energy in) or energy production rate (kWh per kWh/m^2) would be of higher interest to the system owner. Low PR values do not necessarily mean bad system performance, as in high irradiation areas also high ambient (and module) temperatures are common practice. The same system would have a high PR in milder climatic zones.

To be able to reach such high performance ratio values, proper monitoring followed by immediate action should malfunctions occur is necessary. Recently, the participants of Task 13 of IEA-PVPS (International Energy Agency, Photovoltaic Power Systems program) have presented an analytical method for fast and accurate performance analyses (Woyte *et al.*, 2014). This method makes use of a collection of data plots, such as final energy yield versus reference (or expected) energy yield, and linear regression analysis. Depending on the availability of specific data of a PV system, more plots can be made, which makes detection of malfunctions easier and faster, thus optimizing the performance. These methods can also be used for residential systems, with limited data availability (Tsafarakis and Van Sark, 2014). Ransome *et al.* (2005) have introduced another collection of plots and interpretation guidelines based on normalizing the measured performance parameters. In addition, a so-called Sophisticated Verification (SV) method was developed that allows the identification of twelve different loss factors based on seven parameters that should be measured (Ueda *et al.*, 2009). Finally, Pearsall and Atanasiu (2009a) have developed new monitoring guidelines that can be used in spreadsheet form (Pearsall and Atanasiu, 2009b).

When only energy yield data is available, statistical methods to analyze performance can be employed to assess performance on regional or country level (Moraitis and Van Sark, 2014; Nordmann, 2014). Comparison of yields with nearby systems was suggested as a peer-to-peer performance indicator (Leloux *et al.*, 2013).

In this chapter, we will provide a short overview of monitoring practices and parameters, which we illustrate with a few examples.

11.4.2 Monitoring Practice

In general, a grid-connected PV system can be described in four parts, see Figure 11.4.1: (1) modules; (2) module/inverter connections; (3) inverter(s); and (4) inverter/grid connections. In a stand-alone system the solar array is connected to a battery and loads via a charge controller; in this case part 3 is the charge controller, and part 4 the battery. In all these parts, malfunctions may occur. A number of possible malfunctions for grid-connected systems are listed in Table 11.4.1, which can be linked to a specific part in the PV system, see also

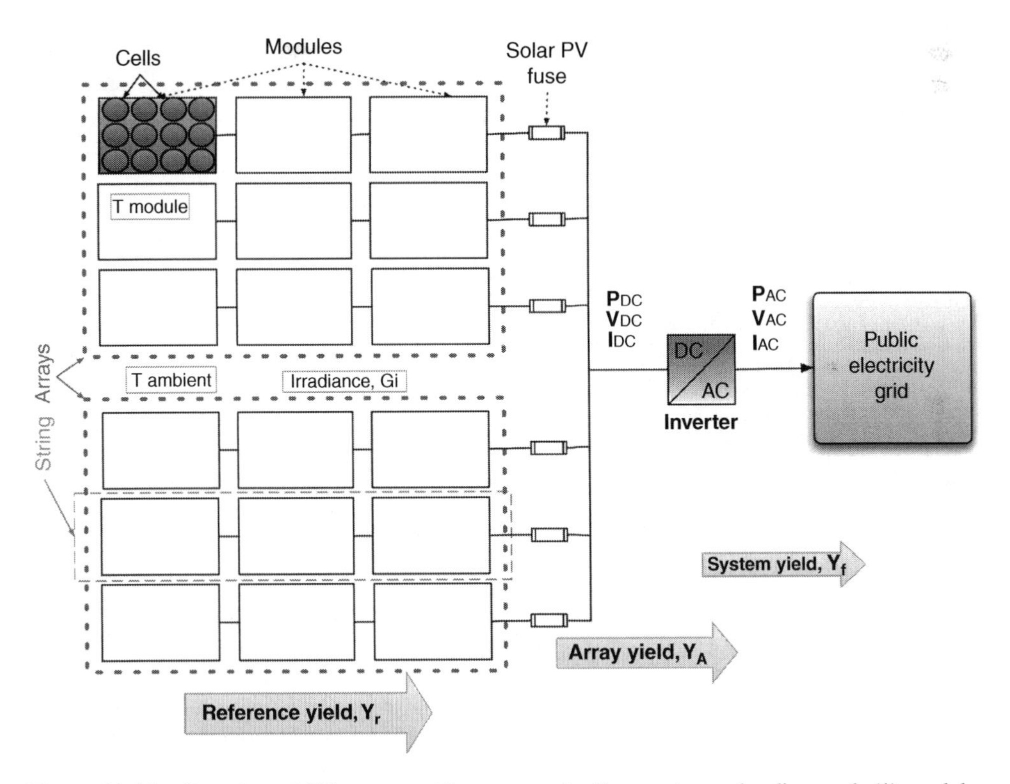

Figure 11.4.1 Overview of PV system with components. Four parts can be discerned: (1) modules; (2) connections between module and inverter; (3) inverter; (4) connection between inverter and public electricity grid. Definitions of string and array are indicated, as well as several measured parameters. Source: (Tsafarakis, 2014)

Table 11.4.1 Examples of malfunctions (after Stettler, 2005)

Energy loss	Module	Module/inverter connection	Inverter	Inverter/grid connection
Constant	Degradation Module overrating Shadowing Module defect	Incorrect connections String defect		Incorrect connections
Changing	Hot modules		Part load behavior Maximum power point tracking Grid outage	
Total blackout		Defect control devices	Defect inverter	Defect control devices

Chapter 11.1 for more details about system failures. Analysis of performance ratio values on different time scales is used to elucidate the origins of malfunctions (Stettler *et al.*, 2005; Drews *et al.*, 2007).

The performance ratio *PR* is defined as the ratio of utilizable AC electricity (at the feed-in meter) or final PV system yield Y_f (in kWh/kWp) to the amount of energy that could be generated if modules were operated under STC (standard test conditions: irradiance 1000 W/m^2, air mass 1.5 spectrum (AM1.5G), and a cell temperature of 25 °C) continuously and without any further losses in the system or reference yield Y_r in (kWh/kWp) (IEC, 1998):

$$PR = \frac{Y_f}{Y_r}, \text{ with } Y_f = \frac{E_{AC}}{P_{STC}} \text{ and } Y_r = \frac{H_{POA}}{G_{STC}} \tag{11.4.1}$$

with E_{AC} the AC energy delivered to the grid in (kWh), P_{STC} the (DC) rated capacity of the PV modules in (kWp), H_{POA} the summed plane-of-array (POA) irradiance in (kWh/m^2), and G_{STC} the STC reference irradiance of 1 kW/m^2 at AM1.5G, and cell temperature of 25 °C. Performance ratio values can be calculated for different periods (annual, monthly, weekly, even daily), which may serve different purposes, such as studying seasonal effects on performance.

Parameters that should be measured (IEC, 1998) for adequate analytical monitoring comprise of PV array output voltage, current and power (V_{DC}, I_{DC}, P_{DC}), utility grid voltage, current and power (V_{AC}, I_{AC}, P_{AC}), ambient and module temperature (T_{amb}, T_{mod}), POA irradiance (G_{POA}), wind speed (S_w), and durations of system outage (t_{outage}). Clearly, for autonomous systems, parameters related to the grid are not required to be measured, while the battery state of charge (SOC) and voltage are relevant parameters.

Measurement of electrical parameters should be done with calibrated energy or power meters, as usually inverter-integrated measurements are not accurate enough (Woyte *et al.*, 2014), or sometimes even show higher power or energy values than actually generated. Nevertheless, system functionality can certainly be tested using inverter-integrated measurements.

On-site measurement of POA irradiance requires irradiation sensors, which are usually too expensive for small PV systems. Pyranometers or crystalline silicon reference cells are commonly used as irradiation sensors. Both are calibrated under indoor and outdoor conditions. Alternatively, on-site irradiation measurement is possible using satellite-based irradiation estimates (Drews *et al.*, 2007), however, this is associated with large short-term inaccuracies, leading to *PR* values with large errors. Also, for residential systems irradiation from ground-based meteorological stations may be used for *PR* determination, while it should be noted that conversion from global horizontal irradiation to POA irradiation requires empirical models that may lead to even greater inaccuracy in *PR* values. Clearly, if one aims for early detection of malfunctions based on *PR*, on-site irradiation monitoring permits much higher certainty (Reich *et al.*, 2012) and better failure response times. Examples of performance of PV systems can be found in the IEA-PVPS Task 13 public database (Task 13, 2014).

Standard graphs that are used in analyzing performance are shown in Figure 11.4.2: (1) a scatter plot of normalized hourly mean array power (P_{AC}/P_{STC}) versus hourly POA irradiance H_{POA}; (2) a bar graph of daily array Y_f and reference yields Y_r for a month (alternatively, a daily *PR*

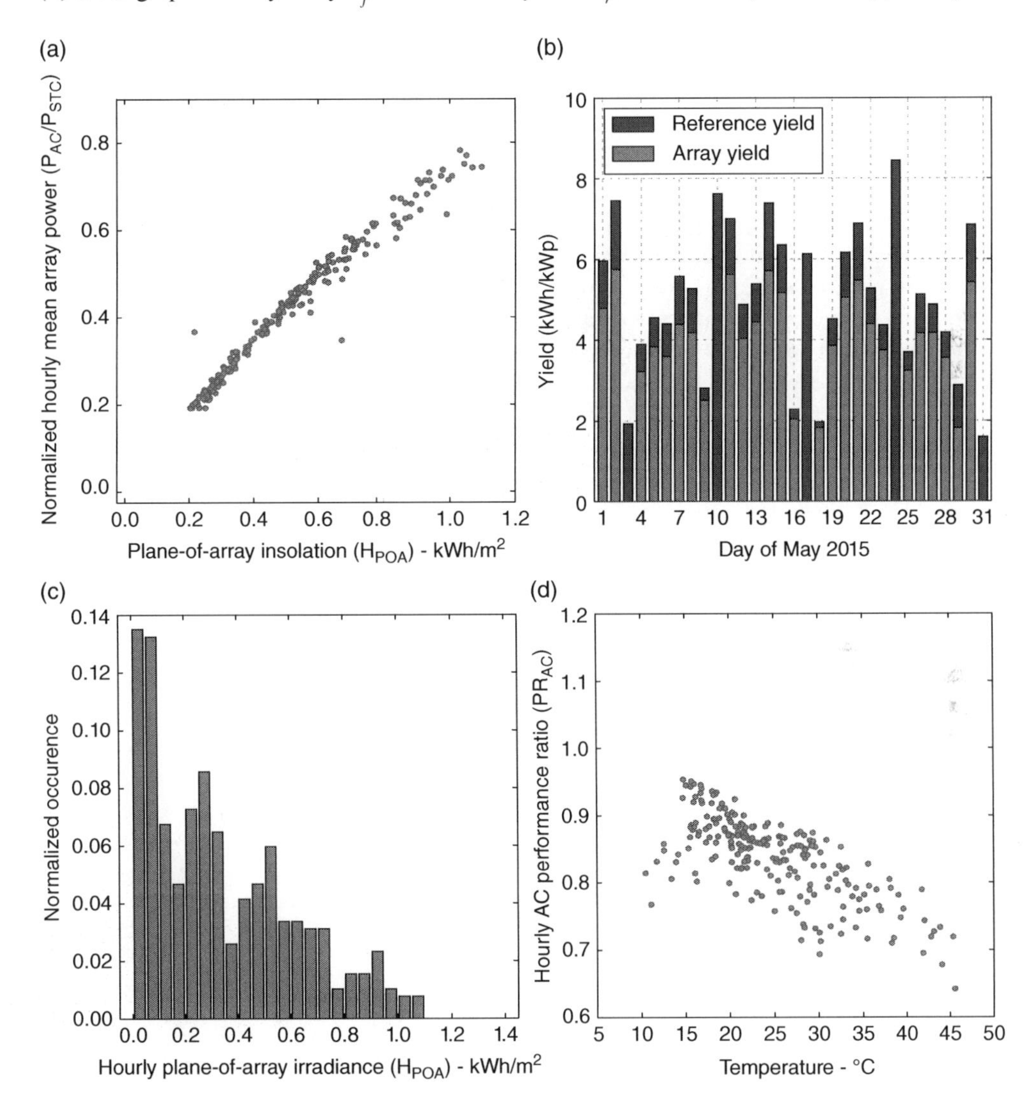

Figure 11.4.2 Standard graphs used in PV performance analysis: (a) normalized hourly mean array power versus hourly POA irradiance; (b) bar graph of daily array (light) and reference (dark) yields for a month; (c) normalized histogram of hourly mean POA irradiance values; and (d) performance ratio versus module temperature. Data (May 2015) are from a system at Utrecht University campus

can be used); (3) a normalized histogram of hourly mean POA irradiance values; (4) performance ratio versus module temperature. Many more are possible, and approaches to show all $1/2\,n(n-1)$ possible graphs for n parameters have been denoted a "stamp collection" (Woyte *et al.*, 2014). Analysis of these graphs reveals any malfunction occurring in the PV system; however, this still requires experts, as automated analysis is not yet sophisticated enough for automatic malfunction detection.

11.4.3 Monitoring Examples

11.4.3.1 Normal Operation

Figure 11.4.3 shows a plot of system yield versus reference yield for a system consisting of 18 crystalline silicon modules with 4.14 kWp total DC capacity installed at the European Academy of Bolzano (EURAC), Italy (Moser, 2014). Data were collected from January to August 2012. The plot shows hourly data, for three different weeks (winter, spring, summer). It is clear that during days with higher solar radiation (spring, summer) the yields are higher (dark and light data points have higher maximum values than shaded ones). Linear regression on weekly data, as suggested by Woyte *et al.* (2014) shows that slopes differ for the three periods: for spring and summer, slopes are smaller than for winter. In winter $PR = 0.92$, while in spring and summer $PR = 0.84$. This is commonly observed in well-functioning systems and is due to the higher ambient and module temperatures in spring and summer.

11.4.3.2 Shading

Figure 11.4.4 shows a plot of system yield versus reference yield for a system consisting of 30 crystalline silicon modules with 3.2 kWp total DC capacity installed at the Development

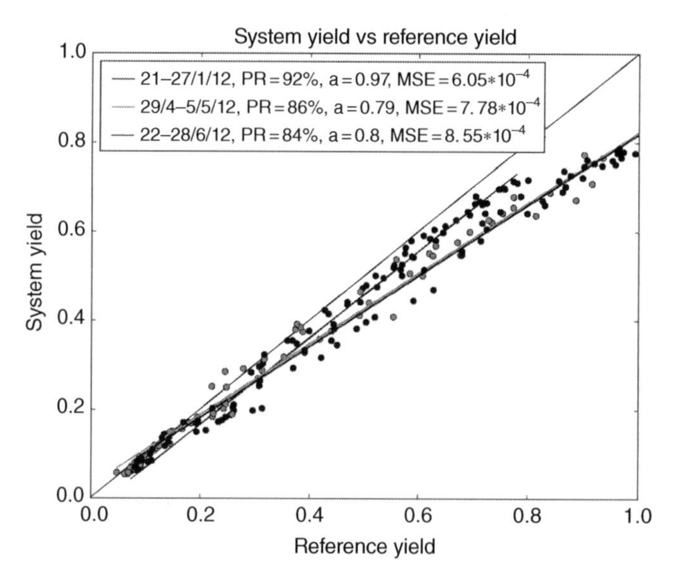

Figure 11.4.3 System yield versus reference yield for a 4.14 kWp PV system. Source: (Tsafarakis, 2014). (*See insert for color representation of the figure*)

Centre of Renewable Energies, Algiers, Algeria (Silvestre *et al.*, 2013). Power as a function of time (for 15-min time intervals) for a clear day without shading on the system is shown in Figure 11.4.4 (a), while a shade is visible in Figure 11.4.4 (b). From the system yield versus reference yield plot, one can infer that the slopes are about equal and the daily PR values are $PR=0.737$ (no shade) and $PR=0.731$ (with shade). From these values alone it is difficult to

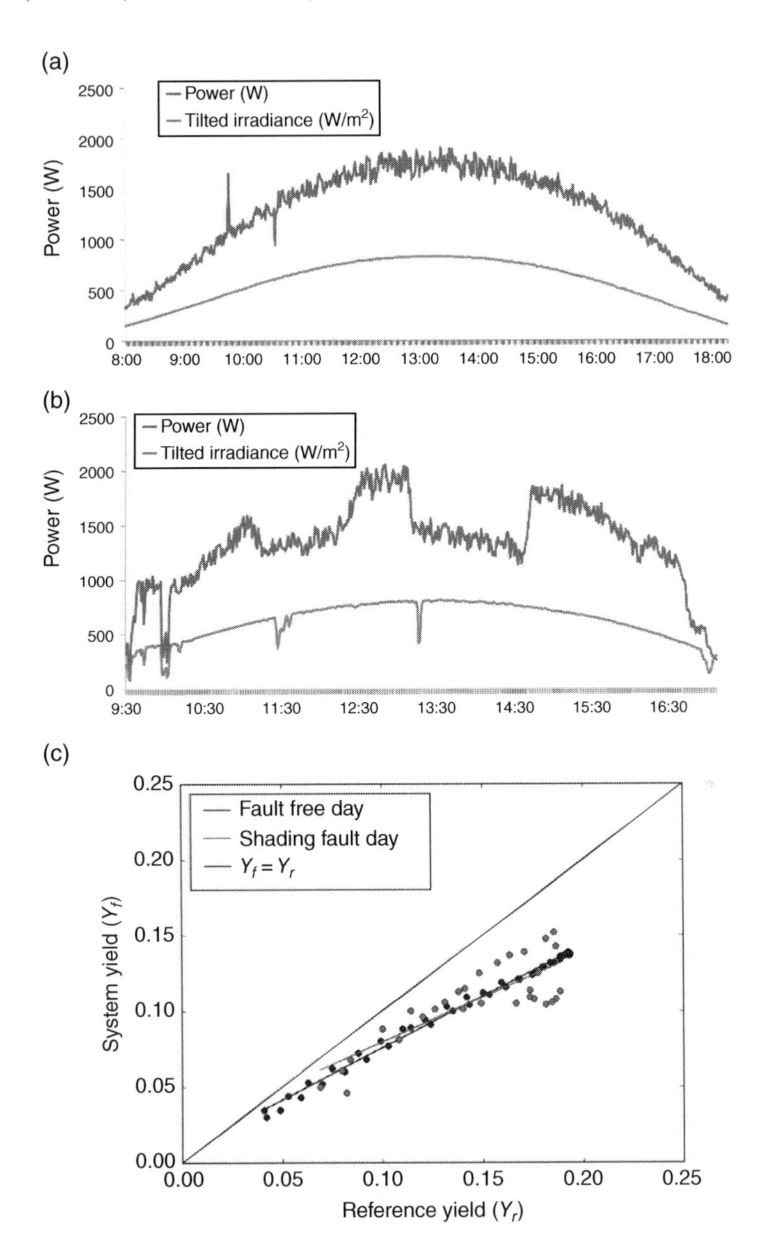

Figure 11.4.4 PV system power and irradiance for a clear day without shade (a), and with shade (b). Panel (c) shows system yield versus reference yield. Source: (Tsafarakis and Van Sark, 2014). (*See insert for color representation of the figure*)

conclude that shading is the cause of a somewhat lower *PR*. However, from the scatter of points around the linear fit, one can infer that something is wrong with the system. The scatter can be quantified using the mean square error: $MSE = 1/2\sum(Y_f - \hat{Y}_f)^2$, where Y_f and $\hat{Y}_f$ are true and predicted values of the final yield, respectively. Predicted yield results from fitting the data and for every reference yield value a final yield prediction is made, and compared to the true (or measured value). In this example, the MSE value for the shading case is 26 times higher than the one for the non-shading case, thus demonstrating that MSE could be an additional metric for performance analysis: plots as in Figure 11.4.4 (c) can be made for daily data and analyzed automatically. Any large MSE value is an indicator for operators to check the data and perform some action to resolve the issue at hand.

11.4.3.3 Statistical Analysis

For small-scale systems, monitoring of energy yield is increasingly being performed using various web tools provided by inverter manufacturers or independent monitoring service companies. Hardware and software are integrated that allow owners to monitor system performance and the production of their system at any time of the day (Moraitis and Van Sark, 2014). For statistical analysis of performance so-called web-scraping techniques can be used to collect performance and system data from the web portals and to organize it in databases for automatic analyses. In this way daily yields (AC and DC) can be analyzed for tens of thousands of systems. Determination of performance ratio is only possible when irradiation data is available. As an attempt to determine the performance ratio variation in a large Dutch data set of systems that were not equipped with an irradiation sensor, hourly global horizontal irradiation data from the Royal Netherlands Meteorological Institute (KNMI) stations was used. Every PV installation was linked to the closest of the 31 weather stations according to geographical coordinates. Total POA irradiation was calculated on a daily basis using a model for every system independently in accordance with the orientation and the tilt of each system (Olmo *et al.*, 1999).

Figure 11.4.5 shows the results of the analysis of 590 systems (average size 10.65 kWp) in the Netherlands for the year 2014, mostly installed between 2011 and 2013: the average value

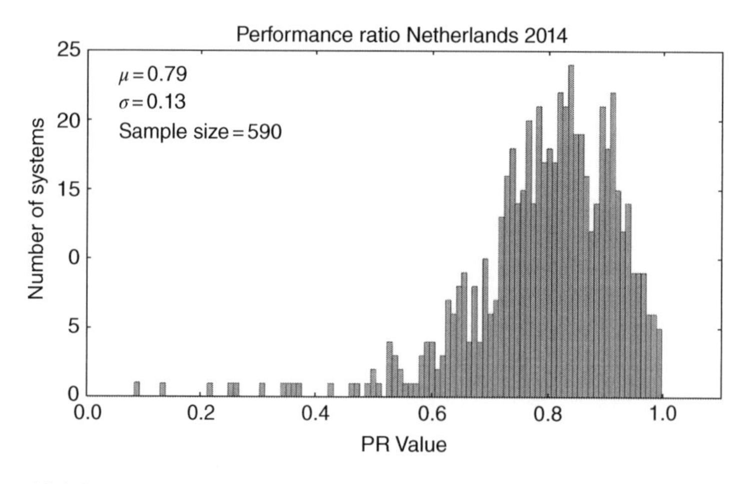

Figure 11.4.5 Distribution of PR values for a sample of 590 Dutch PV systems in 2014

of PR is 0.79 ± 0.13. Some systems clearly show malfunctions, with $PR<0.50$. Systems with $0.6<PR<0.7$ most probably suffer from some kind of shading. About half of the systems perform well with $PR>0.8$. Note that values of PR close to 1 are unrealistic, and these are due to incorrect irradiance values as the distance between irradiance sensor and system can be as large as 30 km. Also, the wide distribution shows that the accuracy of input data (yield, capacity, inverter readings) is unknown. Clearly, statistical approaches offer a rough indication of the performance of systems, and should be interpreted with care. Nonetheless, presenting maps based on geographical information systems is visually interesting to quickly see at which locations system performance may deviate (Moraitis *et al.*, 2015).

11.4.4 Conclusion

Once installed, PV systems need to be monitored in order to be able to assess if systems are delivering the energy as predicted by the installer. Large systems will be equipped with proper monitoring devices complying with IEC standards. This chapter summarized performance-monitoring issues and showed some examples.

Analysis of performance ratios at different time resolution or plotting various parameters versus another allows the detection of malfunctions, and prompt action to solve these leads to highly reliable power production. Smaller, residential systems are usually not monitored with the same attention, and automatic monitoring services are under development that ensure less knowledgeable system owners enjoy carefree PV energy. The introduction of smart meters that also are able to register PV production is particularly of interest in that respect, as these are increasingly being used in the energy management of home energy systems. Algorithms that analyse PV performance could be part of these systems, thus changing current monitoring practices.

Acknowledgments

We would like to gratefully acknowledge D. Moser (EURAC, Italy) and A. Chouder (Development Centre of Renewable Energies, Algeria) for kindly providing monitoring data, and participants in Task 13 of IEA-PVPS for fruitful discussions. This work is financially supported by the Netherlands Enterprise Agency (RVO).

List of Symbols

Symbol	Description	Unit
E_{AC}	AC energy delivered to grid	kWh
G_i	Incident irradiance	(1000) W/m^2
G_{STC}	STC reference irradiance	(1000) W/m^2
H_{POA}	Summed plane-of-array irradiance	kWh/m^2
I_{AC}	AC current	A
I_{DC}	DC current	A
MSE	Mean square error	(kWh/kWp)^2

(Continued)

(*Continued*)

Symbol	Description	Unit
PR	Performance ratio	- (or %)
P_{AC}	AC power	W
P_{DC}	DC power	W
P_{STC}	DC rated capacity of PV system or modules	W (or Watt-peak)
S_w	Wind speed	m/s
t_{outage}	Duration of system outage	s
T_{amb}	Ambient temperature	°C
T_{mod}	Module temperature	°C
V_{AC}	AC voltage	V
V_{DC}	DC voltage	V
Y_A	Array yield	kWh/kWp
Y_f	Final yield	kWh/kWp
Y_r	Reference yield	kWh/kWp

List of Acronyms

Acronym	Description
AC	Alternating Current
AM	Air Mass
DC	Direct Current
IEA-PVPS	International Energy Agency, Photovoltaic Power Systems program
IEC	International Electrotechnical Committee
KNMI	Royal Netherlands Meteorological Institute
MSE	Mean square error
POA	Plane Of Array
PR	Performance Ratio
PV	PhotoVoltaic
SOC	State Of Charge
STC	Standard Test Conditions
SV	Sophisticated Verification
VLS	Very Large-Scale

References

Blaesser, G. and Munro, D. (1995a) Guidelines for the Assessment of Photovoltaic Plants, Document A, Photovoltaic System Monitoring, Report EUR 16338.

Blaesser, G. and Munro, D. (1995b) Guidelines for the Assessment of Photovoltaic Plants, Document B, Analysis and Presentation of Monitoring Data, Report EUR 16339.

Decker, B, and Jahn, U. (1997) Performance of 170 grid connected PV plants in Northern Germany—analysis of yields and optimization potentials. *Solar Energy*, **59**, 127–133.

Drews, A., De Keizer, A.C., Beyer, H.G. *et al.* (2007) Monitoring and remote failure detection of grid-connected PV systems based on satellite observations. *Solar Energy*, **81**, 548–564.

IEA-PVPS (2014) Trends 2014 in photovoltaic applications, Survey Report of Selected IEA Countries between 1992 and 2013, Report IEA-PVPS T1-25:2014, 2014.

IEC (1998) International Electrotechnical Committee, Photovoltaic System Performance Monitoring-Guidelines for Measurement, Data Exchange and Analysis, standard 61724.

Jahn, U. and Nasse, W. (2004) Operational performance of grid-connected PV systems on buildings in Germany. *Progress in Photovoltaics*, **12**, 441–448.

Leloux, J., Narvarte, L., Luna, A., and Desportes, A. (2013) Automatic detection of PV systems failures from monitoring validated on 10,000 BIPV systems in Europe. *Proceedings of the 28th European Photovoltaic Solar Energy Conference*, Paris, France, pp. 4013–4016.

Moraitis, P. and Van Sark, W.G.J.H.M. (2014) Operational performance of grid-connected PV systems. *Conference Record of the 40th IEEE Photovoltaic Specialists Conference,* Denver, Colorado, pp. 1953–1956.

Moraitis, P., Kausika, B.B., and Van Sark, W.G.J.H.M. (2015) Visualization of operational performance of grid-connected PV systems in selected European countries. *Conference Record of the 42nd IEEE Photovoltaic Specialists Conference,* New Orleans, LA, pp. 1–3.

Moser, D. (2014) Data provided from EURAC test field (2014).

Nordmann, T., Clavadetscher, L., and Jahn, U. (2007) PV system performance and cost analysis, a report by IEA PVPS Task 2. *Proceedings of the 22nd European Photovoltaic Solar Energy Conference*, Milan, Italy.

Nordmann, T., Clavadetscher, L., Van Sark *et al.* (2014) Analysis of long-term performance of PV systems, different data resolution for different purposes. Report IEA-PVPS T13-05:2014.

Olmo, F.J., Vida, J., Foyo, I. *et al.* (1999) Prediction of global irradiance on inclined surfaces. *Energy*, **24**, 689–704.

Pearsall, N. and Atanasiu, B. (2009a) Assessment of PV system monitoring requirements by consideration of failure mode probability. *Proceedings of the 24th European Photovoltaic Solar Energy Conference*, Hamburg, Germany, pp. 3896–3903.

Pearsall, N. and Atanasiu, B. (2009b) The European PV system monitoring guidelines – Modernisation under the PERFORMANCE project. *Conference Record of the 34th IEEE Photovoltaic Specialists Conference*, Philadelphia, PA, pp. 256–261.

Ransome, S.J., Wohlgemuth, J.H., Poropat, S., and Aguilar, E. (2005) Advanced analysis of PV system performance using normalised measurement data, *Conference Record of the 31st IEEE Photovoltaic Specialists Conference*, Orlando, FL, pp. 1698–1701.

Reich, N.H., Mueller, B., Armbruster, A. *et al.* (2012) Performance Ratio revisited: are PR>90% realistic?. *Progress in Photovoltaics*, **20**, 717–726.

Silvestre, S., Chouder, A., and Karatepe, E. (2013) Automatic fault detection in grid connected PV systems. *Solar Energy*, **94**, 119–127.

Stettler, S., Toggweiler, P., Wiemken, E. *et al.* (2005) Failure detection routine for grid-connected PV systems as part of the PVSAT-2 project. *Proceedings of the 22nd European Photovoltaic Solar Energy Conference*, Barcelona, Spain, pp. 2490–2493.

Task 13 (2014) PV Performance database, available at: http://iea-pvps.org/index.php?id=1

Tsafarakis, O. (2014) Development of a data analysis methodology to assess PV system performance, MSc thesis. Utrecht University.

Tsafarakis, O. and Van Sark, W.G.J.H.M. (2014) Development of a data analysis methodology to assess PV system performance. *Proceedings of the 29th European Photovoltaic Solar Energy Conference*, Amsterdam, the Netherlands, pp. 2908–2910.

Ueda, Y., Kurokawa, K. Kitimura, K. *et al.* (2009) Performance analysis of various system configurations on grid-connected residential PV systems. *Solar Energy Materials and Solar Cells*, **93**, 945–949.

Van Sark, W.G.J.H.M., Reich, N.H. *et al.* (2012) Review of PV performance ratio development. *Proceedings of the World Renewable Energy Forum*, Denver, Colorado, pp. 4795–4800.

Woyte, A., Richter, M., Moser, D. *et al.* (2014) Analytical monitoring of photovoltaic systems: good practices for monitoring and performance analysis. Report IEA-PVPS T13-03:2014.

11.5

Energy Prediction and System Modeling

Joshua S. Stein
Sandia National Laboratories, Albuquerque, NM, USA

11.5.1 Introduction

Accurate prediction and modeling of the energy that a PV system will produce at a given location are essential for obtaining financing and choosing between different PV technologies for a given project. Modeling of system power under different environmental conditions such as irradiance, temperature, humidity, and so on, is also useful for monitoring of the operation of a PV system to ensure it is functioning as designed, and if not, modeling of energy losses in a PV system can help to identify what is wrong and causing deviations of the expected power.

PV system performance is typically modeled using a computer application that takes as input design information and time series of weather data such as irradiance, temperature, wind speed, precipitation, and so on. Design information covers, among others, location, module and inverter performance parameters, array and module orientations, mounting configuration, horizon map, and wiring design. This information is used in a series of sub-modeling steps to calculate how much energy (kWh) is produced during each time step, which is usually set at 1 hour for annual model prediction and less time for monitoring applications. Such modeling applications include PVSyst, PVSoL, System Advisor Model (SAM), among others.

A PV performance model is actually a collection of submodels where the output of one model provides the inputs to other models downstream. It is informative to think of the flow of energy from the photon to the electron, or sunlight to AC power. Figure 11.5.1 shows this process flow for a grid-connected PV system.

Photovoltaic Solar Energy: From Fundamentals to Applications, First Edition.
Edited by Angèle Reinders, Pierre Verlinden, Wilfried van Sark, and Alexandre Freundlich.

Companion website: www.wiley.com/go/reinders/photovoltaic_solar_energy

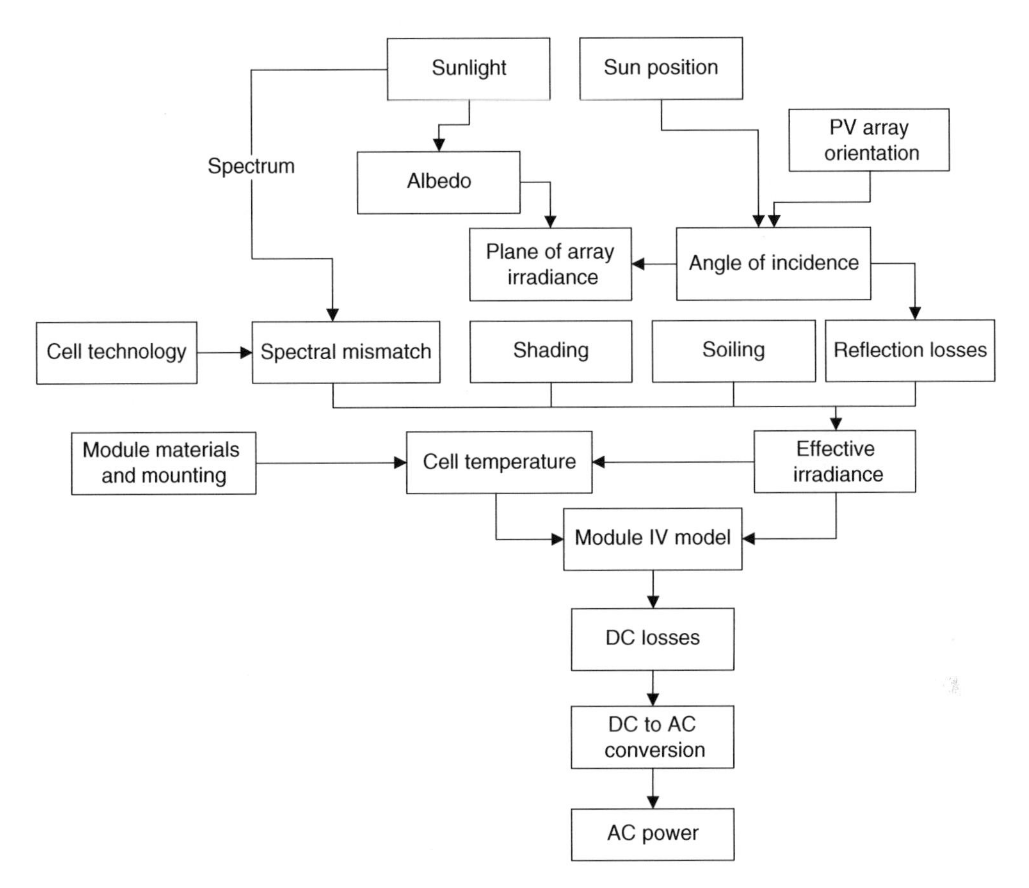

Figure 11.5.1 PV performance modeling process for a grid-connected PV system from sunlight to AC power

11.5.2 Irradiance and Weather Inputs

The first step to modeling a PV system is to collect irradiance and weather data as inputs. Irradiance is the intensity of the sunlight and is expressed in W/m^2 and can be separated into direct and diffuse components. Of all the inputs to PV performance, irradiance is the most directly correlated to power output. Temperature, wind speed, and humidity are also important since they influence the operating temperature of the PV cells and thus their efficiency. Since irradiance and weather vary diurnally and seasonally, it is necessary to measure it relatively frequently (to resolve high frequency variability caused by passing cloud shadows and the daily profile) as well as for an entire year (to resolve seasonal changes in sun position, length of day, and temperature patterns).

To estimate annual energy production from a PV system it is common to use measured irradiance and weather from numerous years to construct a synthetic "typical" year. A common source for such data is Meteonorm, which creates typical annual datasets by stochastically generating hourly and even minute data from monthly mean data available at over 8,300 different stations worldwide (Remund *et al.*, 1998). In the USA these generated years are called

Typical Meteorological Years (TMY). TMYs are constructed by analysis of a long record (e.g., 30 years) of measured irradiance and weather conditions at a set of sites. Data is either measured or more typically modeled (e.g., from satellite or observed sky conditions). The TMY is then constructed by selecting the most typical calendar month from the set available. For example, all of the Januarys in the set are examined and the one judged most typical is selected to be included in the TMY. The selection is based on a weighted average of irradiance, dry bulb, and dew point temperatures. The exact weighting factors have changed with every new version of the TMY (e.g., TMY, TMY2, TMY3, etc.), see Marion and Urban, (1995) and Wilcox and Marion (2008).

11.5.3 Plane of Array Irradiance

Once an irradiance and weather dataset has been selected, the next step is to calculate the plane-of-array (POA) irradiance (in W/m^2). This is the total amount of light hitting the top surface of the PV modules on the array. POA irradiance is calculated by separating it into beam (E_b), sky-diffuse (E_{sd}) and ground reflected (E_g) components:

$$G_{POA} = E_b + E_{sd} + E_g \tag{11.5.1}$$

$$E_b = DNI \times cos\left(AOI\right) \tag{11.5.2}$$

where:

DNI is the direct normal irradiance (in W/m^2)
AOI is the angle of incidence of the DNI onto the array (in degrees)

The *AOI* is calculated using geometry as:

$$AOI = cos^{-1}\left[cos\left(\theta_Z\right)cos\left(\theta_T\right) + sin\left(\theta_Z\right)sin\left(\theta_T\right)cos\left(\theta_A - \theta_{A,array}\right)\right] \tag{11.5.3}$$

where:

θ_Z is the sun zenith angle (degrees)
θ_A is the sun azimuth angle (degrees) (N=0°, E=90°, S 180°, W=270°)
θ_T is the array tilt angle from horizontal (degrees)
$\theta_{A,array}$ is the array azimuth angle (degrees) (N=0°, E=90°, S 180°, W=270°)

The ground reflected irradiance is:

$$E_g = GHI \times albedo \cdot \frac{\left(1 - cos\left(\theta_T\right)\right)}{2} \tag{11.5.4}$$

where:

GHI is the global horizontal irradiance (in W/m^2)
albedo is the relative reflectivity of the ground surface (unitless), which can change with time (e.g., snow cover (Marion *et al.*, 2013))

The sky-diffuse irradiance is estimated using a model. Determining the best model for a site is an active area of research and many models have been proposed. The concepts important to estimating the sky-diffuse irradiance are that the intensity of this component is not isotropic, as is sometimes assumed, but rather there is a slight enhancement or brightening near the sun's disk and near the horizon. One of the most widely used models was proposed by Hay and Davies (1980). It has the following form:

$$E_{sd} = DHI\left[\frac{DNI}{E_a}cos(AOI)+\left(1-\frac{DNI}{E_a}\right)\frac{1+cos(\theta_T)}{2}\right] \tag{11.5.5}$$

where *DHI* is the measured diffuse horizontal irradiance in (W/m^2), and E_a is extraterrestrial irradiance in (W/m^2), which can be estimated from the day of the year, since the Earth's orbit is elliptical (Paltridge and Platt, 1976; University of Oregon Solar Radiation Basics, 2015).

$$E_{sd} = DHI\left[\frac{1+cos(\theta_T)}{2}\right] \tag{11.5.6}$$

This is a simple model that assumes all diffuse light is isotropic. The additional terms and factors in the Hay and Davies model are meant to account for the brightening around the solar disk. Once all these components are determined, G_{POA} is calculated from Equation (11.5.1). There exist many other sky diffuse irradiance models (e.g., Loutzenhiser *et al.*, 2007).

11.5.4 Shading, Soiling, Reflection, and Spectral Losses

The next step is to determine the amount of light available for photovoltaic conversion. This quantity is typically called the "effective irradiance." Effective irradiance is the G_{POA} reduced by any shading losses from near and far objects, transmission losses from dirt accumulation on the module's surface (soiling), reflection losses from the modules' surface, and spectral losses caused by the mismatch between the available spectrum and the absorption spectrum of the PV cell material.

11.5.4.1 Shading Losses

Shading is divided into object (near) and horizon (far) shading. Object shading is from objects or structures that are in the immediate vicinity of the array and will therefore only affect part of the array. Row-to-row shading and shading from nearby buildings and trees are typical examples. Most object shade is usually assumed to remove all of the direct beam irradiance. One exception can be shade from a tree without leaves. Horizon shading is from the effects of the far horizon (e.g., mountains). Horizon shading is characterized by more diffuse shadows. Quantifying the extent and effect of shading is usually done by creating a look-up table listing shading reduction factors as a function of time (e.g., by month and hour of the day). Such tables can be output from instruments that use a fish-eye lens to map the horizon view and all obstructions from the perspective of the array (usually these measurements are made at several locations within the array). These values are then combined with a calculation of sun position in the sky over the year, and times when the sun is obscured by the obstructions are identified. The shading factors are then applied only to the beam irradiance component of the G_{POA}. Deline *et al.* (2013) developed a simplified model to quantify the effect of shading on PV arrays.

11.5.4.2 Soiling Losses

Soiling losses depend on many complicated and site-specific factors such as the composition and physical properties of the regional soil, the prevailing wind direction and speed, the relative humidity, the proximity of industrial activity, and the precipitation patterns (Pavan *et al.*, 2011). Predicting the expected soiling loss for a given site is an active research area and currently such estimates are based on conservative assumptions from measurements made at sites deemed to be similar. Models include soil accumulation periods followed by precipitation cleaning events (e.g., Kimber *et al.*, 2006).

11.5.4.3 Reflection Losses

Reflection losses are due to optical effects at the module surface and depend on the AOI and the module material's optical properties. Reflection losses are expressed as an incident angle modifier (IAM), which is quantified as the ratio between the transmittance at the *AOI* vs. a normal incident angle ($AOI=0°$). The IAM is multiplied by the G_{POA} to adjust for reflection losses. Figure 11.5.2 shows IAM vs. *AOI* measured on a standard PV module with a glass top sheet. Reflection losses begin to become significant when AOI>60°. There are a number of different empirical equations used to fit such data. A simple model used in the PVsyst modeling package is documented in ASHRAE 93-3003:

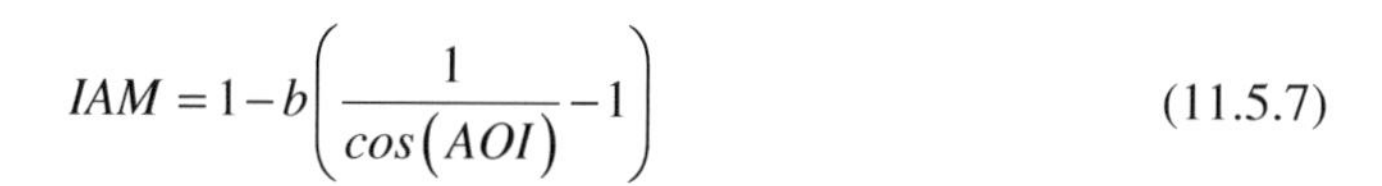

$$IAM = 1 - b\left(\frac{1}{cos(AOI)} - 1\right) \tag{11.5.7}$$

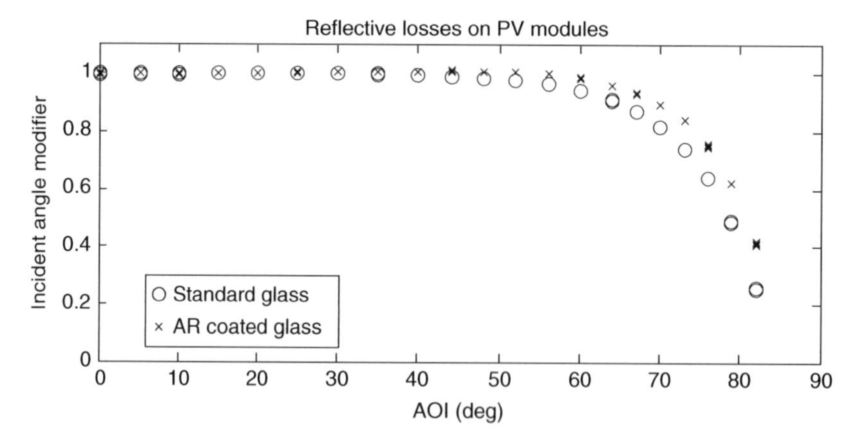

Figure 11.5.2 Measured reflection losses on two modules with different top cover glass

b is a unitless fitting parameter which typically has a value near 0.05 for standard c-Si PV modules. The value of b would be lower for modules with antireflective coatings.

11.5.4.4 Spectral Mismatch Losses

Spectral mismatch occurs when a portion of the available light is not utilized in the PV conversion process because the photon's energy level does not correspond with the available quantum levels to free electron-hold pairs in the cell. This lost portion of the spectrum only contributes to heating the cell. PV cells and modules are rated at a reference spectrum (ASTM G173-03), which is referred to as AM1.5. For different spectra, a spectral mismatch factor is calculated using the actual spectra and the spectral sensitivity or quantum efficiency of the PV device (IEC 60904-7).

11.5.4.5 Effective Irradiance

The effective irradiance (E_e) is the light that is available to be converted into DC current by the PV array. It is calculated by starting with the G_{POA} and adjusting it for shading, soiling, reflection, and spectral losses. One expression for E_e is:

$$E_e = f_1\left\{\left(E_b f_2 + f_d\left(E_{sd} + E_g\right)\right)/E_0\right\}SF \quad (11.5.8)$$

where f_1 is a function to describe the spectral loss, f_2 is a function that describes the reflection loss, f_d is the fraction of the diffuse irradiance used by the module and typically equals 1 for flat plate modules but can be <1 for concentrating PV modules. *SF* is the soiling factor (unitless) which is the fraction of light that is obscured by the soil layer on the PV device. E_0 is the reference irradiance (1,000 W/m^2).

11.5.5 Cell Temperature

The next step is to estimate the cell temperature (T_c) as a function of time. One popular model is based on Faiman (2008) and has the following form:

$$T_c = T_a + \frac{\alpha E_e\left(1-\eta_m\right)}{U_0 + U_1 WS}, \quad (11.5.9)$$

where T_a is air temperature (°C), α is the adsorption coefficient of the module (typical value is 0.9), η_m is the efficiency of the module (typically 0.08 to 0.2), *WS* is the wind speed (m/s). U_0 is the constant heat transfer coefficient (Wm^{-2}°C^{-1}) and U_1 is the convective heat transfer coefficient (W/m^2°C). Typical values for U_0 and U_1 range from 23.5 to 26.5 Wm^{-2} °C and 6.25 to 7.68 Wm^{-3} s °C, respectively. Other model forms are available (e.g., King *et al.*, 2004; Luketa-Hanlin and Stein, 2012).

11.5.6 Module IV Models

In this modeling step, the current-voltage characteristic of a single module is determined. There are two general model types used: equivalent circuit diode models, which evaluate the full I-V curve, and point value models, which only determine key points on the I-V curve (e.g., maximum power point).

11.5.6.1 Equivalent Circuit Diode Models

Equivalent circuit diode models assume that the performance of a PV module can be represented as a simple diode in the circuit shown in Figure 11.5.3. Versions of this basic circuit with more than one diode are also popular.

The current-voltage (*I-V*) characteristic of this circuit is described as:

$$I = I_L - I_0\left[exp\left(\frac{V + IN_sR_s}{nN_sV_T}\right) - 1\right] - \frac{V + IN_sR_s}{N_sR_{sh}} \tag{11.5.10}$$

where I_L is the light current (A), I_0 is the diode reverse saturation current (A), R_s is the series resistance (Ω), R_{sh} is the shunt resistance (Ω), n is the diode ideality factor, Ns is the number of cells in the module, and V_T is the thermal voltage. $V_T = \frac{kT_c}{q}$, where k is Boltzmann's constant (1.381×10^{-23} J/K) and q is the elementary charge (1.602×10^{-19} C). This equation has five module parameters needed to solve for current and voltage (I_L, I_0, n, R_s, and R_{sh}). Several of these parameters vary with irradiance and temperature. When these relationships are defined mathematically, a module performance model is made (e.g., De Soto *et al.*, 2006; Mermoud and Wittmer, 2014). The technical manuals for such performance models should provide these details.

11.5.6.2 Fixed-Point Models

There are several examples of fixed point module and system models. These models are limited in that they only provide selected points on the I-V curve for a module. The Sandia Photovoltaic Array Performance Model (King *et al.*, 2004) predicts I-V values for five points shown in Figure 11.5.4.

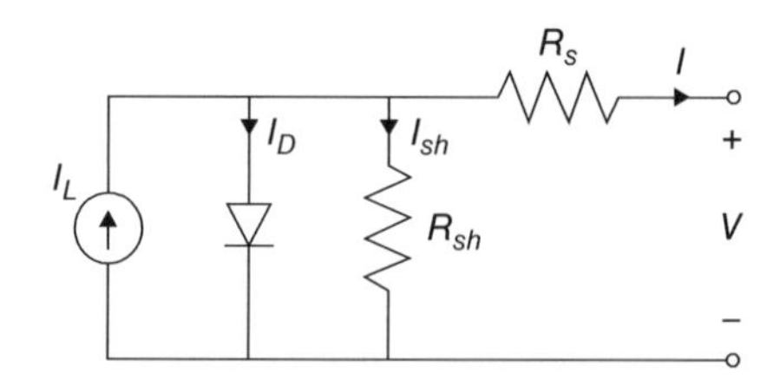

Figure 11.5.3 Single diode equivalent circuit

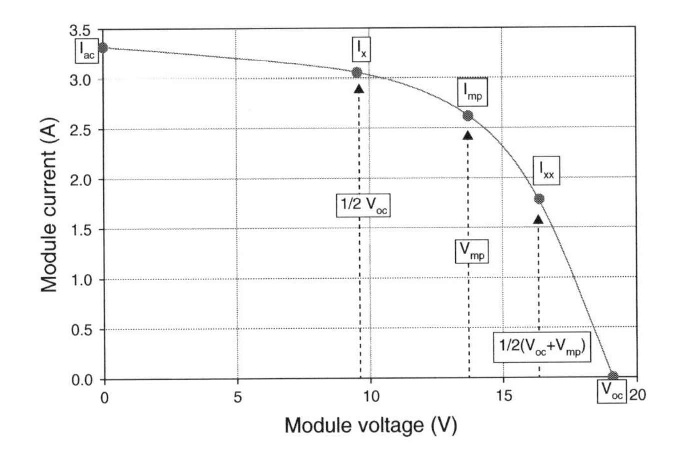

Figure 11.5.4 Five points determined by the Sandia module IV curve model

The Loss Factors Model (Sellner *et al.*, 2012; Stein *et al.*, 2013) can estimate the maximum point, V_{oc} and I_{sc}. Other simpler models such as, for instance, PVWatts are only designed to estimate the maximum power (P_{mp}) and do not resolve the current or voltage separately (PVWatts, 2015).

11.5.7 DC-DC Maximum Power Point Tracking and DC Losses

To optimally obtain power from a PV array the operating voltage must be controlled by a maximum power point tracking (MPPT) algorithm, which continuously perturbs voltage and seeks to maximize power. The maximum power voltage (V_{mp}) varies with irradiance and temperature) and usually MPPT is controlled by the inverter for grid-tied systems or voltage is controlled by a charge controller for off-grid, battery applications. MPPT losses can occur when the MPPT controller cannot rapidly find the MPP. Typically these losses are very low (<0.5%).

DC losses due to resistance in the wiring and interconnections can lead to current and power losses. To minimize these losses, PV designers can connect PV modules in series, thus increasing the voltage while keeping the current fixed to the current at maximum power I_{mp} of a single module. Since resistance losses increase with the square of the current, this configuration keeps these losses low without increasing the costs associated with larger conductor. A consequence of connecting modules in series is that any mismatch between the I-V characteristics of the module population can lead to mismatch losses. For example, in a series-connected string of modules, since the current flowing through all the modules has to be equal, the module with the lowest current will limit the current in the string. Such mismatch can occur from module inconsistencies or from uneven soiling or partial shading. Module manufacturers bin their modules to minimize mismatch and system designers try and avoid string configurations that are affected by partial shading. If these situations cannot be easily avoided, system designers can use module scale power electronics such as microinverters or module-scale DC-DC converters that perform MPPT for each module separately, thus avoiding many of the mismatch losses, but at a higher cost. Most PV performance modeling applications assume that MPPT and DC losses will be estimated outside the software and are entered in as system derate factors.

11.5.8 DC to AC Conversion

Most PV systems are grid-connected and produce AC power. One of the inverter's primary functions is to convert DC power to AC power. This conversion process results in power losses that need to be accounted for in the modeling of PV system performance. These losses are expressed in terms of inverter conversion efficiency, which is equal to P_{AC}/P_{DC}. Inverter efficiency varies with both DC power level and DC voltage and these variations are very manufacturer- and design-specific. Figure 11.5.5 shows an example of this variation for an example inverter.

Inverter performance models aim to represent this complex behavior mathematically. Models are based on measurements made by testing labs that measure efficiency at specific DC power and voltage levels. Model parameters are derived from fitting these curves. The Sandia Photovoltaic Inverter Performance Model (King *et al.*, 2007) is one such model with the following form:

$$P_{AC} = \left\{ \frac{P_{AC0}}{A - B} - C(A - B) \right\} (P_{DC} - B) + C(P_{DC} - B)^2 \tag{11.5.11}$$

where:

$$A = P_{DC0}\left\{1 + C_1\left(V_{DC} - V_{DC0}\right)\right\}$$

$$B = P_{s0}\left\{1 + C_2\left(V_{DC} - V_{DC0}\right)\right\}$$

$$C = C_0\left\{1 + C_3\left(V_{DC} - V_{DC0}\right)\right\}$$

P_{DC} is the DC power (W), V_{DC} is the input voltage (V), V_{DC0} is the DC voltage level (V) at which the AC power rating is achieved at reference conditions, P_{AC0} is the AC power rating (W) at reference conditions, P_{S0} is the DC power (W) required to start the inversion process, or self-consumption by the inverter. The C_1, C_2 and C_3 parameters are fitting coefficients. Other inverter models are also available (e.g., Driesse *et al.*, 2008).

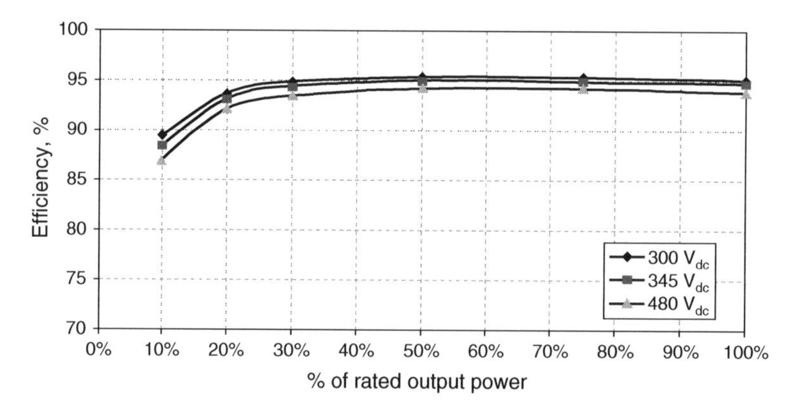

Figure 11.5.5 Example inverter efficiency profile

11.5.9 AC Losses

The final step in modeling the performance of a PV system is to account for any AC losses between the inverter and the final revenue meter that determines how much energy is generated. For small systems (e.g., residential) the meter is directly adjacent to the inverter and AC losses are negligible. However, for large systems it is not uncommon to have an AC distribution system between the inverters and the meter. Transmission-connected, utility-scale systems may have additional transformers. These wires and transformers will introduce losses in the system output that need to be considered.

11.5.10 Modeling of Stand-Alone PV Systems

Stand-alone or "off-grid" PV systems share most of the performance characteristics described in the previous sections except that they operate with additional constraints, equipment, and there is a much tighter integration to the specific equipment (loads) that are included in the system. While the performance of grid-connected PV systems is typically quantified by the number of kWh produced, the performance of stand-alone PV systems is evaluated in their ability to serve loads when required and at a minimal cost. Loads are of two general varieties: primary loads, which must be met immediately by the system, and deferrable loads, which can met within some defined time window. If primary loads can exceed the available power output of the PV system, then energy storage (e.g., battery) or alternate generation (e.g., generator) needs to be included in the system. The main modeling questions for stand-alone PV systems are concerned with module and component selection, system sizing and array wiring decisions. These variables all affect the efficiency and cost of the final system.

One popular software tool that is used to design and size standalone power systems including PV systems is HOMER (HOMER, 2015). This software runs an energy balance calculation at an hourly time interval and is able to compare and evaluate many different system design options to identify the least-cost solutions that meet the load requirements. However, the PV performance model in HOMER is very simplistic. It assumes linear performance with irradiance and a scalar "derate" factor that accounts for all other losses, including thermal, soiling, spectral, mismatch, and wiring losses. Because stand-alone PV systems are usually designed for "worst case" scenarios (e.g., long cloudy periods, maximum loads), there are periods when energy produced by the system is not usable and therefore the need for accurate representations of PV performance at any one instant is unnecessary.

11.5.10.1 Direct-Drive Water Pumping PV System

The simplest stand-alone PV systems are those designed to meet deferrable loads, for example, water pumping applications. These systems usually do not require battery storage, because they store potential energy by filling a tank or reservoir, from which water can flow by gravity when it is needed, regardless of the availability of solar energy. The performance of such a system is quantified by determining how effective the system is at meeting the water demand over time. If the tank runs dry during a period of low irradiance, then there is either a performance or design problem. To prevent performance problems in such an application, it is incumbent upon the designer to size the PV array appropriately. Important variables include daily water need and minimum number of sun hours per day. These are used to estimate a pumping rate. The pumping rate, pumping height, and pipe friction are used to calculate the pumping power required. Next the ampere-hour requirements of the selected DC pump are used along with the pumping power to size the PV array and define its wiring configuration. Other devices, such as a linear current booster (LCB), can be added to reduce the size of the pump needed and extend the pumping time. The details of how to design such a system are beyond the scope of this chapter, but such information is readily available (e.g., Messenger and Ventre, 2010; Häberlin, 2012).

11.5.10.2 Stand-Alone PV system with Battery Storage and Inverter

Linking a PV system to a battery system and inverter can result in a home microgrid that can support a wide range of loads and be designed to operate as if the home were grid connected, with only a few limitations, such as maximum power and energy requirements, and state of charge and depth of discharge limitations of the battery (e.g., Copetti *et al.*, 1993; Manwell and McGowan, 1993; Bajpai and Dash, 2012). For example, some large loads are inappropriate for a PV/battery system, especially if a nonelectric version exists (e.g., clothes dryer, heater, etc.), since it would require the system size to be increased at a cost greatly exceeding replacing the appliance in question. There are a number of basic design options with a PV/battery system. Some systems create separate DC and AC circuits, and segregate loads to each as appropriate (e.g., DC circuit for LED lighting and charging of small electronics like laptops and cell phones). Other systems connect the PV and battery system to an inverter and supply exclusive AC power to the home. To ensure that the batteries are not damaged by overcharging, the system will include a charge controller. These devices come in a number of different varieties. Basically charge controllers monitor battery voltage and temperature, and limit the charging current to protect the battery. When the battery voltage is below a certain threshold, constant current is injected, when the battery voltage reaches the threshold, the charge controller switches to constant voltage mode and the current gradually decreases as the battery becomes fully charged. At this stage, the controller maintains a "float" voltage, which is low enough not to damage the battery, but high enough to compensate for self-discharge. Battery chemistries other than lead-acid require modifications to this charging cycle. Basic charge controllers are designed to work with PV arrays with a voltage level that roughly matches the battery's voltage. This design works well for small systems, but as systems get larger, it makes more sense to connect PV modules in series to boost the array voltage far above the battery voltage and reduce current levels and wiring losses. In this case a more advanced MPPT charge controller is useful. These devices set the array voltage to the optimal MPP while employing a DC-DC converter to supply the battery with a voltage that meets its needs for charging (Salas *et al.*, 2006). Such functionality can also be found in specialized hybrid inverters that can control battery charging as well as inversion to AC. Use of MPPT charge controllers have been shown to reduce the size of both the array and battery needed in certain situations (Chen *et al.*, 2007).

11.5.11 Conclusion

Understanding the performance of a PV system requires knowledge of a number of modeling steps that roughly follow the flow of energy through the system from solar radiation to the delivery of electrical power. A detailed model of these processes helps researchers to identify opportunities for improving the efficiency of the technology or system designs. More accurate performance models increase the predictability and lower the risk for PV investors, all of which results in lower interest rates and easier access to capital for investment in PV. It is important for users of PV performance models to understand the basic algorithms used and the assumptions made in order that the results are not misleading. For example, a PV performance analysis based on historical weather data should not be expected to match measured performance for any given month, since there is considerable variability year to year in monthly weather. Similarly, performance should be expected to vary from predictions during periods of snow, reducing yields when snow covers the array and raising performance when the array is clear but snow on the ground increases albedo. These are just some common examples of confusion about PV performance. Taking the time to identify and understand each of the submodels that are included in a PV performance model is important. It is also important to remember that predicting future performance has two kinds of uncertainties; the first is related to the model's ability to represent the performance under defined conditions. The second is related to not knowing what the future conditions will be. The task of PV performance modeling is to try and reduce each of these uncertainties as much as possible. A good source of more detailed information is the PV Performance Modeling Collaborative (Stein, 2012; Stein and King, 2013; PVPMC, 2015).

List of Symbols and Acronyms

Symbol	Description	Units
albedo	Relative reflectivity of ground surface. Typical values between 0 and 1, for instance 0.2 (grass) and 0.8 (fresh snow)	unitless
AOI	Angle Of Incidence	°
b	Empirical coefficient for ASHRAE IAM model	unitless
C_1, C_2, and C_3	Fitting parameters	unitless
DHI	Diffuse Horizontal Irradiance	W/m^2
DNI	Direct Normal Irradiance	W/m^2
E_a	Extraterrestrial irradiance	W/m^2
E_b	Beam irradiance on a surface	W/m^2
E_e	Effective irradiance (normalized light available to be converted into DC current)	unitless
E_g	Ground-reflected irradiance on a surface	W/m^2
E_{sd}	Sky-diffuse irradiance on a surface	W/m^2
f_1	Empirical function describing spectral loss	function
f_2	Empirical function describing reflection losses	function
f_d	Fraction of the diffuse irradiance used by the module	unitless
GHI	Global Horizontal Irradiance	W/m^2
G_{POA}	Irradiance on the plane-of-array	W/m^2
I	Current	A
I_0	Diode reverse saturation current	A
IAM	Incident Angle Modifier	unitless
I_L	Light current	A

(*Continued*)

(*Continued*)

Symbol	Description	Units
I_{mp}	Maximum power current	A
I_{sc}	Short circuit current	A
k	Boltzmann's constant = 1.381×10^{-23}	J/K
n	Diode ideality factor	unitless
Ns	Number of cells in a module	unitless
P_{AC}	AC power	W
P_{AC0}	AC power rating of inverter	W
P_{s0}	DC power required to start the inversion process, or self consumption by the inverter	W
q	Elementary charge, 1.602×10^{-19} Coulombs	C
R_s	Series resistance	Ω
R_{sh}	Shunt resistance	Ω
SF	Soiling Factor	unitless
T_a		°C
T_c	PV cell temperature	°C
U_0	Constant heat transfer coefficient	Wm^{-2}°C
U_1	Convective heat transfer coefficient	$Wm^{-3}s$°C
V	Voltage	V
V_{DC}	DC input voltage	V
V_{DC0}	DC voltage level at which AC power rating is achieved	V
V_{mp}	Maximum power voltage	V
V_{oc}	Open circuit voltage	V
WS	Wind speed	m/s
η_m	Module efficiency	%
θ_A	Azimuth angle of the sun (degrees East of North)	°
$\theta_{A,array}$	Azimuth angle of the array (degrees East of North)	°
θ_T	Tilt angle of the array from horizontal	°
θ_Z	Zenith angle of the sun	°

Acronyms	Description
LCB	Linear Current Booster
MPP	Maximum Power Point
MPPT	Maximum Power Point Tracking
α	Module Adsorption Coefficient
V_T	Thermal Voltage
TMY	Typical Meteorological Year

References

ASTM G173-03 (2012) Terrestrial Reference Spectra for Photovoltaic Performance Evaluation. American Society for Testing and Materials, West Conshohocken, PA.

Bajpai, P. and Dash, V. (2012) Hybrid renewable energy systems for power generation in stand-alone applications: a review. *Renewable and Sustainable Energy Reviews*, **16** (5), 2926–2939.

Chen, W., Shen, H., Shu, B. *et al.* (2007) Evaluation of performance of MPPT devices in PV systems with storage batteries. *Renewable Energy*, **32** (9), 1611–1622.

Copetti, J.B., Lorenzo, E. and Chenlo, F. (1993) A general battery model for PV system simulation. *Progress in Photovoltaics: Research and Applications*, **1** (4), 283–292.

Deline, C., Dobos, A., Janzou, S. *et al.* (2013) A simplified model of uniform shading in large photovoltaic arrays. *Solar Energy*, **96**, 274–282.

De Soto, W., Klein, S. A. and Beckman, W. A. (2006) Improvement and validation of a model for photovoltaic array performance. *Solar Energy*, **80** (1), 78–88.

Driesse, A., Jain, P. and Harrison, S. (2008) Beyond the curves: modeling the electrical efficiency of photovoltaic inverters. Paper presented at 33rd IEEE Photovoltaic Specialists Conference, San Diego, CA.

Faiman, D. (2008) Assessing the outdoor operating temperature of photovoltaic modules. *Progress in Photovoltaics*, **16** (4), 307–315.

Häberlin, H. (2012) Photovoltaics System Design and Practice. John Wiley & Sons, Ltd, Chichester.

Hay, J.E. and Davies, J.A. (1980) Calculation of the solar radiation incident on an inclined surface. *Proceedings of First Canadian Solar Radiation Data Workshop* (eds J.E. Hay and T.K. Won), Ministry of Supply and Services Canada. Vol. 59.

HOMER, http://homerenergy.com/(accessed Feb. 2015).

IEC 60904-7 (2008) Photovoltaic devices – Part 7: Computation of the spectral mismatch correction for measurements of photovoltaic devices.

Kimber, A., Mitchell, L., Nogradi, S. and Wenger, H. (2006) The effect of soiling on large grid-connected photovoltaic systems in California and the Southwest Region of the United States. IEEE Photovoltaics Specialist Conference, Waikoloa, Hawaii.

King, D.L., Boyson, E.E. and Kratochvil, J.A. (2004) Photovoltaic Array Performance Model. Sandia National Laboratories, Albuquerque, NM, SAND2004-3535

King, D.L., Gonzalez, S., Galbraith, G.M. and Boyson, W.E. (2007) Performance Model for Grid-Connected Photovoltaic Inverters. Sandia National Laboratories, Albuquerque, NM, SAND2007-5036.

Loutzenhiser, P.G., Manz, H., Felsmann C. *et al.* (2007) Empirical validation of models to compute solar irradiance on inclined surfaces for building energy simulation. *Solar Energy*, **81** (2), 254–267.

Luketa-Hanlin, A. and Stein, J.S. (2012) Improvement and validation of a transient model to predict photovoltaic module temperature. World Renewable Energy Forum, Denver, CO.

Manwell, J.F. and McGowan, J.G. (1993) Lead acid battery storage model for hybrid energy systems. *Solar Energy*, **50** (5), 399–405.

Marion, B., Schaefer R., Caine, H. and Sanchez, G. (2013) Measured and modeled photovoltaic system energy losses from snow for Colorado and Wisconsin locations. *Solar Energy*, **97** (0): 112–121.

Marion, W. and Urban, K. (1995) User's manual for TMY2s. Available at: http://rredc.nrel.gov/solar/pubs/tmy2/

Mermoud, A. and Wittmer, B. (2014) PVSYST user's manual. Available at: http://files.pvsyst.com/help/index.html.

Messenger, R.A. and Ventre, J. (2010) Photovoltaic Systems Engineering. CRC Press, New York.

Paltridge, G.W. and Platt, C.M.R. (1976) Radiative Processes in Meteorology and Climatology. Elsevier Scientific Pub. Co., New York.

Pavan, A.M., Mellit, A. and Pieri, D.D. (2011) The effect of soiling on energy production for large-scale photovoltaic plants. *Solar Energy*, **85**, 1128–1136.

PVPMC, PV Performance Modeling Collaborative. Available at: https://pvpmc.sandia.gov (accessed Feb. 2015).

PVWatts, http://pvwatts.nrel.gov/(accessed Feb. 2015).

Remund, J., Salvisberg, E. and Kunz, S. (1998) On the generation of hourly shortwave radiation data on tilted surfaces. *Solar Energy*, **62** (5), 331–344.

Salas, V., Olías, E., Barrado, A., and Lázaro, A. (2006) Review of the maximum power point tracking algorithms for stand-alone photovoltaic systems. *Solar Energy Materials and Solar Cells*, **90** (11), 1555–1578.

Sellner, S., Sutterlueti, J., Ransome, S. *et al.* (2012) Understanding module performance further: validation of the novel loss factors model and its extension to AC arrays. EU PVSEC, Frankfurt, Germany.

Stein, J. (2012) The Photovoltaic Performance Modeling Collaborative. Paper presented ate 38th IEEE PVSC, Austin, TX.

Stein, J.S. and King, B.H. (2013) Modeling for PV plant optimization. Photovoltaics International, Solar Media Ltd. 19th: 101–109.

Stein, J.S., Sutterlueti, J., Ransome, S. *et al.* (2013) Outdoor PV performance evaluation of three different models: single-diode, SAPM and loss factor model. 28th EU PVSEC. Paris, France.

University of Oregon Solar Radiation Basics, available at: http://solardat.uoregon.edu/SolarRadiationBasics.html#Ref1 (accessed Feb. 2015).

Wilcox, S. and Marion, W. (2008) User's manual for TMY3 data sets, *NREL/TP-581-43156.*

11.6

Building Integrated Photovoltaics

Michiel Ritzen[1,2], Zeger Vroon[1,3], and Chris Geurts[3]
[1]*Zuyd University of Applied Sciences, Heerlen, the Netherlands*
[2]*Eindhoven University of Technology, Eindhoven, the Netherlands*
[3]*TNO, Delft/Eindhoven, the Netherlands*

11.6.1 Introduction

Photovoltaic (PV) installations can be realized in different situations and on different scales, such as at a building level. The application of PV installations at the building level fits in with the international tendency to realize energy-efficient and zero-energy buildings (EU, 2010). Application of PV installations on the building level has the following advantages, compared to other energy systems and other applications of PV installations:

- An installation can easily be applied to buildings, because a PV installation is relatively easily connected to the electrical system of a building.
- A PV installation is not based on either dangerous processes or use of dangerous resources.
- A PV installation does not consist of moving parts that need maintenance (except for sun trackers).
- No additional land is needed for a PV installation.
- Transmission losses are minimal because a large part of the generated energy is consumed at the same location.
- PV installations on buildings are less vulnerable to theft and damage.

PV installations at the building level can either be added to the building envelope, which is called Building Added PV (BAPV) (Figure 11.6.1, left), or they can be integrated into the building envelope, called Building Integrated PV (BIPV) (Figure 11.6.1, right). In general, we speak of a BIPV system if the installation is technically and aesthetically integrated, contributing to a homogeneous coverage of the building surface (Sinapsis and Donker, 2013). BIPV (and BAPV) can either be applied in a grid-connected situation (see Chapter 11.1) or be autarkic with storage on-site.

Photovoltaic Solar Energy: From Fundamentals to Applications, First Edition.
Edited by Angèle Reinders, Pierre Verlinden, Wilfried van Sark, and Alexandre Freundlich.

Companion website: www.wiley.com/go/reinders/photovoltaic_solar_energy

Figure 11.6.1 Free-standing and roof mounted Building Added PV in Rotterdam, the Netherlands, 2013 (left) vs Building Integrated PV in Heerlen, the Netherlands, 2013 (right)

BIPV is seen as a necessary step in coping with our energy challenge in the next decades by realizing energy generation with socially accepted solutions. In general, it is assumed that by realizing BIPV the rise of Not on My Roof (NoMyR) opposition by people can be prevented, which is based on the belief that PV is necessary but should be realized further away, comparable with the Not In My BackYard (NIMBY) opposition to wind turbines. With a share of 1 to 3%, BIPV has a relatively small market share compared to BAPV in the total PV market in 2012 (Clover, 2014).

In this chapter we will cover BIPV mainly from a holistic building viewpoint, covering the building design aspects of BIPV, the main regulatory and building codes issues related to the application of PV in the built environment, and conclude with the barriers ahead for large-scale deployment of BIPV.

11.6.2 BAPV vs BIPV

The definition of BIPV is still being internationally discussed. However, the main indicators of a BIPV system that are widely accepted are the following (Kaan and Reijenga, 2004; Schoen *et al.*, 1998):

- BIPV generates electricity.
- BIPV possibly replaces conventional building envelope materials.
- BIPV is aesthetically integrated into the building envelope:
  - The PV components fit into the design grid, or the design grid is based on PV components.
  - Harmony of PV components in the design composition.
  - PV component color fits into the design.
  - The PV technology fits into the design or the design is based on PV technology.

BIPV is seen as one of the four tracks for the realization of large-scale PV deployment, besides higher PV efficiency, a lower market price, and a storage network (Raugei and Frankl, 2009). The possibilities and acceptance of BIPV depend on the local energy situation, the scale of the project, the local culture, the type of financing (Reijenga, 2005), the regulations and governmental incentives, and should be an integral part of the building design to accomplish a successful result.

Figure 11.6.2 Two examples of BIPV on office buildings, contributing to the environmental consciousness impression of the occupants (left: office building in Barcelona, Spain, 2014, right: office building in Westerlo, Belgium, 2011)

The main advantage of BIPV compared to BAPV is that it contributes to an aesthetically more acceptable result. Secondary benefits are the possible material savings, financial savings, and a contribution to the environmental consciousness impression of the building owner or occupant (Figure 11.6.2).

The main disadvantage of BIPV compared to BAPV is that the integration may influence the building physics of the building envelope, increasing the risk of higher operating temperatures and higher levels of relative humidity (Ritzen *et al.*, 2014), affecting the efficiency of the installation and possibly the lifespan of the installation.

The power output and efficiency of a PV installation depend linearly on the operating temperature (Dubey *et al.*, 2013). The efficiency of PV crystalline silicon cells decreases by approximately 0.5% per °C temperature rise (Hasan *et al.*, 2010; Quesada *et al.*, 2012). The operating temperature of a PV module is influenced by the ambient temperature, the thermal properties of the module, the thermal properties of the installation and the insolation (Norton *et al.*, 2011).

Higher operating temperatures can be found in BIPV due to the decreased and/or sealed air gap between PV systems and the underlying envelope layer. This can be particularly problematic for fully building integrated PV (Mei *et al.*, 2009). Norton *et al.* (2011) indicate that temperature rises in BIPV result in losses between 2.2% and 17.0%. Higher operating temperatures can be prevented with cooling of PV systems, based on either air, fluids or Phase Change Materials (PCMs).

Passive back-string ventilation cooling is seen as one of the most effective and easily applicable methods to cool PV systems (Wang *et al.*, 2006; Gan, 2009a, 2009b; Mei *et al.*, 2009; Maturi *et al.*, 2010; Huang, 2011; Norton *et al.*, 2011; Sadineni *et al.*, 2011; Bloem *et al.*, 2012; Petter Jelle *et al.*, 2012; Tyagi *et al.*, 2012). The effect of back-string ventilation cooling depends on factors such as project scale, location, orientation, and inclination and should be an integral part of the BIPV design and realization process.

11.6.3 BIPV Design

In future, the integration of PV modules in the building envelope should be seen as part of a new architectural era. BIPV products and components should not be positioned in the process as merely energy-generating devices but should be seen as a building material as well,

Figure 11.6.3 Two examples of BIPV systems in France (2013) in which the modules are, from a technical point of view, integrated into the building skin, but where the aesthetic integration is lacking

providing the optimal solution between aesthetic considerations and economic considerations. Current economic considerations are mainly related to energy generation, and energy has only economic value when needed.

There are many examples of physical integration of PV in the building envelope that lack an aesthetic integration (Reijenga, 2005), negatively affecting acceptance of BIPV as a sustainable solution for our energy demands (Figure 11.6.3).

To select the best BIPV solution from the wide range of BIPV products for a specific building project, a collaborative system approach should be followed in which different experts cooperate closely, such as architects, installers and the project manager. During the design and building process, different aspects have to be taken into account, such as the architectural design, the location, thermal properties, building typology and function, building installations, user behavior, grid connection, energy storage, passive measures, etc. Within the scope of the definition of BIPV a very wide range of products can be applied.

Currently there are more than 100 different BIPV products developed with different techniques (Jelle and Breivik, 2012; Cerón *et al.*, 2013; Sinapis and Donker, 2013). On different websites such as www.bipv.ch, www.pvnord.org, and www.solarintegrationsolutions.org, and in the publications of Verberne *et al.* (2014), Memari *et al.* (2014), and Sinapis and Donker (2013) an extensive overview of different BIPV products and applications is presented. In general, two categories of BIPV products exist, namely, roof systems and façade systems, that have different sub-categories, such as flat roof, etc., for which different techniques and solutions are suitable (Eiffert and Kiss, 2000; Memari *et al.*, 2014), indicated and visualized in Table 11.6.1.

To select the most appropriate BIPV product and develop a successful BIPV project, the following four aspects of integration have to be taken into account (Schoen *et al.*, 1998; Bahaj, 2003; Kaan and Reijenga, 2004; Reijenga, 2005; Petter Jelle *et al.*, 2012; Memari *et al.*, 2014):

1. *Level of contribution to building aesthetics*: This aspect can have the following characteristics (but not inclusively all of them):
   - It is not visible (e.g., rooftop).
   - It has added architectural elements (e.g., shading devices).
   - It contributes to the aesthetic quality (e.g., in façades/visible roofs).

Table 11.6.1 BIPV categories, sub-categories and techniques. (*See insert for color representation of the figure*)

Category	Sub category	Technique	Example
Roof systems	Flat roof	Roofing material	
Roof systems	Pitched roofs	Opaque modules	
Roof systems	Pitched roofs	Colored cells/ modules	
Roof systems	Pitched roofs	Shingles	 Reproduced with permission by the Department of Geosciences, University of Wisconsin-Madison

(*Continued*)

Table 11.6.1 (*Continued*)

Category	Sub category	Technique	Example
Roof systems	Pitched roofs	Tiles	
Roof systems	Pitched roofs	Skylights/ semitransparent modules	 

Table 11.6.1 (*Continued*)

Category	Sub category	Technique	Example
Façade systems		Curtain walls	 Reproduced with permission by Oskomera
Façade systems		Opaque panels	 Reproduced with permission by Brooks Scarpa Architects.
Façade systems		Semitransparent modules	
Façade systems		Shading devices	

  - It determines the aesthetic quality (e.g., building design based on orientation, inclination and PV technology applied).
  - The PV installation has resulted in a new architectural concept.
2. *Building quality*: This aspect can be described by the following features
  - The BIPV system is aesthetically integrated into the building envelope.
  - The BIPV system architecturally contributes to the building's appearance.
  - The BIPV system consists of colors, materials and composition that fit in and contribute to the building design.
  - The BIPV system contributes to the quality of the urban tissue.
  - The BIPV system contributes to an innovative design.
  - The building design prevents possible theft and damage of the BIPV installation.
  - In the site and building design, shading (by trees, adjacent buildings, and other building components) is minimized complying with the client's expectations.
3. *PV installation quality*: This aspect covers technical factors as described below:
  - The system and its integration are well engineered, and the expected lifetime of BIPV component complies with the client's expectations and the building component it replaces.
  - The orientation and inclination of BIPV installation are optimized within the building project's constraints.
  - The system can easily be incorporated into the design and realization of the building envelope, displacing its costs.
  - All system components (wiring, inverter placement, etc.) are an integral part of the system design and are easily reachable for maintenance, replacement, and cleaning.
  - The PV temperature increase is minimized within the design constraints.
4. *Process quality*: This aspect merely covers the embedding of BIPV in environmental, societal, electrical, and building-related regulations, as described below:
  - The BIPV installation is an integral part of the lifecycle cost of the building project and business case.
  - The BIPV system complies with the relevant electrical regulations and building codes.
  - All actors in the BIPV project are supplied with sufficient information about the project.
  - Minimal additional installation time should be required.

These four aspects with different levels and parameters are indicators of the level of "BIPV-ness" of products and applications, but a general rating framework has yet to be developed.

11.6.4 BIPV Building Aspects, Codes and Regulations

BIPV products are more complex products than regular PV products and regular building envelope components because they are a combination of both. As indicated in the Section 11.6.3, many technical solutions are possible and there are many aspects of integration to be taken into account. In the building process, architects and builders prefer modular elements (Cerón *et al.*, 2013), resulting in a selection of a standard 60-cell module of 1 x 1.6 m^2, but the lack of technical knowledge about sustainable buildings in general among architects is one of the main barriers in the current building process (Attia, 2010). The level of integration of a standard PV module-based installation depends on the mounting structure, which mainly consists of aluminum girders that are placed on the building structure instead of a regular cladding or roofing material. Fitting the dimensions of the modules in the measurement of the complete building demands knowledge and implementation of the product dimensions in the design process and might result in dummies at the edges of roofs and façades. In the case of semi-transparent modules, these can be placed in the glazing frame.

BIPV products have to comply with both existing and pending building codes and regulations and have to comply with the electricity generation regulations. These regulations differ in many cases from country to country, and from application to application (Weller *et al.*, 2010), having a negative effect on BIPV product market introduction in different markets. In the French market this is partly compensated by an extra subsidy related to the level of BIPV-ness of the project (AgentschapNL, 2011).

As BIPV is relatively new on the market, suitable codes and regulations are not always developed yet and BIPV is in a 'grey' area of building legislation, which might cause complications and lengthy building permission procedures, resulting in potentially high administrative fees (Garbe *et al.*, 2012). Weller *et al.* (2010) have investigated the regulations with which different BIPV solutions have to comply, resulting in an overview of seven regulations in Germany. To facilitate the realization of BIPV in Germany, a document that combines the electrical, technical, and building regulations is being developed, the DIN VDE 0216-21. In the Netherlands and other countries that are on track for larger market penetration of BIPV, regulations have either recently been developed or are under development. In the Netherlands, a regulation, NEN 7250, specifies the application of solar energy systems (or complete building elements with PV) as an integrated component of the building envelope (NEN, 2014). Considering Europe, the Construction Product Directive (CPD) was established in 1989 to facilitate the possibility of the application of qualitative building components in the different countries. The CPD is based on six essential aspects: in construction safe and sound, fireproof, not harmful to man and animal, safe to use, low noise, and energy- and cost-efficient (AgentschapNL, 2011).

Aspects such as fire regulation (fire tests are not yet included in BIPV EU standards) (Weller *et al.*, 2010), solar access rights, regulatory instability (Barth *et al.*, 2014), Product Category Rules (PCRs) to develop Environmental Product Declarations, and BIPV-centered financing schemes have not (fully) been taken into account. In a number of European countries demands and criteria for BIPV tests have been developed, focusing on wind loads, mechanical loads, snow loads, fire safety, condensation, temperature behavior, and installation safety. These tests are being implemented by the European Institute for Normalization (CEN) in a European Regulation, but this can be a time-consuming process.

At the time of writing, the situation for large-scale BIPV application is complex due to a combination of a large variety of BIPV products, the large scope of regulations the products have to comply with, the gap between the technology and the facilitating framework, the financial aspect, and the small market penetration. However, in the end, with or without building codes and regulations, the design, realization, and functioning of the system have to result in a safe and sustainable functioning for 25 years (AgentschapNL, 2011).

11.6.5 Outlook

The development of BIPV is still relatively new and its deployment small. Furthermore, there is not yet a clear consensus on the definition of BIPV. The current range of BIPV products offers a wide variety of possibilities for building integration, meeting the clients' expectations, with increasing efficiency and lower prices. The current thresholds for large-scale deployment of BIPV therefore seem not to be technically based, with these lower price levels and market readiness of products, but based more on building process and financing.

To reach a successful large-scale realization of BIPV projects, the building process has to be adapted to the implementation of BIPV. Tools such as a rating framework for BIPV-ness, 3D mapping, an effective framework of legislation and regulations, financing mechanisms and environmental assessment have to be improved or developed to facilitate this adaption.

Only with larger market share will the prices of BIPV products become competitive with BAPV products and regular building envelope components, and only then will we be able to cope with our energy demand in a societal accepted way.

List of Acronyms

Acronym	Description
BAPV	Building Added PV
BIPV	Building Integrated PV
NIMBY	Not In My BackYard - characterization of opposition, based on the belief that wind turbines are necessary but should be realized further away
NoMyR	Not on My Roof - characterization of opposition, based on the belief that PV is necessary but should be realized further away.
PV	Photovoltaic

References

AgentschapNL (2011) Gebouwintegratie zonnestroomsystemen, Praktijkvoorbeelden van succesvolle producten.

Attia, S. (2010) Sizing photovoltaic systems during early design: a decision tool for architects. Paper presented at the SB10, Maastricht, the Netherlands.

Bahaj, A. S. (2003) Photovoltaic roofing: issues of design and integration into buildings. *Renewable Energy*, **28** (14), 2195–2204. DOI 10.1016/s0960-1481(03)00104-6.

Barth, B., Concas, G., Zane, E.B. *et al.* (2014) PV Grid Final Project Report.

Bloem, J.J., Lodi, C., Cipriano, J., and Chemisana, D. (2012) An outdoor test reference environment for double skin applications of building integrated photovoltaic systems. *Energy and Buildings*, **50** (0), 63–73. DOI: 10.1016/j.enbuild.2012.03.023.

Cerón, I., Caamaño-Martín, E., and Neila, F.J. (2013) 'State-of-the-art' of building integrated photovoltaic products. *Renewable Energy*, **58** (0), 127–133. DOI: http://dx.doi.org/10.1016/j.renene.2013.02.013.

Clover, I. (2014) BIPV sector to reach 1.15 GW by 2019, report says. http://www.pv-magazine.com/news/details/beitrag/bipv-sector-to-reach-115-gw-by-2019--says-report_100014922/#ixzz3IHpBOXu8 (accessed 11 June 2014).

Dubey, S., Sarvaiya, J.N., and Seshadri, B. (2013) Temperature dependent photovoltaic (PV) efficiency and its effect on PV production in the world – a review. *Energy Procedia*, **33** (0), 311–321. doi: http://dx.doi.org/10.1016/j.egypro.2013.05.072

Eiffert, P., and Kiss, G. (2000) *Building-Integrated Photovoltaic Designs for Commercial and Institutional Structures*. National Renewable Energy Laboratory report BERL/BK-20-25272 Oakridge, TN.EU Directive 2010/31/Eu of the European Parliament and of the Council of 19 May 2010 on the energy performance of buildings (recast).

Gan, G. (2009a) Effect of air gap on the performance of building-integrated photovoltaics. *Energy*, **34** (7), 913–921. DOI: 10.1016/j.energy.2009.04.003.

Gan, G. (2009b) Numerical determination of adequate air gaps for building-integrated photovoltaics. *Solar Energy*, **83** (8), 1253–1273. DOI: 10.1016/j.solener.2009.02.008.

Garbe, K., Latour, M., and Sonvilla, P.M. (2012) PV legal final report – reduction of bureacratic barriers for succesful PV deployment in Europe.

Hasan, A., McCormack, S. J., Huang, M. J., and Norton, B. (2010) Evaluation of phase change materials for thermal regulation enhancement of building integrated photovoltaics. *Solar Energy*, **84** (9), 1601–1612. DOI: 10.1016/j.solener.2010.06.010.

Huang, M J. (2011) Two phase change materials with different closed shape fins in building integrated photovoltaic system temperature regulation. Paper presented at the World Renewable Energy Congress 2011, Linkoping, Sweden.

Jelle, B.P. and Breivik, C. (2012) State-of-the-art building integrated photovoltaics. *Energy Procedia*, **20** (0), 68–77. DOI: 10.1016/j.egypro.2012.03.009.

Kaan, H. and Reijenga, T. (2004) Photovoltaics in an architectural context. *Progress in Photovoltaics: Research and Applications*, **12**, 395–408.

Maturi, L., Sparber, W., Kofler, B., and Bresciani, W. (2010) *Analysis and monitoring results of a BIPV system in northern Italy. Paper presented at the PVSEC*, Valencia, Spain.

Mei, L., Infield, D.G., Gottschalg, R. *et al.* (2009) Equilibrium thermal characteristics of a building integrated photovoltaic tiled roof. *Solar Energy*, **83** (10), 1893–1901. DOI: 10.1016/j.solener.2009.07.002.

Memari, A.M., Iulo, L.D., Solnosky, R.L., and Stultz, C.R. (2014) Building integrated photovoltaic systems for single family dwellings: innovation concepts. *Open Journal of Civil Engineering*, **4**, 102–119. DOI: http://dx.doi.org/10.4236/ojce.2014.42010

NEN (2014) NEN 7250 Solar energy systems: integration in roofs and facades - Building aspects. NEN.

Norton, B., Eames, P.C., Mallick, T.K., *et al.* (2011) Enhancing the performance of building integrated photovoltaics. *Solar Energy*, **85** (8), 1629–1664. DOI: 10.1016/j.solener.2009.10.004.

Petter Jelle, B., Breivik, C., and Drolsum Røkenes, H. (2012) Building integrated photovoltaic products: A state-of-the-art review and future research opportunities. *Solar Energy Materials and Solar Cells*, **100** (0), 69–96. DOI: 10.1016/j.solmat.2011.12.016.

Quesada, G., Rousse, D., Dutil, Y. *et al.* (2012) A comprehensive review of solar facades. Opaque solar facades. *Renewable and Sustainable Energy Reviews*, **16** (5), 2820–2832. DOi: 10.1016/j.rser.2012.01.078.

Raugei, M. and Frankl, P. (2009) Life cycle impacts and costs of photovoltaic systems: Current state of the art and future outlooks. *Energy*, **34** (3), 392–399. DOI: http://dx.doi.org/10.1016/j.energy.2009.01.001.

Reijenga, T. (2005) PV in architecture. In *Handbook of Photovoltaic Science and Engineering* (eds A. Luque and S. Hegedus). John Wiley & Sons, Ltd., Chichester.

Ritzen, M., Vroon, Z., Rovers, R., and Geurts, C. (2014) Comparative performance assessment of four BIPV roof solutions in the Netherlands. Paper presented at the ICBest 2014, Aachen.

Sadineni, S.B., Madala, S., and Boehm, R.F. (2011) Passive building energy savings: A review of building envelope components. *Renewable and Sustainable Energy Reviews*, **15** (8), 3617–3631. DOI: 10.1016/j.rser.2011.07.014.

Schoen, T., Prasad, D., Toggweiler, P. *et al.* (1998) Status report of Task VII of the IEA program: PV in buildings. *Renewable Energy*, **15** (1–4), 251–256. DOI: 10.1016/s0960-1481(98)00169-4.

Sinapis, K., and Donker, M. v. d. (2013) BIPV Report 2013, State of the art in Building Integrated Photovoltaics. In SEAC (Ed.) Eindhoven: Solar Energy Application Centre.

Tyagi, V.V., Kaushik, S. C., and Tyagi, S. K. (2012) Advancement in solar photovoltaic/thermal (PV/T) hybrid collector technology. *Renewable and Sustainable Energy Reviews*, **16** (3), 1383–1398. DOI: 10.1016/j.rser.2011.12.013.

Verberne, G., Bonomo, P., Frontini, F., *et al.* (2014) BIPV Products for facades and roofs: a market analysis. Paper presented at the PVSEC, Amsterdam, the Netherlands.

Wang, Y., Tian, W., Ren, J. *et al.* (2006) Influence of a building's integrated-photovoltaics on heating and cooling loads. *Applied Energy*, **83** (9), 989–1003. DOI: 10.1016/j.apenergy.2005.10.002.

Weller, B., Hemmerle, C., Jakubetz, S., and Unnewher, S. (2010) *Photovoltaics, Technology, Architecture, Installation*. Birkhauser, Dresden.

11.7

Product Integrated Photovoltaics

Angèle Reinders[1] and Georgia Apostolou[2]
[1] *University of Twente, Enschede, the Netherlands*
[2] *Delft University of Technology, the Netherlands*

11.7.1 Introduction: What Is Product Integrated Photovoltaics?

Since the 1970s, a variety of different products, such as watches, flashlights, chargers, mp3 players, solar lamps, and more recently laptops and key boards, etc., have been released that are powered by photovoltaic solar cells at various levels of integration. However, only since the beginning of this century, the research area associated with evaluations and the improvement of the design and engineering of products with integrated PV cells, has existed under the title of product integrated photovoltaics (PIPV). Several researchers and designers have investigated various aspects of product-integrated PV, among others Randall (Randall *et al.*, 2001; Randall and Jacot, 2003; Randall *et al.*, 2006), Kan (2006), Kan *et al.* (2006), Reich *et al.* (2007; 2008; 2009), Mueller *et al.* (2009a; 2009b), Freunek (Mueller) *et al.* (2013), Veefkind *et al.* (2006), Alsema *et al.* (2005), Reinders and van Sark (2012), Reinders *et al.* (2012), Apostolou *et al.* (2012a; 2012b; 2014a; 2014b; 2015; 2016a; 2016b), however, the final word has not yet been said on this topic with various ties to the wider PV community. Therefore, before discussing several aspects of product integrated PV, it makes sense to define a set of criteria that characterize a product as a product integrated PV (PIPV) product. For these criteria we refer to (Reinders and van Sark, 2012) and (Apostolou and Reinders, 2014) while listing them below:

A product can be defined as product integrated PV by the following criteria:

1. The existence of one or more integrated PV cell(s) on or in a product's surface.
2. The energy generated by the PV cells is used to power the main functions of the product.
3. Energy generated by the PV cells can be stored in batteries or other forms of energy storage.
4. The product is used in a terrestrial application, i.e. satellites are not categorized as PIPV products.
5. The PIPV product can easily be transported.
6. Often the product comprises interaction with an end-user.

Photovoltaic Solar Energy: From Fundamentals to Applications, First Edition.
Edited by Angèle Reinders, Pierre Verlinden, Wilfried van Sark, and Alexandre Freundlich.

Companion website: www.wiley.com/go/reinders/photovoltaic_solar_energy

11.7.2 Application Areas of Product Integrated Photovoltaics

To inform the reader about the vast area of applications of PIPV, we will shortly illustrate each application area of PIPV based on established product categories of PIPV as indicated below:

1. Consumer products (see Figures 11.7.1(a)–(d), 1(g))
2. Lighting products (see Figures 11.7.1(e)–(f), 1(h)–(i))
3. Business-to-business applications (see Figures 11.7.1(j)–(k))
4. Recreational products (see Figure 11.7.1(o))
5. Vehicles and transportation (see Figures 11.7.1(l)–(n))
6. Arts (see Figure 11.7.1(p))

1. *Consumer products with integrated PV.* To the category "consumer products" belong products of daily use with PV cells' power ranging from 0.001 W up to 10 W. PV cells are applied in products like toys with integrated PV cells, PV-powered radios, calculators, solar-powered watches, solar-powered mp3 players, PV headsets, automated lawn mowers, kitchen appliances, mobile phones, PV chargers used in cell phones, which constitutes a big category by itself, lamps, and portable consumer electronics.
2. *Lighting products with integrated PV.* Numerous self-powered lighting products such as flashlights, ambient lights, lamps for bicycles, garden lights, pavement lights, indoor desk lamps, street lighting systems, and other products for lighting of public spaces are commercially available. The power of the lighting products varies between 1 W for one LED and 100 W which could be the power of a street lighting system. A special category of lighting products targets the rural lighting sector in countries with limited grid penetration. These products are often called solar lanterns and are advertised as a means of supporting socio-economic development.
3. *Business-to-business applications with integrated PV.* Similar to consumer and lighting products, PIPV has been also applied in business-to-business (B2B) applications. The power of solar cells of B2B products varies approximately between 10 W to 200 W. Examples are: parking meters, traffic control systems, traffic lights, and trash bins. Nowadays, PV-powered public trash cans, with automated control of trash collection, are available and they are being used successfully in many cities in the USA and Asia. Another well-known B2B application is the PV-powered ventilator that operates in parked cars to decrease the temperature and humidity in the car. PV is also used in products for telecommunication (e.g., radiotelephone systems, microwave, telephone and television repeaters), security, and environmental monitoring.
4. *Recreational products with integrated PV.* Products in the power range from 50 W up to 500 W, which belong to this product group, are the following: PV-powered caravans, motorhomes, campers, tents, solar-powered pond equipment (e.g., pond lights), solar-powered fountains and PV products for watersports (e.g., underwater lens). Nowadays, modern caravans, motorhomes or solar tents are gradually using PV panels attached to their roof, which enables the use of lighting, refrigeration, laptops' charging and entertainment equipment, on any camp site.
5. *Vehicles and transportation.* The vehicles and transportation category includes bikes, boats, cars, and planes; from 200 W power to 1500 W for cars up to several kilowatts for boats and tens of kilowatts for planes. Most of these PV applications are still in the demonstration phase and they are not commercially available yet. Main drivers for product development in this category are contests like the Solar Challenge for racing cars in Australia and for solar-powered boats in The Netherlands, which is why new PV vehicles are designed, realized, and investigated

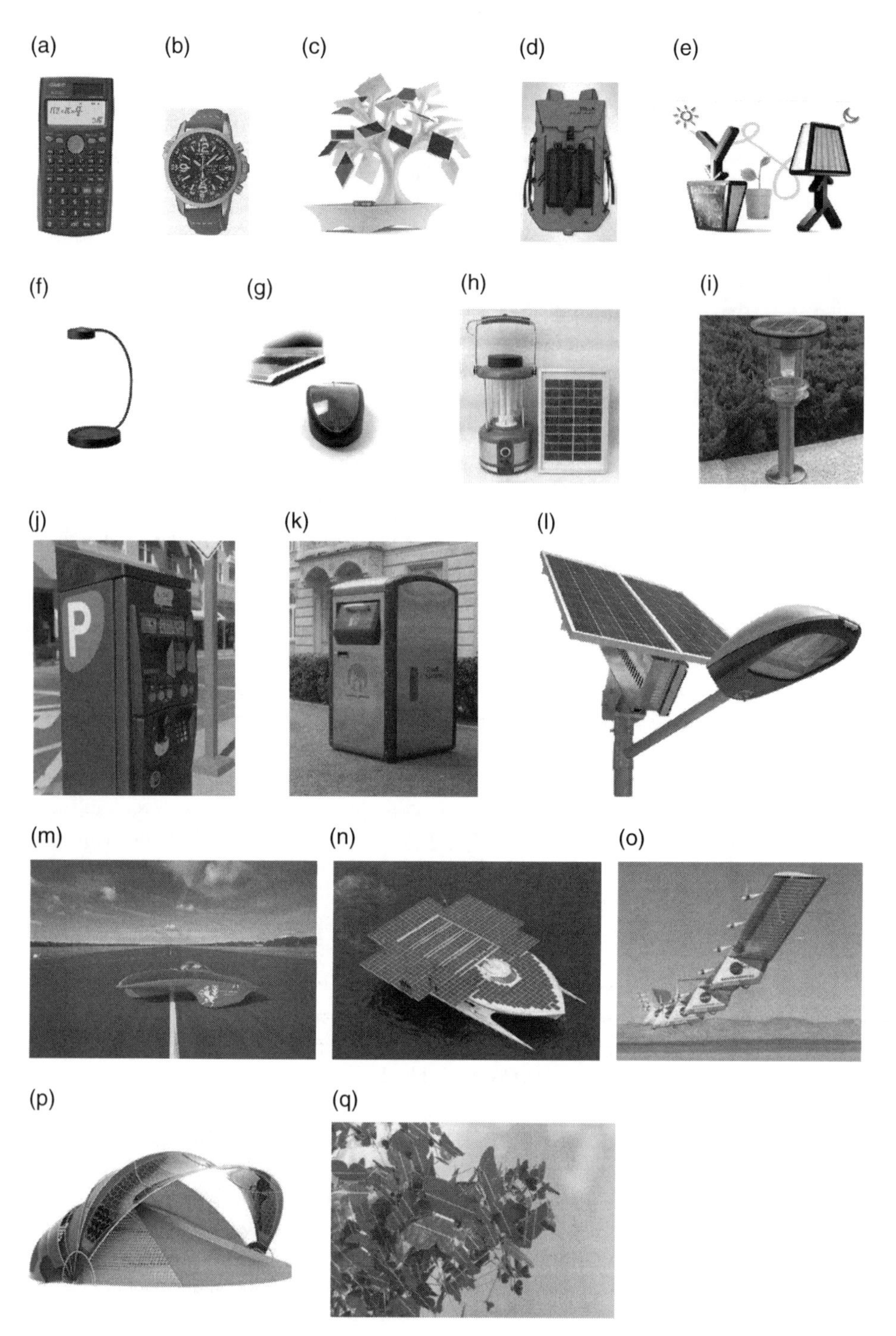

Figure 11.7.1 Photovoltaic products of various product-categories. (a) solar calculator (Sale Stores, 2015), (b) solar watch (Express Watches, 2015), (c) phone charger by Vivien Muller (Muller, 2015), (d) solar-powered bag (Ralph Lauren, 2015), (e) Spark lamp (Spark, 2015), (f) IKEA Sunnan lamp (IKEA, 2015), (g) PC computer mouse Sole-Mio (DDI, 2015), (h) solar lantern (Solar Lantern, 2015), (i) solar garden light (Solar Garden Light, 2015), (j) solar-powered parking meter in Virginia (Matray, 2015), (k) automated trash bin Big Belly (Big Belly, 2015), (l) solar traffic light, (m) Solar-powered car from University of Twente in 2011 (n) Planet Solar Catamaran (PlanetSolar, 2015), (o) Helios solar aircraft (NASA, 2015), (p) solar-powered tent (Harris, 2015), (q) PV-powered chandelier (Renewable Energy Magazine, 2015). (*See insert for color representation of the figure*)

for potential future market implementation. Planes and drones are developed for military or environmental applications or as major innovation projects by individuals.

6. *Arts.* The category 'arts' contains mainly products for decoration and emotional experience. The PV power of art products can vary significantly – from milliWatts (mW) to kiloWatts (kW) – as well as the location of use (indoors and outdoors). Some examples of "arty" PV products are the following: PV jewellery, art for public spaces (e.g. statues, fountains, art constructions for decoration of parks or squares, etc.), and indoor art such as "Virtue of Blue," which is a PV-powered chandelier, and decorative PV lights. In this category combinations of aesthetics and elements of imagination result in a beautiful visual outcome using special PV features of color, flexibility, and light reflections.

11.7.3 Selected Items for Product Integrated Photovoltaics

11.7.3.1 System Design and Energy Balance

A dominant factor in the design of PIPV is the required area on products surfaces that can be covered by solar cells. The typical area of solar cells in PIPV is determined by a trade-off of the internal power consumption of a product, its characteristic run time that results from the user behavior (that is to say the patterns of energy consumption), the available area on the product, the PV technology applied, the storage capacity, and the irradiance conditions in the product's surrounding. For instance, in the case of PV-powered consumer products that are used indoors, area is constrained by the geometries of consumer products, which indirectly implies a very low internal power requirement of these products that can range from 0.005 mW up to a few Watts. Under indoor irradiance conditions of 10 W/m^2, crystalline silicon (c-Si) PV cells perform with an efficiency of 8% or less, see Figure 11.7.2. Assuming a run time similar to the charging time of batteries, an area of solar cells of 10^{-6} m^2 up to several square meters is required to meet the internal power requirement of these products. Under outdoor conditions at STC, PV cells can generate power in the range of 100 W/m^2, using commercially available amorphous silicon PV, up to about 240 W/m^2, using record efficiencies for silicon PV technologies as shown in Figure 11.7.2.

As such, PIPV that is used outdoors can meet the power requirements of outdoor lighting products, vehicles – such as cars, boats, and lightweight planes – portable accommodation – such as campers, caravans, and tents – and business-to-business applications – such as parking machines, traffic control, and public information displays. In these products we find either multi-crystalline silicon or amorphous silicon PV cells in combination with either NiMH or Lithium-ion batteries (Apostolou and Reinders, 2012a). Only a few PIPV products contain PV cells with a capacitor and rarely is solar power used directly only.

The situation is different for PIPV that is mainly used indoors due the rapid decrease of efficiency at low irradiance levels that occur in indoor environments, such as offices and homes. Indoor irradiance consists of a mixture of daylight passing through windows and artificial irradiance originating from various sources such as incandescent lamps, fluorescent lamps, and LEDs. In general, the indoor irradiance decreases as the reciprocal of distance from a window, leading to typical values in the order of 1 to 10 W/m^2. Moreover the spectral distribution of indoor irradiance differs from the spectra of solar terrestrial irradiance. In Figure 11.7.3 the difference between the intensity and spectral distribution of outdoor and indoor irradiance is shown. This indoor spectrum shows peaks on top of a relatively continuous base. These peaks originate from artificial light sources in the specific room where the measurements were taken.

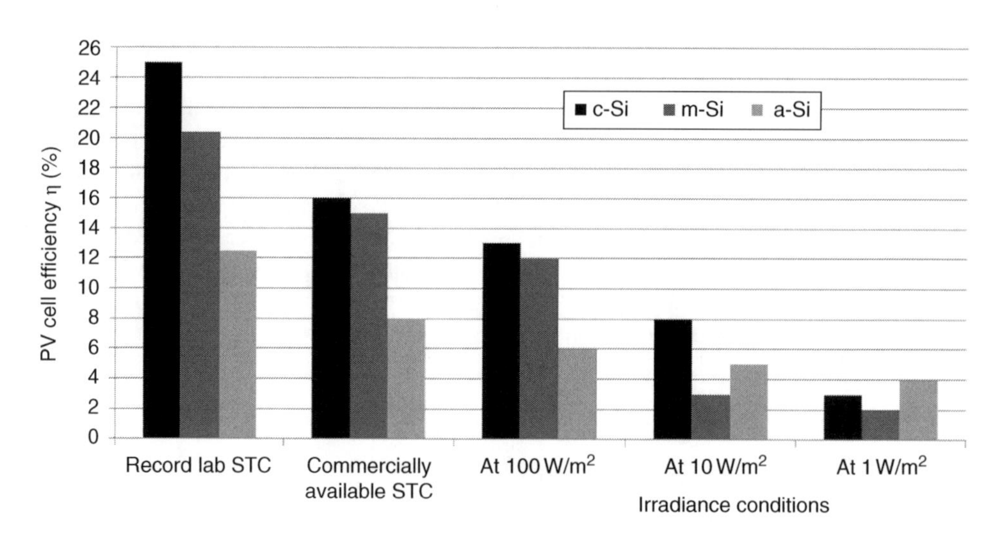

Figure 11.7.2 PV cell efficiencies of c-Si, mc-Si and a-Si at different irradiance conditions respectively at STC, 100W/m², 10W/m² and 1W/m² as stated in the literature until 2014. Source: (adapted from Apostolou *et al.* (2014b); Freunek (Mueller) *et al.* (2013))

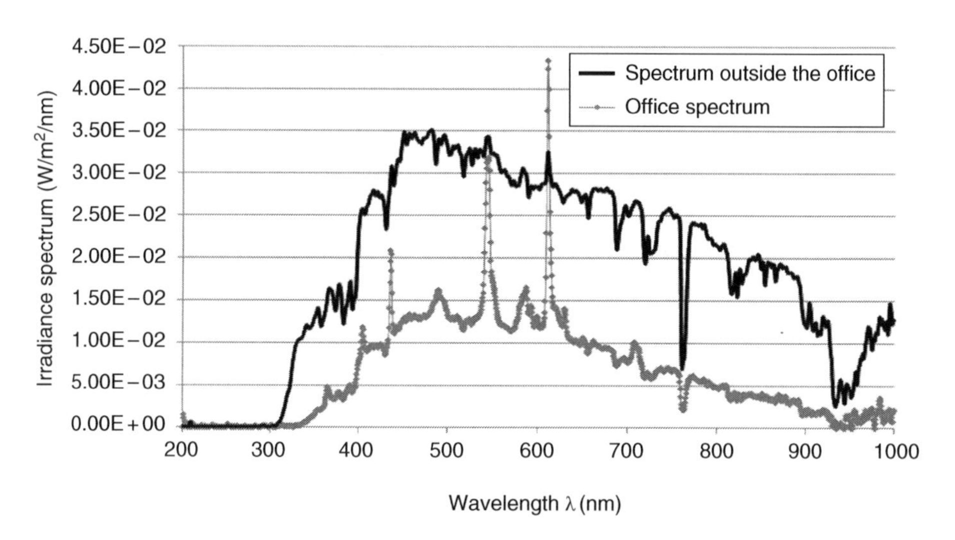

Figure 11.7.3 Global horizontal irradiance measurements indoors under mixed indoor lighting and outside the window in The Netherlands, in January 2014, at a distance of 50 cm inside and outside the window respectively

11.7.3.2 Environmental Impact

"Sustainable" or "ecological" design of products establishes the incorporation of environmental features within the product design aimed at improving the conservational performance of the product, leading to an environmental impact lower than that of the present alternative designs. The eco-design of products as a preventive approach is designed to enhance the products' environmental performance, while preserving their functional properties and providing original chances for the manufacturers, consumers, and society. However, it is quite a new field of research and it is still in progress. The most effective approach to deal with the environmental impact is to use life-cycle analysis (LCA) during the initial phases of design and to identify the extent of the problem by imposing (environmental) priorities and focusing on effective solutions, see also Chapter 13.3. An environmental life-cycle analysis (LCA) is defined as:

> consecutive and interlinked stages of a product system, from the raw material acquisition or the creation of natural resources to the final disposal (ISO 14040:2006, 2006). The main stages of a product's life include: the acquisition of raw materials, the phase before construction, manufacturing, packaging and distribution, use and end of critical life.

An LCA on small PV lighting products was carried out in South-East Asia by Durlinger *et al.* (2012), aimed at evaluating the environmental impact performed by the production, use, and end of critical life of these products. Illumination in rural areas of developing countries is usually provided by candles and oil lamps (kerosene), whereas torches and flashes often powered by car batteries or lead acid batteries are also used as a portable source of lighting. The study shows that solar PV lighting products have a lower environmental effect than the conservative options for lighting in these countries. Batteries' recycling is one way, whereby the eco-friendly profile of the small PV lighting products can be enhanced. Intrinsically, PV lighting products offer an environmentally-friendly and advantageous illumination service for off-grid households. Another important conclusion of this study is the possible enhancement of the environmental profile of solar lighting products by sufficient battery waste control or the use of circuit panels of reduced size.

Another study executed by Flipsen *et al.* (2012) and by Dafnomilis (2012) was a battery life-cycle assessment for mobile phones executed with a focus on cradle-to-grave (CTG) energy and greenhouse gases (GHG) emissions. The study is based on literature data and on a comparison of four different types of smartphones using their original battery and after a reduction of the battery capacity while using a solar cover. Findings of this study indicate that a reduction of 30% of the batteries' capacity will result in a reduction of approximately 30% of the energy consumed during the production of raw materials, manufacturing and CO_2 emissions during the battery's lifetime. From this point of view, solar phones can be an environmentally-friendly solution to oversized batteries – especially for moderate and light users or in countries with sufficient annual sunshine. Besides, they can be an ideal solution for off-grid areas. Results from this study are presented in Figure 11.7.4, where a comparison of CO_2 emissions between two different technologies – a mobile phone without modification and one with smaller battery including a solar cover – for four types of smartphones is shown.

11.7.3.3 User Interaction

Consumers prefer to buy products that not only have a sufficient function, but also because of their looks. They want products with a nice visual appeal, color, material, design, as well as products that satisfy their emotional reactions. This applies also to PIPV products.

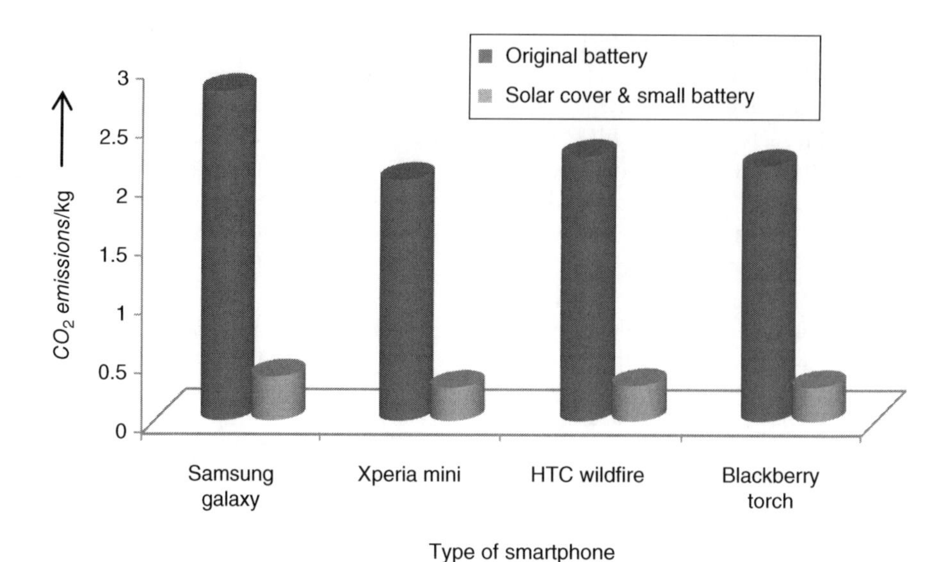

Figure 11.7.4 Comparison of the amount of carbon dioxide emissions required during the life-cycle of four mobile phones, between two different technologies; original battery and solar cover and small battery. Source: Apostolou and Reinders (2014) based on Flipsen *et al.* (2012) and Dafnomilis (2012)

In order to understand how users interact with PV-powered products, Apostolou and Reinders (2015) observed and analysed the behavior of 100 respondents while interacting with six different photovoltaic-powered products in their daily life. The tested PV products were: the IKEA Sunnan lamp, the Waka Waka light, the Waka Waka Power (charger and light), the Little Sun light, the Logitech solar keyboard and the Beurer kitchen weight scale, see Figure 11.7.5. In this study the design of the six tested PV products was analyzed, users' expectations were outlined and the users' opinion of the products' performance after use was addressed.

The results depicted the users' expectations at a time before using the product and their evaluation after the field trial, see Table 11.7.1. It is interesting to observe the difference between the two stages. During the first contact of the user with the product, and before the field trial starts, users criticize the look of the product (e.g., the design, color, materials, size) and they try to predict its function and usefulness. Around 60% of the respondents feel comfortable with the product and consider it "a nice gadget" to use. While their first impression is positive, there are doubts concerning the functionality and performance of the product.

After the field trial, users' feedback, which is in the form of written reports and answered questionnaires, mainly concerns the product's performance; around 40% of the respondents are totally dissatisfied, 38% find the product useless, around 60% find the design of the product of bad or low quality, 54% believe the design of the product is quite simple and can easily be used by everybody, while only 4% find it difficult to use the specific product. 88% of the respondents would not buy the PV product or propose it to a

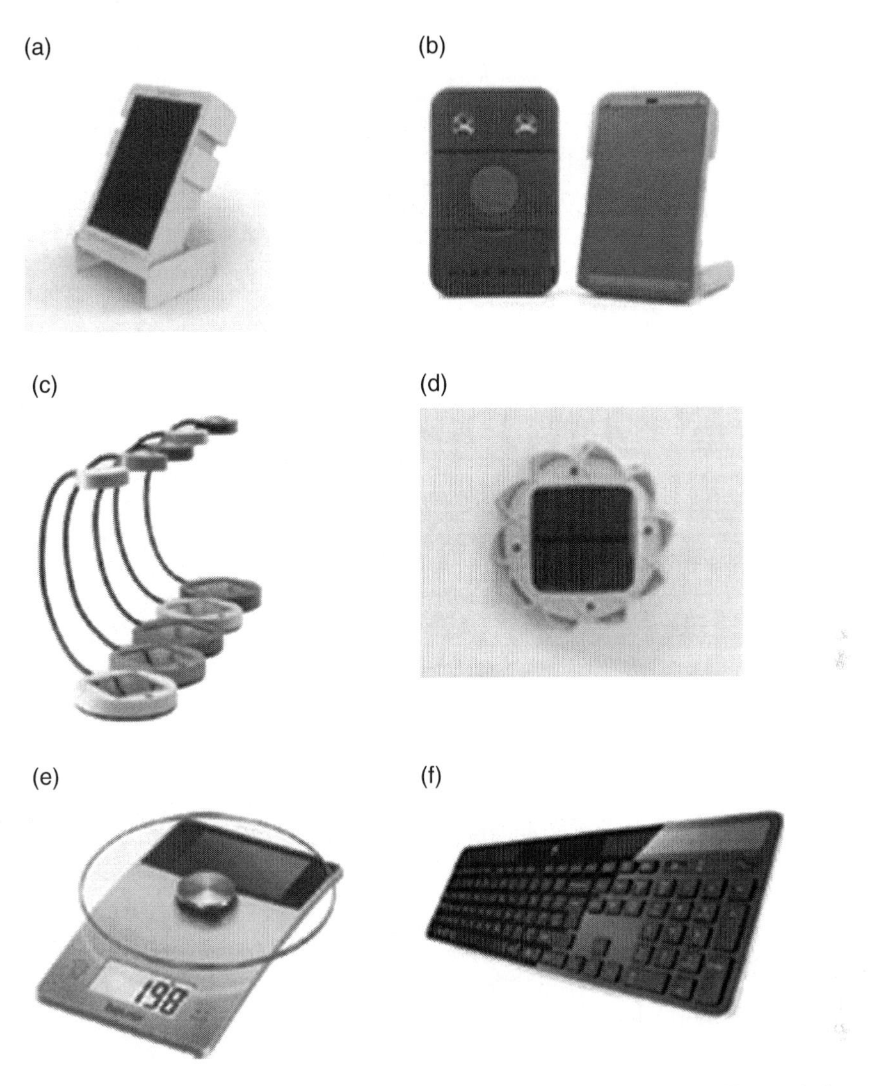

Figure 11.7.5 The tested PV products: (a) Waka Waka light; (b) Waka Waka Power light and charger; (c) Sunnan IKEA lamp; (d) Little Sun light; (e) Beurer kitchen weight scale; and (f) Logitech solar keyboard. Source: (Apostolou and Reinders, 2015, 2016b). (*See insert for color representation of the figure*)

friend and around 70% believe that the price of the PV product does not correspond to its quality and performance. Main results show that the users need more reliable PV products, made with nice materials, that have an interesting design, and perform sufficiently. Consumers are willing to pay money and buy a PV product if it is useful and works properly. Furthermore, it is noticeable that consumers are quite positive about PV products that have an environmentally friendly or a social character, such as, for instance, donations to the developing countries when buying a solar-powered lighting product, such as the Waka Waka or the Little Sun light.

Table 11.7.1 Comparison of six PIPV products according to 100 users' feedback after having used the products. Source: Apostolou and Reinders (2015)

Number of users per product (n)	Sunnan lamp $n=50$	Waka Waka light $n=15$	Waka Waka power $n=4$	Little Sun $n=15$	Beurer kitchen scale $n=10$	Logitech solar keyboard $n=6$
Form	+/−	+	+	+	+/−	+
Compactness	+/−	+	+	+	−	+
Use and repair	+	+	+	+/−	+/−	+
Safety	+	+	+	+	+	+
Solidity	−	+	+	+/−	−	+/−
Price affordable	+	−	−	+/−	−	+/−
Performance outdoors/ indoors	+/−	+/−	+/−	+/−	−	+
Charge capacity	−	+	−	+	−	+
Efficiency	+/−	+/−	+/−	+	−	+
Adjustability	−	−	−	+/−	−	+
Durability	−	+	+	−	−	+
Sustainability	+	+	+	+	+/−	+
Environmentally friendly character	+	+	+	+	+	+

11.7.4 Conclusion

Product-integrated PV is a relatively new field of interest in the PV community with great potential to use photovoltaics in the direct context of energy consumption by products. The minimum benefits are in using autonomously sustainable electricity produced by solar cells and by the reduction of the required battery capacity in mobile products. The big challenges for the nearby future are in improving the users' perception of PIPV as being something beyond gadgets, and in finding good solutions for PIPV for indoor use. In this section we have not yet addressed manufacturing issues of PIPV or design and styling, apart from the aesthetics of PIPV, which will both become more important when the market for PIPV gradually becomes more mature. In the near future you may find new PIPV applications in lighting products, boats and other vehicles, and in indoor sensor networks. Also new categories of PIPV products may be created, because product innovation will continue in this field.

List of Acronym

Acronym	Description
PIPV	Product Integrated PV

References

Alsema, E.A., Elzen, B., Reich, N.H. *et al.* (2005) Towards an optimized design method for PV-powered consumer and professional applications – the Syn-energy project. In *Proceedings of the 20th European Photovoltaic Solar Energy Conference, Barcelona 2005.* Spain, WIP-Renewable Energies, 1981–1984.

Apostolou, G. and Reinders, A.H.M.E. (2012) A comparison of design features of 80 PV-powered products, *27th EU Photovoltaic Solar Energy Conference and Exhibition*, Frankfurt.

Apostolou, G. and Reinders, A.H.M.E. (2014) Overview of design issues in product-integrated PV. *Journal of Energy Technology*, **2** (3), 229–242.

Apostolou, G. and Reinders, A.H.M.E. (2015). Users' interaction with PV-powered products: an evaluation of 6 products by 100 end-users. *Proceedings of the International Photovoltaic Specialists Conference PVSC 42*, June 2015, New Orleans, USA.

Apostolou, G. and Reinders, A.H.M.E. (2016), How do users interact with photovoltaic-powered products? Investigating 100 'lead-users' and 6 PV products, *Journal of Design Research*, **14** (1), 66–93.

Apostolou, G., Verwaal, M., and Reinders, A.H.M.E. (2014) Modeling the performance of product integrated photovoltaic (PIPV) cells indoors, *26th EU Photovoltaic Solar Energy Conference and Exhibition*, Amsterdam.

Bakker, C.A., Eijk, D.J. van, Silvester, S. *et al.* (2010) Understanding and modelling user behaviour in relation to sustainable innovations: the living lab method. In *Proceedings of TMCE 2010 Symposium* (pp. 1–12). TMCE. (TUD).

Big Belly Solar Recycling Bin, ©BigBelly: http://www.bigbelly.com/ about/bigbelly-in-the-wild/?nggpage = 3, accessed July, 2015.

Bringing the butterfly effect to life, PV, ©*Renewable Energy Magazine*: http://www.renewableenergymagazine.com/article/bringing-the-butterfly-effect-to-life, accessed July, 2015.

Dafnomilis, I. (2012) Battery capacity and performance in relation to user profiling in smartphones, Student report on *System Integration Project II (SIP 2),* TU Delft, the Netherlands.

Durlinger, B., Reinders, A.H.M.E., and Toxopeus, M.E (2012) A comparative life cycle analysis of low power PV lighting products for rural areas in South East Asia, *Renewable Energy*, **41**, 96–104.

Express Watches, Seiko Men's Solar Watch SSC081P, ©Seiko:http://www.expresswatches.co.uk/seiko_solar_mens_sports_watch, accessed July, 2015.

Flipsen, B., Geraedts, J., Reinders, A. *et al.* (2012) Environmental sizing of smartphone batteries, *Proceedings of Micro-energy Supplies and Energy Harvesting. Electronics Goes Green 2012, Berlin 2012*, Germany, pp. 1–9.

Freunek (Müller), M., Freunek, M., and Reindl, L.M. (2013) Ideal and empirical maximum efficiencies for indoor photovoltaic devices. *IEEE Journal of Photovoltaics*, **3** (1), 59–64.

IKEA, Sunnan solar charged LED table lamp, ©IKEA : http://www.amazon.com/Ikea-Sunnan-Table-Lamp-Black/dp/B00ATIHUT2, accessed July, 2015.

ISO 14040:2006 (2006) Environmental management – life cycle assessment – principles and framework, International Organization for Standardization, ISO/TC 207/SC 5, Geneva.

Kan, S.Y. (2006) SYN-Energy in solar cell use for consumer products and indoor applications. PhD thesis, Final Report 014-28-213, NWO/NOVEM, Technical University of Delft, Delft, the Netherlands.

Kan, S.Y., Verwaal, M., and Broekhuizen, H. (2006) The use of battery-capacitor combinations in photovoltaic powered products short communication. *Journal of Power Sources*, **162**, 971–974.

Müller, M., Hildebrandt, H., Walker,W.D., and Reindl, L.M. (2009a) Simulations and measurements for indoor photovoltaic devices. In: *Proceedings of the 24th European Photovoltaic Specialist Conference, Hamburg*: WIP, 2009, CD.

Müller, M., Wienold, J., Walker, W.D., and Reindl, L.M. (2009b). Characterization of indoor photovoltaic devices and light. In *Proceedings of 34th IEEE Photovoltaic Specialist Conference. Philadelphia: IEEE*, 2009, 738–743, poster award.

Netcomposites, Helios Solar-Powered Aircraft Crashes, ©NASA.gov :http://www.netcomposites.com/news/helios-solar-powered-aircraft-crashes/1736, accessed July, 2015.

PlanetSolar Turanor, ©BlueBird-Electric.net : http://www.bluebird-electric.net/planetsolar.htm, accessed July, 2015.

PilotOnline.com, Parking's just a swipe away in Virginia Beach, by Margaret Matray : http://hamptonroads.com/2013/03/parkings-just- swipe-away-virginia-beach, accessed July, 2015.

Ralph Lauren, Ralph Lauren Launches Solar-powered Waterproof Backpack for $800, ©ralphlauren.fr:http://inhabitat.com/ralph-lauren-launches-solar-powered-waterproof-backpack-for-800/, accessed July, 2015.

Randall, J.F. (2006) Designing indoor solar products: photovoltaic technologies for AES. PhD thesis, Delft, the Netherlands.

Randall, J.F., Droz, C., Goetz, M. *et al.* (2001) Comparison of 6 photovoltaic materials across 4 orders of magnitude of intensity. *Proceedings of 17th EUPVSEC*, October 2001, Munich, Germany, pp. 603–606.

Randall, J.F. and Jacot, J. (2003) Is AM 1.5 applicable in practice? Modeling eight photovoltaic materials with respect to light intensity and two spectra. *Renewable Energy*, **28**, 1851–1864.

Reich, N.H., Elzen, B., Netten, M.P. *et al.* (2008) Practical experiences with the PV powered computer mouse 'Sole-Mio'. *Components for PV Systems, Engineering and System Integration in Proceedings of 23rd European Photovoltaic Solar Energy Conference*, Valencia, 2008, Spain, pp. 3121–3125.

Reich, N.H., Netten, M.P., Veefkind, M. *et al.* (2007) A solar powered wireless computer mouse: design, assembly and preliminary testing of 15 prototypes. *Proceedings of 22nd European Photovoltaic Solar Energy Conference*, Milan, Italy.

Reich, N.H., Veefkind, M., van Sark, W.G.H.M. *et al.* (2009) A solar powered wireless computer mouse: Industrial design concepts. *Journal of Solar Energy*, **83**, 202–210.

Reinders, A.H.M.E., Diehl, J.C., and Brezet, H. (2012) The power of design: product innovation. In *Sustainable Energy Technologies*. Wiley, Weinheim.

Reinders, A.H.M.E. and van Sark, W.G.H.M. (2012) Product-integrated photovoltaics. In *Comprehensive Renewable Energy* (ed. A. Sayigh). Elsevier, Oxford, Vol. 1, pp. 709–732.

Sale Stores, Casio FX-300ES Scientific Calculator, ©Casio : http://salestores.com/casiofx300es.html, accessed July, 2015.

Solar bonsai Electree, Ecological charger for mobile devices ©Vivien Muller : http://vivien-muller.fr/#electree-original, accessed July, 2015.

Solar Lantern SCL-05, ©Solar-Kingdom.com : http://www.solar-kingdom.com/, accessed July, 2015.

Solar Outdoor Lighting For An Ecofriendly Exterior, Beufl.com, Solar Garden Light: http://beufl.com/solar-outdoor-lighting-for-an-ecofriendly-exterior/, accessed July, 2015.

Solar Powered Mouse—The Sole Mio, photo by Matthijs Netten, ©Delft Design Institute/SynEnergy : http://www.tudelft.nl/en/current/latest-news/article/detail/minister-cramer-test-eerste-muis-op-zonnecellen/, accessed July, 2015.

Solar Spark Lamp, The Solar Spark Lamp Visualizes Energy Use, by Jorge Chapa, 2008, Spark Lamp by Beverly Ng : http://inhabitat.com/solarspark-lamp-by-beverly-ng/, accessed July, 2015.

Solar tent Carpa Para La Playa, ©Emma Harris : http://www.elchiltepe.com/2010/11/carpa-para-la-playa.html, accessed July, 2015.

Veefkind, M., Reich, N.H., and Elzen, B. (2006) The design of a solar powered consumer product, a case study. *Proceedings of Going Green – Care Innovation 2006, International Journal of Automation Austria*, Vienna, Austria.

Part Twelve

PV Deployment in Distribution Grids

12.1

PV Systems in Smart Energy Homes: PowerMatching City

Albert van den Noort
PowerMatching City, DNV GL, Arnhem, the Netherlands

12.1.1 Introduction

The demand for energy is increasing globally, renewable energy sources such as photovoltaic systems (PV) and wind energy are emerging and fossil fuel supply will eventually decrease. Renewable energy sources are becoming more affordable, even for individual households. More consumers are installing PV systems without the help of a subsidy. On the other hand, there is a trend towards an increasing demand for electricity with the rise of electric mobility and heating of households with electric heat pumps. As the energy system traditionally is designed as a top-down, demand-driven system, these developments have consequences. Since the sun is not always shining and the wind is not always available, the generation of renewable energy from these sources is variable. As a consequence, a paradigm change is necessary: the supply-driven system needs to change to a demand-driven system. Flexibility is needed to balance the demand and supply of energy in the system. In PowerMatching City in the Netherlands, it is shown how smart grids or, rather, smart energy systems can support this requirement.

PowerMatching City is an internationally recognized lighthouse project that demonstrates our future energy system in an existing neighborhood in Groningen, the Netherlands. This project is a typical example of what democratizing the energy market means in real life; it gives a practical and replicable example of how a more sustainable energy future can be achieved using existing energy infrastructures. PowerMatching City's aim is to enable the transition from centralized fossil fuel-based power production to decentralized sustainable power production. We therefore have developed and demonstrated a smart energy infrastructure with related customer-focused smart energy services, and have validated the costs and benefits of such a system in practice.

Photovoltaic Solar Energy: From Fundamentals to Applications, First Edition.
Edited by Angèle Reinders, Pierre Verlinden, Wilfried van Sark, and Alexandre Freundlich.

Companion website: www.wiley.com/go/reinders/photovoltaic_solar_energy

Figure 12.1.1 Overview of the Thomsonstraat in Groningen, which is taking part in the PowerMatching City project

In PowerMatching City this is demonstrated in 40 real households (Figure 12.1.1). The participants of PowerMatching City are normal consumers. With the help of a fully integrated smart energy system, they are inspired and empowered to become active energy prosumers, which implies that they become both a consumer as well as a producer of energy. The smart energy system enables them to control where and when they want to buy or sell energy. A home energy management system with smart apps educates them about the status and implications of their energy use, and helps them to save energy and reduce costs. The smart energy system controls the indoor climate in their homes and charges their electric vehicles in time, providing them with an enhanced level of comfort. The user's comfort and freedom of choice are of primary concern and are always guaranteed.

Beside the prosumers, other stakeholders use the smart grid. As such energy suppliers, network operators and prosumers cooperate to actively balance the demand for and supply of energy in the grid. PowerMatching City's main goal is to pinpoint and quantify the added value of a smart energy infrastructure to these stakeholders. The results of these analyses are used to enable and accelerate a large-scale rollout of smart grids.

The project started in 2007 as a three-year EU demonstration project (EU FP6-038576). Phase II (Sept. 2011–Sept. 2014) is part of the Dutch IPIN program. The project is executed by a consortium consisting of DNV GL, RWE/Essent, Enexis, Gasunie, ICT Automatisering and TNO. The following knowledge partners have contributed to the project: Delft University of Technology, Eindhoven University of Technology, and Hanzehogeschool Groningen.

12.1.2 Technology

When people talk about smart grids, they often believe it mainly consists of advanced metering infrastructure (AMI) technologies. PowerMatching City's smart energy system goes beyond AMIs; it is a total solution that combines smart technologies such as AMI, distributed intelligence, grid capacity management, near real-time wholesale processes, and customer energy services in a real-life environment.

The homes in PowerMatching City are equipped with a variety of smart appliances that can be monitored and controlled by the user's smart energy system. These appliances range from smart washing machines to intelligent heating systems with smart thermostats. An advanced home energy management system makes optimal use of the smart infrastructure by giving insight, control, and automation to the residents, combined with an increased level of comfort.

Gas-fueled appliances integrate the gas and electricity systems into a smart energy system on a household level. By using local heat buffers to store any surplus of heat for use at a later moment in time, the system creates flexibility to manage peak loads in electricity demand.

Figure 12.1.2 Remote PV system for PowerMatching City

12.1.2.1 Photovoltaic Panels

Most of the participating households are capable of producing electricity locally using photovoltaic panels installed in their homes. The PV panels in the households were not provided by the project but were bought privately by the participants. Therefore, various types, sizes and efficiencies of PV modules are used in PowerMatching City. The total installed peak power is 55 kWp (on average 1400 Wp per household). When a user has a surplus of locally produced electricity, the smart grid enables this surplus to be shared with other users in the community. The users' smart appliances automatically advance their energy use when there is an abundance of locally produced electricity, and postpone their use when local availability of electricity is low.

Users who do not have solar panels installed in their homes are connected to an array of solar panels that is installed at the office of DNV GL in Groningen (Figure 12.1.2). Using a real-time certificate system, the smart infrastructure guarantees the origin of the produced electricity at every moment in time. Using this mechanism the project explores the effect of PV-produced electricity on the local energy market.

12.1.2.2 Hybrid Heat Pumps

Half of the participating households have hybrid electric heat pumps (Figure 12.1.3) that use a combination of electricity and natural gas to heat their houses and tap water. These houses are equipped with heat buffers that enable decoupling between the moment when heat pumps are operating and the moment when heat is actually used in the homes. When the price of electricity is low, the heat pumps fill the buffers with heat. When the price is high, the heat pumps are turned off and the heat buffers provide heat when necessary. In PowerMatching City two brands of heat pumps are used, both use approximately 1 kW of electricity and generate, depending on outside air temperature, 2–3 kW of heat (Figure 12.1.4).

Figure 12.1.3 Hybrid heat pump system

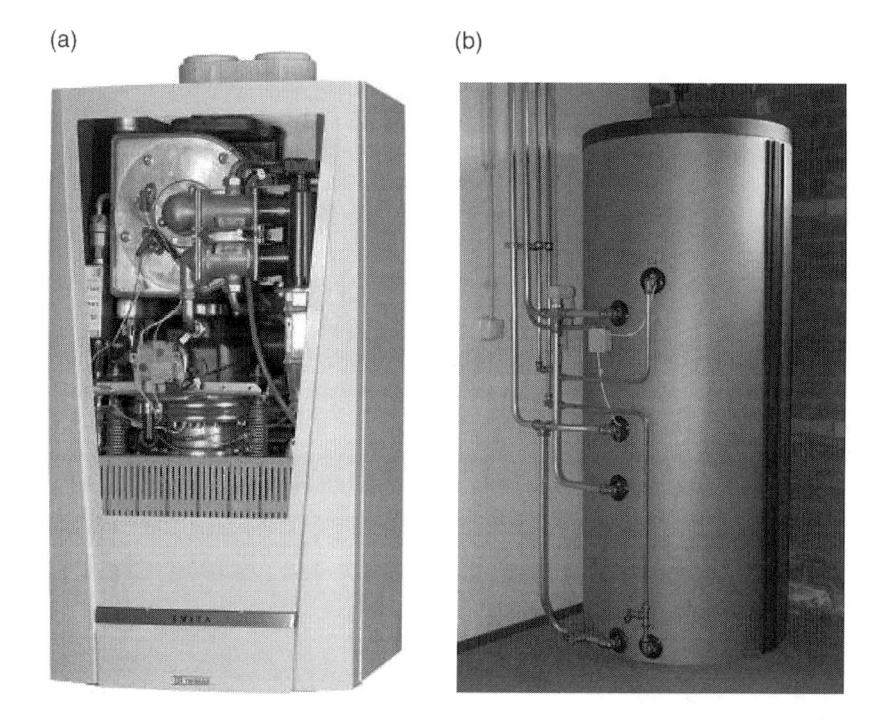

Figure 12.1.4 Heat pumps

12.1.2.3 Micro Combined Heat and Power Units (MicroCHP)

In the other half of the households small-scale generators are deployed that produce both electricity (1 kWe) and heat (6 kWth). Similar to the heat pumps, these natural gas-fired units are equipped with heat buffers, enabling decoupling between electricity production and heat usage in the homes. The microCHP units work exactly opposite to the heat pumps: when the price of electricity is high, the microCHPs start to produce electricity and fill the buffers with heat. When the price is low, the microCHPs stop running and the heat buffers provide heat when necessary. In this way, the supply of electricity by microCHPs can balance the demand for electricity by the heat pumps in PowerMatching City.

12.1.3 Matching Supply and Demand

In PowerMatching City, consumers, energy suppliers and grid operators cooperate to actively balance the demand for and supply of energy in the grid. A local energy market is created by using algorithms as part of the PowerMatcher technology. The basic idea of the PowerMatcher is that each device is represented by a software agent representing the device on an energy market. The agent constructs bid curves where energy supply or demand is plotted versus the price. An aggregator agent collects the bids for a household from the various devices and aggregates them to the bid curve of the household. Subsequently the "neighborhood" aggregator agent collects the bids of various houses, Finally, the auctioneer agent determines the equilibrium price at which supply and demand are equal. This price is communicated back to the devices, which will react to this. This local market system is connected to the national power exchange through an auctioneer service, enabling local and central energy supply systems to supplement each other. The algorithm uses the flexibility required to integrate large amounts of renewable energy into our energy systems, at both a central and a local level. The system can easily be scaled up.

The project uses a multi-goal optimization algorithm to simultaneously balance the needs of the energy supplier, grid operator, and end user, via various agents. This makes it possible to guarantee freedom of choice for consumers while keeping peak loads in the grid within acceptable limits. In the first phase of PowerMatching City (2007–2011), we have shown that this control algorithm is actually working as expected: the grid operator can use it to lower the peak on the network, the supplier can use it to trade on the energy markets, and the prosumer can use it to optimize energy use in their home.

12.1.4 New Energy Services

The second phase of the project (2011–2014) is unique in the way it engages and empowers customers to become active prosumers. In a co-creation process with the participants, the energy supplier, the technology providers and the grid operators, two new customer energy services have been developed that make optimal use of the possibilities of a smart infrastructure. The first service enables users to create a self-supporting energy community. The second service focuses on reducing electricity costs by applying an optimal buying and selling strategy on the local energy market.

Customer interaction is key in the PowerMatching City project. During the course of the project, quarterly information sessions were organized and many surveys were taken, aimed at getting a profound understanding of customer drivers and needs. This information was used to create an interactive energy management system, the Energy Monitor. This Energy Monitor is a tablet-based application that enables the users to interact with the smart grid as well as with each other (Figure 12.1.5). The Energy Monitor provides feedback about the optimum

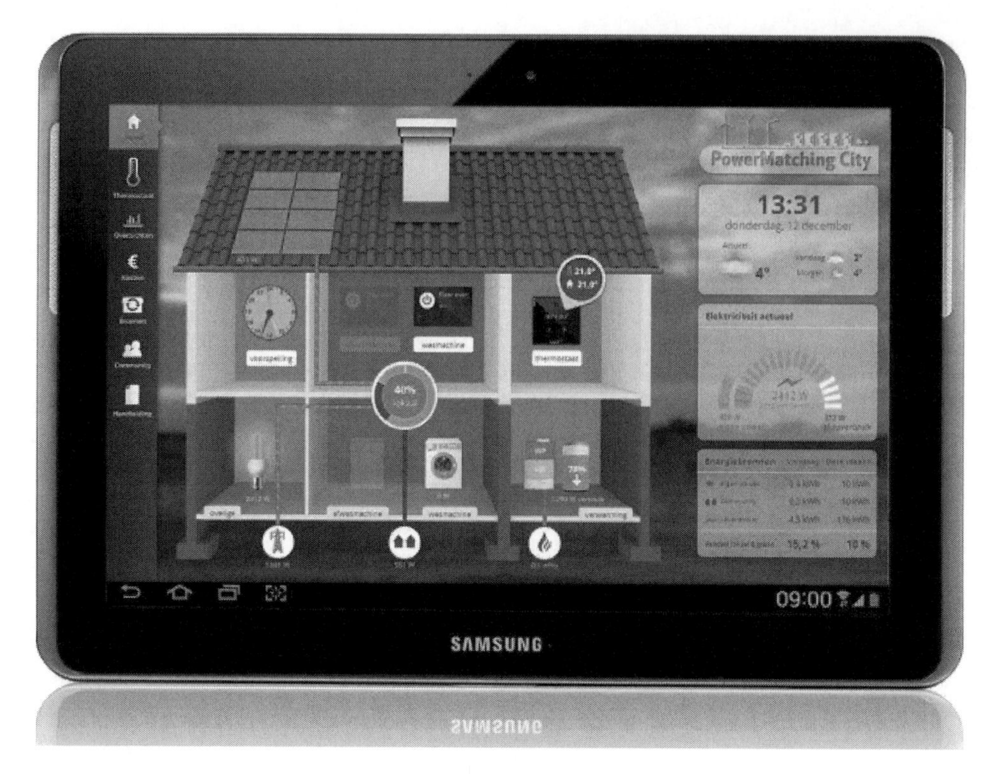

Figure 12.1.5 The Energy Monitor of PowerMatching City. The various energy systems are visually represented as well as their energy production in quantitative data. The PV system is shown on the roof

time to use electricity. By giving suggestions, the app helps users to save energy and to achieve their personal energy goals, although participants indicate that automated control of the heating system is the best way to achieve their energy goals. A digital community service provides them with the opportunity to meet and to start new energy initiatives that go beyond the scope of the current project.

12.1.5 Results and Lessons Learned

One of the most important lessons learned from the project is that it is only possible to fully exploit the opportunities inherent in the smart grid if all parties in the entire energy chain align their efforts. Although separate technologies and optimization goals can be implemented in a rather isolated way, the true strength of the smart energy system lies in the cooperation between the different actors along the value chain. This will optimize the total benefits of the system.

The experiment demonstrated that it is technically feasible to allow demand to follow supply, rather than supply following demand as it is today (Bliek *et al.*, 2010; Bliek *et al.*, 2011). Measurements from the micro-combined heat and power (CHP), the hybrid heat pumps and the charging of electric vehicles all indicate that the system responds quickly to fluctuating demand and maintains comfort levels for the end-user over the long term. This is

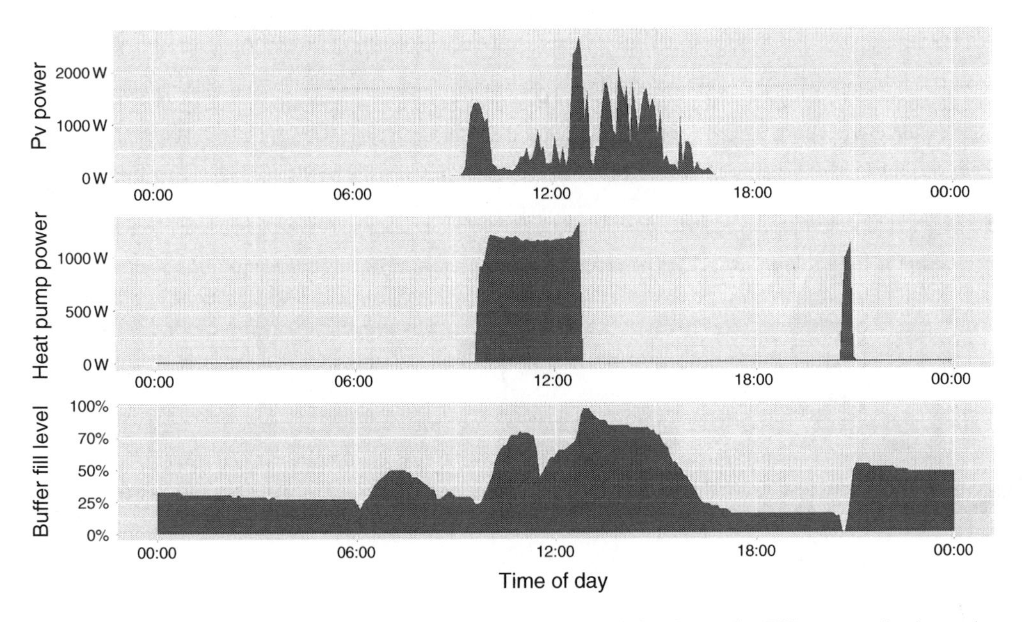

Figure 12.1.6 The interaction between the generated electricity from the PV system (top) on the electric heat pump (middle) and its buffer (bottom). The buffer fill level represents the "state of charge" of the heat buffer. Source: (Wijbenga *et al.*, 2014)

favorable for the smooth integration of renewable wind and solar energy, because the energy supply from these sources is variable. In addition, it is evident that there is a need to design (household) appliances in a different manner. The aim would be to allow appliances to decide for themselves whether to switch on or off, depending on the current electricity rate, for example, when the rate falls because the supply from renewable sources is high. At present, a heat pump is designed to supply heat when the consumer has a need, not when the electricity rate is favorable.

An example of the interaction between the energy service, the heat pump and the generated PV power is given in Figure 12.1.6 (Wijbenga *et al.*, 2014). The behavior of one heat pump follows from a combination of the demand and supply of energy in the neighborhood, the user's comfort setting and the energy service for this household that optimizes the use of renewable energy in the community. Around 6 in the morning the warm water buffer is filled to guarantee user comfort using the auxiliary gas-fired burner. This is due to the heat pump's limitation that it does not work when the outside temperature is below 5 degrees Celsius, occurring in the early morning. Later in the day, the heating by electrical means occurs simultaneously with the peak in solar power and enforces a higher buffer level.

This full-scale trial shows that over time, the role of consumers will fundamentally change (Geelen *et al.*, 2013). Who will supply what service, at which time, and to whom, in the future? Consumers will not only evolve into prosumers, they will also gain an influential voice in the demand and supply of energy and consequently on the energy supply as whole. This even affects the billing process: who will then be billing whom? Because if consumers are "called on" to supply electricity, or to consume it via rate incentives in order to maintain the balance within the power grid, then they should also be able to financially profit from it.

The residents of the 40 smart grid households in Groningen are trying to find a balance between energy saving and home comfort, and to a limited extent become self-supporting as well, while having access to local energy. Based on the measurements in PowerMatching City and combined with a cost-benefit model for the Dutch energy market, the benefits of flexibility for the Dutch consumer market were calculated at between 1 and 3.5 billion euros. These benefits are made up of deferred grid operation costs and the added value for the energy markets (PMC Consortium, 2015).

With the PowerMatching City system, consumers are able to balance personal preferences between green, cheap and local energy. Such freedom of choice for thousands or even all users in a region or country would make most of the current regulatory framework obsolete. To allow users to choose the optimum between green, cheap and local energy on a much larger scale, the adaption of the regulatory framework is a prerequisite. Energy tariffs will need elements that are based on time and location of generation and use. In particular for PV systems, the network is an important driver for consumers to invest in more PV. The topic of novel regulatory frameworks clearly goes beyond the scope of the real-life demonstrations in Groningen, but needs to be resolved for a large-scale rollout of smart grids.

12.1.6 Conclusion

PowerMatching City is a smart energy system demonstration project in Groningen, the Netherlands. The project demonstrates an integrated solution for the energy system of the future. It includes various technologies, ranging from innovative heating systems, PV panels and renewable, distributed generation as well as advanced in-home display systems that provide insight and control to end users. With increasing use of electricity and lowering costs for renewable generation, in particular, PV systems, the different stakeholders of the electricity system need to be prepared for a more decentralized and sustainable energy system. The solution in PowerMatching City provides advantages to the various stakeholders in the energy value chain: end users gain more insight and control via novel smart energy services, grid operators can reduce peak loads on the grid, and utilities can operate the systems as a virtual power plant. In PowerMatching City, we have shown that smart energy systems are technically feasible. The flexibility has economic value and new energy services can be created that meet the needs of the consumers. It is a valuable stepping stone towards the large-scale implementation of smart energy systems that allows a sustainable, affordable and reliable energy system.

List of Acronyms

Acronym	Description
AMI	Advanced Metering Infrastructure
IPIN	Innovatie Programma Intelligente Netten (Dutch subsidy programme 2011–2015)
kW	KiloWatt
kWp	KiloWatt peak; a measure for the peak power of a PV panel.
MicroCHP	Combined heat and power; a device that generates and uses both heat and power. Here we use small generators (1 kW), hence the name microCHP
PV	Photovoltaic

References

Bliek, F., van den Noort, A., Roossien, B. *et al.* (2010) PowerMatching City, a living lab smart grid demonstration. In *Innovative Smart Grid Technologies Conference Europe (ISGT Europe), 2010 IEEE PES*, pp. 1–8.

Bliek, F., van den Noort, A., Roossien, B. *et al.* (2011) The role of natural gas in smart grids. *Journal of Natural Gas Science and Engineering*, **3** (5), 608–616.

Geelen, D., Vos-Vlamings, M., Filippidou, F. *et al.* (2013) An end-user perspective on smart home energy systems in the PowerMatching City demonstration project. In *Innovative Smart Grid Technologies Europe (ISGT EUROPE), 2013 4th IEEE/PES*, pp. 1–5.

PMC Consortium (2015) *End Report PowerMatching City*, available at: http://www.powermatchingcity.nl (accessed 28 June 2015).

Wijbenga, J. P., MacDougall, P., Kamphuis, R. *et al.* (2014) Multi-goal optimization in PowerMatching city: a smart living lab. In *Innovative Smart Grid Technologies Conference Europe (ISGT-Europe), 2014 IEEE PES*, pp. 1–5.

12.2

New Future Solutions: Best Practices from European PV Smart Grid Projects

Gianluca Fulli and Flavia Gangale
European Commission, Joint Research Centre, Institute for Energy and Transport, Petten, the Netherlands

12.2.1 Introduction

In the past few decades, the European energy sector has undergone radical changes. Growing concerns about market competitiveness, climate change, and energy security have led the European Union (EU) to set ambitious policy goals which have called for a major restructuring of the entire energy system.

The EU is on track to meet its 2020 targets for greenhouse gas emissions reduction and renewable energy, and significant improvements have been made to reach the energy efficiency target. In its recent Communication, *A Policy Framework for Climate and Energy in the Period from 2020 to 2030* (European Commission, 2014), the European Commission proposed a 40% emissions reduction target as a milestone for 2030, with a target for the share of renewable energy consumed in the EU of at least 27%.

This newly proposed target for renewables in 2030 will translate into a target of about 45% renewable energy for the electricity sector. The rapid deployment of renewable energy poses challenges for the electricity system, which needs to adapt to increasingly decentralized and variable production. Much of the growth of renewables beyond 2020 is anticipated to come from wind and solar photovoltaic (PV), as these resources are abundant, have a wide applicability and are becoming cost-competitive. Their integration into the grid, however, may result in additional challenges for the secure operation of the electricity system and may require substantial investments in firm generation capacity as well as in additional infrastructure at the transmission and distribution level.

Photovoltaic Solar Energy: From Fundamentals to Applications, First Edition.
Edited by Angèle Reinders, Pierre Verlinden, Wilfried van Sark, and Alexandre Freundlich.

Companion website: www.wiley.com/go/reinders/photovoltaic_solar_energy

Smart grid concepts and technologies, however, can have an important role to play to address the challenges posed by the increased integration of PV into the electricity system (see Chapter 11.1) and to reduce or postpone the need for costly investments in grid reinforcements. In line with the Smart Grids European Technology Platform, by smart grids we mean "electricity networks that can intelligently integrate the behaviour and actions of all users connected to it – generators, customers and those that do both – in order to efficiently deliver sustainable, economic and secure electricity supplies" (European Technology Platform for Electricity Networks of the Future, 2006).

A number of smart solutions already exist, including storage and demand side management. Furthermore, the grid hosting capacity can be enhanced by the provision of grid support services and their integration into the planning and operational stages of the distribution grids, thus further reducing the need for grid reinforcements. PV systems themselves can contribute actively to smooth network operation through the delivery of local grid support services, such as frequency and voltage support.

To address these issues, many research and demonstration projects have been set up in Europe to investigate and demonstrate new technologies, tools and techniques towards the attainment of the European climate and energy goals. Most of these projects were included in the 2014 version of the Joint Research Centre's (JRC) inventory of smart grid projects, the most updated and comprehensive database of smart grid projects in Europe. Interactive and dynamic charts and maps of these projects can be accessed through the JRC website (Joint Research Centre, n.d.). Building on this vast knowledge base, in the following sections we will try to identify the latest trends in PV integration in distribution grids and discuss some of the main challenges addressed by the projects in the database.

12.2.2 Insights from the JRC Smart Grid Inventory

Since 2011, the Joint Research Centre (JRC) of the European Commission has monitored the state of the art of smart grid projects in Europe with a view to assessing current developments and drawing lessons learned. The first comprehensive inventory of smart grid projects in Europe was published in July 2011 (Giordano *et al.*, 2011) and its updates were released in 2013 (Giordano *et al.*, 2013) and in 2014 (Covrig *et al.*, 2014). The inventory has proved to be an important instrument to monitor the direction Europe is taking, to benchmark investments and to accelerate the innovation process.

The latest version of the inventory includes 459 smart grid projects from all 28 European Union countries and others. Through an in-depth analysis of the database, we identified 58 projects with the primary focus on PV integration into the electricity grid. Many other projects in the catalogue address PV integration, but we did not include them in our analysis because they do so only in the more general context of distributed energy resources (DER) integration into the grid. It also needs to be stressed that PV integration projects assessed in this study are not those contemplating mere grid connection/reinforcement actions but those entailing the adoption of smarter electricity network concepts and models.

The number of PV smart grid projects in Europe has significantly grown over the years, reaching a peak in 2012, when 16 new projects began. We can identify a first phase with some sporadic activity until 2008, followed by a sharp increase in the following years. This result seems to be in line with the introduction of the EU climate and energy package in 2009 and

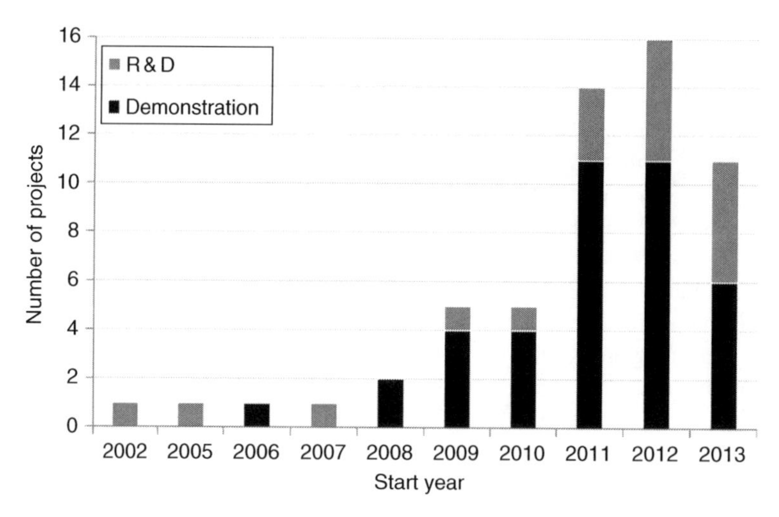

Figure 12.2.1 Number of R&D and demonstration PV smart grid projects starting each year

the resulting need to identify and deploy solutions to integrate a growing share of variable renewable generation into the grid. The drop in the number of projects in 2013, more than to the reduction of PV-specific incentivization schemes, seems to be linked to the cut-off date of the inventory exercise and the delay with which smart grid projects proponents tend to share results of their R&D and demonstration activities, see Figure 12.2.1.

The increasing share of R&D projects from 2009 onwards (from 20% in 2009 to 40% in 2013) suggests that there is still a need to identify and study new solutions to foster PV integration. Presumably, in the future new demonstration projects will test the results of these R&D projects to investigate their suitability for more extensive diffusion.

As for the geographical distribution of investments, PV smart grid projects are not uniformly distributed across Europe, with Italy, France, and Spain accounting for about 60% of the total investments, see Figure 12.2.2. The strong focus of smart grid projects on PV integration in these three countries can be explained by their present level of penetration and by their high photovoltaic potential. Some European countries with comparable potentials have shown less interest but they might be investing in more general DER integration projects. Many projects in the database (38%) are carried out by multinational consortia, revealing strong cooperation links between EU countries.

12.2.3 Main Solutions Investigated by the Projects in the JRC Inventory

Current power systems were designed to connect large generation with controllable and predictable output to large consumers and substations, from where power was then distributed to smaller consumers and to households.

The higher variability of PV generation challenges the previous paradigm and requires new operation and planning practices to ensure adequate control and balancing of generation and demand at all times. To maintain adequate electricity security it becomes crucial to fully exploit and further develop system flexibility by using a mix of different technologies and mechanisms.

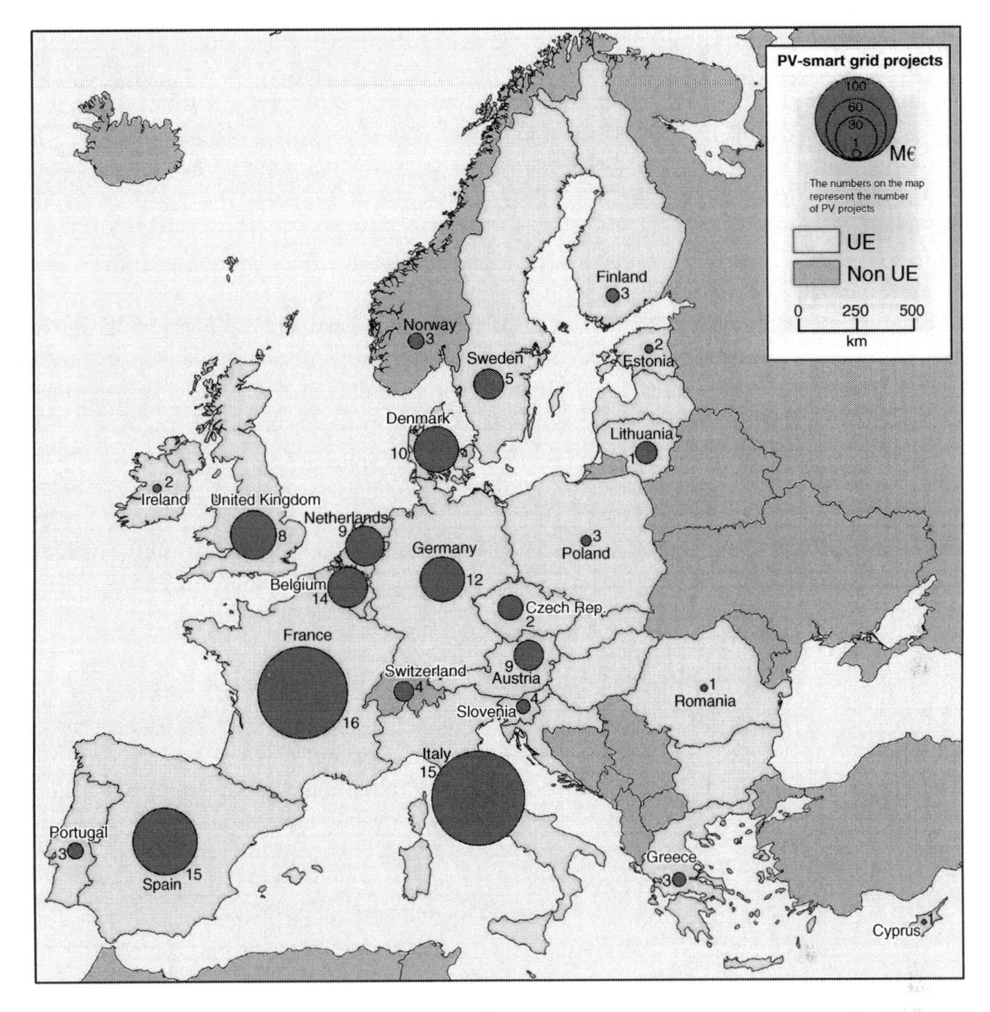

Figure 12.2.2 Investments in R&D and Demonstration PV smart grid projects in Europe (M€), 2002–2013. (*See insert for color representation of the figure*)

Smart and cost-effective solutions already exist to address the challenges associated with weather-dependent resources but they need to be tested and validated in real-life environments before they can be deployed on a large scale. Further research is also needed to prepare the accommodation of an increasing share of variable generation in the low-carbon electricity scenarios.

The analysis of the projects included in the JRC inventory shows that a variety of R&D and demonstration projects have already been carried out in Europe and many more are still ongoing. These projects have investigated a broad variety of solutions, including storage, demand side management and different solutions for grid voltage control.

Energy storage in particular has attracted much interest as a cost-effective solution to accommodate a large share of variable generation and increased demand peaks. Storage provides flexibility by shifting the time between generation and consumption of electricity, thus reducing the need for expensive grid reinforcements. The flexible nature of storage can offer a wide range of services at short notice to help with system balancing.

Different storage solutions have been investigated by the surveyed project as a means to support increased levels of PV generation, including hydrogen, e.g., Myrte (FR), compressed air, e.g., Energy Positive IT 2.0 (FR), flywheel energy storage systems, e.g., Smart ZAE (FR), lithium-ion battery, e.g., Sol-Ion (DE, FR), Nice Grid (FR), Isernia (IT), Millener (FR) and electric vehicles, e.g., Isernia (IT), S2G (CH), Ashton (UK), and Irene (DE), see Box 12.2.1 for a project example.

The recourse to storage solutions by Distribution System Operators (DSOs), however, is still hindered by a lack of experience, high costs, and regulatory constraints.

Demand side management (DSM) solutions have also been widely investigated by the projects in our catalogue. DSM is a source of flexibility on the consumption side and can be used as a solution to support the large-scale integration of PV energy into the grid. Insufficient flexibility may limit the share of PV generation a power system can accommodate. By shifting the flexible part of the demand to when PV sources produce most, network operators can reduce excess generation and the need for increased ramping capacity, thus improving system operation and reliability. While DSM solutions are already common in the industrial sector, they are still quite new in the residential sector. At the household level, DSM solutions require the previous deployment of advanced monitoring and communication infrastructure.

Box 12.2.1 Project Example: Energy Storage

The Nice Grid project (FR) – one of the six demonstrators of the FP7 European project Grid4EU – aims to develop a smart solar district in Carros, in the Nice-Côte d'Azur area. The project focuses on the optimization of PV integration into the low voltage grids by using PV and load forecasts, flexible loads, electric storage, and islanding. It involves about 1,500 residential and commercial customers equipped with the Linky smart meter, 200 solar rooftops and 100 lithium-ion batteries equivalent to 2.7 MWh of storage capacity.

The batteries will be installed at three distinct levels of the electricity distribution network: (1) a 560 kWh/1.1 MW lithium-ion battery at the Carros primary substation, to link the distribution and transmission networks; (2) three 106 kWh/33 kW lithium-ion batteries installed in five medium and low voltage distribution substations, to control peak generation of PV installations and manage peak demand periods, while also allowing for operation in islanded mode and more effective management of energy flows and voltages; and (3) several 4 kWh/4.6 kW lithium-ion batteries installed in volunteer customer homes to facilitate load shedding (Nicegrid.fr, n.d.).

The project includes the design and integration of a local energy management system that communicates with the various local dispersed devices and technologies. The objective is to ensure that the primary substation has sufficient flexibility by controlling the operation of batteries connected to the grid.

By optimizing the charge–discharge cycle of the distributed storage facilities, the system will allow the reduction in the number of distributed generators disconnections due to the infringement of voltage constraints. Furthermore, the area will be able to reduce its load in case of planned outage thanks to the use of demand side management of local loads and purpose-built probabilistic solar power forecast and load power forecast tools (Michiorri *et al.*, 2014).

Box 12.2.2 Project Example: DSM

The *Jouw Energie Project* (NL) which means Your Energy Moment in English, is a demand response demonstration project, which aims at mobilizing flexibility in the use of electricity, giving participants financial and/or emotional incentives to change their use in time. The project involves about 250 residential participants who have been provided with products and services to enable them to choose their preferred times to use electricity. Equipped with a special In-Home Display (IHD) and a smart washing machine, users will be able to choose whether they want to run their washing machine during times when their local sustainable electricity is produced by PV panels or at those when the cost of electricity is low on the wholesale market. Participants can enter their preferences and constraints and the energy computer does the rest autonomously (Giordano *et al.*, 2013).

The IHD provides consumers with information on their energy consumption, weather forecasts and expected production of solar power. The time of use (ToU) tariffs, local production of PV electricity and the price differences on the wholesale market are presented to consumers 24 hours in advance, to allow them to make their choices on the use of the washing machine.

A lot of work and effort has been dedicated to the design of the interface to ensure that the information is presented in an attractive and easily understandable way. In particular, the visualization of the relationship between PV production and household consumption is an important feature to help consumers/prosumers to shift the flexible part of their demand to when PV sources produce more, thus improving the grid stability. The project has developed a comprehensive strategy for consumer engagement, which addresses the individuals and the community to which they belong, using a variety of tools, such as historical comparison, gamification, continuous interaction, use of trusted third parties, social events. More information about the project and its results can be found online on the project website Jouwenergiemoment.nl.

Several projects have investigated the technical and economic viability of DSM solutions to increase PV penetration in the distribution grid. Consumer engagement issues play a crucial role in these projects, and project developers are showing growing attention to the interaction between the new technological solutions and end-users, e.g., the Smart Domo Grid (IT), and the Jouw Energie Moment (NL), see Box 12.2.2 for a project example. Different consumer involvement strategies are being tested, ranging from community engagement, e.g., Ashton (UK) to wider automation, e.g., Millener (FR).

Finally, a variety of *grid voltage control* solutions have been tested by the surveyed PV-smart grid projects as a means to increase PV hosting capacity in the distribution grid, e.g., MorePV2Grid (AT), PVNetDK (DK), POI (IT), MetaPV (AT, BE, DE, SI), DG Demonet - Smart LV Grid (AT), and REserviceS (BE, DE, DK, ES, FI, IE), see Box 12.2.3 for a project example. Several projects tackled the issue of active and reactive power control to keep voltages within the admissible ranges. The aim of the different tested solutions is to limit voltage rises caused by a high local PV power feed-in, hence increasing the share of potential PV penetration.

Several projects have focused on smart inverters for voltage support. Given their capability to control their active and reactive power, PV inverters are well suited to provide different kinds of ancillary services (such as frequency and voltage ride through capabilities), minimizing the need for costly grid reinforcement and operational countermeasures or for PV power curtailment, e.g. PVNetDK, MetaPV.

Box 12.2.3 Project Example: Grid Voltage Control

The *REserviceS* (BE, DE, DK, ES, FI, IE) project intends to establish a reference basis and policy recommendations for future network codes and market design in the area of ancillary services from variable renewable energy technologies. The project delivered technical insights and economic elements to support the establishment of proper market mechanisms and grid code formulations in the EU, and investigated whether ancillary services can generate additional value for network operators by involving grid users, notably wind and solar PV generation (Giordano *et al.*, 2013).

The REserviceS study identified the key grid support services that variable renewable generation can provide – notably frequency and voltage support – and assessed the capabilities of wind and solar PV to provide them. The project also looked at the technical and economic impacts of a large deployment of grid services on transmission and distribution grids.

The analysis showed that the technical and operational capabilities required to provide grid support services are already in place or can be implemented at a reasonable cost. For solar PV in particular, potential technical enhancements for frequency and voltage support services include: the estimation of available power/forecasting, faster and reliable communication and control within the plant, control strategies for portfolios composed of numerous small and medium-sized units, improving the interoperability of different networks, and enhancing compliance to a multitude of non-harmonized grid code requirements.

Finally, the REserviceS simulations of power systems confirmed the economic benefits of utilizing wind and solar PV for frequency support. These benefits increase as the share of variable renewable generation increases (Van Hulle *et al.*, 2014).

More qualitative and quantitative results can be found in the deliverables published on the project website (Reservices-project.eu).

12.2.4 Conclusion

The transition towards a more competitive, secure and sustainable energy system will not be possible without significantly higher shares of renewable energy, including solar photovoltaic.

Technical solutions already exist to address many of the concerns raised by higher penetration rates of PV in the distribution grid but more research and demonstration projects are needed to assess the viability and cost-effectiveness of the new solutions. The JRC is continuously monitoring the state of the art of smart grid projects in Europe by updating the JRC inventory and analyzing current trends and developments.

The analysis of the projects in the inventory has revealed the increasing interest of project developers in new solutions and techniques to improve the grid hosting capacity via non-conventional solutions and to use PV systems as active suppliers of local ancillary services. PV systems are likely to become more active players in the future system operation, supporting the power grid and performing grid-related control functions.

DSOs are the most active stakeholders in this field, and they are taking the lead in developing new and more ambitious demonstration projects. Smaller utilities with fewer resources and other stakeholders, however, can face difficulties in conducting similar tests. Sharing projects' methodologies and results at the European level is therefore of crucial importance to achieve the European energy and climate policy goals in time. To this aim, the JRC will continue to monitor the state of development of smart grid projects in Europe and to disseminate the lessons learned.

List of Acronyms

Acronym	Description
AT	Austria
BE	Belgium
CH	Switzerland
DE	Germany
DER	Distributed Energy Resources
DK	Denmark
DR	Demand Response
DSM	Demand Side Management
DSO	Distribution System Operator
ES	Spain
EU	European Union
FI	Finland
FP7	EU's Seventh Framework Programme for Research
FR	France
IE	Ireland
IHD	In-Home Display
IT	Italy
JRC	Joint Research Centre
NL	The Netherlands
PV	Solar Photovoltaic
R&D	Research and Development
SI	Slovenia
ToU	Time of use
UK	United Kingdom

References

Covrig, C.F. *et al.* (2014) Smart Grid Projects Outlook 2014, JRC Science and Policy Report (JRC Report No. EUR 26651 EN). European Commission, Joint Research Centre, Petten.

European Commission (2014) A policy framework for climate and energy in the period from 2020 to 2030. Communication of the Commission to the European Parliament, the Council, the European economic and social committee and the committee of the regions, COM(2014) 15 final.

European Technology Platform for Electricity Networks of the Future (2006) *The SmartGrids European Technology Platform*. Available at: http://www.smartgrids.eu/ETPSmartGrids. (accessed 17 December 2014).

Giordano,V. *et al.* (2011) Smart grid projects in Europe: lessons learned and current developments, JRC Reference Report (JRC Reference Report No. EUR24856EN). European Commission, Joint Research Centre, Petten.

Giordano, V. *et al.* (2013) Smart Grid projects in Europe: lessons learned and current developments 2012 update, JRC Reference Report (JRC Scientific and Policy Report No. EUR 25815). European Commission, Joint Research Centre, Petten.

Joint Research Centre, n.d. Smart Grid Projects Outlook 2014. Available at: http://ses.jrc.ec.europa.eu/smart-grids-observatory (accessed 20 Feb. 2015).

Jouwenergiemoment, n.d. Available at: http://jouwenergiemoment.nl/ (accessed 22 Dec. 2014).

Michiorri, A. *et al.* (2014) A local energy management system for solar integration and improved security of supply: The Nice Grid project. Paper presented at the 3rd IEEE PES International Conference and Exhibition on Innovative Smart Grid Technologies, Dec. 2012, Berlin, Germany.

Nicegrid, n.d. *Nice Grid: Le Stockage d'énérgie*, n.d. Available at: http://www.nicegrid.fr/nice-grid-energy-storage-10.htm (accessed 19 Dec. 2014).

REserviceS project., n.d. *Final Publication and Recommendations*. Available at: http://www.reservices-project.eu/publications-results/ (accessed 28 Dec 2014).

Van Hulle, F. *et al.* (2014) *Economic Grid Support Services by Wind and Solar PV: A Review of System Needs, Technology Options, Economic Benefits and Suitable Market Mechanisms*. Final publication of the REserviceS project, September 2014. Available at: http://www.reservices-project.eu/wp-content/uploads/REserviceS-full-publication-EN.pdf (accessed 19 Dec. 2014).

Part Thirteen

Supporting Methods and Tools

13.1

The Economics of PV Systems

Matthew Campbell
SunPower Corporation, Richmond, CA, USA

13.1.1 Introduction

The widespread global deployment of PV systems is contingent on reducing the cost of generated electricity to levels that make systems economically competitive without incentives, this is a milestone also referred to as solar "grid-parity." In recent years, substantial progress has been made to this end with significant reductions in all aspects of PV system costs from the solar panel to the electrical system to construction and operation. Fundamental to maximizing cost reduction is a thorough understanding of PV system economic drivers.

How to derive a cost of electricity from a PV system is a complicated topic, depending on a wide variety of variables, some well known and some uncertain because they are based on future assumptions. This section will introduce some key tools to understand PV system economics.

Calculating a PV system's cost of energy is fundamentally different from that of traditional generation, namely, fossil fuel. For example, in a power plant generating electricity through coal or natural gas, the cost to generate electricity today is easily measurable based on the price of fuel, the efficiency of the generating unit, the operating costs, and the depreciation of the facility. For PV systems, the cost of electricity depends on a forecast of the future, since in essence the electricity is being paid for upfront in the higher capital cost of the system. Since there is no fuel, and the operating costs are relatively low, one must take the purchase price and amortize it across the asset's useful life to derive an average cost of electricity.

To be able to financially evaluate PV systems in accordance with existing frameworks, we will introduce basic PV system economic principles, a general taxonomy of PV system economic drivers, and review five of the most important categories, which influence the cost of electricity generated by PV systems.

Photovoltaic Solar Energy: From Fundamentals to Applications, First Edition.
Edited by Angèle Reinders, Pierre Verlinden, Wilfried van Sark, and Alexandre Freundlich.

Companion website: www.wiley.com/go/reinders/photovoltaic_solar_energy

13.1.2 Levelized Cost of Electricity (LCOE)

The levelized cost of electricity (LCOE) is perhaps the most commonly referenced metric of PV system economics. In simple terms, the LCOE gives a levelized (average) cost of electricity generation over the life of the asset. The LCOE is an analytical tool that can be used to compare alternative technologies when different scales of operation, investment, or operating time periods exist, thus facilitating an "apples to apples" comparison of technologies. For example, the LCOE could be used to compare the cost of energy generated by a PV power plant with that of a fossil fuel generating unit or another renewable technology (Short *et al.*, 1995).

The calculation for the LCOE is the net present value of the total life-cycle costs of the PV project divided by the quantity of energy produced over the system's life:

$$\text{LCOE} = \frac{\text{Total Life Cycle Cost}}{\text{Total Lifetime Energy Production}} \tag{13.1.1}$$

The above LCOE equation can be disaggregated for solar generation as follows:

$$\text{LCOE} = \frac{\begin{array}{r}\text{Initial Investment} - \sum_{n=1}^{N} \frac{\text{Depreciation}}{(1+\text{Discount Rate})^{n}} \times (\text{Tax Rate}) \\ + \sum_{n=1}^{N} \frac{\text{Annual Costs}}{(1+\text{Discount Rate})^{n}} \\ \times (1-\text{Tax Rate}) - \frac{\text{Residual Value}}{(1+\text{Discount Rate})^{N}}\end{array}}{\sum_{n=1}^{N} \frac{\text{Initial kWh / kWp} \times (1-\text{System Degradation Rate})^{n}}{(1+\text{Discount Rate})^{n}}} \tag{13.1.2}$$

The costs include the initial capital investment and the lifetime operating expenses, less the end-of-life value discounted into the present value. The lifetime costs are divided by the lifetime energy production to derive the average cost of electricity.

When evaluating the LCOE and comparing other commonly known cost per kWh benchmarks used in the PV industry, it is important to remember that the LCOE is an evaluation of levelized life-cycle energy costs and not the price of energy established under power purchase agreements[1] (PPAs) or by feed-in-tariffs[2] (FITs). These may differ substantially from the LCOE of a given PV technology, as they may represent different contract or incentive durations, the inclusion of incentives such as tax benefits or accelerated depreciation, financing structures, and in some cases, the value of time-of-day production tariffs. One might

[1] A Power Purchase Agreement contract is an agreement between the party that generates the electricity (the seller) and the party that is looking to purchase the electricity (the buyer). In the case of solar, the PPA contract generally covers a duration of 10, 20, or 25 years. A PPA provides the contracted annual revenue stream which can support the financing of a solar project when no FIT exists.

[2] A feed-in-tariff is an economic policy created to promote active investment in and production of new energy sources. Feed-in tariffs typically make use of long-term agreements and pricing tied to costs of production for renewable energy producers. The long-term security of a 10–25 year FIT contract provides a solar power system with the revenue needed to secure financing.

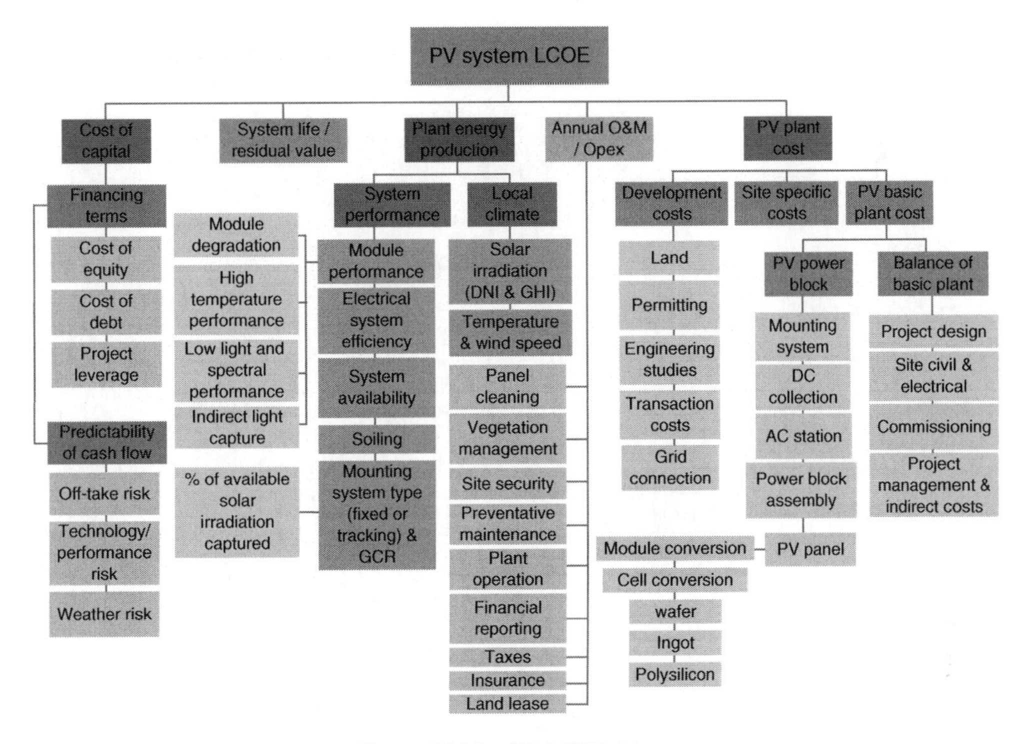

Figure 13.1.1 PV LCOE drivers

consider LCOE to be similar to an automobile's nameplate average fuel efficiency versus what a driver actually realizes, based on their driving patterns (analogous to a FIT or PPA).

The drivers of a system LCOE may be disaggregated into the five key high-level drivers listed in Figure 13.1.1: cost of capital, system life/residual value, plant energy production, annual operating costs, and PV system cost. Under these five categories dozens of variables influence LCOE, and each can have a material influence on system economics.

13.1.3 PV System Cost

The first major driver of PV system economics is the PV system cost. How to reduce the system cost has been the primary focus of PV technology development and innovation during the past three decades. Just 10 years ago a PV system typically may have cost between $5 and $10 per watt of installed peak capacity (Wp), rendering an associated cost of electricity well above traditional sources in most applications. This high capital cost was a commonly-used argument by those skeptical of PV's ability to meaningfully contribute to global electricity production. With the strong support of governments that helped develop the PV industry through incentive programs, however, the industry has achieved cost reductions well beyond the most optimistic predictions made by industry analysts in the not-too-distant past, with large system prices in 2015 as low as $1–2/Wp, a remarkable improvement.

When trying to understand the costs of a PV system it is important to note that there may be 100 or more individual drivers, all of which must be addressed to improve system economics. Often those new to the field search for a "magic bullet" breakthrough that could quickly bring

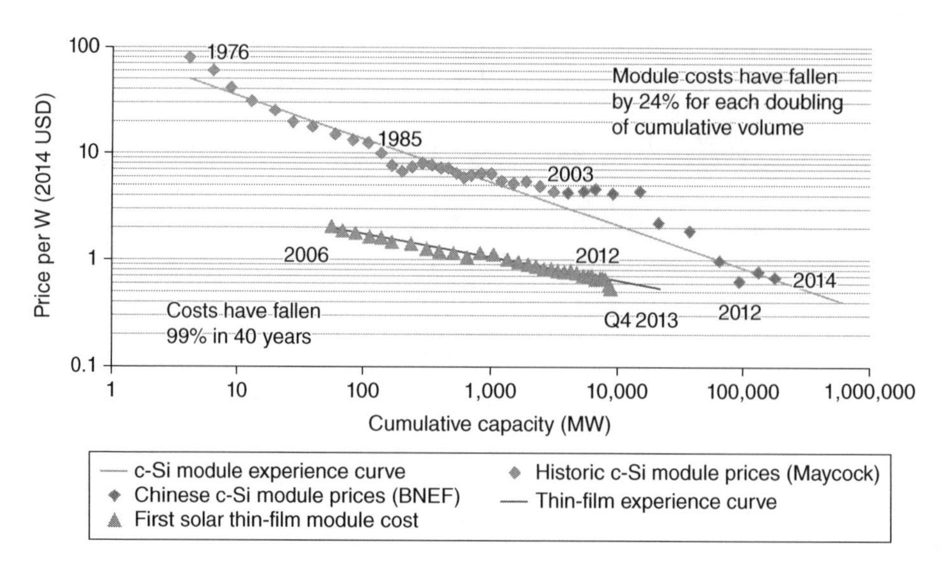

Figure 13.1.2 PV module prices vs. cumulative production. Source: (Bromley, 2015)

down the cost of a PV system. The reality is that methodical improvement on a large number of details will be required to continue progress down a price curve rather than a singular breakthrough invention.

The biggest improvement to system costs has been the steady decline of module prices through a process of "learning by doing," or an experience curve. An experience curve is a graphical representation of the process by which the costs of an activity go down over time through innovation and the benefits of scale, both required by firms in order to be profitable in a given sector. Figure 13.1.2 shows a commonly used measurement of PV module average selling price (ASP) improvements associated with an experience curve. Price improvements follow a logarithmic trajectory as module installations continue their substantial annual growth rate leading to the current low module prices seen in the market. A so-called progress ratio of 76% seen in Figure 13.1.2 means that doubling of cumulative production leads to a decrease in module ASP of 24%.

Similar to module price declines, balance of system (BOS) costs have fallen more than 75% over the past 10 years. Plotting BOS comparisons over time is difficult given different project sizes, scope of work, geographic location, and lack of published data, but improvement has been seen in essentially all areas, from construction to hardware components. Central inverters, for example, once cost more than \$0.50/Wp and can now be purchased for less than \$0.10/Wp in the market.

Other cost considerations in a PV system are site-specific costs such as the type of foundations required, roof repair work, road building, fencing, or other costs which may be unique to a project. Development costs include site acquisition, engineering studies, obtaining permits, and project financing, and are also site-specific. Interconnection costs include costs to connect a system to the electrical grid or to a customer's electrical system. These also vary widely by scope and can range from a low cost tie-in to an existing substation, to an expensive high-voltage transmission line running many kilometres with associated new switchyards and substations.

Table 13.1.1 gives a simplified view of what a large (10 MW+) system cost in 2008, what it may cost at the time of writing in early 2015, and a simplified projection for 2025. As can

Table 13.1.1 Simplified PV system costs, 2008, 2015 and 2025

Generic PV System Costs ($/Wp) @ 10 MW + Project Size						
		2008*		2015**		2025 Estimate***
PV module price	$	4.02	$	0.55	$	0.41
Balance of system (power block & construction)	$	1.77	$	0.68	$	0.50
Interconnection costs	$	0.07	$	0.06	$	0.05
Development costs	$	0.06	$	0.04	$	0.03
Total system price	**$**	**5.92**	**$**	**1.33**	**$**	**0.99**
% of 2008 baseline		*100%*		*22%*		*17%*

*Deutsche Bank, January 2011 Report; also consistent with SunPower projects at that time
**Cost for a generic fixed tilt system built in 2015 with commodity modules in a moderate or low labor cost country – estimate by the author based on project experience
***Estimate based on an extrapolation of 2015 system costs and applying a simple 3% annual improvement rate through experience curve effects

be seen, there has been a remarkable reduction in the cost of PV systems, largely driven by PV module price reductions but also seen across almost all system components. Considering that the first 10 MW system in the world was just built in 2004, it is not surprising that companies quickly learned how to reduce costs on bigger and much more numerous large-scale systems.

PV project costs vary widely depending on scope, project condition and location, but system prices for large ground systems can be found today in the mid $1/Wp range. Although future predictions of system costs are difficult, one could expect BOS costs to conservatively follow an experience curve similar to PV modules. Applying a simple 3% annual progress curve across the plant, one might expect large-scale PV system prices at or below $1/Wp by 2025 if not sooner.

13.1.4 System Energy Production

The second key driver of PV system economics is the system's annual energy production. The energy production of a system is expressed in kilowatt hours generated per rated kilowatt peak of capacity per year (kWh/kWp). The kWh/kWp is a function of:

- the amount of sunshine the project site receives in a year and the local weather (wind, temperature, humidity);
- how the system is mounted and oriented (e.g., flat, fixed tilt, tracking, etc.);
- the spacing between PV panels as expressed in terms of system ground coverage ratio (GCR), more spacing reduces shading losses and improves yield but requires more space;
- the energy harvest of the PV panel (e.g., performance sensitivity to high temperatures, efficiency as affected by low or diffuse light, etc.);
- system losses from soiling, transformers, inverters, and wiring inefficiencies;
- system availability, largely driven by inverter reliability improvements.

Figure 13.1.3 Solar tracking system with robotic cleaning

PV system performance has improved over the past ten years due to a number of factors, including high levels of system availability (99%+for the best plants), improvements in inverter efficiency, the development of modules with improved temperature coefficient, improvements in cleaning (e.g., through the use of robotics, see Figure 13.1.3), and the widespread adoption of trackers for large power plants in high-sunshine markets.

Improved system performance is very valuable in the reduction of LCOE since most of a PV system's upfront and long-term costs are fixed: more energy directly reduces the LCOE.

Solar technology developers and system designers must constantly evaluate the tradeoffs between cost and performance in arriving at the lowest LCOE. Components with a low initial cost may suffer from poor performance and higher maintenance costs over the long-term life of the plant and may contribute to a higher rather than a lower LCOE. Cost reduction and reliability are often in tension in many industries. The key is to innovate and reduce cost without using substandard materials or manufacturing/construction practices, which may impact the long-term performance.

Because all systems in an outdoor environment experience degradation over time, the energy yield of a plant will follow a declining curve during its useful life. The durability of the components versus environmental variables of high temperature, moisture, wind, and UV will influence the rate of average annual performance degradation. Project investors may consider annual degradation rates from 0.25% –2.00% per year (Jordan and Kurtz, 2013) depending on the system, location, and technology type. Annual degradation can have a major impact on system LCOE, since it represents an annual decrease in project revenue (electricity sales).

System yields can range from less than 1000 kWh/kWp for a fixed system in Western Europe to more than 2500 kWh/kWp for a tracking system in the Atacama Desert in Chile. Factoring in a range of degradation rates, the highest-performing systems can generate 3–4 times more electricity over their lifetime than the lowest, leading to a commensurate reduction in electricity cost. One must also consider the time value of money. A kWh and its associated revenue are more valuable today than in the future when computing the LCOE. This value of future energy production is closely tied to the project's cost of capital. As financing costs decline, the value of future energy production goes up.

13.1.5 Cost of Capital

The third driver of PV system economics is the cost of capital. The cost of capital, or effectively the level of interest rate required to finance a system, has a substantial influence on economics, equal in importance to the system cost or performance. Since essentially all of the energy from a system is prepaid in the form of the system price, the cost to finance that asset is critical.

The average cost of financing is linked to the real and/or perceived risk of the asset. Table 13.1.2 gives examples of the potential risks to investors in a PV power plant. The first category of risk is the off-take risk, the risk that the electricity from a plant will be sold for the price and/or volume expected by project investors. Government FIT programs and long-term PPAs with investment grade utilities substantially reduce the off-take risk to an investor in a PV power plant since an associated contract will provide a guarantee on future electricity pricing. Some PV power plants are now being built without the security of long-term off-take contracts; these fall under a merchant power plant designation since the electricity is sold into the wholesale energy market at the prevailing energy prices. In this model there is no certainty of future electricity pricing. Due to the higher risk associated with taking the market price risk, merchant plants require a higher investment return (and thus cost of capital).

The next major category of risk is performance risk, meaning the risk that the system will not generate the expected energy and/or achieve the operating costs in line with original project estimates, therefore reducing expected cash flows. Factors impacting performance could include inaccurate weather predictions or technology underperforming relative to

Table 13.1.2 Project risks influencing financing costs

Risk Category	Risk Driver
Off-take risk	**Risk that the system electricity sale price and/or volume will be lower than planned**
Government FIT	Credit rating of the country providing a FIT guarantee, risk that a government would retroactively change the FIT
PPA	Credit rating of the utility/customer signing PPA agreement
Merchant plant	Price volatility in associated wholesale power market
Performance risk	**Risk of lower than expected cash flows due to weather conditions or system technical issues**
Weather risk	The level of solar resource or other weather variables reduce the system performance below financial model expectations
Technology risk	The technology underperforms relative to expectations
O&M risk	Operations and maintenance costs exceed those forecasted in financial model
Property risk	**Risk of property damage/loss**
Damage	System damaged by severe weather or seismic event
Theft	Theft of system components

system specifications. One of the strongest attributes of PV power plants as a new financial asset class is that the performance can be predicted with a high level of accuracy. The standard deviation of annual solar resource is low, and the performance of a well-constructed PV plant built with quality components is very high and predictable. Well-engineered systems built on sites developed with proper meteorological diligence should have low performance risk.

Similar to other physical infrastructure, PV systems face additional risks from property damage or theft, exchange rate or interest rate changes, inflation, and country risk among others. The collection of these risks will be factored into the cost of capital for a project.

As investor understanding of PV power plants has matured, the PV industry has seen a dramatic reduction in the financing costs of PV systems. PV power plants provide attractive risk-adjusted returns on investment, particularly in the current environment of historically-low interest rates. PV power plants offer bond-like predictable returns with the security of an underlying physical asset that will be capable of generating energy and revenue for decades. The reduction in average PV financing costs from 10% + in past years to less than 6% in some markets today has created a proportionate reduction in LCOE. Innovative financial structures such as the "Yieldco" method of financing have substantially reduced the cost of capital and led to large reductions in PPA prices in the USA, triggering an increase in utility demand for solar. Similar financing mechanisms are likely to be adopted in other markets as investors optimize investment structures to the attributes of PV assets.

13.1.6 System Life

The operational life of a PV system is a fourth key contributor to PV system economics. History has demonstrated high performance from PV panels even with more than 20 years of outdoor exposure. No fundamental barrier to operating a PV plant for more than 30 years exists. In fact, systems will likely operate for 40–80 years with periodic refurbishments. An example of the future life of PV systems can be seen in the experience of other generating assets such as conventional fossil or hydro power plants. For example, in hydropower, the 2GW Hoover Dam project was built in 1936 for a cost of $49 million. After more than 75 years of operation, the system is capable of generating as much energy today as at any time during its existence and has an electricity cost of about $0.02/kWh, less than that of any other generating source in the southwest USA (Brean, 2014). This asset benefits from the fact that the system was long ago paid for and that current generation costs are just those associated with annual operations and maintenance. Similarly other power plants in the USA have a long operational history with 51% of all electricity generation capacity and 73% of coal plants built over 30 years ago (U.S. Energy Information Administration, 2011), as shown in Figure 13.1.4 (note that since this 2011 study a significant amount of wind, solar, and gas has been added to the US power generation fleet).

Since existing power plants provide empirical evidence that power stations can have a long life, the question is whether nuclear, coal, and hydro stations have a fundamental longevity advantage over a PV power plant. The reality is that conventional power plants must contend with a much more difficult operational environment than PV. Steam generation units, industrial pollution, corrosion, and complicated mechanical systems that require constant maintenance are some examples of why conventional power plants face larger operational hurdles than PV, yet continue to operate for more than 50 years. In comparison, PV systems have few moving

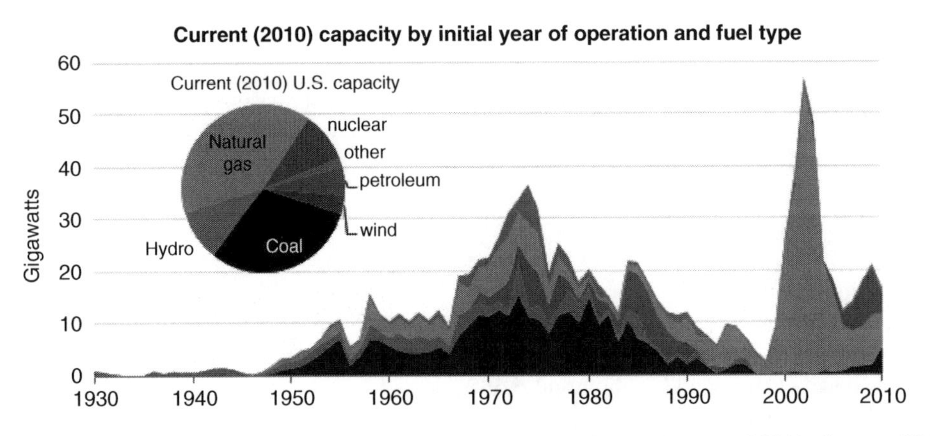

Figure 13.1.4 US power generating capacity by initial year of operation, as of 2011. Source: (U.S. Energy Information Administration, 2011)

parts, use steel structures and cabling components similar to a conventional power plant, and generally have a more benign operating environment. In many dry, sunny regions of the world we should expect properly engineered PV components to last a very long time. The oldest operating solar energy generating system (SEGS) power plants are the SEGS solar thermal trough systems in Southern California, which were built in phases between 1984 and 1990. After more than 25 years of operation the mechanical and motor drive components are in excellent condition and their solar collector systems have been repowered with the latest technology, increasing their output. Although not identical to a PV power plant, many of the field components are similar. Furthermore, PV plants do not require the steam power block and heat transfer fluid that the SEGS plants have, which are more complicated to operate than PV. These solar thermal plants are likely to continue operation for the foreseeable future, which portends well for the potential operational life of PV power plants. After the first 20–25 years of operation, we might expect that the cost of electricity for a PV system may drop to a very low level, with the underlying asset being depreciated and operating costs being relatively low. In some power markets depreciated PV systems could actually have the lowest cost of electricity in the second phase (after years 20–25) of their operation, similar to the experience of Hoover Dam. The conclusion is that PV systems will likely have a long economic life and that should be factored in when calculating the lifetime LCOE of a system.

13.1.7 Annual Operating Costs

The fifth major LCOE cost driver is the annual operating cost of the system. The annual operating cost of a PV system includes the operations and maintenance of the plant (O&M), the cost to administer the project, such as accounting and tax reporting, property insurance, land lease as applicable, site security and property or revenue taxes that may apply. Note that PV plants often have an associated special purpose vehicle (SPV) company for the purposes of project financing, which has an associated annual administrative overhead which contributes to the annual operating expense.

Historically, annual operating costs of a PV system have been low on a relative basis, since they contributed less than 5% to a project's LCOE. However with the substantial declines in PV system prices and the cost of capital, they are becoming increasingly important to manage and reduce. In fact, going forward, operating costs will have almost as much influence as the module or BOS cost. It should also be noted that annual costs tend to go up every year for a specific project due to inflationary effects on labor and material costs, which also increases their relative importance in a low-interest-rate environment.

As PV power plants have scaled up to larger than 100 MW in recent years, substantial improvements to O&M costs have been made. The largest O&M cost in a PV power plant, inverter maintenance, is seeing substantial improvements in operating expense and performance as the technology of central inverters improves. Automation is also providing benefit in areas such as panel cleaning and in the remote operation of a system, which reduces the need for field labor in remote areas. The variable operating costs of PV systems generally compromise less than 15% of a new project's LCOE, far lower than the fuel and maintenance costs which may contribute more than 75% of the cost of a fossil fuel-based generating plant (even excluding the external costs associated with plant emissions or future carbon taxes, should they be implemented).

13.1.8 PV LCOE and Grid Parity

There has been a remarkable decline in the PV LCOE over the past ten years. This has been accomplished through innovation in all aspects of PV systems from the module to the BOS to system performance, financing, and operation. With continued forecasted PV growth and the associated billions of dollars of investment being made in new systems, factories, research and development programs, and financing, it is likely that the PV LCOE will continue to decline. At the time of writing, the LCOE for 10 MW-scale PV systems is on the order of $0.065/kWh–$0.272/kWh depending on the location, type of technology, cost of capital and solar resource (Lazard, 2014). This cost of electricity is lower than some conventional power plants, especially those relying on liquefied natural gas or diesel-based electricity generators, meaning that grid parity has been reached in selected markets. When considering the potential for future declines in PV LCOE as well as the possibility of carbon emission charges, it is clear that PV will become competitive in an increasing number of more common electricity markets which rely on coal and natural gas. This will likely continue the virtuous circle of increased demand which will help enable further reductions in PV cost, thus further triggering demand as prices fall.

In distributed rooftop PV markets, the story is similar with substantial cost reductions experienced and with PV LCOE levels becoming increasingly competitive with commercial and retail rates in many countries. Unsubsidized LCOEs for residential PV systems are estimated to be $0.157–$0.224/kWh in the USA as of 2015, with commercial rooftop LCOEs of $0.107–$0.163/kWh (Lazard, 2014). As distributed PV becomes economically competitive, fundamental changes to traditional utility business models and regulatory structures may be needed. Utility systems consist of substantial fixed costs relating to transmission and distribution as well as the cost of insuring reliability and back-up capacity to insure the integrity of the grid, if utility retail revenue is reduced due to more customers having distributed PV systems, the costs must be spread over the remaining consumption. This may necessitate changes to residential and commercial electricity rates to reallocate how system costs are recaptured.

A further variable impacting retail and commercial energy economics will be lower cost battery storage which is on the not-too-distant horizon. It is likely that retail and commercial electricity markets will experience substantial changes now that the point of grid parity has been reached for some and is in sight for other PV markets, especially when combined with energy storage.

13.1.9 Conclusion

The LCOE is a tool that enables objective PV economic analysis and comparisons with other technologies. The derivation of the LCOE depends on many variables, with the general categories of system cost, cost of capital, annual operating expenses, performance, and system life having the largest influence. The past 30 years have seen remarkable improvement in all of these key LCOE drivers, with new PV systems having a much lower cost, a longer operating life, higher performance, lower operating expense, and lower cost of capital, enabling LCOEs to approach a competitive level with conventional generation sources without incentives. Through continued scale-up and methodical cost reductions, PV systems are likely to become increasingly competitive on LCOE, and over the next 20 years could become one of the most cost-effective electricity generation sources.

List of Acronyms

Acronym	Description
ASP	Average selling price
BOS	Balance-of-system
FIT	Feed-in tariff
GCR	Ground coverage ratio
LCOE	Levelized cost of electricity
O&M	Operations and maintenance of the plant
PPA	Power purchase agreement
SEGS	Solar Energy Generating System
SPV	Special purpose vehicle

References

Brean, H. (2014) Cheaper Hoover Dam power up for grabs. *Las Vegas Review Journal*, September 11, 2014, available at: http://www.reviewjournal.com/news/nevada-and-west/cheaper-hoover-dam-power-grabs (accessed 30 July 2015).

Bromley, H. (2015) PV module makers: tiers and trends. Paper presented at Australian Clean Energy Summit, July 16, Bloomberg New Energy Finance.

Jordan, D.C. and Kurtz, S.R. (2013) Photovoltaic degradation rates: an analytical review. *Progress in Photovoltaics: Research and Applications*, **21**, 12–29.

Lazard (2014) Lazard's Levelized Cost of Energy Analysis – Version 8.0, September 2014, available at: http://www.lazard.com/media/1777/levelized_cost_of_energy_-_version_80.pdf (accessed 30 July 2015).

Short, W., Packey, D., and Holt, T. (1995) A manual for the economic evaluation of energy efficiency and renewable energy technologies. *National Renewable Energy Laboratory*, Report NREL/TP-462-5173, available at: http://www.nrel.gov/docs/legosti/old/5173.pdf (accessed 30 July 2015).

U.S. Energy Information Administration (2011) Age of electric power generators varies widely, available at: http://www.eia.gov/todayinenergy/detail.cfm?id=1830 (accessed 29 August 2015).

13.2

People's Involvement in Residential PV and their Experiences

Barbara van Mierlo
Department of Social Sciences, Knowledge, Technology and Innovation, Wageningen University, the Netherlands

13.2.1 Introduction

In the novel *Solar* by Ian McEwan, Professor Michael Beard accepts a position as the head of a renewable energy research centre and, after the failure with a small-scale wind turbine, ends up developing photovoltaics. He sees the job as an opportunity to increase his income and has no interest in the solar energy technology whatsoever. In the, at times, hilarious descriptions of the professor's lectures all over the world, the readers of the book learn more about his digestion system than a PV system. This is in stark contrast to what PV advocates expect from people with space available for solar panels on their rooftops or balconies. The latter are addressed as potentially interested in PV, willing to consider and buy PV as a benign alternative to fossil fuel-generated electricity. Reduction of costs is often seen as the most essential factor to motivate not only the small group of frontrunners driven by environmental concerns, but to reach a larger group of consumers as well. Once grid parity[1] is reached, the main barrier to social acceptance of photovoltaics is removed.

The first houses with grid-connected photovoltaic systems appeared in the USA in the 1980s as a response to the oil crisis when utilities started experimenting with this new source of electricity. The built environment has remained a major application domain ever since; the physical infrastructure is almost readily available and it consists of a large market of residents. In the past couple of years, the number of solar homes has risen greatly. Estimations for the

[1] Grid parity occurs when a renewable source generates electricity at a cost that is less than or equal to the price of purchasing power from the electricity company or utility.

Photovoltaic Solar Energy: From Fundamentals to Applications, First Edition.
Edited by Angèle Reinders, Pierre Verlinden, Wilfried van Sark, and Alexandre Freundlich.

Companion website: www.wiley.com/go/reinders/photovoltaic_solar_energy

increase in the USA are from 30,000 solar homes in 2006 to 400,000 in 2013 (Wisland, 2014). In Germany, the other world-leading country for solar systems, about 510,000 domestic PV systems were installed in 2014 (Harvey, 2014). The market segment of housing is expected to continue to expand greatly. In its technology roadmap for the role of solar PV in the mitigation of climate change, the International Energy Agency suggests that if the right actions are taken, 20% of the 1700 GW installed globally in 2030 would consist of decentralized residential rooftop systems (International Energy Agency, 2014).

The residents of houses are unavoidably involved in the application of PV in housing in one way or another (van Mierlo, 2002, 2012). Hence, relevant questions are: How are they involved? What motivates them to buy or accept solar panels or a solar house or participate in a collective energy project with photovoltaics? What are their experiences in practice and how should these inform the design of follow-up initiatives, as well as further technology and product development? Unfortunately, the attention paid to the efficiency of solar panels, new generations of the technology, and policy measures by far exceeds the attention paid to the (prospective) users living in houses with solar panels.

This chapter takes stock of the current knowledge and understanding of people's[2] involvement in and experiences of photovoltaics,[3] which is important for the design of PV systems as well as wise market introduction and support. It takes an interest in everybody who is not professionally involved, as a 'prosumer', an active citizen or a gadget hunter, and more. Their diverse roles and related experiences are discussed individually, after which the main insights are synthesized. In this way, the chapter provides an overview of relevant research globally. Suggestions are provided on how to engage in a more effective learning process to improve further socio-technological development.

13.2.2 Residents Purchasing and Owning a PV System

People who own a single house can individually purchase a PV system to be mounted on or integrated into their rooftops or façades if the physical conditions and their financial resources allow it. In this role they are most frequently addressed in PV campaigns and policy. People living in an apartment or rental house are confined to becoming involved in collective initiatives (see Section 13.2.4) or buying a small add-on system. Whether residents consider the purchase of a PV system depends on the characteristics and design of the available PV systems, the characteristics of their houses, their demographic variables, as well as the more general local and national regulation and subsidy schemes.

In most studies on motivations to consider or buy solar systems, the financial aspects are defined as the main or one of the most important characteristics of a PV system, both to adopt and to reject PV (Palm and Tengvard, 2011; Rai *et al.* 2012; Vasseur and Kemp, 2015).[4] Faiers and

[2] The term "people" is used in order to include all private persons who in one way or another are involved with PV; as a user, tenant of a solar house, member of a critical energy collective and more. This general term is used because none of the well-known terms such as consumers, end-users, residents and citizens cover all these types of involvement, as will be shown in this chapter.

[3] With thanks to Tobias Hiemstra, the student-assistant, who did part of the literature search.

[4] Most of these studies that investigate why people purchase PV systems build on the adoption diffusion model of Rogers (1995). It explains the choices and behavior of individuals from the premise that they consciously consider alternative behavioral options and take a rational, conscious decision after weighing pros and cons.

Neame (2006), for instance, investigated why people who had taken energy-saving measures had not bought a PV system despite a grant for it being available. They turned out to be negative about affordability, the payback period, and the level of available grants.

PV systems, unless very small, are capital-intensive, because most expenditures are made upfront. The investment costs consist of the costs of the total system and mounting, installation and grid connection costs. The proportion of the latter, what are called the soft costs, has increased over the years alongside the steep decrease of the costs of the panels. How long it will take for the costs to be earned back depends on the yield and net metering rules and the financial measures to stimulate PV system adoption. These rules and policies have changed and can be expected to continue changing in the near future because of the political struggles around them (Huijben and Verbong, 2013). Net metering,[5] for instance, is criticized as being unfair to non-PV customers who are faced with increasing electricity prices.

An important question is how households themselves define the financial benefit or viability of a PV system, if this is, even for experts, a hard nut to crack. What people regard as financially acceptable is unfortunately not very clear, because most studies tend to ask the respondents about one financial aspect only. This issue is well addressed by Rai *et al.* (2012). Table 13.2.1 shows what financial metrics PV owners in Texas used. The payback period turned out to be considered the most important factor by many more respondents than the net present value. Their perceptions of the payback period, however, varied widely, so their accuracy should be questioned. Reported payback periods ranged from 1.5 to 35 years with the majority reporting a range between 7 and 10 years. Prospective owners might thus face uncertainty regarding the financial viability of the systems.

Interestingly, a study by Leenheer and colleagues (2011) of 2,047 Dutch households about their intention to generate their own power with micro-generation installations concluded that financial aspects were not at all important and environmental concerns were the most important driver. Positive PV characteristics arising from the literature are indeed its environmental benefits and the possibility to be self-sufficient, i.e. to generate one's own power (Claudy *et al.*, 2011; Palm and Tengvard, 2011; Vasseur and Kemp, 2015).

Other influential factors are information and knowledge. The decision-making process is resource-intensive, requiring effort, information and knowledge, in addition to money. In China, the low awareness of the existence and functioning of PV (26% among a representative sample of the residents of a Chinese province) is a major general barrier to adoption

Table 13.2.1 Use of financial metrics, more than one answer allowed ($N=360$)

Financial metric	(%)
Payback period	87
Internal Rate of Return	36
Net Present Value	12
None	7

Source: Rai *et al.* (2012).

[5] With net metering, the surplus PV electricity fed into the grid gets a credit equal to the general electricity price paid to the electricity company. It requires no special metering.

(Yuan *et al.*, 2011). However, people who are aware of the existence of PV are not necessarily interested in it. Vasseur and Kemp (2015) found that Dutch people who reject PV are more of the opinion that there is sufficient information available, than people interested in PV who perceive the available information as insufficient. The literature is rather ambiguous with regard to people's uncertainty before buying a PV system. Rai and colleagues (2012) conclude from a study of 360 PV owners in Texas that they seem to have had little uncertainty or concern about PV at the time of installation. They had spent a significant amount of time in the information search process (which had taken almost 9 months on average), which was mostly spent on understanding the finances and performance of PV systems. Whether uncertainty about financial aspects stops people considering buying PV is unknown.

In the process of decision-making, what are called peer effects play a role. Having solar houses in the vicinity stimulates the purchase of PV systems. Peer effects include both mere imitation and social learning. The latter involves communication with the neighbors about, for instance, trustworthy contractors. Studies based on geographical data, such as ZIP codes and street names, numbers of installations and installation dates show both types of peer effects (Bollinger and Gillingham, 2012; Müller and Rode, 2013). Remarkably, the peer effect was found to be greater with *less* marketing by PV companies (Bollinger and Gillingham, 2012).

The provision of good information would certainly not be a luxury, given that when testing the knowledge of 98 PV owners in Japan, Mukai *et al.* (2011) found that these users had many unrealistic expectations. Among others, 83 respondents were hoping the system would last more than 20 years, 44 respondents hoped for more than 30 years, and 14 respondents wanted more than 40 years. They did not expect it could fail in the meantime either and therefore widely overestimated the lifetime of the inverter and the warranty. Jager (2006) shows that the information and support meetings organized in Groningen for people interested in buying PV systems with the local subsidy had a strong positive effect on the diffusion of PV systems, probably due to the reduction of perceived technical and bureaucratic barriers, and in that way it was essential to help the PV owners to take advantage of the available grant. Another interesting finding is that people who knew more owners of PV systems perceived the bureaucratic procedures as less of a barrier after the support meetings than people who knew fewer PV system owners and stuck to their earlier stance.

13.2.3 Residents Commissioning a House Renovation

A decisive moment to consider a PV system is when people plan to renovate their house. In a noteworthy ethnographic study of why people in Australia chose a certain package of measures in an energy-efficiency enhancing renovation of their house, Judson and Maller (2014) reach some striking findings. The homeowners wanted their houses to provide an example to others: "We looked at all the materials that go into [the renovation] and the energy intensity of those ... it was really important...that my house is a shining beacon." The bulk of the changes brought about by the renovation, however, involved the extensive expansion of interior space with a second story and additional bedrooms and bathrooms. While overall the narratives of environmental sustainability were about incorporating products and technologies into renovations, in fact, something else happened. Accommodating daily life in terms of comfort, cleanliness and convenience took precedence in the home renovations. The authors thus illustrate how changes in the way the house is used in everyday life shapes the renovation practices. The boundary between work and daily routines is becoming less distinct through the availability of information and communication technologies (ICT) and more flexible working arrangements. As this "fluidity" often occurs alongside increases in floor area, dedicated space and more intensive use of space, and increasing numbers of appliances, it is likely to result in increased electrical consumption in the home.

13.2.4 People Receiving a PV System

Many people become the user of a PV system without deciding to buy it. For this phenomenon, I use the term *reception*, because it better covers the way in which they become involved with PV than the popular theoretical term adoption, established by Rogers (1995), which refers to an active, rational, individual decision-making process. Reception occurs when people live in a rental house, where the housing association has decided to adopt PV project-wide. Also buying a solar house, however, is often a matter of reception. In general, when people move to a new house with solar panels, e.g., when project developers have implemented PV in large housing projects, or when they move to an existing house which happens to have PV installed, other characteristics of the house are much more important than the PV system, such as the location, the total price, and the size (van Mierlo, 2002). Exceptions to this general tendency are people who deliberately seek a high performance, energy-neutral or sustainable house.

It may hence happen that the residents are not positively inclined toward the PV system. In a representative sample of Dutch inhabitants, it was found that 4.6% owned a solar house. More than half of them did not have a positive attitude towards PV, but were either neutral or negative about it (Vasseur and Kemp, 2015). Tenants in social housing were found to be indifferent about PV if the yield was used for the property owner's supply (Wheal *et al.*, 2004). This adverse effect can be expected to occur more often with the increase in the number of solar houses and may threaten the social acceptance of PV.

13.2.5 Owners Using a PV System

A recent IEA report (International Energy Agency, 2014) reports the following about the functioning of residential PV systems: "Most common defects were broken interconnections, solder bonds and diodes, or encapsulant discoloration or delamination. Other problems arose because local installers lacked the required skills or the initial design was poor." Other reports point to the inverters as the weakest part of the system, with 10%, 17% and 20% replacements respectively (Jablonska *et al.*, 2005), Kato (2009) in Mukai *et al.* (2011), and Oozeki *et al.* (2010) in (Mukai *et al.* 2011). A Dutch energy consultancy concludes in addition that active maintenance still gets too little attention in the Netherlands, especially for the relatively small residential systems (Goud *et al.*, 2012).

Strikingly little is known about what these technical problems mean from a user's perspective. Rai *et al.* (2012) describe in the aforementioned study that most of the PV owners in Texas were satisfied with the performance, operation, maintenance, and financial attractiveness of their systems after the installations. Almost all stated that installing a PV system was a wise decision. Mukai *et al.* (2011) found that those Japanese PV owners who perceived financial profits from the residential PV system, due to decreasing grid electricity charges or selling PV electricity, tended to be satisfied with the use of the product. However, these PV owners also had a lack of understanding regarding the technical aspects of the PV technology necessary for maintenance, lifetime use, and related financial returns. For this reason, the system's reliability and failure risk should be addressed more to avoid problems that may hamper a sound embedding of PV systems in housing.

Generally, the installation of a PV system seems to stimulate households to reduce their overall electricity consumption and to shift the demand to the peaks of production by the systems. At least, so is reported by the owners themselves (Keirstead, 2007; Rai *et al.*, 2012). In the study by Rai *et al.* (2012) nearly 46% of the PV owners reported that their total electricity consumption is "lower or much lower" after the installation and 34% reported load shifting. Keirstead (2007), who investigated changes in electricity consumption on the basis of the installation of insulation measures, the efficiency of appliances, and the use of efficient lighting, found significant differences to the respondents' pre-PV situation in two measures: the use of a green electricity tariff (more than half of the respondents) and the use of efficient lighting (on average around half of the lighting points). Significant changes in appliance efficiency were not seen, arguably because of the slow turnover in these stocks (75% of respondents had owned their PV systems for less than 2 years). The average of the self-assessed overall energy saving was 5.6%, weighed against certainty. In this study, 43% of the PV owners thought they had saved a little or a lot of energy. The same number of respondents reported some form of load shifting,[6] especially when people were at home during the day or people spending a lot of time using appliances.

However, in a study based on in-depth monitoring of nine low-energy social housing units equipped with PV systems, commissioned in 2004, it is concluded that electricity consumption increased over the 12-month period studied (Palm and Tengvard, 2011). Only one of the nine houses, the one with the highest electricity consumption, showed a slight negative gradient. The three high-energy consumers showed a rise of 3%; the six low-to-medium energy consumers showed an average increase of 34%. This seemed to be due to large screen televisions and computers with broadband connections. Several houses had added a second chest freezer, sometimes placed in the sun room. Others had even exchanged the low energy lighting with what they perceived to be an aesthetically pleasing alternative, albeit less efficient.

Whether and why people tend to save electricity or shift demand after the installation of a PV system hence is not clear. Energy saving as well as rebound effects[7] are observed. Meters, performance monitors, and additional information seem to have an important influence on these behavioral responses and hence may provide a trigger to stimulate electricity saving. The development of this interface between the PV system and the user is bound to be influenced by the recent introduction of smart meters.

Interfaces, such as smart meters, can only be expected to have a positive influence if they seriously engage with people's lifestyles. The commonly supplied smart monitoring devices only generate data about energy and sometimes finances and emissions. Studies in the UK and in the Netherlands show that the recent approaches to smart meters and grids are driven by technical and financial concerns and hardly stimulate an active role of end-users to save energy or shift electricity consumption to times when PV energy is available (Geelen *et al.*, 2013; Pullinger *et al.*, 2014). Pullinger *et al.* (2014) suggest developing and using alternative

[6] Moving household electricity use from hours of collective peak demand, to off-peak hours of the day.

[7] The common understanding of the rebound effect in energy management relates to behavioral responses to energy-saving measures because of which actual savings are less than expected savings or even negative.

feedback devices instead. These should support forms of reflection about the relationships between energy-using practices, personal meanings and values, and options for change, given wider societal constraints. This could be done among other things by disaggregating total electricity use to daily household practices (rather than by appliance), and targeting not only financial and environmental motivations, but meanings of comfort and cleanliness as well. In addition, Geelen *et al.* (2013) recommend designing smart grids that support the interaction between users and the formation of communities for the joint purchase and maintenance of PV systems and local energy management.

13.2.6 Citizens Participating in Collective Initiatives

While in the sections above, the focus was on people's involvement on an individual basis, there are several types of community-based or collective initiatives in which people can engage. These initiatives have proliferated greatly in the past decade. Although most authors claim collective initiatives are relatively new, Dewald and Truffer (2011) show how solar initiatives constituted by local citizen associations were actually the basis of the success of the Feed-in tariff[8] in Germany.

Four main types of collective initiatives exist, distinguished on the basis of the types of networks and internal relations (Schwencke, 2012; Huijben and Verbong, 2013):

1. *Collective buying initiative*: The initiatives to co-ordinate potential purchasers are taken by PV suppliers, regional and local governments, companies for their own employees, or environmental organizations. By collectively buying, installing, maintaining and insuring systems, costs can be reduced as well as the perceived technological risks (Huijben and Verbong, 2013). This type of initiative is to be found in Europe predominantly.
2. *Community shares*: In this model, multiple users, lacking the proper on-site conditions for solar energy or the financial capacity to invest in a full individual PV system on their own house, purchase a portion of their electricity from a facility located off-site. Again costs and risks may be reduced in this way. Examples include both local and national initiatives. Legislation to enable off-site or "virtual net metering" is an important condition for this model.
3. *Third party ownership*: PV companies own and operate customer-sited PV systems and either lease PV equipment or sell PV electricity to the building occupant. Potential advantages for people are the (partial) absence of investments costs and that they do not have to worry about the technological aspects and risks. Third-party PV systems represent a rapidly growing market segment in the USA especially, and are emerging in Europe as well. In California, these systems grew from 9% of residential PV installations during the first quarter of 2009 to 36% of residential PV installations during 2011 for projects tracked in the California Solar Initiative database (Drury *et al.*, 2012). By controlling the numbers of PV adopters using several demographic and voting characteristics of the population, it is concluded that third party PV products are increasing the PV market demand because they reach new customers who are younger, less educated and with a lower income.

[8] A feed-in tariff is a policy mechanism to stimulate investments in renewable energy technologies with defined payments for the generated energy. The tariffs may change over time to follow or stimulate the decrease in investments costs.

4. *Local renewable energy cooperatives*: In this variant of the community shares model, the cooperative (informal or officially registered) conducts several utility-related activities, such as the collective purchase of green energy, the local production of renewable energy, supply to the local community, financing of or participating in renewable energy projects, and supporting energy saving within the community. The wider goals and trading of energy distinguish them from the community shares model. In this model, the energy users are expected to have more autonomy than as a regular customer of the large energy-supplying companies.

Drivers for people to set up a local energy initiative vary from a concern about energy prices, a desire to have control over their own energy supply, a wish to jointly save energy, a concern about the environment, a wish to realize more efficient energy generation, to improve their quality of life, and a dissatisfaction with the energy services of large companies (Oostra and Jablonska, 2014). While collective initiatives theoretically are assumed to help overcome important barriers for the diffusion of residential PV systems, such as risks, high investment costs and uncertainty, recent experiences show this is not self-evident. Regarding the functioning of a collective purchase initiative, problems mentioned are, among others, lengthy periods of time to secure the electrical connection, hidden additional fees, and arduous administrative processes (Adachi, 2009). Schwencke (2012) also reports that after-sales care and warranty conditions need more attention. Regarding community shares, the results of a German study on two projects suggest that acceptance of renewable energy is lower if the co-owners are not part of the same local community and hence do not benefit from the generated income (Musall and Kuik, 2011). Therefore, it is recommended to let the 'community of interest' overlap with the 'community of location'. Regarding the local renewable energy cooperatives, research on their operation and performance is needed to evaluate whether they are indeed radically differently organized than the current unsustainable system of energy provision and whether citizens are actively setting the agenda and making decisions.

Hisschemöller (2012) is quite critical about the existing collective initiatives. On the basis of an evaluation of Dutch initiatives, he states that while they have proven to be powerful in shaping people's mind sets, they tend to follow a step-by-step approach, primarily focusing on the "low hanging fruit." Because they are likely to encounter resistance from powerful energy companies, they face an inherent dilemma:

> They can focus on strengthening their own boundary role [of establishing clear boundaries between the innovation promoted by critical citizens and the regular discourse of the energy institutions, BVM] as a resource for effectively mobilising end users, but then they may lose access to the networks of influence. Or they can avoid the knowledge conflict inherent to innovation by focusing on relatively safe projects (solar PV), but safeguard access to policy networks as a resource for their clientele.
>
> (Hisschemöller and Sioziou, 2013)

13.2.7 Synthesis and Conclusion

The overview of the studies presented above provides insights about the conditions and reasons for people becoming involved with PV or ignoring it, as well as their decision-making processes and underlying values, albeit not unambiguously. The knowledge and understanding deriving from the studies do not lead to easily applicable rules of thumb either for governments regarding the way to support PV system uptake and development, or for marketers on how to put PV systems onto the market, or for researchers and engineers about how to improve the design of PV systems.

The insights also provide reasons for concern:

1. There is evidence that without other positive developments, the decline of costs and prices will not immediately lead to an enthusiastic response by the public, notwithstanding their general support for solar energy (Dewald and Truffer, 2011; Vergragt and Brown, 2012).
2. Little seems to have been learned about the components of a PV system that are essential in the interaction with the (prospective) users, despite more than 20 years of experience in practice. Inverters are still a weak spot and maintenance services are still underdeveloped.
3. With a wider diffusion of PV systems, more people might unwillingly move to a solar home, and hence undermine the social acceptance of PV and the positive peer effect.
4. It seems that a focus on PV as an isolated product hampers realizing the wider goals of climate mitigation and transition to renewable energy provision (Hisschemöller and Sioziou, 2013; Judson and Maller, 2014).

A general insight is the importance of going beyond regarding people as the users at the threshold of purchasing their own PV system, as they are often held to be. The ways in which people are currently involved in PV adoption demonstrate great diversity. People's involvement in the PV-related activities (of buying, maintaining, using, etc.) ranges from situations in which their autonomy is high, to situations in which their autonomy is low, i.e. they are dependent on other actors. Another difference in the types of involvement relates to whether the choice for PV adoption is an isolated one or embedded in a wider process (see Figure 13.2.1).

Figure 13.2.1 shows that in many of those situations people cannot be expected to act as rational individuals knowing the advantages and disadvantages very well in advance, and weighing up the pros and cons before buying a PV system, as is presumed in theoretical models like the renowned and often applied adoption-diffusion model of Rogers (1995). This is because they are often either not fully autonomous in their decision-making or their choice

		Autonomy		
		Low	Middle	High
Embeddedness PV choice	High	New housing projects existing solar homes social housing	Local renewable energy cooperatives	Retrofitting
	Low		Collective buying initiative third party ownership community shares	Individual PV-system

Figure 13.2.1 The diversity of people's involvement with PV in housing

of PV system is part of a much wider process of changing their daily practices. At the one extreme, we find homeowners who buy a PV system (or tenants of rental housing buy a small system). At the other extreme, are people who just happen to become involved with PV, because, for instance, they have bought an attractive house regardless of the solar panels. In the middle, are the diverse collective initiatives that either focus primarily on PV systems or on PV embedded in a local renewable energy provision system. It needs to be acknowledged that PV systems are fusing with people's everyday lives, their values as well as standards, national regulations, and social norms. An integrated perspective on people's involvement in residential PV is needed and may be more effective in stimulating social acceptance.

The developments are in a flux, with new types of collective initiatives emerging in the different countries. The supply of PV is not definite either, the available promoted PV systems vary greatly per project or initiative and will keep on changing in the near future. Hence, although it is suggested that the PV market has become mature in some countries, it would be impossible to claim that clear-cut, regular products have been developed for identifiable market segments. Relevant knowledge and learning experiences are scattered around the globe. They depend on the type of PV system, the ownership relations of the house and the system, the physical infrastructure, the local and national financial support systems and regulations, with regard to quality and price, the policies of utilities, and more. For this reason, it is impossible to draw generic lessons about how PV systems could best be introduced in housing by taking into account people's roles, values and experiences, and how to ensure PV use will contribute to the achievement of societal sustainability goals. It is therefore important to find ways to learn about the PV technology in its endless variety of applications as well as system designs. What can be done is to learn from experiences in a specific situation by explicating them and translating them to other, new geographical and social contexts (van Mierlo, 2012).

Engineers and researchers are recommended to engage in such a learning-by-doing process in the following ways:

1. *Involve solar home residents, future prosumers and the like, in the design and preparation of any large project.* Many other stakeholders claim to represent people's interests, but tend to simplify them and see these only from their own perspective. (Prospective) users' involvement helps to better explore possibilities and alternative options and anticipate constraints and unintended effects.
2. *Actively elicit people's experiences after installation*, to learn from them for the further development and improvement of PV systems. People seldom respond as designers anticipated, hence their personal and active feedback is necessary.
3. *Collaborate with critical social scientists.* They provide integrated theoretical perspectives and research methods to bring about insights into how the involvement of citizens, users, homeowners and others is embedded in people's everyday life and in wider value-driven decisions and how they act as community members in the collective initiatives. Relevant theories include those emphasizing regular social practices, lifestyles, and the embedding of consumption in wider socio-technological systems, some of which have been used in the studies mentioned above. They may help to overcome the limitations of the adoption-diffusion model and offer a more profound understanding of people's involvement with PV in housing and unexpected negative effects, like the rebound effect.
4. *Collaborate with strategic communication specialists.* They will, rather than just promote the PV products, look critically from the outside to the supplying company from a long-term perspective and hence help to keep the PV producers' and suppliers' license to produce.

References

Adachi, C.W.J. (2009) The adoption of residential solar photovoltaic systems in the presence of a financial incentive: a case study of consumer experiences with the renewable energy standard offer program in Ontario (Canada), University of Waterloo.

Bollinger, B. and Gillingham, K. (2012) Peer effects in the diffusion of solar photovoltaic panels. *Marketing Science*, **31** (6), 900–912.

Claudy, M.C., Michelsen, C. and O'Driscoll, A. (2011) The diffusion of microgeneration technologies – assessing the influence of perceived product characteristics on home owners' willingness to pay. *Energy Policy*, **39** (3), 1459–1469.

Dewald, U. and Truffer, B. (2011) Market formation in technological innovation systems—diffusion of photovoltaic applications in Germany. *Industry and Innovation*, **18** (03), 285–300.

Drury, E., Miller, M., Macal, C.M. *et al.* (2012) The transformation of southern California's residential photovoltaics market through third-party ownership. *Energy Policy*, **42** (0), 681–690.

Faiers, A. and Neame, C. (2006) Consumer attitudes towards domestic solar power systems. *Energy Policy*, **34** (14), 1797–1806.

Geelen, D., Reinders, A., and Keyson, D. (2013) Empowering the end-user in smart grids: recommendations for the design of products and services. *Energy Policy*, **61** (0), 151–161.

Goud, J., Alsema,, E., and Nuiten, P. (2012) *Goede voorbeeldprojecten op PV-gebied.* W/E Adviseurs, Utrecht/Eindhoeven.

Harvey, F. (2014) UK should have 10 million homes with solar panels by 2020, experts say, *The Guardian*, 29 January.

Hisschemöller, M. (2012) Local energy initiatives cannot make a difference, unless .... *Journal of Integrative Environmental Sciences*, **9** (3), 123–129.

Hisschemöller, M. and Sioziou, I. (2013) Boundary organisations for resource mobilisation: enhancing citizens' involvement in the Dutch energy transition. *Environmental Politics*, **22** (5), 792–810.

Huijben, J.C.C.M. and Verbong, G.P.J. (2013) Breakthrough without subsidies? PV business model experiments in the Netherlands. *Energy Policy*, **56** (0), 362–370.

International Energy Agency (2014) Technology Roadmap, Solar Photovoltaic Energy, 2014 edition. International Energy Agency, Paris.

Jablonska, B., Kooijman-van Dijk, A., Kaan, H. *et al.* (2005) PV-prive project at ECN: five years of experience with small-scale AC module PV systems. Paper presented at the 20th European Photovoltaic Solar Energy Conference and Exhibition.

Jager, W. (2006) Stimulating the diffusion of photovoltaic systems: a behavioural perspective. *Energy Policy*, **34** (14), 1935–1943.

Judson, E.P. and Maller, C. (2014) Housing renovations and energy efficiency: insights from homeowners' practices. *Building Research & Information*, **42** (4), 501–511.

Keirstead, J. (2007) Behavioural responses to photovoltaic systems in the UK domestic sector. *Energy Policy*, **35** (8), 4128–4141.

Leenheer, J., de Nooij, M. and Sheikh, O. (2011) Own power: motives of having electricity without the energy company. *Energy Policy*, **39** (9), 5621–5629.

Mukai, T., Kawamoto, S., Ueda, Y. *et al.* (2011) Residential PV system users' perception of profitability, reliability, and failure risk: an empirical survey in a local Japanese municipality. *Energy Policy*, **39** (9), 5440–5448.

Müller, S. and Rode, J. (2013) The adoption of photovoltaic systems in Wiesbaden, Germany. *Economics of Innovation and New Technology*, **22** (5), 519–535.

Musall, F.D. and Kuik, O. (2011) Local acceptance of renewable energy: a case study from southeast Germany. *Energy Policy*, **39** (6), 3252–3260.

Oostra, M. and Jablonska, B. (2014) Understanding local energy initiatives and preconditions for business opportunities. *Policy Studies*, **2013**: 2012.

Palm, J. and Tengvard, M. (2011) Motives for and barriers to household adoption of small-scale production of electricity: examples from Sweden. *Sustainability: Science, Practice, & Policy*, **7** (1), 6–15.

Pullinger, M., Lovell, H., and Webb, J. (2014) Influencing household energy practices: a critical review of UK smart metering standards and commercial feedback devices. *Technology Analysis & Strategic Management*, 1–19.

Rai, V., McAndrews, K. and Offshore, W. (2012) Decision-making and behavior change in residential adopters of solar PV. *Proceedings of the World Renewable Energy Forum*, Denver, Colorado.

Rogers, E.M. (1995) *Diffusion of Innovations*. 4th edition. The Free Press, New York.

Schwencke, A.M. (2012) *Energieke BottomUp in Lage Landen – De Energietransitie van Onderaf, Over Vrolijke energieke burgers, Zon- en windcoöperaties, Nieuwe nuts*. Leiden, AS I-Search.
van Mierlo, B. (2002) *Kiem van maatschappelijke verandering : verspreiding van zonnecelsystemen in de woningbouw met behulp van pilotprojecten* [The seed of change in society. Diffusion of solar cell systems in houding by means of pilot projects]. Aksant, Amsterdam.
van Mierlo, B. (2012) Convergent and divergent learning in photovoltaic pilot projects and subsequent niche development. *Sustainability: Science, Practice, & Policy*, **8** (2), 4–18.
Vasseur, V. and Kemp, R. (2015) The adoption of PV in the Netherlands: a statistical analysis of adoption factors. *Renewable and Sustainable Energy Reviews*, **41** (0), 483–494.
Vergragt, P.J. and Brown, H.S. (2012) The challenge of energy retrofitting the residential housing stock: grassroots innovations and socio-technical system change in Worcester, MA. *Technology Analysis & Strategic Management*, **24** (4), 407–420.
Wheal, R., Fulford, D., Wheldon, A. and Oldach. R. (2004) Photovoltaics (PV) in social housing. *International Journal of Ambient Energy*, **25** (1), 12–18.
Wisland, L. (2014) How many homes have rooftop solar? The number is growing… *The Equation* 2015.
Yuan, X., Zuo, J., and Ma. C. (2011) Social acceptance of solar energy technologies in China: end users' perspective. *Energy Policy* **39** (3), 1031–1036.

13.3

Life Cycle Assessment of Photovoltaics

Vasilis Fthenakis
Center of Life Cycle Analysis, Columbia University, New York, USA; Photovoltaic Environmental Research Center, Brookhaven National Laboratory, Upton, NY, USA

13.3.1 Introduction

Life cycle assessment (LCA) is a comprehensive framework for quantifying the environmental impacts caused by material and energy flows in each and all of the stages of the "life cycle" of a product or an activity. It describes all the life stages, from "cradle to grave," thus from raw materials extraction to the end of the product's life. The cycle typically starts from the mining of materials from the ground and continues with the processing and purification of the materials, manufacturing of the compounds and chemicals used in processing, manufacturing of the product, transport, installation if applicable, use, maintenance, and eventual decommissioning and disposal, and/or recycling. To the extent that materials are reused or recycled at the end of their first life into new products, then the framework is extended from "cradle to cradle." The most common metrics used in comparative life cycle evaluations of the environmental impacts of energy systems are Energy Payback Time (EPBT), Energy Return on Energy Investment (EROI), Greenhouse Gas Emissions, Toxic Emissions, Land Use, and Water Use.

13.3.2 Methodology

The methodology employed in PV life cycle assessment (LCA) studies should conform to the ISO Standards 14040 and 14044 (ISO 2006a, 2006b) and should adhere to the IEA PVPS Guidelines (Fthenakis *et al.*, 2011a). The ISO Standards prescribe the four steps for conducting

---

Photovoltaic Solar Energy: From Fundamentals to Applications, First Edition.
Edited by Angèle Reinders, Pierre Verlinden, Wilfried van Sark, and Alexandre Freundlich.

Companion website: www.wiley.com/go/reinders/photovoltaic_solar_energy

an LCA (1) goal and scope definition; (2) life cycle inventory (LCI); (3) life cycle impact assessment (LCIA); and (4) interpretation. The first step calls for a clear definition of the research objective and the systems boundaries used for the analysis. The inventory step quantifies the flows of materials, energy, and emissions in each stage of the life cycle of a PV system. In LCIA, the impact on the environment of energy consumption, resource consumption, pollutant emissions, and greenhouse gas emissions is quantified and their cumulative impact on the environment is estimated. The fourth step of the LCA as defined by the ISO is the interpretation of the results; this is extremely important as different assumptions can result to very divergent results. To this end, the International Energy Agency (IEA) Task 12 (Fthenakis *et al.*, 2011a) adds specificity regarding what major assumptions are valid in stationary PV LCAs and on the interpretation of the results.

13.3.3 Cumulative Energy Demand (CED) during the Life of a PV System

The life cycle stages of photovoltaics involve: (1) the production of raw materials; (2) their processing and purification; (3) the manufacture of solar cells, modules, and balance of system (BOS) components; (4) the installation and operation of the systems; and (5) their decommissioning, disposal, or recycling (Figure 13.3.1). Typically, separate LCAs are undertaken for the modules and the BOS components (inverters, transformers, mounting, supports, and wiring), as the module's technologies evolve more rapidly and entail more options than do the BOS structures.

The life cycle Cumulative Energy Demand (CED) (Frischknecht *et al.*, 2007) of a PV system is the sum total of the (renewable and non-renewable) primary energy harvested from the

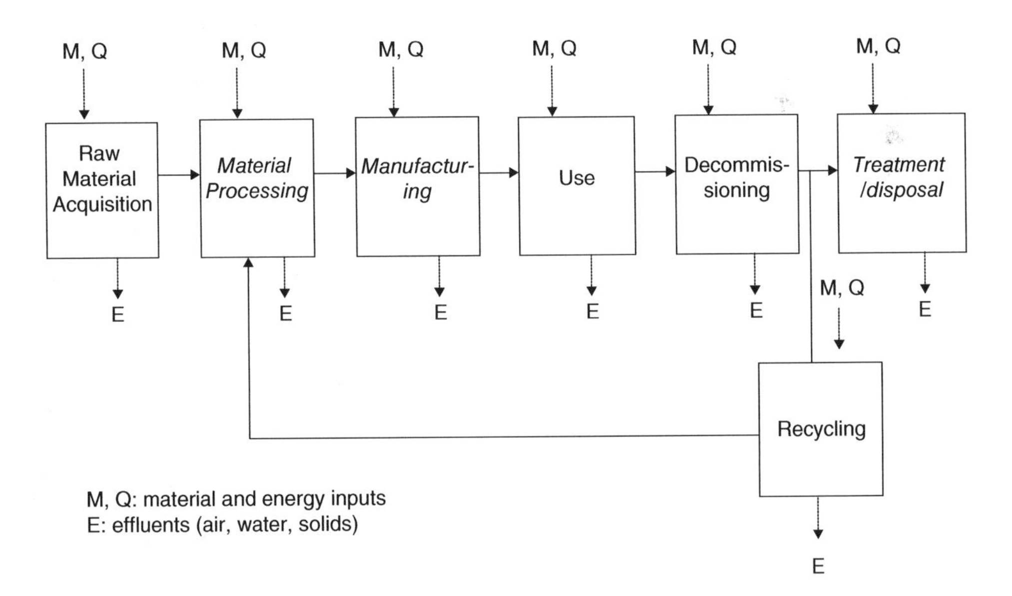

Figure 13.3.1 Flow of the life cycle stages, energy, materials, and wastes for PV systems

geo-biosphere in order to supply the direct energy (e.g., fuels, electricity) and material (e.g., Si, metals, glass) inputs used in all its life cycle stages (excluding the solar energy directly harvested by the system during its operation). Thus,

$$\mathrm{CED}\left[\mathrm{MJ}_{\text{PE-eq}}\right] = E_{mat} + E_{manuf} + E_{trans} + E_{inst} + E_{EOL} \tag{13.3.1}$$

where:

E_{mat} [$MJ_{PE\text{-}eq}$]: Primary energy demand to produce materials comprising PV system
E_{manuf} [$MJ_{PE\text{-}eq}$]: Primary energy demand to manufacture PV system
E_{trans} [$MJ_{PE\text{-}eq}$]: Primary energy demand to transport materials used during the life cycle
E_{inst} [$MJ_{PE\text{-}eq}$]: Primary energy demand to install the system
E_{EOL} [$MJ_{PE\text{-}eq}$]: Primary energy demand for end-of-life management

The life cycle non-renewable Cumulative Energy Demand (NR-CED) is a similar metric in which only the non-renewable primary energy harvested is accounted for; details are given elsewhere (Fthenakis *et al.*, 2011a).

13.3.3.1 Energy Payback Time (EPBT)

Energy payback time (EPBT) is defined as the period required for a renewable energy system to generate the same amount of energy (in terms of equivalent primary energy) that was used to produce (and manage at end of life) the system itself (Fthenakis *et al.*, 2011a).

$$\mathrm{EPBT}\left[\text{years}\right] = \mathrm{CED} / \left(\left(E_{agen} / \eta_G\right) - E_{O\&M}\right) \tag{13.3.2}$$

where:

E_{agen} [MJ_{el}/yr]: Annual electricity generation
$E_{O\&M}$ [$MJ_{PE\text{-}eq}$]: Annual primary energy demand for operation and maintenance
η_G [$MJ_{el}/MJ_{PE\text{-}eq}$]: Grid efficiency, i.e. the average life cycle primary energy to electricity conversion efficiency at the demand side

The annual electricity generation (E_{agen}) is converted into its equivalent primary energy, based on the efficiency of electricity conversion at the demand side, using the grid mix where the PV plant is being installed. Thus, calculating the primary-energy equivalent of the annual electricity generation (E_{agen}/η_G) requires knowing the life cycle energy conversion efficiency (η_G) of the country-specific energy-mixture used to generate electricity and produce materials. The average η_G for the United States of America and Western Europe are respectively approximately 0.30 and 0.31 (Franklin Associates, 1998; Dones *et al.*, 2000).

13.3.3.2 Energy Return on Investment (EROI)

The CED of a PV system may be regarded as the energy investment that is required in order to be able to obtain an energy return in the form of PV electricity. The overall Energy Return On (Energy) Investment may thus be calculated as:

$$EROI_{PE\text{-}eq}\left[MJ_{PE\text{-}eq} / MJ_{PE\text{-}eq}\right] = T / EPBT \tag{13.3.3}$$

where T is the period of the system operation; both T and EPBT are expressed in the same units of time, for example years.

$EROI_{PE\text{-}eq}$ and EPBT provide complementary information. $EROI_{PE\text{-}eq}$ looks at the overall energy performance of the PV system over its entire lifetime, whereas EPBT measures the point in time (t) after which the system is able to provide a net energy return.

Further discussion of the EROI methodology can be found in an IEA PVPS Task 12 report (Raugei *et al.*, 2015). It is noted that some research groups are spreading invalid claims regarding PV as having a low EROI by using outdated data, worst conditions of deployment, and assigning financial burdens to the cumulative energy demand of PV. A critical overview of an on-going debate on this issue can be found in (Carbajales-Dale *et al.*, 2015).

13.3.3.3 Greenhouse Gas (GHG) Emissions and Global Warming Potential (GWP)

The overall Global Warming Potential due to the emission of a number of GHGs along the various stages of the PV life cycle is typically estimated using an integrated time-horizon of 100 years (GWP_{100}), whereby the following CO_2-equivalent factors are used: 1 kg CH_4 = 23 kg CO_2-eq, 1 kg N_2O = 296 kg CO_2-eq, and 1 kg chlorofluorocarbons = 4,600–10,600 kg CO_2-eq. It is noted that the CO_2-equivalent factor of CH_4 has recently been updated to 32; the effect of this update on the reported GHG has not been assessed yet. Electricity and fuel use during the production of the PV materials and modules are the main sources of the GHG emissions for PV cycles, and specifically the technologies and processes for generating the upstream electricity play an important role in determining the total GWP of PVs, since the higher the mixture of fossil fuels is in the grid, the higher are the GHG (and toxic) emissions. For example, during 1999–2002, the GHG emission factor of the average US electricity grid was 676 g CO_2-eq/kWh, that of Germany was 539 g CO_2-eq/kWh, and that of China was 839 g CO_2-eq/kWh (US Department of Energy, 2007).

In this chapter we present the most up-to-date estimates of EPBT, GHG emissions, and heavy metal emissions from the life cycles of the currently commercial PV technologies (e.g., mono-crystalline Si, multi-crystalline Si, and CdTe). Information on the production of solar grade silicon can be found in (Aulich and Schulze, 2002; Woditsch and Koch, 2002) and detailed LCIs can be found in an IEA PVPS Task 12 report (Fthenakis *et al.*, 2011b).

Information and details of energy and materials inventory for CdTe thin-film PV can be found in (Fthenakis *et al.* 2009a; Fthenakis *et al.*, 2009b; Fthenakis *et al.*, 20011b) and detailed assessments of both c-Si and CdTe can be found in (Fthenakis *et al.*, 2008; Kim *et al.*, 2012; Chu *et al.*, 2011; Wild-Scholten *et al.*, 2006; and Wild-Scholten, 2013). Balance of systems data is also included in the IEA PVPS report (Fthenakis *et al*, 2011b). Roof-top BOS data was compiled by Wild-Scholten *et al.* (2006), whereas utility scale, ground-mounted BOS data was compiled by Mason *et al.* (2006).

13.3.4 Results

13.3.4.1 Energy Payback Time and Energy Return on Investment

The energy payback time of photovoltaics has been reduced by almost two orders of magnitude over the last three decades, as material use, energy use, and efficiencies have been constantly improving; this trend is shown in Figure 13.3.2 (Fthenakis, 2012). For example, the cumulative energy demand (CED) energy used in the life cycle of complete rooftop Si-PV systems was 2,700 and 2,900 MJ/m^2, respectively, for multi-, and mono-crystalline Si modules, down from 5,000 and 2,700 in 2006 (Fthenakis *et al.*, 2009a). The EPBT of these systems is 1.8 years for module efficiencies of 13.2% and 14% correspondingly, in rooftop installations under Southern European insolation of 1,700 kWh/(m$^2 \cdot$ yr), with a grid efficiency (η_G) of 0.31 and a performance ratio (PR) of 0.75 (Figure 13.3.3). The corresponding $EROI_{PE\text{-}eq}$ ratio, assuming a 30-year lifetime, is 17. In these estimates, the BOS for rooftop application accounts for 0.3 years of EPBT, but there are different types of roof-top mounting systems with different energy burdens.

For CdTe PV, the primary energy consumption is 850 MJ/m^2 and 970 MJ/m^2, correspondingly based on actual production from First Solar's plant in Frankfurt-Oder, Germany, and Perrysburg, Ohio, USA. The energy burden at the US plant is higher than that in Germany because the former includes R&D and additional administrative functions. For insolation

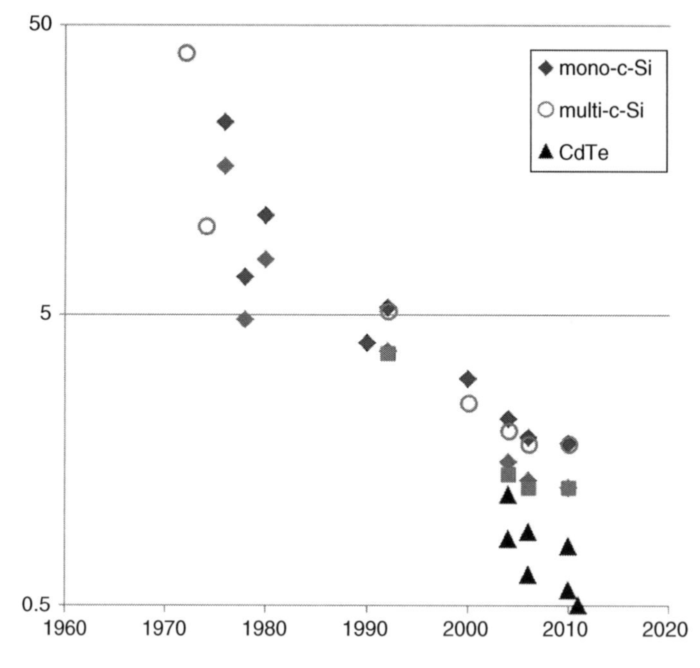

Figure 13.3.2 Historical evolution of Energy Payback Times. Energy payback times of various PV systems were reduced from about 40 years to 0.5 years from 1970 to 2010. The low numbers correspond to insolation of 2,400 kilowatt-hours per square meter per year (US-SW) and the high numbers correspond to insolation of 1,700 kilowatt-hours per square meter per year (Southern Europe)

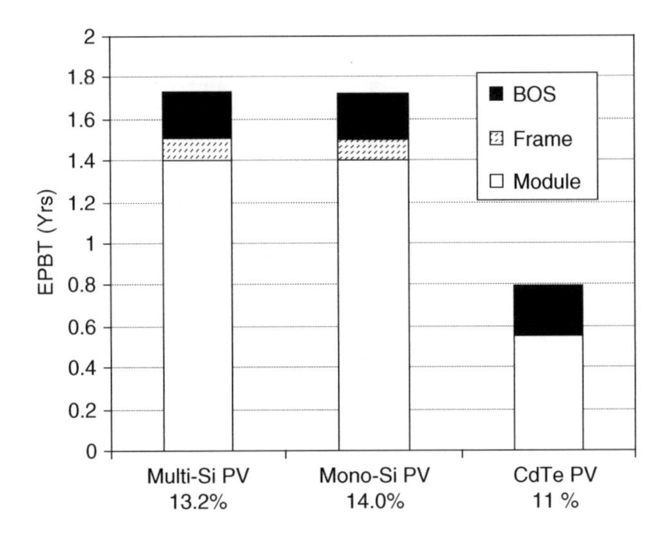

Figure 13.3.3 EPBT of PV systems: Rooftop-installed module with insolation =1,700 kWh/(m^2·yr) and performance ratio = 0.75; European production

levels of 1,700 kWh/(m^2 · yr) and a grid efficiency (η_G) of 0.31, the EPBT and $EROI_{PE\text{-}eq}$ values for the European production are: EPBT = 0.8 yrs, $EROI_{PE\text{-}eq}$ = 37.5; and for the US production: EPBT = 0.9 yrs, $EROI_{PE-eq}$ = 33.

EPBT decreases (and EROI increases) as the solar irradiation levels increases; for example, in Phoenix, Arizona, in the US South-west (latitude optimal irradiation of 2,370 kWh/(m^2 · yr)), the EPBTs of crystalline silicon- and cadmium telluride-PV in fixed tilt ground mount utility installations, respectively, are 1.3 and 0.6 years. In the highest solar irradiation regions (e.g., Northern Chile; irradiation of 4,000 kWh/(m2 · yr) on a 1-axis sun-tracking plane) (Fthenakis *et al.*, 2014), these systems are generating all the CED used in their life cycle in only 0.8 and 0.4 years. With an assumed life expectancy of 30 years, their EROI is in the range of 37–75.

13.3.4.2 Greenhouse Gas Emissions

The GHG emissions of PV systems with mono-crystalline Si modules are in the range of 29 g CO_2-eq./kWh for a rooftop application under Southern European insolation of 1,700 kWh/(m^2 · yr), and a performance ratio[1] (PR) of 0.75 (Figure 13.3.4). Notably, according to these estimates, the BOS for rooftop applications accounts for ~5 g CO_2-eq/kWh of GHG emissions. These calculations were based on the electricity mixture for the current (2009) production of Si feedstock that is powered by hydro-electricity.

The GHG emissions of CdTe PV systems total 17 g CO_2-eq/kWh for irradiation levels of 1,700 kWh/m^2·/yr (Fthenakis *et al.* 2008; Raugei and Fthenakis, 2010; Bhandari *et al.*, 2015). The manufacturing of the module and upstream mining/smelting/purification operations comprises most of the energy and greenhouse burdens. The BOS share was about 5 g CO_2-eq/kWh under 1,700 kWh of insolation, with 14% module efficiency, and performance ratio of 0.75.

[1] Ratio between the DC rated and actual AC electricity output.

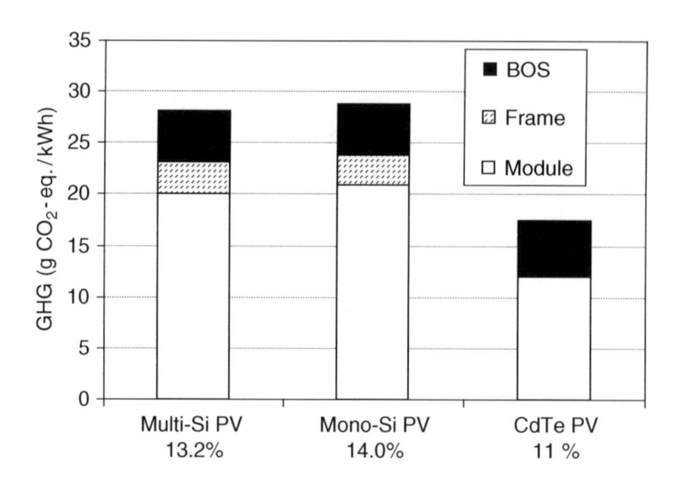

Figure 13.3.4 GHG emissions of PV systems:rooftop-installed module with insolation = 1,700 kWh/(m^2·yr) and performance ratio = 0.75; European production

All the recently published LCA studies found CdTe PV having the lowest EPBT and GHG emissions among the currently commercial PV technologies, as it uses less energy in its material processing and module manufacturing (Raugei and Fthenakis, 2010; Bhandari *et al.*, 2015). This is explained by the lower thickness of the high purity semiconductor layer (i.e., 2–3 um for CdTe versus 15–200 um for c-Si) and by the fact that CdTe module processing has fewer steps and is much faster than c-Si cell and module processing.

Figure 13.3.5 compares GHG emissions from the life cycle of PV with those of conventional fuel-burning power plants, revealing the environmental advantage of using PV technologies. The majority of GHG emissions come from the operational stage for the coal cycle, the natural gas cycle, and the oil-fuel cycle, while the material and device production accounts for nearly all the emissions for the PV cycles. With over 50% contributions, the GHG emissions from the electricity demand in the life cycle of PV are the input with the most impact. Therefore, the LCA results strongly depend on the available electricity mix. The GHG emissions from the nuclear fuel cycle mainly are related to the fuel production, i.e., mining, milling, fabrication, conversion, and the enrichment of uranium fuel. The details of the US nuclear fuel cycle are described elsewhere (Fthenakis and Kim, 2007).

Other comparisons between the life cycles of photovoltaics and conventional power generation cover land use (Fthenakis and Kim, 2009) and water use (Fthenakis and Kim, 2010). Accounting for the land occupation in coal mining, it is shown that the life cycles of PV in the US-SW occupy about the same or less land, and orders of magnitude less water, on an electricity-produced (GWh) basis, than the average coal-based power generation.

13.3.4.3 Toxic Gas Emissions

The emissions of toxic gases (e.g., SO_2, NO_x) and heavy metals (e.g., As, Cd, Hg, Cr, Ni, Pb) during the life cycle of a PV system are largely proportional to the amount of fossil fuel consumed during its various phases, in particular, processing and manufacturing PV materials. Figure 13.3.6 shows estimates of SO_2 and NO_x emissions. Heavy metals may be emitted

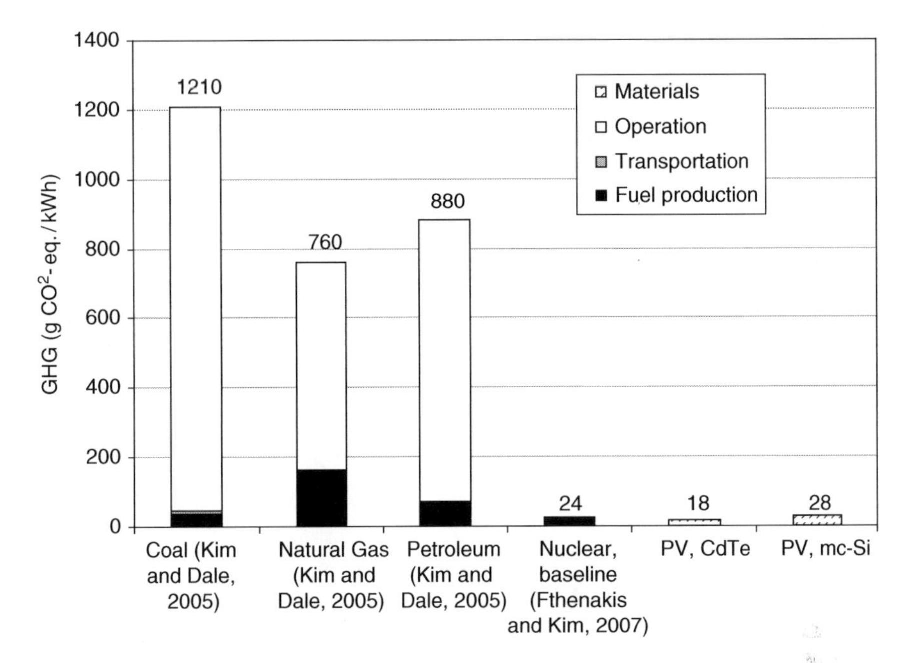

Figure 13.3.5 Comparison of GHG emissions from PV with those from conventional power plants (insolation of 1,700 kWh/(m^2·yr), performance ratio of 0.75

directly from material processing and PV manufacturing, and indirectly from generating the energy used at both stages. For the most part, they originate as trace metals in the coal used.

Direct emissions of cadmium in the life cycle of CdTe PV have been assessed in detail (Fthenakis, 2004). These experiments were designed to replicate average conditions, and the estimated emissions were calculated by accounting for US fire statistics pointing to 1/10,000 houses catching fire over the course of a year in the United States where most houses have wooden frames, by assuming that all fires involve the roof. The indirect Cd emissions from electricity usage during the life cycle of CdTe PV modules (i.e., 0.24 g/GWh) are an order of magnitude greater than the direct ones (routine and accidental) (i.e., 0.016 g/GWh) (Aulich and Schulze, 2002; ISO, 2006a).[2] Cadmium emissions from the electricity demand for each module were assigned, assuming that the life cycle electricity for the silicon-and CdTe-PV modules was supplied by the Union for the Coordination of Transmission of Energy's (European) grid. The complete life cycle atmospheric Cd emissions, estimated by adding those from the usage of electricity and fuel in manufacturing and producing materials for

[2]Indirect emissions of heavy metals result mainly from the trace elements in coal and oil. According to the US Electric Power Research Institute's (EPRI) data, under the best/optimized operational and maintenance conditions, burning coal for electricity releases into the air between 2–7 g of Cd/GWh (EPRI, 2002). In addition, 140 g/GWh of Cd inevitably collects as fine dust in boilers, baghouses, and electrostatic precipitators (ESPs). Furthermore, a typical US coal-powered plant emits per GWh about 1,000 tons of CO_2, 8 tons of SO_2, 3 tons of NOx, and 0.4 tons of particulates. The emissions of Cd from heavy-oil burning power plants are 12–14 times higher than those from coal plants, even though heavy oil contains much less Cd than coal (~0.1 ppm); this is because these plants do not have particulate-control equipment (EPRI, 2002).

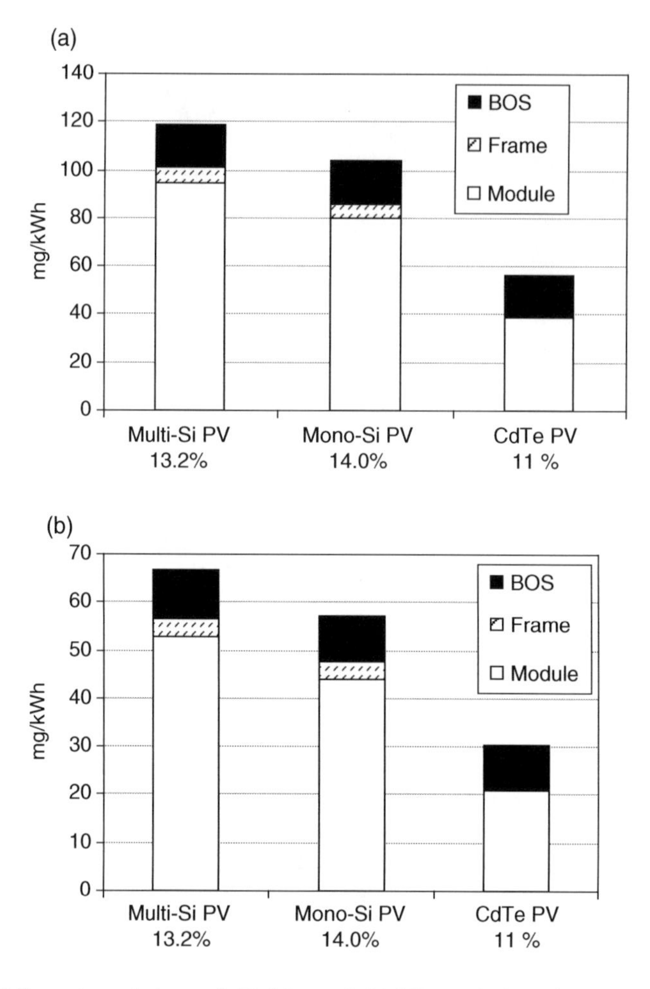

Figure 13.3.6 Life-cycle emissions of (a) SO_2, and (b) NOx emissions from silicon and CdTe PV systems. BOS includes module supports, cabling, and power conditioning. The estimates are based on rooftop-mounted installation, insolation of 1,700 kWh/(m^2·yr), performance ratio of 0.75, lifetime of 30 years, and European production with electricity supply from the UCTE grid

various PV modules and Balance of System (BOS), were compared with the emissions from other electricity-generating technologies (Figure 13.3.7) (Fthenakis *et al.*, 2008). Undoubtedly, displacing the others with Cd PV markedly lowers the amount of Cd released into the air. Thus, every GWh of electricity generated by CdTe PV modules can prevent around 5 g of Cd air emissions if they are used instead of, or as a supplement to, the UCTE electricity grid. In addition, the direct emissions of Cd during the life cycle of CdTe PV are 10 times lower than the indirect ones due to the use of electricity and fuel in the same life cycle, and about 30 times less than those indirect emissions from crystalline photovoltaics. The same applies to total (direct and indirect) emissions of other heavy metals (e.g., As, Cr, Pb, Hg, Ni); CdTe PV has the lowest cumulative energy demand and, consequently, the fewest heavy-metal emissions (Fthenakis *et al.*, 2008). Regardless of the particular PV technology, these emissions are

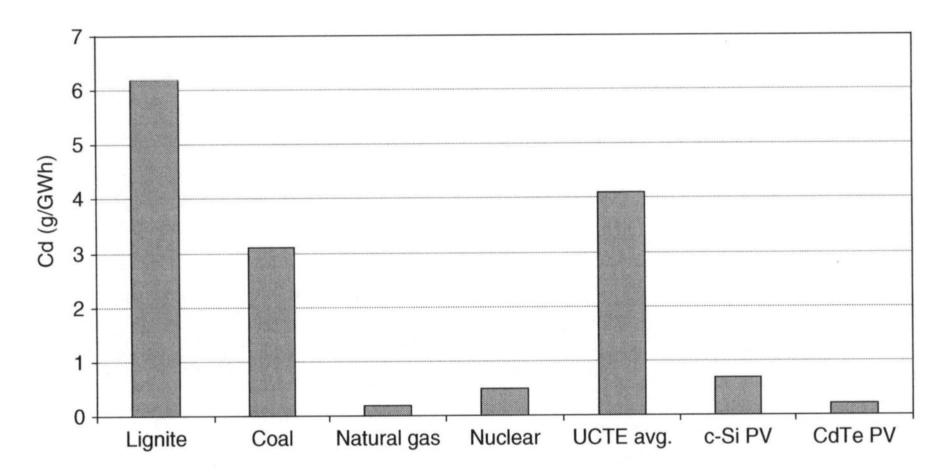

Figure 13.3.7 Life cycle atmospheric Cd emissions for PV systems from electricity and fuel consumption, normalized for a Southern Europe average insolation of 1,700 kWh/(m^2·yr), performance ratio of 0.75, and lifetime of 30 yrs

extremely small compared to the emissions from the fossil fuel-based plants that PV will replace. Furthermore, the external environmental costs of photovoltaics are negligible in comparison to the external costs of fossil fuel life cycles (Fthenakis and Alsema, 2006; Sener and Fthenakis, 2014; Fthenakis, 2015).

Ongoing LCA studies include evaluating replacing building materials with PV in building-integrated photovoltaics (BIPV) (Perez *et al.*, in pressin press), and recycling of PV modules at the end of their useful life (Fthenakis, 2000; Fthenakis and Wang, 2006; Fthenakis, 2012).

13.3.5 Conclusion

This chapter gives an overview of the life cycle environmental performance of photovoltaic (PV) technologies. Energy payback time (EPBT) is a basic metric of this performance; it measures the time it takes for a PV system to generate as much electricity as could be generated by the current electric mix, when using the same amount of primary energy that is used for the production of the PV system. The lower the EPBT, the lower will also be the emissions to the environment, because emissions mainly occur from using fossil-fuel-based energy in producing materials, solar cells, modules, and systems. These emissions differ in different countries, depending on that country's mixture in the electricity grid, and the varying methods of material/fuel processing. Under average US and Southern European conditions (e.g., 1,700 kWh/(m^2·yr)), the EPBT of crystalline Si-, and CdTe-roof-top PV systems, were estimated respectively to be 1.8 and 0.8 years correspondingly. Ground-mounted installations entail a greater amount of BOS and, therefore, have a longer EPBT. Under US SW irradiation (e.g., Phoenix, AZ, 2,370 kWh/(m^2·yr) at a fixed latitude tilt), the EPBT of ground-mounted installations is estimated to be 1.3 years and 0.6 years, respectively, for crystalline silicon and cadmium telluride PV. In the highest solar irradiation regions (e.g., Northern Chile; irradiation of 4,000 kWh/(m^2·yr) on a 1-axis sun-tracking plane) these numbers are reduced to 0.8 and 0.4 years correspondingly.

The environmental impacts of the lifecycle of photovoltaics as assessed by the common metrics of GHG emissions, toxic emissions, and heavy-metal emissions are very small in comparison to those of the power generation technologies they replace.

References

Aulich, H.A. and Schulze, F.-W. (2002) Crystalline silicon feedstock for solar cells. *Progress in Photovoltaics: Research and Applications*, **10**, 141–147.

Bhandari, K., Collier, J., Ellingson, R. and Apul, D. (2015) EPBT and EROI of PV: A systematic review and meta-analysis. *Renewable and Sustainable Energy Reviews*, **47**, 133–141.

Carbajales-Dale, M., Raugei, M., Barnhart, C.J., and Fthenakis V. (2015) Energy return on investment (EROI) of solar PV: an attempt at reconciliation. *Proceedings of the IEEE*, **103** (7), 995–999.

Chu, D., Donahue, P., Fthenakis, V. *et al.* (2011) Life cycle GHG emissions of crystalline-si photovoltaic electricity generation: systematic review and harmonization. *Journal of Industrial Ecology*, **16** (Supplement 1), S122–S135.

Dones, R. *et al.* (2000) *Sachbilanzen von Energiesystemen: Final report Ecoinvent, 2000*. vol. 6. 2003, Swiss Centre for Life Cycle Inventories, PSI, Dübendorf.

Electric Power Research Institute (EPRI) (2002) PISCES data base for US power plants and US coal. EPRI.

Franklin Associates (1998) *USA LCI Database Documentation*. Prairie Village, Kansas.

Frischknecht, R., Jungbluth, N., Althaus, H.-J. *et al.*, (2007) Implementation of life cycle impact assessment methods. Ecoinvent report No. 3, vol. 2.0. Swiss Centre for Life Cycle Inventories, Dübendorf.

Fthenakis, V.M. (2000) End-of-life management and recycling of PV modules, *Energy Policy*, **28** (14), 1051–1058.

Fthenakis, V.M. (2004) Life cycle impact analysis of Cadmium in CdTe PV production. *Renewable and Sustainable Energy Reviews*, **8** (4), 303–334.

Fthenakis, V.M. (2012a) Sustainability metrics for extending thin-film photovoltaics to terawatt levels. *MRS Bulletin*, **37** (4), 425–430.

Fthenakis V.M. (2012b) PV Energy ROI tracks efficiency gains, *Solar Today, American Solar Energy Society (ASES)*, **26** (4), 24–26. Available at: http://www.solartoday-digital.org/solartoday/201206#pg24

Fthenakis, V. (2015) Considering the total cost of electricity from sunlight and the alternatives, *Proceedings of the IEEE*, **103** (3), 283–286.

Fthenakis, V.M. *et al.* (2005) Emissions and encapsulation of cadmium in CdTe PV modules during fires. *Progress in Photovoltaics: Research and Applications*, **13**, 713–723.

Fthenakis, V. and Alsema, E. (2006) Photovoltaics energy payback times, greenhouse gas emissions and external costs: 2004–early 2005 status. *Progress in Photovoltaics: Research and Applications*, **14**, 275–280.

Fthenakis, V., Atia, A., Perez, M. *et al.* (2014) Prospects for photovoltaics in sunny and arid regions: a solar grand plan for Chile, Part I – investigation of PV and wind penetration, *Proceedings of the 40th IEEE Photovoltaic Specialists Conference, Denver, CO*, pp. 1424–1429.

Fthenakis, V., Frischknecht, R., Raugei, M. *et al.* (2011a) Methodology guidelines on life cycle assessment of photovoltaic electricity, 2nd edition. IEA PVPS Task 12, International Energy Agency Photovoltaic Power Systems Programme. Report IEA-PVPS T12-03:2011. Available a: http://www.iea-pvps.org

Fthenakis V.M. and Kim H.C. (2007a) CdTe Photovoltaics: life-cycle environmental profile and comparisons. *Thin Solid Films*, **515**, 5961–5963.

Fthenakis, V.M. and Kim H.C. (2007b) Greenhouse-gas emissions from solar electric- and nuclear power: a life-cycle study. *Energy Policy*, **35**, 2549–2557.

Fthenakis, V.M. and Kim, H.C. (2009) Land use and electricity generation: a life-cycle analysis. *Renewable & Sustainable Energy Reviews*, **13**, 1465–1474.

Fthenakis, V.M. and Kim, H.C. (2010) Life cycle uses of water in U.S. electricity generation. *Renewable & Sustainable Energy Reviews*, **14**, 2039–2048.

Fthenakis, V.M., Kim, H.C. and Alsema, E. (2008) Emissions from photovoltaic life cycles. *Environmental Science & Technology*, **42**, 2168–2174.

Fthenakis V., Kim, H.C., Frischknecht, R. *et al.* (2011b) Life cycle inventories and life cycle assessments of photovoltaic systems. International Energy Agency, Report IEA-PVPS T12-02:2011, October.

Fthenakis, V., Kim, H.C., Held, M., Raugei, M. and Krones, J. (2009a) Update of PV energy payback times and Life-Cycle Greenhouse Gas Emissions, paper presented at 24th European Photovoltaic Solar Energy Conference, Hamburg, Germany, 21–25 September.

Fthenakis, V.M. and Wang, W. (2006) Extraction and separation of Cd and Te from cadmium telluride photovoltaic manufacturing scrap. *Progress in Photovoltaics: Research and Applications*, **14**, 363–371.

Fthenakis, V., Wang, W., and Kim, H.C. (2009b) Life cycle inventory analysis of the production of metals used in photovoltaics. *Renewable and Sustainable Energy Reviews*, **13**: 493–517.

ISO (2006a) ISO 14040 — Environmental Management. Life Cycle Assessment. Principles and Framework. International Organization for Standardization.

ISO (2006b) ISO 14044 — Environmental Management. Life Cycle Assessment. Requirements and Guidelines. International Organization for Standardization.

Kim, H.C., Fthenakis, V., Choi, J.K. *et al.* (2012) Life cycle greenhouse gas emissions of thin-film photovoltaic electricity generation: systematic review and harmonization. *Journal of Industrial Ecology*, **16** (Supplement1), S119–S121.

Mason, J., Fthenakis, V.M., Hansen, T. and Kim, C. (2006) Energy pay-back and life cycle CO_2 emissions of the BOX in an optimized 3.5 MW PV installation. *Progress in Photovoltaics: Research and Applications*, **14**, 179–190.

Perez, M., Fthenakis, V., Kim, H.C., and Pereira, A., in press. Façade-integrated photovoltaics: a lifecycle and performance assessment case study. *Progress in Photovoltaics: Research and Applications*.

Raugei, M., Frischknecht, R., Olson, C. *et al.* (2015) Methodological guidelines on net energy analysis of photovoltaic electricity. IEA-PVPS Task 12, Report T12-XX:2015 (in press). http://www.iea-pvps.org

Raugei, M. and Fthenakis, V. (2010) Cadmium flows and emissions from CdTe PV: future expectations. *Energy Policy*, **38** (9), 5223–5228.

Sener, C. and Fthenakis, V. (2014) Energy policy and financing options to achieve solar energy grid penetration targets: accounting for external costs. *Renewable and Sustainable Energy Reviews*, **32**, 854–868.

US Department of Energy (2007) Energy Information Administration, Form EIA-1605 (2007), OMB No. 19 (1905-0194 Voluntary Reporting of Greenhouse Gases; Appendix F. Electricity Emission Factors.

Wild-Scholten de, M. (2013) Energy payback time and carbon footprint of commercial photovoltaic systems. *Solar Energy Materials & Solar Cells*, **119**, 296–305.

Wild-Scholten de, M.J., Alsema, E., ter Horst, E.W. *et al.* (2006) A cost and environmental impact comparison of grid-connected rooftop and ground-based PV systems. *Proceedings 21st European Photovoltaic Solar Energy Conference, Dresden, Germany*, 4–8 September 2006, pp. 3167–3172.

Woditsch, P. and Koch, W. (2002) Solar grade silicon feedstock supply for PV industry. *Solar Energy Materials & Solar Cells*, **72**, 11–26.

13.4

List of International Standards Related to PV

Pierre Verlinden[1] and Wilfried van Sark[2]
[1] *Trina Solar, Changzhou, Jiangsu, China*
[2] *Copernicus Institute, Utrecht University, The Netherlands*

13.4.1 Introduction

The establishment of standards is an important part of the development of an industry. Preparing new standards is generally a very long procedure based on discussions with many stakeholders, involving research laboratories, testing laboratories, manufacturers, and users. The lists below demonstrate the extent of topics covered by PV standards, including quality, reliability, safety, qualification, design requirements or recommendations, dimensions, characterization and measurement procedures, equipment, material specifications, etc. The standardization of specifications, materials, and procedures has also been critical to reduce the cost of PV manufacturing and deployment, and to ensure the quality of PV components.

The history of PV standards (Ossenbrink, *et al.*, 2012) started in 1978 with a demonstration program supported by the US Department of Energy (DOE) and managed by the Jet Propulsion Laboratory (JPL) of Pasadena. The 1980 original qualification standard for PV modules was called the "Block V Specification" that included a series of qualification tests forming the basis of what today has become the well-known and widely used IEC 61215 Standard. The task of developing standards was taken over by the Institute of Electrical and Electronic Engineers (IEEE) in the 1980s, then by the International Electrotechnical Commission (IEC, 2015a), the Underwriters' Laboratories (UL, 2015) and Semiconductor Equipment and Materials International (SEMI, 2015). Many countries have their own national series of PV-related standards but they are for the most part based on the standards developed by IEC. The UL standards are in general related to the safety of PV components or systems, while the SEMI standards are related to the manufacturing of PV modules.

Photovoltaic Solar Energy: From Fundamentals to Applications, First Edition.
Edited by Angèle Reinders, Pierre Verlinden, Wilfried van Sark, and Alexandre Freundlich.

Companion website: www.wiley.com/go/reinders/photovoltaic_solar_energy

The different groups of experts working on the development of IEC standards are part of a photovoltaic Technical Committee (TC), called TC 82 (IEC, 2015b). An overview of TCs is given by IEC (2015c). The technical committee TC 82 is itself divided into several Working Groups (WG), Project Teams (PT) and Joint Working Groups (JWG):

- WG 1: Glossary
- WG 2: Modules, non-concentrating
- WG 3: Systems
- WG 6: Balance of System Components
- WG 7: Concentrating modules and systems
- WG 8: Photovoltaic (PV) Cells
- PT 62994-1: Environmental Health and Safety (EH&S) Risk Assessment for the sustainability of PV manufacturing – Part 1. General principles and definition of terms
- JWG 1: JCG TC 82/TC 88/TC 21/SC 21A
- JWG 82: TC 21/TC 82 – Secondary cells and batteries for Renewable Energy Storage, managed by TC 21
- JWG 32: Electrical safety of PV system installation, managed by TC 64

13.4.2 IEC Standards Overview

The active and draft IEC standards at the time of writing (August 2015) related to PV are listed in Table 13.4.1. Also listed are Technical specifications (IEC/TS), which are not standards but recommendations. At the bottom of Table 13.4.1 are shown the current standards in preparation or "Proposed New Work" (PNW), which will eventually become a draft, then a standard.

Table 13.4.1 Active and Draft IEC standards and technical specifications

Standard	Title	Remarks
IEC 60891:2009	Photovoltaic devices – Procedures for temperature and irradiance corrections to measured I-V characteristics	Active
IEC 60904-1 Ed. 2.0, 2006	Photovoltaic devices – Part 1: Measurement of photovoltaic current-voltage characteristics	Active
IEC 60904-1 Ed. 3.0	Photovoltaic devices – Part 1: Measurement of photovoltaic current-voltage characteristics	Draft
IEC 60904-1-1 Ed. 1.0	Photovoltaic devices – Part 1-1: Measurement of current-voltage characteristics of multi-junction photovoltaic devices	Draft
IEC 60904-2 Ed. 3.0	Photovoltaic devices – Part 2: Requirements for reference solar devices	Active
IEC 60904-3 Ed. 3.0	Photovoltaic devices – Part 3: Measurement principles for terrestrial photovoltaic (PV) solar devices with reference spectral irradiance data	Active
IEC 60904-4 Ed. 1.0, 2009	Photovoltaic devices – Part 4: Reference solar devices – Procedures for establishing calibration traceability	Active
IEC 60904-5 Ed. 2.0, 2011	Photovoltaic devices – Part 5: Determination of the equivalent cell temperature (ECT) of photovoltaic (PV) devices by the open-circuit voltage method	Active

(Continued)

Table 13.4.1 (*Continued*)

Standard	Title	Remarks
IEC 60904-7 Ed. 3.0, 2008	Photovoltaic devices – Part 7: Computation of the spectral mismatch correction for measurements of photovoltaic devices	Active
IEC 60904-7 Ed. 4.0	Photovoltaic devices – Part 7: Computation of the spectral mismatch correction for measurements of photovoltaic devices	Draft
IEC 60904-8 Ed. 2.0, 2014	Photovoltaic devices – Part 8: Measurement of spectral response of a photovoltaic (PV) device	Active
IEC 60904-8-1 Ed. 1.0	Photovoltaic devices – Part 8-1: Measurement of spectral response of multi-junction photovoltaic (PV) devices	Draft
IEC 60904-9 Ed. 2.0, 2007	Photovoltaic devices – Part 9: Solar simulator performance requirements	Active
IEC 60904-9 Ed. 3.0	Photovoltaic devices – Part 9: Solar simulator performance requirements	Draft
IEC 60904-10 Ed. 2.0, 2009	Photovoltaic devices – Part 10: Methods of linearity measurement	Active
IEC 60904-11 Ed. 1.0	Photovoltaic devices – Part 11: Measurement of initial light-induced degradation of crystalline silicon solar cells and photovoltaic modules	Draft
IEC/TS 60904-12 Ed. 1.0	Photovoltaic devices – Part 12: Infrared thermography of photovoltaic modules	Draft
IEC/TS 60904-13 Ed. 1.0	Photovoltaic devices – Part 13: Electroluminescence of photovoltaic modules	Draft
IEC 61215-1 :2016	Terrestrial photovoltaic (PV) modules – Design qualification and type approval – Part 1: Requirements for testing	Active
IEC 61215-1-1 :2016	Terrestrial photovoltaic (PV) modules – Design qualification and type approval – Part 1-1: Special requirements for testing of crystalline silicon photovoltaic (PV) modules	Active
IEC 61215-1-2 Ed. 1.0	Terrestrial photovoltaic (PV) modules – Design qualification and typ approval – Part 1-2: Special requirements for testing of cadmium telluride (CdTe) photovoltaic (PV) modules	Draft
IEC 61215-1-3 Ed. 1.0	Terrestrial photovoltaic (PV) modules – Design qualification and type approval – Part 1-3: Special requirements for testing of amorphous silicon (a-Si) and microcrystalline silicon (micro c-Si) photovoltaic (PV) modules	Draft
IEC 61215-1-4 Ed. 1.0	Terrestrial photovoltaic (PV) modules Design qualification and type approval – Part 1-4: Special requirements for testing of copper indium gallium selenide (CIGS) and copper indium selenide (CIS) photovoltaic (PV) modules	Draft
IEC 61215-2 :2016	Terrestrial photovoltaic (PV) modules – Design qualification and type approval – Part 2: Test procedures	Active
IEC 61345 (1998-02) Ed.2.0	UV test for photovoltaic (PV) modules	Active
IEC 61646 (2008-05) Ed.2.0	Thin-film terrestrial photovoltaic (PV) modules-Design qualification and type approval	Active
IEC 61683-1999, Ed.1	Photovoltaic systems – Power conditioners – Procedure for measuring efficiency	Active
IEC 61683 Ed. 2.0	Photovoltaic systems – Power conditioners – Procedure for measuring efficiency	Draft

Table 13.4.1 (*Continued*)

Standard	Title	Remarks
IEC 61701 (2011-03) Ed.2.0	Salt mist corrosion testing of photovoltaic (PV) modules	Active
IEC 61724-1998, Ed.1	Photovoltaic system performance monitoring – Guidelines for measurement, data exchange and analysis	Active
IEC 61724-1 Ed. 1.0	Photovoltaic system performance – Part 1: Monitoring	Draft
IEC 61724-3 :2016	Photovoltaic system performance – Part 3: Energy evaluation method	Active
IEC 61725-1997, Ed.1	Analytical expression for daily solar profiles	Active
IEC 61727-2004, Ed.2.0	Photovoltaic (PV) systems – Characteristics of the utility interface	Active
IEC 61730-1 Ed. 2.0	Photovoltaic (PV) module safety qualification – Part 1: Requirements for construction	Active
IEC 61730-2 Ed. 2.0	Photovoltaic (PV) module safety qualification – Part 2: Requirements for testing	Active
IEC 61829 Ed.2.0, 2015	Crystalline silicon photovoltaic (PV) array – On-site measurement of I-V characteristics	Active
IEC 61836-2007, Ed.1	Solar photovoltaic energy systems – Terms, definitions and symbols	Active
IEC 61853-1 (2011-01) Ed.1.0	Photovoltaic module performance testing and energy rating – Part 1: Irradiance and temperature performance measurements and power rating	Active
IEC 61853-3 Ed. 1.0	Photovoltaic (PV) module performance testing and energy rating – Part 3: Energy Rating of PV Modules	Draft
IEC 61853-4 Ed. 1.0	Photovoltaic (PV) module performance testing and energy rating – Part 4: Standard reference climatic profiles	Draft
IEC 62093-2005, Ed.1	Balance-of-system components for photovoltaic systems – Design qualification natural environments	Active
IEC 62093 Ed. 2.0	Balance-of-system components for photovoltaic systems – Design qualification natural environments	Draft
IEC 62108 Ed. 2.0	Concentrator photovoltaic (CPV) modules and assemblies – Design qualification and type approval	Active
IEC 62109-1-2010, Ed.1	Safety of power converters for use in photovoltaic power systems – Part 1: General requirements	Active
IEC 62109-2-2011, Ed.1	Safety of power converters for use in photovoltaic power systems – Part 2: Particular requirements for inverters	Active
IEC 62109-3 Ed. 1.0	Safety of power converters for use in photovoltaic power systems – Part 3: Particular requirements for electronic devices in combination with photovoltaic elements	Draft
IEC PAS 62111-1999, Ed.1	Specifications for the use of renewable energies in rural decentralised electrification	Active
IEC 62116-2014, Ed. 2	Test procedure of islanding prevention measures for utility-interconnected photovoltaic inverters	Active
IEC 62124-2004, Ed.1.0	Photovoltaic (PV) stand alone systems – Design verification	Active

(*Continued*)

Table 13.4.1 (*Continued*)

Standard	Title	Remarks
IEC 62253-2011, Ed.1.0	Photovoltaic pumping systems – Design qualification and performance measurements	Active
IEC/TS 62257-1 (2015) Ed. 3.0	Recommendations for small renewable energy and hybrid systems for rural electrification – Part 1: General introduction to rural electrification	Active
IEC/TS 62257-2 (2015) Ed. 2.0	Recommendations for small renewable energy and hybrid systems for rural electrification – Part 2: From requirements to a range of electrification systems	Active
IEC/TS 62257-3 (2015) Ed. 2.0	Recommendations for small renewable energy and hybrid systems for rural electrification – Part 3: Project development and management	Active
IEC/TS 62257-4 (2015) Ed. 2.0	Recommendations for small renewable energy and hybrid systems for rural electrification – Part 4: System selection and design	Active
IEC/TS 62257-5 (2015) Ed. 2.0	Recommendations for small renewable energy and hybrid systems for rural electrification – Part 5: Protection against electrical hazards	Active
IEC/TS 62257-6 (2015) Ed. 2.0	Recommendations for small renewable energy and hybrid systems for rural electrification – Part 6: Acceptance, operation, maintenance and replacement	Active
IEC/TS 62257-7 (2008) Ed. 1.0	Recommendations for small renewable energy and hybrid systems for rural electrification – Part 7: Generators	Active
IEC/TS 62257-7-1 (2010) Ed. 2.0	Recommendations for small renewable energy and hybrid systems for rural electrification – Part 7-1: Generators – Photovoltaic Generators	Active
IEC/TS 62257-7-3 (2008) Ed. 1.0	Recommendations for small renewable energy and hybrid systems for rural electrification – Part 7-3: Generator set – Selection of generator sets for rural electrification systems	Active
IEC/TS 62257-8-1 (2007) Ed. 1.0	Recommendations for small renewable energy and hybrid systems for rural electrification – Part 8-1: Selection of batteries and battery management systems for stand-alone electrification systems – Specific case of automotive flooded lead-acid batteries available in developing countries	Active
IEC/TS 62257-9-1 (2008) Ed. 1.0	Recommendations for small renewable energy and hybrid systems for rural electrification – Part 9-1: Micropower systems	Active
IEC/TS 62257-9-2 (2006) Ed. 1.0	Recommendations for small renewable energy and hybrid systems for rural electrification – Part 9-2: Microgrids	Active
IEC/TS 62257-9-3 (2006) Ed. 1.0	Recommendations for small renewable energy and hybrid systems for rural electrification – Part 9-3: Integrated system – User interface	Active
IEC/TS 62257-9-4 (2006) Ed. 1.0	Recommendations for small renewable energy and hybrid systems for rural electrification – Part 9-4: Integrated system – User installation	Active
IEC/TS 62257-9-5 (2016) Ed. 3.0	Recommendations for small renewable energy and hybrid systems for rural electrification – Part 9-5: Integrated system – Selection of portable PV lanterns for rural electrification projects	Active
IEC/TS 62257-9-6 (2008) Ed. 1.0	Recommendations for small renewable energy and hybrid systems for rural electrification – Part 9-6: Integrated system – Selection of Photovoltaic Individual Electrification Systems (PV-IES)	Active

Table 13.4.1 (*Continued*)

Standard	Title	Remarks
IEC/TS 62257-12-1 (2015) Ed. 2.0	Recommendations for small renewable energy and hybrid systems for rural electrification – Part 12-1: Selection of self-ballasted lamps (CFL) for rural electrification systems and recommendations for household lighting equipment	Active
IEC/TS 62257-7 Ed. 2.0	Recommendations for renewable energy and hybrid systems for rural electrification – Part 7: Generators	Draft
IEC/TS 62257-7-1 Ed. 3.0	Recommendations for renewable energy and hybrid systems for rural electrification – Part 7-1: Generators – Photovoltaic generators	Draft
IEC/TS 62257-7-3 Ed. 2.0	Recommendations for renewable energy and hybrid systems for rural electrification – Part 7-3: Generator set – Selection of generators sets for rural electrification systems	Draft
IEC/TS 62257-8-1 Ed. 2.0	Recommendations for renewable energy and hybrid systems for rural electrification – Part 8-1: Selection of batteries and battery management systems for stand-alone electrification systems – Specific case of automotive flooded lead-acid batteries available in developing countries	Draft
IEC/TS 62257-9-5 Ed. 3.0	Recommendations for renewable energy and hybrid systems for rural electrification – Part 9-5: Integrated systems – Selection of stand-alone lighting kits for rural electrification	Active
IEC/TS 62257-9-6 Ed. 2.0	Recommendations for renewable energy and hybrid systems for rural electrification – Part 9-6: Integrated system – Selection of Photovoltaic Individual Electrification Systems (PV-IES)	Draft
IEC/TS 62257-12-1 Ed. 2.0	Recommendations for renewable energy and hybrid systems for rural electrification – Part 12-1: Selection of self-ballasted lamps (CFL) for rural electrification systems and recommendations for household lighting equipment	Active
IEC 62446-1 Ed. 1.0	Photovoltaic (PV) systems – Requirements for testing, documentation and maintenance – Part 1: Grid connected systems – Documentation, commissioning tests and inspection	Active
IEC 62446-2 Ed. 1.0	Grid connected PV systems – Part 2: Maintenance of PV systems	Draft
IEC 62446-3 Ed. 1.0	Photovoltaic (PV) systems – Requirements for testing, documentation and maintenance – Part 3: Outdoor infrared thermography of photovoltaic modules and plants	Draft
IEC 62509-2010	Battery charge controllers for photovoltaic systems – Performance and functioning	Active
IEC 62548 Ed. 1.0	Photovoltaic (PV) arrays – Design requirements	Active
IEC 62670-1 Ed.1.0, 2013	Concentrator photovoltaic (CPV) module and assembly performance testing and energy rating-Part 1:Performance measurements and power rating-Irradiance and temperature	Active
IEC 62670-2 Ed. 1.0,2015	Concentrator photovoltaic (CPV) performance testing – Part 2: Energy measurement	Active
IEC 62670-3	Concentrator photovoltaic (CPV) performance testing – Part 3: Performance measurements and power rating	Draft
IEC 62688 Ed. 1.0	Concentrator photovoltaic (CPV) module and assembly safety qualification	Draft

(*Continued*)

Table 13.4.1 (*Continued*)

Standard	Title	Remarks
IEC 62716 Ed.1.0 2013	Ammonia corrosion testing of photovoltaic (PV) modules	Active
IEC TS 62727-2012 Ed.1	Photovoltaic systems – Specification for solar trackers	Active
IEC/TS 62738 Ed. 1.0	Design guidelines and recommendations for photovoltaic power plants	Draft
IEC 62759-1 Ed. 1.0	Transportation testing of photovoltaic (PV) modules – Part 1: Transportation and shipping of PV module stacks	Active
IEC 62782 Ed. 1.0	Dynamic mechanical load testing for photovoltaic (PV) modules	Active
IEC 62787 Ed. 1.0	Concentrator photovoltaic (CPV) solar cells and cell-on-carrier (COC) assemblies – Reliability qualification	Draft
IEC 62788-1-2 Ed. 1.0	Measurement procedures for materials used in photovoltaic modules – Part 1-2: Encapsulants – Measurement of volume resistivity of photovoltaic encapsulation and backsheet materials	Active
IEC 62788-1-4 Ed. 1.0	Measurement procedures for materials used in Photovoltaic Modules – Part 1-4: Encapsulants – Measurement of optical transmittance and calculation of the solar-weighted photon transmittance, yellowness index, and UV cut-off frequency	Active
IEC 62788-1-5 Ed. 1.0	Measurement procedures for materials used in photovoltaic modules – Part 1-5: Encapsulants – Measurement of change in linear dimensions of sheet encapsulation material under thermal conditions	Active
IEC 62788-1-6 Ed. 1.0	Measurement procedures for materials used in photovoltaic modules – Part 1-6: Encapsulants – Test methods for determining the degree of cure in Ethylene-Vinyl Acetate encapsulation for photovoltaic modules	Draft
IEC 62788-2 Ed. 1.0	Measurement procedures for materials used in photovoltaic modules – Part 2: Polymeric materials used for front sheets and backsheets	Draft
IEC 62788-7-2	Measurement procedures for materials used in photovoltaic modules – Part 7-2: Environmental exposures – Accelerated weathering tests of polymeric materials=	Draft
IEC/TS 62789 Ed.1.0, 2014	Specification for concentrator cell description	Active
IEC 62790 Ed.1.0 2014	Junction boxes for photovoltaic modules – Safety requirements and tests	Active
IEC/TS 62804 Ed. 1.0	Test methods for detection of potential-induced degradation of crystalline silicon photovoltaic (PV) modules	Active
IEC 62805-1 Ed. 1.0	Method for measuring photovoltaic (PV) glass – Part 1: Measurement of total haze and spectral distribution of haze	Draft
IEC 62805-2 Ed. 1.0	Method for measuring photovoltaic (PV) glass – Part 2: Measurement of transmittance and reflectance	Draft
IEC 62817-2014 Ed.1	Photovoltaic systems – Design qualification of solar trackers	Active
IEC 62852 Ed. 1.0 2014	Connectors for DC-application in photovoltaic systems – Safety requirements and tests	Active

Table 13.4.1 (*Continued*)

Standard	Title	Remarks
IEC 62891 Ed. 1.0	Overall efficiency of grid connected photovoltaic inverters	Draft
IEC 62892-1 Ed. 1.0	Comparative testing of PV modules to differentiate performance in multiple climates and applications – Part 1: Overall test sequence and method of communication	Draft
IEC 62894 Ed.1.0 2014	Photovoltaic inverters – Data sheet and name plate	Active
IEC/TS 62910 Ed. 1.0	Test procedure of Low Voltage Ride-Through (LVRT) measurement for utility-interconnected photovoltaic inverter	Active
IEC/TS 62915 Ed. 1.0	Photovoltaic (PV) Modules – Retesting for type approval, design and safety qualification	Draft
IEC/TS 62916 Ed. 1.0	Bypass diode electrostatic discharge susceptibility testing	Draft
IEC 62920 Ed. 1.0	EMC requirements and test methods for grid connected power converters applying to photovoltaic power generating systems	Draft
IEC 62925 Ed. 1.0	Thermal cycling test for CPV modules to differentiate increased thermal fatigue durability	Draft
IEC 62938 Ed. 1.0	Non-uniform snow load testing for photovoltaic (PV) modules	Draft
IEC/TS 62941 Ed. 1.0	Guideline for increased confidence in PV module design qualification and type approval	Active
IEC 62979 Ed. 1.0	Photovoltaic module bypass diode thermal runaway test	Draft
IEC 62980 Ed. 1.0	Photovoltaic modules for building curtain wall applications	Draft
IEC/TS 62989 Ed. 1.0	Primary Optics for Concentrator Photovoltaic Systems	Draft
IEC 62994-1 Ed. 1.0	Environmental Health & Safety (EH&S) Risk Assessment for the sustainability of PV module manufacturing – Part 1. General principles and definition of terms	Draft
IEC 62788-5-1 Ed. 1.0	Measurement procedures for materials used in photovoltaic modules – Part 5-1 Suggested test methods for use with edge seal materials (proposed future IEC 62788-5-1)	Draft
IEC 62788-6-2	Measurement procedures for materials used in photovoltaic modules – Part 6-2: Moisture permeation testing with polymeric films (proposed future IEC 62788-6-2)	Draft
IEC/TS 63019 Ed. 1.0	Information model for availability of photovoltaic (PV) power systems	Active
IEC/TS 61724-2 Ed. 1.0	Photovoltaic system performance – Part 2: Capacity evaluation method	Active
IEC/TS 62446-3 Ed. 1.0	Photovoltaic (PV) System – Requirements for testing, documentation and maintenance – Part 3: Outdoor Infrared thermography of PV modules and plants	Draft
IEC 60904-9-1 Ed. 1.0	Photovoltaic devices – Part 9-1: Collimated beam solar simulator performance requirements	Draft
IEC 63027 Ed. 1.0	DC arc detection and interruption in photovoltaic power systems	Draft
IEC 62446-2 Ed. 1.0	Grid connected photovoltaic (PV) systems – Part 2: Maintenance of PV systems (proposed IEC 62446-2)	Draft

13.4.3 Underwriters' Laboratories (UL) Standards

The UL standards related to the reliability and safety of PV components and systems are listed in Table 13.4.2 (status August 2015).

13.4.4 The SEMI Standards

The SEMI standards and specifications related to the manufacturing of PV wafers, cells and modules, as well as the characterization of measurement methods are listed in Table 13.4.3 (status August 2015).

Table 13.4.2 UL standards

Standard	Title
OOI 1279	Solar Collectors
OOI 1699B	PV DC Arc-fault circuit protection
OOI 2579 (248-19)	Low Voltage Fuses – Fuses for Photovoltaic Systems
OOI 2703	Standard for Mounting Systems, Mounting Devices, Clamping/Retention Devices, and Ground Lugs for Use with for Flat-Plate Photovoltaic Modules and Panels
OOI 3703	Solar Trackers
OOI 3730	Photovoltaic Junction Boxes
OOI 4248-18	Fuseholders, Part 18: PV
OOI 4703	PV Wire
OOI 489B	Molded case circuit breakers, MC switches, and circuit breaker enclosures for use with PV systems
OOI 508I	Outline of investigation for Disconnect Switches Intended for use in Photovoltaic systems
OOI 5703	Determination of the Max Operating Temp Rating of PV Backsheet Materials
OOI 6703	Connectors for Use in PV Systems
OOI 6703A	Multi-pole connectors for use in PV Systems
OOI 8703	Concentrator Photovoltaic Modules and Assemblies
OOI 9703	Distributed Wiring Harnesses
OOI 98B	Outline of investigation for enclosed and dead-front switches for use in PV systems
UL 1703	Flat Plate Photovoltaic Modules and Panels
UL 1741	Inverters, Converters, Controllers and Interconnection System Equipment for Use With Distributed Energy Resources
UL 61215	Crystalline silicon terrestrial photovoltaic (PV) modules – Design qualification and type approval
UL 61646	Thin-film terrestrial photovoltaic (PV) modules – Design qualification and type approval
UL 62108	Concentrator photovoltaic (CPV) modules and assemblies – Design qualification and type approval
UL 62109-1	Safety of power converters for use in photovoltaic power systems – Part 1: General requirements

Table 13.4.3 SEMI standards and specifications

Standard	Title	Remarks
SEMI PV1-0211	Test Method for Measuring Trace Elements in Silicon Feedstock for Silicon Solar Cells by High-Mass Resolution Glow Discharge Mass Spectrometry	Active
SEMI PV2-0709E	Guide for Equipment Communication Interfaces (PVECI)	Active
SEMI PV3-1115	Guide for High Purity Water Used in Photovoltaic Cell Processing	Active
SEMI PV4-0311	Specification for Range of 5th Generation Substrate Sizes for Thin Film Photovoltaic Applications	Active
SEMI PV5-1115	Guide for Oxygen (O2), Bulk, Used in Photovoltaic	Active
SEMI PV6-1115	Guide for Argon (Ar), Bulk, Used in Photovoltaic Applications	Active
SEMI PV7-1115	Guide for Hydrogen (H2), Bulk, Used in Photovoltaic Applications	Active
SEMI PV8-1115	Guide for Nitrogen (N2), Bulk, Used in Photovoltaic Applications	Active
SEMI PV9-1115	Test Method for Excess Charge Carrier Decay in PV Silicon Materials by Non-Contact Measurements of Microwave Reflectance After a Short Illumination Pulse	Active
SEMI PV10-0716	Test Method for Instrumental Neutron Activation Analysis (INAA) of Silicon	Active
SEMI PV11-1115	Specifications for Hydrofluoric Acid, Used in Photovoltaic Applications	Active
SEMI PV12-1115	Specifications for Phosphoric Acid Used in Photovoltaic Applications	Active
SEMI PV13-0714	Test Method for Contactless Excess-Charge-Carrier Recombination Lifetime Measurement in Silicon Wafers, Ingots, and Bricks Using an Eddy-Current Sensor	Active
SEMI PV14-1215	Guide for Phosphorus Oxychloride, Used in Photovoltaic Applications	Active
SEMI PV15-1215	Guide for Defining Conditions for Angle Resolved Light Scatter Measurements to Monitor the Surface Roughness and Texture of PV Materials	Active
SEMI PV16-0316	Specifications for Nitric Acid, Used in Photovoltaic Applications	Active
SEMI PV17-1012	Specification for Virgin Silicon Feedstock Materials for Photovoltaic Applications	Active
SEMI PV18-0912	Guide for Specifying a Photovoltaic Connector Ribbon	Active
SEMI PV19-0712	Guide for Testing Photovoltaic Connector Ribbon Characteristics	Active
SEMI PV20-0316	Specifications for Hydrochloric Acid, Used in Photovoltaic Applications	Active
SEMI PV21-1011	Guide for Silane (SiH4), Used in Photovoltaic Applications	Active

(*Continued*)

Table 13.4.3 (*Continued*)

Standard	Title	Remarks
SEMI PV22-1011	Specification for Silicon Wafers for Use in Photovoltaic Solar Cells	Active
SEMI PV23-1011	Test Method for Mechanical Vibration of Crystalline Silicon Photovoltaic (PV) Modules in Shipping Environment	Active
SEMI PV24-1011	Guide for Ammonia (NH3) in Cylinders, Used in Photovoltaic Applications	Active
SEMI PV25-1011	Test Method for Simultaneously measuring Oxygen, Carbon, Boron and Phosphorus in Solar Silicon Wafers and Feedstock by Secondary Ion Mass Spectrometry	Active
SEMI PV26-1011	Guide for Hydrogen Selenide (H2Se) in Cylinders, Used in Photovoltaic Applications	Active
SEMI PV27-0316	Specifications for Ammonium Hydroxide, Used in Photovoltaic Applications	Active
SEMI PV28-0316	Test Methods for Measuring Resistivity or Sheet Resistance with a Single-Sided Noncontact Eddy-Current Gauge	Active
SEMI PV29-0212	Specification for Front Surface Marking of PV Silicon Wafers with Two-Dimensional Matrix Symbols	Active
SEMI PV30-0316	Specifications for 2-Propanol, used in Photovoltaic Applications	Active
SEMI PV31-0212	Test Method for Spectrally Resolved Reflective and Transmissive Haze of Transparent Conducting Oxide (TCO) Films of PV Application	Active
SEMI PV32-0312	Specification for Marking of PV Silicon Brick Face and PV Wafer Edge	Active
SEMI PV33-0316	Specifications for Sulfuric Acid, used in Photovoltaic Applications	Active
SEMI PV34-0213	Practice for Assigning Identification Numbers to PV Si Wafer and Solar Cell Manufacturers	Active
SEMI PV35-0215	Specification for Horizontal Communication Between Equipment for Photovoltaic Fabrication System	Active
SEMI PV36-0316	Specifications for Hydrogen Peroxide, used in photovoltaic applications	Active
SEMI PV37-0912	Guide for Fluorine (F2), Used in Photovoltaic	Active
SEMI PV38-0912	Test Method for Mechanical Vibration of C-SI PV cells in Shipping Environment	Active
SEMI PV39-0513	Test method for in-line measurement of cracks in PV silicon wafers by dark field infrared imaging	Active
SEMI PV40-0513	Test Method for In-Line Measurement of Saw Marks on PV Silicon Wafers by A Light Sectioning Technique	Active
SEMI PV41-0912	Test Method for In-Line, Non-Contact Measurement of Thickness and Variation of Silicon Wafers for PV Applications using Capacitive Probes	Active
SEMI PV42-0314	Test Method for In-Line Measurement of Waviness of PV Silicon Wafers by a Light Sectioning Technique Using Multiple Line Segments	Active

Table 13.4.3 (*Continued*)

Standard	Title	Remarks
SEMI PV43-0113	Test Method for the Measurement of Oxygen Concentration in PV Silicon Materials for Silicon Solar Cells by Inert Gas Fusion Infrared Detection Method	Active
SEMI PV44-0513	Specification for Package protection technology for PV Modules	Active
SEMI PV45-0513	Vinyl Acetate (VA) content test method for Ethylene-Vinyl Acetate (EVA) applied in photovoltaic modules—Thermal Gravimetric Analysis (TGA)	Active
SEMI PV46-0613	Test Method for In-Line Measurement of Lateral Dimensional Characteristics of Square and Pseudo-Square PV Silicon Wafers	Active
SEMI PV47-0513	Specification for Anti-reflective-coated Glass, Used in Crystalline Silicon Photovoltaic Modules	Active
SEMI PV48-0613	Specification for Orientation of Fiducial Marks for PV Silicon Wafers	Active
SEMI PV49-0613	Test Method for the Measurement of Elemental Impurity Concentrations in Silicon Feedstock for Silicon Solar Cells by Bulk Digestion, Inductively Coupled-Plasma Mass Spectrometry	Active
SEMI PV50-0114	Specification for Impurities in Polyethylene Packaging Materials for Polysilicon Feedstock	Active
SEMI PV51-0214	Test Method for In-Line Characterization of Photovoltaic Silicon Wafers by Using Photoluminescence	Active
SEMI PV52-0214	Test Method for In-Line Characterization of Photovoltaic Silicon Wafers Regarding Grain Size	Active
SEMI PV53-0514	Test method for In-line monitoring of flat temperature zone in the horizontal diffusion furnaces	Active
SEMI PV54-0514	Specification for Silver Paste, Used to Contact with N+ Diffusion Layer of Crystalline Silicon Solar Cells	Active
SEMI PV56-1214	Test Method for Performance Criteria of Photovoltaic (PV) Cells and Modules Package	Active
SEMI PV57-1214	Test Method for Current-Voltage (I-V) Performance Measurement of Organic Photovoltaic (OPV) And Dye-Sensitized Solar Cell (DSSC)	Active
SEMI PV58-0115	Specification for aluminum paste, used in back surface field of crystalline silicon solar cells	Active
SEMI PV59-0115	Test Method for Determination of Total Carbon Content in Silicon Powder by Infrared Absorption After Combustion in an Induction Furnace	Active
SEMI PV60-0115	Test Method for Measurement of Cracks in Photovoltaic (PV) Silicon Wafers in PV Modules by Laser Scanning	Active
SEMI PV61-0115	Specification for Framing Tape for PV Modules	Active
SEMI PV63-0215	Specification for Ultra-thin Glasses Used for Photovoltaic Modules	Active
SEMI 4826	Specification for Silicon Wafers for Use as Photovoltaic Solar Cells	Work in progress
SEMI M79-0211	Specification for Round 100 mm Polished Monocrystalline Germanium Wafers for Solar Cell Applications	Active

(*Continued*)

Table 13.4.3 (*Continued*)

Standard	Title	Remarks
SEMI 5427	Specification for front Surface Silver Paste, Used in P-type crystalline silicon solar cells	Work in progress
SEMI 5477	Test Method for Determining B, P, Fe, Al, Ca Contents in Silicon Powder for PV Applications by Inductively-Coupled-Plasma Optical Emission Spectrometry	Work in progress
SEMI 5478	Test Method for Thin-film Silicon PV Modules Light Soaking	Work in progress
SEMI 5564	Test Method for the Measurement of Chlorine in Silicon by Ion Chromatography	Work in progress
SEMI 5644	Terminology for Back Contact PV Cell and Module	Work in progress
SEMI 5648	Test Method for the Integrated Efficiency of Installed PV Components	Work in progress
SEMI 5659	Test Method for C-Si Solar Cell Color	Work in progress
SEMI 5661	Test Method for Electrical Parameters of Bifacial Solar Module	Work in progress
SEMI 5724	Guide for Specifying Quasi Monocrystalline Silicon Wafers used in Photovoltaic Solar Cells	Work in progress
SEMI 5725	Practice for Metal Wrap Through (MWT) Back Contact PV Module Assembly	Work in progress
SEMI 5726	Test Method for Determining the Aspect Ratio of Solar Cell Metal Fingers by Confocal Laser Scanning Microscope	Work in progress
SEMI 5727	Test Method for the Etch Rate of A Crystalline Silicon Wafer by Determining The Weight Loss	Work in progress
SEMI 5728	Test Method for the Wire Tension of Multi-wire Saws	Work in progress
SEMI 5767	Guide for Material Requirements of Internal Feeders Used in Mono-crystal Silicon Growers	Work in progress
SEMI 5768	Specification for Testing Requirements of Electroluminescence Defect Detection System for Crystalline Silicon PV Modules	Work in progress
SEMI 5773	Test Method for Cell Defects in Crystalline Silicon PV Modules by Using Electroluminescence	Work in progress
SEMI 5830	Classification for Electroluminescence Inspection of Crystalline Silicon Photovoltaic Modules	Work in progress
SEMI 5840	Guide for Calibration of PV Module UV Test Chambers	Work in progress
SEMI 5841	Guide for Specifying Low Pressure Horizontal Diffusion Furnace	Work in progress
SEMI 5842	Test Method for Metal-Wrap-Through Solar Cell Via Resistance	Work in progress
SEMI F108-0310	Guide for Integration of Liquid Chemical Piping Components for Semiconductor, Flat Panel Display, and Solar Cell Manufacturing Applications	Active

Acknowledgements

The authors would like to thank TaoYun Xiao, JingJing Gong and Nan Du (Trina Solar) for helping to update the list of standards.

References

IEC (2015a) http://www.iec.ch (accessed 24 August 2015).

IEC (2015b) http://www.iec.ch/dyn/www/f?p=103:7:0::::FSP_ORG_ID:1276 (accessed 24 August 2015).

IEC (2015c) http://www.iec.ch/dyn/www/f?p=103:6:0 (accessed 24 August 2015)

Ossenbrink, H., Müllejans, H., Kenny, R., and Dunlop, E. (2012) Standards in photovoltaic technology, in *Photovoltaic Technology* (ed. W.G.J.H.M. van Sark), vol. 1 in *Comprehensive Renewable Energy* (ed. A. Sayigh), Elsevier, Oxford, pp. 787–803.

SEMI (2015) http://www.semi.org/en/(accessed 24 August 2015)

UL (2015) http://www.ul.com (accessed 24 August 2015).

Index

Photovoltaic Solar Energy: From Fundamentals to Applications, First Edition.
Edited by Angèle Reinders, Pierre Verlinden, Wilfried van Sark, and Alexandre Freundlich.

Companion website: www.wiley.com/go/reinders/photovoltaic_solar_energy